Calculus and Analytic Geometry

*A text for courses of 8 or 9 semester-hours,
with technical emphasis, including conventional topics in
analytic geometry and linear algebra.*

In three parts

CALCULUS
and Analytic Geometry

WILLIAM L. DUREN, JR.   *University of Virginia*

XEROX COLLEGE PUBLISHING   *Lexington, Massachusetts* | *Toronto*

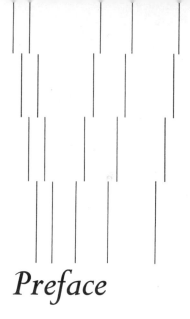

Preface

Why Another Calculus?

There have been a number of changes in the makeup of college calculus classes, calling for a new arrangement of text material, from which shorter courses can be taught as well as the current standard courses of 8 or 9 semester hours. Most important, a need has arisen for a definitive calculus course of 6 semester hours that includes enough multivariable calculus to serve as a prerequisite for such subjects as mathematical statistics, mechanics, and physical chemistry. Many of the new candidates for calculus are in curricula that do not provide time for more than a year of mathematics, and yet they need calculus. Moreover, even students in mathematics, physical sciences, and engineering, whose curricula provide for at least two years of mathematics, now feel the need to cut back the time spent on calculus to 6 semester hours, so as to save the second year for other mathematical subjects whose importance is increasing. These subjects include probability, linear algebra, and computation, which can be included in calculus but are better presented independently.

Two volumes, *Calculus* and *Calculus and Analytic Geometry*, offer a solution to these problems.* They are made up from three semester-course units: Part I, Calculus of Elementary Functions; Part II, Multivariable Calculus; and Part III, Analytic Geometry and Calculus. The full *Calculus and Analytic Geometry*, consisting of all three Parts, will implement the current standard course of 8 or 9 semester hours, with strong technical emphasis. The last third, however, contains the topics that can be subsumed into a new course for the second year or handled in some other way. Thus it is easy to cut back to the 6-semester-hour *Calculus*, Parts I and II. The new selection and arrangement of topics for this one-year course is dictated by a requirement for the student to attain within the first 6 semester hours a complete structure of ideas, including some

---

* These texts are two separate editions published by Xerox College Publishing in 1972. *Calculus and Analytic Geometry* includes the entire *Calculus* book plus a section on analytic geometry and linear algebra.

multivariable calculus. After that, the topics and techniques were chosen for their importance in applications.

Another important advantage of the arrangement in this text is that it fits well into a program of advanced placement. Part I is a one-semester course in elementary functions and introductory calculus, which can be covered in secondary school. This permits the student to enter the second semester of a course based on this text.* Moreover the design of the book favors advanced placement through the entire year for students who qualify.

This effort originated in my service with the Committee on the Undergraduate Program in Mathematics (CUPM). Its report, *A General Curriculum in Mathematics for Colleges* (GCMC), Mathematical Association of America 1965, called for a sequence of semester courses that begins with a calculus, Math 1. Part I is intended to implement the GCMC Math 1. That report suggested two versions of Math 2, preferring a multivariable calculus form. With some license, Part II of this book will implement the GCMC Math 2 (multivariable form). The main discrepancy comes from the fact that in its 1965 report CUPM allowed three semesters for completing calculus, Math 1, 2, 4, whereas this text emphasizes a calculus course that can be completed in two semesters.

A more complete description of the provisions of these texts, called *How to Use These Texts*, follows this preface.

I cannot acknowledge all the sources of ideas in this volume. First of all, I enjoyed years of critical discussion in CUPM, and these left me with an advanced education in college calculus, but I do not now know who taught me what. At the University of Virginia, my colleagues Marvin Rosenblum and Eugene Paige taught an earlier version of this material, which has undergone several metamorphoses. I owe several ideas to E. J. McShane. Younger student colleagues R. E. L. Nelson and Lyle Jenkins assisted with the early manuscript. Early-manuscript preparation was done by Mrs. Jane L. Miller and Sally Duren Heatter. The final manuscript was typed by Mrs. Mildred G. Gibson, whose fantastic accuracy and imaginative involvement made it possible to complete this task.

Outside of the University of Virginia, Don E. Edmondson worked with me on the structure, planning for a formerly projected Part III. George Springer, J. P. King, Donald R. Kerr, Jr., and Ronald Douglas read the manuscript and were most helpful with criticisms and encouragement. Finally, I have a special debt to A. W. Tucker, who started me on this effort years ago and stayed with it to the end as counsel, critic, and source of ideas.

Charlottesville, Virginia

---

* Part I, Sections 1–34, and Part II, 1–7, of this text cover, and are substantially covered by, the Calculus AB of the 1969–70 *Advanced Placement Mathematics* of the College Entrance Examination Board, Princeton, N.J. Also, a year course can be selected from this book to cover the more advanced Calculus BC, designed for a year's advanced placement.

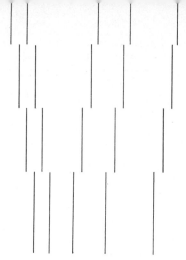

How to Use These Texts

The main effort in writing these texts went into arranging the material so as to provide for flexibility and control of coverage in short courses, as well as full-length ones. To take full advantage of these provisions for budgeting time, one must know something about how the book is put together.

1. *Sections and numbering.* There are 45 class meetings in a 3-semester-hour course. Leaving 11 lessons for review and quiz, this book assumes 34 advancing lessons. So the book is written in sections, each intended for a day's lesson. They are numbered sequentially through each Part to make it easier to compare progress with the calendar.

The section symbol § is used in front of a number when reference to another section is made. For example, a reference to Section 1.3 in Part I would appear as (I, §1.3). If no roman numeral is used, the reference is to a section in the same part of the book.

2. *Review sections and extra sections.* Some additional review sections are provided that are not included in the 34 advancing lessons. They are written primarily for self study.

The optional additional sections (I, §35–38) are not prerequisite to Part II. They can be used, where time allows, to emphasize matters of special interest to students of the social and life sciences.

Some choice is provided in Part II by extra sections (II, §34–39). The polar coordinate sections there should be included in longer courses of 8 or 9 semester hours. They are prerequisite to a good deal of Part III.

3. *Exercises and problems.* There are two types of problem sets. In each section the standard exposition, with proofs of only crucial theorems, leads as directly as possible to the first set of Exercises. After this, in many cases, comes more theoretical exposition and theoretical exercises called Problems. In my own teaching I do not try to present to the class as a whole the material after the first Exercises. I have found, however, that some of the more interested students will get together to work through them. Some or all of these theoretical appendices can be used for a more rigorous, honors type of course. For a complete study by the class as a whole, more time will be required. Some more difficult problems and exercises are marked with an asterisk, for example 23*.

A solutions manual is available to teachers upon request. It is primarily directed to inexperienced teachers and includes some suggestions on teaching certain topics.

4. *Treatment of analytic geometry.* Part I neither assumes a previous knowledge of analytic geometry proper nor includes it. It is based on a knowledge of graphs, as learned in secondary school algebra. This is one of the main economies that permits the completion in one semester of a course in differential and integral calculus of the elementary functions, including the usual transcendental ones. Some analytic geometry, as needed, is presented in Part II. Those who want analytic geometry for its own sake, including transformation of coordinates and the focal properties of conics, will find it in Part III.

5. *Uses of Part I.* For several reasons, this elementary course in differential and integral calculus is accessible to a wider population of students than conventional texts can accommodate. The prerequisites and technical demands are kept to a minimum. Not only is analytic geometry reduced to graphs, but analytic trigonometry is held back until near the end, where only sine, cosine, and tangent, with a limited list of identities, appear. The general chain rule is introduced for theoretical purposes, but the student is asked to use it in differentiation technique only in the forms $Df(cx)$ and Du^n. The methods of integration are less extensive than normally required in year courses. The problems are, on the whole, technically easy.

Therefore Part I can provide a one-semester calculus for students who will transfer in their second semester to a course in probability and statistics, which can use calculus as a prerequisite. It does not require that these students be identified in advance and this is not a "terminal course." If these students later have reason to study more calculus, they can continue with Part II.

The arrangement favors advanced placement, as mentioned in the preface.

6. *Uses of Part II.* The prerequisites of Part II, aside from Part I or its equivalent in any standard text, include no more secondary school mathematics than Part I requires. However, the level of technical skill required is considerably greater. Part II begins with a review and extension of the techniques of differentiation and integration, and brings them up to standard level. This technique-building material may be reduced, if desired, for it is not needed very much in the remainder of Part II. Alternatively, it may be expanded, by including Sections 5, 6, 7 from Part III.

The main body of Part II is a course in multivariable calculus, including a brief and elementary treatment of multiple integrals, II, §8–22 and §29–33.

Within a 3-semester-hour course one must choose between the sections on differential equations, II, §23–28, and the sections on polar coordinates and infinite series, II, §34–39. Polar coordinates should be included in longer courses of 8 or 9 semester hours, since much of Part III depends on it.

7. *Uses of Part III.* Either for year courses of 8 semester hours with technical emphasis and applications appropriate for physical science and engineering, or for three semester courses of 3 hours each, Part III provides a variety of selections. More analytic geometry is found here. The linear algebra chapters may be omitted without disturbing the prerequisites for the remainder of the book.

8. *Quarter courses.* The first 23 sections of Part I compose a 3-quarter hour course in the differential and integral calculus of polynomials. Then Sections I, §24 to II, §7 form another quarter course extending to the transcendental functions with appropriate techniques of integration. The final third of the *Calculus*, II, §8–33, is approximately a quarter course in multivariable calculus, with some differential equations.

9. *Comments on infinite series.* Taylor's polynomial with remainder appears rather

early in Part I. For computing the values of a repeatedly differentiable function, this gives more information than the convergent infinite Taylor's series. Hence, I regard the Taylor's theorem expansion in Part I as an adequate substitute for infinite series in a first course. For those who want infinite series proper, they appear in optional extra sections, II, §37–39.

10. *Comments on differential equations.* Differential equations were once withheld until after calculus was completed, but their immense usefulness has been established again and again. Consider their recently developed applications in biology—I, §38, II, §26, 28. Hence they belong among the selected topics for even a 6-semester-hour course in calculus. I have included about as much as one finds in the first chapter of a book on differential equations, using only naive solution techniques. The emphasis is on applications. The idea of the differential equation begins in Part I with the antidifferentiation, $DF(x) = f(x)$. Thus integration in closed form consists of solving a differential equation. In all of this, the assumption is that the student's first introduction to a rigorous theory of differential equations will occur in a later course.

11. *Mathematics for the following year.* Part III, completing the *Calculus and Analytic Geometry*, was written in recognition of the current practice in many universities of teaching 9 hours of calculus. It is my expectation that universities that use it for 3 semesters will eventually reduce their course called calculus to the 2-semester *Calculus* and teach a second year course, or other courses, presenting probability, linear algebra, numerical analysis and complex variables. Another possibility is an advanced calculus. But, if for no other reason than that of articulation with junior colleges, the unwieldy 3-semester sequences to complete a calculus course should give way to calculus for one year and something else the next.

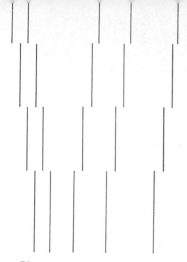

Contents

Part Three | *Analytic Geometry and Calculus*

Chapter **14**   **Rotation in Analytic Geometry**

Chapter **15**   Linear Algebra

Chapter **16**   Applications of Linear Algebra

Chapter **17**   Vector Differential Calculus

Chapter **18**   Techniques of Multiple Integration

Part One
Calculus of Elementary Functions

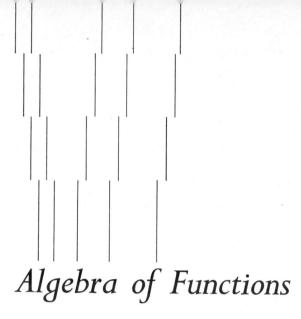

Algebra of Functions

CHAPTER 1

1 Algebra of Functions

We assume a previous knowledge of the real numbers as the numbers that fill up the real number line. They are numbers like 0, $\frac{1}{3}$, 1, 1.32, -6.47, $\pi = 3.14159\ldots$. When we say "number," we mean real number, as contrasted with natural numbers or rational numbers that do not fill up the line, and contrasted on the other hand with complex

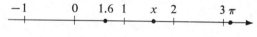

FIGURE 1.1   Real number line.

(imaginary) numbers that are not included in the real number line. A more precise description of the real numbers can be made in terms of their decimal representations (§5.1). We assume also a knowledge of relations, functions, and graphs but review the definitions (§§1.1–1.4). We recall both the definition of a function as a relation, or graph, and the idea of a function as a mapping, or rule that selects a number y when a number x is chosen. Our main objective here is to pass on to the idea of adding and multiplying functions to obtain a sum function and a product function (§§1.5–1.6).

1.1 Graph of a Relation. We recall the idea of an ordered pair of numbers, such as $(2, -1)$, in which the first number is 2 and the second number is -1. When we say that $(2, -1)$ is *ordered*, we mean that $(2, -1)$ is different from $(-1, 2)$. A set of ordered pairs of numbers, selected by any means, is a *relation*. A Cartesian coordinate system in the plane consists of a pair of number scales on straight lines intersecting in the zero point of both scales (Figure 1.2). This point is called *the origin* of coordinates. Every ordered pair of numbers (x, y) then determines a point P in the plane. Conversely, every point P determines by projection onto the axes a unique ordered pair (x, y), called the coordinates of P. In writing the name of the point, we use either a capital letter, such as P, or the ordered pair of coordinates (x, y). The *graph of a relation* consists of all points P, and only those points whose coordinate pairs (x, y) are in the relation.

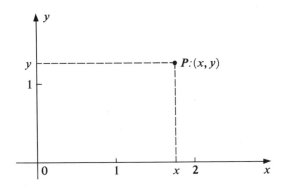

FIGURE 1.2   Cartesian coordinate system.

Some pairs of numbers may be substituted for (x, y) in an equation such as $y = 4/(1 + x^2)$ and produce a true statement. These ordered pairs are said to *satisfy* the equation. For example, this equation is satisfied by $(-1, 2)$, but not by $(2, -1)$. Thus an equation in x and y selects a set of all ordered pairs (x, y) that satisfy the equation. In this way it determines a relation. Hence, for brevity we speak of "the relation" $y = 4/(1 + x^2)$.

We describe the graph of an equation as follows.

Pivotal principle of analytic geometry. A point $P:(x, y)$ is in the graph of an equation if and only if its coordinates satisfy the equation.

We observe that the basic idea of a graph is not a curve, or a path, or a shape. The fundamental idea is that the graph is a *set of points*. When we plot the graph in a coordinate plane by inking the points whose coordinates satisfy the equation, we also say that the points left blank do not satisfy the equation. Hence, a graph both determines a relation and is determined by the relation.

1.2   Plotting Graphs

Example A.   Plot the graph of $y = \dfrac{4}{1 + x^2}$.

Solution.   We choose a few numbers for x, say $x = -5, -4, -3, -2, -1, 0, 1, 2, 3, 4, 5, \ldots$, arranging them in increasing order. We substitute each of these numbers for x in the equation and let the equation determine the number y that goes with it to form a pair (x, y) in the graph. We arrange these pairs in a table, and plot each tabulated point, circling it in the graph. Then, assuming that the coordinates of the other points on the graph vary continuously between these plotted ones, we draw in the remainder of the graph as a curve (Figure 1.3(a)).

Example B.   Plot the graph of $t^2 - u^2 = 1$.

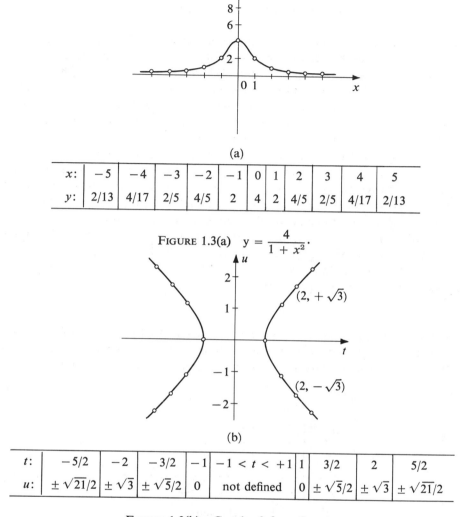

(a)

x:	-5	-4	-3	-2	-1	0	1	2	3	4	5
y:	2/13	4/17	2/5	4/5	2	4	2	4/5	2/5	4/17	2/13

FIGURE 1.3(a) $y = \dfrac{4}{1 + x^2}$.

(b)

t:	$-5/2$	-2	$-3/2$	-1	$-1 < t < +1$	1	3/2	2	5/2
u:	$\pm \sqrt{21}/2$	$\pm \sqrt{3}$	$\pm \sqrt{5}/2$	0	not defined	0	$\pm \sqrt{5}/2$	$\pm \sqrt{3}$	$\pm \sqrt{21}/2$

FIGURE 1.3(b)   Graph of $t^2 - u^2 = 1$.

Solution.   We construct the graph of $t^2 - u^2 = 1$ in similar fashion (Figure 1.3(b)). Here we encounter the difficulty that, since $u^2 = t^2 - 1$, and since there is no real number whose square is negative, there is no pair (t, u) in the relation for which $-1 < t < 1$. This causes the graph to fall into two separated pieces.

1.3   Idea of a Function as a Special Kind of Relation.   There is an important distinction between the two graphs above (Figures 1.3(a) and 1.3(b)). Each graph is a set of points $P : (x, y)$ in the plane but the equation $y = 4/(1 + x^2)$ selects one and only one number y for each number x to form a pair (x, y). No two different points have the same x-coordinates. On the other hand, the equation $t^2 - u^2 = 1$ selects two numbers u for each t greater than 1, $u = +\sqrt{t^2 - 1}$ and $u = -\sqrt{t^2 - 1}$. For $t = 2$ we have two points in the graph, $(2, +3)$ and $(2, -3)$. We call the relation $y = 4/(1 + x^2)$ a *function*, while the relation $t^2 - u^2 = 1$ is not a function. Let us state the formal definition.

DEFINITION.   *A function of x.*   A relation f is a *function* if and only if no two different ordered pairs in (x, y) in f have the same first number x.

For each function f there is a set X of all numbers x that are the first elements of some pair (x, y) in f; this set X is called the *domain* of f. Usually we specify the domain in advance but do not attempt to specify precisely the set of numbers y in the function f.
We only specify a larger set, called the codomain, which includes all of the values y of the function; usually the codomain is the set Y of all real numbers. In describing a function, we often say, "A function f on X to Y." For example, the function $4/(1 + x^2)$ is on the set of all real numbers to the real numbers.
The name of the function as a whole, that is, a set of pairs of numbers, is written in boldface type. According to context, we shall use either boldface f or $f(x)$ to represent the function. The symbol $f(x)$ in ordinary italic type means the number $f(x)$ that the function f, or $f(x)$, assigns to the number x. In each pair, $(x, f(x))$, $f(x)$ is called *the value* of f at x. If we wish to make this distinction in handwriting, we may underline the symbol, or whole formula, with a wavy line; thus f, or $f(x)$, or $x^2 + 3x - 2$, denotes a function, while $f(x)$ or $x^2 + 3x - 2$ denotes a number that is determined when x is specified. While we must always *be able* to make this distinction, we do not always insist upon making it in writing (or printing).

1.4   Idea of a Mapping.   For our purposes it is most useful to think of a function f as a rule that, given a number x, then selects a number $f(x)$ to go with it. The graphical procedure is as follows (Figure 1.4). Given any function, we pick a number x in the

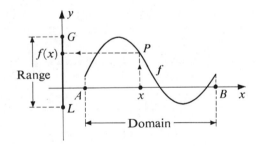

FIGURE 1.4   A function as a mapping.

domain of f. We plot x on the x-axis. Then we proceed from the point x along a line parallel to the y-axis until we intersect the graph of f at P. Then from P we proceed along a line parallel to the x-axis until we intersect the y-axis at a number that is designated by $f(x)$. We say that the function f *maps* the number x on the x-axis onto its *image*, $f(x)$, on the y-axis. The set of all image points y, $y = f(x)$, is called the *range* of f. It is contained in the codomain.
Let us look at the same idea in algebraic interpretation. Suppose that the function is given by a formula such as $f(x) = x^2 - 3x + 1$. Suppose we pick $x = 2$; then this function selects the number $f(2) = (2)^2 - 3(2) + 1 = -1$ as the image of 2, that is, $f(x)$ maps 2 onto -1. Or if $x = b$, this function maps b onto $b^2 - 3b + 1$. Thus this mapping "sends" 2 to -1 and "sends" b to $b^2 - 3b + 1$. In general, f sends its domain *into* its codomain.

Finally, we say that a function f maps its domain set *onto* its range and that the range is the map of the domain. Thus in Figure 1.4 the function maps the interval $[A, B]$ onto the interval $[L, G]$ on the y-axis, and $[L, G]$ is the map of $[A, B]$ under the mapping $f(x)$. Or the function $f(x) = +\sqrt{1 - x^2}$ maps the interval $[-1, 1]$ onto the interval $[0, 1]$.

1.5   Algebra of Functions.   The arithmetic of real numbers is concerned with means of naming particular numbers, as by decimals, and of performing the operations of addition, subtraction, multiplication, and division between specifically named numbers. After that, we recognize that there are many properties of these operations that are shared by all numbers. A suitable language for expressing these general properties of numbers depends on the introduction of a letter symbol, such as x, to stand for any (unspecified) number. In the language of algebra, x is an indefinite pronoun, or more specifically a pronumeral, since x always stands for the name of a number. Thus it is the symbol x as an *unknown*, or *literal number*, or *pronumeral*, or *variable*, or *indeterminate* that enables us to carry over the arithmetic operations, $+$, $-$, $\times$, $\div$, to unspecified numbers and effects the transition from the arithmetic of numbers to the algebra of numbers.*

Turning from numbers to functions, we observe that the formula, or graph, that determines a particular function is like the arithmetic representation of a particular number. But once we have introduced symbols, f, g, ..., for unspecified functions we can proceed to an algebra of functions by defining a sum, $f + g$, a scalar multiplication of f by numbers c, and a multiplication of functions, fg.

DEFINITION (Algebraic operations on functions).   Let f and g be two unspecified functions with the same domain. Then we define functions that are formed by *algebraic operations* as follows.

The value of the *sum $f + g$* at x is $f(x) + g(x)$;
The value of the *scalar multiple cf* at x is $c\, f(x)$;
The value of the *product fg* at x is $f(x) \cdot g(x)$.

In general, a mathematical system with these three operations is called a *linear ring* or an *algebra*.

1.6   Polynomial Algebra.   As our first example, we consider the familiar algebra of polynomial functions, that is, functions like $2x^2 - 3x + 7$. To define polynomials, we start with the special functions **1** and x. The constant function, **1**, is the relation that consists of the pairs $(x, 1)$ for all real numbers x (Figure 1.5). The identity function x is the relation consisting of all pairs (x, x), that is, the relation $y = x$. We recognize also the zero function, **0**, which is the relation consisting of all pairs $(x, 0)$. As a mapping, it sends every number x onto the number 0. The functions **1** and **0** are *constants*. A constant c is defined as a function consisting of all pairs (x, c). That is, the constant c sends every number x to the number c.

---

* The introduction of the essential zero element, the decimal system, and the idea of an algebra of numbers are a great contribution of the Hindu and Arabic civilizations to our own. Their number symbols, calculation schemes, and algebra began to enter the Western world about the seventh century A.D. Even the words *zero*, *cipher*, *algebra*, and *algorithm* (meaning a formal calculation procedure) are from the Arabic.

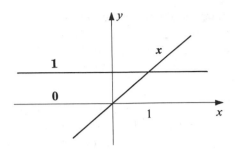

FIGURE 1.5 Zero function, one function, and identity function.

Now we consider the set of all functions we can form from **1** and *x* by a finite sequence of the three linear ring operations. The set of functions so obtained is defined to be the set of polynomials. Thus **1** and *x* are themselves polynomials; **1** + *x* is a polynomial, x^2 is a polynomial, and $4x^3 - 7x^2 + 3x - 2$ is a polynomial.* The zero function, **0**, is a polynomial.

ACADEMIC NOTE. Appendix A provides a review of polynomial algebra, remainder theorem, and synthetic division, with exercises. If such a review is appropriate, the natural place for it is here, before Section 2.

1.7 Exercises

In Exercises 1–6, plot the graphs of the relations given.

1. $y = 2x + 1$.
2. $3y - 2x = 5$.
3. $y^2 = x^2$.
4. $x = 4y^2 + 1$.
5. $x^2 - y + 1 = 0$.
6. $y = x^3 - 5x + 1$.

7. (a) Which of the relations in Exercises 1–6 are functions of *x*?
    (b) For each relation that is a function, find the domain and range.
8. Which of the relations in Exercises 1–6 are functions of *y*?
9. Given $f(x) = 2x + 3$, find $f(2)$, $f(-1)$, $f(-\frac{3}{2})$, $f(0)$, $f(a)$.
10. What is the image of each of the numbers 1, 3, 0, *t* under the mapping $f(x) = x - 1$?
11. Find the numbers for which $f(x) = 0$ if $f(x) = (x + 3)x$.
12. If $f(x) = 2x + 1$ and $g(x) = 3x^2 - 5$, calculate as polynomials, $f + g$, $6f$, and fg.
13. If $f(t) = 1 + t$ and $g(t) = 1 - t$, calculate $f + g$ and fg. Plot the graphs of f, g, $f + g$, fg in the same coordinate plane.
14. For the function $f(x) = (x + 1)(x)(x - 2)$, we observe that the first factor changes sign when *x* increases through -1, the second at $x = 0$, and the third at $x = 2$. We tabulate

	$x < -1$	$x = -1$	$-1 < x < 0$	$x = 0$	$0 < x < 2$	$x = 2$	$2 < x$
$x + 1$	$-$	0	$+$	$+$	$+$	$+$	$+$
x	$-$	$-$	$-$	0	$+$	$+$	$+$
$x - 2$	$-$	$-$	$-$	$-$	$-$	0	$+$
$(x + 1)(x)(x - 2)$	$-$	0	$+$	0	$-$	0	$+$

* In many high school textbooks, x^2 would be called a *monomial*, **1** + *x*, a *binomial*, and the word *polynomial* would be reserved for a sum of several terms. Although this distinction has a historical basis, we find in modern mathematics that the distinction is undesirable, and we make the word *polynomial* cover expressions with a single power term, two power terms, or many power terms.

in the accompanying table the algebraic signs of the factors in the intervals of the x-axis separated by these zeros of the factors. Then in the last line we calculate the sign of $f(x)$ in each interval (column) by the algebraic rule of signs in multiplication. Use this information to sketch the graph of $y = f(x)$.

15. As in Exercise 14, make a table of signs for the function $f(x) = (x + 2)(x - 1)$. Use this information to sketch the graph. At what x is the value of $f(x)$ a minimum?

16. As in Exercise 14, make a table of signs for the function $f(x) = x^2 - x - 2$. Use this information to sketch the graph [First factor].

17. As in Exercise 14, make a table of signs for the function $f(x) = (x + 1)(x)(x - 1) \times (x - 2)$. Use this information to sketch the graph of $f(x)$.

18. Consider that x is increasing through the real number line which is the domain of the function $f(x) = (2x + 3)(x + 1)(x - 1)(x - 2)$. At what points does the sign of $f(x)$ (a) change from negative to positive? (b) change from positive to negative?

19. Show that all constant functions c are polynomials.

20. Is the quotient $(x^2 - 4)/(x + 2)$ a polynomial?

21. Is $\sqrt{1 + x}$ a polynomial? Prove your statement.

22. If y is a function of x, is it true that x is a function of y? Consider $y = x^2$ on the real numbers x.

23. If $fg = 0$, is it true that either $f = 0$ or $g = 0$, as in the algebra of numbers? Consider the functions f and g defined on all real numbers x by

$$f(x) = 0 \text{ if } x \text{ is negative and } f(x) = x, \text{ if } x \text{ is positive or zero};$$

$$g(x) = -x \text{ if } x \text{ is negative and } g(x) = 0, \text{ if } x \text{ is positive or zero.}$$

Graph these functions. Compute the product fg. Is either $f = 0$ or $g = 0$?

2 Idea of an Unknown Function

2.1 Solving Equations for Unknown Functions. Just as the algebra of numbers enables us to solve an equation for an unknown number x, the algebra of functions enables us to solve an equation for an unknown function f. This idea represents a major intellectual step for the student of mathematics. It is an idea that is difficult to grasp, but failure to comprehend it bars the way to some of the most potent applications of calculus. So without attempting to develop a theory of functional equations, that is, equations in which the unknown is a function, let us study a variety of simple examples.

We will begin with an example that is like the numerical equation $2x + 5 = 0$. To solve this equation for the unknown number x, we add -5 to both members and multiply both members by $\frac{1}{2}$. This gives us $2x + 5 = 0$ if and only if $x = -\frac{5}{2}$.

Example A. Find the function $f(x)$ that satisfies the equation $2f + 5x^2 - 1 = 0$.

Solution. We add $-5x^2 + 1$ to both members and multiply both members by $\frac{1}{2}$, which tells us that the unknown function $f = -\frac{5}{2}x^2 + \frac{1}{2}1$.

Example B. Find a function f that satisfies the equation $(2x + 1)f - 2x + 4 = 0$. Is the unknown function a polynomial? Plot its graph.

Solution. As in Example A, we may add $2x^2 - 4$ to both members to find the equivalent equation $(2x + 1)f = 2x^2 - 4$. However, the polynomial $2x^2 - 4$ does not divide by $2x + 1$ in polynomial algebra. We therefore cannot find a solution that is a polynomial. We can proceed to find a solution by the device of going into the real number scale and using the algebra of numbers. The two functions are equal, $(2x + 1)f = 2x^2 - 4$, if and only if for every number x, $(2x + 1)f(x) = 2x^2 - 4$. Now if the number $2x + 1 \neq 0$, we may divide both members by $2x + 1$ to find that the value $f(x)$ is given by

$$f(x) = \frac{2x^2 - 4}{2x + 1} \qquad \text{when} \quad x \neq -\frac{1}{2}.$$

We plot the graph of this function in Figure 1.6, noting especially the behavior of the function near the point where $x = -\frac{1}{2}$, where the function is undefined because the divisor is zero there. The domain of the function is all real numbers except $x = -\frac{1}{2}$.

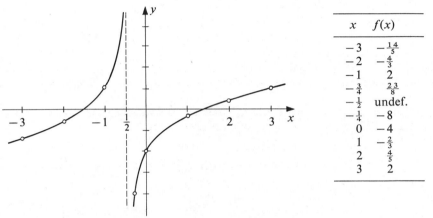

x	$f(x)$
-3	$-\frac{14}{5}$
-2	$-\frac{4}{3}$
-1	2
$-\frac{3}{4}$	$\frac{23}{8}$
$-\frac{1}{2}$	undef.
$-\frac{1}{4}$	-8
0	-4
1	$-\frac{2}{3}$
2	$\frac{4}{5}$
3	2

FIGURE 1.6   The function $\dfrac{2x^2 - 4}{2x + 1}$.

Example C. Find the quadratic polynomial function $f(x) = ax^2 + bx + c$ whose graph contains the three points $(0, 1)$, $(2, 3)$, $(-1, 5)$.

Solution. Here we are selecting from a restricted class of functions, all of which are polynomials of degree 2, one of which satisfies three given conditions. Our solution takes the form of applying "the method of undetermined coefficients."

Since $f(0) = 1$, we know that $1 = 0 \cdot a + 0 \cdot b + c$.
Since $f(2) = 3$, we know that $3 = 4a + 2b + c$.
Since $f(-1) = 5$, we know that $5 = 25a + 5b + c$.

This gives us three equations to be solved simultaneously for a, b, c. However, the first equation gives us directly that $c = 1$, and thus the remaining two equations reduce to

$$\begin{cases} 4a + 2b = 2 & \text{and} \\ 25a + 5b = 4. \end{cases}$$

We proceed by the method of elimination, using the sign $\Leftrightarrow$ to mean "if and only if."

$$\begin{cases} 4a + 2b = 2 & \text{and} \\ 25a + 5b = 4 \end{cases} \Leftrightarrow \begin{cases} 20a + 10b = 10 & \text{and} \\ 50a + 10b = 8 \end{cases} \Leftrightarrow$$

$$\begin{cases} 20a + 10b = 10 & \text{and} \\ 30a = -2 \end{cases} \Leftrightarrow \begin{cases} 60a + 30b = 30 & \text{and} \\ 60a = -4 \end{cases} \Leftrightarrow$$

$$\begin{cases} 30b = 34 & \text{and} \\ 60a = -4 \end{cases} \Leftrightarrow \begin{cases} b = 17/15 & \text{and} \\ a = -1/15 \end{cases}.$$

Hence we have $a = -\frac{1}{15}$, $b = \frac{17}{15}$, and $c = 1$. There the unknown function, $f(x) = -\frac{1}{15}x^2 + \frac{17}{15}x + 1$.

Example D.   Suppose we operate a rooming house of 24 rooms and that the economic conditions are such that the net income is a function of the rental price. We estimate that we could keep all of the rooms rented at \$40 per month each and that we would have one vacancy for each \$5 per month added to the rental; but we would save \$8 per month in maintenance costs for each room left vacant. Find a formula for the function **R**, which represents the net income.

   Solution.   Let r represent the monthly room rental. Then for all rentals for which $r \geq 40$ the number of vacancies is $(r - 40)/5$. Hence the number n of rooms rented is given by $n = 24 - [(r - 40)/5]$. The gross income from the n rooms rented is rn. The total cost of maintaining 24 rooms is not given, but we may assume that there is some fixed cost C for maintaining 24 rooms, so that the total cost is $C - 8[(r - 40)/5]$. The net income from operating the rooming house at a rent r for which $r \geq 40$ is given by

$$R(r) = rn - C + 8\left(\frac{r - 40}{5}\right)$$

$$= r\left(24 - \frac{r - 40}{5}\right) + 8\left(\frac{r - 40}{5}\right) - C.$$

If we combine terms and simplify, we reduce this result to the function

$$R(r) = \frac{-r^2 + 168r - 320}{5} - C$$

for the required net income function on the domain of all numbers r for which $r \geq 40$. We leave to the exercises the sketching of the graph of this function.

2.2   Proportion. In experimental science it is often helpful to describe an unknown function partially by saying, "y is directly proportional to x." This means that for some constant k, $y = kx$. The unknown function of x is then of the form $f = kx$. The use of the word *proportional* is justified by the following reasoning. If (x_1, y_1) and (x_2, y_2) are in the relation $y = kx$, then $y_1 = kx_1$ and $y_2 = kx_2$, so that $k = y_1/x_1$ and $k = y_2/x_2$. Since k is a constant, this implies the proportion $y_1/x_1 = y_2/x_2$. We prefer the formulation $y = kx$, however, to the formulation of the relation as a proportion because the statement $y = kx$ is simpler, more general, and more flexible.

   We also say that "y varies directly as the pth power of x" if $y = kx^p$. We say that "y varies inversely as the qth power of x" if $y = k(1/x^q)$, and we say that "y varies directly as the pth power of x and inversely as the qth power of r" if $y = k(x^p/r^q)$.

Example E.   An electric charge e is attracted to a pole with a force F that is directly proportional to the charge and inversely proportional to the square of the distance r of the charge from the pole. When the charge is 1 coulomb and the distance is 10 cm, the force of attraction is 180 dynes. Find the function describing the attraction.

   Solution.   The statement reads $F = k(e/r^2)$, where k is a constant and F is an unknown function. When $e = 1$ and $r = 10$ then $F = -180$, where the minus sign signifies that the force is directed back toward the point where $r = 0$. Hence we have at this point

$-180 = K(\frac{1}{10})^2$, which implies that $K = -18{,}000 = -1.8(10^4)$. Hence the unknown force function is $F(e, r) = -(1.8)10^4 e/r^2$.

2.3 Difficulties with Division in Function Algebra.

We have encountered some difficulties in function algebra. One is that division $f \div g$ is not always possible even when g is not the zero function. For example, we found (Example B) that the quotient

$$\frac{2x^2 + 1}{2x + 1}$$

is not in the polynomial algebra and that, even when we compute it by numerical division, we must exclude the number where $x = -\frac{1}{2}$ from the domain. So when we divide by a function, we must always pay close attention to the points where the denominator is zero. Unless the numerator is also zero at the same point, we shall find that as x approaches this point, the quotient becomes infinite (in a positive or negative direction depending on the signs of numerator and denominator). See also Section 1.7, Exercise 23.

There is another thing to be noted. The function x is not an unknown function. It is the perfectly definite identity function whose points are $(\dot{x}, x)$. As an equation involving the function x, the statement $2x + 5 = 0$ is simply false. It does not have the "solution" $x = -\frac{5}{2}$. But the equation $2f + 5 = 0$ has the solution $f(x) = -\frac{5}{2}$.

2.4 Recapitulation of Interpretations and Notations for Functions.

The simplest symbol for a function is the single boldface letter f or g. We consider it as the name of the mapping that sends each number x in its domain X to a number y in the codomain Y. It also serves as a variable in algebraic operations involving functions. Single but important functions so denoted are the constant function 1 and the identity function x. When we wish to indicate symbolically the whole system, the domain set X, the codomain Y, and the mapping, we may write $f: X \to Y$.

When functions are built up from the identity function x, we can indicate this by the boldface symbol $f(x)$, as in $f(x) = x^2 - 5x + 4$. When the function x is evaluated at the number x, its value is x. Correspondingly when x is evaluated at x, the value of the function $f(x)$ is the number $f(x)$. The more complicated notation $f(x)$ saves a place to substitute something for x, but means the same as f.

Our first concept of a function as a single-valued relation, or graph, in numbered pairs (x, y) may be indicated by the equation $y = f(x)$, where both sides are numerical variables. This has the advantage of keeping both "independent variable" x and "dependent variable" y in the symbol for the function. We often need this in using functions to describe relations between physical quantities.

We do not try to make one single notation or way of thinking of a function serve all our purposes but we use all of these logically equivalent ones as the context suggests.

2.5 Exercises

In Exercises 1–6, solve for the unknown function $f(x)$. Define its domain carefully, state in each case whether $f(x)$ is a polynomial or not, and plot the graph of $f(x)$.

1. $3f - x^2 + 1 = 0.$ 2. $(x - 1)f - x^2 + 1 = 0.$
3. $xf + x - 1 = 0.$ 4. $xf^2 + (x - 1)f(x) + 4 = 0.$
5. $f^2 + 4 = 0.$ 6. $xf + 10 = 0.$

7. A spring has a force proportional to the distance x that the end is displaced from equilibrium. The force is 10 lb when $x = 5$ in. Express the force as a function of x. Find the force when $x = 12$.

8. Find the second-degree polynomial function $f(x)$ that satisfies the conditions $f(-1) = 4$, $f(0) = 0$, $f(2) = 3$.

9. Find the second-degree polynomial function $f(x)$ that satisfies the conditions $f(-2) = -3$, $f(1) = 8$, $f(4) = -6$. Find $f(3)$.

10. Find the polynomial function $f(x)$ of degree 3 so that $f(0) = 1$, $f'(0) = 0$, $f^2(0) = 0$, $f^3(0) = 6$. Find $f(4)$.

11. A rectangular box to contain 108 cu in. is to be made with a square base. Express the total area of top, sides, and bottom as a function of the length x of a side of the base. State explicitly the domain of the function.

12. A radiator holds 16 qt and is filled with a solution that is 25 percent alcohol. We drain out x qt and replace this by pure alcohol. Express the concentration of the resulting solution as a function of x.

13. Originally automobiles made a 160-mile trip in 8 hours. If the velocity is increased by x mi/hr express the time saved over the original trip as a function of x. Plot the graph. What is the time saved when the speed is increased from 40 mi/hr to 80 mi/hr?

14. A man 6 ft tall walks directly away from a lamp post 12 ft high on which there is a light. Express the length S of his shadow as a function of his distance x from the post.

15. A cylindrical can is to contain 600 cu in. The material for the top costs 3 cents per sq in. and that of the rest costs 1 cent per sq in. Express the cost as a function of the radius.

16. In Example D find a formula for the net income that is valid when $0 \leq r < 40$.

17. Plot the graph of the net income function R in Example D. Estimate from this graph the rental r that gives maximum net income.

18. Newton's law of gravitation asserts that the force of attraction of two masses M and m is proportional to the product of the masses and inversely proportional to the square of the distance r between them. Express this force F by a formula as a function of r, regarding M and m as fixed constants.

19. In Exercise 18, find the explicit formula for F if $M = 2$, $m = 1$, $F(1) = -\frac{1}{10}$.

20. In cgs units the coefficient k in the law of gravitation $F(r) = -k(Mm/r^2)$ has the value 1. Find the mass of the earth from the fact that a mass of 1 gram (g) at the surface of the earth is attracted by the earth with a force of 980 dynes. Take the radius of the earth to be $6.44(10^8)$ cm.

21. The area of the surface of a sphere is proportional to the square of the radius r. If the area is π when $r = \frac{1}{2}$, express the area function $A(r)$ as a formula.

22. Find a formula for the function of r that expresses the ratio of one unit area of the surface of a sphere to the entire surface of the sphere. This is the basic formulation of the *inverse square law*.

3   Integral of a Step Function

3.1   Step Functions.   Functions whose graphs are made up of pieces of straight lines are useful in approximating more complicated functions whose graphs are curved lines. Our first example is a step function s defined on the interval $[2, 12]$. We do not attempt to give an analytic formula for s, but we define it by giving its value at every

x	$2 \leq x < 5$	$5 \leq x < 7$	$7 \leq x < 10$	$10 \leq x \leq 12$
$s(x)$	3	4	1	-1

point in the accompanying table. In words this says that on the interval from 2 to 5 (excluding 5) the value of *s* is the constant 3; on the interval from 5 to 7 (excluding 7) the value of *s* is the constant 4, etc. We plot the graph from these data (Figure 1.7).

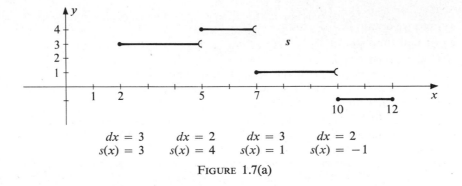

$$dx = 3 \qquad dx = 2 \qquad dx = 3 \qquad dx = 2$$
$$s(x) = 3 \qquad s(x) = 4 \qquad s(x) = 1 \qquad s(x) = -1$$

FIGURE 1.7(a)

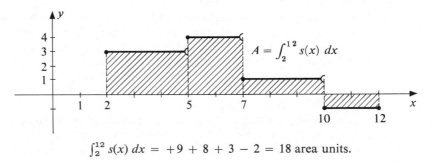

$$A = \int_2^{12} s(x)\, dx$$

$\int_2^{12} s(x)\, dx = +9 + 8 + 3 - 2 = 18$ area units.

FIGURE 1.7(b) Step function and its integral. The half circled ends (———⊂) indicate that the endpoint is not included in this segment.

This is a function on the domain [2, 12]; for at every number *x* in this interval the table assigns a number $s(x)$. The partition points 5, 7, 10 must be allocated to one or the other of the two adjacent subintervals there to avoid giving *s* two values at these points. If this occurred, the table would not properly define a function. We choose to put each partition point in the interval above it, not in the interval below. Thus $s(x) = 4$ when $5 \leq x < 7$ but $s(7) = 1$.

We define a step function in general in the same way; that is, to define any step function *s* on an interval [*a*, *b*], we introduce a finite sequence of partition points, $x_1, x_2, \ldots, x_n$, so that $a < x_1 < x_2 < \cdots < x_n < b$. Then we assign a fixed number as the value of *s* in each of the subintervals $a \leq x < x_1, x_1 \leq x < x_2, \ldots, x_n \leq x \leq b$.

Step functions occur in numerous applications. For example, the postage rate scale is a step function, the income tax rate is a step function. Electric power rates, life insurance rates, and freight rates are usually given as step functions. In quantum physics, quantities like energy often vary as step functions, having discrete energy levels.

3.2 Integral of a Step Function. *Area interpretation.* Let us represent by *dx* (difference in *x*) the length of one of the subintervals in a partition of the interval [*a*, *b*]. For example, if we consider the step function *s* we defined above (Figure 1.7(a)), we see that in the first subinterval $2 \leq x < 5$, $dx = 5 - 2 = 3$. In the second subinterval where $5 \leqq x < 7$, $dx = 7 - 5 = 2$. In the third subinterval, $dx = 10 - 7 = 3$. In the last subinterval, $dx = 12 - 10 = 2$. Now, if for each subinterval we compute $s(x)\, dx$, it will be the area of the rectangle having base of length dx and height $s(x)$. We compute these numbers for each of the subintervals (Figure 1.7(b)). The numbers $s(x)\, dx$ represent the areas of the shaded rectangles in the graph. Finally we add these numbers, $s(x)\, dx$, and we obtain the algebraic sum A of the areas of the rectangles bounded by the graph of s and the x-axis between the limits $x = 2$ and $x = 12$. We write this

$$A = \int_2^{12} s(x)\, dx = 9 + 8 + 3 - 2 = 18 \text{ area units.}$$

Here the symbol $\int$ is the integral sign. It stands for a sum. (It is derived from the old-style long s of early printing.) The numbers 2 below and 12 above are the *limits of the integral* and they indicate that the sum extends over the interval from 2 to 12. After the integral sign we have written $s(x)\, dx$, which indicates that it is a sum of terms of this type. Thus $\int_2^{12} s(x)\, dx$ means a sum of terms of the type $s(x)\, dx$ extending from $x = 2$ to $x = 12$. It represents the net total area of the region underneath the graph of s (shaded in Figure 1.7(b)), where we take account of the negative area below the x-axis.

In general, if s is any step function on an interval $[a, b]$ the integral

$$\int_a^b s(x)\, dx$$

is a number determined by the function s and the two limits a and b. It represents the algebraic sum of the areas of rectangles with bases $[a, x_1[, [x_1, x_2[, [x_2, x_3[, \ldots, [x_n, b]$ and heights $s(x)$ for each of these subintervals. The reversed brackets ([) indicate that the preceding endpoint is not included.

3.3 Integral of a Speed Function.

Suppose we drive a car at the fixed speed of 40 mi/hr for three hours from 2 o'clock to 5 o'clock. Let us first graph this speed function $s(t)$ on the time interval $[2, 5]$ (Figure 1.8). It is a step function with only one step.

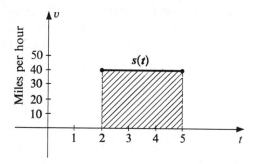

FIGURE 1.8   Single-step speed function.

What does the number $\int_2^5 s(t)\, dt$ represent? Here there is only one term in the sum. For this term, $dt = 5 - 2 = 3$ hr and $s(t) = 40$ mi/hr, so that $s(t)\, dt = (40 \text{ mi/hr}) (3 \text{ hr}) = 120$ mi. It is the distance given by rate $\times$ time that we travel during the interval of time $t = 2$ to $t = 5$. We can still picture it graphically as an area (Figure 1.8), but it is distance in this problem.

Example.   A man started out on a 50-mile hike, getting an early start at 2 A.M. He walked at a steady 3 mi/hr until 5 A.M. Then he speeded up to 4 mi/hr and held this speed until 7 A.M. At 7 he developed a blister and slowed to 1 mi/hr until 10 A.M. Then at 10 he turned back and proceeded at −1 mi/hr until noon. How far was he from home at noon?

Solution.   His speed at any time *t* in the hike is given by a step function *s*(*t*), whose graph is mathematically the same as that of Figure 1.7(a). Physically the axes are *t* (time) and *v* (velocity) scales instead of length scales. The integral

$$L = \int_2^{12} s(t)\, dt = (3)(3) + 4(2) + 1(3) + (-1)(2) = 18 \text{ mi.}$$

It now represents the algebraic sum of the distances he traveled in each of his four different speed intervals. Thus it says that at the end he was 18 mi from home (although he has walked a total distance of 22 mi).

3.4 The Integral in Averages. To compute the average of a finite sequence of numbers, such as 1, 3, 1, 2, 3, −1, 2, we first count the sequence to find that there are 7 numbers in it, then we add the numbers and divide by 7. In this case we get for the average value $\bar{x}$,

$$\bar{x} = \frac{1 + 3 + 1 + 2 + 3 + (-1) + 2}{7} = \frac{11}{7}.$$

When we have a function *f* on an interval [*a*, *b*], this procedure is impossible because we cannot count the infinitely many numbers *x* in the interval [*a*, *b*] to obtain the denominator above. However, for step functions *s* we can get by this difficulty by the aid of the integral and define an average for step functions. Let *s* be a step function on the interval [*a*, *b*]. We define the average $\bar{s}$ as the value of a constant function on [*a*, *b*] whose integral is the same as that of *s*,

$$\int_b^b \bar{s}\, dx = \int_a^b s(x)\, dx.$$

Since for the constant function $\bar{s}$ the integral is the area $\bar{s}(b - a)$ of a rectangle of height $\bar{s}$ and base (*s* − *a*), this leads us to the following.

DEFINITION.   The *average value* $\bar{s}$ of the step function *s* on [*a*, *b*] is given by

$$\bar{s} = \frac{1}{b - a} \int_a^b s(x)\, dx.$$

We observe that the physical units of $\bar{s}$ are those of *s* and not those of *s*(*x*) *dx*.

Example.   Find the mean value of the height of a group of men whose heights are ranked according to the accompanying table:

$5'4'' \leqq h < 5'6''$	$5'6'' \leqq h < 5'8''$	$5'8'' \leqq h < 5'10''$	$5'10'' \leqq h < 6'0''$	$6'0'' \leqq h \leqq 6'2''$
12%	24%	36%	18%	10%

Solution.   We construct a step function *h* representing the height of the man whose percentile rank is *q*. (A man's percentile rank with respect to height is *q* if his height

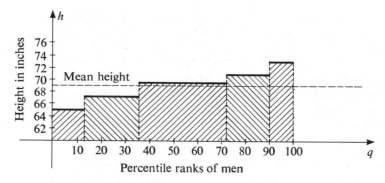

FIGURE 1.9   Step function for heights of men.

equals or exceeds q percent of the men in the group.) Since we know nothing about the distribution of heights of the men in each group, we approximate their heights by a single number in the middle of the heights for the group. For example, for the shortest group, we assign the height $h(q) = 5'5''$ to the percentile ranks up to 12. We represent this approximating step function in Figure 1.9. Now we compute the sum of the areas of the region below these steps. This leads us to make the computation as in the accompanying table.

Interval	$64'' \leqq h < 66''$	$66'' \leqq h < 68''$	$68'' \leqq h < 70''$	$70'' \leqq h < 72''$	$72'' \leqq h \leqq 74''$
$h(q)$	65	67	69	71	73
dq	0.12	0.24	0.36	0.18	0.10
$h(q)\,dq$	7.80	16.08	24.84	12.78	7.30
$\int_0^{1.00} h(q)\,dq$					68.80

Now we find the average height $\bar{h}$ from

$$\bar{h} = \frac{\int_0^{1.00} h(q)\,dq}{1.00 - 0} = \frac{68.8}{1.00} = 68.80 \text{ in.}$$

3.5 Exercises

1. A step function s is defined on [3, 15] by the accompanying table.

x	$3 \leqq x < 5$	$5 \leqq x < 8$	$8 \leqq x < 12$	$12 \leqq x \leqq 15$
$s(x)$	1	0	-2	1

Plot the graph of s and compute the integral $\int_3^{15} s(x)\,dx$. What is the geometric interpretation of the answer?

2. A step function r represents a velocity in feet per second at time t, a point moving on the x-axis. It is given by the accompanying table.

t	$0 \leqq t < 1$	$1 \leqq t < 2$	$2 \leqq t < 3$	$3 \leqq t \leqq 4$
$r(t)$	-16	-32	$+16$	$+32$

Find, by means of an integral, the distance from its starting point to its final point at the end of 4 seconds. What was the average velocity over the interval [0, 4]?

3. A car goes 30 mi/hr for 20 min, then 40 mi/hr for the next 15 min, then 60 mi/hr for the next 40 min. Draw the graph of the step function s that represents this velocity function.
   (a) Compute the integral $\int_0^{75} s(t)\,dt$. What does it represent?
   (b) Compute the average velocity over the interval [0, 75].

4. The federal income tax rate r is a step function r of the taxable income t. For single taxpayers in 1959, the rates were in part given by the schedule herewith, where t is in thousands of dollars.

t	$0 \leq t \leq 2$	$2 < t \leq 4$	$4 < t \leq 6$	$6 < t \leq 8$	$8 < t \leq 10$
r	0.20	0.22	0.26	0.30	0.34

   (a) Graph the step function r.
   (b) Compute the sum $\int_0^{10} r(t)\,dt$. What does it represent?

5. In Exercise 4, find the average value of the tax on a \$10,000 taxable income.

6. In Exercise 4, find total tax and mean value of the tax rate on a taxable income of \$4800.

7. In a French class the final grades in grade points were given by the accompanying table. Represent these data graphically by a step function and compute the average grade.

Number of students	2	4	11	5	3
Grades	-1	0	1	2	3

8. In the study of physics it is found that the energies of atoms often do not vary continuously but occupy a few "discrete" energy levels. Suppose it is found that 40 percent of the atoms in a vessel have energy 1 unit and 60 percent have energy 1.7 units. Draw a graph showing the relation between the atoms and their energies and compute the average value of the energies of the atoms in the vessel.

9. Electric power rates in a certain town are given by the following table in dollars per kilowatt-hour (kWh):

KWh	0 to 10	10 to 50	50 to 100	100 to 500
Rate	0.10	0.05	0.03	0.02

   (a) Plot the rate schedule as a step function $s(p)$, where p is the amount of electricity consumed.
   (b) Find the average rate for a consumer who uses 250 kWh.

10. In Exercise 9, find the value of the integral $\int_0^{250} s(p)\,dp$. What is the meaning of the integral?

11. In Exercise 9, find the average rate for a consumer who uses 500 kWh.

12. Calculate the average value $\bar{s}$ of the step function s on [0, 6] given by Figure 1.10.

13. In Exercise 12, calculate averages $\bar{x}$ of x and $\bar{y}$ of y for all points in the region below the graph of s. Plot the point $(\bar{x}, \bar{y})$. It is called the *center of area* or *centroid* of the shaded figure (Figure 1.10). Ans: $\bar{x} = 97/30$, $\bar{y} = 84/30$.

14. Assume that the region below the graph in Figure 1.10 is a sheet of metal of uniform thickness. Show that if it is suspended at the point $(\bar{x}, \bar{y})$, it will balance. Hence the centroid is also the center of gravity.

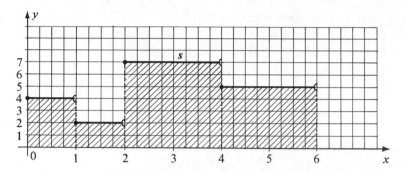

FIGURE 1.10   Step function for Exercise 12.

15. Compute the common arithmetic mean of the numbers 3, 2, 5, -1, 6, -4, 8. Now graph a step function on [0, 7] having 7 steps of heights given by these same numbers with each subinterval of length 1. Compute the average of this step function and show that it is the same as the common arithmetic mean.

16. Show that for every finite sequence of numbers there is a natural way of defining a step function s so that the average value of this step function is the same as the arithmetic mean of the numbers.

17. Prove that for every two step functions r and s,

$$\int_a^b [r(x) + s(x)] \, dx = \int_a^b r(x) \, dx + \int_a^b s(x) \, dx$$

and

$$\int_a^b c \, s(x) \, dx = c \int_a^b s(x) \, dx.$$

18. Prove that for every two step functions r and s, the average of $r + s$ is the sum of the averages of r and s.

19. Prove that for every step function s with average $\bar{s}$ and for every constant k, the average of $s + k$ is $\bar{s} + k$.

20. For the step function s in Exercise 1, define the integral $\int_3^w s(x) \, dx$, where w is any number in the interval [3, 15]. Plot as a function of w the graph of the area below the graph of s from 3 to w.

21. Plot the graph of x^2 on the interval [0, 10]. Consider the area of the region bounded by this curve and the x-axis between the limits $x = 0$ and $x = 10$. Divide the interval [0, 10] into subintervals of length 1 and set up a step function s with 10 steps whose area $\int_0^{10} s(x) \, dx$ approximates the area under the graph of x^2. Compute the area.

22. Plot the graph of x on the interval [0, 10] and as in Exercise 21 approximate it by a step function s with 10 steps of equal base. In this case show how to choose the step function so that its area $\int_0^{10} s(x) \, dx$ is *exactly* the area under the graph of x.

23. For some polynomial $f(x)$ on [0, 10] we connect the points $(0, f(0))$, $(1, f(1))$,..., $(10, f(10))$ by straight-line segments to form a polygon. Then we set up a step function $s(x)$ with 10 steps each of a height that is the average of $f(x)$ at the beginning and end of the interval. For example, in $2 \leq x \leq 3$, $s(x) = \frac{1}{2}[f(2) + f(3)]$. Show that $\int_0^{10} s(x) \, dx$ is exactly the area under the polygon and that it is $\frac{1}{2}f(0) + f(1) + f(2) + \cdots + f(9) + \frac{1}{2}f(10)$.

4 Slopes and Rates

4.1   Directed Line Segments in the Real Line.   Using the idea that the real numbers determine points on the number line, we indicate the association between the point P and the number x by the symbol $P:(x)$. The number x is the length of the directed line segment from the origin $O:(0)$ to P. This length is called the *run* from O to P, written $x = \text{run } \overrightarrow{OP}$. And $-x = \text{run } \overrightarrow{PO}$.

Now, if $P_1:(x_1)$ and $P_2:(x_2)$ are two points in the line, we find that run $\overrightarrow{P_1P_2}$ is given by

$$\text{run } \overrightarrow{P_1P_2} = x_2 - x_1.$$

For in every case (Figure 1.11(a)),

$$\overrightarrow{P_1P_2} = \overrightarrow{P_1O} + \overrightarrow{OP_2} = -\overrightarrow{OP_1} + \overrightarrow{OP_2}.$$

Since run $\overrightarrow{OP_1} = x_1$ and run $\overrightarrow{OP_2} = x_2$, this implies run $\overrightarrow{P_1P_2} = -x_1 + x_2$, as stated. We observe (Figure 1.11(b)) in some particular cases how the signs of x_1 and x_2 enter into the calculation of run $\overrightarrow{P_1P_2}$.

When we have a vertical axis, where points are associated with numbers y, then we call the directed distance from P_1 to P_2 the rise $\overrightarrow{P_1P_2}$. As for the horizontal axis,

$$\text{rise } \overrightarrow{P_1P_2} = y_2 - y_1.$$

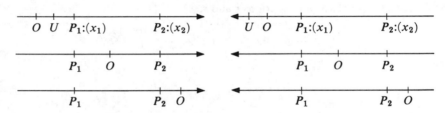

FIGURE 1.11(a)   Directed line segments (six cases).

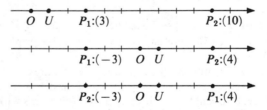

FIGURE 1.11(b)   Calculation of run in three special cases.

4.2   Slope of a Pair of Points in the Plane

DEFINITION.   The *slope* of the pair of points $P_1:(x_1, y_1)$ and $P_2:(x_2, y_2)$ with $x_1 \neq x_2$ (Figure 1.12) is given by the number m,

$$m = \frac{y_2 - y_1}{x_2 - x_1}.$$

This is easily remembered as

$$\text{slope} = \frac{\text{rise}}{\text{run}} = \frac{y_2 - y_1}{x_2 - x_1}.$$

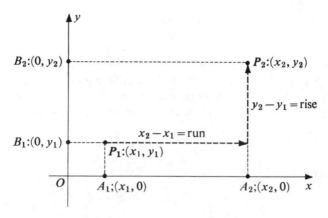

FIGURE 1.12   Slope of two points.

Example A.   Find the slopes of the following pairs of points. Plot showing the run and rise (Figures 1.13, 1.14): (a) $(-5, 2)$ and $(3, 7)$; (b) $(-4, 2)$ and $(6, -3)$.

Solution.

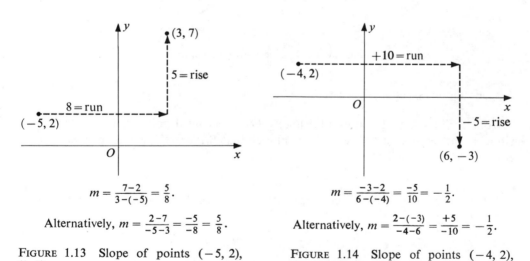

$$m = \frac{7-2}{3-(-5)} = \frac{5}{8}.$$

Alternatively, $m = \frac{2-7}{-5-3} = \frac{-5}{-8} = \frac{5}{8}.$

FIGURE 1.13   Slope of points $(-5, 2)$, $(3, 7)$.

$$m = \frac{-3-2}{6-(-4)} = \frac{-5}{10} = -\frac{1}{2}.$$

Alternatively, $m = \frac{2-(-3)}{-4-6} = \frac{+5}{-10} = -\frac{1}{2}.$

FIGURE 1.14   Slope of points $(-4, 2)$, $(6, -3)$.

Example B.   Find a point P such that the pair of points $P_1:(-3, 5)$ and P shall have slope $-4/7$.

Solution.   Let P have unknown coordinates (x, y). Then the slope of P_1P,

$$\frac{y - 5}{x - (-3)} = -\frac{4}{7}.$$

There are infinitely many points $P:(x, y)$ that satisfy this condition. *One* such point has

rise $P_1P_2 = -4$ and run $P_1P_2 = 7$. That is, $y - 5 = -4$, $y = 1$, and $x + 3 = 7$, or
$x = 4$. To plot this particular solution we start from P_1, lay off the run 7 to point Q
(Figure 1.15), then from Q lay off the rise -4 (down) to point P_2. But any other point
P located in such a way that

$$\frac{\text{rise } RP}{\text{run } P_1R} = -\frac{4}{7}$$

satisfies the condition. This is true whenever the triangle P_1RP is similar to the triangle
P_1QP_2. Hence the set of all points $P:(x, y)$ that offer a solution to this problem lie on a
line through P_1 and P_2. In fact, every point P on this line is a solution except P_1. In this
case the pair of identical points P_1P_1 has no slope because the definition yields

$$m = \frac{y_1 - y_1}{x_1 - x_1} = \frac{0}{0},$$

which is indeterminate.

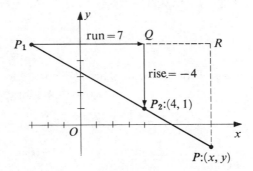

FIGURE 1.15   Line with given slope.

4.3   Equations of Lines

THEOREM (Point-slope equation of a line).   For every fixed point $P_1:(x_1, y_1)$ and fixed
slope m, the set of all points with $P:(x, y)$ such that P_1P has slope m is the line with the
equation

$$y - y_1 = m(x - x_1).$$

Proof.   If $P:(x, y)$ does not coincide with $P_1:(x_1, y_1)$, then P_1P has the required slope
m if and only if

$$\frac{y - y_1}{x - x_1} = m.$$

This equation is true and only if $y - y_1 = m(x - x_1)$. If P coincides with P_1, then this
last equation is still satisfied. This completes the proof. We call m the slope of the
line.

Not all lines have slopes. The equation $x = a$ has a graph that is a vertical straight
line, but it has no slope. Every pair of points, (x_1, y_1) and (x_2, y_2), in it has $x_1 = x_2 = a$
and hence $x_2 - x_1 = 0$. Thus the slope is undefined. If a line has a slope, it has an
equation in *slope form*

$$y = mx + b.$$

The point-slope equation with $(x_1, y_1) = (0, b)$ reduces to this.

THEOREM (Parallel lines).   Two lines that have slopes are parallel if and only if their slopes are equal.

Proof.   By definition, two lines are parallel if they do not intersect or they coincide. If the lines have equal slopes, they have equations of the form

$$y = mx + b_1, \qquad \text{and} \qquad y = mx + b_2.$$

In order that a point (x, y) be in both graphs, its coordinates must satisfy both equations. Then the equation obtained by subtracting the second equation from the first is also true. It is

$$0 = 0 + b_1 - b_2.$$

If $b_1 = b_2$, this equation is true for every (x, y) and the lines coincide. If $b_1 \neq b_2$, this equation is false for every (x, y) and hence no point $P:(x, y)$ is in both lines; so they are distinct parallel lines. Conversely, if the lines have distinct slopes m_1 and m_2, we can prove (Exercises) that they intersect in a single point.

4.4   Slope at a Point Determined by a Limit Process.

We now set out to devise a technique whereby we can give a precise meaning to the slope of a line,

$$f(x) = mx + b,$$

at a point. Let $(x, f(x))$ and $(w, f(w))$ be two distinct points in the graph of the line. Then the slope m of these two points is

$$\frac{f(w) - f(x)}{w - x} = m, \qquad w \neq x.$$

If we attempt to localize this slope at the point $(x, f(x))$ by setting $w = x$, we find that the slope ratio reduces to $0/0$, which is meaningless. To escape this difficulty and give a precise meaning to the slope at the point $(x, f(x))$, we first observe that the slope is well defined when the second point $(w, f(w))$ has $w \neq x$, however close to x we take w. We think of w as variable and let it approach x. Then we inquire whether the function

$$\frac{f(w) - f(x)}{w - x}$$

approaches some number, as w approaches x. Thus we seek to determine the slope at the point where $w = x$, in terms of the trend of slopes when $w \neq x$. Actually, the answer is trivial in this case; for the slopes of pairs of points $(x, f(x))$ and $(w, f(w))$ in the line are all equal to the same number m. Hence we just extend this constant function to have the same value when $w = x$. We describe this symbolically by saying that

$$\lim_{w \to x} \frac{f(w) - f(x)}{w - x} = m.$$

This is our *definition* of the slope at the point $(x, f(x))$. Mathematically, it means that for every prescribed positive number ϵ, however small, there is a number δ such that, when w is near x in the sense that

$$x - \delta < w < x + \delta,$$

then the slope ratio is near m, in the sense that

$$m - \epsilon < \frac{f(w) - f(x)}{w - x} < m + \epsilon.$$

This mathematical definition will become crucial when we undertake to define the slope of a curve at a point. For, in such a case, the slope ratio

$$\frac{f(w) - f(x)}{w - x}$$

is not a constant (§10).

4.5   Instantaneous Velocity in Uniform Motion.

Suppose a point moves on an s-axis with its position $s = f(t)$ given at any time t by the function

$$f(t) = vt + b,$$

where v and b are constants. This function describes a uniform motion on the s-axis, for let us calculate the rate over any interval of time from time t to time u ($u \neq t$). Since $f(u) = vu + b$ and $f(t) = vt + b$, we may subtract and divide by $u - t$. This gives us rate over the time interval $[t, u]$,

$$\frac{f(u) - f(t)}{u - t} = v.$$

Once again, as in the case of the slope of a line (§4.3), we take the limit as u approaches t in order to localize this rate over an interval into an instantaneous rate at time t. We find

$$\lim_{u \to t} \frac{f(u) - f(t)}{u - t} = v.$$

Hence we define this limit v to be the instantaneous velocity at time t. Moreover, since for the motion defined by the function $f(t) = vt + b$ this instantaneous velocity is a constant, we call this motion "a uniform motion" in a straight line.

4.6   Slopes of Ramp Functions.

A ramp function, like a step function, has a graph that consists of a finite number of straight line segments (Figure 1.16).

DEFINITION. *A ramp function $R(x)$ on $[a, b]$ is any function whose graph is constructed as follows. We partition the interval $[a, b]$ by partition points $a < x_1 < x_2 < \cdots < x_n = b$. At each of the partition points we assign values $R(a)$, $R(x_1)$, $R(x_2)$, . . ., $R(x_n) = R(b)$. Then in sequence we connect the vertices $(a, R(a))$ to $(x_1, R(x_1))$, $(x_1, R(x_1))$ to $(x_2, R(x_2))$, . . ., $(x_{n-1}, R(x_{n-1}))$ to $(b, R(b))$ by straight-line segments. This defines $R(x)$ at every point in $[a, b]$.*

In other words, we pick a finite number of vertices (corners) over the interval $[a, b]$ and we connect these (Figure 1.16(a)) in left-to-right sequence by straight-line segments. Thus the graph of a ramp function $R(x)$ is a continuous polygonal path made up of line segments and it is completely determined by the vertices.

Example. Plot the graph of the ramp function $R(x)$ on $[2, 12]$, which is determined by the vertices $(2, 1)$, $(4, 3)$, $(7, -1)$, $(9, 2)$, $(10, 2)$ and $(12, 3)$. Calculate and plot the derived step function $R' = m(x)$, which gives the slope of $R(x)$ at each point.

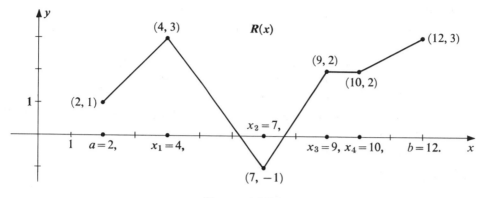

FIGURE 1.16(a).

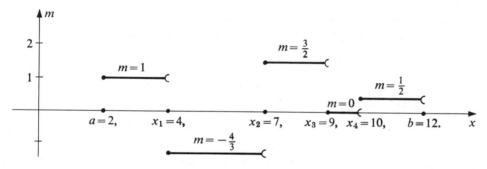

FIGURE 1.16(b)   Ramp function $R(x)$ on [2, 12].

Solution. We plot the graph according to the definition (Figure 1.16(a)). Now in each straight-line segment the slope at each point is a constant equal to the slope of the graph over the interval. We compute the slopes such as

$$m = \frac{R(x_2) - R(x_1)}{x_2 - x_1} = \frac{3 - 1}{4 - 2} = 1$$

in the table below and plot the resulting step function $m(x)$ (Figure 1.16(b)).

x	$2 \leqq x < 4$	$4 \leqq x < 7$	$7 \leqq x < 9$	$9 \leqq x < 10$	$10 \leqq x \leqq 12$
m	1	$-\frac{4}{3}$	$\frac{3}{2}$	0	$\frac{1}{2}$

This step function $R' = m(x)$ is not well defined at the corners because $\lim_{w \to x} [R(w) - R(x)]/(w - x)$ is ambiguous. If we agree to calculate the slope at each corner from the right-hand segment entering the corner, the ambiguity is removed and $R'(x)$ is well defined at each point of the interval $[a, b]$. We call the slope function $R'(x)$ the *derived function* of the ramp function $R(x)$.

We observe that a function $f(x) = mx + b$ whose graph is a straight line is also a ramp function with only one line segment and that its derived function $f'(x) = m$, a constant.

4.7 Exercises

In Exercises 1–6, find slope ratios for each of the pairs of points. Plot the points and show the directed-line segments whose lengths $x_2 - x_1$ and $y_2 - y_1$ appear in the slope ratio.

1. (2, 6) and (−1, 3).
2. (6, 3) and (−7, 3).
3. (−1, 4) and (2, −3).
4. (5, −1) and (5, 4).
5. $(-\frac{2}{3}, \frac{1}{2})$ and $(-\frac{1}{2}, 1)$.
6. (x, y) and (−2, −3).

In Exercises 7–12, plot lines that contain all other points making the given slope with the given point.

7. Through (1, 2) with slope $\frac{4}{7}$.
8. Through (1,2) with slope $-\frac{1}{3}$.
9. Through (2, −3) with slope 5.
10. Through (0, 0) with slope 0.
11. Through (1, 3) with slope 0.
12. Through (1, −3) with slope 0.2.

13. Find equations of the lines in Exercises 7–12 as relations between x and y.
14. Write formulas defining functions $f(x)$ whose graphs are the lines in Exercises 1–6.

In Exercises 15–18, each of the following pairs of numbers gives a time t and a distance s of a moving point from some reference point in the path of the motion. For such a pair of positions, compute the speed over the interval and draw a figure. Times are measured in seconds, distances in feet.

15. (1, 4) to (7, 50).
16. (1, 20) to (8, 6).
17. (4, 6) to (14, 6).
18. $(\frac{3}{2}, \frac{11}{5})$ to $(\frac{14}{3}, \frac{3}{8})$.

19. Assuming that the motions in Exercises 15–18 are uniform, find the instantaneous velocity at each time.

In Exercises 20–23, a point moves along the y-axis at uniform velocity between the designated points. In each case, plot the ramp function $f(t)$ representing the position and the function $f'(t)$ representing its velocity.

20. The point starts from the point where $y = 0$, then in 4 sec it moves to the point where $y = 7$, then in 4 sec it returns to 0.
21. The point starts from the point where $y = 1$, then in 3 sec it moves to the point where $y = 3$, there it remains at rest for 3 sec, then it returns to 0 in 3 sec.
22. The point starts from the point where $y = 10$, then it moves in 2 sec to the point where $y = 0$, then it moves in 2 sec to the point where $y = -5$, then it moves in 2 sec to the point where $y = -6$, then it remains stationary at 6 for 2 sec.
23. The point starts from the point where $y = 0$, then in 4 sec it moves to the point where $y = 7$, then it remains at rest for 3 sec, then it moves to the point where $y = 1$ in 4 sec.

24. Consider the equation $3x + 7y = 5$.
    (a) Solve for y in terms of x.
    (b) Show that this equation represents a line of slope $-\frac{3}{7}$ through the point $(0, \frac{5}{7})$.
25. As in Exercise 24, show that $Ax + By = C$, where $B \neq 0$, is a line of slope $-A/B$ through the point $(0, C/B)$.
26. Show that the graph of $x = 4$ is a straight line but cannot be the graph of a function and has no slope.
27. As in Exercise 26, show that the graph of $x = c$, where c is any fixed number, is a straight line but cannot be the graph of a function and has no slope. Graph the set of all such lines.

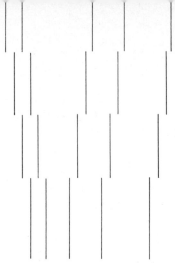

Numerical Values of Functions

CHAPTER 2

Remarks on mathematical theory. The practical man usually has some dislike of theory and distrusts it. In his experience, things may be true in theory but not in practice. These attitudes should not apply to mathematical theory, however. A mathematical theory consists of general statements that are provable, without exception, under conditions established by the hypotheses and axioms. It is efficient and economical thinking to know what is *always* true in given circumstances, without having to rediscover it in each particular instance. The theory of real numbers and that of continuous functions contain many general statements that are widely useful in pure and applied mathematics.

5 Inequalities and Absolute Values

5.1 Our View of the Real Numbers. We have been thinking of the real* numbers as the set of points that completely fill up a continuous line, or number scale (Figure 2.1). However this view, though simple and easily grasped intuitively, is difficult to work with. We shall now consider the real numbers to consist of all the decimal sequences with algebraic sign prefixed, as in $-\pi = -3.141592653589793\ldots$, imbedded in an algebraic system with the operations addition, subtraction, multiplication, and division. Each

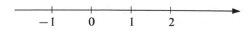

FIGURE 2.1 The real number line.

---

* The word "real" does not have its dictionary meaning as contrasted with "imaginary." Instead, real numbers are the ones which fill a line, in contrast to the complex numbers, which fill a plane, or smaller systems of numbers in the line, which do not fill it.

such sequence is punctuated by the decimal point. We also use three dots at the right to indicate that this is not the end but the sequence continues indefinitely. This decimal is a sum of powers of 10 with coefficients taken from the digits {0, 1, 2, 3, 4, 5, 6, 7, 8, 9}. Thus

$$\pi = 3(10)^0 + 1(10)^{-1} + 4(10)^{-2} + 1(10)^{-3} + 5(10)^{-4} + 9(10)^{-5} + \cdots.$$

We should like to say that two real numbers are equal if and only if they have the same decimal representations. However, in decimals we must make one exception and admit that

$$0.9999\ldots = 1.0000\ldots.$$

In general, any decimal that contains an infinitely long, unbroken string of 9's is equal to one in which the digit preceding the 9's is increased by 1 and then all of the 9's are converted to 0's. Otherwise two real numbers are equal if and only if they have the same decimal sequences.

With this representation, we assume the arithmetic operations of addition, subtraction, and multiplication of any pair of real numbers by means of the standard arithmetic procedure, or "algorithm." Similarly, we assume the division of any real number x by a real number y provided that $y \neq 0$, read "y is not equal to zero." We assume the algebraic properties of the field of real numbers determined by the operations, $+$, $-$, $\times$, $\div$.

5.2 Positive and Negative Numbers.

We distinguish the positive real numbers. They are the nonzero decimals with $+$ attached. We assume that the sum of two positive numbers is a positive number, the product of two positive numbers is a positive number, and (trichotomy) for every real number x exactly one of the statements, x is positive, $-x$ is positive, or $x = 0$, is true. We observe that when we designate the real number by a letter x we cannot tell whether $+x$ or $-x$ is positive, as we can when a number is represented by its own decimal representation. We assume the rules of signs: $+x = x$, $-(-x) = x$, $(-x)y = -xy = x(-y)$, and $(-x)(-y) = xy$. With this basis we can prove the following theorem about real numbers.

THEOREM ON SQUARES OF REAL NUMBERS.   For every real number x, the square x^2 is either positive or zero, and $x^2 = 0$ if and only if $x = 0$.

5.3 Inequalities.

If x and y are real numbers and $x \neq y$, then $x < y$, read "x is less than y," means that $y - x$ is a positive number. Also $x \leq y$ means: $x < y$, or $x = y$. We can prove from this definition and the properties of positive numbers the following properties of inequality in the real numbers. They may also be considered as properties of order.

O1.   (Trichotomy)   If x and y are real numbers then exactly one of the statements, $x < y$, $x = y$, $y < x$, is true.

O2.   (Transitive law)   If $x < y$ and $y < z$, then $x < z$.

O3.   (Addition)   If $x < y$, and z is any real number, then $x + z < y + z$.

O4.   (Multiplication)   If $x < y$, and p is any positive number, then $xp < yp$.

O5.   (Reversal of signs)   If $x < y$, then $-y < -x$.

O6.   (Positive numbers) x is positive if and only if $0 < x$.

O7.   (Negative numbers) x is negative if and only if $x < 0$. Proofs are in the exercises.

In terms of the picture of real numbers as a number line, we see that $x < y$ means that x stands to the left of y on the number line, not that y is numerically bigger than x. For example, $-1000 < 1$.

We define the dual symbol $x > y$, read "x is greater than y," by $x > y$ if and only if $y < x$. Similarly, $x \geqq y$ if and only if $x \leqq y$.

5.4   For Every Epsilon.   Admitting that the decimal name of a real number r is ordinarily infinitely long, we cannot write down all of it or know all of it. For example, nobody knows all of the decimal representation of the number π. We write $\pi = 3.14159\ldots$, meaning that the digits before the three dots are the first six digits in its full decimal name. What does it mean then to say that we know a real number r?

Let us denote by ϵ, Greek epsilon, any one of the set of small positive numbers

$$\{\cdots 10^2, 10^1, 10^0, 10^{-1}, 10^{-2}, 10^{-3}, \ldots, 10^{-k}, \ldots\}.$$

These epsilons are precision measures. Then we say that we know an infinite decimal sequence r if and only if for every ϵ (*specified in advance*) we can exhibit a finite logical procedure that will produce a finite decimal b that differs from r by less than ϵ, that is, $r - \epsilon < b < r + \epsilon$ (Figure 2.2).

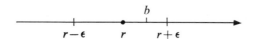

FIGURE 2.2   The finite decimal b differs from r by less than ϵ.

Ideally we would like to think that if we know r then for every integer k specified in advance we can produce the digit in the kth place of the decimal. The description of r by means of epsilon precision measures almost accomplish this, but not quite. There remains the difficulty that arises from the ambiguity

$$1.000\ldots = 0.999\ldots.$$

Moreover, our description of r by approximations that fulfill the "for every ϵ" requirement certainly does not permit us to write down the best approximation to r in terms of a k-place terminating decimal b. We accept these difficulties without attempting to resolve them. Statements beginning with "for every $\epsilon \cdots$" thus occur with great frequency in mathematical language when we are seeking to determine real numbers exactly.

For example, if the real number we seek is $\sqrt{2}$, the square root algorithm of arithmetic enables us to produce the following terminating decimal approximations b.

$b = 1,$ with error less than $\epsilon = 10^0$.

$b = 1.4,$ with error less than $\epsilon = 10^{-1}$.

$b = 1.41,$ with error less than $\epsilon = 10^{-2}$.

$b = 1.414,$ with error less than $\epsilon = 10^{-3}$.

$\cdot\ \cdot\ \cdot\ \cdot\ \cdot\ \cdot\ \cdot\ \cdot\ \cdot\ \cdot\ \cdot\ \cdot\ \cdot\ \cdot\ \cdot\ \cdot\ \cdot\ \cdot\ \cdot\ \cdot$

And so on indefinitely for each ϵ specified in advance.

5.5   Absolute Values.   We define $|x|$, read "the absolute value of x," by $|0| = 0$; and if $x \neq 0$, then $|x|$ is the positive number of the pair $\{x, -x\}$.

5.6   Some Properties of Absolute Value.   A lemma in mathematics is a preliminary statement, usually technical, leading up to the proof of a main result, called a theorem.

LEMMA 1.   For every number x, $-|x| \leq x \leq |x|$.

Proof.   By the definition, if x is positive, $-|x| < x = |x|$; if $x = 0$, then $-|x| = x = |x|$; and if x is negative, then by definition of absolute value $-|x| = x < |x|$. Thus the statement of the lemma is true in every case.

LEMMA 2.   If $d > 0$, then $|x| < d$ if and only if $-d < x < d$.

Proof.   We first prove that if $|x| < d$, then $-d < x < d$. Since by Lemma 1, $-|x| \leq x$, then $-d < x$. Thus $-d < x < d$.

Next we prove that if $d > 0$ and $-d < x < d$, then $|x| < d$. By hypothesis, $x < d$. If x is positive or zero, then by definition $|x| = x$ and hence $|x| < d$. On the other hand, if x is negative, $|x| = -x$, and since by hypothesis $-d < x$, we have $-x < d$ and therefore in this case also $|x| < d$.

LEMMA 3.   If $d \geq 0$, then $|x| \leq d$ if and only if $-d \leq x \leq d$.

The statement and proof of this lemma are analogous to those of Lemma 2 with $\leq$ replacing $<$.

THEOREM (Absolute value of a sum).   For every pair of real numbers x and y, $|x + y| \leq |x| + |y|$.

Proof.   By Lemma 1, $-|x| \leq x \leq |x|$ and $-|y| \leq y \leq |y|$. We add these inequalities and find that

$$-(|x| + |y|) \leq x + y \leq |x| + |y|.$$

Then by Lemma 3 this implies that $|x + y| \leq |x| + |y|$.

THEOREM (Absolute value of a product).   For every pair of real numbers x and y,

$$|xy| = |x| \, |y|.$$

Proof.   (Exercises.)

5.7   Solving Inequalities for an Unknown x.   We now use the properties of absolute value to solve some inequalities involving the unknown number x.

Example A.   Solve the inequality $|2x - 5| < 1$.

Solution.   If this were the equation $|2x - 5| = 0$, which by definition is equivalent to $2x - 5 = 0$, we find the solution $x = \frac{5}{2}$. Since $0 < 1$, this is clearly one solution of the inequality in the example. To find all of the solutions we proceed algebraically as follows.

By Lemma 2, $|2x - 5| < 1$, if $-1 < 2x - 5 < 1$, or if, on our dividing by 2, $-\frac{1}{2} < x - \frac{5}{2} < \frac{1}{2}$. Adding $\frac{5}{2}$ to all members, we find that the inequality is satisfied if $\frac{5}{2} - \frac{1}{2} < x < \frac{5}{2} + \frac{1}{2}$, or if $2 < x < 3$. Thus the complete solution is a *set* of numbers that is an interval in the real number scale centered at $\frac{5}{2}$ and extending $\frac{1}{2}$ unit up and down the scale from this point. We represent this in Figure 2.3 with a heavy black line segment terminated by half circles to indicate that 2 and 3 are not included.

$$0 \qquad 1 \qquad 2 \quad \tfrac{5}{2} \quad 3$$

FIGURE 2.3   The inequality $|2x - 5| < 1$.

Example B.   Solve the inequality $|x - a| < d$, where d is a fixed positive number.

Solution.   We follow the steps of Example A and find that x satisfies the inequality if and only if $a - d < x < a + d$. As before, this solution is a set of numbers in an interval in the scale with center at a and extending d units in each direction (Figure 2.4).

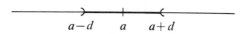

$$a-d \qquad a \qquad a+d$$

FIGURE 2.4   The inequality $|x - a| < d$.

5.8   *Exercises*

1. In the following statements, verify from the definition of absolute value which ones are true and which ones are false:

    (a) $-2 < 1$.

    (b) $|-1| < 0$.

    (c) $|-3| < 2$.

    (d) $2 < |-3|$.

    (e) $0 < |-2|$.

    (f) $|-\frac{1}{2}| < 1$.

2. Plot the intervals of numbers x on the x-axis that are selected by the conditions below:
    (a) $|x| < 1$.   (b) $|x| \leq 2$.   (c) $-2 \leq x < 1$.   (d) $|x - 2| < 1$.

In each of Exercises 3–10, solve each inequality for all of the values of x that satisfy it. Plot the interval of all solutions.

3. $|x - 1| < 4$.

4. $|3 - 2x| < 1$.

5. $|x + 2| < 5$.

6. $|ax + b| < 1$.

7. $|2x - 1| < 0.3$.

8. $|1 + 2x| \leq 1$.

9. $|3x + 5| \leq \frac{1}{2}$.

10. $x^2 \geq 0$.

11. Prove that the equation $(ax - b)^2 = -1$ has no real roots.

12. Prove that the equation $x^2 + x + 1 = 0$ has no real roots. We are working in the real number system! Do not say that "the roots are imaginary" and do not introduce the idea of $\sqrt{-3}$.

13. When ϵ is specified, a finite procedure for determining a finite decimal b approximating a real number r in the interval $[0, 1]$ is as follows. Partition the interval $[0, 1]$ into 10 equal subintervals

$$[0, 10^{-1}[, \ [10^{-1}, 2(10)^{-1}[, \ [2(10)^{-1}, 3(10)^{-1}[, \ldots, [9(10)^{-1}, 1],$$

where the reversed bracket [on the right means that 10^{-1} is excluded. It is included in the next subinterval. We then ask, In what subinterval of the partition is r located? On determining this, we partition this subinterval into 10 smaller subintervals each of length

10^{-2} and repeat the question. On k applications of this decision procedure we find that we have determined the k digits of a number b that approximates r with error less than $\epsilon = 10^{-k}$. Show how this procedure applies when we divide to find $r = 21 \div 47$ as a decimal.

Solve for x the inequalities in Exercises 14–17.

14. $(x - 1)(x - 2) < 0$.   15. $|a(x - 1)| < 1$.

16. $\left|\dfrac{x}{4}\right| < 1$.   17. $\left|\dfrac{x - 2}{10}\right| < 1$.

18. Prove Lemma 3, Section 5.6.
19. Under what conditions does the equals sign hold in the theorem on the absolute value of a sum; that is, what is the relation between x and y when $|x + y| = |x| + |y|$?
20. Prove that if $x > 1$, then $x^2 > x$, but if x is positive and $x < 1$, then $x^2 < x$.

5.9 Problems

T1. Using the properties of positive numbers (§5.2) and the definition of $x < y$ (§5.3) prove the properties of inequalities O1 to O7. (§5.3).
T2. Prove that if $d > 0$, then $|c| > d$ if and only if either $c < -d$ or $c > d$.

In each of the Exercises T3–T6, solve the inequality given.

T3. $|x| > 3$.   T4. $|4x| > 1$.
T5. $|x - 1| > 5$.   T6. $|2x + 3| > 5$.

T7. Prove that $|a + b + c| < |a| + |b| + |c|$.
T8. Prove the theorem on the absolute value of a product (§5.6).
T9. Prove that $|a - b| \leq |a| + |b|$.
T10. Prove that $|a| - |b| \leq |a - b|$.
T11. Solve the inequality $x^2 - x - 6 > 0$.
T12. Solve the inequality $|2x^2 + 3x| < 1$.

6 Problem of the Greatest Number

6.1 Sets. A set S of numbers is any collection of numbers. More workably, S is a set of numbers if we are able to consider any number and decide whether it is included in S or not. If x is included in S, we say that $x \in S$, read "x is an element of S" or "x is in S." We shall use the idea of a set of numbers, or a set of pairs of numbers (§1), or a set of functions (§2), without attempting to make the idea of a set mathematically or logically precise. We use the set relation $S \subset R$, read "Set S is contained in Set R," to mean that for every number x, if $x \in S$ then $x \in R$. "Sets $S = R$" means $S \subset R$ and $R \subset S$. We use braces $\{ \ \}$ to denote sets of numbers described inside the braces. For example, $\{-1, 0, 1, 2, 3\}$ is the set of five numbers named inside the braces. We use the braces also to describe sets by means of a device called the *set builder*. If R denotes the set of real numbers, we shall write $S = \{x \in R \mid$ defining properties$\}$, which will mean that S is the name of a set consisting of elements x that are real numbers and then, after the vertical stroke, included in S if and only if the stated properties are satisfied. For example,

$$S = \{x \in R \mid x > 0 \ \& \ x^2 + x - 2 = 0\}$$

is the set of all real numbers x that are positive and satisfy the equation $x^2 + x - 2 = 0$. In this case $S = \{1\}$. We regard an empty collection of numbers as a set and denote the empty or null set by $\varnothing$. For every set S it is true that $\varnothing \subset S$. An example is $\varnothing = \{x \in R \mid x^2 + 1 = 0\}$.

6.2   Finite and Infinite Sets.

We say that a set is *finite* either if it is an empty set, or if we can count its elements and the count comes to an end with some positive integer n. We often denote the counting assignments by affixing the counting integer as a subscript to the name of the number. Thus $\{x_1, x_2, \ldots, x_n\}$ denotes a counted set of numbers. It is provable that if a set is finite, every counting of it ends with the same integer n, which is called the *number of elements* in the set. An infinite set of numbers is one that is not finite. For example, the set R of real numbers is infinite; so is the set of integers.

Ideally we can always find the maximum number in a finite set of numbers x by comparing them in pairs and throwing out the smaller number until the set is exhausted and only the largest number, x_{max}, remains.

6.3   Infimum and Supremum.

The infinite set $S = \{x \in R \mid -1 < x < 1\}$ is bounded above by $+1$ and below by -1, yet it has no maximum or minimum. For if x is any candidate for the maximum in the set, then $\frac{1}{2}(x + 1)$ is larger and still in the set. Any number b, $b > 1$, is also an upper bound of S, but 1 is the *least upper bound*. No number less than 1 is an upper bound of S. It is obvious that this least upper bound of S is a uniquely determined number that behaves as the maximum of S except that it is not necessarily in S. We call this least upper bound 1 the *supremum* of S. Similarly, -1 is the *infimum* of S.

More generally, any set S of numbers is *bounded* if there exist numbers a and b such that for every number x in S, it is true that $a \leq x \leq b$.

DEFINITION.   If S is a nonempty set of real numbers that is bounded, then the *infimum* of S is a number, inf S, such that for every number x in S, it is true that inf $S \leq x$, and for every lower bound L of S it is true that $L \leq$ inf S. Analogously, the *supremum* of S is a number, sup S, such that for every number x in S, it is true that $x \leq$ sup S, and for every upper bound G of S, it is true that sup $S \leq G$.

In words, inf S is the greatest lower bound of S, and sup S is the least upper bound of S. We can prove from the definition that if the set S includes a maximum number, x_{max}, then $x_{max} =$ sup S; similarly, if x_{min} exists, then $x_{min} =$ inf S. The example of the set $\{x \in R \mid -1 < x < 1\}$ shows that not all bounded sets of real numbers include a maximum or minimum, but the following theorem tells what sets have a supremum and infimum.

THEOREM.   A nonempty set S of real numbers has a unique supremum and a unique infimum if and only if it is bounded.

We illustrate the proof by showing how to construct the infimum of the set $S = \{x \in R \mid x^2 > 12\}$. In words, we are trying to find the greatest lower bound of the set of numbers whose squares exceed 12. By trial and error we find that $3^2 < 12$ and $4^2 > 12$, so 3 is the largest integer that is a lower bound of S (Figure 2.5). We then partition the interval [3, 4] into 10 equal parts by the points {3.0, 3.1, 3.2, ..., 3.9, 4.0}. We find by trial that $(3.4)^2 < 12$ and $(3.5)^2 > 12$, so 3.4 is the largest of these partitioning numbers that is a lower bound of S. We then partition the small interval [3.4, 3.5] into 10 equal parts and find that 3.46 is the largest partition point that is a lower bound of S. We repartition the interval [3.46, 3.47] into 10 parts and find that 3.464 is a lower bound of S but 3.465 exceeds some number in S. Continuing in this way, we uniquely determine each decimal place in an infinite decimal 3.464..., which is the greatest lower bound of S, that is inf $S = 3.464....$

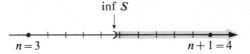

FIGURE 2.5   Infimum of S. The set S is represented by the shaded portion of the line, $S = \{x \in R \mid x^2 > 12\}$.

We see that this procedure can be applied to any nonempty, bounded set S of real numbers to determine inf S and sup S as nonterminating decimals and so complete the proof of the theorem.

6.4 Extrema of Quadratic Functions. It is remarkable that the principle of real squares (§5.2) equips us to resolve many maximum and minimum problems in infinite sets. To avoid excessive words, we say that a number is an *extremum* of a set of numbers if it is either a maximum or a minimum of the set.

Example A.   Investigate for extrema the infinite set

$$Y = \{f(x) \mid f(x) = 2x^2 + 6x - 1, \ x \in R\}.$$

Solution.   We explore the problem by plotting the graph of $f(x) = 2x^2 + 6x - 1$. The graph in Figure 2.6 reveals a minimum point and indicates that there is no maximum.

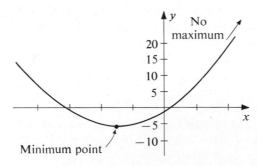

FIGURE 2.6 $f(x) = 2x^2 + 6x - 1$.

Now let us proceed to an exact algebraic analysis. We first rewrite $f(x) + 1 = 2(x^2 + 3x)$ and then we complete the square of terms inside the parentheses by adding $2(\frac{3}{2})^2$, to both sides. We obtain

$$f(x) + 1 + \tfrac{9}{2} = 2(x^2 + 3x + (\tfrac{3}{2})^2), \quad \text{or} \quad f(x) + \tfrac{11}{2} = 2(x + \tfrac{3}{2})^2.$$

Since $(x + \frac{3}{2})^2 \geqq 0$ and attains its minimum value when $x = -\frac{3}{2}$, we see that this form tells us that $f(x)$ attains its minimum value when $x = -\frac{3}{2}$ and that this minimum $f(-\frac{3}{2}) = -\frac{11}{2}$.

We can show mathematically that the set Y of values of $f(x)$ has no maximum, indeed that it is not bounded above, if we prove that for every positive number M, however large, there is a number x for which $f(x) > M$. Given M, we choose x so that

$$x + \frac{3}{2} > \frac{M}{2} + \frac{11}{4} \qquad \text{and} \qquad x + \frac{3}{2} > 1;$$

then

$$f(x) + \frac{11}{2} \geqq 2\left(x + \frac{3}{2}\right)\left(x + \frac{3}{2}\right) > 2\left(\frac{M}{2} + \frac{11}{4}\right)(1) = M + \frac{11}{2}.$$

Hence $f(x) > M$, and $f(x)$ is not bounded above.

6.5 Exercises

In Exercises 1–7, plot the intervals represented by the following set builders.

1. $S = \{x \in R \mid |x - 1| < 2\}$.

2. $S = \{x \in R \mid |x| > 2\}$.

3. $S = \{x \in R \mid x < -2 \ \& \ |x| \leqq 3\}$.

4. $S = \{x \in R \mid x^2 = 0\}$.

5. $S = \{x \in R \mid |2x - 1| < \epsilon\}, \ \epsilon = 2^{-2}$.

6. $S = \{x \in R \mid x^2 \leqq 2\}$.

7. $S = \{x \in R \mid x^2 - x - 7 \leqq 0\}$.

In Exercises 8–17, find the extrema of the sets indicated. If no maximum or minimum exists, state this and prove it.

8.   The set of all numbers y, given by $y = (x + 2)^2 - 3$ for all real numbers x.
9.   The set of all numbers y, given by $y = -2(x - 3)^2 - 8$ for all real numbers x.
10. The set of all numbers y, given by $y = |x - 2|$ for $-3 \leqq x \leqq 3$.
11. The set of all numbers y, given by $y = x^2 - 3x + 4$, for all real numbers x.
12. The set of all numbers y, given by $y = x^2 - 3x + 4$, for $-3 \leqq x \leqq 3$.
13. The set of all numbers y, given by $y = (x - 2)^2 + (x + 1)^2$.
14. The set of all numbers y, given by $y = 2x^2 - 3x + 1$ for $1 < x < 2$.
15. The set of all numbers y, given by $y = x^2 - 2x + 3$ such that $y < 2$.
16. The set of all numbers y, given by $y = ax^2 + bx + c$, $a > 0$, for all real x.
17. The set of all numbers y, given by $y = ax^2 + bx + c$, $a < 0$, for all real x.

18. Find the area of the largest triangle with sides 2 and 5.
19. Find the number that exceeds its square by the greatest amount.
20. At a charge of 20 cents a mile a taxi company finds that it does 800 passenger-miles of business a day. The number of passenger-miles is 40 less for each cent increase in fare. What rate per mile yields the greatest gross receipts?
21. Observe that for any positive numbers x and y, $(\sqrt{x} - \sqrt{y})^2 \geqq 0$. Show that hence the arithmetic mean $(x + y)/2$ always exceeds the geometric mean $\sqrt{xy}$ unless $x = y$.
22. Show that the area A of a rectangle with fixed perimeter p and base x is given by

$$A = -\left(x^2 - \frac{p}{2}x\right).$$

23. In Exercise 22, complete the square of the expression for A and thus find the dimensions of the rectangle of maximum area with the fixed perimeter p.
24. A physical quantity was measured three times in the laboratory and the measurements were a_1, a_2, a_3, not all equal. To determine the best single value x to assign to this quantity we agree that x will be chosen by "the principle of least squares," which says that the

sums of the squares of the errors $x - a_1$, $x - a_2$, $x - a_3$ shall be a minimum. Find x by the principle of least squares, showing that

$$x = \frac{a_1 + a_2 + a_3}{3}.$$

25. In Exercise 20, if operations cost \$50 a day in fixed charges plus 10 cents a mile, what fare rate yields the greatest profit?

26. Find the cylinder of greatest curved surface area that can be inscribed in a given right circular cone (Figure 2.7).

27. In a triangle ABC of altitude H, and base L, a rectangle $abcd$ is to be inscribed so as to have maximum area (Figure 2.8). What is the altitude of the maximal rectangle?

28. Using the fact that the geometric mean never exceeds the arithmetic mean (Exercise 21), show that if a positive number P is to be factored into two factors x and y having minimum sum $x + y$, then $x = y = \sqrt{p}$.

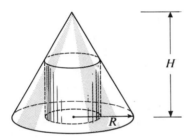   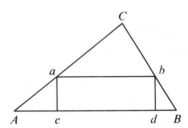

FIGURE 2.7   Exercise 26.   FIGURE 2.8   Exercise 27.

6.6 Problems

T1. Show that if x_1, x_2, x_3, x_4 are positive numbers

$$\sqrt[4]{x_1 x_2 x_3 x_4} \leqq \frac{x_1 + x_2 + x_3 + x_4}{4}.$$

T2. Using Exercise 21, show that if n is an even number, then for any positive numbers

$$\sqrt[n]{x_1 x_2 \cdots x_n} \leqq \frac{x_1 + x_2 + \cdots + x_n}{n}.$$

T3. If $a_1, a_2, \ldots, a_n$ are any n measurements, then show that the number x that minimizes $(x - a_1)^2 + (x - a_2)^2 + \cdots + (x - a_n)^2$ is the average of the numbers $a_1, \ldots, a_n$.

T4. Find a number x that gives the minimum value to $\sqrt{x^2 + 6x + 1}$.

T5. Show that the graph of $2x^2 + 3y^2 = 0$ consists of a single point.

In Problems T6–T12, find inf S and sup S. In each case, decide whether inf S = min S.

T6. $S = \{x \mid -2 < x < 1\}$. T7. $S = \{x \mid |x - 2| < 1\}$.

T8. $S = \{x \mid x^2 < 2\}$. T9. $S = \{y \mid y = -x^2\}$.

T10. $S = \{-1, 3, 4, 6\}$. T11. $S = \left\{ y \mid y = \dfrac{1}{1 + x^2}, x \in R \right\}$.

T12. $S = \left\{ y \mid y = \dfrac{1}{x^2}, x \in R, x \neq 0 \right\}$.

T13. Show that there are no points on the graph of the equation

$$x^2 + y^2 - 2x + 4y + 9 = 0.$$

T14. Prove that if S is bounded, there is only one inf S and only one sup S.

T15. By geometry, find where the point P on the line CD shall be to make the distance $AP + PB$ a minimum (Figure 2.9).

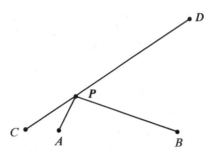

FIGURE 2.9   Problem T15.

T16.* (Difficult) Show that if the inequality Problem T2 is true for an even number n it is true for $n - 1$. Prove the inequality hence for every n.

7 Limits of Infinite Sequences

7.1 Infinite Sequences. An ordered array of numbers

$$s_1, s_2, s_3, \ldots, s_n, s_{n+1}, \ldots$$

is called an *infinite sequence* if it assigns one number s_n to each positive integer n. In shorter notation, we denote the entire sequence by the symbol $\{s_n\}$. The numbers s_n are called *elements* of the sequence. An infinite sequence has a first element s_1 (or s_0) and, after any element s_n, it has a next element s_{n+1}. A sequence is *finite* if it has a last element, after which there is no next element.

For example, if we divide 2 by 7, we obtain an infinite sequence of decimals as partial quotients,

$$s_1 = 0.2, \ s_2 = 0.28, \ s_3 = 0.285, \ s_4 = 0.2857, \ s_5 = 0.28571, \ldots.$$

As we progress through the sequence, these numbers trend closer and closer together. In particular, any two elements s_m and s_n after s_4 differ by less than 0.00001 because they agree to the first four places after the decimal. In general, we call an infinite sequence that behaves like this *convergent*.

DEFINITION. An infinite sequence $\{s_n\}$ is *convergent* if and only if for every positive number ϵ (however small) there is an integer N such that if $m > N$ and $n > N$, then $|s_m - s_n| < \epsilon$.

Just as the elements of the sequence in our example approach closer and closer to the exact quotient $\frac{2}{7}$, we expect that the elements of convergent sequence approach closer and closer to some fixed real number, which we will call the limit of the sequence. Mathematically, we define the limit as follows.

DEFINITION. The number s is the *limit* of the sequence $\{s_n\}$ if and only if for every ϵ there is an integer N such that when $n > N$, $|s_n - s| < \epsilon$.

In symbols this is written

$$\lim_{n \to \infty} s_n = s,$$

or more briefly, $s_n \to s$. In words, we read it "the limit of s_n, as n tends to infinity, is s," or more briefly, "s_n approaches the limit s." We now readily guess the relationship between convergent sequences and limits, which we state as a theorem.

THEOREM (Existence of the limit). For every convergent sequence $\{s_n\}$ of real numbers there is one and only one number s, which is the limit of the sequence.

Proof. We recall (§5.4) that the real number s is known if for every $\epsilon = 10^{-k}$, we can produce a finite decimal b_k that differs from s by less than ϵ. To do this for the limit of a convergent sequence, we take an infinite sequence of epsilons

$$\epsilon_1 = 0.1, \epsilon_2 = 0.01, \epsilon_3 = 0.001, \ldots, \epsilon_k = 10^{-k}, \ldots.$$

Starting with ϵ_1, the hypothesis that $\{s_n\}$ is convergent tells us that there is an integer N_1 such that if $m > N_1$ and $n > N_1$, then $|s_m - s_n| < \epsilon_1$. Thus we see that all of the elements of the sequence beyond the stage N_1 are in an interval $\{a \leqq x < b\}$ of real numbers whose length is 0.1. This interval contains exactly one decimal b_1, which terminates with the first digit after the decimal point. Moreover, for every $n > N_1$, $|b_1 - s_n| < \epsilon_1$. Then we repeat with ϵ_2, and find an integer N_2 such that all of the elements beyond stage N_2 differ by less than ϵ_2. Hence all are in an interval of length 0.01 that is included in the previous one. This smaller interval contains exactly one decimal b_2, which terminates with the second digit after the decimal, and it has the same digits as b_1 except for the last one. In this way we determine step by step every digit of a nonterminating decimal s. Moreover, for every $\epsilon = 10^{-k}$ there is an N_k such that if $n > N_k$, then $|s_n - s| < \epsilon$. That is, the convergent sequence approaches s as a limit.

To finish the proof, we must show that there is only one limit. Suppose there is another limit r, $r \neq s$, such that $s_n \to s$ and also $s_n \to r$. For $\epsilon = |r - s|/3$ there is an N_1 such that $n > N_1$ implies $|s_n - s| < \epsilon$. Also there is an N_2 such that $n > N_2$ implies $|s_n - r| < \epsilon$. If we now take N to be the larger of N_1 and N_2, when $n > N$, we have both $|s_n - s| < \epsilon$ and $|s_n - r| < \epsilon$. Hence, by the triangle inequality (§6),

$$|r - s| = |(s_n - s) - (s_n - r)| \leqq |s_n - s| + |s_n - r| \leqq \tfrac{2}{3}|r - s|.$$

If $r \neq s$, the statement, $|r - s| \leqq \tfrac{2}{3}|r - s|$ is false. Hence only one limit can exist. This completes the proof.

DEFINITION. A sequence *diverges* if and only if it has no limit as n approaches infinity.

7.2 Discussion of the Meaning and Uses of Limits. To start with a familiar example from geometry, we consider the hypotenuse of a right triangle whose sides both have length 1 (Figure 2.10). The length c of the hypotenuse is the physical reality we seek. Geometry tells us that $c^2 = 2$. When we attempt to compute c by the square root process we get an infinite sequence of decimal approximations,

$$s_1 = 1, s_2 = 1.4, s_3 = 1.41, s_4 = 1.414, \ldots,$$

which is convergent to the limit c. Hence in this case the limit of the sequence is the desired reality and we regard the elements, $s_1, s_2, s_3, \ldots$, as inexact approximations to it.

It also works the other way around. For example, for motion (§4.5) it is simpler to think of an instantaneous rate, which can only be realized in the limit, than to think of some one of the actual rates that can be measured over a short interval of time. Here, the only actual rates are the rates obtained by dividing by the elapsed time the distance traveled between two times. We can form an infinite sequence of these actual rates over shorter and shorter intervals of time. The limit of such a convergent sequence gives the simpler idea of the instantaneous rate. In such a case it is the limit that is the approximation to the more complicated actual rates. Many examples of both kinds of application of limits exist in calculus, where the limit s is the exact physical reality and the elements s_n are only approximations to it, and where the elements s_n are the physical realities, which can be approximated conveniently by the limit s.

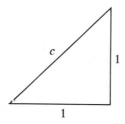

FIGURE 2.10

7.3   Some Stated Theorems on Limits.   We state a few basic properties of limits, deferring their proofs, which are not as difficult as the main theorem on the existence of the limit of a convergent sequence (§7.1).

THEOREM (Algebra of limits).   If for the infinite sequences $\{s_n\}$ and $\{t_n\}$, $s_n \to s$ and $t_n \to t$, then

(a) $s_n + t_n \to s + t$;

(b)   For every constant c, $cs_n \to cs$;

(c) $s_n t_n \to st$;

(d)   If $t \neq 0$, then $\dfrac{s_n}{t_n} \to \dfrac{s}{t}$;

(e) $|s_n| \to |s|$.

THEOREM (Boundedness of convergent sequences).   If $\{s_n\}$ is convergent, then there exist lower and upper bounds of the elements such that for every n

$$a < s_n < b.$$

THEOREM (Convergence of monotone sequences).   If $\{s_n\}$ is bounded and monotone-increasing, that is, if

$$s_1 \leqq s_2 \leqq s_3 \leqq \cdots \leqq s_n \leqq \cdots < b,$$

then the sequence is convergent and has the limit $s = \sup s_n$.

This theorem is an easier version of the main existence theorem (§7.1) applicable when the sequence is monotone-increasing (or monotone-decreasing and bounded below).

7.4 Example. Determine whether the sequence

$$s_1 = 1, \; s_2 = 1 - \frac{1}{2^2}, \; s_3 = 1 - \frac{1}{2^2} + \frac{1}{3^2}, \; \ldots, \; s_n = 1 - \frac{1}{2^2} + \cdots + (-1)^n \frac{1}{n^2}, \cdots$$

converges, and, if it does, compute the limit s with error less then 0.01.

Solution.   We observe that each element is a sum of terms that decrease in absolute value and alternate in sign. We can deduce that under these circumstances the absolute value of the sum is less than the absolute value of the first term. For example, $|s_n| < 1$. More generally, consider $|s_m - s_n|$, $m > n$. After cancelling the terms of s_n, we obtain

$$s_m - s_n = (-1)^n \left[\frac{1}{(n+1)^2} - \frac{1}{(n+2)^2} + \cdots \pm \frac{1}{m^2} \right].$$

Therefore

$$|s_m - s_n| < \frac{1}{(n+1)^2}.$$

Now for any ϵ,

$$|s_m - s_n| < \epsilon \qquad \text{if } \frac{1}{(n+1)^2} < \epsilon, \quad \text{or } n > \sqrt{\epsilon} - 1.$$

Take as N the smallest integer bigger than $\sqrt{\epsilon} - 1$. Then $n > N$ implies that $|s_m - s_n| < \epsilon$. Hence the sequence converges to some limit s.

To compute s with error less than $\epsilon = 10^{-2}$, we find that

$$\frac{1}{(n+1)^2} < 10^{-2} \qquad \text{if } (n+1)^2 > 10^2, \quad \text{or if } n > 9.$$

This gives us $N = 9$. We compute s_{10},

$$s_{10} = 1 - \frac{1}{4} + \frac{1}{9} - \frac{1}{16} + \frac{1}{25} - \frac{1}{36} + \frac{1}{49} - \frac{1}{64} + \frac{1}{81} - \frac{1}{100}$$

$$= 0.81 \cdots = s,$$

with error less than 0.01.

In this example the sum could not be discovered exactly. So we first proved the convergence to some exact sum, s. Then we found that if we go beyond $n = 9$, any element will give s to two places. We then computed s_{10} to evaluate the sum with 2-place accuracy. The same reasoning applies to the sequence

$$\{s_n\} = \left\{ 1 - \frac{1}{2} + \frac{1}{3} - \cdots + (-1)^{n-1} \frac{1}{n} \right\}.$$

An easier type of problem is one in which the exact limit can be guessed.

Example.   Determine whether the sequence

$$s_1 = \frac{5}{1}, \; s_2 = \frac{7}{2}, \; s_3 = \frac{9}{3}, \ldots, \; s_n = \frac{2n+3}{n}, \ldots$$

converges, and evaluate the limit.

Solution. We observe that

$$s_n = \frac{2n + 3}{n} = 2 + \frac{3}{n}.$$

As $n \to \infty$, the term $3/n \to 0$. Hence we guess $s = 2$. Then, rigorously by definition, for any ϵ

$$|s_n - s| = \left|\left(2 + \frac{3}{n}\right) - 2\right| = \left|\frac{3}{n}\right| = \frac{3}{n} < \epsilon, \qquad \text{if} \quad n > \frac{3}{\epsilon}.$$

So

$$\lim_{n \to \infty} \frac{2n + 3}{n} = 2 \qquad \text{exactly.}$$

7.5 *Example.* Determine whether the sequence of powers $\{r^n\}$ of a fixed number r converges, and find the limit if it exists.

Solution. We see that the situation is different when $|r| > 1$ from the case where $|r| < 1$. In fact, we readily see that when $|r| > 1$, the sequence diverges. To prove it rigorously we may write $|r| = 1 + \rho$, where $\rho > 0$. Then, by the binomial theorem,

$$|r|^n = (1 + \rho)^n = (1 + n\rho + \text{positive terms}) > 1 + n\rho.$$

The sequence cannot approach any number s as a limit because $|r|^n > s + 1$ whenever $1 + n\rho > s + 1$. This is true for all elements r^n of the sequences such that $n > s/\rho$.

If $|r| < 1$, we may write $|r| = 1/(1 + \rho)$, where $\rho > 0$. Then we guess that the limit is 0. To prove it, we use the binomial theorem as before to write

$$|r^n - 0| = |r|^n = \frac{1}{(1 + \rho)^n} = \frac{1}{1 + n\rho + \text{positive terms}} < \frac{1}{1 + n\rho}.$$

For any ϵ we have $|r^n - 0| < \epsilon$ if $1/(1 + n\rho) < \epsilon$. This is true if $n > (1 - \epsilon)/\rho\epsilon$. We summarize these results in a theorem.

THEOREM. The $\lim_{n \to \infty} r^n = 0$ if $|r| < 1$, and if $|r| > 1$ the sequence $\{r^n\}$ diverges.

The remaining cases are easily settled. If $r = 1$, the sequence $\{1^n\} = 1, 1, 1, \ldots, 1, \ldots$, which has the limit 1. If $r = -1$, the sequence $\{(-1)^n\} = -1, 1, -1, \ldots, (-1)^n, \ldots$, which can have no limit s because, as n increases, the elements oscillate between $+1$ and -1.

7.6 Estimate for N

Example. For $\epsilon = 10^{-3}$ and $r = \frac{1}{5}$, find explicitly a stage N such that if $n > N$, then $|r^n - 0| < \epsilon$. Find the smallest N.

Solution. As in the general case $\{r^n\}$, for $|r| < 1$, we write $|r| = \frac{1}{5} = 1/(1 + 4)$, so $\rho = 4$. We found that

$$|r^n - 0| < \epsilon \qquad \text{if} \quad n > \frac{1 - \epsilon}{\rho\epsilon} = \frac{1 - 10^{-3}}{4(10^{-3})} > \frac{1}{4(10^{-3})} = 250.$$

Hence, if $N = 250$, $n > N$ implies that

$$\left|\left(\frac{1}{5}\right)^n - 0\right| < 10^{-3}.$$

This estimate, $N = 250$, which was obtained by discarding all except two terms of the binomial expansion of $(1 + \rho)^n$, was adequate to prove that r^n converges to 0. But for computation it is wastefully large. We can find a much smaller N in this case. For

$$\left|\left(\tfrac{1}{5}\right)^n - 0\right| = \frac{1}{5^n} < \frac{1}{10^3} \qquad \text{if} \quad 5^n > 1000.$$

Running through the powers of 5, we find that this is true if and only if $n \geq 5$. Hence we make $N = 4$, and $n > N$ implies that

$$\left|\left(\tfrac{1}{5}\right)^n - 0\right| < 10^{-3}.$$

This is the smallest N that meets the precision requirement.

7.7 Exercises

ACADEMIC NOTE.   These exercises are designed to test understanding rather than to develop proficiency in making proofs of limit statements at this stage.

1. Write out several elements, $s_1, s_2, s_3, s_4, \ldots, s_n, \ldots$ of the infinite sequence $\{s_n\}$ given by $s_n = (3n - 2)/n$.
2. Show why we can guess that the sequence in Exercise 1 converges to the limit 3.
3. (a) In Exercise 1, find N such that if $n > N$, then

$$\left|\frac{3n - 2}{n} - 3\right| < \epsilon.$$

   (b) Find the smallest N if $\epsilon = 10^{-3}$.
4. Given $\epsilon = 10^{-4}$, find the smallest N such that if $n > N$, then $|(1/2)^n - 0| < \epsilon$.
5. Prove that the sequence $\{n\}$, $1, 2, 3, \ldots, n, \ldots$ diverges.
6. We can plot a sequence graphically by plotting the points (n, s_n) for each positive integer n. Plot a graph of the sequence $\{1 - (1/n)\}$.
7. Prove that the sequence

$$s_0 = 1, \; s_1 = 1 + \frac{1}{4}, \; s_2 = 1 + \frac{1}{4} + \frac{1}{4^2}, \ldots, \; s_n = 1 + \frac{1}{4} + \frac{1}{4^2} + \cdots + \frac{1}{4^n}, \ldots$$

   converges to a limit that is less than 1.
8. Investigate for convergence the sequence $\{\sqrt{n}/n\}$.
9. The sequence $\{s_n\}$ given by $s_n = (-1)^n$ does not converge. Form the averages

$$\sigma_n = \frac{s_1 + s_2 + \cdots + s_n}{n}.$$

   Does the sequence of averages $\{\sigma_n\}$ converge? What is the limit?
10. If the elements of the sequence $\{s_n\}$ are not known exactly, but it is known that for every n, $s_n > n$, prove that the sequence diverges.
11. Investigate for convergence the sequence $\{s_n\}$ given by $s_n = (1 - r^n)/(1 - r)$, where r is constant.
12. Show that the sequence $\{s_n\}$ converges, where $s_n = 1 - \frac{1}{2} + \frac{1}{3} + \cdots (-1)^{n-1} \frac{1}{n}$.
13. Compute with error less than 10^{-2} the limit of the convergent sequence $\{s_n\}$, given by

$$s_1 = 1, \; s_2 = -\frac{1}{1 \cdot 2}, \; s_3 = \frac{1}{1 \cdot 2 \cdot 3}, \ldots, \; s_n = (-1)^{n-1} \frac{1}{n!}.$$

   Here "n factorial" means the product of all integers from 1 to n, $n! = 1 \cdot 2 \cdot 3 \cdot \cdots \cdot n$.

8   Continuous Operations and Functions

8.1   Problems of Precision Machining.   As a preparation for the idea of continuity in arithmetic operations, let us consider some problems of precision machining.

Example A.   In cutting a metal rod of prescribed length, the cutting machine is never absolutely accurate. Small errors must be accepted but we can demand that the machined parts be accurate within prescribed precision measures. Suppose we are to cut two metal rods of nominal lengths a and b so that when they are placed end to end their combined length will be $a + b$ with an error that will not exceed 0.001 inches (Figure 2.11). What precision in cutting the rods will assure this accuracy?

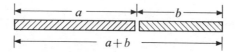

FIGURE 2.11   Cut rods.

Solution.   In this simple problem we can readily see that if each rod is cut with an error not more than $(\frac{1}{2})(0.001)$, their combined lengths will be in error by not more than 0.001 inches. However, let us now solve this problem formally.

We have $\epsilon = 0.001$ prescribed. Let x be the actual length of the a rod and y the actual length of the b rod. We are required to find a second precision measure δ such that if $|x - a| < \delta$ and $|y - b| < \delta$, then $|(x + y) - (a + b)| < \epsilon$. We consider the absolute difference of the sums

$$|(x + y) - (a + b)| = |(x - a) + (y - b)| \leq |x - a| + |y - b| \leq \delta + \delta = 2\delta.$$

Here we used the inequality for the absolute value of a sum (§5.6). From this result we see that, given ϵ, if we choose $\delta = \epsilon/2$, then $|x - a| < \delta$ and $|y - b| < \delta$ implies that

$$|(x + y) - (a + b)| < 2\left(\frac{\epsilon}{2}\right) = \epsilon.$$

Example B.   A rectangular plate of nominal base b and height a inches is to be machined, subject to the requirement that its actual area shall differ from ab square inches by not more than ϵ (Figure 2.12). Find a precision measure δ for the sides that will insure that the required precision ϵ in area is met.

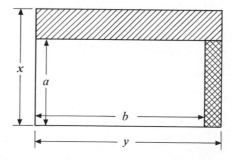

FIGURE 2.12   Cutting a rectangle of prescribed area.

Solution.   Let x be the actual height and y the actual base as machined.   Our problem is to determine a precision δ so that if $|x - a| < \delta$ and $|y - b| < \delta$, then $|xy - ab| < \epsilon$. We shall not try to find the exact largest precision δ that will assure the required area accuracy but we shall use overestimates of the area to find a slightly smaller-than-necessary precision δ.

To do this we shall try to overestimate the area error $|xy - ab|$ in terms of the differences $|x - a|$ and $|y - b|$.   We observe that $|xy - ab| = |xy - ay + ay - ab| = |(x - a)y + a(y - b)|$. Hence by the properties of absolute values of sums and products (§5.6),

$$|xy - ab| \leq |x - a|\,|y| + |a|\,|y - b|.$$

Similarly $|y| = |(y - b) + b| \leq |y - b| + |b|$, so that

$$|xy - ab| \leq |x - a|(|y - b| + |b|) + |a|\,|y - b|.$$

Now if $|x - a| < \delta$ and $|y - b| < \delta$, where δ is as yet undetermined, this implies that

$$|xy - ab| \leq \delta(\delta + |a| + |b|).$$

In the second factor on the right we may make the preliminary requirement that $|\delta| < 1$. Then $|xy - ab| < \epsilon$ if δ is chosen so that $|\delta| < 1$ and

$$\delta(1 + |a| + |b|) < \epsilon.$$

This is a definite solution of the problem.

To take some numbers, let us suppose that $a = 5$ and $b = 10$ in. and the area is required to be 50 sq in. with error not more than $\epsilon = 0.01$. Our solution says that if δ is chosen so that

$$\delta < \frac{0.01}{1 + 5 + 10} = \frac{0.01}{16} = 0.000625 \text{ in.}$$

and we machine the sides to this precision, the area will certainly differ from 50 by less than 0.01 sq in.

8.2   Continuity of Addition and Multiplication.   The solution of the problem of machining the metal plate depends on the continuity of the operation of multiplication in the real numbers. Loosely speaking, this means that the product ab changes very little if a and b are varied slightly.   Mathematically, we state it formally as follows.

DEFINITION.   *Multiplication is continuous* if and only if for every pair of numbers (a, b) and every positive number ϵ, there is a positive δ so that if $|x - a| < \delta$ and $|y - b| < \delta$, then $|xy - ab| < \epsilon$.

DEFINITION.   *Addition is continuous* if for every pair of numbers (a, b) and every positive number ϵ, there is a positive δ so that if $|x - a| < \delta$ and $|y - b| < \delta$, then $|(x + y) - (a + b)| < \epsilon$.

THEOREM.   In the real numbers, addition is continuous and multiplication is continuous.

Proof.   The solutions of the problems of cutting the rods and machining the rectangular plate are proofs of these assertions.

We observe, in passing, the pivotal role played by the relations $|x + y| \leq |x| + |y|$ and $|xy| = |x|\,|y|$ in these proofs.

We see at once that this theorem implies that subtraction is continuous; for

$$x - y = x + (-y) = x + (-1)y,$$

which reduces subtraction to a combination of addition and multiplication. Somewhat similarly, since $y \div x = y(1/x)$, the fact that multiplication is a continuous operation implies that division is continuous, if we show that the reciprocal $1/x$ is continuous.

THEOREM.   For every ϵ and every number a, where $a \neq 0$, there exists a δ such that if $|x - a| < \delta$, then $|(1/x) - (1/a)| < \epsilon$.

Proof.   We will need a lower estimate for $|x|$ in the denominator when $|x - a| < \delta$. In fact,

$$|a| = |(x - a) - x| \leq |x - a| + |x| \leq \delta + |x|.$$

Hence when $|x - a| < \delta$, we have $|x| \geq |a| - \delta$. To be sure that $|x|$ in the denominator does not become zero we will first specify that $\delta < |a|/2$. Then

$$\left|\frac{1}{x} - \frac{1}{a}\right| = \frac{|x - a|}{|a|\,|x|} \leq \frac{\delta}{|a|(|a| - \delta)} \leq \frac{\delta}{|a|(|a|/2)} < \epsilon,$$

if $\delta = \text{minimum } \{|a|/2, \epsilon|a|^2/2\}$. This completes the proof.

Thus all four of the arithmetic operations, $+, -, \times, \div$, are continuous.

8.3   Continuous Functions.   A more general idea is the following:

DEFINITION.   A function f is continuous at a number c in its domain if and only if for every positive number ϵ, there is a number δ such that if $|x - c| < \delta$, then $|f(x) - f(c)| < \epsilon$.

THEOREM.   Every polynomial is a continuous function on the real numbers.

Sketch the proof.   Every value $f(x)$ of a polynomial can be expressed as a finite sequence of additions of real numbers and multiplications by x. We apply the theorems on the continuity of addition and multiplication through each step of this finite sequence of operations *to prove* that the polynomial function is continuous.

8.4   Basic Theorem on Continuous Functions.   We recall that a closed interval $[a, b]$
is the set of numbers $\{x \mid a \leq x \leq b\}$, including both a and b. We shall use the idea of a function as a mapping (§1.4) to state an important theorem on continuous functions that we shall use; but the nonelementary proof will be deferred (§8.7).

THEOREM.   A function f that is continuous at every point x of a closed interval $[a, b]$ maps $[a, b]$ onto a closed interval $[L, G]$.

CONSEQUENCES OF THE THEOREM.   This basic theorem (Figure 2.13) implies that f continuous on $[a, b]$ has a minimum and a maximum. In fact, $L = $ minimum value of f, and $G = $ maximum value of f. Moreover, for every number y, $L \leq y \leq G$, there is some number x in $[a, b]$ where $f(x) = y$. In particular, if $f(a) < 0 < f(b)$, or

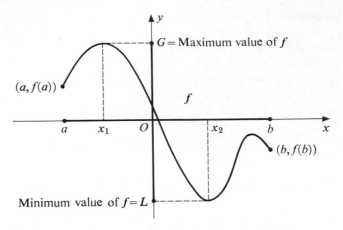

FIGURE 2.13   Values of a continuous function.

$f(b) < 0 < f(a)$, there is some number x where $f(x) = 0$. This last result is the basis of the graphical method of solving equations.

On the other hand, we find that a step function s does not satisfy the definition of a continuous function at the points where the steps up or down occur. If we plot a particular step function (Figure 2.14), we find that the values of this discontinuous function do not form a closed interval of all numbers from L to G but form instead a separated set of numbers on the y-axis.

The function f defined by $f(x) = 1/(x - 1)$ on all real numbers except where $x = 1$ (Figure 2.14) is continuous in the closed interval $[-2, 4]$ except at 1. It does not follow from the theorem that it has a maximum or minimum or that if $f(-2) < 0 < f(4)$ there is a point where $f(x) = 0$. In fact, there is no minimum, no maximum, and no point where $f(x) = 0$.

We have seen other examples where the hypotheses were not satisfied that the function $f(x) = 4/(1 + x^2)$ (Figure 1.2) is continuous on the real numbers. The real numbers are not a closed interval. This function is bounded but has no minimum.

The simple function $f(x) = x$ on the real numbers, though continuous, has no minimum and no maximum. This does not contradict the theorem because the real numbers do not form a closed interval. This function does map any closed interval $[a, b]$ onto the closed interval $[a, b]$.

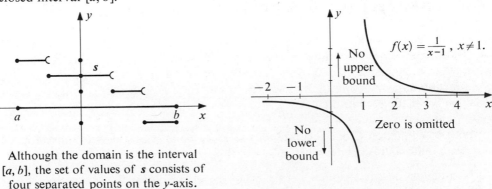

Although the domain is the interval $[a, b]$, the set of values of s consists of four separated points on the y-axis.

(a)

(b)

FIGURE 2.14   Values of discontinuous functions.

8.5 Graphical Solution of Equations. We apply the basic theorem on continuous functions in the familiar method of solving an equation graphically.

Example.   Solve graphically $x^3 + 2x - 1 = 0$.

Solution.   "Solve graphically" means: Use a graph to find the real roots of the equation. We begin by constructing a graph of the function $f(x) = x^3 + 2x - 1$ (Figure 2.15). The function $f(x)$ is continuous because it is a polynomial. Hence by the basic

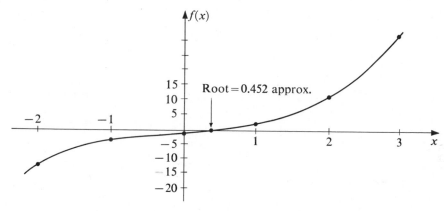

FIGURE 2.15   Graphical solution of $x^3 + 2x - 1 = 0$.

theorem, when we find that $f(0) = -1$ and $f(1) = 2$, we know that there is some number x, $0 < x < 1$, where $f(x) = 0$. This is our root. We narrow the search by plotting additional points in the interval [0, 1] and find that

$$\left. \begin{array}{l} f(0.4) = -0.136 \\ f(x)\ \ =\ \ \ 0 \\ f(0.5) = +1.125 \end{array} \right\} \begin{array}{l} \left.\begin{array}{l} \\ \end{array}\right\} +0.136 \\ \end{array} \right\} 261$$

Therefore the root is between 0.4 and 0.5. If we interpolate in this small table, we find that

$$x \approx 0.4 + \frac{0.136}{0.261}(0.1) = 0.452\ldots.$$

Hence the approximate root is 0.452. Further examination of the graph shows that this is the only root.

8.6 Exercises

1. Three rods of nominal lengths a, b, c are to be cut so that their combined length will be $a + b + c$ with error less than ϵ. What precision in cutting each of the three rods will meet this requirement?
2. A rectangular plate of nominal dimensions 6×9 is to be machined from a brass plate so that the area will be 54 sq in. with error less than 0.1 sq in. What precision in the measures of the sides will assure this accuracy?
3. The number $1/\pi$ is to be computed correctly to the fifth place. How accurately must π be represented to assure the required accuracy in the quotient?
4. When the number x is near zero, the quotient $1/x$ is very sensitive to small errors in x. It is required to compute the quotient with error less than 0.001 when $0.004 < x < 0.0005$. With what precision must x be known?

5. The function $\sqrt{x}$ is continuous when $x > 0$. With what precision may π be represented to insure that $\sqrt{\pi}$ is correct with an error less than $\epsilon = 0.001$?

Hint: $|\sqrt{x} - \sqrt{\pi}| = \dfrac{|x - \pi|}{\sqrt{x} + \sqrt{\pi}} < \dfrac{|x - \pi|}{\sqrt{\pi}} < \dfrac{|x - \pi|}{\sqrt{3}}$.

6. Prove that $\sqrt{x}$ is continuous when $x > 0$.

For the continuous functions described in Exercises 7–14, find the interval $[L, G]$ into which $f(x)$ maps the given interval $[a, b]$.

7. $f(x) = 2x + 3$ on $[a, b] = [-1, 2]$. 8. $f(x) = 5 - 3x$ on $[a, b] = [-1, 2]$.
9. $f(x) = x^2$ on $[a, b] = [-1, 2]$. 10. $f(x) = x^2 - 3x + 5$ on $[a, b] = [-1, 2]$.
11. $f(x) = 1/x$ on $[a, b] = [1, 10]$. 12. $f(x) = x + 1/x$ on $[a, b] = [1, 10]$.
13. $f(x) = 2x^2 - 5x + 11$ on $[a, b] = [-3, 3]$.
14. $f(x) = 11 - x - x^2$ on $[a, b] = [-1, 3]$.

In Exercises 15–18, apply the basic theorem to locate and compute all real roots graphically with error $\epsilon < 0.001$.

15. $x^3 - 5x + 7 = 0$. 16. $x^4 - 2 = 0$.
17. $x^3 + 7x + 5 = 0$. 18. $x^4 + x^2 + 1 = 0$.

19. In Exercises 7–14, use the theorems on the arithmetic operations to show that the functions are continuous as claimed.
20. The number $x^2 + x + 1$ is to be computed with error less than $\epsilon = 0.001$ when x is approximately 3. With what precision must x be known?
21. We observe for the function $f(x) = x + 1/x$ that $f(-1) = -2$ and $f(1) = +2$. Does this imply that there is a number x, $-1 < x < 1$, where $f(x) = 0$?
22. A continuous function f has a positive value at c. Show that there is a neighborhood of c where $f(x) > 0$ for every number in the neighborhood. Similarly, if $f(c)$ is negative, there is a neighborhood of c in which $f(x)$ is negative.

8.7 (Theoretical Section). *Proof of basic theorem on continuous functions* (§8.4).

ACADEMIC NOTE. The proof of the basic theorem on continuous functions is considered to be nonelementary. It is the custom to assume that it is true and go on. Indeed, to most persons encountering these ideas for the first time the conclusion seems obvious, not necessary to prove. For students who may be ready to confront the subtle difficulties the following, rather informally presented, proof is about as simple as possible.

We recall (§6.5) that a set S of real numbers is *bounded* if and only if there are two fixed numbers p and q such that for every y in S, $p \leqq y \leqq q$.

LEMMA 1. If f is continuous at c, then for every positive ϵ there exists a closed interval $[c - \delta \leqq x \leqq c + \delta]$ such that the values of f on this interval are bounded by $f(c) - \epsilon < f(x) < f(c) + \epsilon$.

Proof. (Figure 2.16) By definition of continuity at c (§8.3), for every ϵ there exists a δ_1 such that if $|x - c| < \delta_1$, then $|f(x) - f(c)| < \epsilon$. We first take a positive number $\delta < \delta_1$; then in the closed interval $c - \delta_1 \leqq x \leqq c + \delta_1$ it is still true that $f(c) - \epsilon < f(x) < f(c) + \epsilon$. This completes the proof of Lemma 1. It is important to realize that the closed interval $[c - \delta, c + \delta]$ reaches both back to include a number less than c and ahead to include a number greater than c.

LEMMA 2. A function that is continuous at every point of a closed interval $\{a \leqq x \leqq b\}$ is bounded.

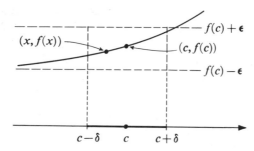

FIGURE 2.16   Local bounds of a continuous function.

Proof. Starting at the left endpoint where $x = a$, we use Lemma 1 to establish that f is bounded on some closed interval $[a, b']$ where $a < b'$. The number b' is not the only right endpoint for closed intervals in which f is bounded. How far to the right can b' be placed so that f is bounded in $[a, b']$? Let $c = \sup \{b' \in [a, b] \mid f$ is bounded in $[a, b']\}$. Then c itself is in $[a, b]$, so f is continuous there. By Lemma 1 there is a closed interval $[c - \delta, c + \delta]$ in which f is bounded. This interval reaches back and intersects the interval $[a, c]$ to include some number less than c (Figure 2.17). Hence f is bounded on the union $[a, c + \delta]$.

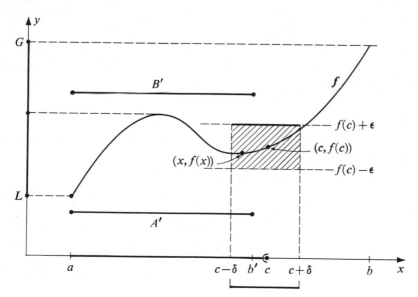

FIGURE 2.17   Extension of a family of intervals of boundedness.

From this we see that c cannot be less than b, for if it were, then the number c is not the *least* upper bound of numbers b' such that f is bounded in $[a, b]$. Moreover, if $c = b$, we see from this also that f is bounded in the interval $[a, b]$. This completes the proof of the lemma.

Proof of the basic theorem. Since, by Lemma 2, f has some lower bound and some upper bound on $[a, b]$, we know that f has a greatest lower bound L and a least upper bound G on $[a, b]$. That is, for all x in $[a, b]$

$$L = \inf f(x) \quad \text{and} \quad G = \sup f(x).$$

We consider the closed interval $[L, G]$ on the y-axis and pick any y' in it so that $L \leqq y' \leqq G$. We wish to show that for some number c in $[a, b]$, $f(c) = y'$. That is, every number in $[L, G]$, including L and G, is a value of f.

Again we start with the left end of $[a, b]$, where $x = a$. If $f(a) = y'$, the conclusion of the theorem is true. But if $f(a) \neq y$, for the sake of definiteness, say $f(a) < y'$, by Lemma 1 there is a closed interval $[a, b']$ on which $f(x)$ has an upper bound less than y'. Let

$$c = \sup \{b' \mid \text{on } [a, b'], f(x) \text{ has an upper bound less than } y'\}.$$

Then one and only one of the three statements,

$$\begin{array}{llll} \text{Case I,} & f(c) < y', & \text{or} \\ \text{Case II,} & f(c) = y', & \text{or} \\ \text{Case III,} & y' < f(c), \end{array}$$

is true.

We can rule out Case I. In this case if $c < b$, then by the argument of Lemma 1 we can extend the interval $[a, c]$ to $[a, c + \delta]$ on which $f(x)$ has an upper bound less than y'. This contradicts the definition of c as the *least* upper bound of the numbers b'. By the same argument, if $c = b$, we can show that on $[a, b]$, $f(x)$ has an upper bound less than y'. Then $G < y'$, so that y' cannot be in the interval $[L, G]$.

We can also rule out Case III. For if Case III is true, since f is continuous at c there is an interval $[c - \delta_2, c + \delta_2]$ on which $y' < f(x)$. This interval reaches back to include a number x', less than c, where $f(x') < y'$. Then for this number x' we have the contradictory statements $f(x') < y'$ and $y' < f(x')$. Hence Case III is impossible. This leaves only Case II and establishes the conclusion of the theorem, that for every y' in $[L, G]$ there is a number c where $f(c) = y'$.

8.8 Problems

In Problems T1–T3, one proves the theorem: If f is continuous on $[a, b]$ and $\epsilon > 0$, there exists a step function s such that

$$|f(x) - s(x)| < \epsilon \text{ on } [a, b].$$

T1. As in Lemma 1 (§8.7), show that there is a closed interval $[a, b']$ with $a < b'$ such that the step function with one step, $s(x) = f(a)$, meets the requirement on $[a, b']$.

T2. Let c be the least upper bound of b' in $[a, b]$ such that there exists a step function s with a finite number of steps, which has $|f(x) - s(x)| < \epsilon$ on $[a, b']$. Prove, as in Lemma 2, that with one more step the conclusion holds on $[a, c]$.

T3. Prove, as in Lemma 2, that $c < b$ is false. Hence, $c = b$ and the theorem is proved.

9 Integral of a Function

9.1 Problem of Exact Area Under a Curve. *Historical sketch.* We have seen (§3) how to calculate the area bounded by the graph of a step function s and the x-axis between the lines $x = a$ and $x = b$. This area is expressed mathematically by the integral $A = \int_a^b s(x)\,dx$ (Figure 2.18). We could calculate this area exactly. Because of the straight-line boundaries it could be resolved into a finite sum of areas of rectangles. When the bounding graph is a curve representing some function f instead of a step function s, it is not possible to define or compute the area of the region below the curve in terms of a

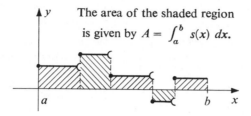

The area of the shaded region is given by $A = \int_a^b s(x)\ dx$.

FIGURE 2.18   Area of a region bounded by a step function.

finite number of areas of elementary figures like rectangles or triangles, and the area determination becomes a major mathematical problem with a long history.

The problem of calculating the *exact* area of a region bounded by a curve was formulated by the Greek geometers as early as the fourth century B.C. Their methods, particularly in the hands of such masters as Eudoxus and Archimedes, yielded many formulas for exact areas of circles and other plane figures, for the areas of curved surfaces such as cones, spheres, cylinders, and for the volumes of solids bounded by curved surfaces. While their methods of calculation have been replaced by more elegant and simple methods of the calculus, which we shall presently study, the approach of the ancient Greeks to the definition of the exact area and to the proof that it exists as a number has not been supplanted by methods of the calculus. There is no way of avoiding the difficulty of these arguments; we cannot understand the fundamental results of the calculus without a comprehension of the exact-area problem. So we accept the difficulty and attack the problem directly.

9.2   Statement of the Exact-Area Problem

DEFINITION.   A function f is increasing if $x_1 < x_2$ implies that $f(x_1) < f(x_2)$.

Let f be an increasing function on the interval $[a, b]$. Consider the region R bounded by the graph of f, the lines $x = a$, $x = b$, and the x-axis (Figure 2.19). The exact-area problem is that of defining the area of R as number and proving that there is one and only one number A that satisfies this definition of the area.

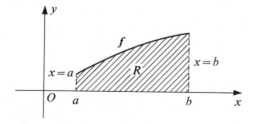

FIGURE 2.19   Region bounded by a curved graph.

9.3   Exploration of the Area Problem.
We are asked to define area, but what do we know out of which we can construct a definition of area? Let us agree on some basic principles of area.

(1)   The area of a region in the plane is a number.

(2) If R and Q are two regions that have areas but have no points in common, then the area of R and Q together is the sum of areas of R and Q (Figure 2.20.) For example, we agree that if R and Q are rectangles that do not overlap, the area of the composite figure consisting of R and Q taken together is equal to the area of R plus the area of Q.

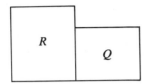

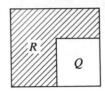

FIGURE 2.20   Regions do not overlap.   FIGURE 2.21   Region Q contained in R.

(3) Conversely, if regions R and Q have area and S is contained in R, then the region R–Q, consisting of all points in R except those in Q, also has an area equal to the area of R minus the area of Q (Figure 2.21).
(4) The rectangle* with sides parallel to the axes has an area equal to the length of the base times the length of the side.

These assumptions are natural. Moreover, the procedures we have used to calculate the area $\int_a^b s(x)\,dx$ of a region bounded by a step function s is consistent with them. It remains to apply these principles of area to the area of a region bounded by a curve.

9.4   Definition of the Integral.   Let f be any bounded function on the interval $[a, b]$, not necessarily continuous. We partition the interval $[a, b]$ into smaller subintervals by introducing subdivision points, $x_0, x_1, x_2, \ldots, x_n$, such that

$$a = x_0 < x_1 < x_2 < \cdots < x_n = b.$$

Then we define two approximating step functions over this partition (Figure 2.22(a)). Let s denote the "understep" function, whose height in each subinterval $s(x) = \inf f(x)$ in that subinterval. Similarly we define the overstep function S of the partition, whose value in each subinterval $S(x) = \sup f(x)$ in that subinterval. We observe, given f, that, the understep and overstep functions, s and S, are uniquely determined by the partition.

Then, by the principles of area, the area A of the region R below the curve must be a number between the integral of the understep function and the overstep function. That is,

$$\int_a^b s(x)\,dx \leq A \leq \int_a^b S(x)\,dx.$$

It is intuitively evident (Figure 2.22(b)) that we can repeatedly refine the partition of the base $[a, b]$ so as to make the areas $\int_a^b s(x)\,dx$ and $\int_a^b S(x)\,dx$ come closer and closer together.

To define a precise number A, we recall that we must demonstrate an explicit procedure that will determine the digit in each decimal point of A. That is, for every ϵ, $\epsilon = 0.1$, $\epsilon = 0.01$, $\epsilon = 0.001, \ldots$, we must show how to calculate an area that differs from A by less than ϵ. We can do this if we can exhibit a sufficiently refined partition of $[a, b]$ on

---

* Actually all we mean is that it is a parallelogram. The content given by base times length of the other side is not necessarily the true area in the Euclidean sense.

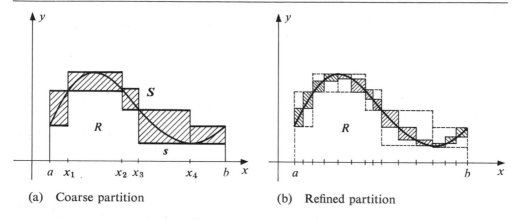

(a) Coarse partition   (b) Refined partition

FIGURE 2.22   Approximating step functions.

which the difference of areas between understep and overstep functions is less than ϵ (Figure 2.22(a), shaded rectangles). When this is done, the area of the understep function (or overstep function) gives A to the prescribed number of decimal places. We formalize this procedure for computing an exact area in a definition.

DEFINITION OF THE INTEGRAL OF A FUNCTION.   Let f be a bounded function on the interval $[a, b]$. For every partition $\{a, x_1, x_2, \ldots, b\}$ of this interval there is a uniquely determined understep function s and overstep function S defined by $s(x) = \inf f(x)$ and $S(x) = \sup f(x)$ in each subinterval of the partition. The function f is *integrable* over $[a, b]$ if and only if there is one and only one number A such that, for all partitions of $[a, b]$,

$$\int_a^b s(x)\, dx \leqq A \leqq \int_a^b S(x)\, dx.$$

If f is integrable, then the *integral* of f over $[a, b]$ is defined by

$$\int_a^b f(x)\, dx = A.$$

This completes the definition.

While we have interpreted the integral

$$\int_a^b f(x)\, dx$$

as the exact area of the region below the curve, we have found (§4), that it has other useful interpretations. For this reason we follow historical precedent and give it the neutral name "integral," implying vaguely "a whole that is the sum of its parts."

The definition does not guarantee that every bounded function is integrable. We now prove for an important class of bounded functions, those that are increasing in the interval $[a, b]$, that the exact integral does exist.

9.5   Procedure for Establishing Exact Area for Increasing Functions

Example.   Let f be an increasing function on the interval $[1, 7]$ having $f(1) = 2$ and $f(7) = 10$. Assuming that there is one number A that represents the area of the region R below f, find a procedure that will compute A with an error less than ϵ, where $\epsilon = 0.1$.

Solution.   The problem will be solved if we construct an understep function *s* and an overstep function *S* so that the error requirement

$$\int_1^7 S(x)\,dx - \int_1^7 s(x)\,dx < 0.1$$

is satisfied. For in this case, both of the areas $\int_1^7 s(x)\,dx$ and $\int_1^7 S(x)\,dx$ will differ from *A* by less than 0.1.

We partition the base [*a*, *b*] into a large number of small subintervals of equal width *w*. We construct (Figure 2.23) step functions *S* and *s* over this partition. The difference

$$\int_1^7 S(x)\,dx - \int_1^7 s(x)\,dx$$

is the sum of the areas of the shaded rectangles, which we will call the difference rectangles. We do not have to compute their areas individually. At the right (Figure 2.23) we stack up the difference rectangles to make one rectangle of base *w* whose altitude $f(7) - f(1) = 10 - 2 = 8$. Hence the sum of the areas of the difference rectangles is 8*w* and therefore

$$\int_1^7 S(x)\,dx - \int_1^7 s(x)\,dx = 8w.$$

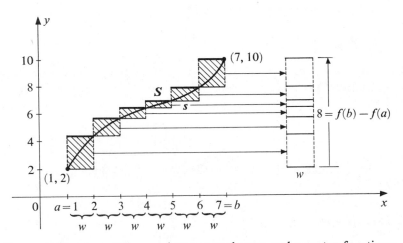

FIGURE 2.23   Area difference between understep and overstep functions.

The error requirement will be met if we make $8w < 0.1$ or $w < (0.1)/8$ or 0.0125. Thus the solution of the problem is to partition the base interval into equal subintervals of width less than 0.0125. Then the understep function *s* will have an area $\int_1^7 s(x)\,dx$, which can be computed arithmetically, and this area will differ from *A* by less than 0.1. This completes the problem.*

We can adapt this procedure immediately to prove the following theorem.

THEOREM.   Every increasing function $f(x)$ on an interval [*a*, *b*] is integrable.

The proof follows the procedure of the example. For any ϵ, $\epsilon = 10^{-n}$, we choose *w* so that $|f(b) - f(a)|\, w < \epsilon$. Then with this *w* as the width of subintervals in a partition of

---

* This procedure is not the most efficient one in actual practice. It will require 480 subintervals in the partition of the base, but it is a construction that can be carried out, and we can be sure that it meets the accuracy requirements.

[a, b], we construct the overstep function S and understep function s. Now $\int_a^b S(x) \, dx$ and $\int_a^b s(x) \, dx$ agree to n decimal places; for, as in the example,

$$\int_a^b S(x) \, dx - \int_a^b s(x) \, dx = w|f(b) - f(a)| < \epsilon = 10^{-n}.$$

Thus the process determines one and only one real number A by giving it to any prescribed number of decimal places, and

$$\int_a^b s(x) \, dx \leqq A \leqq \int_a^b S(x) \, dx.$$

Hence the process defines the exact integral $\int_a^b f(x) \, dx$ for any increasing function.

9.6   Computing the Exact Area in a Particular Case.

We now use these methods to compute one exact area bounded by a curve. As a preliminary we work out some algebraic inequalities.

Let u and x be real numbers such that $0 < u < x$. We factor the difference of two cubes, writing

$$x^3 - u^3 = (x - u)(x^2 + xu + u^2).$$

We estimate the second factor $x^2 + xu + u^2$ above and below. If we replace x in $x^2 + xu + u^2$ by the smaller number u, we get the smaller number $u^2 + u^2 + u^2 = 3u^2$. Similarly, if we replace u in $x^2 + xu + u^2$ by the larger number x we get the larger number $x^2 + x^2 + x^2 = 3x^2$. Hence

$$3u^2 < x^2 + xu + u^2 < 3x^2.$$

Now since $u < x$, $x - u$ is positive, and therefore if we multiply the inequalities by $x - u$, we find

$$3u^2(x - u) < (x^2 + xu + u^2)(x - u) < 3x^2(x - u)$$

or

(1)
$$u^2(x - u) < \frac{x^3}{3} - \frac{u^3}{3} < x^2(x - u).$$

We shall use this special algebraic result in solving the following problem.

Example.   Compute the exact area bounded by the graph of x^2, the lines $x = 1$, $x = 5$, and the x-axis.

Solution.   We construct the graph of x^2 (Figure 2.24). We partition the base interval [1, 5] into smaller subintervals and consider a typical subinterval from u to x, where $u < x$. The width of this interval is $dx = x - u$, and the height of the overstep is x^2. Hence we obtain the area of the rectangle below the overstep, $x^2(x - u) = S(x) \, dx$; and similarly, the area below the understep, $u^2(x - u) = s(x) \, dx$. Recalling our inequality (1) above, we find that we can relate these areas as follows:

(2)
$$s(x) \, dx = u^2(x - u) < \frac{x^3}{3} - \frac{u^3}{3} < x^2(x - u) = S(x) \, dx.$$

Now consider the entire partition of the interval [1, 5] with partition points 1, x_1 $x_2, \ldots, x_n = 5$. The formula (2) can be applied to each subinterval, and we may then

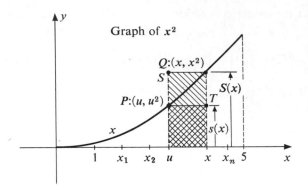

FIGURE 2.24   Single steps of $s(x)$ and $S(x)$ for x^2.

add the inequalities that we get in this manner. We find that for every partition

$$\int_1^5 s(x)\, dx < \frac{x_1^3 - 1}{3} + \frac{x_2^3 - x_1^3}{3} + \cdots + \frac{5^3 - x_n^3}{3} < \int_1^5 S(x)\, dx.$$

All of the terms in the middle cancel in pairs except $(-1/3) + (5^3/3)$. Since $(-1/3) + (5^3/3) = 124/3$, it now turns out that for *every* understep function s and over-step function S

$$\int_1^5 s(x)\, dx \leqq \tfrac{124}{3} \leqq \int_1^5 S(x)\, dx.$$

But x^2 is an increasing function on $[1, 5]$, and thus we know (§9.5) that there is only one number that is thus bracketed by $\int_1^5 s(x)\, dx$ and $\int_1^5 S(x)\, dx$ for every s and S. Therefore $124/3$, or $41.33 \cdots$ area units, is the exact area required by the problem. In notation we say that the integral $\int_1^5 x^2\, dx = \tfrac{124}{3}$.

9.7 *Exercises*

1. For the function x^2 on $[1, 5]$, partition the interval $[1, 5]$ into subintervals of length 1. Find the understep function s and overstep S. Compute: (a) $\int_1^5 s(x)\, dx$; (b) $\int_1^5 S(x)\, dx$.

2. For the finer partition with $dx = \tfrac{1}{2}$ in Exercise 1, compute: (a) $\int_1^5 s(x)\, dx$; (b) $\int_1^5 S(x)\, dx$.

3. If we wish to use the understep function $\int_1^5 s(x)\, dx$ to compute the integral $\int_1^5 x^2\, dx$ to two places after the decimal, what width of partition intervals will assure this accuracy?

4. Show that the exact integral

$$\int_a^b x^2\, dx = \frac{b^3 - a^3}{3} \qquad \text{and that} \qquad \int_a^b 2x^2\, dx = \frac{2}{3}(b^3 - a^3).$$

5. (a) Calculate the exact areas of

$$\int_0^b x^2\, dx, \qquad \text{and} \qquad \int_0^3 (2x^2 + 5)\, dx.$$

(b) Draw the graphs and show the regions whose areas are represented by the integrals.

In Exercises 6–13 below draw the graph and show the region whose exact area is represented by the integral.

6. $\displaystyle\int_1^4 x^3\, dx.$ 7. $\displaystyle\int_0^1 1\, dx.$

8. $\displaystyle\int_1^5 (x^2 - 4)\, dx.$ 9. $\displaystyle\int_{-1}^4 (1 - x)\, dx.$

10. $\int_0^1 (1 - x^2)\, dx.$ 11. $\int_1^3 (2x^2 - 3x + 4)\, dx.$

12. $\int_{-1}^4 x^2\, dx.$ 13. $\int_0^3 (x^2 - 2x)\, dx.$

14. By the method of the Example (§9.6) compute the exact area $\int_1^5 x\, dx.$

15. Show that in Figure 2.24 the area below the curved arc PQ is $(x^3 - u^3)/3.$

16. (a) As in Exercise 1, using steps of length 1, find s and S for x^3 on $[1, 5]$.
     (b) Calculate

$$\int_1^5 s(x)\, dx \qquad \text{and} \qquad \int_1^5 S(x)\, dx.$$

17. (a) Draw a figure like Figure 2.24 for the exact area $\int_1^5 x^3\, dx.$
     (b) Show in this figure that

$$s(x)\, dx < \frac{x^4 - u^4}{4} < S(x)\, dx.$$

18. Using the result of Exercise 17, calculate the exact area $\int_1^5 x^3\, dx.$

19. Show that for every constant c and every function f that is increasing or decreasing (see Exercise 4)

$$\int_a^b c\, f(x)\, dx = c \int_a^b f(x)\, dx.$$

20. Show that if f and g are two increasing functions, then

$$\int_a^b [f(x) + g(x)]\, dx = \int_a^b f(x)\, dx + \int_a^b g(x)\, dx.$$

Both the understep and overstep functions are extreme cases for approximating an integral. It is obvious that a better approximation can be computed by a rule called the *trapezoidal rule*. A step function t is defined for each x in $[a, b]$ by the average

$$t(x) = \tfrac{1}{2}[s(x) + S(x)]$$

and the approximate area A' is computed by

$$A' = \int_a^b t(x)\, dx.$$

Exercises 21–28 below refer to this trapezoidal rule.

21. In Exercise 1, compute t and the integral $\int_1^5 t(x)\, dx.$
22. Show that for any function and any partition,

$$\int_a^b t(x)\, dx$$

can be represented as a sum of areas of trapezoids.
23. Show that for $\int_a^b x\, dx$, the trapezoidal rule gives the exact integral.
24. Show that for the function f and the partition

$$a_0 = x_0 < x_1 < x_2 < \cdots < x_n = b,$$

the trapezoidal rule gives

$$\int_a^b t(x)\, dx = \frac{b - a}{n} \left[\frac{1}{2} f(a) + f(x_1) + f(x_2) + \cdots + f(x_{n-1}) + \frac{1}{2} f(b) \right].$$

25. Use the formula of Exercise 24 to compute $\int_1^6 x^2\, dx$ by the trapezoidal rule for a partition having 10 equal subintervals.
26.* Show that the trapezoidal rule gives the average value of the function f over the interval $[a,b]$ (§4.4) to be

$$\bar{f} = \frac{1}{n} \left[\frac{1}{2} f(a) + f(x_1) + \cdots + f(x_{n-1}) + \frac{1}{2} f(b) \right],$$

and that the units in which $\bar{f}$ is measured are those of $f(x)$, not area units.

27. In Exercise 26, justify the appearance of the endpoints with reduced weight $\frac{1}{2}$.
28. The average value of a function f over the interval $[1, 9]$ is given for a certain partition to be $\bar{f} = 7.2$. Compute $\int_1^9 t(x)\, dx$ for this same function and partition.

9.8 Problems

We shall compute the exact area $\int_a^b x^n\, dx$ bounded by x^n, $x = a$, $x = b$, and the x-axis, where $0 < a < b$, by generalizing Example 9.6.

T1. Verify that $x^{n+1} - u^{n+1} = (x - u)(x^n + x^{n-1}u + \cdots + xu^{n-1} + u^n)$.
T2. Show that (1) of Section 9.6 can be generalized to

$$u^n(x - u) < \frac{x^{n+1} - u^{n+1}}{n + 1} < x^n(x - u).$$

T3. Apply the result of Problem T2 to each subinterval of a partition $a, x_1, x_2, \ldots, x_n, b$ of $[a, b]$ to show that for every such partition

$$\int_a^b s(x)\, dx < \frac{b^{n+1} - a^{n+1}}{n + 1} < \int_a^b S(x)\, dx.$$

T4. Finally, show that by definition the exact area

$$\int_a^b x^n\, dx = \frac{b^{n+1} - a^{n+1}}{n + 1}.$$

In Problems T5–T8, compute the exact areas represented by the integrals.

T5. $\displaystyle\int_1^7 x^3\, dx.$ T6. $\displaystyle\int_0^{10} x^2\, dx.$

T7. $\displaystyle\int_2^4 x^5\, dx.$ T8. $\displaystyle\int_0^1 x^9\, dx.$

T9. Calculate the exact area between the curves $y = 3x - 2$ and $y = x^2$.

10 Limits and Tangent Lines

10.1 Limit of a Function. Let us now formalize the idea of the limit of a function f as x approaches c, written

$$\lim_{x \to c} f(x) = k.$$

We recall that the idea we need is that of a number k that represents the localized trend of values $f(x)$ near c. This may and may not be the same as $f(c)$. Indeed f may not be defined at c (§4.3). Also we recall (§5.4) that to determine the number k as a decimal we must be able to produce a finite decimal that agrees with k to any prescribed number n of decimal places. So the characterization of k will necessarily begin with "for every ϵ," where epsilon is some one of the precision measures

$$\{\cdots 10, 1, 10^{-1}, 10^{-2}, 10^{-3}, \ldots, 10^{-n}, \cdots\}.$$

Finally we must make sure that f has values at numbers x arbitrarily close to c.

We say that c is *approachable* in the domain of f if every δ-neighborhood of c, $\{|x - c| < \delta\}$, includes numbers x, $x \neq c$, in the domain of f. For example, the function $1/x$ is defined when $x \neq 0$. Although $1/x$ is not defined at 0, the number 0 is approachable in the domain of $1/x$.

DEFINITION (Limit of a function).   Let c be a number that is approachable in the domain of the function f. Then

$$\lim_{x \to c} f(x) = k$$

if and only if for every ϵ there is a positive δ such that if x is in the domain of f and $|x - c| < \delta$, $x \neq c$, then

$$|f(x) - k| < \epsilon.$$

We observe at once that the definition does not give instructions on how to find the limit k and does not imply that a limit exists. It only provides a test that enables us to decide for any number k whether it is the limit of f at c or not.

Example.   Calculate $\lim_{x \to 1} (4x^2 - 4)/(x - 1)$, where the function is defined for all numbers x except $x = 1$.

Solution.   Here $c = 1$ and $f(x) = (4x^2 - 4)/(x - 1)$, $x \neq 1$. We must find the number k that is the limit of $f(x)$ as x approaches 1. We cannot evaluate $f(x)$ at 1 because there the formula gives $f(1) = 0/0$, which does not determine any number. However, for any x near 1 but not equal to 1, we may divide twice by $x - 1$ and find that

$$f(x) = \frac{4x^2 - 4}{x - 1} = 4\frac{(x - 1)(x + 1)}{(x - 1)} = 4(x + 1) = 4(x - 1) + 8.$$

The last expression on the right shows that $|f(x) - 8| = 4|x - 1|$ when $x \neq 1$ (Figure 2.25).

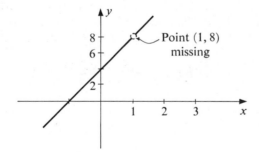

FIGURE 2.25 $f(x) = \dfrac{4x^2 - 4}{x - 1}$.

This shows that if $|x - 1| < \delta$, where δ is not yet determined,

$$|f(x) - 8| < 4\delta.$$

Then for any prescribed ϵ, if we choose δ so that $4\delta \leq \epsilon$, we shall have met the requirement that $|f(x) - 8| < \epsilon$. Thus, in the form of the definition, we can say that for every ϵ we choose $\delta = \epsilon/4$. Then if $|x - 1| < \delta$, it is true that $|f(x) - 8| < \epsilon$. Hence we have shown that

$$\lim_{x \to 1} \frac{4x^2 - 4}{x - 1} = 8.$$

10.2 The Slope of a Curve at a Point. We now undertake to apply the idea of a limit to the problem of finding the slope of the tangent line to the graph of a function f at a point $(x, f(x))$ in the graph. We observe intuitively (Figure 2.26) that a curve ordinarily has a different slope at each point so that the slope of the curve depends both on the coordinates $(x, f(x))$ and on the function f itself. Therefore it is appropriate to use the notation $f'(x)$ for the slope of f at the point $(x, f(x))$. Assigning this notation does not define the slope of the curve, but once we have defined this slope $f'(x)$ mathematically we can define the tangent line to the curve to be the line through $(x, f(x))$ with slope $f'(x)$.

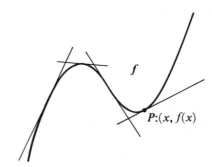

FIGURE 2.26   Tangent lines to a curve.

Example. If $f(x) = 3x^2$, defined on all real numbers, find the slope of the graph at the point $(1, 3)$.

Solution. The first object of the problem is to define the slope of the curve in this special case. As we did for the straight line (§5.3), we calculate the slope for two points on the graph, one point P for which $(x, f(x)) = (1, 3)$ and another point $Q : (w, f(w))$ on the curve near $(1, 3)$. We compute the slope of PQ from the ratio

$$\text{slope } PQ = \frac{f(w) - f(x)}{w - x}$$

and arrange the results in the accompanying table. This table suggests that we use the limit as w approaches 1 to define the slope, and it also suggests that the limit is 6. Let us carry this out.

x	w	$f(x)$	$f(w)$	*Slope*
1	1.1	3	$3(1.1)^2$	6.3
1	1.01	3	$3(1.01)^2$	6.03
1	1.001	3	$3(1.001)^2$	6.003
1	1	3	3	Undefined
1	0.999	3	$3(0.999)^2$	5.997
1	0.99	3	$3(0.99)^2$	5.97
1	0.9	3	$3(0.9)^2$	5.7

We plot a portion of the graph of $3x^2$ near $P : (1, 3)$ and draw the line PQ from the fixed point P to a variable point $Q : (w, f(w))$ near P on the curve (Figure 2.27). We

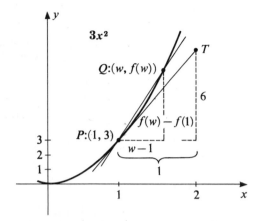

FIGURE 2.27   Slope of a curve at a point.

cannot compute the slope of the curve at P from the formula

$$\text{slope} = \frac{f(w) - f(1)}{w - 1} = \frac{3w^2 - 3}{w - 1}$$

because when $w = 1$ this formula gives the indeterminate result $0/0$. However, we can take the limit

$$\lim_{w \to 1} \frac{f(w) - f(1)}{w - 1} = \lim_{w \to 1} \frac{3w^2 - 3}{w - 1}$$

as the slope of the curve at P.

To do this we proceed as in the example on page 59. We divide twice by $w - 1$, which gives

$$\frac{f(w) - f(1)}{w - 1} = \frac{3w^2 - 3}{w - 1} = \frac{3(w - 1)(w + 1)}{w - 1} = 3(w + 1) = 3(w - 1) + 6,$$

or

$$\frac{f(w) - f(1)}{w - 1} - 6 = 3(w - 1).$$

This equation shows that

$$\lim_{w \to 1} \frac{f(w) - f(1)}{w - 1} = 6;$$

for (§10.1) given any ϵ, we take $\delta = \epsilon/3$ and if $|w - 1| < \delta$, then

$$\left| \frac{f(w) - f(1)}{w - 1} - 6 \right| < \epsilon.$$

10.3   Numerical Derivative.

DEFINITION (Numerical derivative).   The number $f'(x)$ is the numerical derivative of $f(x)$ at x if and only if

$$\lim_{w \to x} \frac{f(w) - f(x)}{w - x} = f'(x).$$

We have not defined the tangent line to a graph at a point. We do this as follows.

DEFINITION (Tangent line).   The tangent line to the graph of f at the point $(x, f(x))$ is the line through the point $(x, f(x))$ that has slope $f'(x)$ at that point.

In xy-coordinates let us find the equation of the tangent line to the graph of $\bar{f}(x)$ at the point $(a, f(a))$, where $x = a$. By definition, the slope of this tangent line is $f'(a)$. Hence (§5.2, and point-slope equation of a line) the equation of the tangent line is

$$y = f(a) + f'(a)(x - a).$$

10.4   Calculation of the Slope of a Curve.

Example.   Find the slope of the graph $3x^2 - 2x + 1$ at the point $(x, f(x))$ and plot several tangent lines.

Solution.   We have defined the slope of the curve $f(x)$ as the number $f'(x)$ determined by

$$f'(x) = \lim_{w \to x} \frac{f(w) - f(x)}{w - x},$$

where in this case $f(x) = 3\dot{x}^2 - 2x + 1$.

We begin by writing down the slope formula for this function. It says that

$$\frac{f(w) - f(x)}{w - x} = \frac{(3w^2 - 2w + 1) - (3x^2 - 2x + 1)}{w - x}$$

if $w \neq x$. We carry out the indicated division by $w - x$. In order to do so, we first rearrange the pairs of terms—those that come from $3x^2$, those that come from $-2x$, and those that come from the term 1.

Thus

$$\frac{f(w) - f(x)}{w - x} = \frac{(3w^2 - 3x^2)}{w - x} - \frac{(2w - 2x)}{w - x} + \frac{(1 - 1)}{w - x} = 3w + 3x - 2.$$

We divide this quotient again by $w - x$, finding a quotient 3 and a remainder $6x - 2$, so that our slope formula can be rewritten

$$\frac{f(w) - f(x)}{w - x} = (6x - 2) + 3(w - x).$$

Now we take the limit as w approaches x and find that

$$f'(x) = \lim_{w \to x} \frac{f(w) - f(x)}{w - x} = \lim_{w \to x} [(6x - 2) + 3(w - x)] = 6x - 2.$$

Thus our solution of the problem shows that the slope of the curve $f(x) = 3x^2 - 2x + 1$ at the point $(x, f(x))$ is given by $f'(x) = 6x - 2$.

Now we return to the example and plot tangent lines to the graph by making a table of three columns for values of $x, f(x), f'(x)$. The first two numbers give the coordinates of the point and the last, $f'(x)$, gives the slope of the tangent line through the point. To plot the tangent lines we have only to plot the line through the given point with given slope as we did for straight lines (§5). We do all this for the function $f(x) = 3x^2 - 2x + 1$ (Figure 2.28).

x	$f(x)$	$f'(x)$
-3	34	-20
-2	17	-14
-1	6	-8
0	1	-2
1	2	4
2	9	10
3	22	16

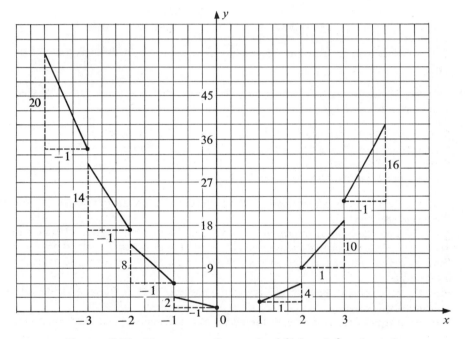

FIGURE 2.28   Tangents to the graph of $f(x) = 3x^2 - 2x + 1$.

10.5   Algebra of Limits.   We now return to the general definition of limits (§10.1) and suppose that we have two functions f and g and that

$$\lim_{x \to c} f(x) = k \qquad \text{and} \qquad \lim_{x \to c} g(x) = l.$$

What can we say about $\lim_{x \to c} (f(x) + g(x))$?   About $\lim_{x \to c} f(x)g(x)$?   About $\lim_{x \to c} af(x)$?   About $\lim_{x \to c} f(x)/g(x)$?   These questions are essentially answered by the continuity of the sum, product, quotient (§8) in conjunction with the definition of the limit (§10.1). We leave proofs to the exercises but state the results in a theorem.

THEOREM (Algebra of limits).   If f and g are two functions on the same domain and c is approachable in this domain and $\lim_{x \to c} f(x)$ exists and $\lim_{x \to c} g(x)$ exists, then

(a) $\lim_{x \to c} [f(x) + g(x)] = \lim_{x \to c} f(x) + \lim_{x \to c} g(x)$;

(b)   if a is any number, $\lim\limits_{x \to c} af(x) = a \lim\limits_{x \to c} f(x)$;

(c) $\lim\limits_{x \to c} f(x)g(x) = \left(\lim\limits_{x \to c} f(x)\right)\left(\lim\limits_{x \to c} g(x)\right)$;

(d)   if $\lim\limits_{x \to c} g(x) \neq 0$, then $\lim\limits_{x \to c} \dfrac{f(x)}{g(x)} = \dfrac{\lim\limits_{x \to c} f(x)}{\lim\limits_{x \to c} g(x)}$.

10.6 Exercises

In Exercises 1–8 find the numerical derivative $f'(x)$ at $(x, f(x))$ as in Section 10.4.

1. $f(x) = x^2 + 3x$.
2. $f(x) = x^2 - x + 1$.
3. $f(x) = -x^2 + 1$.
4. $f(x) = 3x + 5$.
5. $f(x) = 1 + 3x - x^2$.
6. $f(x) = x^4$.
7. $f(x) = ax^2 + bx + c$.
8. $f(x) = x^3 - 5x + 7$.

9. In Exercise 2, without drawing the graph of $f(x)$, plot tangent lines at the points where $x = -3, -1, 0, 1, 3$.

In Exercises 10–13, compute $f'(x)$, plot tangent lines at the indicated points, and then draw the graph of $f(x)$ using the tangent lines as guides.

10. $f(x) = x^2 + 3x$, points where $x = -5, -3, 0, 2$. (See Exercise 1.)
11. $f(x) = 1 + 3x - x^2$, points where $x = -2, 1/2, 2, 4$. (See Exercise 5.)
12. $f(x) = x^3 - 8x$, points where $x = -3, -1, 0, 1, 3$.
13. $f(x) = -x^2/4$, points where $x = -4, -2, 0, 2, 4$.

14. Find the slopes of the tangent lines to the parabola x^2 at the points where $x^2 = 3$. Plot these tangent lines.
15. As in Section 4.5, a point moves on an s-axis with its position s given at time t by $x = -t^2 + 1$. Find the instantaneous velocity when $t = 2$; when $t = 0$; when $t = 1$.
16. Find the equations of the tangent lines to the graph of f in Exercise 10 at each of the indicated points.
17. Find the equations of the tangent lines to the graph of $f(x)$ in Exercise 11 at each of the indicated points.
18. Find equations of tangent lines to the graph of f in Exercise 12 at each of the indicated points.

In Exercises 19–22, calculate the indicated limits showing that the definition 10.1 is satisfied.

19. $\lim\limits_{x \to 2} \dfrac{x^2 - 4}{x - 2}$.
20. $\lim\limits_{x \to 1} \dfrac{x^2 - 1}{x + 1}$.

21. $\lim\limits_{x \to 0} \dfrac{4x^2 + 5x}{x}$.
22. $\lim\limits_{x \to 2} \dfrac{3x^2 - 2x - 8}{x - 2}$.

23. Is there a number k such that:

(a) $\lim\limits_{x \to 0} \dfrac{4x}{x^2} = k$?

(b) $\lim\limits_{x \to 1} \dfrac{|x - 1|}{x - 1} = k$?

Prove that your answers are correct.

24. For the following functions calculate the numerical derivative $f'(x)$ at $(x, f(x))$.
(a) $f(x) = x$; (b) $f(x) = x^2$; (c) $f(x) = x^3$.

10.7   Existence of Limit.   In practice we need a criterion that will insure that

$$\lim_{x \to c} f(x)$$

exists as a real number. With this information we can approximate the limit by evaluating f near c. It is easy to see what the form of such a criterion might be. If the trend of values $f(x)$ for x near c is to cluster arbitrarily close together, then there should be a number k that represents the localized trend. We now state this mathematically.

THEOREM (Existence of a limit).   If the number c is approachable in the domain of the function f, and if for every ϵ there is a δ such that for every pair of numbers x and x' in the δ-neighborhood of c it is true that $|f(x) - f(x')| < \epsilon$, then there exists some number k such that $\lim_{x \to c} f(x) = k$.

Method of proof.   We take a decreasing sequence of epsilons,

$$\epsilon_1 = 10^{-1}, \epsilon_2 \doteq 10^{-2}, \epsilon_3 = 10^{-3}, \ldots, \epsilon_n = 10^{-n}, \ldots.$$

Starting with ϵ_1, we find by hypothesis of the theorem that in some δ_1-neighborhood of c all of the values $f(x)$ agree, with error less than ϵ_1. That is, as decimals they agree to one place after the decimal point. Any one of these values of $f(x)$ then gives k down one place after the decimal point. Next we take ϵ_2, and the hypothesis tells us that there is a δ_2-neighborhood, contained in the preceding δ_1-neighborhood, where all of the values $f(x)$ differ by less than 10^{-2}. Hence all of these values $f(x)$ agree to two places after the decimal point and thus they give the second digit of k after the decimal. Proceeding in this way, we find that if the hypothesis of the theorem is true, we can produce the digit in each decimal place of k in the sense that we mean when we say that the number k is known (§10.3). It is then easy to review these steps to find that the number k, so determined, satisfies the definition of $\lim_{x \to c} f(x) = k$.

10.8   Problems

We will prove the statements in the theorem on the algebra of limits (§10.5). For a model, let us prove the theorem on the product (c).

Proof.   Let

$$\lim_{x \to c} f(x) = k \quad \text{and} \quad \lim_{x \to c} g(x) = l.$$

Now we multiply the numbers $f(x)$ and $g(x)$ when $f(x)$ is near k and $g(x)$ is near l. By the continuity of multiplication (§8.2), for every ϵ there is a positive number n such that when $|f(x) - k| < \eta$ and $|g(x) - l| < \eta$ then $|f(x)g(x) - kl| < \epsilon$. On the other hand, since $\lim_{x \to c} f(x) = k$, there is a number δ_1 such that if $|x - c| < \delta_1$, then $|f(x) - k| < \eta$. Similarly, there is a number δ_2 such that if $|x - c| < \delta_2$, then $|g(x) - l| < \eta$. Finally we choose δ to be the smaller of δ_1 and δ_2. Then if $|x - c| < \delta$, both $|f(x) - k| < \eta$ and $|g(x) - l| < \eta$, and hence $|f(x)g(x) - kl| < \epsilon$. This completes the proof.

T1.   Prove statement (a) in the algebra of limits (§10.5).
T2.   Prove statement (b) in the algebra of limits.
T3.   Prove statement (d) in the algebra of limits.
T4.   Prove that if $f(x)$ is continuous at x, then $\lim_{w \to x} f(w) = f(x)$.
T5.   Prove that if x is in the domain of f and $\lim_{w \to x} f(w) = f(x)$, then f is continuous at x.
T6.   Prove that f cannot have two different limits

$$\lim_{x \to c} f(x) = k, \quad \text{and} \quad \lim_{x \to c} f(x) = k_2.$$

T7.   Prove that if f has a limit, $\lim_{x \to c} f(x) = k$, then the hypothesis of the theorem on the existence of a limit must be satisfied (§10.7).

10R   Summary and Review: Foundations of Calculus

ACADEMIC NOTE.   This is an optional extra section, not essential to the development of the ideas or techniques of calculus. Besides serving as a general review, it may be helpful in assessing the state of one's preparation for the analytical program of calculus that follows.

10.1R   Arithmetic and Algebraic Foundations of Calculus.

The theory of limits, which is essential to calculus, requires that we know more about real numbers than about the decimal representations and elementary operations. We must also be able to operate with inequalities (§§5.1–5.3) and absolute values (§§5.5–5.6), particularly in the solving of inequalities for x (§5.7). These inequalities are often associated with small positive precision measures $\epsilon, \delta, \ldots$ (§5.4). For we cannot write down *all* of the digits of the infinite decimal that represents a real number r. We say that we know r if on demand we can produce the digit in any prescribed place, the kth place. What does this mean precisely (§5.4)?

For a set S containing infinitely many real numbers, even if the set S is bounded above and below, we cannot expect that there is a least and a greatest number in it; but if the set is not empty and is bounded, we find that the structure of the nonterminating decimals insures that there is a least upper bound, or supremum, of S in the real number system (not necessarily in S), written sup S. Also there is in the real number system a greatest lower bound, or infimum, of S, written inf S (§6.3). The numbers sup S and inf S appear frequently in calculus (§8).

We must know more about a single function $y = f(x)$ than that it is a set of pairs of numbers (x, y) in which no two different pairs have the same first element, and from this be able to construct the graph of any function given by an equation like $y = -4x^3 + x + 2$ on $[-10, 10]$. We must know precisely what a continuous function is (§8.3) and understand the basic theorem (§8.4), which says (§§ 8.4–8.7) that a continuous function $f(x)$ on $[a, b]$ is bounded, and attains its maximum value G, its minimum value L, and every value of y, $L \leqq y \leqq G$.

Certain discontinuous functions, such as step functions, become important in calculus as simple functions that approximate more complicated ones (§§ 5.1, 8.8, 9.4, 9.5). The related ramp functions (§5.5) are continuous but made up of broken-line segments and, like step functions, are not represented by single algebraic formulas involving a finite number of operations. They also are important in the numerical program of calculus as approximations, as in the trapezoidal rule (§9.7, Exercise 20).

When we pass from arithmetic to algebra, we imbed the individual numbers of arithmetic in an algebraic system with algebraic operations, and this is implemented and emphasized by introducing the symbol x for an unknown number in equations and formulas. We now make an analogous transition from that of an individual function $y = f(x)$, thought of as a graph, to an algebra of functions, in which the unknown function is represented by the symbol $f(x)$, or just f, usually thought of as a mapping (§§1, 2, 3). We add, subtract, and multiply functions. This algebra of functions is exemplified by polynomials (§1.6), which have some algebraic techniques of their own, such as the explicit algorithms for multiplication and division, and the remainder and factor theorems

(Appendix A). The relationship between algebraic operations with functions and the corresponding operations with numbers is given by the familiar evaluation principle (Appendix A, 1.2), which we used to check calculations in elementary algebra. Division by functions is restricted by the exclusion of division by zero, not only by the zero function but also by any function that has a value zero at some point (§2.3). In the latter case we can work around the difficulty, though it is still present.

10.2R   The Two Fundamental Limits of Calculus.

We studied the integral of a function $f(x)$ over an interval $[a, b]$ written $\int_a^b f(x)\, dx$, first for step functions (§3), which did not require any limit-theoretic ideas, and then for more general functions (§9) in connection with the problem of finding the exact area of a region bounded by a curve. We carried out this exact-area procedure numerically in a few cases (§9.6) but we seek other procedures for evaluating integrals numerically, first approximate ones (§9.5). Other exact integration methods are yet to be studied, involving the algebra of functions. Besides area, other physical interpretations of the integral are possible (§§3.3, 3.4).

The other fundamental limit of calculus is that of the derivative $f'(x)$, defined abstractly (§4.3), by the limit

$$\lim_{w \to x} \frac{f(w) - f(x)}{w - x}.$$

It applies to a line or a curve $y = f(x)$ (§§10.2, 10.4) and may be interpreted as the slope of the tangent line to the curve at the point $(x, f(x))$, or more simply as the slope of the curve. When the independent variable has significance as physical time and the dependent variable $f(x)$ represents a distance, the derivative $f'(x)$ can be thought of as representing an instantaneous velocity (§4.5).

All of this involves the idea of the limit of a function itself that requires a precise definition (§10.1). We defined the limit but did not develop the theory of limits in detail, only the fundamental algebraic operations (§10.5). Moreover, while our definition of the exact integral $\int_a^b f(x)\, dx$ involved a limit-like procedure (§9.4), we did not succeed in stating the definition of the limit of a function so that it covers the procedure for defining the exact integral. In spite of this gap, we think of the exact integral $\int_a^b f(x)\, dx$ as a "limit of a sum."

10.3R   Self Quiz on Fundamentals

1. Define a function (§1.3).
2. Define $\lim_{x \to c} f(x) = k$ (§10.1).
3. What does it mean to say that we "know" a nonterminating decimal (§6.4)?
4. What does it mean to say that $\lim_{x \to c} f(x)$ exists (§10.7)?
5. Prove that there exists a real number q such that $4 \div 7 = q$ (§6.3).
6. Define: The function $f(x)$ is continuous at the point where $x = c$ (§8.3).
7. Let f be a continuous function on the interval $[0, 1]$ having $f(0) = -1$ and $f(1) = 3$. How do we know that there is a number c between 0 and 1 where $f(c) = 0$? Would this necessarily be true if f is replaced by a step function s (§8.4)?
8. If $s(x)$ is a step function on the interval $[a, b]$, what is meant by the integral $\int_a^b s(x)\, dx$ (§3.2)?
9. Define $\int_a^b f(x)\, dx$ if f is an increasing function on $[a, b]$, or if f is continuous (§9.5).
10. Define the numerical derivative $f'(x)$ at a point $(x, f(x))$ in the graph of f if f is a function on $[a, b]$ (§10.3).

11. State some geometric and mechanical interpretations of $f'(x)$ (§4).
12. The graph of $y = f(x)$ is ordinarily a curve. Define mathematically the tangent line to such a curve at the point $(x, f(x))$ (§10.3).
13. What does the statement

$$E = \{x \in R \mid |x - c| < \epsilon\}$$

mean? Describe this set by showing it graphically on the number line.
14. What number r is represented by

$$r = \sup \{x \in R \mid x^2 < 3\}?$$

Is r in the set?

10.4R Miscellaneous Exercises

1. Find an unknown function $f(x)$ that is a polynomial of degree not more than 2 and for which $f(1) = 5, f(-1) = -1, f(3) = 11$. Is $f(x)$ uniquely determined?
2. Plot the graph of the function $f(x)$ given, where $x \neq 0$, by $f(x) = x/|x|$, with $f(0) = 1$.
3. Solve the following inequalities for x and represent the solutions graphically in the number line.
   (a) $|x - 5| < 2$; (b) $|x + 5| > 2$; (c) $x^2 + x + 1 < 2$.
4. Calculate $f'(3)$ from the definition, where $f(x) = (x^2 - 4x + 5)/12$. Find the equation of the tangent line to the graph at the point where $x = 3$ and plot this tangent line.
5. Calculate the exact integral $\int_1^5 (x^2 - 3x + 1)\, dx$ (§9.6).
6. We know that if $f(x) = ax^2 + bx + c$, then $f'(x) = 2ax + b$ (§10.6, Exercise 7). Find an unknown function g whose derivative at x is given by $g'(x) = -5x + 7$. Is g uniquely determined?
7. With the additional information that $g(0) = 8$ in Exercise 6, determine g uniquely.
8. A baseball is thrown vertically upward at time $t = 0$. At any time t seconds after that, its height y is given by $y = -16.1t^2 + 50t$.
   (a) What is the instantaneous velocity at the start, at $t = 0$?
   (b) What is the velocity 4 sec later?

In Exercises 9–11, a point moves uniformly along the y-axis *between the vertex positions given*. Plot the motion $y = f(t)$ in the ty-plane as a ramp function. Find the instantaneous velocity $v = f'(t)$ and plot its graph in a tv-plane.

9. The point starts from $(0, 0)$, then moves uniformly to $(2, 5)$, then uniformly to $(4, 0)$.
10. The point starts from $(0, 0)$, then moves uniformly to $(2, 5)$, then $(4, 0)$, then $(6, 5)$, then $(8, 0)$.
11. The point starts from $(1, 5)$, then moves uniformly to $(2, 8)$, then to $(3, 8)$, then to $(4, 6)$, then to $(5, 1)$, then to $(6, -3)$.

12. Show that the equation $(y + 2)/x = 5$ represents a straight line with one point missing. What point is missing?
13. Prove that the slope of any pair of points $P:(a, b)$, $Q:(c, d)$ is the same whether it is computed with $(x_1, y_1) = (a, b)$ and $(x_2, y_2) = (c, d)$ or reversed.
14. The function $f(x) = mx + b$ has a straight line graph with slope m. What is the significance of the number b?
15. Factor $f(x) = x^3 + x^2 - 2x$ and construct its graph by multiplication using the table of signs (§1.7, Exercise 14).
16. The velocity of a point moving on the y-axis is given for time t by $v = t^3 + t^2 - 2t$. Find the time intervals for when the point is moving up and when it is moving down.
17. Use synthetic division to express the polynomial

$$f(x) = x^3 + 6x^2 - 3x + 5$$

as a polynomial in powers of $x - 1$ (§2.5).

18. Is it true in the algebra of functions that if $fg = 0$, then either $f = 0$ or $g = 0$? Consider the ramp functions in which f joins $(-1, 0)$ to $(0, 0)$ to $(1, 3)$ and g joins $(-1, 4)$ to $(0, 0)$ to $(1, 0)$.

19. Compute as a polynomial by synthetic division

$$\frac{f(x) - f(a)}{x - a}, \qquad \text{where} \quad f(x) = x^3 + 3x - 7.$$

Why does the division always produce the remainder 0?

20. Find the maximum and minimum values (if they exist) of $1/(x^2 + x + 1)$.

21. Solve graphically $x^3 + 5x - 10 = 0$.

22. Prove that for all real numbers x, y, it is true that $2xy \leqq x^2 + y^2$.

23. Find sup $\{x \in R \mid x^2 + 2x + 2 < 3\}$.

24. Find the overstep function $S(x)$ and understep function $s(x)$ of $f(x) = 1/x$ on $[1, 5]$, using steps of width 1. Calculate $\int_1^5 S(x)\, dx$ and $\int_1^5 s(x)\, dx$ and find an approximate value of $\int_1^5 f(x)\, dx$.

25. Compute from the definition $f'(x)$ if $f(x) = 3x^4 - 1$.

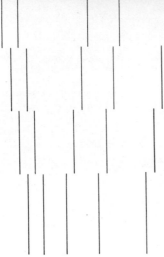

Differential Calculus of Polynomials

CHAPTER 3

11 Idea of the Derivative Function

11.1 Numerical Derivatives of Polynomials. Let us compute the numerical derivative $f'(x)$ for a polynomial function $f(x)$ at the point $(x, f(x))$. We recall (§10) that, regarding $(x, f(x))$ as a fixed point and $(w, f(w))$ as a variable point on the graph, the numerical derivative is given by

$$f'(x) = \lim_{w \to x} \frac{f(w) - f(x)}{w - x}.$$

Example. (a) Compute the numerical derivative at $(x, f(x))$ of the polynomial $f(x) = 4x^3 - 7x^2 - 10x + 6$. (b) Find the slope of the tangent line to the graph of $f(x)$ at several points, plot these tangent lines and then draw the graph of $f(x)$. (c) Set $m = f'(x)$ and, regarding $f'(x)$ as a function, plot its graph in separate xm-coordinates.

Solution. We form the differential quotient

$$\frac{f(w) - f(x)}{w - x} = \frac{(4w^3 - 7w^2 - 10w + 6) - (4x^3 - 7x^2 - 10x + 6)}{w - x}.$$

We rearrange this expression so as to associate like powers of w and x. This gives us

$$\frac{f(w) - f(x)}{w - x} = 4\left(\frac{w^3 - x^3}{w - x}\right) - 7\left(\frac{w^2 - x^2}{w - x}\right) - 10\left(\frac{w - x}{w - x}\right).$$

Proceeding to the limit, we get

$$f'(x) = \lim_{w \to x} \frac{f(w) - f(x)}{w - x}$$

$$= 4 \lim_{w \to x} \left(\frac{w^3 - x^3}{w - x}\right) - 7 \lim_{w \to x} \left(\frac{w^2 - x^2}{w - x}\right) - 10 \lim_{w \to x} \left(\frac{w - x}{w - x}\right),$$

where in the last expression we have applied the theorem on the algebra of limits (§10.6), first in distributing the limit operation term by term and then in factoring out the coefficients 4, −7, and −10. We recognize the three limits in the resulting expression as the derivatives of x^3, x^2, and x. We have previously calculated these derivatives (§10.7, Exercise 10) and we found them to be $3x^2$, $2x$, and 1, so that

$$f'(x) = 4(3x^2) - 7(2x) - 10(1) = 12x^2 - 14x - 10.$$

This completes Part (a).

We learn from this solution that, through the algebra of limits, the limit process involved in the calculation of a derivative of a polynomial can be applied one term at a time and that the problem is thereby reduced to a calculation of the derivative of a term of the form x^n. Thus if we calculate once and for all the derivative of x^n, we can find the derivative of any polynomial without going back to the epsilon-delta definition of the limit.

(b) Before carrying out Part (b) of the problem, it is well to remind ourselves that the function notation $f(x)$ enables us to eliminate many words. We do not have to say, "Substitute 2 for x in the polynomial $f(x) = 4x^3 - 7x^2 - 10x + 6$ and obtain the number -10." Instead, we write only $f(2) = -10$. Similarly, in the formula $f'(x) = 12x^2 - 14x - 10$, when we write $f'(2) = 10$, we mean that we have substituted 2 for x in the formula for $f'(x)$ and we get the number 10.

Also to make a number of these substitutions into polynomial formulas it is easier to use synthetic division (Appendix A). We exhibit below the synthetic division to calculate $f(2)$ and $f'(2)$.

$$
\begin{array}{rrrr|l}
4 - 7 - 10 + 6 & & & & \underline{2} \\
+ 8 + 2 - 16 & & & & \\
\hline
4 + 1 - 8 & \underline{-10} = f(2).
\end{array}
\qquad
\begin{array}{rrr|l}
12 - 14 - 10 & & & \underline{2} \\
+ 24 + 20 & & & \\
\hline
12 + 10 & \underline{+ 10} = f'(2).
\end{array}
$$

Thus we make a table of values of x, $f(x)$, $f'(x)$ (Figure 3.1), plot the points $(x, f(x))$, and through each one construct the line with slope $f'(x)$. Then we finish the graph of $f(x)$, using these tangent lines as guides. For example, at the point where $(2, f(2)) = (2, -10)$, we plot the tangent line whose slope is given by $f'(2) = +10/1$. This completes Part (b).

(c) We plot the graph of $m = 12x^2 - 10x - 10$ on the same x-scale but we use slope units on the vertical scale instead of y (Figure 3.2). We recognize that the values of the

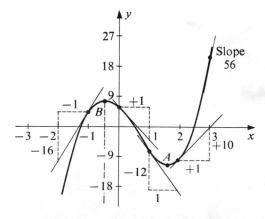

FIGURE 3.1   Graph of $y = 4x^3 - 7x^2 - 10x + 6$ with tangent lines.

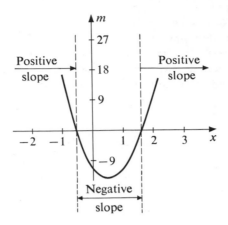

FIGURE 3.2   Graph of the derivative $m = 12x^2 - 10x - 10$.

numerical derivative at each point $(x, f(x))$ generate a new polynomial function $12x^2 - 10x - 10$ associated with the original polynomial $4x^3 - 7x^2 - 10x + 6$. We observe that at the turning points A and B on the graph of $f(x)$, where $f(x)$ changes from decreasing to increasing or from increasing to decreasing, the values of x are the same as those where the graph of $f'(x)$ crosses the x-axis; that is, $f'(x)$ changes from negative to positive, or positive to negative. This completes the solution of the example.

11.2   Derivative Functions and Operators.   In general, a function $f(x)$ is said to be *differentiable* if at each point $(x, f(x))$ of its graph it has a numerical derivative $f'(x)$.

DEFINITION (Derivative function).   If $f(x)$ is a differentiable function, the *derivative function $f'(x)$* of $f(x)$ is the function whose value at each number x in the domain of $f(x)$ is the numerical derivative $f'(x)$.

We also find it convenient to indicate the derivative function by means of an operator D.

DEFINITION.   The operator D is defined on the domain of all differentiable functions $f(x)$ by $Df(x) = f'(x)$.

11.3   Formulas for Differentiation.   To *differentiate* a function $f(x)$ means to find the derivative function $f'(x)$. We have found that we can differentiate a polynomial function largely by algebraic calculation without having to resort to the direct calculation of the limit

$$f'(x) = \lim_{w \to x} \frac{f(w) - f(x)}{w - x}.$$

We will use the operator symbol D and introduce single-letter symbols u, v, etc., for functions to simplify the formulas. A first basic set of formulas is the following. We memorize them.

Let u and v be differentiable functions and c a constant function; then

I. $D(u + v) = Du + Dv,$
II. $Dcu = cDu,$ $\Big\}$ (Linearity)

III. $Duv = uDv + vDu$, (Product Rule)

IV. $Dc = 0$,

V. $Du^n = nu^{n-1} Du$, where n is a positive integer,

V(b). $Dx^n = nx^{n-1}$, where n is a positive integer.

11.4 Linear Operator. Rules I and II, asserting that $D(u + v) = Du + Dv$ and $Dcu = cDu$, express the fact that D is a *linear operator* on differentiable functions.

This property of linearity obviously simplifies the algebra involving the operator D, but we observe that D does not preserve multiplication of functions as well. Duv is not equal to $(Du)(Dv)$.

Let us now use Rules I–V before we prove them.

11.5 Calculation of Derivatives. We have previously computed $Dx = 1$ (§10.7, Exercise 10). Also by IV, $D1 = 0$. These are the derivatives of the functions 1 and x, which generate polynomials (§2.2).

Example A. Compute $D(5x^3 - 7x^2 + 2x - 9)$.

Solution. By Rule I,

$$D(5x^3 - 7x^2 + 2x - 9) = D(5x^3) + D(-7x^2) + D(2x) + D(-9).$$

Then by Rule II, this becomes $5Dx^3 - 7Dx^2 + 2Dx - 9D1$.

Now by Rule V(b), $Dx^3 = 3x^2$, $Dx^2 = 2x$, $Dx = 1$; and by Rule IV, $D(1) = 0$. Therefore, finally

$$D(5x^3 - 7x^2 + 2x - 9) = 5(3x^2) - 7(2x) + 2(1) - 9(0) = 15x^2 - 14x + 2.$$

Example B. Compute $D(2x + 1)^7$.

Solution. This is in the form u^7, with $u = 2x + 1$. First by Rule V,

$$D(2x + 1)^7 = 7(2x + 1)^6 D(2x + 1)$$
$$= 7(2x + 1)^6 (2(1) + 0)$$
$$= 14(2x + 1)^6.$$

We observe that the result would be wrong if we forgot the last factor Du in V. It is called the *Chain rule factor*. Here $Du = D(2x + 1) = 2$, which gave us the coefficient 14 in the final answer. Failure to incorporate the chain rule factor Du would have given the erroneous result $7(2x + 1)^6$.

11.6 Differentiation with Respect to Other Variables. In most cases, the letter we use for the independent variable in a function $f(x)$, or $f(t)$, or $z^3 - 4$ is clear enough. So it is sufficient to write D to indicate differentiation with respect to that variable. However, in some cases it is desirable to emphasize the variable with respect to which D differentiates. We do this by affixing a subscript to D. Thus $D_t f(t) = f'(t)$ indicates differentiation with respect to t, $D_z(z^3 - 4) = 3z^2$ with respect to z, and $D_x f(x) = f'(x)$ denotes differentiation with respect to x.

Example.   Compute $D_t(3t - x)^4$.

Solution.   By Rule V,

$$D_t(3t - x)^4 = 4(3t - x)^3 D_t(3t - x)$$
$$= 4(3t - x)^3(3 - 0)$$
$$= 12(3t - x)^3.$$

11.7   Proofs of the Formulas for Differentiation.   The first three formulas express the relationship of the derivative operator D to the basic operations in the algebra of functions (§2.1). We leave the proofs of I and II to the exercises and prove the product rule, III, here. We need a preliminary theorem.

LEMMA.   If $u(x)$ has a numerical derivative $u'(x)$ at x then

$$\lim_{w \to x} u(w) = u(x).$$

Proof.   We easily prove that $\lim_{w \to x} (w - x) = 0$. Then

$$\lim_{w \to x} [u(w) - u(x)] = \lim_{w \to x} (w - x) \frac{u(w) - u(x)}{w - x}$$
$$= \lim_{w \to x} (w - x) \lim_{w \to x} \frac{u(w) - u(x)}{w - x}$$
$$= 0 \cdot u'(x) = 0.$$

In this we used the theorem on a limit of a product (Theorem 10.5(c)). Hence $\lim_{w \to x} u(w) = u(x)$.

Now we return to the proof of III. By definition of the derivative, the value of Duv at x is given by

$$\lim_{w \to x} \frac{u(w)v(w) - u(x)v(x)}{w - x}.$$

We add and subtract $u(w)v(x)$ to the numerator and find that

$$\frac{u(w)v(w) - u(x)v(x)}{w - x} = \frac{u(w)v(w) - u(w)v(x)}{w - x} + \frac{u(w)v(x) - u(x)v(x)}{w - x}$$
$$= u(w) \frac{v(w) - v(x)}{w - x} + v(x) \frac{u(w) - u(x)}{w - x}.$$

Hence, applying the algebra of limits (§10.6), we find that at x

$$Duv = \lim_{w \to x} u(w) \lim_{w \to x} \frac{v(w) - v(x)}{w - x} + v(x) \lim_{w \to x} \frac{u(w) - u(x)}{w - x}.$$

By the lemma, the first factor on the right is $u(x)$. Hence

$$Duv = u(x)v'(x) + v(x)u'(x) \qquad \text{at any} \quad x,$$

or

$$Duv = uDv + vDu,$$

which was to be proved.

We now prove V by mathematical induction. The formula $Du^n = nu^{n-1} Du$ is true when $n = 1$. In that case it says that $Du^1 = (1)Du^{1-1} Du = (1)u^0 Du = Du$, since

$u^0 = 1$. Next, we prove that if it is true for some exponent n, it is true for the next exponent $n + 1$. By Rule III,

$$Du^{n+1} = D(u^n)(u) = u^n \, Du + u \, Du^n.$$

Now if Rule V is true for some integer n, in the last term we may replace Du^n by $nu^{n-1} \, Du$. This gives $Du^{n+1} = (n + 1) \, u^n \, Du$, which says that if Rule V is true for some integer n, it is true for the next integer $n + 1$.

Hence Rule V is true for every positive integer n. It is true for $n = 1$ and hence for the next exponent $n = 2$. Then, since it is true when $n = 2$, it is true for the next exponent $3, \ldots$, and for every integer n.

We easily prove Rule V(b) from V. In fact, by Rule V,

$$Dx^n = nx^{n-1} \, Dx = nx^{n-1}(1) = nx^{n-1}.$$

11.8 Exercises

In Exercises 1–6, compute the derivative of the polynomials. Make a table of values of $(x, f(x)), f'(x)$. Plot several points $(x, f(x))$ and tangent lines. Then plot the graph of the function. Find the coordinates of the stationary points.

1. $f(x) = x^2 - 3x + 2.$
2. $f(x) = x^3 + 7x + 1.$
3. $f(x) = -x^2 + 3x - 5.$
4. $f(x) = x^4 + x^2 + 1.$
5. $f(x) = x^3 - 7x + 1.$
6. $f(x) = x^3.$

7. Differentiate the following functions if they represent polynomials. Indicate which ones are not polynomials.
    (a) $x^2 + 1$, (b) $x^2 - \sqrt{3}$, (c) $10^{\sqrt{5}}$, (d) $1/x$, (e) $x^{-1/2}$, (f) $\sqrt{x + 1}$,
    (g) $2x^2 - \pi x + 3.$

In Exercises 8–11, find the equation of the tangent line to each of the graphs at the point indicated. Sketch the curve and the tangent line.

8. $f(x) = x^3 - 5x + 2$ at $(3, f(3)).$
9. $f(x) = x^6$ at $(1, f(1)).$
10. $f(x) = (x - 1)^3$ at $(1, f(1)).$
11. $f(x) = (3x + 1)^4$ at $(0, f(0)).$

In Exercises 12–25, calculate the derivatives indicated using Rules I–V.

12. $D(3x^2 + 4x - 9).$
13. $D(1 - x - x^2).$
14. $D_t(2t^2 - 3t + 4).$
15. $D_z(z^2 - 1).$
16. $D_t(t^2x - ty - z).$
17. $D_u u.$
18. $D_t 0.$
19. $D_x(ax^2 + bx + c).$
20. $D(2x - 1)^3.$
21. $D(2x - 1)^3(x + 4)^2.$
22. $Dx^2(3x + 1)^4.$
23. $D(2x - 1)^5(3x + 2)^7.$
24. $D[7(1 - x)^3 + 5(1 - x)^2 - 10]$
25. $D(1 - x - x^2)^4.$

In Exercises 26–28, calculate the derivative in three ways and compare the results (1) by Rules I–V directly; (2) by first carrying out the indicated operations and then applying Rules I–IV; (3) by the definition of the derivative as a limit.

26. $D(3x + 4)^2.$
27. $D(2x - 1)(3x + 5).$
28. $D(3x - 5)^5$ at the point where $x = 0.$

29. Prove Rule I.
30. Prove Rule II.
31. Prove Rule IV.

11.9   Problems

T1.   Verify by synthetic division that

$$\frac{w^n - x^n}{w - x} = w^{n-1} + xw^{n-2} + x^2w^{n-3} + \cdots + x^{n-1}.$$

Hint: The synthetic division begins

$$1 + 0 + 0 + \cdots + 0 - x^n \underline{\lfloor x}$$
$$\frac{+ x}{1 + x} \qquad\qquad\qquad \lfloor \text{Remainder}$$

T2.   Verify by a second division that

$$\frac{w^n - x^n}{w - x} = nx^{n-1} + (w - x)Q,$$

where

$$Q = w^{n-2} + 2xw^{n-3} + 3x^2w^{n-4} + \cdots + (n - 1)x^{n-2}.$$

T3.   Use T2 to prove Rule V(b) by another method.

12   Interpretations and Applications of Derivatives

12.1   Evaluation Theorem of Differential Calculus.   Differential calculus is the study of derivatives of functions: the theory, interpretation, calculation techniques, and applications of derivatives.   The basic definition of the numerical derivative $f'(x)$ of $f(x)$ at x was given by the limit

$$f'(x) = \lim_{w \to x} \frac{f(w) - f(x)}{w - x}.$$

The interpretations and applications usually come from this definition.

On the other hand, we now see an entirely different aspect of differential calculus, the derivative function $f'(x)$ can be calculated, for all polynomials at least, without recourse to limits, by applying a few algebraic rules, I–V (§11).   This leads to a view of differential calculus as algebra operating within the algebra of functions (§1).   We have already established the connection between these two aspects of differential calculus but because of the importance of this idea let us record it here as a theorem.

THEÓREM (Evaluating theorem of differential calculus).   If the function $f(x)$ has a derivative function $f'(x)$, then the numerical derivative of $f(x)$ at $(x, f(x))$ is the value $f'(x)$ of the derivative function $f'(x)$ at x.

Proof.   We defined the derivative function $f'(x)$ to be that function whose value at x is the numerical derivative $f'(x)$.   Then we proved the Rules I–V in terms of this definition.   Hence the theorem is already proved.

12.2   The Sign of the Derivative.   It is graphically obvious that a function that has a positive derivative at x is increasing at x and a function that has a negative derivative at x is decreasing at x (Figure 3.3).   Let us construct an analytic description of these facts.

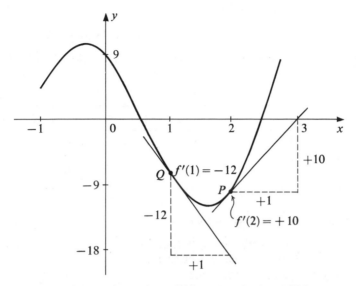

FIGURE 3.3   Points where $f'(x) > 0$ and where $f'(x) < 0$.

DEFINITION.   The function $f(x)$ is *locally increasing* at x if there is some δ-neighborhood of x in which

$$0 < \frac{f(w) - f(x)}{w - x}$$

for all numbers w in the neighborhood. $f(x)$ is *locally decreasing* if there is some δ-neighborhood of x in which

$$\frac{f(w) - f(x)}{w - x} < 0.$$

It follows from this definition that if $f(x)$ is locally increasing at x, and if z and w are near x, and $z < x < w$, then $f(z) < f(x) < f(w)$.

THEOREM.   If $f'(x) > 0$, then $f(x)$ is locally increasing at x. If $f'(x) < 0$, then $f(x)$ is locally decreasing at x.

Proof.   We will prove the statement for $f'(x) > 0$. By the definition of the derivative and that of a limit, for every ϵ there is a δ such that if $|w - x| < \delta$, then

$$f'(x) - \epsilon < \frac{f(w) - f(x)}{w - x} < f'(x) + \epsilon.$$

Now since this is true for *every* ϵ, it is true in particular if $\epsilon < \frac{1}{2}f'(x)$. When δ is determined to meet this precision we know, from the left member of the inequality above, that if $|w - x| < \delta$, then

$$0 < \frac{1}{2}f'(x) < \frac{f(w) - f(x)}{w - x}.$$

This completes the proof.

Example.   For the function $f(x) = 4x^3 - 7x^2 - 10x + 6$ (Example 11.1), find the set of points where $f(x)$ is increasing and the set where $f(x)$ is decreasing.

Solution.   We differentiate to find that

$$f'(x) = 12x^2 - 14x - 10 = 2(2x + 1)(3x - 5).$$

Using the techniques for constructing graphs by multiplication (§1.6), we tabulate the signs of $f'(x)$.

	$x < -\frac{1}{2}$	$x = -\frac{1}{2}$	$-\frac{1}{2} < x < \frac{5}{3}$	$x = \frac{5}{3}$	$\frac{5}{3} < x$
$2x + 1$	$-$	0	$+$	$+$	$+$
$3x - 5$	$-$	$-$	$-$	0	$+$
$f'(x)$	$+$	0	$-$	0	$+$

With the aid of the theorem above, we see that:

when $c < -\frac{1}{2}$, $f(x)$ is locally increasing;
when $-\frac{1}{2} < x < \frac{5}{3}$, $f(x)$ is locally decreasing;
when $\frac{5}{3} < x$, $f(x)$ is locally decreasing.

This completes the example.

12.3   Geometric Interpretation of the Derivative.   We recall the geometric interpretation of the derivative.   The numerical derivative $f'(x)$ of $f(x)$ at $(x, f(x))$ is the slope of the tangent line to the graph at $(x, f(x))$ (§10.4).   Alternatively, $f'(x)$ is the slope of the curve at $(x, f(x))$ (§10.3).

12.4   Instantaneous Velocity.   We turn now to a mechanical interpretation of the derivative.   Let $y = f(t)$ represent the position (t, y) of a point moving on the y-axis t seconds after the start of the motion.   Let $(t, f(t))$ be a fixed position at time t and let $(s, f(s))$ be a position at variable time s and t.   Then the ratio

$$\frac{f(s) - f(t)}{s - t}$$

represents the velocity of the moving point over the time interval from t to s.   To localize this velocity at time t, we take the limit as s approaches t and *define* this limit $f'(t)$ to be the instantaneous velocity at time t.

DEFINITION.   For the motion $y = f(t)$,

$$\textit{instantaneous velocity at time } t = \lim_{s \to t} \frac{f(s) - f(t)}{s - t} = f'(t).$$

Example.   Find the instantaneous velocity at any time t of a point mass that falls freely on the y-axis according to the relation $y = -16t^2$, y in feet, t in seconds.

Solution.   Here we have $f(t) = -16t^2$ and the derivative $f'(t) = -32t$.   According to the above interpretation of the value of the derivative, the instantaneous velocity at time t is $-32t$ ft/sec.

We plot the graph of this velocity relation $v = -32t$ (Figure 3.4).

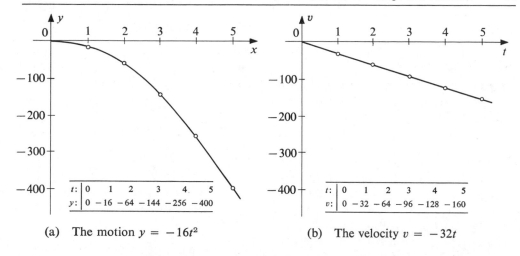

t:	0	1	2	3	4	5
y:	0	-16	-64	-144	-256	-400

(a)   The motion $y = -16t^2$

t:	0	1	2	3	4	5
v:	0	-32	-64	-96	-128	-160

(b)   The velocity $v = -32t$

FIGURE 3.4   Instantaneous velocity function.

12.5   Acceleration.   Another mechanical interpretation of the derivative is acceleration. For the motion on the y-axis given by $y = f(t)$, we define a velocity function by $v(t) = f'(t)$ in accordance with the instantaneous velocity interpretation of $f'(t)$ (§12.3). Then we make the following definition.

DEFINITION.   The *instantaneous acceleration* α at time t of a point moving with velocity $v(t)$ is given by $\alpha = v'(t)$.

An alternative form of this definition involves the idea of the second derivative. We introduce the notation $f^{(2)}(t)$ for the "second derivative," $D_t[D_t f(t)]$, and represent this twice-applied differentiation operator by D_t^2, so that $D_t^2 f(t) = D_t[D_t f(t)] = f^{(2)}(t)$. Then at time t, $\alpha = f^{(2)}(t)$.

Example.   Find the acceleration of the freely falling point mass of Example 12.3.

Solution.   The position function is given by $f(t) = -16t^2$. By definition, the acceleration function is the second derivative, acceleration = $D_t^2(-16t^2) = D_t(-32t) = -32$ feet per second, per second. Thus the acceleration is a constant.

12.6   Local Rate of Change of a Function.   The idea of instantaneous velocity can be generalized to any relation between variables x and y given by a differentiable function $f(x)$. Let $(x, f(x))$ be a fixed point in the graph and let $(w, f(w))$ be a nearby variable point in the graph. Then we can think of the ratio

$$\frac{f(w) - f(x)}{w - x} = \frac{\text{change in } f(x)}{\text{change in } x}$$

as the rate of change of $f(x)$ with respect to x between the points $(x, f(x))$ and $(w, f(w))$. We can localize this at the fixed point $(x, f(x))$ by passing to the limit as w approaches x. We thus define the quantity "the local rate of change of $f(x)$ with respect to x" by means of the derivative.

Example.   What is the rate of change of the area of a circle with respect to the radius?

Solution.   The area A of a circle is given by $A = \pi r^2$. We compute the derivative $D_r \pi r^2 = 2\pi r$. Thus the rate of change of the area of a circle with respect to the radius is $2\pi r$ area units per unit length of the radius. For example, when the radius is 2 feet, the rate of change of its area is given by $2\pi(2) = 4\pi = 12.57$ square feet per foot of increase in the radius.

12.7   Local Scale Factor in a Mapping.   We recall the idea of a function f as a mapping (§1.4) in which each number x on the x-axis is mapped onto the number $f(x)$ on the y-axis. Similarly, the number w is mapped onto $f(w)$. We now compare the length of the interval $[x, w]$ on the x-axis with the length of the interval $[f(x), f(w)]$ on the y-axis by forming the ratio

$$\frac{f(w) - f(x)}{w - x}.$$

Assuming that f is differentiable, we define the quantity $E(w, x)$ by

$$\frac{f(w) - f(x)}{w - x} - f'(x) = E(w, x).$$

This can be rewritten in the form

$$f(w) - f(x) = [f'(x) + E(w, x)](w - x).$$

This shows that the length of the interval $[x, w]$ is multiplied by the scale factor $[f'(x) + E(w, x)]$ when it is mapped on the y-axis by f. We can localize this scale factor at x by taking the limit as w approaches x. By definition of the derivative, $\lim_{w \to x} E(w, x) = 0$, so that, in the limit, the local scale factor of the mapping becomes $f'(x)$.

Example A.   Find the local scale factor of the mapping $f(x) = 2x$.

Solution.   We plot the graph of $2x$ (Figure 3.5(a)). The length of the interval $[x, w]$ is $w - x$ and it maps onto $[2x, 2w]$, whose length $2w - 2x = 2(w - x)$. This says that the image of the interval $[x, w]$ is twice as long as $[x, w]$. The local scale factor at every point is therefore 2. This agrees with our definition; for $f'(x) = 2$.

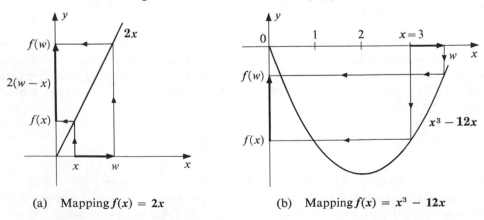

(a)   Mapping $f(x) = 2x$ (b)   Mapping $f(x) = x^3 - 12x$

FIGURE 3.5   Local scale factor.

Example B.   Find the local scale factor of the mapping $f(x) = x^3 - 12x$ at points (a) where $x = 0$; (b) where $x = 2$; (c) where $x = 3$ (Figure 3.5(b)).

Solution.   We compare the length of the map interval $f(w) - f(x)$ with the length of the interval $w - x$ by the ratio

$$\frac{f(w) - f(x)}{w - x} = \frac{(w^3 - 12w) - (x^3 - 12x)}{w - x} = (3x^2 - 12) + (w + 2x)(w - x),$$

where the last expression may be found by division or by subtracting the derivative $f'(x) = 3x^2 - 12$ from the left member. The scale factor for the interval $[x, w]$ is then $(3x^2 - 12) + (w + 2x)(w - x)$. If we proceed to the limit as w approaches x the last term, $E(w, x) = (w + 2x)(w - x)$, approaches zero and the local scale factor at x is given by $f'(x) = 3x^2 - 12$. (a) At the point where $x = 0$, the local scale factor is -12. (b) At the point where $x = 2$, the local factor is given by $f'(2) = 0$. (c) At the point where $x = 3$, the local scale factor is given by $f'(3) = 15$. We observe what this signifies geometrically (Figure 3.5(b)).

12.8   Marginal Quantities in Economics.

Let $C(q)$ be a function that relates two economic quantities q and $C(q)$.

DEFINITION.   The economic quantity, *marginal value* of $C(q)$ at q, is defined by the value of the derivative $C'(q)$.

Example.   The total cost $C(q)$ of producing q units of a certain good is given by

$$C(q) = 10 + 15q - 6q^2 + q^3.$$

Calculate the marginal cost.

Solution.   By the preceding definition, we obtain the marginal cost at q units,

$$C'(q) = 15 - 12q + 3q^2.$$

In words, it is the rate of change of the total cost per unit change in production, or, approximately, the marginal cost at the production level q is the cost of producing one more unit when the production is already q units.

12.9   Exercises

1. A point is moving along a line with uniform velocity 40 ft/sec. Express the distance y as a function $f(t)$ of time and compute $D_t f(t)$. What is the interpretation of this derivative?
2. In Exercise 1, express the instantaneous velocity v as a function $v = g(t)$, and compute $D_t g(t)$. What is the interpretation of this derivative?
3. A weight is thrown vertically upward, and its distance $y(t)$ above the ground is given t seconds thereafter by the function $y(t) = -16t^2 + 96t$. Compute the derivative $D_t(-16t^2 + 96t)$ and plot the graph of this derivative function. What is the interpretation of the derivative?
4. For the motion of Exercise 3:
    (a) What was the velocity at time $t = 0$? Up or down?
    (b) When did the weight stop instantaneously?
    (c) What was the velocity 4 seconds after the start? Up or down?
    (d) When did the weight return to ground level?
    (e) What was the velocity at the instant of return to ground level?

5. A freely falling body near the earth's surface has a distance y above the earth's surface given by

$$y = -(\tfrac{1}{2})gt^2 + v_0 t + y_0,$$

where g is a gravitational constant, and v_0 and y_0 are constants. (g is $+32.2$ ft/sec^2.) We neglect the resistance of air.
    (a) Compute $D_t y$ and interpret this derivative.
    (b) Let $v = D_t(-\tfrac{1}{2})gt^2 + v_0 t + y_0)$ and compute $D_t v$. Interpret this derivative.
    (c) Find the instantaneous velocity at the time $t = 0$ sec.
    (d) Find the height y at the time $t = 0$ sec.

6. Regarding Exercise 5:
    (a) During what interval of time was the function $y = -(\tfrac{1}{2})gt^2 + v_0 t + y_0$ increasing?
    (b) During what interval was it decreasing?
    (c) When did the body reach maximum height?

7. What is the rate of change of the volume of a cube with respect to the length of an edge when the edge is 4 inches long?

8. What is the growth rate of a population P given at time t years by the function $P = $ **$2400t + 30{,}000$**? Is this a constant *percent* rate of increase?

9. Plot the marginal cost curve $C' = 15 - 12q + 3q^2$ in Example 12.8. For what productions was the marginal cost decreasing? Increasing? At what production level q did it reach its minimum value?

10. In Example 12.8 and Exercise 9, suppose q represents the daily production of fiberglass power boats that sell for 12, meaning \$12,000, each. At what daily production level will the producer begin to encounter diminishing returns on each boat produced and sold? If he exceeds this figure, will his total profits still increase?

11. In Exercise 10, at what production level will the unit cost of each additional boat produced exceed the sale price so that additional production will decrease the total profit?

12. A factory's cost C dollars of producing q tons of its product daily is given by $C = 20 + 60q - 0.075q^2$. What is the rate of change of C with respect to q when it produces 10 tons daily? What is the marginal cost?

13. The heat radiated from a hot object varies approximately as the fourth power of the absolute temperature. Find the percent rate of increase of radiated heat per degree increase in temperature when the temperature $T = 800$.

14. A point moves along a line with distance y feet given in terms of t seconds by the function $y = t^3 + 6t^2 + 8t$.
    (a) Find the instantaneous velocity at time t.
    (b) During what intervals of time was y increasing as t increased? Decreasing?
    (c) Describe the motion starting with time $t = 0$ seconds.

15. (a) Find the set of points where $f(x) = x^3 - 3x^2 - 9x + 3$ is increasing.
    (b) Is decreasing.
    (c) Find the stationary points, where $f'(x) = 0$. Plot the graph.

16. Solve the following problem in two ways: by ordinary algebra and by calculus. The Centigrade temperature scale T is mapped onto the Fahrenheit scale θ by the function $\theta = (\tfrac{9}{5})T + 32$. What is the local scale factor of the mapping at each temperature T?

17. Find the local scale factor at the points indicated for the mapping $f(x) = x^3 - x$ of the x-axis onto the y-axis: (a) at $(0, 0)$; (b) at $(1, 0)$; (c) at $(-2, -6)$.

18. Find the local scale factor of the mapping $4x^2 - 1$ at points where $x = -2$, $x = 0$, $x = 1$.

19. If (Exercise 16) the Fahrenheit scale is mapped onto the Centigrade scale, what is the scale factor?

20. In Exercise 17, suppose we consider the mapping of the y-axis onto the x-axis near $(1, 0)$ induced by $f(x)$. What is the local scale factor of this inverse mapping at the point where $y = 0$?

21. In general, if $f'(x) \neq 0$, show that $f(x)$ also determines a mapping of the y-axis onto the x-axis for y near $f(x)$. What is the scale factor of this inverse mapping?

22. Prove that if $f(x)$ is locally increasing at x and w is sufficiently near x, if $w > x$, then $f(w) > f(x)$, and if $w < x$, then $f(w) < f(x)$ (compare §9.2).

13   Mean Value Theorem

13.1   Interior Maximum and Minimum Points of a Function.   A number c is an *interior point* of a set X of real numbers if there is a neighborhood of c, contained in X, that includes numbers both less than c and greater than c. For example, $\frac{1}{3}$ is an interior point of the set $X = [0, 1]$ but 0 is not an interior point and 1 is not.

Let us consider an interior point c of the domain of $f(x)$, where $f'(c) \neq 0$. Can this be a maximum point or a minimum point of $f(x)$? Suppose $f'(c) > 0$; then (§12.2) the function is locally increasing at c, which implies that there is a neighborhood of c including numbers z and w in the domain of $f(x)$ such that $z < c < w$ and $f(z) < f(c) < f(w)$. Hence $f(c)$ is neither a maximum nor a minimum (Figure 3.6). Alternatively, suppose that $f'(c) < 0$. Then $f(x)$ is locally decreasing at c, which implies that there is a neighborhood of c including numbers z and w such that $z < c < w$ and $f(z) > f(c) > f(w)$. (Point d in Figure 3.6.) Since the only remaining possibility for a differentiable function at an interior point c is $f'(c) = 0$, these arguments prove the following theorem.

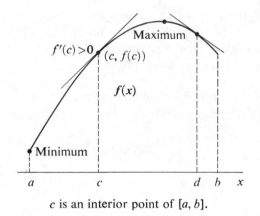

c is an interior point of $[a, b]$.

FIGURE 3.6   Interior points where $f'(x) \neq 0$.

THEOREM (On interior maxima and minima).   If $f(x)$ has a maximum or minimum value at an interior point c of its domain and if $f(x)$ is differentiable at c, then $f'(c) = 0$.

An interior point c where $f'(c) = 0$ is called a *stationary point*.

We readily see that the hypothesis that c is an interior point cannot be left out and still permit us to draw the conclusion that $f'(c) = 0$. For example, the simple function $f(x) = x$ on $[0, 1]$ reaches its maximum at 1 and its minimum at 0. Its derivative function is $f'(x) = 1$, and nowhere is $f'(x) = 0$. This does not contradict the theorem, since neither 0 nor 1 is an interior point of $[0, 1]$ (Figure 3.7).

Another simple example where $f'(x) \neq 0$ at the minimum point is given by the ramp function $f(x) = |x|$ on the real number line. The function has a minimum at the interior

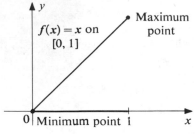
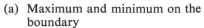

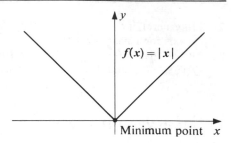

(a) Maximum and minimum on the boundary

(b) Function not differentiable at minimum

FIGURE. 3.7   Maximum and minimum points where $f'(x) \neq 0$.

point 0, but $f'(0) \neq 0$. In fact (Figure 3.7), the function is not differentiable at 0. Hence this does not contradict the theorem either.

13.2 Mean Value Theorem. We draw the graph of a function $f(x)$ on a closed interval $[a, b]$ (Figure 3.8). We connect the points $A:(a, f(a))$ and $B:(b, f(b))$ by a straight line with slope

$$m = \frac{f(b) - f(a)}{b - a}.$$

We ask the question: Is there a point z between a and b where $f'(z) = m$? That is, is there a point where the tangent line is parallel to the secant AB? This question is answered affirmatively by the following theorem, whose proof we leave to the exercises, assuming the difficult basic theorem on continuous functions (§8.4).

THEOREM OF THE MEAN.   If $f(x)$ is continuous on the closed interval $[a, b]$ $a \neq b$, and has a numerical derivative $f'(x)$ at every interior point, then there is an interior point

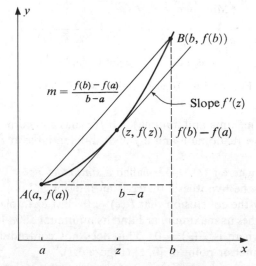

FIGURE 3.8   Theorem of the mean. There is a point $(z, f(z))$ between A and B where the slope $f'(z)$ is equal to the slope of the line AB.

z at which

$$\frac{f(b) - f(a)}{b - a} = f'(z).$$

13.3   Functions Increasing Over an Interval.   The theorem of the mean is a valuable tool to deduce information about a function on an entire interval from local information about the function at each point.

For example, we recall that, by definition, a function $f(x)$ is increasing on the interval $[a, b]$ if for any two distinct points $(x, f(x))$ and $(w, f(w))$ in its graph

$$0 < \frac{f(w) - f(x)}{w - x}.$$

THEOREM (Increasing functions).   If $f(x)$ on $[a, b]$ satisfies the conditions of the theorem of the mean and $0 < f'(x)$ at each point, then $f(x)$ is increasing on $[a, b]$.

COMMENT: We know (§12.2) that $f(x)$ is locally increasing at each x, but this theorem says that we can conclude more: that, dropping the word "locally," we can say that $f(x)$ is increasing on the entire interval.

Proof.   Let x and w be any two points of the interval $[a, b]$. Since $f(x)$ is differentiable in $[a, b]$ and therefore continuous (§11.9, Lemma), we can apply the theorem of the mean to the interval $[x, w]$. It says that there is a point z between x and w where

$$f'(z) = \frac{f(w) - f(x)}{w - x}.$$

Since $0 < f'(x)$ at every point of $[a, b]$ by hypothesis, we know in particular that at the point z, $0 < f'(z)$. Hence

$$0 < f'(z) < \frac{f(w) - f(x)}{w - x},$$

which completes the proof.

Example.   Find the maximum and minimum points of $x^4 - 4x + 11$ on the real numbers.

Solution.   We differentiate $f'(x) = 4x^3 - 4$, which factors into $f'(x) = 4(x - 1) \times (x^2 + x + 1)$. Completing the square in the last factor we find

$$f'(x) = 4[x - 1][(x + \tfrac{1}{2})^2 + \tfrac{3}{4}].$$

The last factor and the factor 4 are everywhere positive. Hence the derivative changes sign with the factor $x - 1$. So we find that when $x < 1$, $f'(x) < 0$, and when $1 < x$, $0 < f'(x)$. Hence by the above theorem on increasing functions, $f(x)$ is decreasing when $x < 1$ and increasing when $1 < x$. This tells us that the minimum occurs at the point $x = 1$ (Figure 3.9).

Hence the minimum value of $f(x)$ is given by $f(1) = 8$. There are no maximum points of $f(x)$; every other point, except 1, is an interior point where $f'(x)$ exists and $f'(x) \neq 0$.

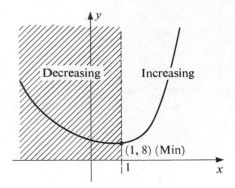

FIGURE 3.9 $f(x) = x^4 - 4x + 11$.

Thus the theorem on interior maxima and minima implies that no maximum can exist. This completes the problem.

13.4   The Simplest Differential Equation.   Let us set ourselves the problem of finding all functions $f(x)$ that have the zero function as derivative, $Df(x) = 0$. This is a problem in which the answer is an unknown function; moreover it is a problem in which the unknown function is described in terms of its derivative. As such, it is a very simple example of a *differential equation*.

We observe that we must specify the domain of the unknown function; for the derivative will not determine the domain. So we restrict the problem to the following: Find all differentiable functions $f(x)$ on the fixed interval $[a, b]$ such that $Df(x) = 0$. Now let x be any number in $[a, b]$. We apply the theorem of the mean to the interval $[a, x]$. It says that there is a number z between a and x such that

$$\frac{f(x) - f(a)}{x - a} = f'(z).$$

But since $f'(x) = 0$ at every point, it follows in particular that $f'(z) = 0$. Hence for every x in $[a, b]$

$$\frac{f(x) - f(a)}{x - a} = 0.$$

Therefore for every x in $[a, b]$, $f(x) = f(a)$. This tells us first that $f(x)$ is a constant. It also tells us that the constant is $f(a)$. But since $f(a)$ was not specified, we may assign any real number C as the value $f(a)$ of $f(x)$ at a and obtain the solution $f(x) = C$. We state the result of this problem in a theorem.

THEOREM (On functions with zero derivative).   Every constant function $f(x) = C$ is a solution of the differential equation $Df(x) = 0$ and the only continuous solutions on an interval $[a, b]$ are these constants.

We now state a related theorem, leaving the proof to the exercises.

THEOREM (On functions with the same derivative).   If $f(x)$ is a function having $Df(x) = f'(x)$ on $[a, b]$, then every other differentiable function $(g)x$ having $Dg(x) = f'(x)$ on $[a, b]$ is of the form $g[x] = f(x) + C$, where C is an arbitrary constant.

13.5 Exercises

In Exercises 1–6, find all of the stationary points of the functions (see theorem on interior maxima and minima).

1. $f(x) = x^2 + 1$ on $[0, 1]$.
2. $f(x) = x^2 + 3x - 4$ on $[0, 1]$.
3. $g(t) = -32 + t^2$ on $[-1, 1]$.
4. $p(u) = (u - 2)^2(3u + 1)^3$ on $[-3, 3]$.
5. $(x^2 - x - 2)^6$ on the real numbers.
6. $(2x + 1)^3 - 4(2x + 1)^2 + 8$ on the reals.

In Exercises 7–14, find the intervals where the function decreases.

7. $f(x) = x^2 + x + 1$.
8. $f(t) = t^3 + t$.
9. $f(t) = (t - 1)^3(2t + 5)^4$.
10. $f(x) = ax^2 + bx + c, a > 0$.
11. $f(x) = ax^2 + bx + c, a < 0$.
12. $f(x) = x^3 - 5x + 7$.
13. $f(x) = (x - a)^3(x - b)^5, a \neq b$.
14. $f(t) = (t^2 - 1)^8$.

15. Using the results of Exercises 10 and 11, show that a quadratic function has a minimum but not a maximum when $0 < a$, and has a maximum but not a minimum when $a < 0$.
16. A differentiable function $f(x)$ on $[a, b]$ has $f(a) = f(b) = 0$. Show that there is a stationary point somewhere in the interval $[a, b]$.
17. Rolle's theorem says that if $f(x)$ is continuous on $[a, b]$ and differentiable at interior points, and $f(a) = f(b)$, then there is an interior point c where $f'(c) = 0$. Show that Rolle's theorem is a special case of the theorem of the mean.
18. A balloon went up, wandered around, and running out of gas, returned to earth at the same point from which it took off. Assuming that it had a velocity at all times, prove that there was a time when its vertical component of velocity was zero; a time when its north-south component of velocity was zero; and a time when its east-west component of velocity was zero. Was there a time when the balloon stopped?

In Exercises 19–22, we only assume that the cars had velocities (positive, negative, or zero) at every time. We do not assume that their velocities are continuous functions of t.

19. I passed a red car on the turnpike. Exactly two hours later, and 60 miles farther down the road, it passed me. Prove that at some time its velocity was exactly 30 miles per hour.
20. In Exercise 19, some time later I passed the red car again. Prove that some time in that interval our speeds were equal.
21. I passed a blue car with an Idaho license. I stopped for lunch and later noticed that I passed it again. Prove that at least twice in the interval between the first and second passing we had exactly the same speeds.
22. In Exercise 21, assuming that we both had accelerations at every time, prove that there was at least one instant when we had equal accelerations.

23. At a certain production level, A's company can either increase or decrease its production. His marginal profit is negative.
    (a) Prove that A can make a larger profit.
    (b) To make his maximum profit can we determine whether A should increase or decrease his production?

In Exercises 24–29, find all the functions on the real numbers that satisfy the given differential equations at every point.

24. $Df(x) = 1$.
25. $Df(x) = 5x^2$.
26. $Df(x) = nx^{n-1}$.
27. $Dh(t) = -gt, g$ constant.
28. $D[f(x) - x^2 + 1] = 0$.
29. $Df(x) = x^n$.

30. (a) Find a solution of $Df(x) = 2x$ for which $f(1) = 3$.
    (b) Prove that it is the only solution having $f(1) = 3$.
31. Determine an integer n for which there is a solution of the differential equation $Dp(x) = nx^{n-1}$ having $p(0) = 1$ and $p(2) = 17$. Find the solution and show that it is the only one.

32. One solution of the differential equation $Df(x) = g(x)$ is $f(x) = x^2 + 1$. Find all solutions and find the function $g(x)$.

33. Prove the theorem on functions with the same derivative (§13.4).

34. A set X consists of all points in the interval $[0, 5]$ with 3 deleted. Find a function $g(x)$ such that $Dg(x) = 0$ on X and $g(1) = -1$, $g(4) = 1$. Is $g(x)$ a constant? Does this contradict the theorem on functions with zero derivative?

35. A point moves on the y-axis with position at time t seconds given by $y(t) = t^3 - 6t^2 + 9t - 8$. After $t = 0$, at what times t is it moving up? Moving down? At what times does the point stop momentarily?

36. A point moves on the x-axis with position at time t seconds given by $x(t) = t^4 - 8t^3 + 18t^2 - 32t + 6$. At what times is its velocity increasing? Decreasing?

37. A function $f(x)$ maps a set of numbers on the x-axis onto a set on the y-axis.
    (a) What is the condition on the local scale factor $f'(x)$ when $f(x)$ magnifies intervals near x?
    (b) When $f(x)$ reduces intervals near x?

In Exercises 38–41, calculate the limits by applying the evaluation theorem of differential calculus.

38. $\lim\limits_{w \to 1} \dfrac{w^2 - 1}{w - 1}$.

39. $\lim\limits_{t \to a} \dfrac{2t^2 - 2a^2}{t - a}$.

40. $\lim\limits_{z \to 0} \dfrac{(z^2 + z + 1) - 1}{z}$.

41. $\lim\limits_{z \to -1} \dfrac{4z^3 + 4}{z + 1}$.

42. The local rate of change of a certain function $f(x)$ is always equal to the value of the function, $f(x)$. Express this relationship as a mathematical equation. Can $f(x)$ be a polynomial?

13.6 Problems

Problems T1–T7 constitute a proof of the theorem of the mean. (See Figure 3.8 for notation.)

T1. Show that the equation of the line AB is $y = m(x - a) + f(a)$.

T2. Define a new function $g(x)$ on $[a, b]$ by

$$g(x) = m(x - a) + f(a) - f(x).$$

Show that $g(x)$ represents the vertical distance from the graph of $f(x)$ to the line AB at x.

T3. Show that $g(x)$ is continuous on $[a, b]$.

T4. Show that $g'(x) = m - f'(x)$ at all interior points.

T5. Verify that $g(a) = 0 = g(b)$.

T6. Show that the result of Problem T5 together with the basic theorem on continuous functions (§8.4) implies that $g(x)$ has a minimum value at some interior point z of $[a, b]$.

T7. Show that the result of Problems T6 and T4 together with the theorem on interior maxima and minima implies that $f'(z) = m$. This proves the theorem of the mean.

14 Taylor's Theorem

14.1 Successive Derivatives of Polynomials.

We consider the polynomial

$$f(x) = 7x^4 - 8x^3 + 5x^2 - 11x + 4.$$

We differentiate it once and find

$$f'(x) = 28x^3 - 24x^2 + 10x - 11.$$

Now we differentiate $f'(x)$, denoting the second-order derivative by $f^{(2)}(x)$, and we find

$$f^{(2)}(x) = 84x^2 - 48x + 10.$$

We differentiate $f^{(2)}(x)$, denoting the third-order derivative obtained by $f^{(3)}(x)$, and we find

$$f^{(3)}(x) = 168x - 48.$$

We repeat this process and find

$$f^{(4)}(x) = 168.$$

We differentiate $f^{(4)}(x)$ and find

$$f^{(5)}(x) = 0.$$

Now since $f^{(4)}(x)$ is a constant, it follows that not only is $f^{(5)}(x) = 0$ but all derivatives of higher order are also zero,

$$f^{(5)}(x) = f^{(6)}(x) = f^{(7)}(x) = \cdots = 0.$$

Let us apply the same successive differentiation to the polynomial $f(x) = x^n$. We find

$$f'(x) = nx^{n-1}, f^{(2)}(x) = (n - 1)nx^{n-2},$$

$$f^{(3)}(x) = (n - 2)(n - 1)nx^{n-3}, \ldots, f^{(n)}(x) = (1)(2) \cdots (n - 1)nx^0 = n!.$$

The symbol $n!$, read "factorial n," is defined for integers when $n \geqq 1$ by

$$n! = (1)(2)(3) \cdots (n),$$

and when $n = 0$ arbitrarily by $0! = 1$. If we continue, we find that all derivatives of order $n + 1$ and higher are zero.

14.2   A Polynomial Expanded in Powers of $x - a$.   If we know the value of a polynomial $f(x)$ and all of its numerical derivatives at the single point where $x = a$, can we write down a formula for $f(x)$ that gives its values at every point?

Solution.   We will try to represent $f(x)$ as a polynomial in power of $x - a$.

$$f(x) = b_0 + b_1(x - a) + b_2(x - a)^2 + b_3(x - a)^3 + b_4(x - a)^4 + \cdots + b_n(x - a)^n,$$

where the numbers $b_0, b_1, b_2, \ldots, b_n$ are unknown coefficients to be determined in terms of the known values $f(a), f'(a), f^{(2)}(a), f^{(3)}(a), \ldots, f^{(n)}(a)$.

We differentiate $f(x)$ successively and find

$$f'(x) = 0 + b_1 + 2b_2(x - a) + 3b_3(x - a)^2 + 4b_4(x - a)^3 + \cdots$$
$$+ nb_n(x - a)^{n-1},$$

$$f^{(2)}(x) = 0 + 0 + 2b_2 + (2)(3)b_3(x - a) + (3)(4)b_4(x - a)^2 + \cdots$$
$$+ (n - 1)(n)b_n(x - a)^{n-2},$$

$$f^{(3)}(x) = 0 + 0 + 0 + (1)(2)(3)b_3 + (2)(3)(4)b_4(x - a) + \cdots$$
$$+ (n - 2)(n - 1)(n)b_n(x - a)^{n-3},$$

$$f^{(4)}(x) = 0 + 0 + 0 + 0 + (1)(2)(3)(4)b_4 + \cdots$$
$$+ (n - 3)(n - 2)(n - 1)(n)b_n(x - a)^{n-4},$$

$$\vdots$$

$$f^{(n)}(x) = 0 + 0 + 0 + 0 + 0 + \cdots + n! \, b_n.$$

We now set $x = a$ in each formula. All of the terms involving powers of $x - a$ vanish and we have left

$$f(a) = b_0, f'(a) = b_1, f^{(2)}(a) = 2! \, b_2, f^{(3)}(a) = 3! \, b_3, f^{(4)}(a) = 4! \, b_4, \ldots, f^{(n)}(a) = n! \, b_n.$$

That is, for each integer k,

$$b_k = \frac{f^{(k)}(a)}{k!}, \qquad k = 0, 1, 2, \ldots, n.$$

This expresses $b_0, b_1, b_2, \ldots, b_n$ in terms of the known values of $f(x)$ and its derivatives at a.

We observe that if $f(x)$ is a polynomial of degree n, then the derivative $f^{n+1}(x)$ and all derivatives of higher order must be zero. Hence the given values $f(a), f'(a), f^{(2)}(a), \ldots, f^{(n)}(a)$ for a polynomial of degree n must have $f^{(n)}(a) \neq 0$ but $f^{(n+1)}(a) = 0, f^{(n+2)}(a) = 0, \ldots$. Thus the procedure above for determining the coefficients $b_0, b_1, \ldots, b_n$ stops automatically with b_n. We now write down the following formula for $f(x)$.

TAYLOR'S* EXPANSION

$$f(x) = f(a) + \frac{f'(a)}{1!} (x - a) + \frac{f^{(2)}(a)}{2!} (x - a)^2 + \frac{f^{(3)}(a)}{3!} (x - a)^3$$

$$+ \frac{f^{(4)}(a)}{4!} (x - a)^4 + \cdots + \frac{f^{(n)}(a)}{n!} (x - a)^n.$$

Example. Find a formula for the unknown function $f(x)$ that satisfies the conditions $f(2) = 3$, $f'(2) = -4$, $f^{(2)}(2) = -9$, $f^{(3)}(2) = 6$, and $f^{(4)}(2) = f^{(5)}(2) = \cdots = 0$ for all derivatives of order higher than 3.

Solution. We apply Taylor's series with $a = 2$ and expand $f(x)$ in powers of $x - 2$. We find at once that

$$f(x) = 3 - 4(x - 2) - \frac{9}{2!} (x - 2)^2 + \frac{6}{3!} (x - 2)^3,$$

$$= 3 - 4(x - 2) - \frac{9}{2} (x - 2)^2 + (x - 2)^3.$$

This completes the problem.

14.3 Cut-off Taylor's Expansion. We observe that if x is near a, near enough so that $|x - a| < 1$, then the powers $(x - a)^2$, $(x - a)^3, \ldots$ in Taylor's expansion grow smaller and smaller as we go to terms of higher degree. Moreover, the factors $1/k!$ of $f^{(k)}(a)$ also decrease very rapidly as k increases. For example, $1/10! = 0.00000025\ldots$. Hence if we regard Taylor's expansion as a formula for the number $f(x)$ when x is near a we may cut off the terms of higher degree and have left terms that give $f(x)$ approximately.

If we cut off all except the first term of Taylor's expansion, we get the first approximation $y = f(a)$, which is exactly equal to $f(x)$ when $x = a$. If we cut off all except the

---

* For Brook Taylor (1685–1731), Cambridge University mathematician, who was a contemporary of Isaac Newton.

first two terms, we obtain a better approximation to $f(x)$, namely

$$y = f(a) + f'(a)(x - a).$$

This looks familiar! It is the equation of a straight line through the point $(a, f(a))$ with slope $f'(a)$, the tangent line to the graph of $f(x)$ (§10.4). When we do this, the terms we have cut off form a remainder, which we will denote by $R_1(x, a)$ for "remainder after terms of order 1," for a polynomial $f(x)$ of degree n,

$$R_1(x, a) = \frac{f^{(2)}(a)}{2!}(x - a)^2 + \cdots + \frac{f^{(n)}(a)}{n!}(x - a)^n.$$

Graphically (Figure 3.10), the remainder is the difference at x between $f(x)$ and the y-coordinate of the tangent line.

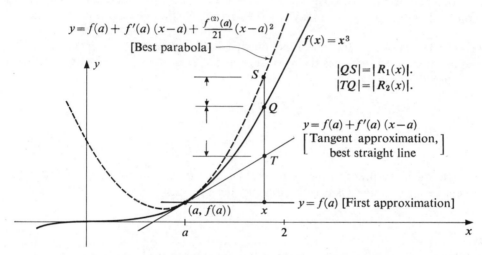

FIGURE 3.10   Cut-off Taylor's expansion as approximation, the example $f(x) = x^3$, with $a = 1$.

This suggests that a still better approximation to $f(x)$ near $(a, f(a))$ would be the Taylor's expansion to terms of order 2, namely

$$y = f(a) + f'(a)(x - a) + \frac{f^{(2)}(a)}{2!}(x - a)^2.$$

This is a parabola that agrees with $f(x)$ in having not only the same $f(a)$ and the same slope of the tangent $f'(a)$, but also the same second derivative $f^{(2)}(a)$. Just as the tangent line is the best-fitting straight line to the graph of $f(x)$ at $(a, f(a))$, this cut-off Taylor's series is the best fitting parabola to the graph of $f(x)$, and it is a closer approximation than the tangent line. (See dotted graph in Figure 14.3.) The remainder of Taylor's expansion cut off after terms of order 2 is then $R_2(x)$, given by

$$R_2(x) = \frac{f^{(3)}(a)}{3!}(x - a)^3 + \cdots + \frac{f^{(n)}(a)}{n!}(x - a)^n.$$

See Figure 3.10, where $R_2(x) = QS$, and compare with $R_1(x) = TS$. This indicates graphically that the Taylor's series cut off after terms of order 2 is a better approximation than the tangent line, which is Taylor's series cut off after terms of order 1.

Clearly we can cut off the Taylor's series after terms of order 3, or still higher order, but in so doing we pay for increasing accuracy with increasing complication. We call these Taylor's polynomials.

14.4 Taylor's Theorem with Exact Remainder. We already know that a slight modification of the Taylor's series cut off after terms of order 1 can make it exact. In fact, there is a number z between a and x such that exactly

$$f(x) = f(a) + f'(z)(x - a).$$

This is the mean value theorem (§13.2). To make this exact, we have replaced the number a in $f'(a)$ by a certain number z, about which we know only that it is between a and x. We can do a similar thing to convert the Taylor's polynomial to terms of order n into an exact formula, variously known as the extended theorem of the mean, or Taylor's polynomial with remainder, or Taylor's theorem, or Taylor's expansion.

TAYLOR'S THEOREM.   If $f(x)$ is a function that has derivatives $f'(x), f^{(2)}(x), \ldots, f^{(n)}(x)$ on some interval that includes the numbers a and x, then there is a number z between a and x such that exactly

$$f(x) = f'(a) + \frac{f'(a)}{1!}(x - a) + \frac{f^{(2)}(a)}{2!}(x - a) + \cdots$$

$$+ \frac{f^{(n-1)}(a)}{(n-1)!}(x - a)^{n-1} + \frac{f^{(n)}(z)}{n!}(x - a)^n.$$

We will defer the proof to the exercises. The last term containing $f^{(n)}(z)$ evaluated at the undetermined point z is called the remainder of order n, denoted by $R_n(x, a)$,

$$R_n(x, a) = \frac{f^{(n)}(z)}{n!}(x - a)^n.$$

All other terms on the right are completely known when we know

$$x, a, f(a), f'(a), \ldots, f^{(n-1)}(a).$$

Example.   An unknown function $f(x)$ has

$$f(\pi) = -1, f'(\pi) = 0, f^{(2)}(\pi) = 1, f^{(3)}(\pi) = 0,$$

and it is known that $|f^{(4)}(x)| < 1$. Compute $f(3)$ and find a measure of the precision of the computed value of $f(3)$.

Solution. We write down Taylor's theorem in powers of $x - \pi$ with remainder of order 4:

$$f(x) = -1 + \frac{0}{1!}(x - \pi) + \frac{1}{2}(x - \pi)^2 + \frac{0}{3!}(x - \pi)^3 + \frac{f^{(4)}(z)}{4!}(x - \pi)^4,$$

$$= -1 + \frac{1}{2}(x - \pi)^2 + \frac{f^{(4)}(z)}{24}(x - \pi)^4,$$

where z is some number between π and x. We substitute $x = 3$ and compute an approximate value of $f(3)$ from the Taylor's polynomial terms without the remainder. This gives us

$$f(3) \cong -1 + \tfrac{1}{2}(3 - \pi)^2 \cong -1 + 0.5(-0.1416)^2$$

$$\cong -1 + 0.010024 = -0.989976.$$

Now how precise is this decimal as a value of $f(3)$? We can use the remainder term to find out, remembering that $|f^{(4)}(z)| < 1$. The error,

$$|R_4(3, \pi)| = \left| \frac{f^{(4)}(z)}{24} (3 - \pi)^4 \right| < \frac{1}{24}(-0.1416)^4 < 0.00002.$$

Thus we know that the computed value of $f(3)$ is correct within two digits in the fifth decimal place. That is, we are justified in writing $f(3) = -0.9900$ as the best four-place decimal value of $f(3)$ but we cannot be sure of the digit in the fifth place. This completes the problem.

This example indicates the way in which Taylor's theorem gives us polynomial approximations to functions* that can be used to compute their values accurately; and the remainder term gives us a bound on the possible error. This enables us to assert in this instance that the computed value is correct to four places, as would be necessary in computing a four-place table. Logarithm and trigonometric tables are computed in this manner.

14.5   Exercises

In Exercises 1–8, produce the Taylor's expansion of the function in the indicated powers:

1. $f(x) = x^2 - 3x + 7$ in powers of $x - 4$.
2. $f(t) = t^4 - 3t^3 + 5t - 9$ in powers of $t - 3$.
3. $f(x) = x^3 - 9x + 4$ in powers of $x + 2$.
4. $f(w) = 3(2w + 1)^5$ in powers of w.
5. $f(t) = (2t + 1)^2 t^3$ in powers of t.
6. $f(z) = 6(3z - 5)^4$ in powers of $z + 1$.
7. $f(x) = x^7$ in powers of x.
8. $f(x) = (1 - x)^6$ in powers of $x + 4$.

In Exercises 9–13, find a formula for the unknown function that satisfies the given conditions.

9. $f(x): f(4) = 1, f'(4) = 1, f^{(2)}(4) = 2, f^{(3)}(4) = 6, f^4(x) = 0$.
10. $g(z): g(0) = 2, g'(0) = -3, g^{(2)}(0) = \frac{1}{2}, g^{(3)}(0) = 2, g^{(4)}(0) = g^{(5)}(0) = \cdots = 0$ for all derivatives of order 4 or higher.
11. $s(t): s(-1) = 2, s'(-1) = 0, s^{(2)}(-1) = 4, s^{(3)}(t) = 0$.
12. $f(x): f^{(k)}(1) = k!$ for $k = 0, 1, \ldots, n$ and $f^{(n+1)}(x) = 0$.
13. $f(x): f(0) = f'(0) = \cdots = f^{(n-1)}(0) = 0, f^n(0) = n!, f^{(n+1)}(x) = 0$.

In Exercises 14–17, find the best first-degree approximations and the best quadratic approximations to the given functions near the indicated points. In each case sketch the graph of $f(x)$ and the approximating line and parabola (Figure 3.10).

14. $f(x) = (x - 1)^3$ at $(2, 1)$.
15. $g(t) = 4(t + 1)^2 - 3$ at $(0, 1)$.
16. $v(t) = t^3 - 3t + 5$ at $(1, 3)$.
17. $c(p) = (p + 1)^4$ at $(-1, 0)$.

18. A rocket moves in a vertical line according to an unknown position function $y(t)$. By sightings and subsequent curve fitting it was found that the best-fitting second-degree polynomial to the motion at the time $t = 5$ sec was $y(t) \cong 2 + 0.3(t - 5) + 0.06(t - 5)^2$. Find the velocity and acceleration of the actual motion at time $t = 5$ seconds.

---

* The function that actually furnished our data here was **cos** x.

19. An unknown motion-function $y(t)$ is to be expanded in a Taylor's expansion in powers of $(t - t_0)$. To determine the velocity and acceleration of the motion at t_0, what coefficients in the Taylor's expansion must be known?

20. An unknown function $f(x)$ has $f(1) = 1.632$ and $f'(1) = 4$, and it is known that $|f^{(2)}(x)| < 1$. Compute $f(1.02)$ and show that the error is less than 0.0002.

21. If we compute the values of the quadratic $0.657 - 9.321t + 0.362t^2$, using the tangent line at the point where $t = 1$ as an approximation, in what neighborhood of 1 will the results be accurate with error less than 0.01?

22. A certain familiar function $f(x)$ has

$$f(0) = f'(0) = f^{(2)}(0) = \cdots = f^{(n)}(0) = \cdots = 1,$$

and it is known that in the interval $0 \leq x \leq 1$ the absolute value of every derivative is less than 3. If we wish to compute $f(1)$ accurately to four places after the decimal point, how many terms of Taylor's expansion in powers of $x - 0$ must be used?

23. Show that if $f^{(2)}(x)$ is positive near a, the tangent line approximation to $f(x)$ always gives an answer that is too small, but if $f^{(2)}(x)$ is negative, the tangent approximation is too large.

24. If we divide the polynomial

$$f(x) = f(a) + f'(a)(x - a) + \cdots + \frac{f^{(n)}(a)}{n!}(x - a)^n$$

by $x - a$ to obtain a quotient $Q_1(x)$ and remainder r_0, what is the remainder? Compare with the remainder theorem (Appendix A).

25. (a) In Exercise 24, if we divide the first quotient $Q_1(x)$ by $x - a$, we obtain a second quotient $Q_2(x)$ and remainder. What is the remainder?
    (b) As we continue to divide successive quotients $Q_2(x)$, $Q_3(x) \cdots$ by $x - a$, what is the remainder after n divisions by $x - a$?

26. Using the division procedure of Exercises 24 and 25, expand $x^3 - 7x - 11$ in powers of $x + 2$. Compare this with the results obtained by Taylor's expansion.

14.6 (Theoretical). *Proof of Taylor's Theorem.* We illustrate by the case $n = 3$. We consider the remainder after terms of order 2.

$$R_3 = f(x) - f(a) - \frac{f'(a)}{1!}(x - a) - \frac{f^{(2)}(a)}{2!}(x - a)^2.$$

To express this remainder as a term of the form $R_3 = (C/3!)(x - a)^3$, where C is to be determined, we replace the number a by a variable t whose domain is the interval $[a, x]$ and consider the function of t defined by the difference

$$g(t) = f(x) - f(t) - \frac{f'(t)}{1!}(x - t) - \frac{f^{(2)}(t)}{2!}(x - t)^2 - \frac{C}{3!}(x - t)^3.$$

We verify that $g(x) = 0$ and we let C be determined so that $g(a) = 0$. Then the function $g(t)$ satisfies the hypotheses of the theorem of the mean over the interval $[a, x]$, where it has the slope

$$\frac{g(x) - g(a)}{x - a} = 0.$$

Hence at some interior point z it is true that $g'(z) = 0$.

We compute $g'(t)$ using the derivative of the product, $Duv = u\,Dv + v\,Du$, and find that, after some simplifying cancellations,

$$g'(t) = -\frac{f^{(3)}(t)}{2!}(x - t)^2 + \frac{C}{2!}(x - t)^2.$$

Hence $g'(z) = 0$ if and only if

$$C = f^{(3)}(z).$$

With this value of C and $g(a) = 0$, it follows that when we set $t = a$ we have

$$0 = f(x) - f(a) - \frac{f'(a)}{1!}(x - a) + \frac{f^{(2)}(a)}{2!}(x - a)^2 - \frac{f^{(3)}(z)}{3!}(x - a)^3,$$

which is Taylor's theorem for the case $n = 3$ with the remainder

$$R_3 = \frac{f^{(3)}(z)}{3!}(x - a)^3.$$

14.7   Problems.   Proof of Taylor's Theorem

T1.   Repeat the proof of the theorem of the mean, case $n = 1$, in the notation of Section 14.6
T2.   Prove the theorem of the mean for the case $n = 2$ by the method of Section 14.6.

Complete the proof of Taylor's theorem in the general case in Problems T3–T7, first defining

$$g(t) = f(x) - (f)t - \frac{f^{(2)}(t)}{2!}(x - t)^2 \cdots - \frac{f^{(n-1)}(t)(x - t)^{n-1}}{(n - 1)!} - \frac{C}{n!}(x - t)^n.$$

T3.   Show that $g(x) = 0$ and define C so that $g(a) = 0$.
T4.   Verify that the theorem of the mean now implies that for some number z between a and x, $g'(z) = 0$ (Rolle's theorem).
T5.   Compute

$$g'(t) = -\frac{f^{(n)}(t)}{(n - 1)!}(x - t)^{n-1} + \frac{C}{(n - 1)!}(x - t)^{n-1}.$$

T6.   Show that $C = f^{(n)}(z)$.
T7.   Set $t = a$ in $g(t)$ and obtain Taylor's theorem.

15   Uses of the Second Derivative

15.1   Deviation of the Curve from the Tangent Line.

Taylor's theorem to terms of order 2 (§14.4) says that

$$f(x) = f(a) + f'(a)(x - a) + \frac{f^{(2)}(z)}{2}(x - a)^2,$$

where z is some number between a and x. We recall that the first two terms on the right give us the tangent line to the graph at $(a, f(a))$. Hence, subtracting, we see that the difference

$$f(x) - [f(a) + f'(a)(x - a)] = \frac{f^{(2)}(z)}{2}(x - a)^2,$$

represents the deviation at x of the graph of $f(x)$ from the tangent line. This is the distance TQ in Figure 3.10.

Thus if $f^{(2)}(x)$ is positive in an interval that includes a and x, we know that $Q:(x, f(x))$ is above the tangent line (Figure 3.11(a)); and if $f^{(2)}(x)$ is negative in an interval that includes a and x, the point $(x, f(x))$ is below the tangent line (Figure 3.11(b)).

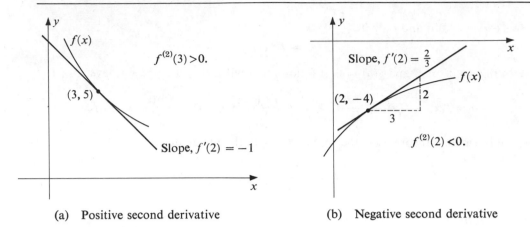

(a)   Positive second derivative   (b)   Negative second derivative

FIGURE 3.11   Geometric interpretation of the second derivative.

Example.   Sketch the graph of a polynomial $f(x)$ in a neighborhood of $(3, f(3))$: (a) Given that $f(3) = 5, f'(3) = -1$, and $f^{(2)}(3) = 15$; (b) given that $f(2) = -4, f'(2) = \frac{2}{3}$, and $f^{(2)}(2) = -8$.

    Solution.   (a) We plot the tangent line (Figure 3.11(a)) through the point $(3, 5)$ with slope $f'(3) = -1$. Since $f^{(2)}(3)$ is positive, the continuous function $f^{(2)}(x)$ is positive in some neighborhood of 3 (§8.6, Exercise 22). Hence we know from Taylor's theorem that in this neighborhood the graph of $f(x)$ deviates upward from the tangent line, as we show it in the figure. We cannot make any graphical use of the fact that the second derivative is precisely 15.

    (b) Similarly (Figure 3.11(b)), we plot the tangent line through $(2, -4)$ with slope $\frac{2}{3}$. Taylor's theorem says that since $f^{(2)}(x)$ is negative near 2, the graph of $f(x)$ deviates downward from the tangent line, so we draw a portion of the curve through $(2, -4)$ below the tangent line and curving downward. Again, we cannot represent graphically the fact that $f^{(2)}(2) = -8$ precisely.

15.2   Rotation of the Secant.   We now consider the direction of rotation of a secant line AP joining a fixed point $A:(a, f(a))$ and variable point $P:(x, f(x))$ on the graph of $f(x)$. The slope of this secant line is a function $m(x)$ defined when $x \neq a$ by

$$m(x) = \frac{f(x) - f(a)}{x - a},$$

and when $x = a$ by the slope of the tangent, $m(a) = f'(a)$. We think of P as moving along the graph as x increases through a (Figure 3.12). Guided by the direction of deviation of the graph of $f(x)$ from the tangent (Figure 3.11), we readily guess the following theorem.

THEOREM (Rotation of the secant).   If $f^{(2)}(x)$ is positive in an interval that includes the point where $x = a$, then the slope function $m(x)$ is increasing in the interval; and if $f^{(2)}(x)$ is negative, the slope function $m(x)$ is decreasing.

    We consider the geometric interpretation. The slope $[f(x) - f(a)]/(x - a)$ is the slope of the secant line AP (Figure 3.12). When this slope increases with x, the secant line AP

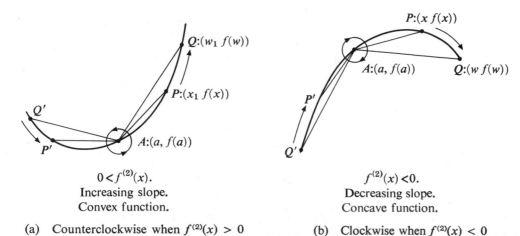

$$0 < f^{(2)}(x).$$
Increasing slope.
Convex function.

$$f^{(2)}(x) < 0.$$
Decreasing slope.
Concave function.

(a)   Counterclockwise when $f^{(2)}(x) > 0$
(b)   Clockwise when $f^{(2)}(x) < 0$

FIGURE 3.12   Rotation of a secant line.

rotates counterclockwise (Figure 3.12(a)) about A, but when the slope decreases, the secant line AP rotates clockwise about A as x increases (Figure 3.12(b)).

Proof of the theorem.   We shall calculate the derivative of the slope function $m(x)$ at a point x by the definition

$$m'(x) = \lim_{w \to x} \frac{m(w) - m(x)}{w - x}.$$

If $x \neq a$, by the definition of $m(x)$ the differential quotient

$$\frac{m(w) - m(x)}{w - x} = \frac{\dfrac{f(w) - f(a)}{w - a} - \dfrac{f(x) - f(a)}{x - a}}{w - x}.$$

Simplifying the fraction on the right, we find that

$$\frac{m(w) - m(x)}{w - x} = \frac{1}{w - a}\left[\frac{f(w) - f(x)}{w - x} - \frac{f(x) - f(a)}{x - a}\right].$$

Hence, applying the algebra of limits (§10.6) and the definition of the derivative of $f(x)$, we find that if $x \neq a$,

$$m'(x) = \lim_{w \to x} \frac{1}{w - a}\left[\frac{f(w) - f(x)}{w - x} - \frac{f(x) - f(a)}{x - a}\right] = \frac{1}{x - a}\left[f'(x) - \frac{f(x) - f(a)}{x - a}\right].$$

We algebraically simplify the formula for $m'(x)$ and find

$$m'(x) = \frac{f(a) - f(x) - (a - x)f'(x)}{(a - x)^2}.$$

Finally, we apply Taylor's theorem to terms of order 2 to the numerator and find that there is a number z between a and x such that

$$m'(x) = \frac{\frac{1}{2}f^{(2)}(z)(a - x)^2}{(a - x)^2} = \frac{1}{2}f^{(2)}(z).$$

If $x = a$, we may calculate (Exercises) that $m'(a) = \frac{1}{2}f^{(2)}(a)$.

These results demonstrate that if $f^{(2)}(x)$ is positive for every x in an interval including a, then in this interval $m'(x)$ is positive. Hence $m(x)$ is an increasing function in the interval. When $f^{(2)}(x)$ is negative we prove the analogous statement by applying this argument to $-f(x)$. This completes the proof.

15.3   Convex and Concave Functions

DEFINITION.   A function $f(x)$ is (locally) *convex* at the point $(a, f(a))$ if in some neighborhood of a the difference quotient

$$\frac{f(x) - f(a)}{x - a}$$

is increasing for all x, $x \neq a$. And $f(x)$ is (locally) *concave* if the difference quotient is decreasing in some neighborhood of a with a deleted.

This is another geometric description. The use of the word *convex* is justified because for a convex function the region above the graph is like a convex lens. Similarly, if $f(x)$ is concave the region above the graph is like a concave lens. (See Figure 3.13.)

(a)   Convex   (b)   Concave

FIGURE 3.13   Convex and concave functions.

We have seen (§15.2) that near a point where $f^{(2)}(x)$ is positive, the function is convex, and that near a point where $f^{(2)}(x)$ is negative, the function is concave.

Example.   Determine where the graph of $f(x) = (x - 1)^3 - 5(x - 1) + 4$ is locally convex and where it is concave. Show this graphically.

Solution.   Differentiating twice, we obtain $f'(x) = 3(x - 1)^2 - 5$ and $f^2(x) = 6(x - 1)$. From the expression for $f^{(2)}(x)$, we see that if $x < 1$, then $f^{(2)}(x)$ is negative and the function is (locally) concave. On the other hand, if $1 < x$, then $f^{(2)}(x)$ is positive and the function is locally convex. This completes the problem except for the graph.

We tabulate the three functions $f(x), f'(x)$, and $f^{(2)}(x)$ at values of x that increase from -3 to 5 (Figure 3.14). We plot the curves in the usual way, using the points $(x, f(x))$ and slopes $f'(x)$. We indicate in the graph points where $f^{(2)}(x) < 0$ by nearby minus signs, and points where $0 < f^{(2)}(x)$ by nearby plus signs. We observe the change in the character of the graph at the point $Q:(1, 4)$, where the second derivative changes sign from negative to positive. To the left, where $f^{(2)}(x)$ is negative, the function is concave. To the right, where $f^{(2)}(x)$ is positive, the function is convex.

15.4   Points Where the Second Derivative is Zero.   We have analyzed what happens when $f^{(2)}(x)$ is positive or negative, so we now ask what the nature of the graph is near

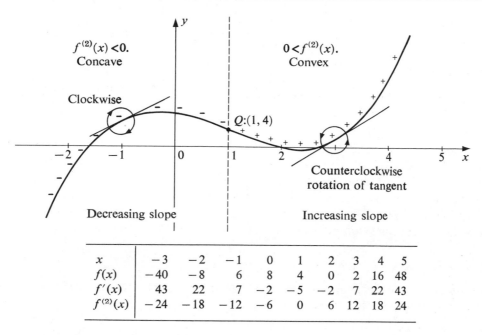

x	−3	−2	−1	0	1	2	3	4	5
$f(x)$	−40	−8	6	8	4	0	2	16	48
$f'(x)$	43	22	7	−2	−5	−2	7	22	43
$f^{(2)}(x)$	−24	−18	−12	−6	0	6	12	18	24

FIGURE 3.14   The second derivative and convexity. $f(x) = (x − 1)^3 − 5(x − 1)$ + 4. The minus signs indicate points where $f^{(2)}(x) < 0$; plus signs where $0 < f^{(2)}(x)$.

a point where $f^{(2)}(x) = 0$. First, we suppose that $f^{(2)}(x) = 0$ at every number x in the domain.

THEOREM (On zero second derivative).   If $f(x)$ is a function on an interval of numbers in which $f^{(2)}(x) = 0$ at every point, then if $(a, f(a))$ is any point in the graph,

$$f(x) = f(a) + f'(a)(x − a),$$

and its graph is a straight line.

Proof.   We apply Taylor's theorem to terms of order 2. There is a number z between a and x such that

$$f(x) = f(a) + f'(a)(x − a) + \frac{f^2(z)}{2}(x − a)^2.$$

But since by hypothesis $f^{(2)}(x) = 0$ at every point, we know that $f^{(2)}(z) = 0$. Hence the last term drops out and the formula for $f(x)$ is that given in the conclusion of the theorem. This completes the proof.

Let us now consider what happens when $f^{(2)}(x) = 0$ at one point. We have seen one example in the point Q (Figure 3.14), where $x = 1$. The direction of curvature reverses at this point because $f^{(2)}(x)$ changes sign (§1.6) at 1. We call such a point an *inflection point*.

DEFINITION.   An inflection point of a function $f(x)$ on $[a, b]$ is an interior point s of $[a, b]$, where $f^{(2)}(s) = 0$ and $f^{(2)}(x)$ changes sign.

Example.   Find the inflection points of

$$f(x) = 3x^5 - 10x^3 - 60x + 12$$

on the real number scale.

Solution.   We differentiate twice and find

$$f^{(2)}(x) = 60x^3 - 60x.$$

We set $f^{(2)}(x) = 0$ and solve for the roots $x = -1$, $x = 0$, $x = +1$. This enables us to factor $f^{(2)}(x) = 60(x + 1)(x)(x - 1)$. We can analyze the signs of $f^{(2)}(x)$ by means of a table of signs of its factors (§1.6). We find: If $x < -1$, then $f^{(2)}(x)$ is negative; if $-1 < x < 0$, then $f^{(2)}(x)$ is positive; if $0 < x < 1$, then $f^{(2)}(x)$ is negative; and if $1 < x$, then $f^{(2)}(x)$ is positive. Thus $f^{(2)}(x)$ changes sign at each point where $f^{(2)}(x) = 0$. Hence by definition the inflection points of $f(x)$ are $-1, 0, 1$ and only these points.

15.5   Exercises

In Exercises 1–5, sketch a portion of the graph of the function near the indicated point, using the information about the second derivative.

1. $f(2) = 3, f'(2) = -\frac{4}{3}, f^{(2)}(2) = -8.$
2. $f(-1) = 4, f'(-1) = 6, f^{(2)}(-1) = +10.$
3. $f(1) = 0, f'(1) = 0, f^{(2)}(1) = -7.$
4. $f(5) = 1, f'(5) = 0, f^{(2)}(5) = +11.$
5. $f(3) = -4, f'(3) = -2, f^{(2)}(2) = +6.$

In Exercises 6–11, find in what intervals of the real number scale $f^{(2)}(x) < 0$ and in what $0 < f^{(2)}(x)$. Also find the inflection points of each function.

6. $f(x) = x^2 - 3x + 2.$ 7. $f(x) = -2x^2 + 3x - 4.$
8. $f(x) = 2x + 1.$ 9. $f(x) = x^3 + 2x + 4.$

10. $g(x) = (x - 1)^3 - 10(x - 1)^2 + 5(x - 1) + 8.$
11. $p(t) = t^4 + 2t^3 - 12t^2 + 4t - 9.$
12.   For the function $f(x) = x^4$, show that the point where $f^{(2)}(x) = 0$ is not an inflection point.
13.   Prove that, at an inflection point, the graph of a function always crosses the tangent line.
14.   A point moves along a line with position y at time t given by

$$y = -2t^3 + 4t^2 + 5.$$

Find at what time the acceleration was zero.
15.   Prove that the parabola $f(x) = ax^2 + bx + c$ opens upward if $0 < a$ and downward if $a < 0$.
16.   A point moves on the x-axis. It starts at time $t = 2$ from the point where $x = 1$ with velocity 3 ft/sec. Its acceleration is always zero. Express its position at any time t by a function $x = f(t)$.
17.   As in Exercise 16, a point moves along the x-axis always with zero acceleration, starting at time $t = t_0$ from x_0 with velocity v_0. Find a function $x = f(t)$ that gives the position of the point at any time.
18.   Prove that if $f^{(2)}(x) > 0$ in an interval $[a, b]$, then every secant line that joins two points on the graph is above the graph between the points.
19.   (Newton's method)   A differentiable function $f(x)$ has $f(a) < 0$ and $f(b) > 0$, and $f'(b) \neq 0$ (Figure 3.15). Hence it has a root r. This root is estimated by finding the number x_1 where the tangent line at $B:(b, f(b))$ intersects the x-axis. Show that if

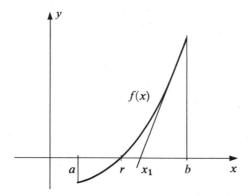

FIGURE 3.15   Newton's method (Exercise 19).

$f^{(2)}(x) > 0$ in $[a, b]$, then there is an error of approximation

$$|x_1 - r| < \frac{1}{2}\frac{f^{(2)}(z)}{f'(b)}(b - a)^2,$$

where z is a number between a and b.

20.   If $P(t)$ is the population of a city at time t, what does d^2P/dt^2 mean?

21.   If $C(q)$ is the cost of the producing of units of goods, what does d^2C/dq^2 mean in terms of marginal cost?

16   Maxima and Minima

16.1   Maximum-Minimum Problem.   We now return to the problem of finding the maximum and minimum point of a function f on a domain X of numbers (§7). We study the problem in the large, that is, we say that $(c, f(c))$ is a *maximum point* of f on X if for every number x in X,

$$f(c) \geqq f(x).$$

Similarly $(c, f(c))$ is a *minimum point* if for every number x in the domain X,

$$f(c) \leqq f(x).$$

It is possible to study local maxima and minima, calling for a weaker conclusion than the problem in the large stated above. For example, a point $(c, f(c))$ is a local maximum point if there is a *neighborhood* of c such that for every number x in this neighborhood $f(c) \geqq f(x)$. The maximum point of f in the large must be a local maximum but a local maximum is not necessarily the solution of the problem in the large.

Let us visualize this situation by an imaginary surveyor's problem. The Albemarle County surveyor is given the problem of finding the highest point in Albemarle County. Suppose he merely explores until he finds a smooth hilltop and determines its altitude to be 765 feet. Is he justified in calling this a solution to his problem? Obviously it is a local maximum but obviously it may not be the highest point in Albemarle County. His location of this hilltop and determination of its altitude is of some use, however. To carry out his original assignment he could locate all of the local maxima and determine their altitudes. Then presumably there are only a finite number of these, so he can compare their altitudes and select the highest point in the county from among the local maxima. He may ignore the infinitely many hillside points in the interior of the county.

We use a mathematical formulation of this method as our basic calculus method to solve the problem of maxima and minima.

16.2   Calculus Tools for Solving Maximum-Minimum Problems.

Instead of attemping to formulate and solve a general problem, we rely on a collection of calculus tools, which aid in the solution of maximum and minimum problems when artfully chosen and skillfully applied. Let us list some of our principal tools.

(1)   The theorem on interior maxima and minima (§13.1) enables us to eliminate from consideration all interior points where $f'(x)$ exists and $f'(x) \neq 0$. This logically leaves only the set of critical points to consider, where critical points are defined as follows.

> DEFINITION.   A number c is a critical value for f on the domain X if and only if:
>   (a) c is not an interior point of X; or
>   (b) f is not differentiable at c; or
>   (c) $f'(c) = 0$.
> A point $(c, f(c))$ is a critical point if c is a critical value.

(2)   The basic theorem on continuous functions (§8.4) assures us that if f is continuous on a domain that is a closed interval $[a, b]$, it has a maximum and a minimum point. Knowing this, we have only to find the maximum and minimum in the set of critical points, usually a finite set, so that the max and min can be found by simple comparison of values of f.

(3)   For a differentiable function f, the mean value theorem (§13.2) often enables us to use information about the sign of $f'(x)$ to show that $f(x)$ increases in X when $x \leqq c$ and decreases when $x \geqq c$. Then $(c, f(c))$ is a maximum (§13.3, Example).

(4)   For a twice-differentiable function f, Taylor's theorem to terms of order 2 (§15.1) may enable us to use information about the sign of $f^{(2)}(x)$. Taylor's theorem for expanding $f(x)$ in powers of $x - c$ says that

$$f(x) = f(c) + f'(c)(x - c) + \tfrac{1}{2}f^{(2)}(z)(x - c)^2,$$

where z is some number between x and c. At an interior critical point, $f'(c) = 0$, so this reduces to

$$f(x) - f(c) = \tfrac{1}{2}f^{(2)}(z)(x - c)^2.$$

If we know that $f^{(2)}(x)$ is everywhere negative, we know also that $f(x) - f(c) \leqq 0$, hence that $f(c) \geqq f(x)$, and that $(c, f(c))$ is the maximum point.

(5)   Convexity (§15.3) can often be applied when f is not known to be differentiable; for the definition of convexity does not involve the derivative. It replaces the information that $f^{(2)}(x) \geqq 0$.

There are many other analytic, graphical, algebraic, and geometric tools that can be applied to maximum-minimum problems of a function, notably the algebraic methods based on the principle of real squares, $x^2 \geqq 0$ (§6).

16.3   Techniques for Maximum-Minimum Problems.

We now proceed by examples.

Example A.   Find the maximum and minimum points of $f(x) = x^3 - 3x + 5$ on $[-3, 3]$.

Solution.   The function is everywhere differentiable and the only points of the domain that are not interior points are the endpoints, -3 and 3. Now we calculate $f'(x) = 3x^2 - 3$ and set $f'(x) = 0$. This gives us $3x^2 - 3 = 0$, which we solve, and we find that the roots are $x = -1$ and $x = +1$. Thus the set of critical values is $\{-3, -1, 1, 3\}$. If there is a maximum or a minimum, it must occur at one of these four points. Now the basic theorem on continuous functions (§8.4) tells us that since $f(x)$ is continuous on the closed interval $[-3, 3]$, $f(x)$ has a maximum and a minimum point. We compute the values of $f(x)$ at the critical values and we find that the maximum and minimum points are included in the set of four points in the table.

x	-3	-1	1	3
$f(x)$	-13	7	3	23

Hence $(-3, -13)$ is the minimum point and $(3, 23)$ is the maximum point. This completes the problem.

We observe that the theorem on interior maxima and minima (§13.1) does not assure us that the points $(-1, 7)$ and $(1, 3)$ where $f'(x) = 0$ are maximum or minimum points. In fact, they are not. We plot the graph (Figure 3.16).

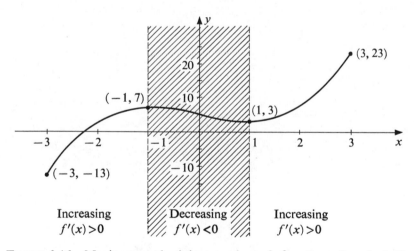

FIGURE 3.16   Maximum and minimum values of $x^3 - 3x + 5$ on $[-3, 3]$.

Example B.   Find the maximum and minimum points of $x^3 + 5x + 1$ on the entire real number line.

Solution.   The real number line is not a closed interval, so the basic theorem on continuous functions is not available to tell us whether maximum and minimum points exist or not. However, since every point in the real number line is an interior point, and since $x^3 + 5x + 1$ is everywhere differentiable, the only critical values of x are those where $f'(x) = 0$. We differentiate and find

$$D(x^3 + 5x + 1) = 3x^2 + 5,$$

which is everywhere positive. Hence there are no critical values of x and therefore no maximum or minimum points.

Example C.   Find the maximum and minimum points of

$$f(x) = x^3 + 3x^2 - 24x + 8$$

on the nonnegative real numbers.

Solution.   The domain, $\{x \in R \mid 0 \leq x\}$, has only one endpoint, $x = 0$. We can differentiate everywhere to find

$$f'(x) = 3x^2 + 6x - 24$$

and set $f'(x) = 0$. This quadratic equation has two roots, -4 and $+2$, but only 2 is in the domain of the function. Hence the set of critical values is $\{0, 2\}$. We factor $f'(x) = 3(x + 4)(x - 2)$ and observe that in the domain, where $0 \leq x$, the factor $3(x + 4)$ is positive. Hence $f'(x)$ is negative if $0 \leq x < 2$ and $f'(x)$ is positive if $2 < x$. Thus by the mean value theorem, $f(x)$ decreases from 0 to 2 and increases when $2 < x$. Therefore $(2, -20)$ is the minimum point of $f(x)$.

If there is a maximum point, it must be the other critical point $(0, 8)$. Indeed $(0, 8)$ is a local maximum, since we know that $f(x)$ is decreasing there. However, we guess that since $f(x)$ increases when $2 < x$, $f(x)$ may exceed 8 at some point where $2 < x$. By trial and error we discover that $f(4) = 34$, which proves that $(0, 8)$ is not the maximum. We know now that there is no maximum point, since we have exhausted the set of critical points without finding one. We plot the graph of this function (Figure 3.17). This completes the solution of Example C.

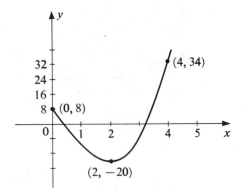

FIGURE 3.17 $x^3 + 3x^2 - 24x + 2$ on $\{0 \leq x\}$.

Let us now consider a more difficult problem where we cannot readily solve the equation $f'(x) = 0$.

Example D.   Find the maximum and minimum points of
$$f(x) = x^4 + 2x^3 + 12x^2 - 40x + 11$$
on the positive real numbers, $\{0 < x\}$.

Solution.   There are no endpoints and no points where $f(x)$ is not differentiable. Hence, the only critical values of x are those where $f'(x) = 0$. We differentiate and set $f'(x) = 0$, which gives us the equation

$$4x^3 + 6x^2 + 24x - 40 = 0.$$

Before attempting to solve this difficult cubic equation, we differentiate again to find

$$f^{(2)}(x) = 12x^2 + 12x + 24.$$

We observe from this formula that $f^{(2)}(x)$ is positive when $0 < x$. (Indeed, completing the square would tell us that it is positive for every real number x.) Hence, $f'(x)$ is an increasing function over the domain where $0 < x$. We calculate a few points on the graph of $f'(x)$ and find a partial table (see accompanying table and Figure 3.18). This

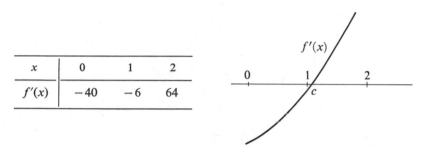

x	0	1	2
$f'(x)$	-40	-6	64

FIGURE 3.18   Equation $f'(x) = 4x^3 + 6x^2 + 24x - 40 = 0$.

shows by use of the basic theorem on continuous functions that there is a number c between 1 and 2 where $f'(c) = 0$. Moreover, since $f'(x)$ is everywhere increasing, this is the only root of $f'(x) = 0$ and therefore the only critical value of x.

We cannot directly apply the basic theorem on continuous functions to assure us that there is a maximum and a minimum because the domain is not a closed interval. However, we may expand $f(x)$ by Taylor's theorem in powers of $x - c$ with second-order remainder (§15.1). Since $f'(c) = 0$, this becomes

$$f(x) = f(c) + \tfrac{1}{2}f^{(2)}(z)(x - c)^2,$$

where z is some number between x and c. Putting in the expression for the second derivative gives us

$$f(x) - f(c) = \tfrac{1}{2}(12z^2 + 12z + 24)(x - c)^2.$$

The right member is never negative, whatever the numbers x, c and z may be, and hence for every x,

$$f(x) \geqq f(c).$$

This proves that $(c, f(c))$ is the minimum point. There is no maximum point, since we have exhausted the set $\{c\}$ of critical points without finding a maximum. This completes the solution of the problem except for the numerical determination of c, where we might use a computer. We can estimate graphically that $c = 1.25$, approximately.

We have not studied any example in which critical points arise because the derivative fails to exist there. This can be demonstrated with simple ramp functions of the sort that we studied earlier (§5.5).

Example E.   Find the maximum and minimum points of the ramp function f on $[0, 10]$ whose graph joins the vertices $(0, 1)$ to $(3, -2)$ to $(6, 4)$ to $(8, 1)$ to $(10, 4)$.

Solution.   We plot the graph in Figure 3.19 and observe that at all points between the vertices the derivative has a definite value, which is the slope of the line segment there. This slope is nowhere zero. Hence the only critical points are the endpoints (0, 1) and (10, 4) and the vertices (corners), where the slope $f'(x)$ is ambiguous because it has one slope to the left of the vertex and another to the right. We examine the values of f at these critical points and find that (6, 4) and (10, 4) are maximum points and (3, −2) is the minimum point. This completes the problem.

The calculus method is hardly necessary here, since direct graphical analysis is sufficient.

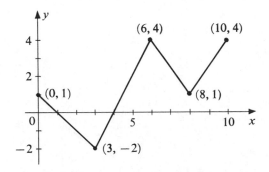

FIGURE 3.19   Critical points of a ramp function.

16.4 Exercises

By inspection, in Exercises 1–6 find the maximum and minimum points of the functions on the intervals indicated.

1. $f(x) = 3x + 1$ on [0, 4].
2. $f(x) = 1 − x$ on [−1, 1].
3. $f(x) = x^2$ on [−1, 1].
4. $f(x) = 2$ on the real numbers.
5. $f(x) = x^7 + 5x^6 + 71x^5 + 20x^2 + 8$ on [0, 1].
6. $f(x) = x^2$ on [0 < x < 1].

Find all the critical points of each of the functions in Exercises 7–11.

7. $f(x) = 2x^2 − 3x + 4$ on [−1, 1].
8. $f(x) = 2x + 3$ on the real number scale.
9. $f(x) = \begin{cases} 1 + x \text{ on } [−1, 0], \\ 1 − x \text{ on } [0, 1], \text{ a ramp function.} \end{cases}$
10. $f(x) = 4$ on the real number scale.
11. $f(x) = (x − 1)(x + 2)$ on [−3, 1].

Investigate for maximum and minimum points each of the functions in Exercises 12–23.

12. $f(x) = −x^2 + 3x + 5$ on [0, 4].
13. $f(x) = x^3 − 42x + 1$ on [−5, 5].
14. $f(x) = x^2 + 6x − 2$ on [−1, 1].
15. $f(x) = x^2 + 6x − 1$ on [−4, 4].
16. $f(x) = 2x^3 + 3x^2 − 12x + 5$ on [−3, 3].
17. $f(x) = x^4 + 4x^2 − 20x + 4$ on the real numbers.
18. $f(x) = x^4 − 4x$ on the real numbers.
19. $f(x) = x^4 + 4x^3 + 4x^2 − 7$ on [−3, 1].
20. $f(x) = x^3$ on all real numbers.
21. $f(x) = x^5 − 5x + 20$ on [−2, 2].
22. Find the maximum and minimum points of the ramp function f on [0, 10] whose graph connects the vertices (0, 1) to (3, −4) to (5, 0) to (7, −1) to (10, 4).

23. Find the maximum and minimum points of a function that is defined by the two formulas $f(x) = -3x + 4$ when $x \leq 0$ and $f(x) = 2x + 4$ when $x > 0$.

Exercises 24–28 on local maxima and minima require the following theorem. The proof is left for the theoretical problems.

THEOREM (The second-derivative test). For the function f on $[a, b]$:

Case I.   If $f'(c) = 0$ and $f^{(2)}(c) > 0$, then $(c, f(c))$ is a local minimum point;
Case II.   If $f'(c) = 0$ and $f^{(2)}(c) < 0$, then $(c, f(c))$ is a local maximum point.

Investigate for local maxima and minima by use of the second-derivative test in Exercises 24–28.

24. $f(x) = x^3 - 3x^2 - 3x + 7$ on the real numbers.
25. $f(x) = x^2 - 5x + 1$ on the real numbers.
26. $f(x) = 1 - 6x - x^2$ on the real numbers.
27. $f(x) = (x - 1)^2(2x + 1)^2$ on the real numbers.
28. $f(x) = 3x^4 - 8x^3 + 6x^2 - 24x + 11$ on the real numbers.

29. Show that, although $(0, 0)$ is a minimum point for the function $f(x) = x^4$ on the real numbers, the second-derivative test fails so to establish.

16.5   Problems

T1. Prove that if $f^{(2)}(c) > 0$ and $f'(c) = 0$, then there is a neighborhood of c in which $f'(x)$ is increasing.
T2. Show that in the neighborhood of c determined by Problem 1 $(c, f(c))$ is a minimum point for $f(x)$.
T3. Prove that if $f'(c) = 0$ and $f^{(2)}(c) < 0$, then $(c, f(c))$ is a local maximum point of $f(x)$.

Problems T1–T3 complete the proof of the theorem on the second derivative test (§16.4).

T4. Prove that if f on $[a, b]$ is convex, then the maximum point occurs at an endpoint, at a or b. (Compare §16.3, Example D.)
T5. Prove that if every point of $[a, b]$ is a critical point of a differentiable function f on $[a, b]$, f is a constant.
T6. In Problem T5, construct a step function to show that the conclusion is not necessarily true if we leave out the hypothesis that f is differentiable at every point of $[a, b]$.

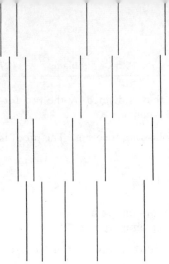

Antiderivatives and Integrals

CHAPTER

17 Antiderivatives

17.1 Definition of Antiderivative. We say that the differentiable function F on the domain X is *an antiderivative* of f on X if and only if $DF = f$. We also introduce the symbol D^{-1} for "an antiderivative of ...," defined by $D^{-1}f = F$ if and only if $DF = f$.

For example, $D^{-1}x^7 = \frac{1}{8}x^8 + 10$; for

$$D\tfrac{1}{8}(x^8 + 10) = \tfrac{1}{8}Dx^8 + D10 = \tfrac{1}{8}(8x^7) + 0 = x^7.$$

Any constant will serve equally as well as **10**.

$$D^{-1}x^7 = \tfrac{1}{8}x^8 + C; \qquad \text{for} \quad D(\tfrac{1}{8}x^8 + C) = \tfrac{1}{8}(8x^7) + 0 = x^7.$$

Thus a formula for the antiderivatives of f has an arbitrary constant added on the end.

17.2 The Differential Equation $DF = f$. We consider f as a known function and we want to find an unknown function F that satisfies the differential equation

$$DF = f$$

on some interval $[a, b]$. For example, the differential equation

$$DF(x) = 3x^2$$

has the solution $F(x) = x^3$, but it has other solutions as well. $F(x) = x^3 + C$ is also a solution for any constant C. If we know one solution of a differential equation of the form $DF = f$, can we find all of them? The following theorem answers the question.

THEOREM. If F is one solution of the differential equation $DF = f$ on an interval $[a, b]$, then every other solution is of the form $F + C$, where C is an arbitrary constant.

Proof. Let G be another solution on the interval, so that $DG = f$. Subtracting, and applying the linearity of D, we find that $D[G - F] = 0$. Then we apply the theorem on functions with zero derivatives (§13.4) and find that the only functions represented by the difference $G - F$ on $[a, b]$ are constants. This immediately implies the conclusion of the theorem.

We plot $f(x) = 3x^2$ and a number of its antiderivatives $F(x) = x^3 + C$ (Figure 4.1). The antiderivatives fill the portion of the plane for which $a \leq x \leq b$; and through each such point there is only one graph of an antiderivative. When one point on the graph is specified, an *initial condition* exists.

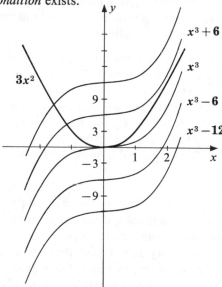

FIGURE 4.1 Antiderivatives of $f(x) = 3x^2$.

Example. Find the antiderivative of $f(x) = x^2 + 3x - 2$ on $[0, 10]$, whose graph includes the point $(4, 7)$.

Solution. By trial and error, using the rules for derivatives, we find that *an* antiderivative of $x^2 + 3x - 2$ is $\frac{1}{3}x^3 + \frac{3}{2}x^2 - 2x$. The theorem says that every antiderivative of $f(x)$ is given by $F(x) = \frac{1}{3}x^3 + \frac{3}{2}x^2 - 2x + C$, where C is an arbitrary constant. For *the* antiderivative we are trying to find, we are given $F(4) = 7$. Hence, subtracting $x = 4$, we find $7 = (\frac{1}{3})4^3 + (\frac{3}{2})4^2 - (2)4 + C$, which determines that $C = \frac{91}{3}$. Thus the required antiderivative is $F(x) = \frac{1}{3}x^3 + \frac{3}{2}x^2 - 2x + \frac{91}{3}$. This completes the problem.

17.3 Some Antiderivative Formulas. If u and v are functions that have antiderivatives $D^{-1}u$ and $D^{-1}v$, then

$$D^{-1}(u + v) = D^{-1}u + D^{-1}v \quad \text{and} \quad D^{-1}Cu = CD^{-1}u.$$

We readily verify that all antiderivatives of the functions 0, 1, x^n on an interval of numbers are given by

$$D^{-1}0 = C,$$

$$D^{-1}1 = x + C,$$

$$D^{-1}x^n = \frac{x^{n+1}}{n+1} + C,$$

where C is an arbitrary constant and n is a positive integer.

These formulas permit us to find the antiderivatives of any polynomial. For example,

$$D^{-1}[x^4 - 7x^3 + 8x^2 - 11x + 4] = D^{-1}x^4 - 7D^{-1}x^3 + 8D^{-1}x^2 - 11D^{-1}x + 4D^{-1}1$$

$$= \frac{x^5}{5} - 7\frac{x^4}{4} + 8\frac{x^3}{3} - 11\frac{x^2}{2} + 4x + C.$$

Example. If a car in a performance test moves with uniformly accelerated motion, increasing its speed at 6 ft/sec², starting from rest at the point when $x = 0$ at time $t = 0$, what is its position $c(t)$ at any time thereafter?

Solution. First let $v(t)$ be the velocity function. We are given $Dv(t) = 6$ and $v(0) = 0$. We find the antiderivative $v(t) = D^{-1}6 = 6t + C_1$ and set $t = 0$ to determine C_1. This gives $v(0) = 0 = 6(0) + C_1$. Therefore, $C_1 = 0$; and $v(t) = 6t$.

The velocity function is the derivative of the position function $x(t)$. That is, $Dx(t) = 6t$ and $x(0) = 0$. Since $D^{-1}6t = 6(t^2/2) + C_2$, we have $x(t) = 3t^2 + C_2$, where C_2 is to be determined by $x(0) = 0$. We are given $x(0) = 0 = 3(0) + C_2$, which implies that $C_2 = 0$. Hence $x(t) = 3t^2$. This is the solution of the problem.

17.4   Differential Equation of Area.

We have previously (§9.4) defined and calculated areas of regions bounded by curved graphs, using the method of sums of areas of small rectangles. The idea of the differential equation provides a different approach to this problem. We make the preliminary assumption that there is an area of the region bounded by the graph of a continuous function f, the x-axis, a vertical line at a, and a vertical line at x. We shall think of x as a variable (Figure 4.2(a)).

For each position x in the domain of f there is an area $A(x)$ of the region swept out behind the moving vertical line at x. Under our assumption, an area function $A(x)$ is thereby defined on the domain of f such that $A(a) = 0$. We assume natural properties of this area as we have done previously (§9.3). Let us calculate the derivative of the area function. Here, for simplicity, we shall treat the case that f is increasing and we shall later remove it (Exercises).

The difference $A(w) - A(x)$ is the area of the region below the graph of f and between the vertical lines at x and at w (Figure 4.2(b), shaded region). When $x < w$, this region includes the rectangle $xwRP$, whose height is $f(x)$, and is included in the rectangle $xwQS$, whose height is $f(w)$ and base is $w - x$. Hence the areas are related by the inequalities

$$f(x)(w - x) \leqq A(w) - A(x) \leqq f(w)(w - x).$$

Dividing by the positive number $w - x$, we find

$$f(x) \leqq \frac{A(w) - A(x)}{w - x} \leqq f(w).$$

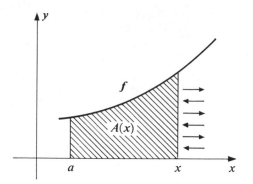

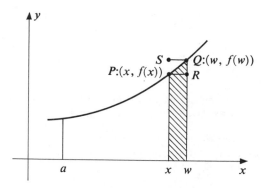

(a)   We think of x moving, sweeping out an area $A(x)$ of the shaded region behind the vertical line at x

(b)   The area of the shaded region is $A(w) - A(x)$. For this increasing function on $[x, w]$,

$$\inf f(u) = f(x) \text{ and } \sup f(u) = f(w)$$

FIGURE 4.2   Differentiation of the area function.

We now proceed to the limit as $w \to x$ and find

$$\lim_{w \to x} f(x) \leqq \lim_{w \to x} \frac{A(w) - A(x)}{w - x} \leqq \lim_{w \to x} f(w),$$

which becomes

$$f(x) \leqq \lim_{w \to x} \frac{A(w) - A(x)}{w - x} \leqq f(x),$$

where on the right we used the fact that f is continuous. Therefore $A'(x) = f(x)$. In other words, $A(x)$ satisfies the differential equation $DA(x) = f(x)$ with the initial condition $A(a) = 0$. Dropping the preliminary assumption that $A(x)$ exists for each x, we can now define the area to be the value $A(x)$, provided that we can find an antiderivative $A(x)$ of $f(x)$.

Example.   Find the area bounded by x^2, the x-axis, and the lines $x = 1$ and $x = 5$ (§9.6, Example).

Solution by antiderivatives.   The area function $A(x)$ satisfies the differential equation $DA(x) = x^2$ with initial condition $A(1) = 0$. We use the rules for antiderivatives to find $D^{-1}x^2 = (x^3/3) + C$, so that $A(x) = (x^3/3) + C$. Evaluating at 1, we have $0 = A(1) = \frac{1}{3} + C$. Therefore C is determined to be $C = -\frac{1}{3}$ and $A(x)$ becomes $A(x) = (x^3/3) - (1/3)$. We now complete the solution by evaluating,

$$A(5) = \frac{5^3}{3} - \frac{1}{3} = \frac{124}{3}.$$

This is the required area.

In general, if $F(x)$ is any antiderivative of $f(x)$, then the area of the region below the graph of $f(x)$ between the limits $x = a$ and $x = b$ is given by $F(b) - F(a)$ (Exercises).

17.5   Exercises

In Exercises 1–12, calculate the indicated antiderivatives, $F(x) = D^{-1}f(x)$.

1. $D^{-1}(x^2 - x + 1)$.

2. $D^{-1}8$.

3. $D^{-1}(1 - x)$.

4. $D^{-1}(x^3 + 7x + 2)$.

5. $D^{-1}(x^3 + 11x^2 - 7x + 4)$.
6. $D^{-1}D^{-1}(6x + 4)$.
7. $D^{-1}D^{-1}(x^2 + 8x - 1)$.
8. $D^{-1}(ax + b)$.
9. $D^{-1}(2x + 3)^7$.
10. $D^{-1}(1 - x)^{10}$.
11. $D^{-1}(3x^2 - 2x + 4)$, for which $F(1) = 4$.
12. $D^{-1}0$, for which $F(7) = -1$.

In Exercises 13–18, find every solution of the differential equation, with initial condition where given, on the real numbers.

13. $DF(x) = x^2 + 1$.
14. $DH(t) = 1 - t + t^2$.
15. $Dg(u) = u^3 - 2u$, $g(0) = 4$.
16. $D^{(2)}p(z) = 2z + 1$, $p(1) = 1$, $p'(1) = -1$.
17. $D^{(2)}f(x) = 4$, $f(0) = 3$, $f'(5) = 7$.
18. $Dg(z) = (2z + 1)^6 - 4$, $g(-\frac{1}{2}) = 2$.

19. A moving point had position $x = 10$ at $t = 3$ sec, and its velocity was given by $v(t) = 10t^2 - 40t$. Where was it at $t = 5$ sec?
20. A car is accelerating uniformly from a stationary start at the rate of 4 ft/sec². How far from the start was it after one minute? How fast was it going at that time?

In Exercises 21–26, find the area of the region below the graph of the given functions between the indicated limits.

21. Below $f(x) = x^3$ between $x = 0$ and $x = 5$.
22. Bounded by $f(x) = x^2 - 7x + 5$ between 0 and 6.
23. Bounded by $f(t) = t^3 - 7t + 1$ between -1 and $+1$.
24. Bounded by $f(x) = x + 1$ between 1 and x. Plot the area function $A(x)$.
25. Bounded by $f(z) = 1 - z - z^2$ between a and b.
26. Bounded by $f(x) = (2x - 1)^6$ between 0 and 4.

27. Let $A(x)$ be the area of the region bounded by $f(x) = 2x - 3$ between 1 and x. Find the number e such that $A(e) = 1$.
28. For a certain manufacturer, the marginal cost $C'(q)$ of producing one additional unit at the production level of q units is given by $C'(q) = 50 - 0.1q$, where $0 \leq q \leq 200$. Find the total cost of producing 200 units.
29. A moving point on a line has velocity $v(t) = t^2 - 10t + 24$ on [0, 10] sec. When in this 10-second interval will it be farthest from the starting point? What is the greatest distance?
30. In Exercise 29, find the approximate times when the point passed back through the starting point.
31. In Exercise 29, what is the acceleration at the time of maximum velocity?
32. A velocity function $v(t)$ on the interval [0, 10] seconds is a ramp function connecting the vertices $(0, 1)$, $(3, -2)$, $(6, 4)$, $(8, 11)$, $(10, 0)$. Find the position function $y(t)$ for which $y(0) = 1$. Plot $y(t)$. Describe the motion in terms of the acceleration.
33. In Exercise 29, what was the total distance traveled by the moving point in the 10 seconds?
34. The local scale factors in mapping $f(x)$ of the x-axis into the y-axis are given by x^2. What magnification of the interval [1, 2] does this mapping produce?

17.6 Problems

T1. The function $f(x) = x^2 + 1$ is defined on a domain X consisting of two disjoint intervals [0, 1] and [3, 4]. Find all antiderivatives $D^{-1}f(x)$ on X.
T2. Prove that if F is any antiderivative of the continuous function f on $[a, b]$, then the area below f between a and b is given by $F(b) - F(a)$.

T3.   Remove the restriction that f is increasing (§17.4) and prove that if f is continuous, then the derivative of the area function $A(x)$ at x is

$$A'(x) = f(x).$$

Figure 4.3 suggests the modification of the argument.

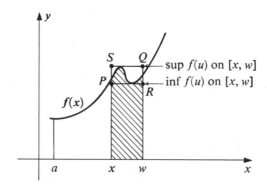

FIGURE 4.3   A hint for Problem T3.

T4.   Show that we can find an antiderivative of $f(x)$ by repeated differentiation, by expanding the unknown antiderivative in a Taylor's series.

T5.   Derive the formula for an antiderivative $F(x)$ of x^n in the manner of T4 by expanding $F(x)$ in powers of $x - 0$.

T6.   By expanding $F(x)$ in powers of $x - 1$ by Taylor's theorem, find $F(x)$ such that $F'(x) = 1/x$ and $F(1) = 0$.

T7.   Find a Taylor's theorem representation of $F(x)$ if $F'(x) = F(x)$ and $F(0) = 1$.

18   Kinetics and Dynamics

18.1   Kinetics.   Let us recall briefly the mathematical representation of motion. The study of pure motion of a tracer point, which has no physical mass or energy, is called *kinetics*. The mathematical representation of an instant of time is simply a real number t. The representation of the duration of an interval of time is a difference $t_2 - t_1$.

When a point P "moves" on a line, it occupies a fixed point on the line given by the coordinate x at time t. Hence we speak of the *position* as the pair of numbers (t, x). The position of the moving point is given every time by a position function $x(t)$ on some domain T of time. The values of the derivative function, $v(t) = x'(t)$, give the velocities at each time t (§12.4) and the values of the second derivative $x^{(2)}(t)$ give the accelerations (§12.5). The point is "moving" at time t if $x'(t) \neq 0$ and is at least momentarily "stationary" if $x'(t) = 0$. It is "at rest" if $x'(t) = 0$ in some interval so that $x(t) = C$ in that interval. Thus the position function $x(t)$ in kinetics is the mathematical representation of a motion.

These ideas may be extended to motions in the plane where the tracer point P occupies the point (x, y) in the plane at time t. Hence a position is a triple of numbers (t, x, y), and the position function that describes the motion becomes a pair of functions $(x(t), y(t))$.

Example A.   A point P moves on the x-axis with the acceleration $a = 12t - 30$, starting when $t = 0$ at the point where $x = 6$ with initial velocity 24 ft/sec. Find the motion and describe it.

Solution. Let $x(t)$ represent the position function. Since $D^2x(t) = 12t - 30$, we apply one antiderivative to find

$$Dx(t) = 12\left(\frac{t^2}{2}\right) - 30t + C_1.$$

This is the velocity function. We find that $C_1 = 24$ from the initial velocity $x'(0) = 24$. Now we have

$$Dx(t) = 6t^2 - 30t + 24.$$

We apply an antiderivative operator again and find

$$x(t) = 6\left(\frac{t^3}{3}\right) - 30\left(\frac{t^2}{2}\right) + 24t + C.$$

We know that $x(0) = 6$, and hence that $C_2 = 6$. This gives us the position function

$$x(t) = 2t^3 - 15t^2 + 246 + 6,$$

which represents the motion (Figure 4.4).

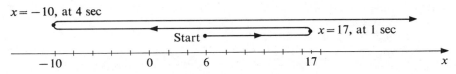

FIGURE 4.4   Motion of Example A.

To describe the motion, we consider first the velocity function

$$x'(t) = 6(t - 1)(t - 4).$$

This tells us immediately that the moving point reverses direction of motion at times $t = 1$ sec and $t = 4$ sec. It starts (Figure 4.4) at position $(0, 6)$ and moves in the positive direction for one second to the position where $(1, x(1)) = (1, 17)$. Then, reversing, it moves in the negative direction for three seconds reaching a position where $(4, x(4)) = (4, -10)$. Then it turns back in the positive direction and, with increasing velocity, it takes off for infinity. This ends our problem.

18.2   Mathematical Representation of a Particle.   When we ascribe a positive numerical mass m to the moving point, we call it a *particle*. It is not enough to give the number m alone to describe the particle because many particles may have the same mass; nor is it enough to say where the point-mass is located in the line at any one time. In mechanics, a particle is known if we know its mass and its position at every time t. Hence a mathematical representation of a particle moving on a line is a pair $(m, x(t))$ consisting of a positive number m and position function $x(t)$.

In terms of this definition, we can express the velocity, acceleration, momentum, kinetic energy, potential energy, and law of motion of the particle in a field of force. At time t the velocity is $x'(t)$; the acceleration, $x^{(2)}(t)$; the momentum, $mv = mx'(t)$. For the kinetic energy, we have $\frac{1}{2}mv^2 = \frac{1}{2}m[x'(t)]^2$.

18.3   Dynamics.   Newton's (second) law of motion states that if $F(x)$ is the force acting on the particle at point x, then the position function must satisfy the differential equation

$$mD^2x(t) = F(x),$$

which says, in words, "mass times acceleration equals applied force."

The principal problem of *dynamics* is to solve this differential equation for the unknown position function $x(t)$ when the force function $F(x)$ and one initial position and velocity are known.   The problem cannot even be understood without the idea of an unknown function.

From the physical point of view, such a procedure may be interpreted as predicting, from the law of motion and the initial position $x(0)$ and initial velocity $x'(0)$, the position $x(t)$ and velocity $x'(t)$ at every time thereafter.   For example, if a ball is thrown vertically upward in the earth's gravitational field at 40 ft/sec from the level $x = 0$ at time $t = 0$, then the law of motion predicts its entire motion thereafter, so long as the only applied force is the earth's gravitation.

18.4   Freely Falling Particle.   A very simple case is the particle in a constant gravitational field whose force function is the constant $F(x) = -mg$, where g is the acceleration due to gravity (Figure 4.5).   Then the differential equation expressing the law of motion becomes

$$mD^{(2)}[y(t)] = -mg,$$

where we denote the motion by $y(t)$ to indicate that it is a vertical line.   We may divide both sides by m to reduce the equation of motion to

$$D^2y(t) = -g,$$

showing that in this case the numerical value of the mass does not affect the motion.*

FIGURE 4.5   Force of gravity.

We take one antiderivative of both sides to find

$$D^{-1}D^2y(t) = Dy(t) = D^{-1}(-g) = -gt + C.$$

Then we take another antiderivative of $Dy(t) = -gt + C_1$ to find

$$y(t) = -g\frac{t^2}{2} + C_1t + C_2,$$

where C_1 and C_2 are arbitrary constants.

---

* This is one observation that Galileo is supposed to have made when he dropped balls of different weights from The Leaning Tower of Pisa.

Example. (a) Find the motion of a freely falling particle that is projected upward at a velocity of 40 ft/sec from a height of 25 ft. (b) At what time does it reach maximum height and what is the maximum height? (c) When does it reach ground level and with what velocity?

Solution. (a) The motion of any freely falling particle of any mass is given by the solution

$$y(t) = -g\frac{t^2}{2} + C_1 t + C_2.$$

We are given that $y(0) = 25$ and $y'(0) = 40$. We set $t = 0$ in this solution and we find $25 = 0 + 0 + C_2$. Differentiating once, we return to $y'(t) = -gt + C_1$ and set $t = 0$ to find that $40 = 0 + C_1$. Hence the motion is

$$y(t) = -g\frac{t^2}{2} + 40t + 25.$$

(b) To find the maximum point, we set $y'(t) = 0$ and find $-gt + 40 = 0$. Hence the time $t = 40/g$ is a critical value of t. This enables us to factor $y'(t) = -g[t - (40/g)]$. The result gives signs of $y'(t)$ as in the accompanying table.

$t < \dfrac{40}{g}$	$t = \dfrac{40}{g}$	$t > \dfrac{40}{g}$
$y'(t) > 0$	$y'(t) = 0$	$y'(t) < 0$
	0	
+	velocities	−

Hence a maximum is reached when $t = 40/g$ sec. The maximum height is $y(40/g) = (800/g) + 25$ ft. If $g = 32.2$ ft/sec^2, then the time to maximum is approximately $t = 1.24$ sec and height 49.8 ft.

(c) At ground level $y(t) = 0$. We set

$$-g\frac{t^2}{2} + 40t + 25 = 0$$

and solve the quadratic equation for two roots

$$t = -\frac{40}{g} - \frac{10\sqrt{16 + (g/2)}}{g}, \quad \text{and} \quad t = -\frac{40}{g} + \frac{10\sqrt{16 + (g/2)}}{g}.$$

The second is the only time that occurs after the start, $t = 0$. So we may approximate, the particle reaches ground level at about time $t = 3.04$ sec and its velocity $-gt + 40 = -57.5$ ft/sec. This completes the problem.

18.5 Exercises

In Exercises 1–6, find the unknown position function of the moving point and describe the motion with a diagram like that of Figure 4.4.
1. $D^2 x(t) = 6$, $x(0) = 10$, $v(0) = -5$.
2. $D^2 x(t) = 12t - 30$, $x(0) = 18$, $v(0) = 36$.

3. $D^2x(t) = 6t - 12$, $x(0) = -4$, $v(0) = -15$.
4. $D^2y(t) = 12 - 6t$, $y(0) = 10$, $v(0) = -9$.
5. $D^2x(t) = 6t - 2$, $x(0) = 0$, $v(0) = 1$.
6. $D^2z(t) = 2t - 2$, $z(0) = -5$, $z'(0) = -2$.
7. Referring to Example A (§18.1), describe the change in motion that occurs at the time $t = \frac{5}{2}$ sec, when the acceleration is zero.
8. A motion in an xy-plane has $D^2x(t) = a$, $x(0) = x_0$, $x'(0) = 0$, $D^2y(t) = c$, $y(0) = y_0$, $y'(0) = 0$. Show that the motion is in a straight line and find the equation of the line.
9. A particle is thrown up from the ground level with a velocity of 64.4 ft/sec.
    (a) Find the position function that predicts its position for every time thereafter.
    (b) Find the time at maximum height and find the maximum height.
10. A particle of mass 4 is released with zero velocity at a height of 400 ft. When does it reach the ground and with what velocity?
11. A baseball is thrown down from the top of the Washington Monument, height 555 ft, with an initial velocity -32 ft/sec. With what speed did it hit the ground after a catch attempt failed?
12. A particle of mass m is released with velocity v_0 at a height s_0 at time t_0. Find the motion.
13. A freely falling particle falls from a height of h ft, starting with velocity 0. Show that it reaches the ground with velocity $-\sqrt{2gh}$.
14. A mortar shoots vertically upward a pellet, which strikes the ground in 8 sec. What was the muzzle velocity?
15. A jet airplane under constant thrust can reach flying speed in R ft of runway. If the thrust of the engines can be doubled without increased weight, what fraction of the takeoff run R will be required to reach flying speed?
16. In Exercise 15, if the doubled thrust can be had at a cost of 15 percent more total weight, what fraction of the original takeoff run R will be required?
17. Two cars are powered with like engines but one car weighs 50 percent more than the other. Assuming that the engine provides a constant driving force, what do we find to be the ratio of the times required by the two cars to reach 60 mi/hr from a standing start?
18. The two cars of Exercise 17 were allowed to race for 10 sec under the same driving from their engines. At the end of 10 sec, what was the ratio of their kinetic energies? Of their momenta?
19. The weight of a body is the force that gravity exerts on it. Find the mass of a body that weighs 100 lb.
20. A body weighing 100 lb starts from rest and is driven without friction on a horizontal line by a force $F(t)$, which varies with time according to the graph (Figure 4.6). Find the unknown position function $x(t)$ and graph it. What is the meaning of the portion of the graph after time 8 sec when $F(t)$ becomes and remains zero?
21. As in Exercise 20, a body of 100 lb is driven on a horizontal line by a force $F(t)$, given by the graph (Figure 4.7). Plot the graph of the velocity function $v(t)$ of the moving mass. What is the significance of the point $(4, x(4))$ where the force changes sign?

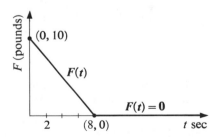

FIGURE 4.6   Force in Exercise 20.

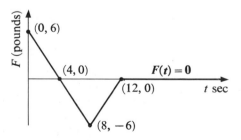

FIGURE 4.7   Force in Exercise 21.

Exercises 22–26 are concerned with Atwood's machine, a simple device in which two masses M and m are connected by a string over a pulley, which we assume to be weightless and frictionless (Figure 4.8). To move, the system must move the total mass $M + m$, but the net force to move them is the difference, $-(M - m)g$.

22. Write the equation of motion for Atwood's machine and show that the motion is not independent of the numerical values of the masses.
23. An Atwood's machine has $M = 9$ and $m = 7$. The heavier mass was set in motion upward at $+4$ ft/sec when $t = 0$ from a height of 8 ft. Find the subsequent motion as a function $y(t)$.
24. An Atwood's machine has $M = m$ and the M mass is given velocity $+4$ ft/sec (upward) from a height 0. Find the subsequent motion. Is it independent of the numerical values of the two equal masses?
25. On the Atwood's machine with $M = m$, a uniform force of 1 lb is applied upward for 1 sec. After that, the force is zero. Find the motion $y(t)$. How does the total mass affect the answer?
26. Find the general solution for the case $M \neq m$ of the motion of an Atwood's machine where the M mass starts with velocity v_0 at height y_0 at time t_0.
27. Two masses m_1 and m_2 slide frictionlessly on a horizontal plane (Figure 4.9) and are connected by strings over a pulley to mass m_3, starting with velocity $v = 0$ at a point where the x-coordinate of m_1 is -10.
    (a) Find the motion.
    (b) Find the force of tension in the string between m_1 and m_2; between m_2 and m_3.

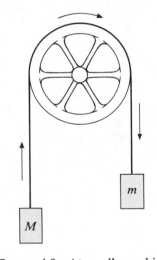

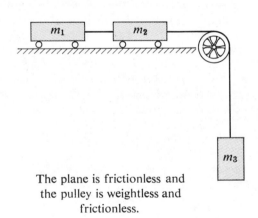

The plane is frictionless and the pulley is weightless and frictionless.

FIGURE 4.8   Atwood's machine.   FIGURE 4.9   Exercise 27.

18.6   Problems

T1. Prove that for any solution x of the equation of motion of the freely falling particle

$$\frac{1}{2} m(x')^2 + mgx = E,$$

where E is a constant and $x' = D_t x$.

T2. In Problem T1, the function $V = mgx$ is called the potential energy at x. Show that $-D_x V$ is the force acting on the particle at the point x. Then show that T1 says that on any motion of a freely falling particle:

Kinetic energy + potential energy = constant.

This is the *law of conservation of energy* for the freely falling particle.

T3.   Write the three equations of motion for $x(t)$, $y(t)$, $z(t)$ that give the x, y, and z coordinates at any time t of a freely falling particle in three-dimensional space. We assume that the only applied force is a constant force of gravity $-mg$ in the vertical direction (usually taken to be the z-direction).

T4.   In Problem T3, do the equations of motion imply that the only motion is in a vertical line?

T5.   Solve the equations of motion in T3 for $x(t)$, $y(t)$, $z(t)$.

T6.   Newton's first law of motion states that if the applied force has a value $F = 0$ everywhere, the particle moves in a straight line with constant velocity. Show that the second law (§18.3) implies this, provided that the motion is continuous and time varies continuously in an interval of real numbers.

19   Fundamental Theorem of Integral Calculus

19.1   Definition of the Integral as a Sum.   We repeat the definition (§9.5) of the integral because of its importance. Let f be a bounded function on the interval $[a, b]$. We recall that a function is bounded if the set of all its values $\{f(x)\}$ is contained in some interval $[L, G]$. Let

$$\{a = x_1, x_2, x_3, \ldots, x_n = b\}$$

be any partition of the interval $[a, b]$ into smaller subintervals. We shall define an understep function s and an overstep function S for this partition (Figure 4.10(a)). For every number u in the subinterval $[x_i, x_{i+1}]$, we define

$$s(u) = \inf\,\{f(x) \mid x_i \leqq x \leqq x_{i+1}\}$$

and

$$S(u) = \sup\,\{f(x) \mid x_i \leqq x \leqq x_{i+1}\}.$$

This reads that $s(u)$ is a constant in the subinterval and is the infimum (§7.5) of the values $f(x)$ in that subinterval. Similarly, $S(u)$ is the supremum of the values $f(x)$ in that subinterval (Figure 4.10(b)). Also, for any subinterval $[x_i, x_{i+1}]$, we define $du = x_{i+1} - x_i$. Then the area of the under-rectangle on any subinterval of the partition is $s(u)\,du$, while the area of the over-rectangle is $S(u)\,du$ (Figure 4.10(a)). Next we sum the areas of all

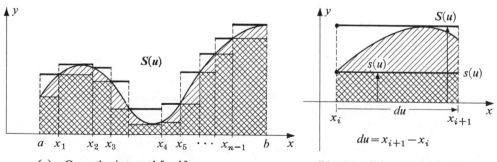

(a)   Over the interval $[a, b]$ (b)   Detail in one subinterval

The area under the curve $\int_a^b f(u)\,du$ is enclosed between the sum $\int_a^b s(u)\,du$ of the areas of the lower rectangles and the sum $\int_a^b S(u)\,du$ of the areas of the upper rectangles.

Figure 4.10   Definition of the integral.

the over-rectangles, denoting this sum by $\int_a^b S(u)\, du$, and we sum the areas of the under-rectangles, denoting the result by $\int_a^b s(u)\, du$. For any partition there is a number A enclosed between these sums,

$$\int_a^b s(u)\, du \leqq A \leqq \int_a^b S(u)\, du.$$

Now we will refine the partition by introducing more and more partition points. If for all possible fine partitions there is one and only one number A enclosed between the under sums and upper sums, this number A is the *integral* of f from a to b and is denoted by

$$A = \int_a^b f(x)\, dx.$$

Also we say that f is *integrable* in $[a, b]$ if the integral exists.

19.2 Conditions for Integrability. We have proved (§9.3) that a function f that is increasing in the interval $[a, b]$ is integrable. It is then trivial to prove that a decreasing function is integrable. Combining these cases into one word, we say that $f(x)$ is *monotonic* in $[a, b]$ if it is either increasing in $[a, b]$ or decreasing in $[a, b]$. More generally, now we can prove (Exercises) the following theorem.

THEOREM. A bounded function f on $[a, b]$ whose graph consists of a finite number of increasing or decreasing arcs is integrable on $[a, b]$ (Figure 4.11).

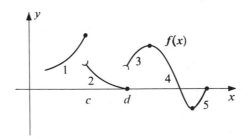

FIGURE 4.11   A function with a finite number of increasing or decreasing arcs. There are five monotonic arcs in the graph.

One important consequence of this theorem is that polynomials are integrable.

THEOREM. Every polynomial, $f(x)$, on a bounded and closed interval $[a, b]$ is integrable.

Proof. Let the degree of the polynomial be n. Then the degree of its derivative $f'(x)$ is $n - 1$. When we set $f'(x) = 0$ and solve, we get at most $n - 1$ interior critical points. The two endpoints included, the function has at most $n + 1$ critical points; these are joined by n pieces of the graph, which are either increasing because $f'(x) > 0$ or decreasing because $f'(x) < 0$. Thus the graph of a polynomial consists of a series of increasing or decreasing arcs joined at critical points, and there are at most n of these arcs. This proves that a polynomial is integrable.

We also state without proof a classical theorem that gives a different condition for integrability.

THEOREM.   Every continuous function f on a closed interval $[a, b]$ is integrable.

We could also use this theorem to establish the integrability of polynomials, since a polynomial is continuous.

19.3   Fundamental Theorem of Integral Calculus

THEOREM.   If f is integrable on the closed interval $[a, b]$ and if F is an antiderivative such that for every x in $[a, b]$, $F'(x) = f(x)$, then

$$\int_a^b f(x)\, dx = F(b) - F(a).$$

Proof.   We partition $[a, b]$ with the partition points

$$\{a = x_1, x_2, x_3, \ldots, x_{i-1}, x_i, \ldots, x_n = b\}$$

and set up the understep function $s(x)$ and overstep function $S(x)$, by the definition of the integral

$$\int_a^b s(x)\, dx \leq \int_a^b f(x)\, dx \leq \int_a^b S(x)\, dx;$$

and $\int_a^b f(x)\, dx$ is the only number that is enclosed between $\int_a^b s(x)\, dx$ and $\int_a^b S(x)\, dx$ for every partition of $[a, b]$.

Now we apply the theorem of the mean (§13.2) to the difference $F(x_i) - F(x_{i-1})$ on each subinterval of the partition. The theorem of the mean says that, since $F'(x) = f(x)$,

$$F(x_i) - F(x_{i-1}) = f(c_i)(x_i - x_{i-1}),$$

where c_i is between x_{i-1} and x_i. We define a new intermediate step function $t(x)$ by giving $t(x)$ the constant value $f(c_i)$ on each subinterval, $i = 1, 2, 3, \ldots, n$. Then on the ith subinterval

$$\inf f(u) \leq f(c_i) \leq \sup f(u).$$

Hence

$$s(x) \leq t(x) \leq S(x),$$

so that

$$\int_a^b s(x)\, dx \leq \int_a^b t(x)\, dx \leq \int_a^b S(x)\, dx.$$

Let us examine more closely the sum $\int_a^b t(x)\, dx$. By the way in which $t(x)$ was defined in the ith subinterval,

$$t(x)\, dx = f(c_i)(x_i - x_{i-1}) = F(x_i) - F(_{i-1}).$$

Hence, if we sum these terms over the entire partition we see that $\int_a^b t(x)\, dx$ is equal to the sum

$$[F(x_1) - F(a)] + [F(x_2) - F(x_1)] + \cdots$$
$$+ [F(x_i) - F(x_{i-1})] + \cdots + [F(b) - F(a)].$$

All of the terms of this sum cancel in pairs except $-F(a)$ from the first bracket and $F(b)$ from the last. Hence, for any partition,

$$\int_a^b t(x)\, dx = F(b) - F(a).$$

This shows that for every partition

$$\int_a^b s(x)\,dx \leq F(b) - F(a) \leq \int_a^b S(x)\,dx.$$

Since $f(x)$ is integrable, there is only one number, $\int_a^b f(x)\,dx$, thus enclosed between $\int_a^b s(x)\,dx$ and $\int_a^b S(x)\,dx$ for every partition. Therefore,

$$\int_a^b f(x)\,dx = F(b) - F(a).$$

This completes the proof of the fundamental theorem.

19.4   Application of the Fundamental Theorem to the Calculation of Integrals

Example A.   Calculate

$$\int_2^{12} (x^3 - 3x^2 + 7x - 5)\,dx.$$

Solution.   We find the antiderivative

$$F(x) = \frac{x^4}{4} - x^3 + 7\frac{x^2}{2} - 5x + C.$$

Since the polynomial is integrable, the fundamental theorem applies.   This gives us

$$\int_a^b (x^3 - 3x^2 + 7x - 5)\,dx = F(12) - F(2) = 3900.$$

This completes the problem.   We observe that the constant C always cancels in the difference $F(b) - F(a)$, so that it need not have been included in this problem.

Example B.   Find the area enclosed between the curves $x^2 - x + 1$ and $5 + 3x - x^2$.

Solution.   We begin by plotting the graphs of the two functions in the same coordinate plane (Figure 4.12).   The area in question turns out to be the area below the curve $5 + 3x - x^2$ and above the curve $x^2 - x + 1$, extending from the x-coordinate of the point of intersection at P to the x-coordinate of the point of intersection at Q.   Accordingly, we solve simultaneously to find the x-coordinates of the points of intersection.   At points of intersection

$$x^2 - x + 1 = 5 + 3x - x^2 \qquad \text{if and only if} \quad x^2 - 2x - 2 = 0.$$

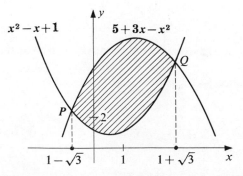

FIGURE 4.12   Area bounded by two curves.

This quadratic equation has the roots $x = 1 - \sqrt{3}$ and $x = 1 + \sqrt{3}$, which are the x-coordinates of P and Q.

Therefore the area A of the region between the two curves can be found by subtracting the area below $x^2 - x + 1$ from that below $5 + 3x - x^2$; that is,

$$A = \int_{1-\sqrt{3}}^{1+\sqrt{3}} (5 + 3x - x^2) \, dx - \int_{1-\sqrt{3}}^{1+\sqrt{3}} (x^2 - x + 1) \, dx.$$

Considering the definition of the integral, we see that this can be written as a single integral

$$A = \int_{1-\sqrt{3}}^{1+\sqrt{3}} [(5 + 3x - x^2) - (x^2 - x + 1)] \, dx = \int_{1-\sqrt{3}}^{1+\sqrt{3}} (4 + 4x - 2x^2) \, dx.$$

An antiderivative of $4 + 4x - x^2$ is

$$F(x) = 4x + 2x^2 - (2x^3/3).$$

Hence by the fundamental theorem,

$$A = F(1 + \sqrt{3}) - F(1 - \sqrt{3}).$$

We calculate by substitution,

$$F(1 + \sqrt{3}) = 4(1 + \sqrt{3}) + 2(1 + \sqrt{3})^2 - \frac{2(1 + \sqrt{3})^3}{3} = \frac{16 + 12\sqrt{3}}{3}$$

and

$$F(1 - \sqrt{3}) = 4(1 - \sqrt{3}) + 2(1 - \sqrt{3})^2 - \frac{2(1 - \sqrt{3})^3}{3} = \frac{16 - 12\sqrt{3}}{3}.$$

This gives us

$$A = F(1 + \sqrt{3}) - F(1 - \sqrt{3}) = 8\sqrt{3} \text{ area units.}$$

This completes our problem.

19.5   A Convenient Symbol.   We define the symbol $F(x)|_a^b$ to mean that

$$F(x)\Big|_a^b = F(b) - F(a).$$

Thus

$$(x^2 - x + 1)\Big|_1^2 = (2^2 - 2 + 1) - (1^2 - 1 + 1) = 3 - 1 = 2.$$

Example.   Calculate $\int_{-1}^{2} (x^2 + 2x - 1) \, dx$.

Solution.   By the fundamental theorem,

$$\int_{-1}^{2} (x^2 + 2x - 1) \, dx = \left(\frac{x^3}{3} + x^2 - x\right)\Big|_{-1}^{2} = \left[\frac{2^3}{3} + 2^2 - 2\right]$$

$$- \left[\frac{(-1)^3}{3} + (-1)^2 - (-1)\right]$$

$$= \frac{14}{3} - \frac{5}{3} = \frac{9}{3} = 3.$$

This completes the problem.

19.6 Exercises

In Exercises 1–10, calculate the integrals by use of the fundamental theorem.

1. $\int_2^6 (x - 2)\, dx.$ 2. $\int_0^4 (x^2 - 3)\, dx.$

3. $\int_{-1}^1 (x^3 + 3x)\, dx.$ 4. $\int_{-1}^1 (x^2 + 2)\, dx.$

5. $\int_{1-\sqrt{2}}^{1+\sqrt{2}} (2 - u)\, du.$ 6. $\int_{-5}^0 (t^3 - 1)\, dt.$

7. $\int_a^b (cx + d)\, dx.$ 8. $\int_a^b (ct + d)\, dt.$

9. $\int_3^5 (u^3 - 3u^2 + 15u - 76)\, du.$ 10. $\int_2^{-4} x^2\, dx.$

In Exercises 11–20, find the areas of the regions bounded by the given curves. In each case draw a sketch and shade the region whose area is calculated. Indicate by plus and minus signs areas that are positive and those that are negative.

11. The region bounded by the x-axis and the curve $x^2 + 3x - 5$ between $x = -1$ and $x = 3$.
12. The region below $1 + 4x - x^2$ between 0 and 4.
13. The region between $1 + x - x^2$ and $2x - 5$.
14. The region between x^2 and **10**.
15. The region bounded by the x-axis, $x^2 - 6x + 8$, $x = 0$, and $x = 6$.
16. The region bounded by **0**, $x^3 - 4x$, $x = -2$, $x = 2$.
17. The region between $x^2 - x + 6x$ and $-x^2 + 11x - 10$.
18. The region between x^3 and $4x$.
19. The region between $x^2 - 2x + 3$ and $5 - x^2$.
20. The region between x^2 and $3x + b$. Find for what values of b such a region exists.

21. Verify that for every three numbers a, b, c,

$$\int_{-c}^c (ax^3 + bx)\, dx = 0.$$

Explain this phenomenon.

22. Show that

$$F(x)\Big|_a^b + F(x)\Big|_b^c = F(x)\Big|_a^c.$$

23. Show that

$$[F(x) + G(x)]\Big|_a^b = F(x)\Big|_a^b + G(x)\Big|_a^b.$$

24. Show that for any numbers a, b, c, d,

$$F(x)\Big|_a^b + F(x)\Big|_b^c + F(x)\Big|_c^d + F(x)\Big|_d^a = 0.$$

25. Evaluate $(x^3 - x^2 - x - 2)\big|_0^2$.
26. Calculate the sum of the absolute values of the areas bounded by **0** and $x^3 - 4x^2 + 3x$ between 0 and 3.
27. A point P moves on the x-axis so that its velocity $v(t) = t^3 - 15t^2 + 50t$. Calculate the total distance traveled in 10 seconds, starting from $x = 0$ when $t = 0$. Also, how far from the starting point was P after 10 seconds?
28. Calculate $\int_3^6 f'(x)\, dx$, given that the graph of $f(x)$ includes the points $(0, 4)$, $(3, 5)$, $(6, -2)$, $(8, -9)$.
29. Calculate $\int_1^x (t^3 - 1)\, dt$. Observe that the result involves x but not t.

30. Show that

$$\int_0^x (t^2 - 2t + 5)\, dt = \int_0^x (z^2 - 2z + 5)\, dz.$$

31. Solve for x, $\int_1^x (u - 1)\, du = 4$.

19.7 Problems

T1. Prove that a decreasing function on $[a, b]$ is integrable (§9.3).
T2. Prove that if $f(x)$ increases on $[a, c]$ and decreases on $[c, b]$, it is integrable ($a < c < b$). *Hint:* As in Section 9.3, stack up the rectangles for $\int_a^b [S(x) - s(x)]\, dx$ in one pile whose height is $|f(c) - f(a)| + |f(b) - f(c)|$.
T3. Extend T2 to prove the theorem (§19.2) that a bounded function $f(x)$ on $[a, b]$, whose graph consists of a finite number of increasing or decreasing arcs, is integrable on $[a, b]$.
T4. Prove that if $f(x) \leq g(x)$ for every x in $[a, b]$ and both functions are integrable, then

$$\int_a^b f(x)\, dx \leq \int_a^b g(x)\, dx.$$

T5. Prove that if $f(x)$ is integrable on $\{a \leq x \leq b\}$ and if $\sup f(x) = G$ on $[a, b]$ and $\inf f(x) = L$ on $[a, b]$, then

$$L(b - a) \leq \int_a^b f(x)\, dx \leq G(b - a).$$

T6. Prove that if $f(x)$ is integrable on $[a, b]$, there is a number y such that

$$\int_a^b f(x)\, dx = y(b - a).$$

Show that $\inf f(x) \leq y \leq \sup f(x)$.
T7. Prove that if $f(x)$ is continuous on $[a, b]$, there is a point c between a and b such that

$$\int_a^b f(x)\, dx = f(c)(b - a).$$

T8. Prove that if the graph of $f(x)$ on $\{a \leq x \leq b\}$ consists of a finite number of increasing or decreasing arcs, then $|f(x)|$ is integrable and $\int_a^b f(x)\, dx \leq \int_a^b |f(x)|\, dx$.
T9. Prove that if $f(x)$ is integrable and $|f(x)| < M$ on $[a, b]$, then

$$\left| \int_a^b f(x)\, dx \right| \leq M|b - a|.$$

T10. Prove the following theorem. The bounded function f is integrable on the interval $[a, b]$ if and only if for every ϵ there exists a partition of $[a, b]$ such that

$$\int_a^b S(x)\, dx - \int_a^b s(x)\, dx < \epsilon,$$

where S is the overstep function and s is the understep function for f on the partition (see §10.2).

20 Calculation of Antiderivatives

20.1 General Chain Rule. We recall that if we take a differentiable function u and substitute it for x in x^n to get u^n, and then differentiate u^n, we get $D_x u^n = nu^{n-1} D_x u$ (§11.3, Rule V). This is the derivative $D_u u^n$ multiplied by the derivative $D_x u$ to form the chain $(u^n)'\, Du$.

More generally, let $u(x)$ be a function on a domain X and let f be a function such that the values $u(x)$ are in the domain of f. We can then substitute $u(x)$ into f. The *composite function $f(u)$* on the domain X is defined at any x to have the value $f[u(x)]$.

Now let us differentiate $f(u)$ with respect to x.

THEOREM (Chain rule). If $f(u)$ is a composite function on the domain X of numbers x, and both f and u are differentiable, then $f(u)$ is differentiable with respect to x, and its derivative is given by VI.

$$D_x f(u) = f'(u)D_x u.$$

Another notation for the chain rule is $D_x f(u) = D_u f \cdot D_x u$. This says: "First differentiate $f(u)$ with respect to u and multiply by the factor $D_x u$."

Proof of chain rule. We shall compute the numerical derivative of $f(u)$ at a point x in the domain X. Since $u(x)$ is differentiable, we know (§11.9, Lemma) that $\lim_{w \to x} [u(w) - u(x)] = 0$; that is, u is continuous at x. Hence, by the definition of the derivative,

$$\lim_{w \to x} \frac{f[u(w)] - f[u(x)]}{u(w) - u(x)} = f'[u(x)].$$

There is a slight difficulty here; for the dividend difference is not defined if it happens that $u(w) = u(x)$. We shall return to this case later, and we proceed now with the ordinary case.

We are required to calculate

$$D_x f[u(x)] = \lim_{w \to x} \frac{f[u(w)] - f[u(x)]}{w - x}.$$

To do this we multiply and divide by $u(w) - u(x)$ to find that

$$D_x f[u(x)] = \lim_{w \to x} \frac{f[u(w)] - f[u(x)]}{u(w) - u(x)} \cdot \frac{u(w) - u(x)}{w - x}$$

$$= f'[u(x)]D_x u(x).$$

This completes the proof except for the difficulty that occurs when $u(w) = u(x)$. This is easily removed by simply defining the divided difference to be equal to its limit at $u(x)$ in this case; that is,

$$\frac{f[u(w)] - f[u(x)]}{u(w) - u(x)} = f'[u(x)]$$

when $u(w) = u(x)$. Then the proof is still valid when $u(w) = u(x)$.

The product $f'(u)D_x u$ is called a *chain*, and the factor $D_x u$ is called the *chain factor*. At this stage, we are interested in the general chain rule only as a theoretical tool, so we work no problems with it.

20.2   Use of the Integral Sign as an Antiderivative Operator. The fundamental theorem (§19.3) reveals a close connection between antiderivatives, $F(x) = D^{-1}f(x)$, and integrals, $\int_a^b f(x)\, dx$. Hence, it is natural to use a symbol for the antiderivatives of $f(x)$ that resembles an integral and becomes an integral on evaluation at a and b, as in $F(x)|_a^b$. Thus the traditional symbol for an antiderivative operator is $\int$, without limits. It becomes a symbol for an integral when evaluation at a and b is indicated thus, $\int_a^b$.

Consequently, in keeping with tradition, let us now adopt the symbol $\int$ in place of D^{-1}, so that

$$D \int u = u \qquad \text{and} \qquad \int Du = u + c.$$

Moreover, a look at such a formula as V (§11.3), which says that $Du^{n+1} = (n+1)u^n \, Du$, tells us that we want to say that the antiderivative of $u^n \, Du$ is $u^{n+1}/(n+1) + C$. This is a more general and powerful statement than the special case we have used so far. The special case (§17.3) says that the antiderivative of x^n is $x^{n+1}/(n+1) + C$, which is included in the one above, since if $u = x$, then $Du = Dx = 1$. Now if we apply the antiderivative operator $\int$ to $u^n \, Du$, the formula that results is

$$\int u^n \, Du = \frac{u^{n+1}}{n+1} + C, \qquad \text{provided that } n \neq -1.$$

This looks even more like an integral, though it is a function, while the related integral, $\int_a^b u^n \, du$, is a number obtained from it by the evaluation $\int u^n \, Du \big|_a^b$.

In general, in our antiderivative formulas we shall always apply the operator $\int$ to a chain $f(u) \, Du$, thus $\int f(u) \, Du$. We shall always keep the chain factor Du present even when it is not needed. This will cause no difficulty when the chain factor is not needed; when $u = x$, $Du = Dx = 1$, and the formula $\int f(u) \, Du$ reduces to $\int f(x)$. But, in our going the other way from $\int f(x)$, to remember to restore the chain factor Du when x is replaced by a function u is likely to induce errors, which can be avoided by keeping Du always present in our formulas.

When we write $\int f(u) \, Du$, the chain $f(u) \, Du$ is also called *the integrand*. For it is the function whose antiderivative we seek to find in order to evaluate an integral $\int_a^b f(u) \, du$ from the limit $x = a$ to the limit $x = b$.

20.3   Notation.   The more usual notation for the antiderivative is $\int f(u) \, du$, just like $\int_a^b f(u) \, du$, but without limits of integration. We use Du in the antiderivative formula instead of du to emphasize that it is not an integral but is the antiderivative of a chain $f(u) \, Du$. The evaluation of the integral $\int_a^b f(x) \, dx$ by the fundamental theorem requires that we first find the antiderivative $F(x) = \int f(x) \, Dx$. Hence, finding the antiderivative is part of this integration procedure. It has become customary to speak of finding the antiderivative $\int f(x) \, Dx$ as "integrating $f(x)$," though it is actually not integration at all.

20.4   Rules for Antiderivatives.   Let us now restate the formulas for derivatives (§11.3) and alongside, the corresponding antiderivative formulas, with the integral sign placed before the Roman numeral to indicate that it is the antiderivative version.

DERIVATIVES	ANTIDERIVATIVES
I. $D(u + v) = Du + Dv.$	$\int$ I. $\int (Du + Dv) = \int Du + \int Dv.$
II. $D\,cu = c\,Du.$	$\int$ II. $\int c\,Du = c \int Du.$

III. $D\ uv = u\ Dv + v\ Du.$

IV. $DC = 0.$

V. $Du^n = nu^{n-1}\ Du.$

VI. $Df(u) = f'(u)\ Du.$

$\int$ III. $\int u\ Dv = uv - \int v\ Du.$

$\int$ IV. $\int 0\ Du = C$ (constant).

$\int$ V. $\int u^n\ Du = \dfrac{u^{n+1}}{n+1} + C,$
where n is an integer and $n \neq -1$ in $\int V.$

$\int$ VI. $\int f'(u)\ Du = f(u) + C.$

Formula $\int$ III is called *integration by parts*. Formula VI is the chain rule and $\int$ VI is called *integration by substitution*.

Example A.   Compute the antiderivative $\int (2x - 3)^4\ Dx.$

Solution.   We would be wrong if we wrote down $(2x - 3)^5/5$ as the result; for if we check this by differentiation we get, using formula V with $u = 2x - 3$ and $Du = 2\ Dx,$

$$D\left[\frac{(2x - 3)^5}{5} + C\right] = \left(\frac{1}{5}\right) 5(2x - 3)^4\ 2\ Dx + 0 = 2(2x - 3)^4\ Dx,$$

which is wrong by a factor of **2**, where **2** is the chain factor Du in the differentiation.

We can escape this difficulty by making appropriate substitutions to transform $\int (2x - 3)^4\ Dx$ into $\int u^4\ Du$, to which $\int V$ applies directly. We let $u = 2x - 3$ so that $Du = 2\ Dx$, and solve this last equation for Dx in terms of Du. We get $Dx = \frac{1}{2}\ Du.$ Then we substitute these expressions into the integrand, which gives us

$$\int (2x - 3)^4\ Dx = \int u^4 \left(\frac{1}{2}\ Du\right) = \frac{1}{2}\int u^4\ Du = \frac{1}{2}\frac{u^5}{5} + C,$$

where the last result was obtained by applying $\int$ V. It was essential in the preceding step to be able to remove the factor $\frac{1}{2}$ outside of the integrand by $\int$ II.

Now, resubstituting $2x - 3$ for u to go back to the variable x, we find

$$\int (2x - 3)^4\ Dx = \frac{1}{2}\frac{u^5}{5} + C = \frac{1}{10} (2x - 3)^5 + C.$$

This is the correct result, as may be verified by differentiation.

Let us work another example to illustrate this important point.

Example B.   Calculate $\int (1 - 5x)^7\ Dx.$

Solution.   Let $u = 1 - 5x$; then $Du = -5\ Dx$ and therefore $Dx = -\frac{1}{5}\ Du.$ With these expressions now,

$$\int (1 - 5x)^7\ Dx = \int u^7 \left(-\frac{1}{5}\ Du\right) = -\frac{1}{5}\int u^7\ Du = -\frac{1}{5}\frac{u^8}{8} + C = -\frac{1}{40} (1 - 5x)^8 + C.$$

Finally, let us take an example where this procedure fails to give us an antiderivative.

Example C.   Calculate $\int (1 + x + x^2)^{15}\ Dx.$

Attempted Solution. Let $u = 1 + x + x^2$; then $Du = (1 + 2x) Dx$ and $Dx = Du/(1 + 2x)$. Now we have

$$\int (1 + x + x^2)^{15}\, Dx = \int u^{15}\, \frac{Du}{1 + 2x}.$$

Here $\int$ II fails to permit us to remove the function $1/(1 + 2x)$ outside the integral sign; for it is not a constant factor. This having failed, our procedure stops; for it does not reduce the indefinite integral to $\int u^{15}\, Du$, where $\int$ V would apply. We could integrate this by the extremely tedious procedure of expanding $(1 + x + x^2)^{15}$ by Taylor's theorem into a polynomial in x of degree 30, and then finding the antiderivative of this. We shall not carry out this procedure.

20.5   Integration by Parts

Example D.   Calculate $\int x(2x - 3)^4\, Dx$.

Solution.   The integrand $x(2x - 3)^4\, Dx$ is a product of two functions, which suggests the use of the formula $\int$ III, $\int u\, Dv = uv - \int v\, Du$. We first try to make $x(2x - 3)^4\, Dx = u\, Dv$. We do this by choosing $u = x$ and $Dv = (2x - 3)^4\, Dx$. Now we calculate the derivative Du, and we find v by calculating an antiderivative of Dv, which we have already done in Example A. We have $Du = Dx$ and $v = \frac{1}{10}(2x - 3)^5$. Then, knowing u, v, Du, and Dv, we apply $\int$ III. This gives us

$$\int x(2x - 3)^4\, Dx = x\left(\frac{1}{10}(2x - 3)^5\right) - \int \frac{1}{10}(2x - 3)^5\, Dx$$

$$= \frac{x}{10}(2x - 3)^5 - \frac{1}{10}\int (2x - 3)^5\, Dx.$$

The last indefinite integral can be found as in Example A and the result is

$$\int x(2x - 3)^4\, Dx = \frac{x}{10}(2x - 3)^5 - \frac{1}{10}\left[\frac{1}{12}(2x - 3)^6\right] + C$$

$$= \frac{x}{10}(2x - 3)^5 - \frac{1}{120}(2x - 3)^6 + C.$$

This completes Example D. We can check the result by verifying that its derivative is the original integrand.

Example E.   Calculate $\int x^5(4 - 3x)^2\, Dx$.

Solution.   We apply integration by parts, letting $u = (4 - 3x)^2$ and $Dv = x^5\, Dx$. Then $Du = 2(4 - 3x)^1(-3)\, Dx$ and $v = x^6/6$ so that the given form becomes

(1)
$$\int x^5(4 - 3x)^2\, Dx = (4 - 3x)^2\, \frac{x^6}{6} - \int \frac{x^6}{6}(2)(4 - 3x)(-3)\, Dx$$

$$= (4 - 3x)^2\, \frac{x^6}{6} + \int x^6(4 - 3x)\, Dx.$$

This leaves us with a new antiderivative form $\int x^6(4 - 3x)\, Dx$ to calculate. We observe that the exponent of $(4 - 3x)$ has been reduced by 1 and that one more application of

$\int$ III will eliminate this factor so we now let $u = 4 - 3x$ and $Dv = x^6 \, Dx$, which gives us $Du = (-3) \, Dx$ and $v = x^7/7$. Now applying $\int$ III, we find that

$$\int x^6(4 - 3x) \, Dx = (4 - 3x)\frac{x^7}{7} - \int \frac{x^7}{7}(-3) \, Dx = (4 - 3x)\frac{x^7}{7} + \frac{3}{7}\int x^7 \, Dx,$$

or

(2) $$\int x^6(4 - 3x) \, Dx = (4 - 3x)\frac{x^7}{7} + \frac{3}{7}\frac{x^8}{8} + C.$$

The last antiderivative form $\int x^7 \, Dx$ is calculated by use of $\int$ V. Now we replace the antiderivative $\int x^6(4 - 3x) \, Dx$ in (1) by its value, which we calculated in (2), and we get

$$\int x^5(4 - 3x)^2 \, Dx = (4 - 3x)^2 \frac{x^6}{6} + \left[(4 - 3x)\frac{x^7}{7} + \frac{3}{56}x^8 + C\right]$$

$$= (4 - 3x)^2 \frac{x^6}{6} + (4 - 3x)\frac{x^7}{7} + \frac{3}{56}x^8 + C.$$

This is our final result.

COMMENT ON EXAMPLE E. This is a complicated two-stage procedure. Its success depends on eliminating the factor $(4 - 3x)^2$ from the integrand by two successive differentiations. When we set $u = (4 - 3x)^2$ and integrated by parts, we still had the factor $(4 - 3x)$ in the integrand. Then on repeating with $u = 4 - 3x$ the second time, our last term $\int v \, Du$ no longer contained the factor $4 - 3x$, and we could find its antiderivative explicitly as we did in the last step.

We could have made the calculation by a different choice of u and Dv. At the start in $\int x^5(4 - 3x)^2 \, Dx$ we could have chosen $u = x^5$ and $Dv = (4 - 3x)^2 \, Dx$. Let us carry out one stage of this calculation. We have $Du = 5x^4$ and $v = -\frac{1}{9}(4 - 3x)^3$ so that

$$\int x^5(4 - 3x)^2 \, Dx = x^5\left(-\frac{1}{9}(4 - 3x)^3\right) - \int -\frac{1}{9}(4 - 3x)^3 5x^4 \, Dx$$

$$= -\frac{1}{9}x^5(4 - 3x)^3 + \frac{5}{9}\int x^4(4 - 3x)^3 \, Dx.$$

We have reduced the exponent of x by 1, but we still have the factor x^4. We can eventually eliminate this factor and compute directly the last integral of the form $\int v \, Du$, but it will take four more stages of rule $\int$ III with $u = x^4$, then $u = x^3, \cdots$ to eliminate the factor that is a power of x. The two-stage calculation we used in the first solution, starting with $u = (4 - 3x)^2$, is more economical.

The use of integration by parts is an art that requires practice. We cannot always make the best choice of u and Dv in our first attempt. We cannot be sure at the outset that it will succeed at all. It is a method that perhaps can best be described as a trial-and-error procedure.

20.6 Exercises

Calculate the derivatives in Exercises 1–4.

1. $D(3x - 4)^5$.

2. $D[(1 - 2x)^2 + 5(1 - 2x)^3]$.

3. $D\dfrac{(6x - 1)^7}{8}$.

4. $D(4x - 9)^0$.

In Exercises 5–14, calculate the indicated antiderivatives.

5. $\int x^2 \, Dx.$

6. $\int (x - 1) \, Dx.$

7. $\int (x^3 - 3x + 4) \, Dx.$

8. $\int (x - 2)^3 \, Dx.$

9. $\int (6x - 5)^{18} \, Dx.$

10. $\int 4(3 - 7x)^{10.} \, Dx.$

11. $\int \frac{(7 - x)^2}{10} \, Dx.$

12. $\int [(x - 1)^5 - 7x + 4] \, Dx.$

13. $\int \frac{4 - x^2}{2 + x} \, Dx.$

14. $\int \left[7 \frac{(3 - 2x)^2}{3} - 4(6x - 2) + 1 \right] Dx.$

In Exercises 15–22, use the fundamental theorem of calculus to evaluate the integrals.

15. $\int_0^{3/2} (2x - 3)^4 \, dx,$ (Example A).

16. $\int_1^2 (x - 2)^3 \, dx.$

17. $\int_0^{1/5} (1 - 5x)^7 \, dx,$ (Example B).

18. $\int_{1/5}^{1/5} (1 - 5x)^3 \, dx.$

19. $\int_0^1 x(2x - 3)^4 \, dx,$ (Example D).

20. $\int_0^1 (6x - 5)^{18} \, dx.$

21. $\int_0^{4/3} x^5(4 - 3x)^2 \, dx,$ (Example E).

22. $\int_4^8 \frac{(7 - x)^2}{10} \, dx.$

In Exercises 23–28, calculate, using integration by parts.

23. $\int x(2x - 1)^7 \, Dx.$

24. $\int x(5 - 2)x^3 \, Dx.$

25. $\int x^8(2x - 1) \, Dx.$

26. $\int (3x - 4)^5 x^2 \, Dx.$

27. $\int (4x - 2)^3(1 - x)^2 \, Dx.$

28. $\int (1 - x)^3(4x - 5)^4 \, Dx.$

29. Assuming that f is a function having a first derivative f' and a second derivative f'', show that $\int xf'' \, Dx = xf' - f + C.$

30. Find the area of the region bounded by $(3x - 2)^4$ and the x-axis from the limit $x = 0$ to $x = 1$.

31. Find the area of the region bounded by the graphs of $2(x - 4)^3$ and $8x - 32$.

32. Find the area of the region bounded by the graph of $x(3x + 1)^3$, the x-axis, and the limits $x = 0$ and $x = 1$.

33. Find the distance from the starting point of a point that moves with velocity $v = (3t - 1)^4$ ft/sec between the times $t = 1$ sec and $t = 2$ sec.

21   Algebraic Properties of Integrals

21.1   Indefinite Integrals.   The integral $\int_a^b f(u) \, du$ is called a *definite integral* if we think of the limits a and b as fixed numbers. It is a number. On the other hand, if we integrate $f(x)$ between the fixed limit a and a variable upper limit x in $[a, b]$, we form a function.

DEFINITION.   The function $F(x)$ on $[a, b]$ whose value at any x is given by

$$F(x) = \int_a^x f(u) \, du$$

is an *indefinite integral* of $f(x)$ in $[a, b]$.

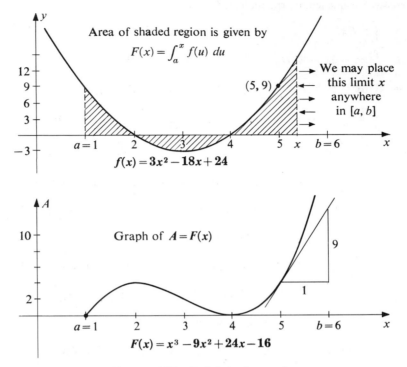

FIGURE 4.13   Indefinite integral.

We may picture an indefinite integral (Figure 4.13) either as a variable area below the graph of $f(x)$ swept out behind a moving ordinate at x, or as a related function $F(x)$ with its own graph in xA-coordinates.

Example.   For the function $f(x) = 3x^2 - 18x + 24$ on $[1, 6]$, find the explicit polynomial formula for the indefinite integral $F(x)$ given by

$$F(x) = \int_1^x (3u^2 - 18u + 24) \, du.$$

Solution.   We find the antiderivative

$$\int (3u^2 - 18u + 24) \, Du = 3\left(\frac{u^3}{3}\right) - 18\left(\frac{u^2}{2}\right) + 24u = u^3 - 9u^2 + 24u.$$

Then by the fundamental theorem,

$$F(x) = \int_1^x (3u^2 - 18u + 24) \, du = (u^3 - 9u^2 + 24u)\Big|_1^x$$

$$= (x^3 - 9x^2 + 24x) - ((1)^3 - 9(1)^2 + 24(1))$$

$$= x^3 - 9x^2 + 24x - 16.$$

Thus the required polynomial formula is $F(x) = x^3 - 9x^2 + 24x - 16$. We plot the graphs of $f(x)$ and the indefinite integral, $F(x)$ (Figure 4.13).

21.2   Dummy variables in a sum.   In the calculation of $\int_1^x (3u^2 - 18u + 24)\, du$ above, we observe that the variable u did not appear in the final result, which depended only on the limits 1 and x and the integrand function $3x^2 - 18x + 24$. We can retrace the definition of the integral $\int_a^b f(u)\, du$ as a sum (§19.1) to see that this is true in general.

Under these circumstances, the variable of summation u is called a *dummy variable*. When it suits our purposes we will express the fact that the dummy variable does not appear in the numerical value of the integral by writing

$$\int_a^b f(u)\, du = \int_a^b f, \quad \text{and} \quad F(x) = \int_a^x f.$$

21.3   Linearity of the Integral.   We prove (Exercises) from the definition of the integral as a sum that if $f(x)$ and $g(x)$ are two integrable functions on $[a, b]$ and c is a constant, then

$$\int_a^b cf(x)\, dx = c \int_a^b f(x)\, dx,$$

and

$$\int_a^b [f(x) + g(x)]\, dx = \int_a^b f(x)\, dx + \int_a^b g(x)\, dx.$$

In terms of the indefinite integral, the same result may be stated

$$\int_a^x cf = c \int_a^x f, \quad \text{and} \quad \int_a^x (f + g) = \int_a^x f + \int_a^x g.$$

These formulas express the fact that $\int_a^x$ is a linear operator on functions as the derivative operator D is.

21.4   When $f(x)$ is Negative.   We have thus far set up all of our integrals with $a < b$ and we have used positive differences $dx = x_{i+1} - x_i$, where $x_i < x_{i+1}$, in evaluating the area elements $f(x)\, dx$ that entered our sums. Under these circumstances, the element of area $f(x)\, dx$ is negative when $f(x)$ is negative.

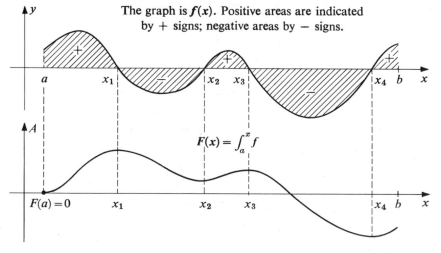

FIGURE 4.14   Positive and negative contributions to an integral. The upper graph is $y = f(x)$. Positive areas are indicated by plus signs and negative areas by minus signs. Compare with the graph of the indefinite integral $A = F(x)$ below.

Hence, in any interval (Figure 4.14) where $f(x)$ is negative, the contribution to the total area bounded by this portion of the graph is negative. It is associated with a region below the x-axis.

The integral $\int_a^b f(x)\,dx$ (Figure 4.14, upper) is the algebraic sum of positive and negative areas

$$\int_a^b f(x)\,dx = \int_a^{x_1} f + \int_{x_1}^{x_2} f + \int_{x_2}^{x_3} f + \int_{x_3}^{x_4} f + \int_{x_4}^b f,$$
$$= (+) + (-) + (+) + (-) + (+).$$

In the case pictured, the net algebraic sum, $\int_a^b f(x)\,dx$, is negative.

The corresponding behavior of the graph of the indefinite integral (Figure 4.14, lower) is that $F(x)$ increases whenever $f(x)$ is positive and attains a maximum as at x_1, at a point where $f(x)$ becomes negative. Then $F(x)$ decreases so long as $f(x)$ is negative. When, as at x_2, $f(x)$ becomes positive again, the value of $F(x)$ begins to increase, and so on.

21.5   Additivity of Integrals.   We now permit "integrating in the negative direction" by a simple definition.

DEFINITION.   If $a < x$,

$$\int_x^a f = - \int_a^x f.$$

With this definition and the definition of the integral itself we can prove (Exercises) the following theorem (Figure 4.15).

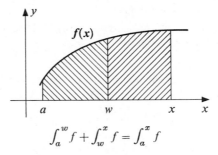

$$\int_a^w f + \int_w^x f = \int_a^x f$$

FIGURE 4.15   Additivity of integrals.

THEOREM (Additivity of integrals).   If $f(x)$ is integrable on $[a, b]$ and if x and w are any numbers in the interval, then

$$\int_a^x f = \int_a^w f + \int_w^x f.$$

A consequence of this theorem is that $\int_a^a f = 0$.

21.6   Total Variation.   There are problems in which it is not desirable to have the integral represent the algebraic sum of positive and negative areas. Instead, we may wish to find the sum of the absolute values of all areas represented by the integral

$\int_a^b f(u)\, du$ over subintervals, without cancelling positive areas against negative ones. This type of sum is called the *total variation* of the indefinite integral $F(x)$, and defined mathematically by

$$\int_a^b |f(x)|\, dx,$$

where the absolute value $|f(x)|$ insures that it is never negative.

For example, if we want the total variation of the area $F(x)$ in the Example of Section 21.1 over [1, 6] we examine the graph of $f(x)$ (Figure 4.13) and observe that $|f(x)| = f(x)$ in the intervals [1, 2] and [4, 6] but $|f(x)| = -f(x)$ in [2, 4]. So

$$\int_1^6 |f(x)|\, dx = \int_1^2 f + \int_2^4 (-f) + \int_4^6 f$$
$$= (x^3 - 9x^2 + 24x)\Big|_1^2 - (x^3 - 9x^2 + 24x)\Big|_2^4 + (x^3 - 9x^2 + 24x)\Big|_4^6$$
$$= 4 - (-4) + 20 = 28.$$

We contrast this with the net area given by the integral of $f(x)$ over [1, 6],

$$A = \int_1^6 f = (x^3 - 9x^2 + 24x)\Big|_1^6 = 20.$$

By additivity this result, 20, is formed of the algebraic sum

$$A = \int_1^2 f + \int_2^4 f + \int_4^6 f = 4 + (-4) + 20 = 20.$$

21.7   Derivative of the Indefinite Integral

THEOREM.   If $f(x)$ is integrable in $[a, b]$ and if $F(x) = \int_a^x f$, then at any point x where $f(x)$ is continuous, $F'(x) = f(x)$.

Proof.   We must show that

$$\lim_{w \to x} \frac{F(w) - F(x)}{w - x} = f(x).$$

By the additivity (§21.5), $F(w) - F(x) = \int_x^w f$. Now we take an understep function and overstep function each with a single step in the interval $[x, w]$ (Figure 4.16). It follows from the definition of the integral that

$$(w - x) \inf f(u) \leq \int_w^x f \leq (w - x) \sup f(u)$$

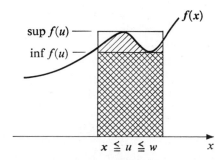

FIGURE 4.16   Single step overstep and understep functions.

for all u in $[x, w]$. Assuming for simplicity that $x < w$, we divide by $w - x$. This gives us

$$\inf f(u) \leqq \frac{F(w) - F(x)}{w - x} \leqq \sup f(u).$$

Now, if we let w approach x, at any point where $f(x)$ is continuous (Exercises),

$$\lim_{w \to x} \inf f(u) = \lim_{w \to x} \sup f(u) = f(x).$$

We then have from the inequality above

$$f(x) \leqq \lim_{w \to x} \frac{F(w) - F(x)}{w - x} \leqq f(x).$$

Therefore $F'(x) = f(x)$, which completes the proof of the theorem.

Example.   For the indefinite integral $F(x)$ defined by

$$F(x) = \int_1^x (3u^2 - 18u + 24)\, du,$$

calculate $F'(x)$ and, in particular, $f(5)$ and $F'(5)$.

Solution.   The theorem says that $F'(x) = 3x^2 - 18x + 24$. We also computed (§21.1, Example) explicitly $F(x) = x^3 - 9x^2 + 24x - 16$, whose derivative is $F'(x) = 3x^2 - 18x + 24$, as predicted by the theorem. In particular, $f(5) = 9$ and $F'(5) = 9$, so that if we take the value $f(5)$ and plot it as a slope at the point $(5, F(5))$, it is the slope of the tangent line to $F(x)$ at that point (Figure 4.13, lower).

This theorem, taken in conjunction with the theorem on integrability of continuous functions, assures us of the existence of an antiderivative of a continuous function on an interval $[a, b]$, whether we can find an algebraic expression for it or not.

21.8   Integration by Inspection of the Graph.   We consider the integral

$$\int_{-a}^a f(x)\, dx,$$

where $f(x) = x^3 - 7x$. The integrand function has the property that for every number x, $f(-x) = -f(x)$. Hence the graph is antisymmetric about the origin (Figure 4.17).

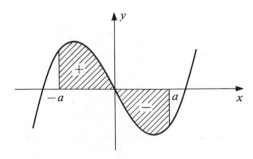

FIGURE 4.17   The area $\int_{-a}^a (x^3 - 7x)\, dx$.

Hence if $f(x)$ bounds a region above the x-axis on the positive x side of the y-axis, it bounds a congruent region below the x-axis on the negative x side. This is expressed by the equation

$$\int_{-a}^{0} (x^3 - 7x)\, dx = -\int_{0}^{a} (x^3 - 7x)\, dx.$$

Hence, when we calculate

$$\int_{-a}^{a} = \int_{-a}^{0} + \int_{0}^{a},$$

the two integrals on the right cancel, and therefore without calculation we see that for any number a

$$\int_{-a}^{a} (x^3 - 7x)\, dx = 0.$$

The detailed proof can be completed by referring to the definition of the integral as a sum (§19.1).

21.9 Exercises

In Exercises 1–10, find the explicit polynomial formula for the indefinite integrals. Check by differentiating the resulting polynomial.

1. $\int_{1}^{x} (t + 1)\, dt.$

2. $\int_{-1}^{x} (u^2 - 1)\, du.$

3. $\int_{-1}^{t} (z^2 - z + 1)\, dz.$

4. $\int_{x}^{1} (2v - 3)\, dv.$

5. $\int_{0}^{x} (3t - 2)^5\, dt.$

6. $\int_{-2}^{x} t(2t + 5)^6\, dt.$

7. $\int_{a}^{x} (u - a)\, du.$

8. $\int_{a}^{z} (a - t)^5\, dt.$

9. $\int_{b/a}^{u} z^2 (az - b)\, dz.$

10. $\int_{-x}^{x} 3t\,(5 - 2t)^8\, dt.$

In Exercises 11–16, sketch the graph of the integrand and show the region whose area is represented by the integral. Indicate which areas are positive and which negative. Calculate the integrals.

11. $\int_{0}^{4} (2x - 5)\, dx.$

12. $\int_{-3}^{3} (u^2 - u - 2)\, du.$

13. $\int_{-3}^{3} (z^3 - 5z)\, dz.$

14. $\int_{-1}^{4} t(2t - 5)^3\, dt.$

15. $\int_{-4}^{4} (t + 1)(t - 1)(t - 3)\, dt.$

16. $\int_{-a}^{a} (x^4 - a^4)\, dx.$

In Exercises 17–20, calculate the total variation of the area $F(x)$ over the interval given by the limits of the integrals.

17. Exercise 11. 18. Exercise 12.
19. Exercise 15. 20. Exercise 16.

21. A point moves on the y-axis with its velocity at every time t, $0 \le t \le 10$ given by $v(t) = (t - 1)(t - 3)$. Find both the distance from the starting point at the end of the 10 seconds and the total distance traveled.

22. Solve for x the equation $\int_{0}^{x} u^2\, du = 11.$

23. Solve for x the equation $\int_{0}^{x} (2u - 1)^2\, du = 9.$

24. Solve for x the equation $\int_{x}^{x+2} u\, du = 0.$

In Exercises 25–30, simplify the expressions involving integrals, using the algebraic properties of integrals, and then evaluate.

25. $\int_0^4 [x^2 - (2x - 1)^7]\, dx + \int_0^4 (2x - 1)^7]\, dx.$

26. $\int_1^5 x[1 + x(2x - 1)^7]\, dx + \int_5^1 x^2[1 + (2x - 1)^7]\, dx.$

27. $\int_0^1 xf(x)\, dx + \int_0^2 x[1 - f(x)]\, dx + \int_1^2 x[1 + f(x)]\, dx.$

28. $2\int_0^{1/2} x(2x - 1)^7\, dx - \int_0^{1/2} (2x - 1)^7\, dx.$

29. $\int_a^x f = x^2 - a^2 - 2\int_a^x f.$

30. $\int_0^1 (7 - x)^5\, dx + \int_1^3 (7 - x)^5\, dx - \int_5^3 (7 - x)^5\, dx.$

31. $\int_0^2 (x^2 - x - 2)(x - 2)^n\, dx.$

32. Investigate for maximum and minimum points the function $f(x)$ on $\{x \geq 1\}$ defined by

$$f(x) = \int_1^x (u - 3)^{15}\, du.$$

21.10   Problems

T1.   Prove from the definition of the integral that if $f(x)$ is an integrable function and c is a constant, then

$$\int_a^b cf(x)\, dx = c\int_a^b f(x)\, dx.$$

T2.   Prove from the definition of the integral that if $f(x)$ and $g(x)$ are integrable on $[a, b]$,

$$\int_a^b [f(x) + g(x)]\, dx = \int_a^b f(x)\, dx + \int_a^b g(x)\, dx.$$

T3.   Prove from the definition of the integral that if $f(x)$ is integrable on $[a, b]$ and x and w are two numbers in $[a, b]$, then

$$\int_a^x f = \int_a^w f + \int_w^x f \qquad \text{(Figure 4.15)}.$$

T4.   Show that if $f(x)$ is a differentiable, increasing function on $[a, b]$, then the total variation of $f(x)$ is $f(b) - f(a)$.

T5.   Prove that if for every value of x, $f(-x) = f(x)$, then

$$\int_{-a}^a f(x)\, dx = 2\int_0^a f(x)\, dx.$$

T6.   Prove that for every integrable function $f(x)$

$$\int_{-a}^a [f(x) - f(-x)]\, dx = 0.$$

T7.   Let $f(x)$ be a continuous function. Prove that

$$D\int_x^a f = -f(x).$$

T8.   Prove that if f is continuous and u is differentiable,

$$D_x \int_a^u f = f(u)u'.$$

Since a continuous function $f(x)$ is integrable (§19.2), we can use the antiderivative $G(x) = \int_a^x f(u)\, du$ (§21.7) to prove another version of the fundamental theorem of integral calculus.

Exercises T9–T12. Complete a proof of the following theorem.

FUNDAMENTAL THEOREM OF INTEGRAL CALCULUS. If $f(x)$ is continuous on the interval $[a, b]$, then it has an antiderivative. If $F(x)$ is any antiderivative, then

$$\int_a^b f(x)\, dx = F(b) - F(a).$$

T9.   Denote by $G(x)$ the special antiderivative (§21.7) given at x by $G(x) = \int_a^x f(u)\, du$. Show that $G(a) = 0$.

T10. Let $F(x)$ be any antiderivative of $f(x)$. By differentiating $G(x) - F(x)$, show that $G(x) - F(x)$ is a constant C.

T11. Prove that, in T10 $C = -F(a)$, hence that $G(x) = F(x) - F(a)$.

T12. Use T11 to prove that

$$\int_a^b f(x)\, dx = F(b) - F(a).$$

This completes the proof.

22   Integration as a Process of Summation

22.1   Resumé of Integration as a Process of Summation.

We have seen (§4) how various physical quantities may be represented by discrete sums of a finite number of terms formed from step functions. We represented these by the sum, or integral, notation $\int_a^b s(x)\, dx$, meaning "the sum of terms of the type $s(x)\, dx$ extending over the interval from the point where $x = a$ to the point where $x = b$." Then we saw (§9 and §19.1) how we can take a continuously varying function $f(x)$ on the interval $[a, b]$, partition the interval into subintervals of length dx, approximate $f(x)$ by a step function $s(x)$ on this partition, form the finite sum $\int_a^b s(x)\, dx$, then refine the partition, and by a limit process obtain a number $\int_a^b f(x)\, dx$ called the definite integral of $f(x)$ over the interval $[a, b]$. Thus, for applying it, we still think of $\int_a^b f(x)\, dx$ as a sum of terms of the type $f(x)\, dx$.

To calculate $\int_a^b f(x)\, dx$, we can often find an antiderivative $F(x) = \int f(x)\, Dx$ of $f(x)$ and use it in the fundamental theorem (§19.3) to evaluate $\int_a^b f(x)\, dx$ without either limits or tedious sums. Alternatively, or when we cannot find an antiderivative, we can compute $\int_a^b f(x)\, dx$ numerically to any desired accuracy by use of approximating step functions. Once we have succeeded in representing a physical quantity by a definite integral $\int_a^b f(x)\, dx$ with explicit integrand function $f(x)$, the problem is set up. The task of finding the numerical value of the quantity is essentially solved, whether we choose to proceed via the fundamental theorem or via numerical integration with approximating step functions.

We now consider a variety of techniques for setting up integrals to represent quantities in geometry and mechanics.

22.2   Areas of Regions Bounded by Two Curves

Example A.   Find the area of the region bounded by the curves $y = x^2 - 4$ and $y = 2x - 1$.

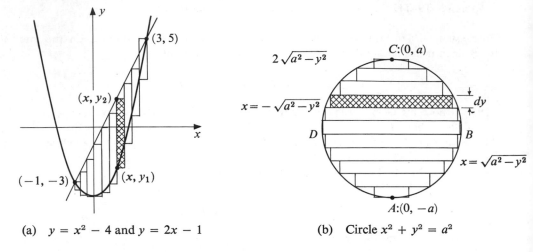

FIGURE 4.18   Regions bounded by curves.

Solution.   We plot the graphs and solve simultaneously to find that the points of intersection have coordinates $(-1, -3)$ and $(3, 5)$. The region whose area we calculate is bounded by the straight line $y = 2x - 1$ above and the curve $y = x^2 - 4$ below (Figure 4.18(a)). We partition the interval $[-1, 3]$ into smaller intervals of length dx and form rectangles with these subintervals as base, extending from the lower curve to the upper one. The altitude of such a rectangle is the difference $(2x - 1) - (x^2 - 4)$ of the y-coordinates of the upper and lower curve. The area of the rectangle is $(-x^2 + 2x + 3)\,dx$. The sum of areas of this type is, in the limit,

$$A = \int_{-1}^{3} (-x^2 + 2x + 3)\,dx.$$

The problem is set up. For this function we have the antiderivative $-(x^3/3) + x^2 + 3x$. Hence

$$A = \int_{-1}^{3} (-x^2 + 2x + 3)\,dx = -\frac{x^3}{3} + x^2 + 3x \,\Big|_{-1}^{3}$$

$$= (-9 + 9 + 9) - \left(-\frac{1}{3} + 1 - 3\right) = \frac{61}{3} \text{ area units.}$$

Example B.   Set up an integral to represent the area of the circle $x^2 + y^2 = a^2$.

Solution.   We plot the circle (Figure 4.18(b)). Let us choose to slice it into rectangles of height dy and base stretching the left arc of the circle to the right. To find the length of this base we solve the equation for x in terms of y and find that we have two functions as solutions,

$$x = -\sqrt{a^2 - y^2} \qquad \text{and} \qquad x = +\sqrt{a^2 - y^2}.$$

The function with the minus sign gives the left arc of the circle ADC and the plus sign gives the right arc ABC.

The length of the base of a typical rectangle is then

$$+\sqrt{a^2 - y^2} - (-\sqrt{a^2 - y^2}) = 2\sqrt{a^2 - y^2}.$$

Its area is $2\sqrt{a^2 - y^2}\,dy$ and thus the area A of the whole circle becomes in the limit

$$A = \int_{-a}^{a} 2\sqrt{a^2 - y^2}\,dy.$$

The problem is set up as an integral. We have as yet found no antiderivative of $\sqrt{a^2 - y^2}$, and we cannot use the fundamental theorem until we do. However, using the fact from geometry that the area of the circle is πa^2, we know that

$$\int_{-a}^{a} \sqrt{a^2 - x^2}\,dx = \frac{\pi}{2}\,a^2.$$

22.3 Volumes by Parallel Slices

Example A. Find the volume of a pyramid with square base of side b, and height h.

Solution. In the figure (Figure 4.19), the height $|OV| = h$. We slice the pyramid with planes parallel to the base and distance dz apart. Choosing one of these planes at height z above the base, we form a square slab of side s and height dz (shaded in the figure). The total volume V is a sum of volumes $dV = s^2\,dz$ of such slabs extending from the base, where $z = 0$, to the apex, where $z = h$. Thus the exact volume is given by an integral

$$V = \int_{0}^{h} s^2\,dz,$$

where s is a function of z.

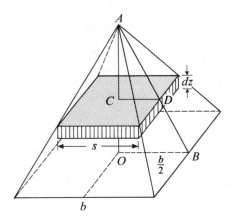

FIGURE 4.19   Volume of a pyramid.

To find s as a function of z, we use ratios from the similar triangles ACD and AOB. Since $|OC| = z$, we have $|CA| = h - z$. Then

$$\frac{|CD|}{|CA|} = \frac{|OB|}{|OA|}$$

gives

$$s = |CD| = \frac{s}{2} = \frac{b}{h}\,(h - z).$$

With this expression for s, the volume integral becomes

$$V = \int_0^h \frac{b^2}{h^2} (h - z)^2 \, dz = \frac{b^2}{h^2} \left(\frac{h^3}{3}\right) = \frac{1}{3} b^2 h.$$

22.4   Volumes of Revolution

Example.   Find the volume of the solid generated by revolving about the y-axis the region above the curve $y = x^2$ and below the line $y = 9$ (Figure 4.20(a)).

Solution.   We partition the altitude segment OT into small subintervals of length dy and approximate the boundary curve $y = x^2$ by vertical segments in each subinterval. This gives a series of thin rectangles of area $2x$, dy whose sum is approximately the area of the generating region for the solid (Figure 4.20(a)). Now if we rotate the region, we

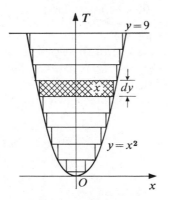
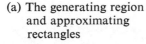
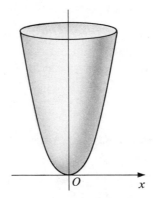
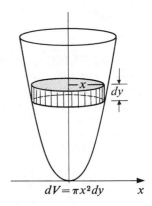

(a) The generating region   (b) The solid of revolution   (c) A typical slice
and approximating
rectangles

FIGURE 4.20   Volumes by summing slices.

generate the solid whose volume we are to calculate (Figure 4.20(b)). Each of the approximating rectangles rotates into a thin cylinder (Figure 4.20(c)) of radius x and height dy. We call this a "horizontal slice" of the solid. Its volume $dV = \pi x^2 \, dy$. Since x is measured from the y-axis to the boundary curve, the value of x at height y is given by $x = \sqrt{y}$. Hence the volume dV of the slice at height y is

$$dV = \pi (\sqrt{y})^2 \, dy = \pi y \, dy.$$

We sum the volumes dV from the bottom, where $y = 0$, to the top, where $y = 9$, and get the volume

$$V = \int_0^9 \pi y \, dy.$$

We can regard this as the exact volume if we think of this sum as the integral resulting from passing to the limit as the number of slices increases indefinitely and dy approaches zero.

We calculate the exact volume V by the fundamental theorem, as follows:

$$V = \pi \int_0^9 y \, dy = \pi \frac{y^2}{2} \Big|_0^9 = \frac{81\pi}{2} \text{ volume units.}$$

22.5   Force Due to Water Pressure

Example.   Find the total force F exerted by water pressure on an equilateral triangle of side 10 ft located in a vertical plane on the wet face of a dam when the top edge is horizontal and 4 ft below the surface (Figure 4.21).

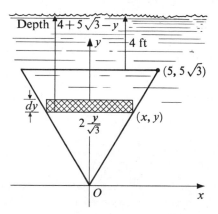

FIGURE 4.21   Triangle on the wet face of a dam.

Solution.   The pressure, in pounds per square foot, due to water at depth h ft is equal to the weight of a column of water h ft high and 1 square ft in horizontal section. Since water weighs 62.5 lb/cu ft, this pressure is $62.5h$ lb/sq ft.

We choose coordinates having origin at the bottom vertex of the triangle. Then the equation of one slanted side is $y = \sqrt{3}\, x$. We slice the triangle into thin horizontal strips so that the water pressure on each strip is approximately constant. We find from the equation $y = \sqrt{3}\, x$ that a strip at distance y above the vertex 0 has width $2y/\sqrt{3}$ and is $(4 + 5\sqrt{3} - y)$ feet above the surface. Hence the pressure at this level is $62.5(4 + 5\sqrt{3} - y)$ lb/sq ft. The area of the strip is $(2y/\sqrt{3})\, dy$. Therefore, the force element the water exerts on the face of the strip is

$$62.5(4 + 5\sqrt{3} - y)\left(\frac{2y}{\sqrt{3}}\right) dy.$$

This is the force element that we sum over the entire face of the triangle to obtain the total force F on the triangle. It is given by the integral

$$F = \int_0^{5\sqrt{3}} 62.5(4 + 5\sqrt{3} - y)\left(\frac{2y}{\sqrt{3}}\right) dy.$$

Simplifying this, and applying the fundamental theorem, we have

$$F = \frac{125}{\sqrt{3}} \int_0^{5\sqrt{3}} [(4 + 5\sqrt{3})\, y - y^2]\, dy$$

$$= \frac{125}{\sqrt{3}} \left[(4 + 5\sqrt{3})\, \frac{y^2}{2} - \frac{y^3}{3}\right]_0^{5\sqrt{3}} = 18{,}638 \text{ lb}.$$

22.6   Work.   The work done by a constant force of F pounds in moving a body through s feet is Fs foot-pounds. Thus a force of 10 pounds applied to lift a weight through 3 feet does 30 ft-lb of work. This is elementary when the force and distance are constant but when the force is a continuously varying function of the distance we need integration.

Example.   Compute the work done in stretching a spring through 10 inches, starting from equilibrium, if the spring reacts with a force of 25 pounds when it has been stretched 1 inch from equilibrium (Figure 4.22).

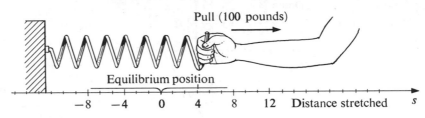

FIGURE 4.22   Stretching a spring.

Solution.   The equilibrium position is that in which the spring rests when no stretching or compressing force acts on it. According to Hooke's law, the force F exerted by a spring is proportional to the distance s through which it is stretched, that is $F = ks$, where k is a constant. Since $F = 25$ when $s = 1$, we can evaluate k and find that $k = 25$. So Hooke's law for this spring is $F(s) = 25s$.

We partition the interval $[0, 10]$ into short intervals of length ds, approximating $F(s)$ by a step function that has constant values in each subinterval. Then the work element to stretch the spring through one of the small subintervals is of the form $F(s)\,ds$. We sum these elements of work over the interval $[0, 10]$, which, after passing to the limit, gives the exact work W as the integral

$$W = \int_0^{10} F(s)\,ds = \int_0^{10} 25s\,ds = 25\,\frac{s^2}{2}\bigg|_0^{10} = 1250 \text{ in.-lb.}$$

22.7 Exercises

In Exercises 1–7, represent the areas as integrals.

1.   The area of the region between the curves $y = x^2$ and $y = x$.
2.   The area of the region between the curves $x = 3y - y^2$ and $x + y = 3$.
3.   The area of the region between the curves $y = x^2$ and $x = y^2$.
4.   The area of the region between the curves $y = 2x^2$ and $y = x^2 + 2x + 3$.
5.   The area of the region enclosed by the graph of $|x| + |y| = 1$.
6.   The area of the region enclosed by the graph of $|x|^{1/2} + |y|^{1/2} = 1$.
7.   The area of the region enclosed by $(y - x^2 + 1)(y + x^2 - 1) = 0$.

In Exercises 8–15, set up the volumes as integrals and compute by the fundamental theorem.

8.   The volume of a pyramid of height h whose base is a right triangle having sides a and b (Figure 4.23).
9.   The volume cut out of a tree of radius 2 ft by a horizontal cut halfway through and a cut at an angle of 45 deg to the same diameter (Figure 4.24).
10.   The volume enclosed between two cylinders of radius a whose axes intersect at right angles. (Observe that a plane slice parallel to both axes is a square.)
11.   The volume of the cone generated by revolving about the y-axis the region above the line $y = 3x$ and below the line $y = 7$.

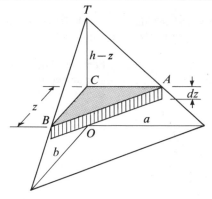

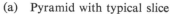

(a)   Pyramid with typical slice

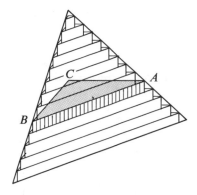

(b)   Slices stacked to fill pyramid

FIGURE 4.23   Exercise 8.

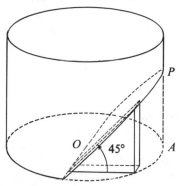

FIGURE 4.24   Wedge cut from a tree, Exercise 9.

12.   The volume of a cone whose base is a circle of radius r and altitude h.

13.   The volume of the solid generated by revolving about the y-axis the region bounded by $x + y = 2$, $y = 0$ and $y = 1$.

14.   The volume of the spherical ball generated by revolving the region bounded by the circle $x^2 + y^2 = r^2$ about the y-axis.

15.   The volume of the spherical cap cut off from the spherical ball of Exercise 14 by a plane at distance h from the center, where $0 \leqq h < r$.

16.   Find the mass of a cube of side s cm if the density ρ varies linearly in horizontal sections from 10 gm/cm³ at bottom to 3 gm/cm³ at top.

17.   Find the force due to water pressure on one face of a square of side 10 ft immersed vertically in water with the top edge in the surface.

18.   Find the force on the square in Exercise 17 if the top edge is horizontal, 12 ft below the surface.

19.   Find the force due to water pressure on the immersed portion of the region above the curve $y = x^2 - 10$ if the water level is at $y = 0$, and x and y units are feet.

20.   Find the work required to stretch a spring from equilibrium to 1 ft if a force of 10 lb is required to stretch the spring 1 in. from equilibrium.

21.   Find the work done by a vertical spring in lifting a weight of 100 lb as far as it can from a position 12 in. below equilibrium, assuming that a force of 25 lb is required to stretch the spring 1 in.

22.   Find the work done by a force that uniformly accelerates a particle of mass m on a smooth horizontal plane from rest to 40 ft/sec in 10 sec.

23. In Exercise 22, show that the work required to speed the mass up to 40 ft/sec is independent of the manner of acceleration and independent of the time required to reach the terminal speed.

24. An irregularly shaped vessel (Figure 4.25) holds water to a variable depth z. The area of a horizontal section at depth z is $A(z)$. Represent the volume of water at depth h as an integral.

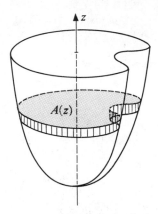

FIGURE 4.25   Exercise 24.

25. In Exercise 24, the water is flowing out from a hole in the bottom at a constant rate c in.^{3}/sec. Show that when the water is h in. deep the surface is falling at the rate $c/A(h)$ in./sec.

26. A vertical cylindrical tank with diameter 8 ft and depth 12 ft is to be pumped full of water flowing through a pipe in the bottom from a lake 24 ft below the bottom. Find the work required. *Hint*: Sum the work elements required to lift each thin horizontal slice of water in the full tank from the lake into position in the tank.

27. Find the work required to fill an open hemispherical vat of diameter 10 ft with water from a lake 30 ft below the bottom.

28. Solve Exercise 26, assuming that the tank is filled through a pipe that pours the water into it from the top.

29. With the tank in Exercise 26 as a source and initially full, a smaller vertical cylindrical tank of diameter 5 ft and depth 6 ft is to be filled with water pumped therefrom. If we assume that the bottom of the small tank is 28 ft above the top of the source tank, what is the work required?

23   Integration as an Averaging Process

23.1   Average Value of a Function

Example.   Find the average value of the function $f(x) = x^2 - 3x + 7$ over the interval $[-1, 3]$.

Solution.   We have seen how to define and compute the average value of a step function $s(x)$ over an interval $[a, b]$ (§4.4). We defined the average value $\bar{s}$ of a step function $s(x)$ to be

$$\bar{s} = \frac{\int_a^b s(x)\, dx}{\int_a^b 1 \cdot dx} = \frac{1}{b-a} \int_a^b s(x)\, dx.$$

Now we have a continuous function $f(x)$ and we define the average value $\bar{f}$ of $f(x)$ to be

$$\bar{f} = \frac{\int_a^b f(x)\,dx}{\int_a^b dx} = \frac{1}{b-a} \int_a^b f(x)\,dx.$$

In the particular case of the example, the mean value $x^2 - 3x + 7$ is given by

$$\bar{f} = \frac{1}{3-(-1)} \int_{-1}^3 (x^2 - 3x + 7)\,dx$$

$$= \frac{1}{4} \left(\frac{x^3}{3} - \frac{3x^2}{2} + 7x \right)\Bigg|_{-1}^{3} = \frac{37}{6}.$$

The average value $\bar{f}$ appears graphically as the height of the single-step (constant) function, which has the same area over the interval $[a, b]$ as $f(x)$ does (Figure 4.26),

$$\int_a^b \bar{f}\,dx = \int_a^b f(x)\,dx.$$

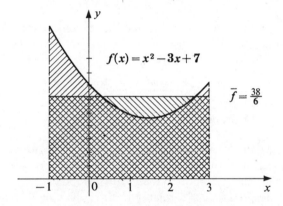

FIGURE 4.26   Mean value of a function.

23.2   Average Value of the Derivative.   We calculate the average value of the derivative $f'(x)$ on the interval $[a, b]$. It turns out, by the fundamental theorem, that

$$\bar{f}' = \frac{\int_a^b f'(x)\,dx}{b-a} = \frac{f(b) - f(a)}{b-a}.$$

That is, the slope of the secant line joining the points $(a, f(a))$ and $(b, f(b))$ on the graph of $f(x)$ is exactly equal to the average value of the derivative $f'(x)$ over the interval $[a, b]$. The mean value theorem (§13.2) says that $f(b) - f(a) = (b-a)f'(z)$, where z is some unspecified number between a and b. The average value of the derivative gives us a new formulation that is more truly a mean value theorem.

THEOREM (Integral form of Mean Value).   If $f(x)$ on $[a, b]$ is continuous and has an integrable derivative $f'(x)$, then

$$f(b) - f(a) = (b-a)\bar{f}',$$

where $\bar{f}'$ is the average value of $f'(x)$ on $[a, b]$.

23.3   Weighted Average of x

Example.   A set of different weights $\{w_1, w_2, w_3, \ldots, w_n\}$ is distributed on a bar, itself weightless, at distances $x_1, x_2, x_3, \ldots, x_n$ from one end. Find the average value $\bar{x}$ of x in the weight distribution (Figure 4.27).

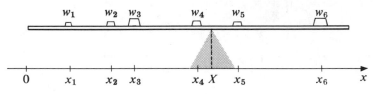

FIGURE 4.27   Weights on a bar.

Solution.   Is it enough to compute the average

$$\frac{x_1 + x_2 + \cdots + xn}{n}?$$

This assigns to each x where a weight is located the same importance as any other. It is clear that we should take the difference in size of weights into account in the averaging process. We do this with the weighted average

$$\bar{x} = \frac{x_1 w_1 + x_2 w_2 + x_3 w_3 + \cdots + x_n w_n}{w_1 + w_2 + w_3 + \cdots + w_n}.$$

This is our solution to the problem.

We observe that the units of measure in this quotient are the same as those of x because the weight units cancel out in the division. We may interpret $\bar{x}$ as locating the single point at which the bar can be supported and be in balance.

There is another interpretation of $\bar{x}$ related to the balancing of a lever. We treat $\bar{x}$ as unknown and require that the sum of the moments

$$(x_1 - \bar{x})w_1 + (x_2 - \bar{x})w_2 + \cdots + (x_n - \bar{x})w_n = 0,$$

where the "moment" of any weight w_i is its distance $x_i - \bar{x}$ from the point $\bar{x}$ multiplied by w_i, that is, $(x_i - \bar{x})w_i$. The equilibrium principle for balancing the beam with weights requires that the sum of the moments is zero. If we solve this equation for $\bar{x}$, we get the same average value $\bar{x}$ that we got before.

23.4   Weighted Average of x in a Continuous Distribution.

We may extend these ideas to a continuous distribution of mass along a line, using the integral of a step function as a link. Let $\rho(x)$ be a function on the interval $[a, b]$ that represents the density (mass per unit length) at x of a distribution of mass along the x-axis (Figure 4.28). We approximate this density function by a step function $s(x)$. Then the area of the rectangle below each step, $s(x)\, dx$, represents a mass that we may think of as located at x in the interval $[x, x + dx]$. Thus we may regard the step function as describing a finite number of masses distributed along the line and the average value of x in this n-particle distribution

$$\bar{x} = \frac{\int_a^b x[s(x)\, dx]}{\int_a^b s(x)\, dx}.$$

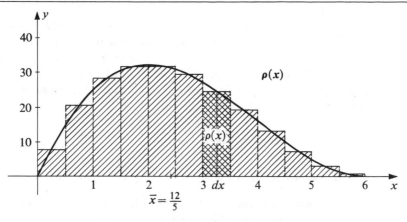

FIGURE 4.28   Continuous distribution of mass and approximating step function.

By refinement of the partition, the finite sums in the numerator and denominator converge to exact integrals, leading us to define the weighted average value of x in the continuous distribution by

$$\bar{x} = \frac{\int_a^b x \, \rho(x) \, dx}{\int_a^b \rho(x) \, dx}.$$

The denominator represents the total mass in the distribution.

Example.   Find the average value of x in the distribution of mass along the x-axis given by the density $\rho(x) = x^3 - 12x^2 + 36x$ on $[0, 6]$.

Solution.   We may plot the graph of the density (Figure 4.28). Then $\bar{x}$ is given by the average

$$\bar{x} = \frac{\int_0^6 x(x^3 - 12x^2 + 36x) \, dx}{\int_0^6 (x^3 - 12x^2 + 36x) \, dx} = \frac{12}{5}.$$

23.5   Standard Deviation.   The weighted average value of x in a distribution in a line is called the *mean* and denoted by μ. Two distributions in a line may have the same mean and the same total mass but one may be more concentrated near the mean and the other more dispersed (Figure 4.29). A measure of the dispersion about the mean is

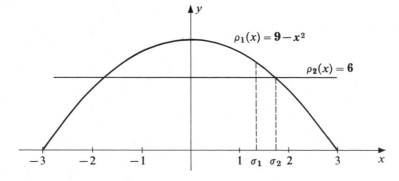

FIGURE 4.29   Two distributions with the same mean.

the average value of $(x - \mu)^2$ in the distribution. For $(x - \mu)^2$ is never negative, and therefore any mass m at distance $x - \mu$ from the mean contributes the positive number $(x - \mu)^2 m$ to the calculation, whether $x - \mu$ is positive or negative. Moreover, it contributes a big number (heavy penalty) to the average of $(x - \mu)^2$ when $|x - \mu|$ is large. We denote the average value of $(x - \mu)^2$ by σ^2. In any distribution in the line segment $[a, b]$ given by the density function $\rho(x)$, it is

$$\sigma^2 = \frac{\int_a^b (x - \mu)^2 \rho(x) \, dx}{\int_a^b \rho(x) \, dx}.$$

The positive number σ is called the standard deviation. It is measured in the units of x.

Example.   Find the standard deviations of the two distributions in the interval $[-3, 3]$ given by $\rho_1(x) = 9 - x^2$ and $\rho_2(x) = 6$ (Figure 4.29).

   Solution.   We verify that both distributions have total mass 36 and both have the mean $\mu = 0$. We calculate σ^2 for each one. For $9 - x^2$

$$\sigma_1^2 = \frac{\int_{-3}^3 (x - 0)^2 (9 - x^2) \, dx}{36} = \frac{\frac{9}{5}(36)}{36} = \frac{9}{5}.$$

Hence for this distribution, $\sigma_1 = \frac{3}{5}\sqrt{5}$.
   For the distribution given by $\rho_2(x) = 6$, we have

$$\sigma_2^2 = \frac{\int_{-3}^3 (x - 0)^2 (6) \, dx}{36} = 3, \quad \text{or} \quad \sigma_2 = \sqrt{3}.$$

23.6   Center of Mass

Example.   A flat sheet of metal of uniform density ρ is cut in the shape of the region bounded by the parabola $y = x^2$, the x-axis between lines $x = 0$ and $x = 5$. Find the center of mass of this piece of metal (Figure 4.30).

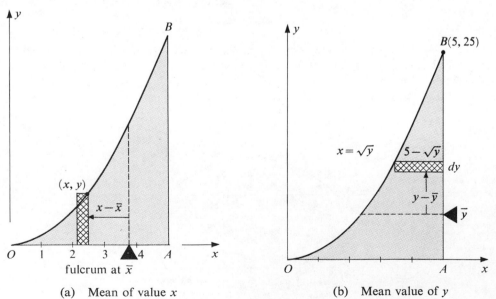

(a)   Mean of value x (b)   Mean value of y

FIGURE 4.30   Archimedes' balance.

Solution.   We follow Archimedes' line of reasoning. We slice the metal into thin strips parallel to the *y*-axis so that *x* is nearly constant in each strip. The mass *dm* of a typical strip is given by $dm = \rho x^2\, dx$. The turning moment about $\bar{x}$, as yet undetermined, is $(x - \bar{x})\, dm$. The metal will balance over a fulcrum placed at $\bar{x}$ if the sum of the moments

$$\int_0^5 (x - \bar{x})\, dm = 0.$$

Since $dm = \rho x^2\, dx$, if we solve this for $\bar{x}$, we find that

$$\bar{x} = \frac{\int_0^5 x(\rho x^2\, dx)}{\int_0^5 \rho x^2\, dx} = \frac{15}{4}.$$

Similarly, to obtain the *y*-coordinate of the center of mass, we slice the metal into thin strips parallel to the *x*-axis. The *y*-coordinate, $\bar{y}$, of the balance point is given by the requirement that

$$\int_0^{25} (y - \bar{y})\, dm = 0.$$

Since the horizontal strip at height *y* has length $5 - \sqrt{y}$, its mass is $\rho(5 - \sqrt{y})\, dy$. Hence

$$\bar{y} = \frac{\int_0^{25} y\, dm}{\int_0^{25} dm} = \frac{\int_0^{25} y\, \rho(5 - \sqrt{y})\, dy}{\int_0^{25} \rho(5 - \sqrt{y})\, dy} = \frac{\int_0^{25} (5y - y^{3/2})\, dy}{\int_0^{25} (5 - y^{1/2})\, dy}.$$

We cannot calculate these integrals by the fundamental theorem since we do not know antiderivatives for $y^{1/2}$ and $y^{3/2}$. As it stands, we can only calculate $\bar{y}$ numerically by approximating step functions. We omit this calculation.

23.7   Integral Form of Mean Value Theorem.   By the fundamental theorem of calculus, if $f'(x)$ is integrable,

$$f(x) - f(a) = \int_a^x f'(u)\, du.$$

Dividing both sides by $x - a$, we find that this statement is equivalent to the statement that the slope of the secant of the graph of *f* over the interval $[a, x]$

$$\frac{f(x) - f(a)}{x - a} = \frac{\int_a^x f'(u)\, du}{\int_a^x du}$$

is equal to the average value of the derivative $f'(x)$ over the interval. This is more truly a "mean value theorem" than the original statement (§13.2), which said that

$$\frac{f(x) - f(a)}{x - a} = f'(z),$$

where *z* was an unspecified number between *a* and *x*.

23.8   *Exercises*

In Exercises 1–3, find the average value of the function indicated.

1. x^2 on the interval $[-2, 2]$.
2. $x^3 - 7x + k$ on the interval $[-a, a]$.
3. $(2x - 5)^3$ on the interval $[0, 5]$.

4. A point P moves on the x-axis with position at any time t given by the function

$$x(t) = -t^2 + 3t + 4.$$

Find the average velocity: (a) in the time interval $[0, 3]$; (b) in the interval $[0, t]$.

In Exercises 5–8, find the mean and standard deviations about the mean of the distributions given by the density function ρ.

5. $\rho(x) = 2x$ on the interval $[0, 1]$.
6. $\rho(x) = 3x^2$ on the interval $[-1, 1]$.
7. $\rho(x) = |x|$ on the interval $[-a, a]$.
8. $\rho(y) = y^2 + y + 1$ on the y-interval $[-1, 3]$.

In Exercises 9–14, find as integrals the coordinates of the center of mass.

9. A thin sheet of uniform density in the xy-plane bounded by $x = 5$, $y = 0$, $y = 3x$.
10. A thin sheet of uniform density bounded below by $4y = x^2$ and above by $y = 5$.
11. A thin sheet of uniform density in the xy-plane bounded by $y = x^2$ and $y = x$.
12. The solid generated by the revolving about the y-axis of the region above the curve $y = x^2$ and below the line $y = 9$ (Figure 4.20 (a)).
13. The solid in Exercise 12, assuming that the density varies as a function of y, $\rho = 1 - \frac{1}{2}y$.
14. The spherical cap cut off from a ball of radius r by a plane at distance h from the center, $0 \leqq h < r$ (§22.7, Exercise 15).
15. Masses are placed on a horizontal bar at distances from the end given by the table below.

m	2	3	1	1	3	1
x	-1	0	2	3	5	6

Find the average value of x and the standard deviation.

16. Scores of subjects on a psychological test were tabulated.

Number of subjects	3	5	13	10	5
Score	0	1	2	3	4

Find the average score and the standard deviation.

17. Prove that the integral form of the mean value theorem can be expressed by

$$f(x) - f(a) = (x - a) \int_0^1 f'[a + t(x - a)]\, dt.$$

18. Show that the integral form of the mean value theorem implies that

$$f(x) - f(a) = (x - a)f'[z],$$

where z is a number between a and x (§13.2).

19. How is the mean of a distribution in x changed (a) when x_0 is added to every value of x? (b) when every value of x is multiplied by 2?

20. How are the mean and standard deviation transformed when we introduce a new scale in the x-axis by the transformation of scale $x = ax' + b$?

21. Prove that in any distribution if $\bar{x}^2$ represents the average value of x^2 in the distribution $\sigma^2 = \bar{x}^2 - \mu^2$.

22. Data from an experiment came in the form of a step function, $s(t)$, with steps of one-second duration as follows.

Interval of time	[0, 1)	[1, 2)	[2, 3)	[3, 4]	[4, 5)	[5, 6)
$s(t)$	2	-1	1	0	-2	1

It was decided to smooth the data, replacing $s(t)$ by the average value of s in the one second preceding t. Compute and plot the smoothed data $\bar{s}(t)$ on the same graph with a plot of $s(t)$.

23.* A step function is defined on unit intervals centered at $0, 1, \ldots, n$ so that its value at k is the kth binomial coefficient in the expansion of $(p + q)^n$, where $p + q = 1$. Show that these equations are true for the mean and standard deviation:

$$\mu = np, \qquad \sigma = \sqrt{npq}.$$

Plot these step functions for several values of n.

24. Prove the following form of the mean value theorem, which is different only in notation (§23.2).

$$f(a + h) - f(a) = \int_0^1 f'(a + th)\, dt = h\bar{f}',$$

where $\bar{f}'$ is the average value of f' on the interval $[a, a + h]$.

23R   Summary and Review: Three Programs for Calculus

ACADEMIC NOTE.   This is an extra lesson, not an essential part of the sequence that builds up the calculus. It is written more to be read for understanding than to be studied for mastery. The list of miscellaneous problems will serve to review the polynomial calculus as it has been developed in Chapters 3 and 4.

23.1R   Three Programs for Calculus: Three Patterns of Thought. *The theoretical program of calculus* consists of the study of two fundamental limits. First, there is the integral as a limit of a sum of elementary areas, or volumes, leading to an exact area or volume to be associated with a geometric figure. This pattern of thought was brought to a high state of perfection by the ancient Greeks, especially by Eudoxus and Archimedes in the fourth and third centuries B.C., and was refined and generalized in modern times, notably by Riemann (German, 1826–1866). We have studied this integration aspect of the fundamental theory of calculus (§§4, 9, 19, 21, 22, and especially 9.6). The other half of the theoretical program defines another precise numerical limit, called *the derivative*, which localizes a slope over an interval, or a velocity over an interval of time at a single point (§§5, 10, 12, 13, 14, 15, 18). The basic geometric problem is concerned with the question, Precisely what do we mean by a tangent line to a curve; or equivalently, what do we mean by the slope of a curve at a point? The analogous question in mechanics is, What do we mean by an instantaneous velocity at a point in space and time? As in integration, the theoretical answer involves a sophisticated limit concept that is difficult to apply in practice.

The *analytical program* of calculus is a modern invention, due independently to Isaac Newton (1642–1727) and Leibniz (1646–1716). It begins with an algebra of functions $f(x)$ (§§1, 2, 3) and develops a method of using the algebra of functions to calculate a derived function $f'(x)$ (§11) from which the numerical derivative can be found by evaluation (§12) without recourse to the defining limits, except for a few simple functions that generate the function algebra. This is the analytical program for differential calculus. Remarkably, the same function algebra provides a calculation program for the apparently unrelated integral calculus. By the fundamental theorem of calculus (§19), we can evaluate the limit of a sum, $\int_a^b f(x)\,dx$, which gives an exact area or volume, if we can solve the *differential equation* $DF(x) = f(x)$ for an unknown function $F(x)$, called the antiderivative. The exact limit, $\int_a^b f(x)\,dx$, can then be calculated by evaluation of $F(x)$; in fact,

$$\int_a^b f(x)\,dx = F(b) - F(a).$$

If we can carry this through successfully, it is very much easier than the fundamental limit processes. We have begun the implementation of these two aspects of the analytical program by applying them in the algebra of polynomial functions,

$$f(x) = a_0 x^n + a_1 x^{n-1} + \cdots + a_n,$$

(§§2, and 11–22). This is possible because a polynomial always has a derived function $f'(x)$, which is a polynomial, and an antiderivative $\int f(x)\,Dx$, which is a polynomial. Moreover, only five fundamental derivatives must be calculated from definition, giving the differentiation rules I, ..., V, which with their inverse statements $\int$ I, ..., $\int$ V serve to calculate derivatives and antiderivatives of all polynomials by simple algebraic techniques. The rest of our study was concerned with interpretations and applications of the derivative and integral (especially §§12, 15, 16, 18, 22, 23).

The third program of calculus is the *numerical program*. In most applications of calculus leading to a numerical derivative or integral, we do not require the exact result. We cannot ordinarily write the exact result as a decimal even if it is theoretically available. For a number of reasons it may not be possible or desirable to carry out exactly either the theoretical program or analytical program when we have a particular problem. In such cases, we can retain the *ideas* of the theoretical and analytic programs but go back to the prelimit stages of the fundamental definitions of integral and derivative to calculate ordinary finite arithmetic sums and divided differences that give a close enough approximation to the exact integral or derivative. Another way to implement the numerical program is to replace the function involved by an approximating step function (§§4, 5). Still another technique of the numerical program is to replace the function by an approximating polynomial, as in Taylor's theorem (§14). The simpler analytical program for polynomials can then be applied to obtain numerical results.

We have so far not developed the numerical program extensively. At the earliest possible stage we will want to bring machine computation into it to reduce the formidable obstacle of large scale arithmetic computation that the program involves.

23.2R   Differential Equation Pattern.   We have seen above that when the analytical program for integral calculus is applied, the differential equation $DF(x) = f(x)$ for the unknown antiderivative $F(x)$ is always involved in the calculation stage after the quantity to be calculated has been set up by an ideal limit-of-a-sum argument as an integral $\int_a^b f(x)\,dx$ (viz. §21). This is not the only possible pattern of thought for solving problems of integral calculus in the analytical program.

An alternative is to omit the summation procedure as a conceptual link in the solution and instead to introduce a differential equation directly into the physico-mathematical formulation of the problem. This seems to be a more natural way to think in problems of motion, flow, and growth (§18) than the more geometric pattern of the limit-of-a-sum, which we use in problems having an apparent static character, like those of area and volume, in the manner of Eudoxus and Archimedes, or Riemann. Historically, the differential equation pattern might be called the Newtonian pattern. We applied it to the area of a region below a graph (§17.4) and there found it to be roughly equivalent to the limit-of-a-sum pattern.

We recapitulate the synthesis and calculation of area by the differential equation pattern because of its importance. We imagine a variable region below the curve $y = f(x)$ bounded on the right by a variable ordinate at x (Figure 4.31). We assign to this an intuitively

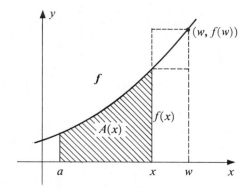

FIGURE 4.31   Differential equation of area.

conceived area function $A(x)$. Then, using some basic assumptions (§9.3) about what an area ought to be, we prove that such an area function must satisfy the differential equation $DA(x) = f(x)$ and have at some initial point at which $x = a$, the value $A(a) = 0$. Then if we can exhibit a continuous solution $A(x)$ of this differential equation and show that it is the only one satisfying $A(a) = 0$, we *define* the area by this function and the argument becomes rigorous. The equivalence of the differential equation pattern to the limit-of-a-sum pattern emerges from the theorem that if $f(x)$ is continuous, a solution of the differential equation is given at x by

$$A(x) = \int_a^x f(u)\, du.$$

To show that this is the only solution, we must appeal to that simplest, yet fundamentally difficult, differential equation $DF(x) = 0$ (§13.4).

We observe that the two patterns are not quite completely equivalent as we developed them; for the fundamental theorem in the limit-of-a-sum approach (§19.3) did not require that $f(x)$ be continuous. In practice, however, we may ordinarily take our choice of the two approaches. Under proper handling they are both rigorous arguments, the apparently less rigorous aspect of the differential equation pattern being due only to a relative delay of the stage at which one makes it rigorous. A competent problem solver must be able to apply both arguments; for many problems can be solved more readily in one pattern than the other.

23.3R   Self Quiz

1.   Define a numerical derivative of $f(x)$ at $(x, f(x))$.
2.   What is a derivative function (§11.2)?
3.   How is the derivative calculated when the derivative function is known (§12.1)?
4.   State several physical interpretations of the derivative: slope, velocity, rate of change, local scale factor, marginal quantity (§12).
5.   What does the sign of the derivative mean (§12.2)?
6.   State the mean value theorem (§13.2).
7.   State Taylor's theorem. What does it do for us (§14)?
8.   What does it mean when we say that the derivative operator D is linear (§11.4)?
9.   State the fundamental theorem of integral calculus.
10. What is an antiderivative? Do antiderivatives always exist (§17.4)?
11. Describe accurately integration by parts (§20.5).
12. Explain why the dummy variable u does not affect the value of $\int_a^x f(u)\,du$ (§21.2).
13. Outline the algebraic properties of indefinite integrals.
14. Describe several applications of the integral (§§22, 23).

23.4R   Exercises (Review)

1.   Find the slope of the tangent line and plot the tangent to the graph of $x[(x - 2)x + 3] - 1$ at the point where $x = 2$.
2.   Find an equation of the tangent line in Exercise 1.
3.   Plot the tangent line and sketch a portion of the graph of $y = (2 - 3x)^7$ near the point where $x = 1$, using the information from the second derivative.
4.   Use a cut-off Taylor's series to evaluate approximately $(2 - x)^7$ at the point where $x = 1.13$.
5.   Solve the differential equation $DF(x) - 4(x - 2)^3 = 0$ with the initial condition that the unknown function $F(x)$ has the value $F(1) = 5$.
6.   Evaluate $\int_0^1 x(2x + 3)^5\,dx$.
7.   Without evaluating, show that $\int_0^2 \sqrt[3]{x - 1}\,dx = 0$.
8.   Find by integration the volume of a pyramid whose base is a square having sides of length 2 and whose apex is located 5 units above one vertex of the base.
9.   Solve the differential equation $DF(x) = 0$ on a domain consisting of two intervals $\{x \mid -1 \leq x \leq 1\}$ and $\{x \mid 2 \leq x \leq 4\}$ given that $F(0) = 1$ and $F(3) = -1$. Is $F(x)$ a constant?
10. A function $f(x)$ is defined on the entire real line by

$$f(x) = \int_0^x \frac{2t - 1}{t^2 + 1}\,dt.$$

Show that $f(x)$ attains its minimum value at the point where $x = \frac{1}{2}$.

11. (a) Find the area of the region enclosed between the graphs of $y = x^3 - x$ and $y = 7x$, $x \geq 0$. (b) Calculate so that every part of the region has positive area.
12. A weight is projected into the air at time $t = 0$ with initial velocity 96 ft/sec upward. Find its height as a function of time. Find its mean height for the time it remains above ground level.
13. The function $y = x^2$ maps points on the x-axis onto points on the y-axis. A small interval of length h centered at the point where $x = 3$ is mapped into an interval on the y-axis of length k. Find the limiting ratio $\lim_{h \to 0} k/h$. Hence find k approximately in terms of h when h is small.
14. Show that the function $f(x) = x^3 - 9x + 5$ on all real numbers is piecewise monotonic, and find the intervals of x in which it is increasing and the intervals in which it is decreasing.

15. For the function in Exercise 14, find the interval over which the function is convex. In moving along the x-axis in the positive direction, at what point does the tangent line to the graph cease to rotate clockwise and begin to rotate counterclockwise?

16. A factory produces x units of a product that costs $C(x)$ dollars and sells for $S(x)$ dollars, where $C(x)$ and $S(x)$ are functions on $\{x \geq 0\}$. Under suitable mathematical assumptions show that maximum profit is attained when marginal sales equals marginal cost. State the assumptions that make it possible to deduce this result.

17. Find the rate of change of y with respect to x when $x = -1$ if

$$y = (x^2 + x + 1)^{15}.$$

Is y increasing or decreasing?

18. Find the maximum and minimum points of the function

$$f(x) = 2x^3 + 3x^2 - 36x + 7$$

on the closed interval $[-5, 5]$.

19. Water weighs 62.5 lb/cu ft. Find the force due to water pressure on the ends of a horizontal vat if the water is 5 ft deep and the ends are shaped like the graph of $y = |x|$.

20. A car is driven along a hilly road against a variable force due to gravity, which we shall call $f(x)$. Starting with $x = 0$, show that the total work $W(x)$ done against this force satisfies the differential equation $DW(x) = f(x)$.

21. Investigate for maximum and minimum points the function

$$f(x) = 2x^3 + 3x^2 + 6x + 5$$

on the real number line.

22. Differentiate $(2x - 3)^7(4x^2 + 1)^5$ and find the numerical derivative at the point where $x = 0$.

23. Find the average value of $3x^2$ in the interval $[-5, 5]$.

24. Investigate for maximum and minimum points $f(x) = x^2 - 2x - 2$ (a) on $[-1, 3]$; (b) on $[1, 2]$; (c) on $[-1, 1]$; (d) on all real numbers; (e) on all negative numbers.

25. What number exceeds its square by the greatest possible amount?

26. Find x_0 so that $\int_1^2 (x - x_0)(x^2 - 1)\, dx = 0$. What is a physical interpretation of this result?

27. Find the mean value of x for the distribution in the interval $[0, 4]$ given by the density function $\rho(x) = x^2$. Also find the standard deviation from the mean.

28. Find x so that $(x - a)^2 + (x - b)^2$ is a minimum.

29. Investigate for maximum and minimum points $|x - 2|$ on the real numbers.

In Exercises 30–32, investigate for local maxima and minima by use of the second derivative test.

30. $f(x) = x^3$ on the real numbers.
31. $f(x) = x^3 - 42 + 1$ on $[-5, 5]$.
32. $f(x) = (3x - 1)^4$ on all real numbers.

33. Investigate for maximum and minimum points the function $F(x)$ defined on all real numbers by

$$F(x) = \int_1^x \left(1 - \frac{3t + 1}{t^2 + 3}\right) dt.$$

Note: We do not yet know how to calculate the integral.

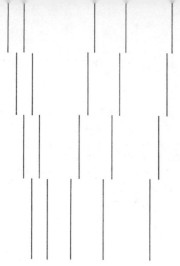

Powers and Transcendental Functions

CHAPTER 5

24. Derivatives and Integrals of Negative Powers

24.1 The Derivative of $1/x$. The function $f(x) = 1/x$ is defined for every x, $x \neq 0$, by dividing 1 by x. Let us compute the derivative. By definition the numerical derivative $f'(x)$ at x is given by

$$f'(x) = \lim_{w \to x} \frac{f(w) - f(x)}{w - x} = \lim_{w \to x} \frac{(1/w) - (1/x)}{w - x}.$$

We simplify the slope ratio as follows

$$\frac{(1/w) - (1/x)}{w - x} = \frac{-(w - x)/wx}{w - x} = \frac{-1}{wx} = -\frac{1}{x^2} + \frac{w - x}{wx^2}.$$

We find the last expression by dividing $-1/wx$ by $w - x$ to get the quotient $1/wx^2$ and remainder $-1/x^2$. Now

$$f'(x) = \lim_{w \to x} \left[-\frac{1}{x^2} + \frac{w - x}{wx^2} \right] = -\frac{1}{x^2}.$$

Example. Find the slope of the tangent line to $1/x$ at the point where $x = -2$, and plot this tangent line.

Solution. The derivative at -2 is given by evaluating $f'(x) = -1/x^2$ at -2. $f'(-2) = -\frac{1}{4}$. This is the slope of the tangent line to $1/x$ at $(-2, -\frac{1}{2})$. We plot this on the graph of $1/x$ (Figure 5.1).

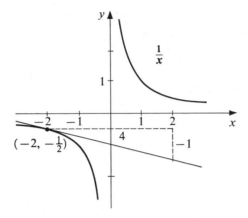

FIGURE 5.1   Tangent line to $1/x$.

24.2   Extension of Rule V to Negative Powers.   We recall some laws of exponents from algebra,

$$x^m x^n = x^{m+n};$$

$$\frac{x^m}{x^n} = x^{m-n}, \qquad x \neq 0;$$

$$(x^m)^n = x^{mn},$$

which are valid for negative and zero as well as positive integers n when we define

$$x^0 = 1 \qquad \text{and} \qquad x^{-m} = \frac{1}{x^m}, \qquad x \neq 0.$$

Examples

$$x^{-2} x^{-3} = x^{-5}, \qquad \frac{x^{-2}}{x^{-3}} = x^{-2-(-3)} = x^{-2+3} = x, \qquad \text{and} \qquad (x^{-2})^{-3} = x^6.$$

We calculate the derivative of u^{-m}, where u is a differentiable function and $u \neq 0$. By applying the derivative of $1/u$ (§24.1), the chain rule, and Rule V, with the laws of exponents, we find

$$\boldsymbol{D u^{-m}} = D \frac{1}{u^m} = D \left(\frac{1}{u}\right)^m = m \left(\frac{1}{u}\right)^{m-1} D \left(\frac{1}{u}\right) = m \left(\frac{1}{u}\right)^{m-1} \left(-\frac{1}{u^2}\right) \boldsymbol{Du} = -\boldsymbol{m u^{-m-1} \, Du}.$$

The result is precisely the same as the formula for $\boldsymbol{Du^n}$ in the case $n = -m$. Hence we may restate Rule V to include negative as well as positive integral exponents.

V. $\boldsymbol{Du^n = n u^{n-1} \, Du}$, if n is any integer.

$\int$ V. $\displaystyle\int \boldsymbol{u^n \, Du} = \frac{\boldsymbol{u^{n+1}}}{\boldsymbol{n+1}} + \boldsymbol{C}$, if n is any integer except $n = -1$.

We observe that in $\int$ V we must exclude $n = -1$ because in this case the denominator, $n + 1$, is zero.

Example.   Differentiate $1/(3x - 5)^2$.

   Solution. $1/(3x - 5)^2 = (3x - 5)^{-2}$ and by Rule V,

$$D(3x - 5)^{-2} = -2(3x - 5)^{-3} D(3x - 5) = -2(3x - 5)^{-3}(3)$$
$$= -6(3x - 5)^{-3}, \quad \text{or} \quad -6/(3x - 5)^3.$$

This is valid at any point except where $3x - 5 = 0$. Hence we specify $x \neq 5/3$.

24.3 Quotient Rule.   If u and v are differentiable functions, then at any point where $v(x) \neq 0$,

$V_a.$ $D\dfrac{u}{v} = \dfrac{v\,Du - u\,Dv}{v^2}.$

We derive this by differentiating $u(1/v)$ (Exercises).

Example A.   Differentiate

$$\frac{x^2 - 3x + 1}{(x^2 + 1)^2}.$$

   Solution.   The denominator is nowhere zero, so the quotient rule applies. It gives us

$$D\frac{x^2 - 3x + 1}{(x^2 + 1)^2} = \frac{(x^2 + 1)^2(2x - 3) - (x^2 - 3x + 1)(2)(x^2 + 1)2x}{(x^2 + 1)^4}$$
$$= \frac{-2x^3 + 9x^2 - 2x - 3}{(x^2 + 1)^3}.$$

Example B.   By linearity and Rule $\int$ V,

$$\int (2u^2 - 3u + 4 + 10u^{-2} - 7u^{-3})\, Du$$
$$= 2 \int u^2\, Du - 3 \int u\, Du + 4 \int Du + 10 \int u^{-2}\, Du - 7 \int u^{-3}\, Du$$
$$= 2 \left(\frac{u^3}{3}\right) - 3 \left(\frac{u^2}{2}\right) + 4u + 10 \left(\frac{u^{-1}}{-1}\right) - 7 \left(\frac{u^{-2}}{-2}\right) + C$$
$$= \frac{2}{3} u^3 - \frac{3}{2} u^2 + 4u - 10^{-1} + \frac{7}{2} u^{-2} + C.$$

Example C.   Calculate the area below the curve $7/(2x - 3)^2$ between the limits $x = 2$ and $x = 5$.

   Solution.   We observe that the function $7/(2x - 3)^2$ is decreasing in the interval $[2, 5]$, hence integrable. The point $x = \frac{3}{2}$, where the denominator $(2x - 3)^2 = 0$, is outside the interval $[2, 5]$ and will cause no difficulty. The area in question is then

$$A = \int_2^5 \frac{7}{(2x - 3)^2}\, dx.$$

To apply the fundamental theorem, we first find an antiderivative,

$$\int \frac{7}{(2x - 3)^2}\, Dx.$$

We let $u = 2x - 3$, compute $Du = 2\,Dx$, solve to find that $Dx = \frac{1}{2}\,Du$, and substitute this into the integrand. It becomes

$$7\int \frac{1}{u^2}\left(\frac{1}{2}\,Du\right) = \frac{7}{2}\int u^{-2}\,Du = \frac{7}{2}\left(\frac{u^{-1}}{-1}\right) + C$$

by Rule V. We now resubstitute $u = 2x - 3$ and find the antiderivative,

$$\int \frac{7}{(2x-3)^2}\,Dx = -\frac{7}{2}(2x-3)^{-1} + C.$$

With this antiderivative we apply the fundamental theorem, which gives us

$$A = \int_2^5 \frac{7}{(2x-3)^2}\,dx = -\frac{7}{2}(2x-3)^{-1}\Big|_2^5 = -\frac{7}{2}\left(\frac{1}{7}-\frac{1}{1}\right) = 3.$$

This completes the problem.

24.4   Restoring the Missing Antiderivative.

We recall again that Rule $\int$ V does not yield an antiderivative for x^{-1}. For, to apply $\int$ V, we would have $\int x^{-1}\,Dx = (x^0/0) + C$, which has no meaning. Does this mean that x^{-1} has no antiderivative? Certainly not. For consider the indefinite integral

$$\int_1^x \frac{1}{u}\,du.$$

The function $1/u$ is integrable over any interval of positive numbers since $1/u$ is a decreasing function (§9.3). Hence the indefinite integral defines a function on the set of all positive numbers x, which can be computed by approximating sums to any required accuracy. Moreover, $1/u$ is continuous when $u > 0$ (§8.2). When we differentiate the function of x, defined by this integral, with respect to x, the numerical derivative is $1/x$ (§21.6). Thus this indefinite integral is an antiderivative of $1/x$. It remains only to give this antiderivative a suitable name. Following tradition, we call it "the natural logarithm of x," $\ln x$. It does not matter for our immediate purposes that this function is a logarithm to some base; our only reason for introducing it is to obtain an antiderivative of $1/x$ that is computable.

DEFINITION (The natural logarithm).   We define (Figure 5.2) the function $\ln x$ at each positive real number x to have the value

$$\ln x = \int_1^x \frac{1}{u}\,du.$$

We observe that $\ln 1 = 0$.

This gives us immediately a pair of new derivative-antiderivative formulas (§20.4).

$$D \ln u = \frac{1}{u}\,Du \qquad \text{and} \qquad \int \frac{1}{u}\,Du = \ln u + C.$$

We record these formulas in a slightly more general form that permits us to differentiate $\ln |u|$ at places where the value of u is negative, so that $\ln u$ does not exist (proofs in theoretical exercises, §24.6).

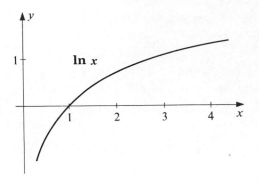

FIGURE 5.2   The natural logarithm.

VII. $D \ln |u| = \dfrac{1}{u} Du$,   and $\displaystyle\int$ VII. $\displaystyle\int \dfrac{1}{u} Du = \ln |u| + C,$

at every point x, where u is differentiable and $u(x) \neq 0$.

With these new rules, we can find an antiderivative of any integral power of x. For example,

$$\int [2x^{-3} - 7x^{-2} + 9x^{-1} + 4 + 7x - 12x^2] \, Dx$$

$$= \left(\frac{x^{-2}}{-2}\right) - 7\left(\frac{x^{-1}}{-1}\right) + 9 \ln x + 4x + 7\left(\frac{x^2}{2}\right) - 12\left(\frac{x^3}{3}\right) + C.$$

Example.   Calculate

$$\int_1^2 \frac{x^3 + 3x - 4}{2x - 1} \, dx.$$

Solution.   Whenever we have an integrand that is a quotient of two polynomials, we *always* perform the indicated division until we have a remainder in which the numerator is of *lower* degree than the denominator. Thus, dividing in this problem, we find

$$\int \frac{x^3 + 3x - 4}{2x - 1} \, Dx = \int \left[\frac{x^2}{2} + \frac{x}{4} + \frac{13}{8} - \frac{19}{8}\frac{1}{2x - 1}\right] Dx$$

$$= \frac{1}{2}\left(\frac{x^3}{3}\right) + \frac{1}{4}\left(\frac{x^2}{2}\right) + \frac{13}{8}x - \frac{19}{8}\int \frac{1}{2x - 1} \, Dx.$$

The integrand in the last term is in the form $1/u$ with $u = 2x - 1$. This requires that $Du = 2 \, Dx$ and hence that $Dx = \frac{1}{2} Du$. Substituting this into the integrand, we find

$$\int \frac{1}{u}\left(\frac{1}{2} Du\right) = \frac{1}{2}\int \frac{1}{u} Du = \frac{1}{2}\ln u = \frac{1}{2}\ln(2x - 1) + C.$$

Now returning with this to the original problem, we have

$$\int_1^2 \frac{x^3 + 3x - 4}{2x - 1} \, dx = \frac{x^3}{6} + \frac{x^2}{8} + \frac{13}{8}x - \frac{9}{16}\ln(2x - 1) \Big|_1^2$$

$$= \left(\frac{8}{6} + \frac{4}{8} + \frac{13}{8}(2) - \frac{9}{16}\ln 3\right) - \left(\frac{1}{6} + \frac{1}{8} + \frac{13}{8} - \frac{9}{16}\ln 1\right)$$

$$= \frac{29}{12} - \frac{9}{16}\ln 3, \quad \text{since} \quad \ln 1 = 0.$$

We can complete the evaluation by taking $\ln 3 = 1.099$ from Table 2 in the Appendix.

24.5 Exercises

In Exercises 1–12, perform the indicated differentiations.

1. $D\dfrac{1}{x^2}$.

2. $D\dfrac{1}{3x+5}$.

3. $D\left[-\dfrac{5}{x^2}+\dfrac{1}{x}+2-7x+4x^3\right]$.

4. $D\left[\dfrac{1}{x^2}-\dfrac{1}{(2x-3)^3}\right]$.

5. $D\ln ax$.

6. $Dx\ln|2x-1|$.

7. $D\dfrac{\ln 2x}{3x+5}$.

8. $D\dfrac{\ln 3x}{x^2}$.

9. $D\dfrac{\ln x}{x}$.

10. $D(\ln x)^2$.

11. $D\dfrac{1}{\ln x}$.

12. $D\dfrac{x}{1+\ln|x|}$.

In Exercises 14–23, find the indicated antiderivatives.

14. $\displaystyle\int\dfrac{1}{x^3}\,Dx$.

15. $\displaystyle\int\dfrac{95}{2x^2}\,Dx$.

16. $\displaystyle\int\left(\dfrac{1}{x^2}-\dfrac{1}{x^3}\right)Dx$.

17. $\displaystyle\int cx^{-5}\,Dx$.

18. $\displaystyle\int(2x^{-3}+4x^{-2}-3x^{-1}+13+5x^2)\,Dx$.

19. $\displaystyle\int\dfrac{5\,Dx}{x}$.

20. $\displaystyle\int\dfrac{1}{3x-4}\,Dx$.

21. $\displaystyle\int\dfrac{1}{1-x}\,Dx$.

22. $\displaystyle\int\dfrac{x^2+x-3}{x-2}\,Dx$.

23. $\displaystyle\int\dfrac{x^3}{2x+4}\,Dx$.

In Exercises 24–29, evaluate the definite integrals.

24. $\displaystyle\int_2^4\dfrac{1}{3x^2}\,dx$.

25. $\displaystyle\int_{-1}^{-2}\dfrac{x-1}{x^2}\,dx$.

26. $\displaystyle\int_0^{1/2}\dfrac{1}{u}\,du$.

27. $\displaystyle\int_1^4\dfrac{t^2-1}{t+4}\,dt$.

28. $\displaystyle\int_1^{0.1}\dfrac{1}{x}\,dx$.

29. $\displaystyle\int_0^x\dfrac{1}{u+1}\,du$.

30. Find the slope of the tangent line to $1/x^2$ at the point where $x=-2$, and plot this tangent line.

31. Find the slope of the tangent line to $\ln 2x$ at the point where $x=\frac{1}{2}$, and plot this tangent line.

32. Find the maximum and minimum points of $(1/x)+2x$ on $\{x>0\}$.

33. Find the maximum and minimum points of $x\ln x$ on $\{x>0\}$

34. A quantity of gas in a cylinder is compressed by a piston (Figure 5.3). It requires 25 pounds of force on the piston to hold it so that the gas occupies a cylinder 12 inches

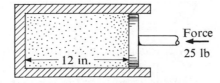

FIGURE 5.3   Exercise 34.

long. The gas obeys Boyle's law, PV = constant. How much work is done in compressing the gas into a cylinder 2 inches long?

35. The force of gravity on a body varies inversely as the square of the distance from the center of the earth. If you weigh 200 pounds at the surface of the earth, how much work must be done against earth's gravity to rocket you 2000 miles high, assuming the earth's radius to be 4000 miles?

36. How much work must be done against the earth's pull on a 200-pound man to send him to the moon 240,000 miles away, without account being taken of the moon's gravitational pull?

37. A body on the moon's surface, 1080 miles from its center, weighs 0.165 times as much as it does on the earth's surface. Taking account of the moon's pull, calculate how much work must be done to send a 200-pound man to the moon.

38. In Exercise 37, how much work must be done to send the astronaut back to earth after landing on the moon?

39. Find the unknown function $f(t)$ if $Df(t) = 1/t$ and $f(2) = 0$.

40. Find the area enclosed between the graphs of the equations $xy = 1$ and $x + y = 4$.

24.6 Problems

T1. Let y be a constant. Show that $D \ln xy = D \ln x$. What does this imply about the difference $\ln xy - \ln x$? Evaluate the difference by setting $x = 1$. State the theorem proved by these steps

T2. Show that $f(x) = (ax + b)/(cx + d)$, where a, b, c, d are constants, defined on the real numbers except where $cx + d = 0$, has no critical points.

T3. If $f(x) = 1/x$, calculate the derivatives of all orders. Show that the derivative of order n is given by

$$f^{(n)}(x) = \frac{(-1)^n n!}{x^{n-1}}.$$

T4. Carry out a formal expansion of $1/x$ by the Taylor's expansion formula (§14.2) in powers of $x - 1$ and show that there is no last term.

T5. Show that if $\ln |x|$ is defined on all real numbers x, $x \neq 0$, then $D \ln |x| = (1/x) Dx$.

T6. Use T5 to show that

$$\int \frac{1}{u} Du = \ln |u| + C, \qquad u \neq 0.$$

25 Maximum and Minimum Problems with Side Conditions

25.1 **Most Economical Can.**   Many problems of finding a maximum or minimum of some quantity, subject to a side condition that holds a related quantity constant, can be reduced to a problem of finding the maximum or minimum of a function involving negative powers of x. Let us consider an example.

Example. A cylindrical can with closed top is to contain V cu in., where V is a constant (Figure 5.4). Find the dimensions that require the least metal.

  Solution.   We first assume that minimum metal use will be achieved with a can of minimum area of top, bottom, and sides. Then let x be the radius of the base and y the height of the can. The circular top and bottom each have area πx^2 and the side has area $2\pi xy$.

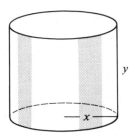

FIGURE 5.4   Cylindrical can.

The can having been cut along a vertical line, the cylindrical surface of the side opens out into a rectangle of base $2\pi x$ and height y. We obtain as the volume, $V = \pi x^2 y$. Hence the problem requires that

$$2\pi x^2 + 2\pi xy = \text{minimum},$$

subject to the side condition $\pi x^2 y = V$. We can eliminate one of the variables by means of the side condition. It appears easier to eliminate y, so we solve the side condition for y as a function of x,

$$y = \frac{V}{\pi x^2}.$$

and substitute the result into the area formula. This gives the area as a function $A(x)$,

$$A(x) = 2\pi x^2 + 2\left(\frac{V}{x}\right)$$

on the domain of all positive numbers x. For any x, the area $A(x)$ given by this function is automatically the area of a can of volume V.

Our problem is now set up for the application of calculus methods (§16). The problem is to find the minimum of the function $A(x)$. The only critical values of x are those for which $A'(x) = 0$, since the domain contains only interior points and there are no points where $A'(x)$ fails to exist. We differentiate and find by Rule V that

$$A'(x) = 4\pi x - 2\left(\frac{V}{x^2}\right).$$

We set the derivative equal to zero.

$$A'(x) = \frac{4\pi x^3 - 2V}{x^2} = 0.$$

The only root, therefore the only critical value, is

$$x = \sqrt[3]{\frac{V}{2\pi}}.$$

If we assume from geometric considerations that the area has a minimum, this x must give the minimum.

To determine by calculus techniques that this critical value actually gives a minimum to the area $A(x)$, we can follow several routes. We look at two of them. Let us first use the second derivative.

The second derivative is given by

$$A^{(2)}(x) = 4\pi + 4\left(\frac{V}{x^3}\right).$$

This function is positive throughout the domain of $A(x)$. Hence by Taylor's theorem with second-order remainder (§14.4), the critical point must give the minimum for all x.

We can also use the first derivative to derive the same information. We write the first derivative at x in factored form

$$A'(x) = \frac{4\pi}{x^2}\left(x^3 - \frac{V}{2\pi}\right).$$

The first factor is everywhere positive, so the sign of $A'(x)$ depends on the sign of $[x^3 - (V/2\pi)]$. If $x < \sqrt[3]{V/2\pi}$, we see that $A'(x)$ is negative; and if $x > \sqrt[3]{V/2\pi}$, $A'(x)$ is positive. Hence $x = \sqrt[3]{V/2\pi}$ gives a minimum to $A(x)$ (Figure 5.5).

It remains to determine y. This can be done by substituting $x = \sqrt[3]{V/2\pi}$ into the side condition $y = V/\pi x^2$, but a better way is to find the ratio y/x from this expression. Dividing both sides by x, we have

$$\frac{y}{x} = \frac{V}{\pi x^3}.$$

Substituting the critical value for x gives $y/x = 2$. Therefore the relative dimensions of the can for minimum metal use are height = diameter. This completes the problem.

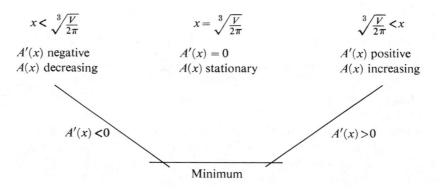

FIGURE 5.5   Proof that the critical point gives a minimum.

25.2   Maximum Utility. An Econometric Problem

Example.   An unsentimental farmer derives from the purchase of land satisfaction in proportion to the number of acres of land he can purchase; at the same time he derives satisfaction from the purchase of higher education for his many children in proportion to the number of years of education he can purchase. If land costs \$200 per acre and education costs \$1000 per year, and he has not more than \$20,000 to spend, how should he apportion his money for maximum satisfaction?

Solution.   Let x be the number of acres of land he purchases and y be the number of years of education he purchases. Let U represent his amount of satisfaction (technically "utility" in economics language). Then the relation

$$U = Cxy,$$

where C is an unknown positive constant, represents the given condition that his satisfaction is jointly proportional to the number of acres of land and the number of years of education he can purchase. We have the side condition expressed by the limitation of his total purchase given by

$$200x + 1000y \leq 20{,}000.$$

Instead of working with this inequality, we shall let A be the number of dollars he actually spends and write the budget condition as an equality with $\$A$ spent. This gives us

$$200x + 1000y = A.$$

We solve this equation of y in terms of x and find that

$$y = \frac{(A - 200x)}{1000}.$$

Then we substitute this expression for y in the satisfaction relation $U = Cxy$ and find that

$$U = Cx \left(\frac{A - 200x}{1000}\right),$$

or

$$U(x) = \frac{C}{1000} (-200x^2 + Ax).$$

This is the function on the set of numbers for which $x \geq 0$, which we will try to make a maximum.

The critical values of x are the endpoint 0 and the values of x, where $U'(x) = 0$. We calculate the derivative

$$U'(x) = \frac{C}{1000} (-400x + A),$$

set $U'(x) = 0$, and find that the only other critical point is given by $x = A/400$. We test the signs of $U'(x)$ and find that if $x < A/400$, then $U(x)$ is increasing, and if $x > A/400$, then $U(x)$ is decreasing. So $x = A/400$ gives the maximum value to $U(x)$. Returning to the equation $200x + 1000y = A$, we find that when $x = A/400$, then $y = A/2000$. Thus for maximum satisfaction the farmer should purchase education and land in the ratio $y/x = 5$ and the measure of his satisfaction will then be

$$U = \frac{CA^2}{800{,}000}.$$

Now clearly U increases as A increases so that maximum satisfaction is attained when $A = \$20{,}000$ and the farmer spends all $\$20{,}000$ to purchase 50 acres of land and 10 years of education for his children. This is the solution of the problem. We observe that the ratio y/x does not depend on that constant C or on the amount, $\$A$, that the farmer spends.

25.3 Exercises

1. The sum of two numbers is 25. Find the numbers if their product is a maximum.
2. What is the smallest number that can be obtained by adding a positive number to its reciprocal?
3. Separate 24 into two parts so that the sum of the squares of the parts shall be a minimum.
4. Find the volume of the largest open-top box that can be made from a rectangular sheet of tin 24 in. long by 16 in. wide by cutting equal squares from the corners as in the figure and bending up the remaining pieces to form the sides (Figure 5.6).
5. A strip of metal 20 in. wide is to be bent to form a gutter of rectangular cross section (Figure 5.7). Find the depth and width of the base for maximum capacity.

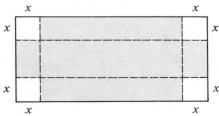

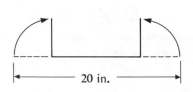

FIGURE 5.6   Exercise 4.   FIGURE 5.7   Exercise 5.

6. The frictional resistance to flow of liquid in an open channel is proportional to the area of the wetted surface. Show that for minimum resistance in moving a given volume of water in a rectangular channel the width should be twice the depth.

7. Find the dimensions of a rectangle of minimum perimeter that has an area of 1000 sq ft.

8. A rectangular box to contain 108 cu in. is to be made with a square base. Find the dimensions that will give a minimum area of top, bottom, and sides.

9. Find the dimensions of the rectangle of maximum area that will have perimeter (a) 400 cm; (b) P cm.

10. A rectangular plot of 10,000 sq ft is to be fenced. If the north-south fences cost $2 per ft and the east-west fences cost $3 per ft, find the dimension N-S and E-W that will require minimum fencing cost.

11. In Section 25.2, Example, the farmer's wife also derived satisfaction from the purchase of land and education that was jointly proportional to the number of units of each purchased, but being more sentimental, she derived 100 times as much satisfaction from a unit of education as from a unit of land. How would she apportion the money for maximum satisfaction?

12. (a) Solve the Example of Section 25.2 under the assumption that land costs $p per acre and education $q per year.
    (b) Show that regardless of the prices p and q, the farmer will always derive maximum satisfaction if he spends half of his money on land and half on education.

13. A rectangular plot of land with undetermined dimensions is to be fenced with a fixed budget for fencing. The back and front fences cost $p per ft and the side fences cost $q per ft. Show that the maximum enclosed area is always attained when the shape of the plot is such that the cost of the back and front fences is the same as the cost of the side fences. Is there a relation between this problem and the preceding one?

14. The intensity of light varies directly as the intensity of the source and inversely as the square of the distance from the source. Two lights are 16 ft apart and one source has 4 times the intensity of the other. At what point on the line between them is the combined intensity from the two lights a minimum?

15. The strength of a rectangular beam is proportional to the (horizontal) width and to the square of the (vertical) depth. What are the dimensions of the strongest beam that can be cut from a log of radius r?

16. A cylindrical vessel with open top is to contain V cu in. Find the dimensions that require the least metal.

17. A cylindrical vessel with open top is to contain V cu in. The material for the bottom costs 3 times as much as the material for the vertical wall. Find the relative dimensions that will permit minimum cost.

18. (a) Prove that for every pair of positive numbers x and y such that $x + y = 2L$, $xy \leq L^2$ and $xy = L^2$ if and only if $x = y$.
    (b) Show from this that $\sqrt{xy} \leq (x + y)/2$ and that the equals sign holds only when $x = y$. Thus the geometric mean never exceeds the arithmetic mean.

19. A trucking company finds that its mileage costs (fuel, tires, etc.) are proportional to the

square of the running speed and that the driver and capital costs are proportional to the time required to make a trip. For a certain trip at 40 mi/hr where the speed limit was 50, the mileage costs are $10 and the driver and capital costs are $40. Find the speed at which the total trip costs are a minimum.

20. In an irreversible chemical reaction, one molecule of sodium hydroxide reacts with one molecule of ethyl acetate to form one molecule of alcohol. The reaction proceeds at a rate proportional to the product of the concentrations of hydroxide and acetate remaining. Find the concentrations at which the formation of alcohol proceeds with minimum speed with an initial 0.60 gram mols of hydroxide and 0.30 gram mols of acetate.

21. An apartment building is to be constructed so that the ground floor alone will contain a space 200 ft by 200 ft. The ground costs $5000 per year. Construction and maintenance costs are $7500 per floor per year plus $500 per floor per year for each floor added above the ground floor. On account of elevators, usable space decreases by 100 sq ft in all floors each time a floor is added above the ground floor. The owner estimates that he can rent the usable space for 10 cents per sq ft. What is the height of building that will yield maximum profit?

26   General Powers and Logarithmic Differentiation

26.1   Inverse of ln x. Since **ln x** is defined for all positive numbers x and has the derivative $1/x$, which is everywhere positive, it is a strictly increasing continuous function (Figure 5.8). We shall use the following plausible theorem, postponing its proof (§26.10).

THEOREM (One-to-one property of **ln x**).   For every real number y, there is one and only one positive number x such that ln $x = y$.

DEFINITION.   The number e is defined by ln $e = 1$.

26.2   Logarithmic Property. So far we have not needed to know that **ln x** is a logarithm. Let us now investigate this question. First we define a logarithm.

DEFINITION.   A *logarithm* is any continuous function $f(x)$ on the positive real numbers such that $f(xy) = f(x) + f(y)$.

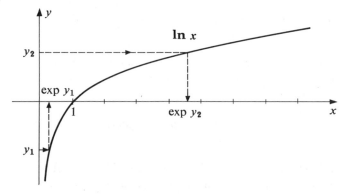

FIGURE 5.8   Inverse of $y = \ln x$.

THEOREM.   **ln** x is a logarithm.

Proof.   **ln** x is continuous, since it is differentiable (§24.4). Let y be a constant. Then

$$D \ln xy = \frac{1}{xy} D xy = \frac{y}{xy} = \frac{1}{x} = D \ln x.$$

Since $D \ln xy = D \ln x$, **ln** xy differs from **ln** x by a constant (§17.2); that is, **ln** $xy =$ **ln** $x + C$. Since ln 1 = 0, we set $x = 1$ and get ln $y = 0 + C$. With C thus evaluated, we find that

$$\ln xy = \ln x + \ln y,$$

which completes the proof of the theorem.

26.3   Antilogarithm Function.

We can use the preceding theorem to define a function that is an inverse of the logarithm.

DEFINITION.   The function **exp** y, read "exponential y," on all real numbers y to positive numbers x is defined by exp $y = x$ if and only if ln $x = y$.

THEOREM.   **exp** y is differentiable and D_y **exp** $y =$ **exp** y.

Proof.   From the definition, $x =$ exp y and $y =$ ln x have the same graphs and therefore the same tangent lines at (x, y). We know that **ln** x is differentiable and has the slope $1/x$ at each positive number x. The equation of the tangent line to $y =$ ln x at (x, y) is then $Y - y = 1/x(X - x)$, where (X, Y) are the coordinates of a variable point on the tangent line. So the *same line* has the slope x with respect to Y, since its equation can be written $X - x = x(Y - y)$. Thus, since $x =$ exp y has the same tangent as $y =$ ln x, its slope is given by D_y exp $y = x$. But by definition, $x =$ exp y. Hence D_y **exp** $y =$ **exp** y. This completes the proof.

26.4   Logarithmic Differentiation

THEOREM.   If u is a positive-valued function such that **ln** u is differentiable, u is itself differentiable and $Du = u \, D \ln u$.

Proof.   exp y is differentiable (§26.3). By hypothesis of the theorem, **ln** u is differentiable. Therefore by the chain rule, VI, **exp** (**ln** u) is differentiable. But **exp** (**ln** u) = u; hence u is differentiable. Now we differentiate **ln** u by Rule VII, $D \ln u = 1/u \, Du$. This gives $Du = u \, D \ln u$, as stated in the conclusion of the theorem.

26.5   General Powers.

For every integer n, it follows from the definition of a logarithm that ln $x^n = n \ln x$ (Proof: Exercises). We now extend this property to all real numbers p with a definition.

DEFINITION.   Let p be any real number; then the function x^p is defined at each positive number x by ln $x^p = p \ln x$.

This is a valid definition, since $p \ln x$ is determined when x and p are given. Then the theorem on the one-to-one property of the logarithm (§26.1) says that x^p is determined uniquely.

THEOREM (General laws of exponents). For every positive real number x and for any real numbers p and q:

$$x^p x^q = x^{p+q},$$

$$\frac{x^p}{x^q} = x^{p-q},$$

$$(x^p)^q = x^{pq},$$

$$x^{1/q} = \sqrt[q]{x}.$$

Proof of $x^p x^q = x^{p+q}$. By definition (§26.5), x^p is defined by $\ln x^p = p \ln x$, x^q by $\ln x^q = q \ln x$. Since $\ln x$ is a logarithm, $\ln x^p x^q = \ln x^p + \ln x^q = p \ln x + q \ln x = (p + q) \ln x = \ln (x^{p+q})$, where the last step follows again from the definition. Now since $\ln x^p x^q = \ln x^{p+q}$, the one-to-one property of $\ln x$ (§26.1) implies that $x^p x^q = x^{p+q}$. This completes the proof. We leave the proof of the remaining statements in the theorem to the exercises.

26.6 Derivatives of General Powers.
Let u be any differentiable function having only positive values. Then (§26.5) we define u^p for any real power p by $\ln u^p = p \ln u$. Differentiating both sides of this defining equation, we find

$$D \ln u^p = \frac{D u^p}{u^p} = p \frac{D u}{u}.$$

Therefore $D u^p = p\, u^{p-1}\, Du$. Thus we do not need a new differentiation formula for general powers, but we extend Rule V as follows:

For every differentiable, positive-valued function u and every real power p,

V_a. $D u^p = p\, u^{p-1}\, Du$, $\int V_a$. $\int u^p\, Du = \dfrac{u^{p+1}}{p+1} + C$,   if $p \neq 1$.

Example A. Differentiate $\sqrt{u}$, $u(x) > 0$.

Solution. $\sqrt{u} = u^{1/2}$. By the generalized power rule, V,

$$D u^{1/2} = \frac{1}{2} u^{1/2 - 1}\, Du = \frac{1}{2} u^{-1/2}\, Du, \quad \text{that is,} \quad D\sqrt{u} = \frac{Du}{2\sqrt{u}}.$$

The importance of the square root function is so great that we record this as a special case of Rule V.

V_b. $D\sqrt{u} = \dfrac{Du}{2\sqrt{u}}$. $u(x) > 0$. $\int V_b$. $\int \sqrt{u}\, Du = \dfrac{2u^{3/2}}{3} + C$.

Example B. Evaluate

$$\int_1^2 \frac{1}{\sqrt{2x - 1}}\, dx.$$

Solution.

$$\int \frac{1}{\sqrt{2x - 1}}\, Dx = \int (2x - 1)^{-1/2}\, Dx.$$

We let $u = 2x - 1$, then $Du = 2 Dx$. So $Dx = \frac{1}{2} Du$. Then our antiderivative transforms into

$$\int (2x - 1)^{-1/2} Dx = \int u^{-1/2} \left(\frac{1}{2} Du \right) = \frac{1}{2} \int u^{-1/2} Du = \frac{1}{2} \frac{u^{-(1/2)+1}}{-\frac{1}{2} + 1} + C$$

$$= u^{1/2} + C = \sqrt{2x - 1} + C.$$

Then, applying the Fundamental theorem, we obtain

$$\int_1^2 \frac{1}{\sqrt{2x - 1}} \, dx = \sqrt{2x - 1} \Big|_1^2 = \sqrt{3} - 1.$$

26.7   *Example A.*   Integrate $\int x^3 \ln x \, Dx$.

Solution.   We use integration by parts (§20.5) with

$$u = \ln x, \qquad Dv = x^3 \, Dx,$$

$$Du = \frac{1}{x} Dx, \qquad v = \frac{x^4}{4}.$$

Then

$$\int x^3 \ln x \, Dx = \frac{x^4}{4} \ln x - \int \frac{x^4}{4} \left(\frac{1}{x} Dx \right)$$

$$= \frac{x^4}{4} \ln x - \frac{1}{4} \int x^3 \, Dx$$

$$= \frac{x^4}{4} \ln x - \frac{1}{12} x^4 + C.$$

Example B.   Integrate $\int (\ln x)^2 \, Dx$.

Solution.   We use integration by parts with

$$u = (\ln x)^2, \qquad\qquad Dv = Dx,$$

$$Du = 2(\ln x) \left(\frac{1}{x} \right) Dx, \qquad v = x.$$

Then

$$\int (\ln x)^2 \, Dx = x(\ln x)^2 - \int x(2 \ln x) \frac{1}{x} Dx$$

$$= x(\ln x)^2 - 2 \int \ln x \, Dx.$$

We can complete this by another integration by parts applied to $\int \ln x \, Dx$ (page 173, Exercise 26).

26.8   Common and Natural Logarithms.   The common $\log_{10} u$ is defined for positive numbers u by

$$\log_{10} u = y \qquad \text{if and only if} \qquad u = 10^y.$$

Taking the natural logarithm of both sides of the last equation, we have $\ln u = y \ln 10$. That is,

$$\log_{10} u = \frac{1}{\ln 10} \ln u \quad \text{and} \quad D \log_{10} = \frac{1}{\ln 10} \frac{Du}{u}.$$

Thus the common logarithm, with base 10, is expressed in terms of the natural logarithm. The conversion factor is

$$\frac{1}{\ln 10} = \log_{10} e = 0.4343\ldots.$$

In differentiation, the more convenient logarithm is the natural logarithm, since its derivative formula does not have the awkward coefficient $1/\ln 10$. But in computation with decimals, the common logarithm is much easier to handle. So we use both logarithms, reserving the base 10 logarithm for computation.

26.9   Exercises

1. Differentiate $\sqrt{3x}$.
2. Find the slope of the tangent line to $y = \sqrt{3x}$ at the point where $x = 2$. Plot this tangent line.
3. (a) Differentiate $\sqrt[3]{4x^2}$.
    (b) Find the numerical derivative at the point where $x = 3$.

In Exercises 4–9, differentiate.

4. $D\sqrt{3x + 5}$.

5. $D\dfrac{1}{\sqrt{x}}$.

6. $D\sqrt{\dfrac{1 - x}{1 + x}}$.

7. $Dx^{\sqrt{2}}$.

8. $D\sqrt[3]{4x - 5}$.

9. $D(2x - 1)^{3/5}$.

In Exercises 10–17, use logarithmic differentiation to calculate the indicated derivatives.

10. De^x.

11. $D10^x$.

12. $D\dfrac{\sqrt{x}}{\sqrt{2x + 1}}$.

13. Db^{x+1}.

14. De^{-x^2}.

15. $De^{\ln x^2}$.

16. $D \log_{10} (2x - 3)$.

17. $D\sqrt[3]{x^2 + 1}$.

In Exercises 18–25, integrate.

18. $\displaystyle\int \frac{Dx}{\sqrt{x}}$.

19. $\displaystyle\int \frac{Dx}{\sqrt{2x}}$.

20. $\displaystyle\int_0^1 \frac{dx}{\sqrt{5x + 4}}$.

21. $\displaystyle\int_1^e \frac{1}{x} dx$.

22. $\displaystyle\int_e^{e^2} \frac{1}{x} dx$.

23. $\displaystyle\int_1^4 \frac{dx}{x^{3/2}}$.

24. $\displaystyle\int_0^1 (p + 1)x^p \, dx$.

25. $\displaystyle\int \frac{\ln x}{x} Dx$.

26. Use integration by parts to show that

$$\int \ln x \, Dx = x(\ln x - 1) + C.$$

In Exercises 27–42, integrate.

27. $\int (\ln x)^2 \, Dx.$

28. $\int x \ln x \, Dx.$

29. $\int \dfrac{\ln x}{x^2} \, Dx.$

30. $\int x^2 \ln x \, Dx.$

31. $\int \left(\dfrac{2}{x} + \ln x \right) Dx.$

32. $\int \left(x^2 - 2x + \dfrac{1}{x} \right) \ln x \, Dx.$

33. $\int \dfrac{Dx}{x \ln x} \cdot$

34. $\int \ln (2x + 1) \, Dx.$

35. $\int \dfrac{(\ln x)^5}{x} \, Dx.$

36. $\int \ln (x^2) \, Dx.$

37. $\int_{1/e}^{e} \ln x \, dx.$

38. $\int_{1}^{e} (\ln x)^2 \, dx.$

39. $\int_{0}^{e-1} \dfrac{x^2 + x + 1}{x + 1} \, dx.$

40. $\int_{1/2}^{e} \ln 2x \, dx.$

41. $\int_{1}^{2} \left(x - \dfrac{1}{x} + \ln x \right) dx.$

42. $\int_{2}^{3} \dfrac{1}{(x - 1)^2} \, dx.$

43. Find the unknown function $f(x)$ such that

$$(x + 1) \, Df(x) = 1 \quad \text{and} \quad f(1) = 2.$$

44. Find the unknown function $f(x)$ such that

$$D^2 f(x) = -\frac{1}{x^2}, \quad f(1) = 0, \quad \text{and} \quad f(e) = 1.$$

26.10 (Theoretical) Proof of One-to-One Property of ln x.   The function **ln** x is defined for all positive numbers x by the indefinite integral

$$\ln x = \int_{1}^{x} \left(\frac{1}{u} \right) du.$$

This implies that the derivative $D \ln x = 1/x$ is positive, so that **ln** x is an increasing function over any interval $\{a \leqq x \leqq b\}$ (§13.3). Moreover **ln** x is continuous, since it is differentiable. Hence by the basic theorem on continuous functions (§8.4), the function **ln** x maps any closed interval $[a, b]$ of positive numbers in one-to-one fashion onto the closed interval $[\ln a, \ln b]$.

We now prove that for any real number y there is a logarithm $\ln 2^n$ and a logarithm $\ln 2^{-m}$ such that $\ln 2^{-m} < y < \ln 2^n$. If y is a (large) positive number, we can construct an understep function $s(x)$ for $1/x$ such that

$$N < \int_{1}^{2^n} s(u) \, du < \int_{1}^{2^n} \frac{1}{u} \, du = \ln 2^n$$

(See Figure 5.9). We partition the interval $[1, 2^{n+1}]$ by partition points

$$2^0 = 1, 2^1, 2^2, \ldots, 2^n, 2^{n+1}, \ldots.$$

In each subinterval, we define $s(x)$ to be the value of $1/x$ at the right-hand end. Thus in the interval $[2^n, 2^{n+1}]$, $s(x) = 1/2^{n+1}$ and for this interval

$$dx = 2^{n+1} - 2^n = 2^n(2 - 1) = 2^n.$$

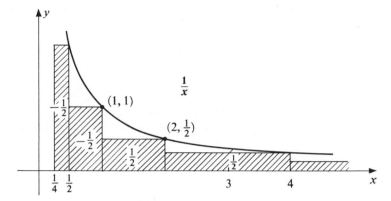

FIGURE 5.9   Lower estimate of the logarithm.

For this subinterval the area of the rectangle,

$$s(x)\, dx = \frac{1}{2^{n+1}}(2^n) = \frac{1}{2}.$$

is less than the area under the curve. Hence

$$\int_1^{2^n} s(u)\, du = \frac{1}{2} + \frac{1}{2} + \frac{1}{2} + \cdots < \int_1^{2^n} \frac{1}{u}\, du = \ln 2^n.$$

Therefore, given a large number y, we can choose n large enough so that $\frac{1}{2} + \frac{1}{2} + \frac{1}{2} + \cdots$ to n terms exceeds y. Then $\ln 2^n$ exceeds y.

Similarly, if y is positive but less than 1 (Figure 5.9), we can underestimate **ln** x by an understep function whose integral is a sum of terms.

$$\left(-\frac{1}{2}\right) + \left(-\frac{1}{2}\right) + \left(-\frac{1}{2}\right) + \cdots$$

Hence we can find a 2^{-m} such that $\ln 2^{-m} < y$. Thus every real number y is bracketed between two logarithms,

$$\ln 2^{-m} < y < \ln 2^n.$$

Hence the basic theorem on continuous functions applied to **ln** x over the interval $[2^{-m}, 2^n]$ assures us that there is a number x in this interval where $\ln x = y$. Finally, since **ln** x is strictly increasing, there is only one such number x. This completes the proof of the theorem (§26.1).

THEOREM.   For every real number y there is one and only one number x such that $y = \ln x$.

27   Numerical Calculations

It is particularly easy to apply the analytic program of calculus to polynomial functions, on account of their simple algebraic construction. But when we begin to include transcendental (nonalgebraic) functions, which are defined only by limit processes, we need some accurate numerical methods that can be applied in a finite number of steps. The same is true as well of functions given only by tables of empirical data. We here investigate some numerical methods for calculating $\int_a^b f(x)\, dx$ and for solving equations, $f(x) = 0$.

27.1 Simpson's Rule. We can calculate the approximate numerical value of $\int_a^b f(x)\,dx$ in terms of three ordinates

$$y_0 = f(a), \qquad y_1 = f\left(\frac{a+b}{2}\right), \qquad y_2 = f(b),$$

at the ends of the interval $[a, b]$ and at the midpoint, $c = (a + b)/2$ (Figure 5.10). Assuming that the function is convex, we have the exact integral as a number between the area A_T of the trapezoid below the chord joining $(a, f(a))$ to $(b, f(b))$ and the area A_M below the tangent at midpoint. Hence we average these two areas. Since the tangent at midpoint is a better approximation to the curve than the chord, it is plausible to give the tangent area twice the weight of the chord area,

$$A = \tfrac{1}{3}(A_T + A_M + A_M).$$

The result is (Exercises)

$$A = \frac{b-a}{6}(y_0 + 4y_1 + y_2).$$

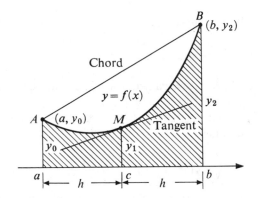

FIGURE 5.10   Simpson's rule.

This gives Simpson's rule,

$$\int_a^b f(x)\,dx \approx \frac{b-a}{6}\left[f(a) + 4f\left(\frac{a+b}{2}\right) + f(b)\right],$$

where $\approx$ means "approximately equal to."

Example. Calculate ln 3 by Simpson's rule and compare the result with the trapezoidal rule (§9.7).

Solution. We may take $a = 1$, $c = 2$, $b = 3$, $b - a = 2$, and calculate

$$f(a) = 1, \quad f\left(\frac{a+b}{2}\right) = \frac{1}{2}, \quad f(b) = \frac{1}{3}.$$

Then, by definition of the logarithm and Simpson's rule,

$$\ln 3 = \int_1^3 \frac{1}{x}\,dx \approx \frac{2}{6}\left[1 + 4\left(\frac{1}{2}\right) + \frac{1}{3}\right] = \frac{10}{9}.$$

The trapezoidal rule, using the same three ordinates, calculates the area below the chords joining the endpoints to the midpoint to be

$$\ln 3 = \int_1^3 \frac{1}{x}\, dx \approx \frac{1}{2}\left(1 + \frac{1}{2}\right)(1) + \frac{1}{2}\left(\frac{1}{2} + \frac{1}{3}\right)(1) = \frac{7}{6}.$$

We may compare the two results with $\ln 3 = 1.09861$, taken from an accurate table of the natural logarithm. The error of Simpson's rule is 0.01250, remarkably accurate for so coarse a partition. The error of the trapezoidal rule, with the same information, is 0.07806, about six times as large.

27.2   Error Estimate in Simpson's Rule. It is not good enough to say that the Simpson's rule formula gives the "approximate value" of the integral. For purposes of calculation we must know that the error does not exceed some definitely calculable bound. We may find this by deriving Simpson's rule mathematically. To do so, we expand $f(x)$ by Taylor's theorem (§14.4) in powers of $x - c$, with the fourth-order remainder $[f^{(4)}(z)/4!](x - c)^4$, where z is a number between $c = (a + b)/2$ and x. Then we integrate the Taylor's polynomial and overestimate the integral of the remainder (§27.6, Proof). The result of this is the following.

THEOREM. (Simpson's rule).   If f has continuous derivatives up to order 4 in the interval $[a, b]$, then the integral of f is given approximately by

$$\int_a^b f(x)\, dx = \frac{b - a}{6}\left[f(a) + 4f\left(\frac{a + b}{2}\right) + f(b)\right].$$

The error,* ϵ, of Simpson's rule is bounded by

$$|\epsilon| \leq \frac{(b - a)^5}{720} \sup |f^{(4)}(x)| \quad \text{in}\quad [a, b].$$

COROLLARY.   Simpson's rule is exact if $f(x)$ is a polynomial of degree 3, or less.

27.3   Improvement of the Accuracy by Partitioning. At the cost of more tedious computation, we can partition $[a, b]$ into n smaller subintervals and apply Simpson's rule to each one. We start with $x_0 = a$ to produce the partition,

$$[x_0, x_2], [x_2, x_4], \ldots, [x_{2n-2}, x_n],$$

where $x_{2n} = b$. Then we evaluate f at the odd-numbered midpoints of each interval, x_{2k-1}, as well as at the even-numbered endpoints, $[x_{2k-2}, x_{2k}]$, setting $y_i = f(x_i)$, $i = 0, 1, 2, \ldots, 2n$. Then adding the results of Simpson's rule for each subinterval, we find

$$\int_a^b f(x)\, dx = \approx \frac{b - a}{6}(y_0 + 4y_1 + 2y_2 + 4y_3 + 2y_4 + \cdots + 2y_{2n-2} + 4y_{2n-1} + y_{2n}).$$

---

* Closer, but more complicated, estimation than that of Taylor's theorem shows that, by best estimate,

$$|\epsilon| \leq \frac{(b - a)^5}{2880} \sup |f^{(4)}(x)| \quad \text{in}\quad [a, b].$$

If d is the length of the maximum subinterval, then $|\epsilon|$, the sum of the errors in the n-subintervals, does not exceed n times the maximum error,

$$|\epsilon| < n \left(\frac{d^5}{720}\right) \sup f^{(4)}(x), \qquad a \leqq x \leqq b.$$

Example.   We are to compute $\ln 5 = \int_1^5 (1/x)\, dx$ by Simpson's rule, using a partition with subintervals of length $\frac{1}{6}$. Find an upper bound to the error $|\epsilon|$.

Solution. $n = 30$. $f^{(4)}(x) = 24x^{-5}$. Hence $\sup f^{(4)}(x) = 24$ on $[1, 5]$. Then for the total error $|\epsilon|$,

$$|\epsilon| < (30) \frac{(\frac{1}{6})^5}{720} (24) = \frac{1}{7776} < 0.00014.$$

27.4   Use of Common Logarithms for Evaluating the Exponential Function.

We recall that the common logarithm of N, written $\log N$, is defined by

$$\log N = z \qquad \text{if and only if} \quad N = 10^z.$$

Calculus provides a suitable foundation for this definition, not possible in elementary algebra. For we have defined 10^z for every real number z (§26.5) and proved (§26.10) that the exponential equation $10^z = N$ has one and only one solution for each positive real number N.

In terms of these definitions, we can then prove that for all positive numbers M and N

$$\log MN = \log M + \log N, \qquad \log \frac{M}{N} = \log M - \log N,$$

$$\log N^p = p \log N, \qquad \log \sqrt[q]{N} = \frac{1}{q} \log N,$$

where p and q are any real numbers.

We may write any real number N uniquely in scientific notation $N = a(10^k)$, where $1 \leqq a < 10$. The integer k is called the *characteristic* of $\log N$ and $\log a$ is called the *mantissa*. It is given in computed tables (Table 2 of the Appendix).

While printed tables of the exponential function are available, it is usually more convenient to use common logarithms for its evaluation, especially for exponentials b^x, where $b \neq e$.

Example A.   Compute $N = 3.72\, K^{-1.32}$, where $K = 21.62$.

Solution.   Using the properties of logarithms, we find $\log N = \log 3.73 - 1.32 \log 21.62$. Then from Table 2 (Appendix),

$$\log N = 0.5705 - 1.32(1.3349) = (10.5705 - 10) - 1.7621 = 8.8084 - 10.$$

Taking the antilog yields $N = 0.06433$.

Example B.   Solve the exponential equation, $2^x = 0.362$.

Solution.   The equation is equivalent to $x \log 2 = \log 0.362$ or $x(0.3010) = 9.5587 - 10$. We want the right-hand side expressed as a simple negative number. Subtracting, we find $x(0.3010) = -0.4413$. Then the solution is

$$x = \frac{-0.4413}{0.3010} = -1.469.$$

Note that we *divide*, as called for by the algebraic operations; the fact that the numbers are logarithms does not mean that we replace division by subtraction.

Example C.   Find the *p*H of a chemical solution if the hydrogen ion concentration $[H^+] = 2(10^{-5})$.

Solution.   By definition, the *p*H is the number x defined by the exponential equation $[H^+] = 10^{-x}$. In this case we have $2(10^{-5}) = 10^{-x}$. Hence, $\log 2 - 5 \log 10 = -x \log 10$. That is, $0.3010 - 5.0000 = -x$. Hence, $x = 4.699 = p$H. We notice that here as in the preceding example it suits our purposes to write the logarithm as a negative number, -4.6990, not $5.6990 - 10$.

27.5   Newton's Method.   We can solve equations $f(x) = 0$, even when f is a transcendental function, by repeatedly applying a numerical approximation due to Newton. We first graph $y = f(x)$ and find that its values change sign between $x = a$ and $x = b$ (Figure 5.11). Then we estimate the unknown true root r by a number x_0 between a and b. At x_0 we calculate $f'(x)$. If $f'(x_0) \neq 0$, then the tangent line crosses the x-axis at the point x_1 given by (Exercises)

$$x_1 = x_0 - \frac{f(x_0)}{f'(x_0)}.$$

The number x_1 is our second estimate of the true root r.

How accurate is x_1? We expand $f(r)$ in powers of $r - x_0$ by Taylor's theorem, with second-order remainder (§14.4). Since $f(r) = 0$, this gives

$$0 = f(x_0) + f'(x_0)(r - x_0) + \frac{f^{(2)}(z)}{2!}(r - x_0)^2,$$

where z is a number between x_0 and r. Since $f'(x_0) \neq 0$, we may divide by $f'(x_0)$. Recalling Newton's formula for x_1, we know that

$$x_1 - r = \frac{1}{2}\frac{f^{(2)}(z)}{f'(x_0)}(r - x_0)^2.$$

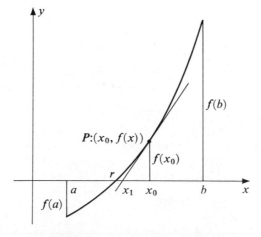

FIGURE 5.11   Newton's method.

Consider the least upper bound, sup $|f^{(2)}(x)|$ on $[r, x_0]$. Then

$$|x_1 - r| \leq \frac{1}{2} \frac{\sup |f^{(2)}(x)|}{|f'(x_0)|} (r - x_0)^2$$

gives an upper bound to the error $\epsilon = |x_1 - r|$. Each of the following conditions tends to produce a small error. The second derivative $f^{(2)}(x)$ is small, so that the curve is nearly straight; $|f'(x_0)|$ is large, so that the tangent line is steep, or the original estimate x_0 is close, so that $|x_0 - r| < 1$.

Example.   Compute by Newton's method $\sqrt[3]{2}$ and estimate the error $|x_1 - r|$.

Solution.   We solve the equation $x^3 - 2 = 0$.   Here $f(x) = x^3 - 2$, $f'(x) = 3x^2$, and $f^{(2)}(x) = 6$. We find that the graph of $y = x^3 - 2$ crosses the x-axis between 1.2 and 1.3.   Estimating $x_0 = 1.25$, we find by Newton's method

$$x_1 = 1.25 - \frac{(1.25)^3 - 2}{3(1.25)^2} = 1.25 - \frac{(1.25)^3 - 2}{3(1.25)^2} = 1.25 + 0.0106 = 1.2606.$$

Since we know that $|x_0 - r| < 0.05$ and that $f^{(2)}(x) < 6(1.3) = 7.8$, we find that the error of the second approximation, x_1, is bounded by

$$|x_1 - r| < \frac{1}{2} \frac{7.8}{3(1.25)^2} (0.05)^2 < 0.0027.$$

27.6 Exercises

In Exercises 1–3, compute $\int_a^b f(x)\, dx$ by the trapezoidal rule and by Simpson's rule, where $c = (a + b)/2$.

1. $f(a) = 2, f(c) = 1, f(b) = 1.5, b - a = 1$.
2. $f(a) = 1.60, f(c) = 1.63, f(b) = 1.62, b - a = 0.1$.
3. $f(a) = 1.30, f(c) = -1.93, f(b) = -2.01, b - a = 0.4$.
4. (a) Using the partition points $a = 1, c = 2, b = 3$, calculate the integral $\int_1^3 x^2\, dx$ by the trapezoidal rule.
    (b) Calculate the integral by Simpson's rule with the same partition points.
    (c) Calculate the integral exactly by the fundamental theorem.
5. A function $f(x)$ is given by a table

x	1.40	1.52	1.64	1.84	2.04
$f(x)$	2.37	2.43	2.52	3.02	3.64

Compute $\int_{1.40}^{2.04} f(x)\, dx$ by Simpson's rule.

In Exercises 6–9, compute to four-place accuracy using common logarithms:

6. $(0.08240)^{-1/3}$.   7. $(356.8)^{-1.1}$.
8. $P = 200(1.04)^{-10}$.   9. $(2.367)^{1.572}$.

10. Evaluate $7.31e^{-6.24t}$, when $t = 0.341$.
11. Evaluate $100e^{0.62t}$, when $t = 45$.
12. Use Newton's method to find the real root of the equation $x^3 - 5 = 0$, accurate to two places after the decimal.

13. Starting with the estimate $x = 1$ for the root of the equation $e^x = x + 2$, apply one step of Newton's method and estimate the error of the resulting second approximation to the root.

In Exercises 14–17, solve the exponential equations.

14. $2^x = 7$.
15. $3(10^x) = 5$.
16. $3^x = 0.4624$.
17. $4(2^{3x}) = 0.7852$.

In Exercises 18–21, the methods of integration that we have considered so far will not give the integral by the fundamental theorem. Compute them by Simpson's rule.

18. $\pi = 4 \int_0^1 \frac{dx}{1 + x^2}$.
19. $\pi = 4 \int_0^1 \sqrt{1 - x^2}\, dx$.
20. $\int_0^1 \exp\left(-\frac{x^2}{2}\right) dx$.
21. $\int_0^1 \sqrt{t}\, e^{-t}\, dt$.

22. Not remembering the arithmetic process for computing square root, we guess the square root of k to be x_1. We divide to find k/x_1. If x_1 was too small, then k/x_1 is too large. So we average them to find a second approximation $x_2 = \frac{1}{2}[x_1 + (k/x_1)]$. Apply Newton's method to the problem and show that this is exactly what Newton's method does.
23. Estimate the error of the approximate square root x_2 in Exercise 22.
24. Using Newton's method as a guide, devise a cube root process analogous to the square root process of Exercise 22. Then devise a process for computing nth roots of positive numbers.
25. Compute $\ln 0.6$ correct to five places using Simpson's rule. Be careful about the sign!

27.7   Proof of Simpson's Rule.   To prove Simpson's rule, we expand f in powers of $x - c$, by Taylor's theorem with fourth-order remainder. Let $x - c = t$ and denote by h the difference,

$$h = \tfrac{1}{2}(b - a) = c - a = b - c.$$

Then, Taylor's theorem says that on the interval $-h \leq t \leq h$,

$$f(x) = f(t + c) = f(c) + \frac{f'(c)}{1!} t + \frac{f^{(2)}(c)}{2!} t^2 + \frac{f^{(3)}(c)}{3!} t^3 + \frac{f^{(4)}(z_1)}{4!} t^4.$$

In the last term, $f^{(4)}$ is evaluated at some unknown point z_1 between c and t. Then, integrating, we find that exactly

$$\int_a^b f(x)\, dx = \int_{-h}^h f(t + c)\, dt = 2hf(c) + 0 + \frac{2f^{(2)}(c)}{2!} \frac{h^3}{3} + 0 + R_4,$$

where the remainder R_4 is given by

$$R_4 = \int_{-h}^h \frac{f^{(4)}(z_1)}{4!} t^4\, dt.$$

In this formula for $\int_a^b f(x)\, dx$, we recognize $y_1 = f(c)$, but we must replace the term containing $f^{(2)}(c)$ by some combination of the ordinates y_0, y_1, y_2. To do this, we use the same Taylor's expansion to compute y_0 by setting $t = -h$, and we compute y_2 by setting $t = h$. Adding the two results and multiplying by $h/3$, we find

$$\frac{h}{3}(y_0 + y_2) = \frac{2}{3}f(c)h + 0 + 2\frac{f^{(2)}(c)}{2!} \frac{h^3}{3} + 0 + S_4.$$

The remainder, S_4, is given by

$$S_4 = \frac{f^{(4)}(z_0) + f^{(4)}(z_2)}{4!} \frac{h^5}{3},$$

where z_0 is some number between a and c and z_2 is a number between c and b. If we solve this for the second derivative term, we find

$$\frac{2 f^{(2)}(c)}{2!} \frac{h^3}{3} = \frac{h}{3} (y_0 - 2y_1 + y_2) - S_4.$$

We substitute this expression for the second derivative term in the expression for $\int_a^b f(x)\, dx$ and, after simplifying, we find that exactly

$$\int_a^b f(x)\, dx = \frac{2h}{6} [y_0 + 4y_1 + y_2] - S_4 + R_4.$$

We now estimate the remainder terms, which all contain a factor involving $f^{(4)}$. We find that

$$|R_4| \leqq \frac{\sup |f^4(x)|}{4!} \int_{-h}^h t^4\, dt = \frac{2h^5}{5!} \sup |f^{(4)}(x)|.$$

Also

$$|S_4| \leqq \frac{2 \sup |f^{(4)}(x)|}{4!} \frac{h^5}{3} = \frac{2h^5}{5!} \left(\frac{5}{3}\right) \sup |f^{(4)}(x)|.$$

Hence the total error $|\epsilon|$ in Simpson's rule is

$$|\epsilon| = |-S_4 + R_4| \leqq |S_4| + |R_4| \leqq \frac{2h^5}{5!} \left(\frac{8}{3}\right) \sup |f^{(4)}(x)| = \frac{(b-a)^5}{720} \sup |f^{(4)}(x)|$$

in $[a, b]$. This completes the proof.

28 Exponential Function

28.1 Definition of the General Exponential Function. If b is any positive number, we can define a positive-valued function b^x on *rational* numbers x by elementary algebraic operations. Thus if x is the natural number 3, b^3 means $(b)(b)(b)$, if $x = \frac{1}{4}$, $b^{1/4}$ means the positive fourth root $\sqrt[4]{b}$, and if $x = p/q$, where p and q are natural numbers $b^{p/q} = \sqrt[q]{b^p}$. Finally, b^{-x} means $1/b^x$, and consistent with this, $b^0 = 1$.

Our purpose here is to extend the definition of the function b^x over all real numbers x. For this purpose we use the properties of **ln** x.

DEFINITION.   The positive-valued function b^x is defined on all real numbers x by saying that $u = b^x$ if and only if $\ln u = x \ln b$.

This is a proper definition of b^x; for, given x and b ($b > 0$), then $x \ln b$ is determined, and by the one-to-one property of the logarithm function (§26.1), the value of u is uniquely determined and is a positive number. Moreover, for every positive b, $b^0 = 1$. The definition agrees with the elementary algebraic definition wherever the algebraic definition applies.

We plot the graphs of several exponential functions with different bases (Figure 5.12).

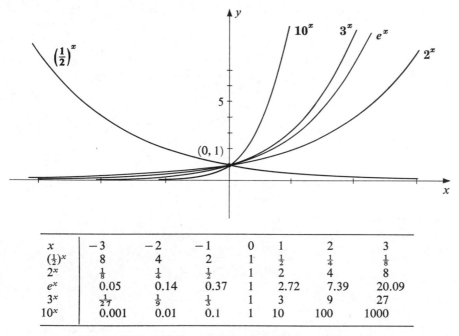

x	-3	-2	-1	0	1	2	3
$(\frac{1}{2})^x$	8	4	2	1	$\frac{1}{2}$	$\frac{1}{4}$	$\frac{1}{8}$
2^x	$\frac{1}{8}$	$\frac{1}{4}$	$\frac{1}{2}$	1	2	4	8
e^x	0.05	0.14	0.37	1	2.72	7.39	20.09
3^x	$\frac{1}{27}$	$\frac{1}{9}$	$\frac{1}{3}$	1	3	9	27
10^x	0.001	0.01	0.1	1	10	100	1000

FIGURE 5.12   Exponential functions with several bases.

Example.   Calculate a table of values for 3^x.

Solution.   We may use common logarithms, since they also have the one-to-one property. To calculate $y = 3^x$, we first find $\log y = x \log 3 = x(0.4771)$. Then for each x we find the antilog of $x \log 3$ (§27.4, Example B). Carrying out these calculations we obtain the following table.

x	-1.5	-1.00	-0.5	0	$+0.5$	$+1.0$	$+1.5$
$\log y$	-0.7157	-0.4771	-0.2386	0.0000	0.2386	0.4771	0.7157
$\log y$	$9.2843 - 10$	$9.5229 - 10$	$9.7614 - 10$	0.0000	0.2386	0.4771	0.7157
y	0.1924	0.3333	0.5773	1.000	1.732	3.000	5.196

From such a table of values of (x, y), we can plot the graph of 3^x (Figure 5.12).

28.2   Differentiation of the General Exponential Function.   We can easily differentiate the functions b^x by means of logarithmic differentiation (§26.4). In fact,

$$Db^x = b^x \, D \ln b^x = b^x \, D(x \ln b) = b^x \ln b.$$

In particular, if $b = e$ so that $\ln b = \ln e = 1$, we have $De^x = e^x$ (§26.1).

This is an interesting property of e. The number e is that base b for which the general exponential function b^x is a self-derivative function. Since the base e simplifies the differentiation of the exponential function, we ask whether all exponential functions cannot be represented with the base e. This is easily seen to be true. We write

$$b^x = e^{kx},$$

where k is an undetermined constant. This is true when $\ln b^x = \ln e^{kx}$, that is, if and only if $k = \ln b$. Thus the composite function e^u, where u is a differentiable function, includes all exponential functions with any base when $u = kx$, with k appropriately chosen. It includes many other functions as well; so we write the derivative and anti-derivative formula, using the chain rule (Rule VI, §20.1) in the form

VIII. $De^u = e^u\, Du$; $\int$ VIII. $e^u\, Du = e^u + C.$

Example A.   Differentiate the following functions.

$$\text{(a)} \quad e^{-x}; \qquad \text{(b)} \quad e^{3x-5}; \qquad \text{(c)} \quad x^2 e^{-ax}.$$

Solution.   (a) $De^{-x} = e^{-x}\, D(-x) = -e^{-x}.$
(b) $De^{3x-5} = e^{3x-5}\, D(3x-5) = 3e^{3x-5}.$
(c)   By the product rule III,

$$\begin{aligned}
Dx^2 e^{-ax} &= x^2\, De^{-ax} + (Dx^2)(e^{-ax}) \\
&= x^2 e^{-ax}(-a) + (2x)(e^{-ax}) \\
&= e^{-ax}(-ax^2 - 2x).
\end{aligned}$$

This completes the problem.

Example B.   Find the slope of the tangent line to the graph of $2e^{(1/3)x}$ at the points where $x = -2$, $x = 0$, $x = 2$. Plot these tangent lines and the graph of the function.

Solution.   With $f(x) = 2e^{(-1/3)x}$, we compute by Rule VIII the derivative $f'(x) = (-2/3)e^{(-1/3)x}$. We find the y-coordinates of the points on the graph by evaluating $f(x)$, and we find the slopes of the tangent lines by evaluating $f'(x)$. For the three points in the example, we tabulate the results of these calculations, which we make either with a table of values of e^x and e^{-x} (Table 3 of the Appendix) or with a table of common logarithms (§27.4, Example B). We plot the three points and their tangent lines (Figure

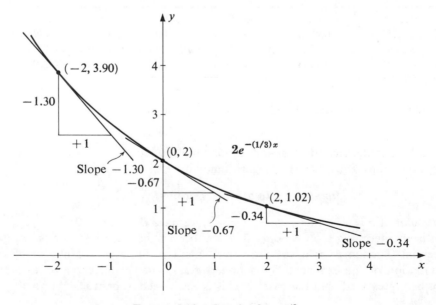

FIGURE 5.13   Graph of $2e^{-x/3}$.

5.13). After we have plotted these points and tangent lines, we sketch the graph of $f(x)$, remembering that $2e^{(-1/3)x}$ is positive for all x.

x	$f(x) = 2e^{-(1/3)x}$	$f'(x) = -\frac{2}{3}e^{(-1/3)x}$
-2	$2e^{2/3} = 3.90$	$-\frac{2}{3}e^{2/3} = -1.30$
0	$2e^0 = 2.00$	$-\frac{2}{3}e^0 = -0.67$
2	$2e^{-2/3} = 1.02$	$-\frac{2}{3}e^{-2/3} = -0.34$

28.3   The Behavior of e^{-kt} as t increases.   If $k > 0$, the value of e^{-kt} approaches zero as t increases. We express this symbolically by saying that if $k > 0$,

$$\lim_{t \to \infty} e^{-kt} = 0.$$

This means that for every ϵ there exists a number T such that if $t > T$, then $|e^{-kt}| < \epsilon$. Let us prove that this is true. Take $\epsilon = 10^{-N}$. We wish to find T such that if $t > T$, then $e^{-kt} < 10^{-N}$. Since $\log x$ is a strictly increasing function, this is true if and only if $\log e^{-kt} < \log 10^{-N}$, that is, if $-kt \log e < -N$. Solving for t, we find that the condition is satisfied if

$$t > \frac{N}{k \log e}.$$

The number on the right is a T, as required by the definition. In case t represents time, we see that, for a fixed ϵ, the larger the factor k is, the more quickly e^{-kt} becomes less than ϵ. We plot e^{-kt} for three values of k to show graphically how k affects the rapidity with which e^{-kt} approaches zero (Figure 5.14).

Example A.   The potential in an electric circuit is 100 volts before a switch is thrown and after that, its transient values are given by $100e^{-5t}$, t in seconds. How long after the switch is thrown does the potential drop below 0.001 volts?

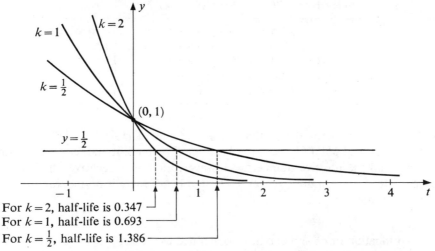

For $k = 2$, half-life is 0.347
For $k = 1$, half-life is 0.693
For $k = \frac{1}{2}$, half-life is 1.386

FIGURE 5.14 e^{-kt} as an exponential decay. The words, "decays," "damps out," "attenuates," are used to describe what e^{-kt} does as t increases. The bigger k is, the faster e^{-kt} decays and the shorter its half-life.

Solution. $100e^{-5t} < 0.001$ if and only if $\log 100e^{-5t} < \log 0.001$; that is, if $-5t \log e + 2 < -3$. Simplifying, we find that this is true if and only if $t > 1/\log e = 2.303$ sec. Hence the voltage drops below the required level in 2.303 seconds.

Example B.   A quantity whose value at time t is represented by Ae^{-kt}, where $k > 0$, is said to *decay exponentially*. In scientific application, it is customary to measure the rapidity of this decay by the "half-life." The half-life is the time t required for Ae^{-kt} to decrease to half its initial value when $t = 0$. Find k for a quantity that decays exponentially and has a half-life of 12 years.

Solution.   The initial value of Ae^{-kt} when $t = 0$ is A. Since the quantity has half-life 12, we have

$$Ae^{-k(12)} = \tfrac{1}{2}A \quad \text{or} \quad e^{-12k} = 0.5.$$

We solve this exponential equation for the unknown k by use of common logarithms and find

$$-12k \log e = -0.3010, \quad \text{whence} \quad k = 0.0578.$$

That is, $Ae^{-0.0578t}$ has half-life 12.

Example C.   Investigate for maximum and minimum points the function $f(x) = x^2 e^{-x}$ on the real numbers.

Solution.   The derivative is defined everywhere by the product rule and Rule VIII.

$$f'(x) = x^2(-e^{-x}) + (2x)e^{-x} = -x(x-2)e^{-x}.$$

We set this derivative equal to zero and find that the only critical values of x are 0 and 2. We now investigate the sign of $f'(x)$ to find where $f(x)$ is increasing and where it is decreasing. Remembering that e^{-x} is always positive, we tabulate the work (§12.2). We construct a table of values and draw the graph (Figure 5.15). Since $f(0) = 0$ and for

Interval	$f'(x) = (-1)(x)(x-2)(e^{-x})$	$f(x)$
$x < 0$	$f'(x) = (-)(-)(-)(+) = -$	Decreasing
$x = 0$	$f'(x) = (-)(0)(-)(+) = 0$	Critical
$0 < x < 2$	$f'(x) = (-)(+)(-)(+) = +$	Increasing
$x = 2$	$f'(x) = (-)(+)(0)(+) = 0$	Critical
$2 < x$	$f'(x) = (-)(+)(+)(+) = -$	Decreasing

every $x, f(x) \geq 0$, the point $(0, 0)$ is the minimum point. The signs of the derivative show that the other critical point $(2, 4e^{-2})$ is a local maximum point, in fact a maximum point for $f(x)$ over all the positive numbers. However, there is a negative number x, where $f(x)$ exceeds $4e^{-2}$, for example, $f(-2) = 30$, and hence there is no maximum point over the entire real number scale.

28.4   Identification of e^x and exp x.   We defined **exp** x as the inverse of **ln** x by saying that $y = \exp x$ if and only if $x = \ln y$ (§26.3). On the other hand, we defined e^x by $y = e^x$ if and only if $\ln y = x \ln e$ (§28.1). But since $\ln e = 1$, this reduces to $\ln y = x$, the same relation as that which defined **exp** x. Hence **exp** $x = e^x$.

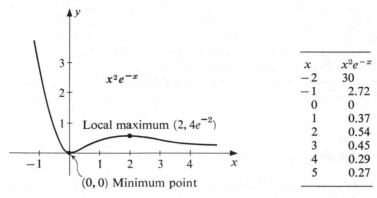

x	$x^2 e^{-x}$
-2	30
-1	2.72
0	0
1	0.37
2	0.54
3	0.45
4	0.29
5	0.27

FIGURE 5.15   Maximum and minimum points of $x^2 e^{-x}$.

28.5   Exercises

1.   Plot the graphs of the power function x^2 and the exponential function 2^x on the same coordinate axes. Observe that they are entirely different functions.

In Exercises 2–11, differentiate the exponential functions

2. e^{-2x}.

3. 3^{-2x+1}.

4. $x e^{-x}$.

5. $x^2 2^x$.

6. $1 - e^{-2t}$.

7. $100 e^{-0.1t}$.

8. $(1 + x^2) e^{3x}$.

9. $\frac{1}{2}(e^x + e^{-x})$.

10. $\dfrac{e^{3x+1}}{x^2 + 1}$,

11. $\sqrt{e^{kx}}$.

In Exercises 12–19, find the indicated antiderivatives.

12. $\displaystyle\int e^{-2x}\, Dx$

13. $\displaystyle\int \frac{e^x}{e^x + 1}\, Dx.$

14. $\displaystyle\int (e^x + e^{-x})^2\, Dx.$

15. $\displaystyle\int 2^x\, Dx.$

16. $\displaystyle\int \frac{e^{2x}}{e^x + 1}\, Dx.$

17. $\displaystyle\int \frac{Dx}{e^x + 1}.$

18. $\displaystyle\int_{-a}^{a} \frac{a}{2} (e^{x/a} + e^{-x/a})\, dx.$

19. $\displaystyle\int_0^1 x e^{2 \ln x}\, dx.$

20.   Use integration by parts (§20.5) to calculate $\int_0^1 x e^x\, dx$.
21.   A particle falls vertically down with velocity v given by $v = -4e^{-2t}$, t in seconds.
       (a) Find the acceleration at time t.
       (b) Describe the motion as t increases.
       (c) If it comes to rest, at what time does it do so?
22.   Find the half-life of each of the following decay functions: (a) $10e^{-0.7t}$; (b) e^{-2t+1}; (c) $(\frac{1}{2})^{t/a}$.
23.   A point moves along the s-axis with distances from 0 given at time t seconds by $s = 10(1 - e^{-t/2})$.
       (a) Find the velocity and acceleration at times $t = 2, 4, 8$ sec.
       (b) Describe the motion.
24.   Find the slope of the tangent to the curve $y = e^{-2x}$ at the point where $x = 1$. Plot.
25.   Find the slope of the tangent to the graph of $y = x - e^{-x}$ at the point where $x = \ln 2$. Plot.

26. Investigate to find where the function $xe^{-x/2}$ is increasing and where it is decreasing.
27. Investigate for maxima and minima xe^{-x} on the real numbers with $x \geq 0$.
28. Investigate for maxima and minima $1 - x - e^{-x}$ on the real numbers.
29. If $f(t) = Ae^{-kt}$, show that the numbers $f(0), f(a), f(2a), f(3a), \ldots, f(na) \cdots$ form a geometric progression, and find the constant ratio.
30. Find the coordinates of the points on the graph of $y = \frac{1}{2}(e^x - e^{-x})$, where the tangent has slope 2.

29 Exponential Growth

29.1 Differential Equation of Growth or Decay.

A frequently occurring problem in science concerns an unknown function $y(t)$ about which it is known that "the rate of change of y is proportional to y," or in symbols,

$$D_t y = ky,$$

where k is a constant. We readily verify that one solution is the function given by $y = e^{kt}$. We should like to find all solutions, which are necessarily continuous functions y since they all have a derivative, $y' = ky$, at every value of t. To answer this question, we let y denote any solution and calculate the derivative of the quotient (Rule V_a),

$$D \frac{y}{e^{kt}} = \frac{e^{kt} Dy - y De^{kt}}{(e^{kt})^2} = \frac{e^{kt}(ky) - y(ke^{kt})}{e^{2kt}} = 0.$$

This is valid, since both numerator and denominator are differentiable, and we know that the denominator e^{kt} is nowhere equal to zero. Hence on any interval,

$$\frac{y}{e^{kt}} = C,$$

where C is a constant (§13.4). Therefore all solutions of the differential equation $Dy = ky$ are of the form $y = Ce^{kt}$. The constant k is called the growth rate.

29.2 The Compound Interest Law.

Example. The principal of $100 is invested at compound interest so as to grow at all times at a rate proportional to the accumulated amount. In one year the amount increases by 4 percent. Find the amount accumulated at any time t, and particularly in 20 years, and find the growth rate.

Solution. Let the function A represent the amount accumulated at time t, $t \geq 0$. Then the description of the growth law in mathematical terms is

$$D_t A = kA,$$

where k is an unknown constant growth rate. Moreover we know that $A(0) = 100$ and $A(1) = 104$. We know (§29.1) that all solutions are of the form $A(t) = Ce^{kt}$. Since $A(0) = 100$, we have $A(0) = 100 = Ce^0 = C$. Hence $A(t) = 100e^{kt}$. Using common logarithms, we find that

$$\log A(t) = \log 100 + (k \log e)t.$$

Since $A(1) = 104$,

$$\log 104 = \log 100 + (k \log e)(1).$$

Hence the constant $k \log e$ is given by

$$k \log e = \log 104 - \log 100 = 0.0170,$$

and at any time t,

$$\log A(t) = 2.0000 + (0.0170)t.$$

This expression contains no unknown constants and gives us a very convenient calcula-tion formula for $A(t)$. When $t = 20$ years, we have

$$\log A(20) = 2.000 + (0.0170)(20) = 2.3400,$$

whence, from the log table,

$$A(20) = \$218.6.$$

We are required to find also the growth rate k, which we can do by returning to the equation above for $k \log e$,

$$k \log e = k(0.4343) = 0.0170,$$

which tells us that the growth rate $k = 0.03922$ per year.

We can also use the value of k to express $A(t)$ in exponential form,

$$A(t) = 100e^{0.03922t},$$

but this is less convenient for calculation than the linear formula for $\log A(t)$ that we used above.

29.3 Alternative Method of Solution of the Growth Equation.

We can solve the differential equation of growth

$$Dy = ky$$

in a different way, which does not require us to guess one solution. The method is called *separation of variables*. We assume that $y \neq \mathbf{0}$ so that we may divide both sides by y and find that the growth equation, with this restriction, is equivalent to

$$\frac{1}{y} Dy = k.$$

Or, since differentiation is with respect to t, $Dt = \mathbf{1}$, and we can write the equivalent equation

$$\frac{1}{y} Dy = k \, Dt.$$

Each member is now a chain in the sense of the chain rule (§20.1) and we can assert that their antiderivatives

$$\int \frac{1}{y} Dy = \int k \, Dt$$

differ at most by a constant, which implies that

$$\ln |y| = kt + c,$$

where c is a constant. That is,

$$|y| = e^{kt+c} = e^c e^{kt} = |Ce^{kt}|,$$

where we have introduced a new constant C, positive or negative, such that $|C| = e^c$. This gives us once more the general solution $y = Ce^{kt}$ for any C, $C \neq 0$. We readily verify that in the excluded exceptional case where $C = 0$ we still have the solution defined by $y = 0$. So $y = Ce^{kt}$ is a solution for every real number C, and these are the only solutions.

29.4 Exercises

1. Certain government bonds may be purchased for $18.75 and are redeemable for $25 ten years later. What is the implied growth rate, and what is the value when $t = 5$?

2. A sugar in solution decomposes at a rate proportional to the amount of sugar still present. If the amount of sugar decreases from 25 lb to 10 lb in 3 hr, find a function that represents the amount of sugar at any time t.

3. Strontium 90 is a radioactive substance (prominent in fallout from hydrogen bomb explosions) that decays at a rate proportional to the amount still present. It has a half-life of 28 years (§28.3). If 10 g of strontium 90 is present initially, find the amount present at any time t years thereafter.

4. Radium disintegrates at a rate proportional to the amount remaining. The half life is 1590 years. Find the function that represents the amount of radium present at any time t out of an original amount A.

5. (Radiocarbon dating.*) Carbon 14 is a rare radioactive form of carbon that has a half-life of about 5600 years. It is believed that the proportion of carbon 14 in the atmosphere has remained constant for many thousands of years. Growing plants take up new carbon 14 in the constant atmospheric proportion. But after the death of the plant, with no new carbon coming in, the proportion of carbon 14 decreases according to the law of radioactive decay. An old coffin was found in a burial mound and the wood of the coffin was found to contain carbon 14 in 0.72 times the atmospheric proportion (Figure 5.16). How many years ago was the tree cut to make the coffin?

6. Under the action of a frictional brake the velocity of a moving body decreases at a rate proportional to the velocity. If the velocity decreases from 50 ft/sec to 45 ft/sec in 2 sec, find a function that gives the velocity at any time t after the brake is applied.

7. The intensity of light transmitted by a plate of glass decreases at a rate proportional to the thickness of the glass. Find a function expressing the percent of original intensity transmitted by a glass of any thickness x if the intensity decreases 50 percent in going through 8 in. of the glass. What percentage of the incident light is transmitted through 12 in. of the glass?

8. The rate at which water is flowing through an opening in the bottom of a vertical cylindrical tank is proportional to the depth of the water remaining. If the depth decreases from 10 ft to 5 ft in 10 minutes, what is the function that represents the depth at any time t minutes? When, if ever, is the tank completely emptied?

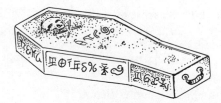

FIGURE 5.16   Exercise 5.

---

* W. F. Libby received the Nobel Prize for this idea of radiocarbon dating. *See* "Radiocarbon dating," *Encyclopedia Brittanica*.

9. An electrical condenser discharges through a resistance, losing voltage at a rate proportional to the voltage remaining. If the initial voltage of 100 volts decreases to 50 volts in 3 seconds, what is the function representing the voltage at the condenser at any time t?

10. A porous material dries in open air at a rate proportional to the moisture content. If a wet towel hung in the wind loses half its free moisture in the first 40 min, when will it have lost 95 percent?

11.* A large, light body falls under the action of gravity in air. Find a function that represents the velocity at any time t if there is a retarding force due to air resistance proportional to the velocity, and if $v = 0$ when $t = 0$, $v = 50$ when $t = 2.16$, and $v = 75$ when $t = 4.32$. Transform this differential equation into one of the same form as the preceding ones.

12. A 100-gallon tank of brine contains 50 lb of dissolved salt. Fresh water is run into the tank at 3 gal per minute, the brine overflowing at the same rate. The solution is kept uniform by stirring. How much salt remains in the tank after one hour?

13. (Cooling of a hot body). See Figure 5.17. The rate of cooling of a hot body is proportional to the difference between its temperature and that of the surrounding medium. It takes a can of hot liquid 20 min to cool from 90° to 80° in air that is at 60°. How long will it take to cool to 65°?

Cool air at 60°

Can of hot liquid at 90° initially

FIGURE 5.17   Exercise 13.

14. (Growth of a bacterial colony.) With plenty of available food and plenty of room, a colony of bacteria grows at a rate proportional to the number of bacteria present at any time. There are 10^5 bacteria present at first and the number doubles every 40 min. Find the growth rate and the number of bacteria in the colony after t min. What length of time is required to have 10^7 bacteria? *Note*: The number of bacteria is a whole number but we find it convenient to measure it by a continuous function $P(t)$.

15. Show that a radioactive decay solution with given half-life can be written by inspection. For example, for strontium 90 (Exercise 3) the amount remaining $A(t)$ at time t out of an original A_0 is given by

$$A(t) = A_0(\tfrac{1}{2})^{t/28}.$$

The exponent $t/28$ represents the time in half-life units.

16. (Parachute jumper.) A parachutist of mass m opens his parachute when he is falling at the rate of $-V$ ft/sec. His parachute exerts a force in the opposite direction to his velocity, which is proportional velocity with coefficient $-k$. Using Newton's law of motion (§18.3), find the equation of motion in terms of the velocity function $v(t)$. Solve for $v(t)$. Show that the rate of drop approaches $-mg/k$ ft/sec, independent of the initial drop-rate $-V$.

17. The principal of $A invested at ordinary annually converted compound interest at the rate i per year becomes in t years $A(1 + i)^t$. Prove this. Then find the annual growth rate k so that $(1 + i)^t = e^{kt}$. This proves that ordinary, annually converted, compound interest is exactly equivalent to "continuously converted" exponential growth at an annual growth rate slightly less than i.

18. In Exercise 17, compute k when $i = 0.04$. Compare Section 29.2.

19. In Exercise 5, if the percentage of carbon 14 can be measured only to an error of 2×10^{-3}, how far back can carbon-14 dating discriminate between dates 1000 years apart?

20. If money is invested at compound interest with annual growth rate 4 percent, what length of time will be required to double the original principal?

21. If money is invested at compound interest with annual growth rate 4 percent, what ordinary compound interest rate, compounded annually, will give the same amount at every time?

22. At what growth rate will money invested at compound interest double in exactly 20 years? At what annually compounded rate?

23. What principal invested today will accumulate to $1000 in 25 years at an annual growth rate of $4\frac{1}{2}$ percent?

29.5 (A theoretical section.)   Calculation of e. We can use Taylor's theorem (§14.4) to calculate e to as many decimal places as we wish. We recall Taylor's theorem applied to a function $f(x)$, whose value, and the values of its derivatives of every order, are known at 0. Taylor's theorem says in this case that there exists a number z between 0 and x such that exactly

$$f(x) = f(0) + \frac{f'(0)}{1!} x + \frac{f^{(2)}(0)}{2!} x^2 + \cdots + \frac{f^{(n-1)}(0)}{(n-1)!} + \frac{f^{(n)}(z)}{n!} x^n.$$

If $f(x) = e^x$, we get at once

$$f(x) = f'(x) = f^{(2)}(x) = \cdots = f^{(n-1)}(x) = f^{(n)}(x) = e^x,$$

so

$$f(0) = f'(0) = f^{(2)}(0) = \cdots = f^{(n-1)}(0) = e^0 = 1.$$

Hence Taylor's theorem tells us that

$$e^x = 1 + \frac{x}{1!} + \frac{x^2}{2!} + \frac{x^3}{3!} + \cdots + \frac{x^{n-1}}{(n-1)!} + e^z \frac{x^n}{n!},$$

where z is some number between 0 and x. Applying this when $x = 1$, with $n = 12$, we get

$$e = 1 + \frac{1}{1!} + \frac{1}{2!} + \frac{1}{3!} + \cdots + \frac{1}{11!} + \frac{e^z}{12!},$$

where z is some number between 0 and 1. We know that e^x is increasing on the interval $[0, 1]$. This tells us that since $0 < z < 1$, $e^z < e^1$. Moreover, we can estimate that

$e < 4$ (§26.10), and so obtain the remainder term $e^z/12! < 4/12!$. Recalling the definition $n! = 1 \cdot 2 \cdot 3 \cdot 4 \cdots n$, we compute

$$1 = 1.00000000 \qquad \frac{1}{6!} = 0.00138889$$

$$\frac{1}{1!} = 1.00000000 \qquad \frac{1}{7!} = 0.00019841$$

$$\frac{1}{2!} = 0.50000000 \qquad \frac{1}{8!} = 0.00002480$$

$$\frac{1}{3!} = 0.16666667 \qquad \frac{1}{9!} = 0.00000276$$

$$\frac{1}{4!} = 0.04166667 \qquad \frac{1}{10!} = 0.00000028$$

$$\frac{1}{5!} = 0.00833333 \qquad \frac{1}{11!} = 0.00000003$$

$$\text{Sum} = 2.71828184.$$

This sum represents e with an error less than $4/12!$, which is less than 0.000000001. However, there were some rounding-off errors in the eighth place after the decimal in our calculations above. Hence the eighth digit after the decimal is not reliable, and we cut back to 7 places to state that

$$e = 2.7182818\ldots,$$

correct to 7 places after the decimal.

29.6 Problems

T1. How many terms of Taylor's expansion would be required to calculate e correct to 10 places?

T2. Find the Taylor's expansion for e^x in powers of $x - 1$.

T3. Knowing the value of e to 7 places, compute the value of $e^{1.2}$ to at least 6 places.

30 Improper Integrals

30.1 Infinity Elements at the Ends of the Real Number Line. Of the several different ideas of infinity in mathematics, we can use one of the simpler ones in elementary calculus. We introduce two artificial numberlike elements, which we denote by $-\infty$ and $+\infty$, read "minus infinity" and "plus infinity," and place them at the ends of the real number line, so that for every real number x we have $-\infty < x < +\infty$. This works well enough when only the relation $<$ is involved, but when we try to add or multiply an infinity element with real numbers it turns out that the algebraic properties of real numbers must be contradicted. Hence we cannot consistently include $-\infty$ and $+\infty$ in the real numbers as an algebraic system with relation $<$ and operations $+$ and $\times$. In particular, it is meaningless to write $|x - \infty|$.

On the other hand, if we use only the relation $<$, we can define some new limits involving $-\infty$ and $+\infty$. We say that x is in a neighborhood $\mathcal{N}$ of $+\infty$ if for some (large) number N, $N < x < +\infty$. Similarly, x is in a neighborhood of $-\infty$ if for some (large) N, $-\infty < x < -N$.

We may generalize the definition of limit. Let c be a number, or $+\infty$, or $-\infty$, which is approachable in the domain of the function f.

DEFINITION. If h is a real number, then

$$\lim_{x \to c} f(x) = h$$

if and only if for every ϵ there is a neighborhood $\mathcal{N}$ of c such that if x is in $\mathcal{N}$, then $|f(x) - h| < \epsilon$.

This definition includes the ordinary definition as well as $\lim_{x \to +\infty} f(x)$ and $\lim_{x \to -\infty} f(x)$ but does not let us say that $\lim_{x \to c} f(x) = +\infty$. Our definition still says that only numbers can be limits. If there is no number h that is the limit of $f(x)$, then the limit does not exist.

The function $f(x) = e^{-kx}$, where $k > 0$, is an example of a function for which $\lim_{x \to +\infty} f(x) = 0$ (§28.3). We also illustrate this definition with the graph of the function $f(x) = (3x + 1)/(x - 1)$ on $\{-\infty < x < 1\}$. For this function $\lim_{x \to -\infty} f(x) = 3$ (Figure 5.18).

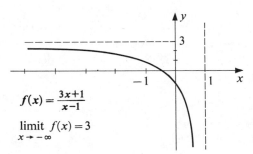

FIGURE 5.18   Limit of a function as x approaches minus infinity.

30.2 Limit Statements Involving Infinities. Again let c be an element (a number, or $+\infty$, or $-\infty$) that is approachable in the domain of the function f.

DEFINITION. The function f *becomes infinite* at c if and only if for every positive number M (however large) there is a neighborhood $\mathcal{N}$ of c such that if x is in $\mathcal{N}$, then $|f(x)| > M$.

For example, the function $(3x + 1)/(x - 1)$ becomes infinite at 1 (Figure 5.18).

THEOREM. (a) If f becomes infinite at c, then $\lim_{x \to c} f(x)$ does not exist and

$$\lim_{x \to c} 1/f(x) = 0.$$

(b) If $\lim_{x \to c} f(x) = 0$, then $1/f(x)$ becomes infinite at c.

Proof. (a) $\lim_{x \to c} f(x)$ does not exist; for if f had a limit h, then in some neighborhood $\mathcal{N}$ of c, $h - \epsilon < f(x) < h + \epsilon$. This contradicts the hypothesis that f becomes infinite at c. Also, by definition, if f becomes infinite at c, then for every ϵ there is a neighborhood $\mathcal{N}$ of c where $|f(x) - 0| > 1/\epsilon$. Taking reciprocals, we find that in this neighborhood

$$\left| \frac{1}{f(x)} - 0 \right| < \epsilon.$$

This proves that

$$\lim_{x \to c} \frac{1}{f(x)} = 0.$$

(b) If $\lim_{x \to c} f(x) = 0$, then for every M take ϵ so that $1/\epsilon > M$. By hypothesis there exists a neighborhood $\mathcal{N}$ of c in which $|f(x) - 0| < \epsilon$. In this same neighborhood,

$$\left| \frac{1}{f(x)} \right| > \frac{1}{\epsilon} > M,$$

so $1/f(x)$ becomes infinite at c.

The following statements, with indications of proof, describe the behavior of certain special functions involving infinities. In all of them, $k > 0$.

1. $\ln |x|$ becomes infinite at $+\infty$. (Proof: §26.10)
2. $\lim_{x \to +\infty} e^{-kx} = 0$. (Proof: §28.3)
3. e^{kx} becomes infinite at $+\infty$. (Proof: §30.2, Theorem (b), and No. 2 above)
4. x^k becomes infinite at $+\infty$.
   (Proof: $y = x^k$ if and only if $\ln y = k \ln x$. Since $\ln x$ becomes infinite at $+\infty$, $\ln y$ and hence y do also.)
5. $\lim_{x \to c} x^{-k} = 0$. (Proof: No. 4 above and §30.2, Theorem (a))
6. x^{-k} becomes infinite at 0, where all values of x are positive. (Proof: $y = x^{-k}$ means $\ln y = -k \ln x$. Near 0 the values of $\ln x$ are large negative numbers (§26.8). Hence $\ln y$ becomes infinite and hence y does also.)
7. $\lim_{x \to 0} x^k = 0$, where all values of x are positive. (Proof: No. 6 above and §30.2 Theorem (a))

30.3   Improper Integrals. Integration to Infinity. The definition (§9.4) of the integral $\int_a^b f(x)\, dx$ applies when the interval $[a, b]$ is bounded and the function itself is bounded on $[a, b]$. In certain cases, we can carry out integration over an unbounded interval, or integrate an unbounded function. Such integrals are called *improper integrals*.

We consider first a specific problem.

Example A.   Find the amount of work that must be done against the pull of the earth's gravity to send a 175-pound man into outer space completely beyond the earth's gravitation (§24.5, Exercises 35–38). We are computing the work to lift the astronaut alone, excluding his vehicle and equipment.

Solution.   We shall measure distances from the center of the earth in a unit of 1000 nautical miles. The force of the earth's gravity is $F = -k/x^2$, and when $x = 4$ we are given that $F = -175$. Hence $k = 2800$, and the force of gravity on the astronaut's body at any height x is given by $F = -2800/x^2$ lb. The work $W(s)$ to raise the astronaut from the height 4 to a height s is given by the definite integral

$$W(s) = 2800 \int_4^s \frac{1}{x^2}\, dx = -\left. \frac{2800}{x} \right|_4^s = 700 - \frac{2800}{s}.$$

(See Figure 5.19.)

By our definition of the limit,

$$\lim_{s \to +\infty} -\frac{2800}{s} = 0.$$

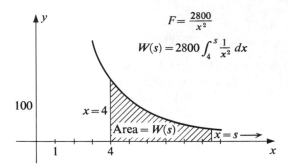

FIGURE 5.19   Work against gravity.

Hence $\lim_{s \to +\infty} W(s) = 700$. Thus the amount of work to shoot the astronaut "to infinity" is not infinite, but is 700 kilomile-pounds. This completes the problem.

It is natural to use the symbol $2800 \int_4^\infty (1/x^2)\, dx$ to represent this amount of work. We now generalize this idea to define $\int_a^\infty f(x)\, dx$.

DEFINITION.   The symbol $\int_a^\infty f(x)\, dx$ is said to represent a *convergent improper integral* if and only if $\lim_{s \to +\infty} \int_a^s f(x)\, dx$ exists. In that case, its numerical value is defined by

$$\int_a^\infty f(x)\, dx = \lim_{s \to +\infty} \int_a^s f(x)\, dx.$$

Example B.   Determine whether $\int_1^\infty (1/\sqrt{x})\, dx$ is convergent, and if so, evaluate it.

Solution.   By definition,

$$\int_1^\infty \frac{1}{\sqrt{x}}\, dx = \lim_{s \to +\infty} \int_a^s \frac{1}{\sqrt{x}}\, dx$$

if and only if that limit exists. We integrate and find that

$$\int_1^s x^{-1/2}\, dx = \frac{1}{2} x^{1/2} \Big|_1^s = \frac{1}{2}(\sqrt{s} - 1).$$

Hence

$$\lim_{s \to +\infty} \int_1^s x^{-1/2}\, dx = \lim_{s \to +\infty} \tfrac{1}{2}(\sqrt{s} - 1).$$

This limit does not exist (§30.2, No. 4). Hence $\int_1^{+\infty} (1/\sqrt{x})\, dx$ is not convergent. This completes the example.

Examples A and B show that, even when $\lim_{x \to +\infty} f(x) = 0$, $\int_a^{+\infty} f(x)\, dx$ converges in some cases and not in others. We must treat each problem individually.

DEFINITION.   The symbol $\int_{-\infty}^b f(x)\, dx$ represents a *convergent improper integral* if and only if $\lim_{s \to -\infty} \int_s^b f(x)\, dx$ exists. In that case, its numerical value is defined by

$$\int_{-\infty}^b f(x)\, dx = \lim_{s \to -\infty} \int_s^b f(x)\, dx.$$

Finally, we say that the improper integral $\int_{-\infty}^{+\infty} f(x)\,dx$ is convergent if and only if both $\int_{-\infty}^{b} f(x)\,dx$ and $\int_{b}^{+\infty} f(x)\,dx$ are convergent for every number b; and we define its value by

$$\int_{-\infty}^{+\infty} f(x)\,dx = \int_{-\infty}^{b} f(x)\,dx + \int_{b}^{+\infty} f(x)\,dx.$$

30.4 Integration of Functions that Become Infinite. We now turn to the problem of defining integrals $\int_{a}^{b} f(x)\,dx$ for functions $f(x)$ that become infinite at either a or b. Since such a function is unbounded in the interval $[a, b]$, all of the methods we have previously used to establish the existence of an exact integral $\int_{a}^{b} f(x)\,dx$ fail (§§9.4, 9.5) because there can be no overstep function $S(x)$ and understep function $s(x)$ that enclose the graph of $f(x)$ between them. Hence we resort to a device similar to that we used to define the improper integral $\int_{a}^{\infty} f(x)\,dx$.

Example A. Give a meaning to $\int_{0}^{1} (1/\sqrt{x})\,dx$, and if this gives a convergent process, compute a value for it (Figure 5.20).

Solution. The function $1/\sqrt{x}$ is defined in the *domain* $[0 < x \leq 1]$. It becomes infinite at 0, which is not in the domain of $1/\sqrt{x}$ but is approachable in this domain. Following Cauchy (French mathematician, 1789–1857), we pick any number s such that $0 < s < 1$ and compute the ordinary integral (area of shaded region, Figure 5.20(a))

$$\int_{s}^{1} \frac{1}{\sqrt{x}}\,dx = 2\sqrt{x}\,\Big|_{s}^{1} = 2 - 2\sqrt{s}.$$

Then we calculate the limit

$$\lim_{s \to 0+} \int_{s}^{1} \frac{1}{\sqrt{x}}\,dx = \lim_{s \to 0+} (2 - 2\sqrt{s}) = 2,$$

where $s \to 0^{+}$ means that we let s approach 0 from the positive side so that s will always be in the domain $[0 < x \leq 1]$. Using this result, we say that the integral of the unbounded function is convergent and that its value

$$\int_{0}^{1} \frac{1}{\sqrt{x}}\,dx = \lim_{s \to 0+} \int_{s}^{1} \frac{1}{\sqrt{x}}\,dx = 2.$$

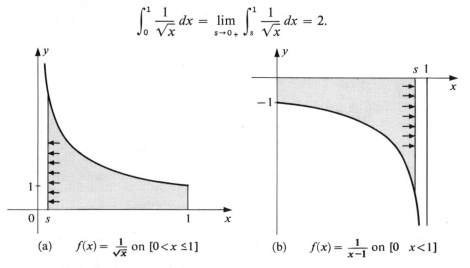

(a) $f(x) = \dfrac{1}{\sqrt{x}}$ on $[0 < x \leq 1]$ (b) $f(x) = \dfrac{1}{x-1}$ on $[0 \ \ x < 1]$

Figure 5.20   Functions which become infinite.

We generalize from this example to define improper integrals of unbounded functions $f(x)$ in the two cases, stating the definition so that it contains instructions for computing.

DEFINITION (Case I). If $f(x)$ is bounded and integrable on every interval $[s, b]$, where $a < s < b$, and $f(x)$ becomes infinite at a, then the improper integral $\int_a^b f(x)\, dx$ is *convergent* if and only if $\lim_{s \to a^+} \int_a^b f(x)\, dx$ exists, and in this case

$$\int_a^b f(x)\, dx = \lim_{s \to a^+} \int_s^b f(x)\, dx.$$

DEFINITION (Case II). If $f(x)$ is bounded and integrable in every interval $[a, s]$, where $a < s < b$, and $f(x)$ becomes infinite at b, then the improper integral $\int_a^b f(x)\, dx$ is *convergent* if and only if $\lim_{s \to b^-} \int_a^s f(x)\, dx$ exists, and in that case

$$\int_a^b f(x)\, dx = \lim_{s \to b^-} \int_a^s f(x)\, dx.$$

Example B. Determine whether the improper integral $\int_0^1 [1/(x - 1)]\, dx$ converges, and if it does, evaluate it.

Solution. We plot the graph of $1/(x - 1)$ on $[0, 1]$ (Figure 5.20(b)). The function becomes infinite at 1. Elsewhere it is integrable in the ordinary sense. For any s such that $0 < s < 1$, we compute the ordinary integral

$$\int_0^s \frac{1}{x - 1}\, dx = \ln |x - 1| \Big|_0^s = \ln |s - 1| - \ln |-1| = \ln |s - 1|.$$

But $\lim_{s \to 1^-} \ln |s - 1|$ does not exist (§30.2). The improper integral does not converge. This completes the problem.

30.5 Exercises

1. $\displaystyle \int_1^\infty \frac{1}{x^3}\, dx.$

2. $\displaystyle \int_1^\infty \frac{1}{x^4}\, dx.$

3. $\displaystyle \int_2^\infty \frac{1}{(2x - 1)^2}\, dx.$

4. $\displaystyle \int_{-\infty}^0 \frac{1}{(2x - 1)^2}\, dx.$

5. $\displaystyle \int_0^\infty e^{-x}\, dx.$

6. $\displaystyle \int_0^\infty \exp \frac{-x}{\sqrt{2}}\, dx.$

7. $\displaystyle \int_{-\infty}^\infty \exp \frac{-|x|}{\sqrt{2}}\, dx.$

8. $\displaystyle \int_{-\infty}^0 e^{2x}\, dx.$

9. $\displaystyle \int_1^\infty x^{-1/3}\, dx.$

10. $\displaystyle \int_{-\infty}^{-1} \frac{1}{\sqrt{1 - x}}\, dx.$

11. $\displaystyle \int_1^\infty x^{-p}\, dx$, where $p > 1$.

12. $\displaystyle \int_1^\infty x^{-p}\, dx$, where $0 \leq p \leq 1$.

13. $\displaystyle \int_0^{1/2} \frac{1}{\sqrt{2x - 1}}\, dx.$

14. $\displaystyle \int_0^8 x^{-2/3}\, dx.$

15. $\displaystyle \int_0^{1/2} \frac{1}{2x - 1}\, dx.$

16. $\displaystyle \int_0^1 x^{-p}\, dx,\ 0 \leq p \leq 1.$

17. $\displaystyle \int_0^1 x^{-p}\, dx,\ p > 1.$

18. $\displaystyle \int_1^2 \left(\frac{a}{x - 1} + \frac{b}{2 - x} \right) dx.$

19. An astronaut weighs 200 lb at the earth's surface. Find the amount of work necessary to lift him free of the earth's gravity from a height of R thousand miles.
20. Find the energy that must be given a particle of mass m at the earth's surface to let it escape from the earth's gravitation.
21. If the kinetic energy of the particle in Exercise 20 is $\frac{1}{2}mv^2$, what must the velocity be for it to escape from the earth's gravity? This is called the *escape velocity*.
22. According to Coulomb's law, two like electric charges q_1 and q_2 repel each other with an inverse-square-law force $F = k(q_1 q_2 / r^2)$, where r is the distance between them and k is a constant. The units are so arranged that $F = 1$ when $q_1 = q_2 = 1$ and $r = 1$. Compute the work W done to bring the charges from a distance very far apart to a distance x apart. Why is it true that the derivative $D_x W = -F$?
23.* A particle slides on a horizontal plane losing speed because of friction. It starts with a velocity V ft/sec and its velocity decreases at a rate proportional to v. The particle slows down to $\frac{1}{2}V$ in 10 sec. How far will it go if it is allowed to slide forever?

31   Trigonometric Functions

31.1   Geometry of the Circle.   We introduce the trigonometric functions **sin *u*, cos *u*, tan *u*** by the conventional geometric definition, making use of the distance formula

$$|P_1 P_2| = \sqrt{(x_2 - x_1)^2 + (y_2 - y_1)^2}.$$

The graph of the relation $x^2 + y^2 = r^2$ is a circle of radius r (Figure 5.21). We take a variable point $P:(x, y)$ in this circle and draw the radius $\overrightarrow{OP}$. We think of $\overrightarrow{OP}$ as rotating about O, sweeping out behind it an area A (shaded region) that is proportional to the central angle θ, starting with P at the position $Q:(1, 0)$. This area is considered positive if OP rotates counterclockwise, negative if $\overrightarrow{OP}$ rotates clockwise from $\overrightarrow{OQ}$. We take the radian measure of angle, which assigns $\theta = 2\pi$ for a complete counterclockwise rotation sweeping out the area $A = \pi r^2$, which is the area of the complete circular disk. With radian measure of angle, we then have the proportion

$$\frac{A}{\pi r^2} = \frac{\theta}{2\pi}.$$

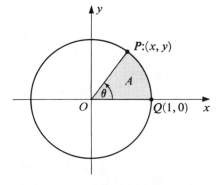

FIGURE 5.21   Circle and central angle.

This implies that the area of a circular sector with central angle θ in radians is given by

$$A = \tfrac{1}{2}r^2\theta.$$

If we introduce the arc length s of the arc QP instead of the area of the sector OQP, then s is also proportional to the central angle, θ. For radian measure of angle this proportion is expressed by

$$s = r\theta.$$

31.2   Geometric Definition of the Trigonometric Functions.   Given a number u, plausibly we can rotate $\overrightarrow{OP}$ (Figure 5.21) until the radian measure of the angle through which it rotates is exactly u. Then the coordinates (x, y) of P are determined by u. That is, x and y are functions of u. We shall assume tentatively that this is true, although it is not obvious how we can compute x and y in every case, for example, $u = 0.1350$ radians. We shall have justified this assumption mathematically when we show ultimately how to compute x and y for every number u.

DEFINITION.   Let (x, y) be the coordinate of the point P on the circle $x^2 + y^2 = r^2$, determined by rotating the radius $\overrightarrow{OP}$ through an angle of radian measure u from the initial position $\overrightarrow{OQ}$, where Q has coordinates $(1, 0)$. Then the trigonometric functions **sin u, cos u, tan u** are defined for each real number u by

$$\sin u = \frac{y}{r}, \qquad \cos u = \frac{x}{r}, \qquad \tan u = \frac{y}{x}, \qquad \text{when} \quad x \neq 0.$$

It follows from these definitions that for every u

$$\sin^2 u + \cos^2 u = 1, \qquad \text{and} \qquad \tan u = \frac{\sin u}{\cos u}, \qquad \text{provided that} \quad \cos u \neq 0.$$

Moreover the values, for clockwise (negative) rotations, are given by

$$\sin(-u) = -\sin u, \qquad \cos(-u) = \cos u, \qquad \text{and} \qquad \tan(-u) = -\tan u,$$

provided that $\cos u \neq 0$.

31.3   Addition Theorems for Cosine.   We must find the sine and cosine of the sum of two angles $u + v$. For **sin u** and **cos u** are not linear functions; that is, $\sin (u + v)$ is not $\sin u + \sin v$. On the unit circle, we rotate $\overrightarrow{OP}$ (Figure 5.22) through u radians, then continuing, for v more radians, terminating with the ray $\overrightarrow{OQ}$ (Figure 5.22(a)). Since $r = 1$, by definition of sine and cosine the coordinates of points S, P, Q are $S:(1, 0)$; $P:(\cos u, \cos v)$; $Q:(\cos (u + v), \sin (u + v))$. By the distance formula the distance $|SQ|$ is given by

$$|SQ|^2 = [\cos (u + v) - 1]^2 + [\sin (u + v) - 0]^2 = -2 \cos (u + v) + 2,$$

where to get the last form we use the fact that Q is on the circle.

Now we rotate the axes clockwise through $+u$ radians, giving P the coordinates $(1, 0)$ in the new coordinates (Figure 5.22(b)). The new coordinates of Q are now

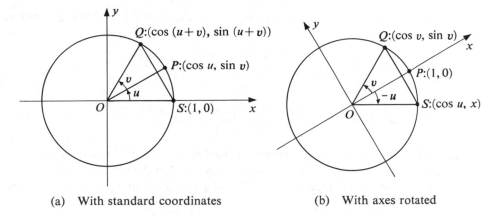

(a)   With standard coordinates   (b)   With axes rotated

Figure 5.22   Addition theorem for sine and cosine.

$(\cos v, \sin v)$ and the coordinates of S are $(\cos u, -\sin u)$. The distance $|SQ|$ is invariant under this change of coordinates, and in the new coordinates,

$$|SQ|^2 = (\cos v - \cos u)^2 + (\sin v + \sin u)^2$$
$$= -2(\cos u \cos v - \sin u \sin v) + 2.$$

Equating the two values of $|SQ|^2$, we find that

$$\cos(u + v) = \cos u \cos v - \sin u \sin v.$$

This is the addition theorem for the cosine.

31.4   Addition Theorems for Trigonometric Functions.   The following is a list of the addition theorems for the trigonometric functions. All are provable from the formula for $\cos(u + v)$ taken with the negative angle relations (§31.2). (Proofs. Exercises.)

(1) $\cos(u + v) = \cos u \cos v - \sin u \sin v.$

(2) $\cos(u - v) = \cos u \cos v + \sin u \sin v.$

(3) $\sin(u + v) = \sin u \cos v + \cos u \sin v.$

(4) $\sin(u - v) = \sin u \cos v - \cos u \sin v.$

(5) $\tan(u + v) = \dfrac{\tan u + \tan v}{1 - \tan u \tan v}.$

(6) $\tan(u - v) = \dfrac{\tan u - \tan v}{1 + \tan u \tan v}.$

(7) $\cos 2\theta = \cos^2 \theta - \sin^2 \theta = 2\cos^2 \theta - 1 = 1 - 2\sin^2 \theta.$

(8) $\sin 2\theta = 2 \sin \theta \cos \theta.$

(9) $\cos^2 \dfrac{\theta}{2} = \dfrac{1 + \cos \theta}{2}.$

(10) $\sin^2 \dfrac{\theta}{2} = \dfrac{1 - \cos \theta}{2}.$

(11) $\tan \dfrac{\theta}{2} = \dfrac{\sin \theta}{1 + \cos \theta} = \dfrac{1 - \cos \theta}{\sin \theta}.$

31.5   Differentiation of sin u and cos u at 0.   By the definition of the derivatives **sin′ u** and **cos′ u**,

$$\sin' 0 = \lim_{\theta \to 0} \frac{\sin (0 + \theta) - \sin 0}{\theta - 0} = \lim_{\theta \to 0} \frac{\sin \theta}{\theta}$$

and

$$\cos' 0 = \lim_{\theta \to 0} \frac{\cos (0 + \theta) - \cos 0}{\theta - 0} = \lim_{\theta \to 0} \frac{\cos \theta - 1}{\theta}.$$

We calculate these two limits by a process of squeezing the quantities between upper and lower estimates whose limits we can evaluate simply. On the unit circle, the coordinates of P are $(\cos \theta, \sin \theta)$ (Figure 5.23). If θ is a positive acute angle, we have the inequalities:

Area of triangle QOP < area of sector QOP < area of triangle OQT.

The area of the sector QOP is $\frac{1}{2}r^2\theta$ (§31.1). The areas of the triangles are $\frac{1}{2}(1) \sin \theta$ and $\frac{1}{2}(1) \tan \theta$. Hence the inequalities can be expressed

$$\tfrac{1}{2} \sin \theta < \tfrac{1}{2}\theta < \tfrac{1}{2} \tan \theta.$$

One result of this is that $\sin \theta < \theta$.
Dividing these numbers by $\frac{1}{2} \sin \theta$ and remembering that $\tan \theta = \sin \theta / \cos \theta$, we have

$$1 < \frac{\theta}{\sin \theta} < \frac{1}{\cos \theta}.$$

Inverting these fractions and changing their signs, we have

$$-1 < -\frac{\sin \theta}{\theta} < -\cos \theta.$$

We add 1 to all members and find that

$$0 < 1 - \frac{\sin \theta}{\theta} < 1 - \cos \theta = \frac{1 - \cos^2 \theta}{1 + \cos \theta} = \frac{\sin^2 \theta}{1 + \cos \theta} < \frac{\theta^2}{1},$$

where to get the last inequality we made use of the fact that $\sin \theta < \theta$ and the fact that $1/(1 + \cos \theta) < 1$. By virtue of these inequalities we have

$$0 < 1 - \frac{\sin \theta}{\theta} < \theta^2.$$

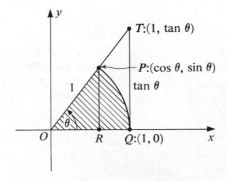

FIGURE 5.23   Geometric inequalities.

We now let θ approach 0, squeezing the middle expression to zero. This gives us

$$\lim_{\theta \to 0} \frac{\sin \theta}{\theta} = 1.$$

We can use the same inequalities to prove that

$$0 < \frac{1 - \cos \theta}{\theta} < \theta.$$

Hence, again letting θ approach 0 gives us

$$\lim_{\theta \to 0} \frac{1 - \cos \theta}{\theta} = 0.$$

We state these results as a preliminary conclusion to be used for differentiating **sin u** and **cos u** in general.

LEMMA.   The derivatives of **sin u** and **cos u** at zero are given by $\sin' 0 = 1$, $\cos' 0 = 0$.

31.6   Derivatives of sin u and cos u at any Point.   By definition of the derivative

$$D_u \cos u = \lim_{v \to 0} \frac{\cos (u + v) - \cos u}{v}.$$

Using the addition theorem for $\cos (u + v)$, we find that

$$\frac{\cos (u + v) - \cos u}{v} = \frac{\cos u \cos v - \sin u \sin v - \cos u}{v}$$

$$= -\cos u \left(\frac{1 - \cos v}{v}\right) - \sin u \left(\frac{\sin v}{v}\right).$$

Using the lemma (§31.5), we find that

$$D_u \cos u = \lim_{v = 0} \left[-\cos u \left(\frac{1 - \cos v}{v}\right) - \sin u \left(\frac{\sin v}{v}\right)\right]$$

$$= - \cos u \cos' 0 - \sin u \sin' 0 = -\sin u.$$

Similarly we find that

$$D_u \sin u = \lim_{v \to 0} \frac{\sin (u + v) - \sin u}{v} = \cos u.$$

We state these results as a theorem.

THEOREM (Differentiation of sine and cosine).   The functions **sin u** and **cos u** are differentiable for all real numbers u. If **u** is a differentiable function of x, their derivatives and antiderivatives are given by:

IX. $D \cos u = - \sin u \, Du.$ $\int$ IX. $\int \sin u \, Du = - \cos u + C.$

X. $D \sin u = \cos u \, Du.$ $\int$ X. $\int \cos u \, Du = \sin u + C.$

Using the relation **tan u** $= $ **sin u/cos u**, we can differentiate **tan u**.

THEOREM (Differentiation of the tangent function). The function **tan u** is differentiable wherever cos $u \neq 0$, and its derivative and corresponding antiderivative are

XI. $D \tan u = \dfrac{1}{\cos^2 u} Du,$ $\int$ XI. $\displaystyle\int \dfrac{Du}{\cos^2 u} = \tan u + C,$

where u is any differentiable function of x.

31.7 Examples of Differentiation

Example A.   Find the derivative of the function $f(x) = \frac{1}{2}(1 - \cos 2x)$. Find the slope of the tangent line to the graph at the point where $x = \pi/6$.

Solution.

$$D \tfrac{1}{2}(1 - \cos 2x) = \tfrac{1}{2}(D1 - D \cos 2x) = \tfrac{1}{2}[0 - (-\sin 2x\, D\, 2x)] = \sin 2x.$$

Set $x = \pi/6$; then $f(\pi/6) = \frac{1}{4}$ and $f'(\pi/6) = \sqrt{3}/2$. Thus at the point $(\pi/6, \frac{1}{4})$ the slope of the tangent line is $\sqrt{3}/2$.

Example B.   Differentiate **$\sin^2 x \cos x$**.

Solution. $\sin^2 x$ means $(\sin x)^2$. To differentiate $\sin^2 x$, we observe that it has the power form u^2, for which $Du^2 = 2u\, Du$. Thus

$$D \sin^2 x = 2 \sin x\, D \sin x = 2 \sin x \cos x.$$

Similarly,

$$D \cos^3 x = 3 \cos^2 x\, D \cos x = -3 \cos^2 x \sin x.$$

Finally, **$D \sin^2 x \cos^3 x$** is in the product form **$D\, uv$**, for which **$D\, uv = u\, Dv + v\, Du$**, so that

$$D \sin^2 x \cos^3 x = \sin^2 x\, D \cos^3 x + \cos^3 x\, D \sin^2 x$$
$$= \sin^2 x(-3 \cos^2 x \sin x) + \cos^3 x(2 \sin x \cos x)$$
$$= -3 \cos^2 x \sin^3 x + 2 \sin x \cos^4 x.$$

This completes the problem.

Example C.   Differentiate $\ln \left| \dfrac{1 + \sin x}{\cos x} \right|.$

Solution.   This is in the form $D \ln |u| = Du/u$, where $u = (1 + \sin x)/\cos x$. Hence we must calculate by the quotient formula Rule V_a,

$$D \left(\frac{1 + \sin x}{\cos x} \right) = \frac{\cos x\, D(1 + \sin x) - (1 + \sin x)\, D \cos x}{\cos^2 x}$$

$$= \frac{\cos x (\cos x) - (1 + \sin x)(-\sin x)}{\cos^2 x} = \frac{1 + \sin x}{\cos^2 x}$$

Putting this back into the derivative of $\ln |u|$, we have

$$D \ln \left| \frac{1 + \sin x}{\cos x} \right| = \frac{\cos x}{1 + \sin x} \cdot \left(\frac{1 + \sin x}{\cos^2 x} \right) = \frac{1}{\cos x}.$$

This also gives us the antiderivative

$$\int \frac{Du}{\cos u} = \ln \left| \frac{1 + \sin u}{\cos u} \right| + C.$$

31.8   Exercises

In Exercises 1–10, differentiate.

1. $3 \sin 2x.$
2. $\tan u.$
3. $1 - \cos 2x + \sin 2x.$
4. $\cos^2 x.$
5. $1 - \tan 2x.$
6. $\sqrt{1 - \cos 2x}.$
7. $\sin^2 \dfrac{x}{2}.$
8. $\ln |\tan \theta|.$
9. $\dfrac{1 + \cos 2x}{\sin 2x}.$
10. $x \sin \dfrac{1}{x}, x \neq 0.$

In Exercises 11–16, find the indicated antiderivatives.

11. $\int \sin 2x \, Dx.$
12. $\int (1 - \cos 2x) \, Dx.$
13. $\int (1 - \sin x + 3 \sin 2x) \, Dx.$
14. $\int \dfrac{dx}{\cos^2 3x}.$
15. $\int -\tan x \, Dx = \ln |\cos x| + C.$
16. $\int \dfrac{\cos x}{1 + \sin x} \, Dx.$

17. Explain why the following two different-appearing antiderivatives are the same.

$$\int \sin x \cos x \, Dx = \int \sin x (\cos x \, Dx) = \int u \, Du = \frac{\sin^2 x}{2} + C.$$

$$\int \sin x \cos x \, Dx = -\int \cos x (-\sin x \, Dx) = -\int u \, Du = -\frac{\cos^2 x}{2} + C.$$

18. Evaluate $\int_{-\pi/4}^{\pi/4} \sin x \cos x \, dx.$
19. Evaluate $\int_0^{\pi} (1 - \cos 2x) \, dx.$
20. Integrate $\int_0^t \sin 2x \, dx.$
21. Show that if $y = A \cos kt + B \sin kt$, then $D_t^2 \, y + k^2 \, y = 0.$
22. Find formulas for $D^n \cos x$ for all n.
23. Find formulas for $D^n \sin x$ for all n.

In Exercises 24–30, we use the definitions $\cot x = 1/\tan x$, $\sec x = 1/\cos x$.

24. Show that $D \tan u = \sec^2 u \, Du.$
25. Show that $D \sec u = \sec u \tan u \, Du.$
26. Differentiate $\sec x \tan x.$
27. Differentiate $\sec^3 ax.$
28. Find the antiderivative $\int \sec^2 3x \, Dx.$
29. Find the antiderivative $\int \tan^2 ax \, Dx.$
30. Find $D_t \ln |\tan x|.$
31. Using Formulas (9), (8), Section 31.4, show that

$$\int \cos^2 x \, Dx = \frac{x}{2} + \frac{1}{4} \sin 2x + C = \frac{1}{2} (x + \sin x \cos x) + C.$$

32. Find all x where $D \sin x = 0$.
33. Find all x where $D \cos 2x = 0$.
34. If $f(x) = \cos x$, what are $f'(\pi/6)$, $f^{(2)}(\pi/6)$.
35. If $f(x) = \cos x$, find $f(\theta)$ if $f'(\theta) = \frac{1}{3}$.

32 Periodic Functions

32.1 Graphs of the Trigonometric Functions. Using the geometry of the isosceles right triangle with acute angles $\pi/4$ and the right triangle with angles $\pi/6$ and $\pi/3$ (Figure 5.24), we can evaluate the trigonometric functions at intervals of $\pi/6$ or $\pi/4$.

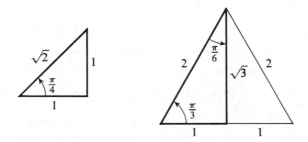

FIGURE 5.24   Special right triangles.

We plot on the unit circle points whose coordinates are calculated in this manner (Figure 5.25). We plot graphs of **sin x** and **cos x** using these points. We can plot these graphs with approximately equal units on the two axes by choosing one unit in y equal to $\pi/3$ on the x-axis (Figures 5.26, 5.27). In plotting the graph of **tan x** (Figure 5.28), we must take account of the fact that at the series of angles

$$\cdots -\frac{3\pi}{2}, -\frac{\pi}{2}, \frac{\pi}{2}, \frac{3\pi}{2}, \cdots,$$

where $\cos x = 0$, **tan x** is not defined. It becomes infinite at these points.

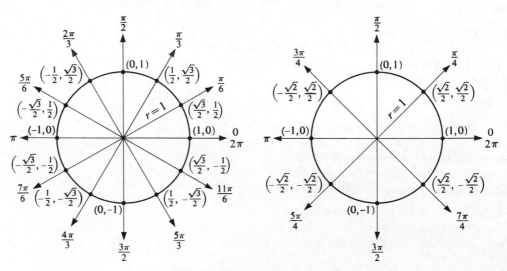

Figure 5.25   Special angles on unit circle.

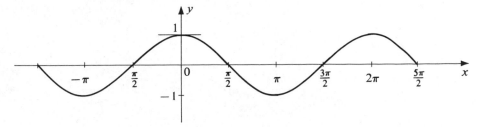

FIGURE 5.26   Graph of $y = \cos x$.

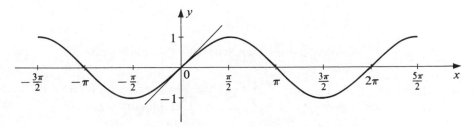

FIGURE 5.27   Graph of $y = \sin x$.

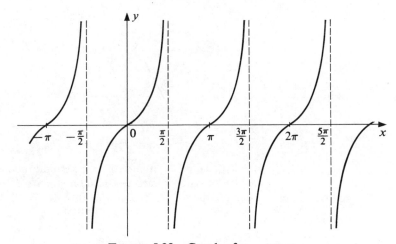

FIGURE 5.28   Graph of $y = \tan x$.

32.2   Periodic Property.   On account of the fact that as the ray $\overrightarrow{OP}$ rotates through a complete revolution, 2π radians, the ray $\overrightarrow{OP}$ returns to its original position, it follows that for any angle θ,

$$\sin(\theta + 2\pi) = \sin\theta, \qquad \cos(\theta + 2\pi) = \cos\theta, \qquad \tan(\theta + 2\pi) = \tan\theta.$$

(See Figure 5.29.) Thus the graphs of these trigonometric functions can be constructed for any interval of length of 2π and then the rest of the graph can be made by repeating this pattern over and over (Figures 5.26, 5.27, 5.28). We say that the trigonometric functions are *periodic* with period 2π.

More generally, we may make the following definition.

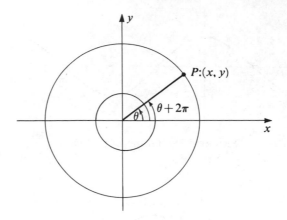

ᵻɪɢᴜʀᴇ 5.29   Periodic property of rotation.

DᴇꜰɪɴɪᴛɪᴏN.   A function $f(x)$ defined on all real numbers is *periodic with period T*, where $T \neq 0$, if and only if for every x, $f(x + T) = f(x)$. (See Figure 5.30.)

In the study of geometric trigonometry the principal role of the trigonometric functions was in solving triangles; but in calculus and its applications in dynamics, it is periodicity of the trigonometric functions that gives them their importance.

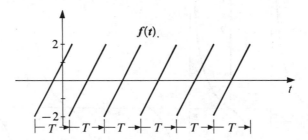

Fɪɢᴜʀᴇ 5.30   A periodic function with period T. In this case the periodic function $f(t)$ is discontinuous, having one discontinuity in each period interval. We do not attempt now to give an analytic representation of $f(t)$, that is, a formula describing it.

32.3 Problem of Simple Harmonic Motion.   A mass particle of mass m is driven by a spring that acts along a horizontal x-axis with force F, $F = -kx$, where $k > 0$. The mass is free to move along the x-axis subject only to the force of the spring (Figure 5.31). We stretch (or compress) the spring to place the particle at the point where $x = b$. Then the system is released with zero initial velocity at time $t = 0$ seconds. Find the motion thereafter.

We can predict intuitively that the particle will move back and forth forever along the line between the points where $x = b$ and $x = -b$, and that the time to execute one cycle and return to the initial point is a constant, T seconds. We predict also that the larger the mass m is for a particular spring, the slower the motion will be and the longer the period T will be. On the other hand, the stiffer the spring, the more rapid the motion, and the shorter the period T.

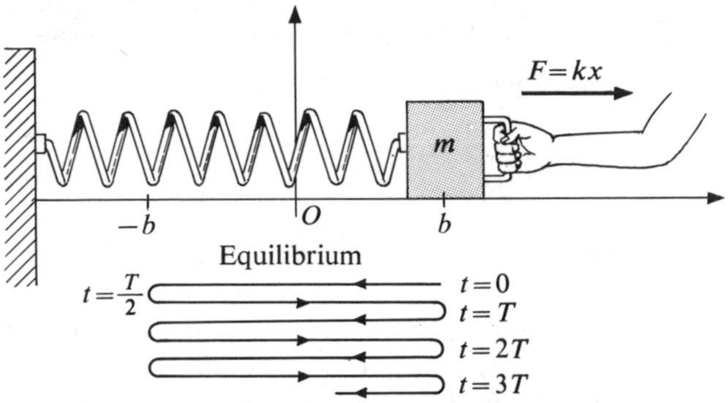

FIGURE 5.31   Spring and simple harmonic motion.

We now proceed to solve the problem mathematically. The unknown motion is a function $x(t)$ that must satisfy Newton's law of motion (§18.3), $mD_t^2 x = -kx$, or

$$(1) \qquad D_t^2 x + \frac{k}{m} x = 0,$$

and the initial conditions

$$(2) \qquad x(0) = b \qquad \text{and} \qquad x'(0) = 0.$$

We find some solutions by inspection, observing that

$$D_t^2 \cos \sqrt{\frac{k}{m}} \, t = -\frac{k}{m} \cos \sqrt{\frac{k}{m}} \, t$$

so that $\cos \sqrt{(k/m)} \, t$ is one solution. Similarly, $\sin \sqrt{(k/m)} \, t$ is another solution of the differential equation (1). We verify furthermore that, for all values of the constants A and B, the linear combination

$$x(t) = B \cos \sqrt{\frac{k}{m}} \, t + A \sin \sqrt{\frac{k}{m}} \, t$$

is a solution.

We apply the initial conditions to determine the coefficients B and A. We set $t = 0$ and find that, since $\cos 0 = 1$ and $\sin 0 = 0$ and $x(0) = b$, $B = b$. Then we differentiate once and find that

$$x'(t) = -B \sqrt{\frac{k}{m}} \sin \sqrt{\frac{k}{m}} \, t + A \sqrt{\frac{k}{m}} \cos \sqrt{\frac{k}{m}} \, t.$$

Again we set $t = 0$, and since $x'(0) = 0$, we find that $A = 0$. Hence a solution satisfying the initial conditions is

$$(3) \qquad x(t) = b \cos \sqrt{\frac{k}{m}} \, t.$$

This function describes the motion.

Let us see what information about the motion we can deduce from this solution. We know that if we add a complete revolution, 2π radians, to an angle θ we have $\cos\theta = \cos(\theta + 2\pi)$. Applying this property to $\cos\sqrt{(k/m)}t$ in (3), we have

$$\cos\sqrt{\frac{k}{m}}\,t = \cos\left(\sqrt{\frac{k}{m}}\,t + 2\pi\right) = \cos\sqrt{\frac{k}{m}}\left(t + 2\pi\sqrt{\frac{m}{k}}\right).$$

This tells us that if we add $2\pi\sqrt{m/k}$ seconds to any time t, the function $x(t) = b\cos\sqrt{(k/m)}t$ returns to its value at time t. Thus the motion is *periodic* and its *period* $T = 2\pi\sqrt{m/k}$ sec.

Since the maximum value attained by the cosine factor in the motion (3) is 1, the maximum distance x that the particle is displaced from equilibrium during the oscillation is given by the absolute value of the coefficient b. Thus $|b|$ is the *amplitude* of the oscillation.

Finally, with a particular driving spring, we observe that when the particle is made heavier, the velocity of the motion,

$$x'(t) = -\sqrt{\frac{k}{m}}\,b\sin\sqrt{\frac{k}{m}}\,t,$$

decreases in proportion to $1/\sqrt{m}$, and the period $T = 2\pi\sqrt{m/k}$ sec is longer in proportion to $\sqrt{m}$. On the other hand, with a fixed particle mass, as the spring is made stiffer the motion speeds up in proportion to $\sqrt{k}$, and the period of oscillation is shorter in proportion to $1/\sqrt{k}$. We predicted this intuitively but had no way of measuring precisely the effect on the period caused by increasing m or k.

32.4   Sinusoidal Functions

DEFINITION.   A trigonometric function of the form $b\sin(pt + \phi)$ or $b\cos(pt + \phi)$, where, b, p, ϕ are constants and $p \neq 0$, is *sinusoidal*. The number ϕ radians is called the *initial phase* and $|b|$ is called the *amplitude*.

THEOREM.   A sinusoidal function $b\sin(pt + \phi)$ or $b\cos(pt + \phi)$ is periodic and its period $T = 2\pi/p$.

Proof.   Exercise.

When the formula for a sinusoidal function is written $b\sin(2\pi\omega t + \phi)$ or $b\cos(2\pi\omega t + \phi)$, the period becomes $T = 1/\omega$ seconds per cycle. The reciprocal, ω cycles per second, written ω sec^{-1}, is called the *frequency*. It is the number of complete cycles in a second. For example, ordinary household alternating electric current is usually "60-cycle," meaning that its frequency $\omega = 60$ sec^{-1}. If it is 110-volt current then 110 is the amplitude, so that in its simplest form, household alternating current has the instantaneous emf given by $E = 110\sin(120\pi t + \phi)$. The phase ϕ is usually of no significance in this case because it does not matter at what stage in the cycle the current is turned on.

When we know the numbers b, p, ϕ, we can sketch the graph of any sinusoidal function $b\sin(pt + \phi)$ by appropriate adaptation of the graph of $\sin t$ (Figure 5.32).

Any linear combination of sinusoidal functions with the same period is a sinusoidal function. We study the technique by an example.

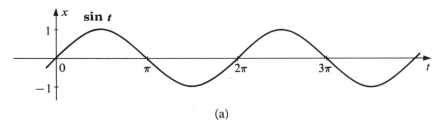

(a)

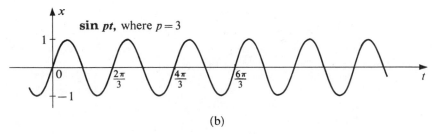

(b)

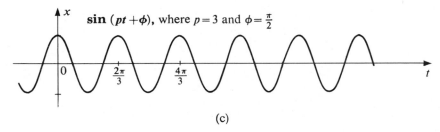

(c)

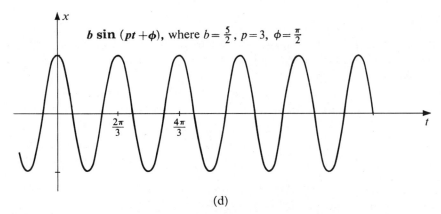

(d)

(a) **sin *t***.
(b) First stage. Frequency modulation. The period of **sin *pt*** is $2\pi/p$. Here $p = 3$.
(c) Second stage. Phase shift, one quarter cycle to left, giving **sin (*pt* + φ)**, where $p = 3$ and $\phi = \pi/2$.
(d) Final stage. Amplitude modulation to amplitude b, giving b **sin (*pt* + φ)**, where $b = 5/2$, $p = 3$, $\phi = \pi/2$.

FIGURE 5.32   Graph of $\frac{5}{2}$ **sin (3*t* + π/2)**, constructed in three stages, starting with **sin *t***.

Example.   Show that **2 cos 3t − 5 sin 3t** is a sinusoidal function. Find the amplitude, period, and initial phase.

Solution.   For the coefficients $(2, -5)$, we compute $\sqrt{2^2 + (-5)^2} = \sqrt{29}$. We multiply and divide the expression by $\sqrt{29}$, obtaining

$$2 \cos 3t - 5 \sin 3t = \sqrt{29} \left[\frac{2}{\sqrt{29}} \cos 3t - \frac{5}{\sqrt{29}} \sin 3t \right].$$

This maneuver converts the coefficients $(2, -5)$ into a pair $(\cos \phi, \sin \phi)$, where $\cos \phi = 2/\sqrt{29}$, $\sin \phi = -5/\sqrt{29}$ (Figure 5.33). That is, $\phi = \arctan(-\frac{5}{2})$. This gives us

$$2 \cos 3t - 5 \sin 3t = \sqrt{29} \left[\cos 3t \cos \phi - \sin 3t \sin \phi \right] = 29 \cos (3t + \phi),$$

where the last step comes from the addition theorem for the cosine (§31.3). Thus **2 cos 3t − 5 sin 3t** is transformed into the formula $\sqrt{29} \cos (3t + \phi)$ for a cosine sinusoidal function, with amplitude $\sqrt{29}$, phase $\phi = \arctan(-\frac{5}{2})$, and period $2\pi/3$, the same as that of **cos 3t** and **sin 3t**. This completes the problem.

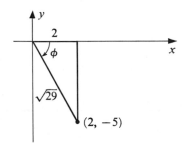

FIGURE 5.33   Determination of phase angle.

32.5   Exercises

In Exercises 1–6, find the period and amplitude and sketch the graphs of the periodic functions. Find the frequency of each function.

1.   **cos 2t.**
2.   **3 sin t/2.**
3.   **− sin t.**
4.   **3 sin ((t/2) + (π/2)).**
5.   **5 sin (12πt + φ).**
6.   **−2 cos (2πt + (π/4)).**

7.   Construct the graph of **b cos (pt + φ)** from **cos x.** (See Figure 5.32.)
8.   Show that if $f(t)$ is periodic with period T, it is periodic with period $2T$, or $3T$, or nT for any positive integer n. The least period of a periodic function, if it exists, is called the minimum period.
9.   Show that **tan t** is periodic and find its minimum period.
10. Show that **1 − cos t** is periodic and find its period. Is it sinusoidal?

In Exercises 11–18, find which functions are periodic and for these find a period and the minimum period if it exists.

11.   **t − sin t.**
12.   **1/sin t.**
13.   **10.**
14.   **$t^2 + 2t - 1$.**

15.   **cos 2t − cos 3t.**
16.   **$f'(t)$**, where $f(t)$ is periodic with period T.
17.   **$F(t)$**, where $F(t) = \int_0^t (1 - \cos 2u) \, du$.
18.   **$\sin^2 t$.**

In Exercises 19–22, express the linear combinations as sinusoidal functions. Find the period, amplitude, and initial phase.

19.   **sin *t* + cos *t*.**

20.   **3 cos *t* − 4 sin *t*.**

21.   ***a* sin *pt* + *b* cos *pt*.**

22. $\textbf{sin } 2t + \textbf{sin}\left[2t - \dfrac{\pi}{2}\right].$

23.   Express **sin *t*** as a cosine sinusoidal function with appropriate phase shift.

24.   Show that any sinusoidal function can be expressed in both sine and cosine forms.

Exercises 25–30 are concerned with finding the motion $x(t)$, and properties of the motion, of a particle of mass m driven by a spring with spring-constant k.

25.   Find the motion $x(t)$ when $m = 2$, $k = 8$, $x(0) = 5$ and initial velocity $x'(0) = 0$. Find the period and amplitude. Plot a graph of the motion in tx-coordinates.

26.   Find the motion $x(t)$ when $m = 2$, $k = 8$, $x(0) = 0$ and $x'(0) = -6$ in./sec. Find the period and amplitude. Plot a graph of the motion in tx-coordinates.

27.   In Exercise 25, the motion was started by stretching the spring until $x(0) = 10$ and again the particle was released with $x'(0) = 0$. What was the change in the period due to this change in initial conditions?

28.   Find when and where the velocity is a maximum in the motion of Exercise 25.

29.   Show mathematically that the motion $x(t) = b \cos \sqrt{(k/m)}t$ stops momentarily when $x = b$ or $x = -b$.

30.   In Exercise 25 with other conditions the same, what (heavier) mass is required to get a period twice as long as in Exercise 25?

33   Inverse Trigonometric Functions

33.1   Geometric Construction of θ when sin θ is given

Example A.   Construct geometrically all rotations θ for which $\sin \theta = \frac{1}{3}$.

Solution.   We recall the definition of $\sin \theta$, which says that if the point $P:(x, y)$ on the circle $x^2 + y^2 = r^2$ is determined by the rotation θ, then $\sin \theta = y/r$ (Figure 5.34(a)). In this problem we may take $r = 3$ and then, since $\sin \theta = \frac{1}{3}$, $y = 1$. We plot the line

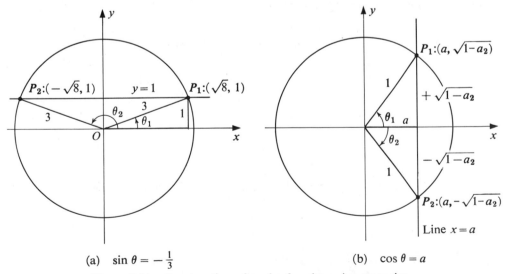

(a) $\sin \theta = -\dfrac{1}{3}$ (b) $\cos \theta = a$

Figure 5.34   Construction of angles for given sine or cosine.

$y = 1$, and find it intersects the circle $x^2 + y^3 = 3^2$ in two points $P_1:(\sqrt{8}, 1)$ and $P_2:(-\sqrt{8}, 1)$. The terminal rays $\overrightarrow{OP_1}$ and $\overrightarrow{OP_2}$ determine infinitely many rotations θ, which have $\sin \theta = \frac{1}{3}$. There are two basic ones, θ_1 and θ_2 (Figure 5.34), for which $0 \leq \theta < 2\pi$. All the others can be formed from these by adding or subtracting complete revolutions. Thus all rotations θ for which $\sin \theta = \frac{1}{3}$ are given by

$$\theta = \theta_1 + 2n\pi \quad \text{or} \quad \theta = \theta_2 + 2n\pi, \quad \text{where } n \text{ is any integer.}$$

This completes the problem.

Example B.   Construct all rotations θ for which $\cos \theta = a$, where $|a| \leq 1$.

Solution.   We plot a unit circle with center at O (Figure 5.34(b)). This amounts to choosing $r = 1$. Since $\cos \theta = x/r$ in general, and since in this case we have $\cos \theta = a$, we plot the line $x = a$. Since $|a| \leq 1$, the line intersects the circle in points $P_1:(a, \sqrt{1-a^2})$ and $P_2:(a, -\sqrt{1-a^2})$. The rays $\overrightarrow{OP_1}$ and $\overrightarrow{OP_2}$ are the only terminal rays of rotations θ for which $\cos \theta = a$. All rotations θ for which $\cos \theta = a$ are found from two basic ones, θ_1 and θ_2, by

$$\theta = \theta_1 + 2n\pi \quad \text{or} \quad \theta = \theta_2 + 2n\pi, \quad \text{where } n \text{ is any integer.}$$

We observe that in the special case that if $|a| = 1$, the two rays $\overrightarrow{OP_1}$ and $\overrightarrow{OP_2}$ coincide, and $\theta_1 = \theta_2$. This completes the problem.

33.2   Inverse Trigonometric Relations.   We have seen that there are infinitely many numbers y such that $\cos y = x$ when $|x| \leq 1$. We define the symbol $\cos^{-1} x$ to mean any one of them. That is,

$$y = \cos^{-1} x \quad \text{if and only if} \quad x = \cos y.$$

This inverse relation does not define a function since the relation does not select a unique number y for each number assigned to x. We plot the graph of this relation (Figure 5.35(a)) and see that it is not a function.

Example A.   Find $\cos (\sin^{-1} \frac{1}{3})$ and $\tan (\sin^{-1} \frac{1}{3})$.

Solution.   We could use tables to find all values of $\sin^{-1} \frac{1}{3}$ and then determine from the tables the cosine and tangent of these rotations. However, the geometric construction of θ when $\sin \theta$ is given (§33.1, Example A) suggests a simpler technique. We construct the two terminal rays $\overrightarrow{OP_1}$ and $\overrightarrow{OP_2}$ corresponding to the relation $\theta = \sin^{-1} \frac{1}{3}$ (Figure 5.34(a)) and compute the x-coordinates, $\sqrt{8}$ and $-\sqrt{8}$. Then, we read off from the figure $\cos (\sin^{-1} \frac{1}{3}) = +\sqrt{8}/3$; and $\tan (\sin^{-1} \frac{1}{3}) = \pm 1/\sqrt{8} = \pm \sqrt{2}/4$. This completes the problem.

Example B.   Find $\sin (\cos^{-1} a)$ and $\tan (\cos^{-1} a)$, where $|a| \leq 1$.

Solution.   We construct the terminal rays for $\theta = \cos^{-1} a$ as before (§33.1, Example B) and compute $b = \pm \sqrt{1-a^2}$. Then we read off

$$\sin (\cos^{-1} a) = \pm \sqrt{1-a^2} \quad \text{and} \quad \tan (\cos^{-1} a) = \frac{\pm \sqrt{1-a^2}}{a}.$$

This completes the problem. The ambiguity of sign is unavoidable so long as we define the symbol $\cos^{-1} a$ so that is is multiple-valued.

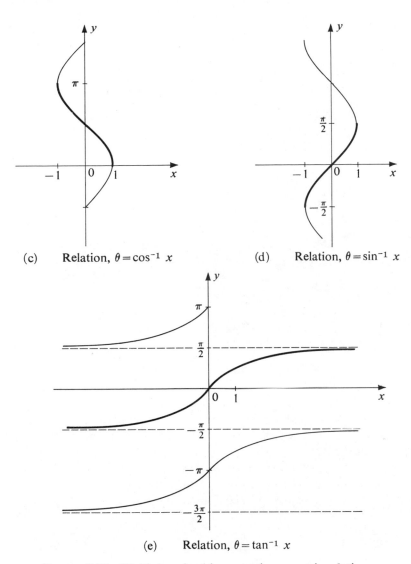

(c)   Relation, $\theta = \cos^{-1} x$ (d)   Relation, $\theta = \sin^{-1} x$

(e)   Relation, $\theta = \tan^{-1} x$

FIGURE 5.35   Multiple valued inverse trigonometric relations.

33.3 The Functions arccos x, arcsin x, arctan x.   It is desirable to have inverses of the trigonometric functions that are properly functions, that is, single-valued. To define such inverse trigonometric functions, we must restrict arbitrarily the range of the inverse trigonometric relations so that they will become single-valued. We do this as follows.

DEFINITION.   *The inverse trigonometric functions.*

We define the functions:

(a)   **arccos x** on $[-1, 1]$ by $y = $ arccos x if and only if $x = \cos y$ and $0 \leqq y \leqq \pi$;
(b)   **arcsin x** on $[-1, 1]$ by $y = $ arcsin x if and only if $x = \sin y$ and $-\pi/2 \leqq y \leqq \pi/2$;
(c)   **arctan x** on $[-1, 1]$ by $y = $ arctan x if and only if $x = \tan y$ and $-\pi/2 \leqq y \leqq \pi/2$.

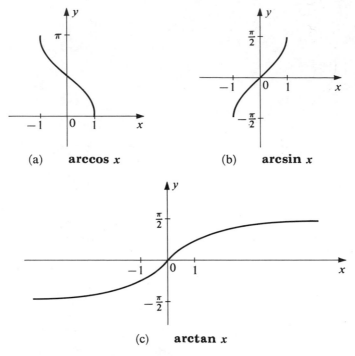

(a)   **arccos** x (b)   **arcsin** x

(c)   **arctan** x

FIGURE 5.36   Inverse trigonometric functions.

We observe that the inequalities restricting y in these definitions serve to pick out one and only one value of y, called the *principal value*. We show this in the graphs of the inverse relations (Figure 5.35(a), (b), (c)) by means of the heavy black portion of each graph, which are the graphs of the inverse trigonometric *functions*. We repeat the graphs of the inverse trigonometric functions separately (Figure 5.36). When the inverse functions are defined in this way we can compute without ambiguity of sign such combinations as:

$$\sin(\arccos a) = +\sqrt{1 - a^2}, \qquad \tan(\arccos a) = +\frac{\sqrt{1 - a^2}}{a},$$

$$\cos(\arcsin x) = +\sqrt{1 - x^2}, \qquad \tan\left(\arcsin \frac{1}{3}\right) = \frac{\sqrt{2}}{4}.$$

$$\tan\left(\arccos - \frac{1}{2}\right) = -\sqrt{3}, \qquad \sin\left(\arccos - \frac{1}{2}\right) = +\frac{\sqrt{3}}{2}.$$

33.4   Differentiation of the Inverse Trigonometric Functions.   By a reasoning similar to that for the exponential function (§26.3), we know that $y = $ **arccos** x is differentiable; that is, it has a derivative $D_x y$. For, by definition, it has the same graph and the same tangent lines as $x = $ **cos** y, for which we can calculate $D_y x$. The operation of forming the inverse is to interchange x and y, thereby inverting all slopes. Hence, wherever $D_y x \neq 0$,

$$D_x y = \frac{1}{D_y x}.$$

To find the derivative $D_x y$ for $y = \textbf{arccos } x$, we differentiate with respect to x the equivalent relation $x = \cos y$, and find $1 = -\sin y\, D_x y$. Since $\sin y = \sin(\textbf{arccos } x)$, we use the right-triangle device (§33.3) to read off $\sin y = \sqrt{1 - x^2}$. Hence

$$D_x \textbf{ arccos } x = -\frac{1}{\sqrt{1 - x^2}} \qquad \text{wherever} \quad |x| < 1.$$

Similar calculations can be applied to $D_x \textbf{ arcsin } x$ and $D_x \textbf{ arctan } x$.

We record these results with their corresponding antiderivative formulas.

XIII. $D \textbf{ arccos } u = -\dfrac{Du}{\sqrt{1 - u^2}},$ $\displaystyle\int$ XIII. $\displaystyle\int \dfrac{Du}{\sqrt{1 - u^2}} = \begin{cases} -\textbf{arccos } u + C, \text{ or} \\ \textbf{arcsin } u + C_1 \end{cases}$

$$\text{where } |u| \leq 1.$$

XIV. $D \textbf{ arcsin } u = \dfrac{Du}{\sqrt{1 - u^2}},$ $\displaystyle\int$ XIII$_\text{a}$. $\displaystyle\int \dfrac{Du}{\sqrt{a^2 - u^2}} = \textbf{arcsin } \dfrac{u}{a} + C,$

$$|u| \leq |a|.$$

XV. $D \textbf{ arctan } u = \dfrac{Du}{1 + u^2},$ $\displaystyle\int$ XV. $\displaystyle\int \dfrac{Du}{1 + u^2} = \textbf{arctan } u + C.$

$$\int \text{XV}_\text{a}. \quad \int \dfrac{Du}{a^2 + u^2} = \dfrac{1}{a} \textbf{ arctan } \dfrac{u}{a} + C,$$

$$\text{for all real numbers } u.$$

33.5 Example of Differentiation and Integration

Example A.   Compute $D \textbf{ arcsin } \frac{3}{5}x$.

Solution.   From XIV with $u = \frac{3}{5}x$ so that $Du = \frac{3}{5}$, we find

$$D \textbf{ arcsin } \tfrac{3}{5}x = \frac{\frac{3}{5}}{\sqrt{1 - (\frac{3}{5}x)^2}} = \frac{3}{\sqrt{25 - 9x^2}},$$

valid where $9x^2 < 25$.

Example B.   Differentiate $f(x) = x \textbf{ arctan } 2x$ and investigate this function for maxima and minima on all real numbers x.

Solution.   We combine the product rule for differentiation (III) with the rule XV for differentiating $\textbf{arctan } u$. This gives

$$f'(x) = x\frac{1}{1 + (2x)^2} \cdot 2 + (1) \textbf{ arctan } 2x = \frac{2x}{1 + 4x^2} + \textbf{arctan } 2x.$$

Since $f(x)$ is everywhere differentiable, a maximum or minimum can occur only where $f'(x) = 0$. The equation

$$\frac{2x}{1 + 4x^2} + \textbf{arctan } 2x = 0$$

has only the solution $x = 0$, since both terms are positive when $x > 0$ and both are negative when $x < 0$. The table of signs

$x < 0$	$f'(x) < 0$
$x = 0$	$f'(x) = 0$
$x > 0$	$f'(x) > 0$

shows that $x = 0$ is a minimum point and that there is no maximum (§16.3). This completes the problem.

Example C. Evaluate

$$\int_{-2}^{2} \frac{dx}{\sqrt{4 - x^2}}.$$

Solution. We use the rule $\int XV_a$ to find the antiderivative

$$\int \frac{Dx}{\sqrt{4 - x^2}} = \arcsin \frac{x}{a}.$$

Hence

$$\int_{-2}^{2} \frac{dx}{\sqrt{4 - x^2}} = \arcsin \frac{x}{2}\Big|_{-2}^{2} = \arcsin 1 - \arcsin(-1)$$

$$= \frac{\pi}{2} - \left(-\frac{\pi}{2}\right) = \pi.$$

33.6 Exercises

In Exercises 1–8, construct geometrically all rotations θ satisfying the conditions given.

1.	$\sin \theta = \frac{3}{5}$.	2.	$\cos \theta = -\frac{4}{5}$.
3.	$\tan \theta = 1$.	4.	$\tan \theta = -1$.
5.	$\cos \theta = 0$.	6.	$\tan \theta = a$.
7.	$\cos \theta = x$.	8.	$\sec \theta = a$.

In Exercises 9–16, evaluate the trigonometric functions by use of the geometric construction of the angle given.

9.	$\sin \arccos \frac{1}{2}$.	10.	$\sin(\arcsin x)$.
11.	$\tan \arccos(-1)$.	12.	$\tan \arccos a$.
13.	$\cos\left(\arctan \frac{x}{a}\right)$.	14.	$\sin\left(\arccos - \frac{1}{\sqrt{2}}\right)$.
15.	$\arctan\left(\tan \frac{3\pi}{4}\right)$.	16.	$\arcsin(\cos x)$.

17. Using the addition theorem for $\tan(\theta + \phi)$ (Section 31.4), show that

$$\arctan \frac{1}{2} + \arctan \frac{1}{3} = \frac{\pi}{4}.$$

18. Since $-\arccos x$ and $\arcsin x$ have the same derivative on $[-1, 1]$ (XIII and XIV), they must differ by a constant. What is this constant difference, $\arcsin x - (-\arccos x)$?

In Exercises 19–24, perform the indicated differentiation.

19.	$D \arccos 2x$.	20.	$D \arctan \frac{x}{4}$.
21.	$D x^2 \arcsin x$.	22.	$D \arctan 2x$.
23.	$D \arctan \frac{x + 1}{4}$.	24.	$D \arccos\left(\frac{a - x}{a}\right)$.

In Exercises 25–28, find the indicated antiderivatives.

25. $\int \dfrac{Dx}{\sqrt{1 - 4x^2}}.$ 26. $\int \dfrac{Dx}{4 + x^2}.$

27. $\int \dfrac{Dx}{1 + 9x^2}.$ 28. $\int \sqrt{1 - 4x}\, Dx, \quad x < \tfrac{1}{4}.$

In Exercises 29–32, evaluate the definite integrals.

29. $\displaystyle\int_0^{1/2} \dfrac{dx}{\sqrt{1 - x^2}}.$ 30. $\displaystyle\int_{-1}^{1} \dfrac{dx}{\sqrt{1 - x^2}}.$

31. $\displaystyle\int_0^1 \dfrac{dx}{1 + x^2}.$ 32. $\displaystyle\int_0^1 \dfrac{dx}{x^2 + x + 1}.$

33. Find the length of arc cut from a circle of radius a by a line h units distant from the center.

34. Find the area of the region below the curve $(a^2 - x^2)y^2 = 1,\ y \geqq 0.$

35. Find the area of the region enclosed between the curves $y = 8/(x^2 + 4)$ and $4y = x^2.$

36.* A movie screen 12 ft high has its lower edge 6 ft above the viewer's eye. At what horizontal distance does the screen subtend the greatest angle at the viewer's eye?

34 Computation and Applications of Sine and Cosine

34.1 Numerical Establishment of Sine and Cosine. Our definition of **sin** x and **cos** x (§31.2) was a geometric one that did not permit us to calculate sin x and cos x when x is given as a number. On the basis·of this geometric definition, however, we found that if such functions exist, they must satisfy what we will call the *harmonic relation*.

Harmonic Relation

$D \cos x = -\sin x,\qquad D \sin x = \cos x,$
$\cos 0 = 1,\qquad \text{and}\qquad \sin 0 = 0.$

LEMMA 1. If **cos** x and **sin** x are defined on the interval $[0, 1]$ and satisfy the harmonic relation, then

$$\sin^2 x + \cos^2 x = 1.$$

Proof. Using the harmonic relation, we calculate the derivative

$$D(\sin^2 x + \cos^2 x) = 2 \sin x \cos x - 2 \cos x \sin x = 0.$$

Hence $\sin^2 x + \cos^2 x = C$, constant on the interval $[0, 1]$. Since sin $0 = 0$ and cos $0 = 1$, we find that $C = 1$. This proves the lemma.

We now recall Taylor's theorem (§14.4), which says that the values of a repeatedly differentiable function $f(x)$ in an interval $[0, 1]$ are exactly given by

$$f(x) = f(0) + \frac{f'(0)}{1!} x + \frac{f^{(2)}(0)}{2!} x^2 + \cdots + \frac{f^{(n)}(0)}{n!} x^n + R_{n+1},$$

where the remainder term $R_{n+1} = [f^{(n+1)}(z)/(n + 1)!]x^{n+1}$, evaluated at some unspecified number z between 0 and x.

LEMMA 2.   If **cos** x and **sin** x are functions on the interval $[0, 1]$ that satisfy the harmonic relation, then their Taylor's expansion are

$$\cos x = 1 - \frac{x^2}{2!} + \frac{x^4}{4!} + \cdots + (-1)^{n/2} \frac{x^n}{n!} + R_{n+1} \qquad (n \text{ even}),$$

$$\sin x = x - \frac{x^3}{3!} + \frac{x^5}{5!} + \cdots + (-1)^{(n-1)/2} \frac{x^n}{n!} + S_{n+1} \qquad (n \text{ odd}).$$

Proof.   Exercise 6, page 225.

LEMMA 3.   The remainder terms R_{n+1} and S_{n+1} in the Taylor expansions of $\cos x$ and $\sin x$ on the interval $[0, 1]$ satisfy the inequalities

$$|R_{n+1}| \leq \frac{1}{(n+1)!} \qquad \text{and} \qquad |S_{n+1}| \leq \frac{1}{(n+1)!}.$$

Proof.   In the Taylor expansion with $f(x) = \cos x$, $|f_{n+1}(z)| \leq 1$, since it is either $|\sin z|$ or $|\cos z|$, and Lemma 1 implies that for every z, $|\sin z| \leq 1$ and $|\cos z| \leq 1$. Also $|x^n| \leq 1$, since $0 \leq x \leq 1$. This gives $|R_{n+1}| \leq 1/(n+1)!$. The estimate, $|S_{n+1}| \leq 1/(n+1)!$, is similar. This completes the proof.

We can now turn the argument around. We use the Taylor expansions to define **cos** x and **sin** x. We use the Taylor expansions (Lemma 2) with remainder estimates (lemma 3) to compute $\cos x$ and $\sin x$ to any prescribed accuracy. For example, to compute $\cos x$ to 5-place accuracy we have only to carry the expansion to n terms, where n is chosen so that $|R_{n+1}| < 10^{-5}$. This is true if $n \geq 9$ (§29.5). Hence the Taylor polynomial

$$1 - \frac{x^2}{2!} + \frac{x^4}{4!} - \frac{x^6}{6!} + \frac{x^8}{n!}$$

computes $\cos x$ to 5-place accuracy for all x in $[0, 1]$. For an x near zero even fewer terms would be required (Exercises). We verify by differentiation of the expansions themselves that the functions defined by these relations satisfy the harmonic relation. We summarize these results in a theorem.

THEOREM (Existence of the trigonometric functions).   The functions defined by the expansions

$$1 - \frac{x^2}{2!} + \cdots + (-1)^{n/2} \frac{x^n}{n!} + R_{n+1} \qquad (n \text{ even}),$$

and

$$x - \frac{x^3}{3!} + \cdots + (-1)^{(n-1)/2} \frac{x^n}{n!} + S_{n+1} \qquad (n \text{ odd}),$$

and the inequalities

$$|R_{n+1}| \leq \frac{1}{(n+1)!}, \qquad |S_{n+1}| \leq \frac{1}{(n+1)!}$$

are respectively **cos** x and **sin** x on the interval $[0, 1]$.

Returning to the geometric definitions of the trigonometric functions (§31.2), we find that it is enough to compute $\sin \theta$, $\cos \theta$, $\tan \theta$ on the interval $[0, \pi/4]$, which is contained in the interval [0, 1], on which our Taylor's expansion applies. All other values can be found from these. For, from their values in [0, $\pi/4$], we can compute the trigonometric functions in the interval [$\pi/4$, $\pi/2$] by the complementary relations

$$\sin \theta = \sin \left(\frac{\pi}{2} - \theta\right), \qquad \cos \theta = \cos \left(\frac{\pi}{2} - \theta\right).$$

Then for θ in the second quadrant [$\pi/2$, π] we can use the supplementary relations

$$\sin \theta = \sin (\pi - \theta), \qquad \cos \theta = -\cos (\pi - \theta).$$

With the trigonometric functions computed on [0, π] we can use the negative angle relations

$$\sin (-\theta) = -\sin \theta, \qquad \cos (-\theta) = \cos \theta,$$

to get them on the interval [$-\pi$, 0]. This gives us **sin θ, cos θ, tan θ** on the interval [$-\pi$, π], except where **tan θ** is undefined. Their values for all other angles are easily derived from these by adding or subtracting complete revolutions, using the periodic property.

34.2   Tangent Line to a Trigonometric Graph

Example.   Find the slope of the tangent line to the graph of $f(x) = \frac{1}{2}(1 - \cos 2x)$ at the point where $x = \pi/6$. Plot the tangent line and the graph.

Solution.

$$D\frac{1}{2}(1 - \cos 2x) = \sin 2x.$$

Thus

$$f\left(\frac{\pi}{6}\right) = \frac{1}{4} \qquad \text{and} \qquad f'\left(\frac{\pi}{6}\right) = \frac{\sqrt{3}}{2}.$$

At the point ($\pi/6$, $\frac{1}{4}$), we plot the line of slope $\sqrt{3}/2$ (Figure 5.37). We can plot the graph of $f(x) = \frac{1}{2}(1 - \cos 2x)$ by first plotting the sinusoidal function $-\frac{1}{2}\cos 2x$ with amplitude $\frac{1}{2}$ and period π. Then add $\frac{1}{2}$ to it, which moves its graph up $\frac{1}{2}$ unit.

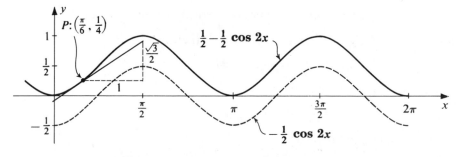

FIGURE 5.37   Tangent to $\frac{1}{2}(1 - \cos 2x)$.

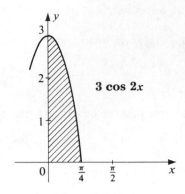

FIGURE 5.38   Area bounded by **3 cos 2x** on [0, $\pi/4$].

34.3   Area Below a Trigonometric Graph

Example.   Find the area below the graph of 3 cos 2x from 0 to $\pi/4$ (Figure 5.38).

Solution.   This area is given by the integral

$$A = \int_0^{\pi/4} 3 \cos 2x \, dx.$$

We find an antiderivative

$$\int 3 \cos 2x \, Dx = \tfrac{3}{2} \int \cos 2x \, (2 \, Dx) = \tfrac{3}{2} \sin 2x.$$

With this antiderivative we apply the fundamental theorem and compute the area

$$A = \int_0^{\pi/4} 3 \cos 2x \, dx = \frac{3}{2} \sin 2x \Big|_0^{\pi/4} = \frac{3}{2} \sin \frac{\pi}{2} = \frac{3}{2} \text{ area units.}$$

34.4   A Maximum-Minimum Problem

Example.   Find the maximum and minimum values of

$$f(x) = 1 + \cos x + \tfrac{1}{2} \cos 2x \quad \text{on } [-\pi, \pi].$$

Solution.   The function is continuous and therefore has a maximum and a minimum on the closed interval $[-\pi, \pi]$. We calculate

$$f'(x) = -\sin x - \sin 2x = -\sin x - 2 \sin x \cos x.$$

The critical values include those where $f'(x) = 0$, that is,

$$\sin x + 2 \sin x \cos x = 0$$

We solve this equation by factoring,

$$\sin x \, (1 + 2 \cos x) = 0,$$

which is true if and only if either

$$\sin x = 0 \quad \text{or} \quad 1 + 2 \cos x = 0.$$

The first equation has the roots $x = -\pi, 0, \pi$. The second equation,

$$\cos x = -\frac{1}{2},$$

has the roots $x = -2\pi/3, 2\pi/3$. The set of all critical values of x is then

$$\left\{-\pi, -\frac{2\pi}{3}, 0, \frac{2\pi}{3}, \pi\right\},$$

including the endpoints of the interval. We compute

$$f(-\pi) = \frac{1}{2}, f\left(-\frac{2\pi}{3}\right) = 1, f(0) = \frac{5}{2}, f\left(\frac{2\pi}{3}\right) = 1, f(\pi) = \frac{1}{2}.$$

Since the maximum and minimum exist, and are included in this set, we see that the maximum occurs at 0 and is $f(0) = \frac{5}{2}$, and minima occur at $-\pi$ and π, where $f(-\pi) = f(\pi) = \frac{1}{2}$.

34.5   A Rate Problem

Example.   A wheel of radius 2 ft is revolving clockwise around a fixed axis at 20 rpm. How fast is a point on the circumference moving horizontally when it is 1 ft below the horizontal line through the axis? (See Figure 5.39.)

Solution.   We choose coordinates so that the origin is at the axis of the wheel and the point $P:(x, y)$ starts from $(0, 1)$ when $t = 0$. Then the function $\theta = -(2\pi/3)t$ gives the rotation at any time t seconds in radians. We wish to express the horizontal coordinate as a function $x(t)$. Then the derivative $x'(t)$ gives the required horizontal velocity at any time t.

In this problem, $x(t) = 2 \cos \theta$. Hence, $x'(t) = -2 \sin \theta\, D_t\theta$. When the moving point P is 1 ft below the axis, $y = -1$ and $\sin \theta = -\frac{1}{2}$. We know $D_t\theta = -2\pi/3$ rad/sec. Hence at any instant t when $y = -1$ we have

$$x'(t) = -2\left(-\frac{1}{2}\right)\left(-\frac{2\pi}{3}\right) = -\frac{2\pi}{3} \text{ ft/sec.}$$

This completes the problem. The minus sign means that P is moving to the left at the prescribed instant (Figure 5.39).

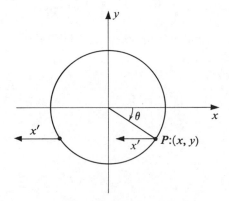

FIGURE 5.39

34.6 Pendulum Problem. Find the motion of a simple pendulum, of length l, that is displaced from equilibrium to an angle β and released with zero initial velocity (Figure 5.40).

Solution. The pendulum is constrained to move in a circle $x^2 + y^2 = l^2$. The only force that affects the motion is the component of gravity tangent to the circle. This is $-mg \sin \theta$, where θ is the angular displacement from equilibrium and m is the mass (Figure 5.40). The law of motion (§18.3) says that

$$mD^2s = -mg \sin \theta,$$

where differentiation is with respect to time t. Since $s = l\theta$, this becomes

$$D^2\theta = -\frac{g}{l} \sin \theta,$$

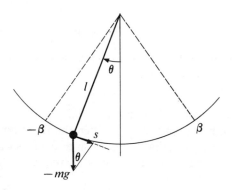

FIGURE 5.40   Forces on pendulum.

with $\theta(0) = \beta$ and $\theta'(0) = 0$. This differential equation of motion can be approximated by a simpler one when θ is small. We replace $\sin \theta$ by θ. The error of this approximation can be found by Taylor's expansion (Lemma 2), which says that

$$\sin \theta = \theta - \cos z \frac{\theta^3}{3!}, \qquad \text{where} \quad 0 \leq z \leq \theta.$$

Since $|\cos z| \leq 1$, this tells us that when $|\theta| < 0.1$ rad,

$$|\sin \theta - \theta| < \frac{\theta^3}{3!} < 0.00017 \text{ rad}.$$

The approximate equation then becomes

$$D^2\theta + \frac{g}{l} \theta = 0, \qquad \theta(0) = \beta, \qquad \theta'(0) = 0,$$

and this is the equation of simple harmonic motion (§32). The solution that satisfies the initial conditions is

$$\theta = \beta \cos \sqrt{\frac{g}{l}} \, t.$$

It represents a periodic oscillation with amplitude $|\beta|$ and period $P = 2\pi\sqrt{l/g}$ sec.

This leaves open the questions of whether the exact motion is periodic with some period T, and if so, whether the period P is a good approximation to the exact period T (Exercises).

34.7 Exercises

1. Find the slope and equation of the tangent line to the graph of $f(t) = 2 \cos x/3$ at the point where $x = \pi$. Plot the tangent line.
2. Find the area of the region under the graph of **sin x** between successive points where it crosses the x-axis.
3. Find the area of the region bounded by the graph of **cos x** from 0 to π. Explain why the result is zero.
4. Compute sin 0.5 and cos 0.5, with error less than 0.0001.
5. Compute sin 3° with error less than 0.0001.
6. Verify the expansions of Lemma 2.
7. In the derivation of $D \cos x$ (Section 31), we found that $|(\sin \theta/\theta) - 1| < \theta$. Show that, in fact,
$$\left| \frac{\sin \theta}{\theta} - 1 \right| < \frac{\theta^2}{6}.$$
8. How many terms of the expansion of **sin x** are needed to compute sin 50° correct to five places?
9. If $x < 0.1$, how many terms of the expansion are needed to compute cos x with fifth-place accuracy?
10. If sin 5° is computed from $x - (x^3/3!)$, obtain an estimate of the error.
11. Compute tan $\frac{1}{2}$ to three-place accuracy.
12. Find the motion of a pendulum, of length 100 cm, that is displaced through an angle $\beta = 0.1$ rad and released. What is the period? Amplitude?
13. In Exercise 12, how long would the pendulum have to be to have a period twice as long? In general, how does the period of the (approximate) pendulum vary with l, m, and β?
14. Find the maximum and minimum of **sin x** $-$ **cos x** on $[-\pi, \pi]$, *if they exist*.
15. Find the maximum and minimum of **sin 2x** $+$ **2 sin x** on $[-\pi, \pi]$, *if they exist*.
16. Find the maximum and minimum of **sin x/cos x** on the real numbers, *if they exist*.
17. An unseen flying saucer, being tracked by a rotating radar antenna, is moving on a straight-line path whose nearest approach to the tracking station is 4 miles due north of it. At the instant that the antenna is pointed at N 60° E, the rate of rotation of the antenna is $-\frac{1}{2}$ rad/sec. How fast is the flying saucer moving?
18. The functions x and y are related by
$$x \cos y + y \cos x = 1.$$
    Find x' when $x = \pi/3$, $y = 0$, and $y' = -1$.
19. A point $P:(x, y)$ moves around the unit circle (Figure 5.41) so that the radius OP rotates

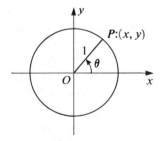

FIGURE 5.41   Exercise 19.

at the constant angular rate $D\theta = 2$ rad/sec, starting at $(1, 0)$ at time $t = 0$.
(a) Find the rates of horizontal and vertical motion of P at any time t.
(b) Find x' and y' for a value of $\theta = 5\pi/6$.

20.   A ball whirls in a vertical plane at the end of a string of length r, maintaining a constant angular rate of one revolution per second starting at the bottom of the circular path when $t = 0$. Find functions that give the horizontal and vertical components of position and of velocity at any time t thereafter.

21.   An open gutter with sides at angle θ to the horizontal is to be made from a long piece of sheet metal 12 inches wide by bending up one-third of the sheet on each side (Figure 5.42). Find the angle θ for maximum capacity.

FIGURE 5.42   Exercise 21.

22.   A kite is 80 ft above the ground with 100 ft of string out. It is moving horizontally away at 5 ft/sec from the boy who holds the string.
(a) At what rate is the string being paid out?
(b) At what rate is the inclination of the string to the horizontal changing?

23.* Consider the exact differential equation of the pendulum

$$\theta'' = -\frac{g}{l}\sin\theta, \qquad \theta(0) = \beta, \qquad \theta'(0) = 0.$$

Multiply both sides by $2\theta'$ and prove that

$$(\theta')^2 = 2\frac{g}{l}(\cos\theta - \cos\beta).$$

24.* Using the result of Exercise 23, prove that the angular velocity is zero when $\theta = \pm\beta$. Then prove that the exact motion oscillates from $\theta = \beta$ to $\theta = -\beta$, and return, with some period T seconds.

25.* Show that the period of the pendulum is given by

$$T = \sqrt{\frac{2l}{g}}\int_{-\beta}^{\beta}\frac{d\theta}{\sqrt{\cos\theta - \cos\beta}}.$$

This integral cannot be evaluated by the fundamental theorem because the integrand has no elementary antiderivative.

34R   The Transcendental Functions (A review)

ACADEMIC NOTE.   This review of Sections 24–34 is not a necessary part of the development of calculus. It is designed to be read rather than studied.

34.1R   Missing Antiderivatives. We found (§24.2) that every integral power x^n, including those with negative exponents, has a derivative $Dx^n = nx^{n-1}$, and this enabled us to write down an antiderivative

$$\int x^n\, Dx = \frac{x^{n+1}}{n+1} + C,$$

except when $n = -1$, which makes the denominator zero. Seeking the missing anti-derivative $\int 1/x \, \mathbf{D}x$, we found that the theory had already provided an answer. It tells us that since $1/x$ is decreasing on every interval of positive numbers x, it is integrable. Indeed, it is continuous. Under these conditions, the derivative with respect to x of the indefinite integral is

$$D \int_1^x \frac{1}{u} \, du = \frac{1}{x} \qquad (\S 17.4).$$

This enabled us to define a new function $\ln x$ by giving its value at each positive x as

$$\ln x = \int_1^x \frac{1}{u} \, du.$$

It was computable and had the right antiderivative $\mathbf{D} \ln x = 1/x$, so we adjoined it to our algebra of functions (§24). The negative powers themselves expanded our capability to solve max-min problems with side conditions (§25).

A closer study of $\ln x$ produced some results not directly related to its definition as an antiderivative of x^{-1}. It turned out to be a logarithm, in fact, to be a constant multiple of the ordinary base 10 logarithm $\log x$ (§§26.2, 26.7). The base of the logarithm $\ln x$ turned out to be a number e defined by $\ln e = 1$ (§§26.1, 26.7), which we later computed to be $e = 2.71828 \ldots$. We also found that $\ln x$ had an inverse function $\exp x$, which had the rather surprising property that it was its own derivative, $\mathbf{D} \exp x = \exp x$ (§26.3). Also technically, it turned out that $\exp x = e^x$ (§28.4). This justified its name as an exponential function.

We found that the self-derivative function $\exp x$ had remarkable applications in the solution of the differential equation $\mathbf{D}f(t) = kf(t)$, which expresses the law of growth, or compound interest law. All solutions of this equation turned out to be given by $f(t) = C \exp kt$ (§29).

The functions $\ln x$ and $\exp x$ were our first *transcendental* functions in calculus, that is, functions whose values $f(x)$ are not producible from x, and the real numbers, by a finite sequence of algebraic operations of addition, subtraction, multiplication, division, and extraction of roots. This fact makes transcendental functions difficult, and requires that we establish a firm theory for each one. All such functions involve limits both in their definition and evaluation (§§26, 28). Previously computed tables, log tables, etc. are useful in their evaluation. We also use Taylor's theorem (§§29.5, 34) for numerical evaluation.

We used $\ln x$ and $\log x$ in computation, also to solve exponential equations like $2^x = 3$ (§§26.7, 26.8).

Another new function that we introduced via $\ln x$ was the power function x^p for all real powers p (§§26.5, 26.6). It is an algebraic function when p is a rational number, for example, $x^{1/2} = \sqrt{x}$, but is transcendental when p is irrational.

The trigonometric functions and their inverses: $\sin x$, $\cos x$, $\tan x$, $\arcsin x$, $\arccos x$, $\arctan x$ completed our list of "elementary" transcendental functions (§§31.2, 33, 34). We introduced all of them into the analytic program of calculus. Typically, it was this analytic program itself, or an important application, or both, that generated the specifications for some missing function that then had to be constructed. These elementary functions by no means exhaust the possibilities. The further we go in the analytic program of calculus (§23R), the more of these "higher" transcendental functions we must introduce. This ultimately becomes so complicated that we return to fundamentals and develop the numerical program for calculus systematically.

We associated other techniques and applications with the transcendental functions. Among them were improper limits and integrals (§30) involving the ordinal infinities, ∞ and −∞. These, in terms of linear order, were end-elements adjoined to the real number line, but which could not be assimilated into the algebraic structure of real numbers. We defined such symbols as

$$\lim_{x \to \infty} f(x) \quad \text{and} \quad \int_a^\infty f(x)\, dx,$$

and studied their properties.

An application of the trigonometric function emerges in their role as periodic functions (§§32, 34). Here we found that $x = \sin kt$ and $x = \cos kt$ are solutions of the differential equation of simple harmonic motion $D^2x + k^2x = 0$, representing periodic oscillations of a spring or (approximately) a pendulum near its equilibrium point. The power of the ideas of calculus in this context is striking.

The geometric definition of the trigonometric functions by way of the circle $x^2 + y^2 = r^2$ is simple and classical (§31.2), but a computable definition emerges from the expansion by Taylor's theorem (§34).

We studied other methods of numerical computation that provide definite answers to many problems. A very accurate method of numerical integration is Simpson's rule (§27.1). Newton's method (§27.5) provides us with a method of successive approximation converging to the real roots of an equation $f(x) = 0$. As in Simpson's rule, an estimate of the error is essential. Finally, the time-honored table of common logarithms is not to be neglected for computation in calculus, especially for exponentials and exponential equations (§27.4).

34.2R Power Series Representations.   We have used the expansions of transcendental functions in powers of $x - a$ by Taylor's theorem as a means of representing them explicitly and establishing them as computable. We recall that Taylor's theorem says that a function $f(x)$ that has derivatives of all orders at the point where $x = a$ can be represented at a nearby x by

$$f(x) = f(a) + \frac{f'(a)}{1!}(x - a) + \frac{f^{(2)}(a)}{2!}(x - a)^2 + \cdots + \frac{f^{(n)}(a)}{n!}(x - a)^n + R_{n+1},$$

with the remainder

$$R_{n+1} = \frac{f^{(n+1)}(z)}{(n+1)!}(x - a)^{n+1},$$

and z is an unspecified number between a and x. This common mode of representation of transcendental functions as "extended polynomials" brings them together. So we list here some of their Taylor's expansions, where in each case the remainder, R_{n+1}, is to be evaluated by Taylor's theorem.

$$\exp x = 1 + \frac{x}{1!} + \frac{x^2}{2!} + \cdots + \frac{x^n}{n!} + \cdots.$$

$$\ln x = \frac{x - 1}{1} - \frac{(x - 1)^2}{2} + \frac{(x - 1)^3}{3} + \cdots + (-1)^{n-1}\frac{(x - 1)^n}{n} + \cdots.$$

$$x^p = 1 + \frac{p}{1!}(x - 1) + \frac{p(p - 1)}{2!}(x - 1)^2 + \cdots$$
$$+ \frac{p(p - 1)\cdots(p - n + 1)}{n!}(x - 1)^n + \cdots.$$

$$\sin x = x - \frac{x^3}{3!} + \frac{x^5}{5!} + \cdots + (-1)^{(n-1)/2}\frac{x^n}{n!} + \cdots, \quad (n \text{ odd}).$$

$$\cos x = 1 - \frac{x^2}{2!} + \frac{x^4}{4!} - \cdots + (-1)^{n/2}\frac{x^n}{n!} + \cdots, \quad (n \text{ even}).$$

$$\tan x = x + \frac{x^3}{3} + \frac{2x^5}{15} + \frac{17x^7}{315} + \cdots.$$

$$\arcsin x = x + \frac{x^3}{6} + \frac{1.3}{2.4}\frac{x^5}{5} + \cdots + \frac{1.3\cdots(n-2)}{2.4\cdots(n-1)}\frac{x^n}{n} + \cdots, \quad (n \text{ odd}).$$

$$\arccos x = \frac{\pi}{2} - \arcsin x.$$

Infinitely many other transcendental functions have power series representations like this. In each of the expansions above, we can prove that for a suitably restricted closed interval of values of x and for any preassigned precision measure ϵ, there is a number N of terms of the Taylor's expansion that will give the numerical value of $f(x)$, with error $|R_{N+1}| < \epsilon$.

The Taylor's expansion for the power function x^p is in fact the binomial series. It reduces to the $n+1$ terms of the binomial theorem when $p = n$, a positive integer. To see this, we observe that

$$x^p = [1 + (x - 1)]^p$$

and expand this binomial in powers of $x - 1$ by the binomial theorem. The result is precisely the same as that given by Taylor's theorem. But Taylor's theorem is valid also when p is any real number, not necessarily a positive integer, provided that we include the remainder term.

34.3R   Self Quiz

1. Define the function x^{-n} for negative integer exponents $-n$, being careful to include in the definition the domain of the function (§24.1).
2. At what value of x does the formula $y = x^{-n}$ fail to define a number y? What is the behavior of the graph of $y = x^{-n}$ near this point (§24.1)?
3. Under suitable assumptions about $f(x)$, what is the derivative $D \int_a^x f(u)\, du$? State the "suitable assumptions" (§21.7).
4. How can we exhibit an antiderivative of x^{-1} (§24.4)?
5. Define $\ln x$ (§24.4). Remember to specify the domain.
6. What is the inverse of $\ln x$ (§28.1)? What is the derivative of this inverse (§28.2)?
7. Define the number e (§26.1).
8. What is the role of e in connection with the function $\ln x$ (§26.7)? With the function $\exp x$ (§28.4)?
9. What is the compound-interest law (§29.2)?
10. Define x^p for all real numbers p. Why is the domain restricted to positive numbers x (§26.5)?
11. What does it mean to say that $f(x)$ becomes infinite at the point where $x = c$ (§30.2)?
12. What do we mean by the symbols $+\infty$ and $-\infty$? They have nothing to do with counting sets that have infinitely many elements. That is an entirely different idea. Are $+\infty$ and $-\infty$ numbers? Do the symbols $0 \cdot \infty$ and ∞/∞, have any meaning (§30.1)?
13. Define $\lim_{x \to \infty} f(x) = b$ (§30.2).
14. Define $\int_a^\infty f(x)\, dx$ (§30.3).
15. What is meant by saying that the integral $\int_a^\infty f(x)\, dx$ converges? Or that $\int_a^\infty g(x)\, dx$ does not exist (§30.3)?

16. What is the radian measure of angle (§31.1)?
17. Define sin θ, cos θ, tan θ (§31.2).
18. $\lim_{\theta \to 0} \sin \theta/\theta = k$. What is k? The answer depends on what angle measure we choose for θ (§31.5).
19. Define **arcsin** x, **arccos** x, **arctan** x, being careful to eliminate multiple values (§33.3).
20. Write down formulas for the derivatives of the inverse trigonometric functions (§33.4).
21. Since the derivatives of the inverse trigonometric functions are algebraic functions, it is obvious that the inverse trigonometric functions could have been defined by indefinite integrals as **ln** x was. Write explicit integral formulas for these alternative definitions.
22. What is a periodic function? the period of a periodic function (§32.2)?
23. What is a sinusoidal function? What are the explicit formulas for its period and amplitude? What differential equation does it satisfy (§32.4)?

34.4R   Miscellaneous Exercises

In Exercises 1–12, differentiate the functions.

1. $\ln (2x - 1)^3$.

2. $\dfrac{1 - \cos 2x}{\sin 2x}$.

3. $\ln \left(\dfrac{x + 1}{x - 1}\right)$.

4. $\cos (\arcsin x^2)$.

5. $x^{\sqrt{3}}$.

6. $\dfrac{1}{a} \arctan \dfrac{x}{a}$.

7. $\ln |x + \sqrt{x^2 - 1}|$.

8. $\displaystyle\int_1^x \dfrac{dt}{t - 2}$.

9. $\sqrt{x^2 + 2x + 1}$.

10. $\dfrac{1}{\sqrt{x^2 + a^2}}$.

11. $\sqrt{\dfrac{x - 1}{x + 1}}$.

12. $e^{-x^2/2}$.

In Exercises 13–18, compute the integrals by using the fundamental theorem, if they converge.

13. $\displaystyle\int_0^{1/2} \dfrac{dx}{\sqrt{1 - x^2}}$.

14. $\displaystyle\int_{-a}^a \dfrac{dx}{x^2 + a^2}$.

15. $\displaystyle\int_0^\infty \dfrac{dx}{1 + x^2}$.

16. $\displaystyle\int_1^2 \dfrac{dx}{\sqrt{x - 1}}$.

17. $\displaystyle\int_0^\infty \dfrac{dx}{(x + 1)^2}$.

18. $\displaystyle\int_{-1}^1 \dfrac{dx}{x^2}$.

19. For the curve $y = x^{3/2}$, find the slope of the tangent at $(1, 1)$. Sketch the curve and plot the tangent line.
20. For the curve $y = \arccos x$, find the slope of the tangent at the point where $x = \frac{1}{2}$, and plot this tangent line.
21. Find the equation of the tangent line to $y = \ln x$ at the point where $x = e$. At what point does this tangent line cross the x-axis?
22. Show that $y = te^{2t}$ satisfies the differential equation

$$D^2y - 4Dy + 4y = 0.$$

23. Show that if $y = \frac{1}{2}(e^x + e^{-x})$, then $\sqrt{1 + y'^2} = y$.
24. Compute sin 0.2 and cos 0.2 correct to 4 decimal places.

In Exercises 25–32, compute π by calculus methods.

25. Verify that for every number x,

$$\int_0^x \frac{du}{1 + u^2} = \arctan x.$$

26. Prove that $\arctan \frac{1}{2} + \arctan \frac{1}{3} = \pi/4$.

27. By polynomial division, or geometric progressions, or Taylor's theorem (§14.4), show that for every natural number n,

$$\frac{1}{1 + t^2} = 1 - t^2 + t^4 + \cdots + (-1)^n t^{2n} - (-1)^n \frac{t^{2n + 2}}{1 + t^2}.$$

28. Using Exercises 25 and 27, show that

$$\arctan x = x - \frac{x^3}{3} + \frac{x^5}{5} + \cdots + (-1)^n \frac{x^{2n + 1}}{2n + 1} + R_{2n + 2},$$

where

$$R_{2n + 2} = (-1)^{n + 1} \int_0^x \frac{t^{2n + 2}}{1 + t^2} \, dt.$$

29. Prove that when $x > 0$,

$$|R_{2n + 2}| \leqq \int_0^x t^{2n + 2} \, dt = \frac{x^{2n + 3}}{2n + 3}.$$

30. If $x = \frac{1}{2}$ and if $x = \frac{1}{3}$, what is the smallest integer n so that $|R_{2n + 2}| < \frac{1}{4}(10^{-4})$.

31. Use Exercises 28 and 30 to evaluate $\arctan \frac{1}{2}$ and $\arctan \frac{1}{3}$ with error less than $\frac{1}{4}(10^{-4})$.

32. Use Exercises 31 and 26 to compute π with error less than 10^{-4}.

33. A particle at P (Figure 5.43) with negative unit charge is at distance x cm from an infinitely long wire with positive charge of uniform density μ per cm distributed along it. The particle at P is attracted to particles in the wire by a force inversely proportional to the square of the distance between them (Coulomb's law). Show that the total force F of attraction of the wire on the particle at P is given by $F = -\pi\mu k/x$.

34. (a) In Exercise 33, show that the work V done against the force F when the particle is moved from the distance 1 to the distance x from the wire is given by $V = \pi\mu k \ln x$. This is called a *logarithmic potential*.

    (b) What is the result of calculating the work V_0 done against the attraction of the wire by the negative particle's being moved from the wire to a distance x away from it?

35. Calculate the improper integral $\int_0^1 dx/\sqrt{1 - x^2}$.

36. Show that

    (a) $\lim\limits_{x \to \infty} \arctan 2x = \dfrac{\pi}{2}$;   (b) $\lim\limits_{x \to \infty} \arctan \dfrac{x}{a} = -\dfrac{\pi}{2}$,   if $a > 0$.

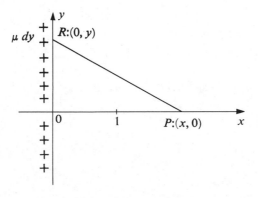

FIGURE 5.43   Exercise 33.

37. Show that

$$\int_{-\infty}^{\infty} \frac{dx}{1 + 2x^2} = \frac{\pi}{\sqrt{2}}.$$

38.* A rectangular box with dimensions x, y, z is to be constructed with fixed surface area S and fixed height $z = 4$. Find the dimensions for maximum volume.

34.5R   Problems

We prove in Problems T1–T4 that when e is defined (§26.1) as the number such that $\ln e = 1$, then

$$\lim_{n \to \infty} \left(1 + \frac{1}{n}\right)^n = e.$$

This was the historical definition of e.

T1. Prove the following equalities—

$$\ln \left(1 + \frac{1}{n}\right)^n = n[\ln (n + 1) - \ln n] = n \int_n^{n+1} \frac{dx}{x} = n \int_1^{1 + (1/n)} \frac{dt}{t}.$$

where in the last integral we have changed the scale on the x-axis by substituting $x = nt$.

T2. In the interval $1 \leqq t \leqq 1 + (1/n)$, we draw the graph of $1/t$ with understep and overstep functions such that

$$\frac{1}{1 + (1/n)} < \frac{1}{t} < 1.$$

From this show that

$$\left(\frac{1}{1 + (1/n)}\right) \frac{1}{n} < \int_1^{1 + (1/n)} \frac{dt}{t} < (1) \frac{1}{n}.$$

T3. The results of T1 and T2 imply that

$$\frac{1}{1 + (1/n)} < \ln \left(1 + \frac{1}{n}\right)^n < 1.$$

From this prove that

$$\lim_{n \to \infty} \ln \left(1 + \frac{1}{n}\right)^n = 1.$$

T4. Show that

$$\left(1 + \frac{1}{n}\right)^n = \exp \left[\ln \left(1 + \frac{1}{n}\right)^n\right],$$

and use this and T3 to prove that

$$\lim_{n \to \infty} \left(1 + \frac{1}{n}\right)^n = e.$$

What properties of **exp** x are used in the argument?

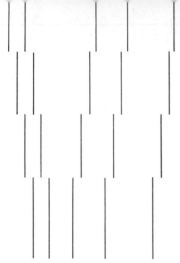

Applications in Economics and Life Sciences

CHAPTER **6**

This chapter depends only on Sections 1–34.

35   Calculus in Economics*

In economics there are many maximum and minimum problems to which calculus applies. We may wish to find the minimum point of the function which represents the cost, or we may wish to find the maximum point of the profit function. Typically the functions involved are not known by explicit formulas, so we use the device of postulating that they have derivatives, and proceed with the analysis on this basis.

35.1   Theory of the Business Firm

Example A.   A business firm produces y units per week of its product at a total cost $C(y)$, which is a function whose value depends on the number of units produced. Its weekly total revenue from the sale of this product is $R(y)$. Find the production level y for maximum profit.

   Graphical Solution.   The total profit is also a function of the production level y. We represent it by $T(y)$, defined for all positive numbers y. The definition of profit is the excess of revenue over cost, which is expressed mathematically by

$$T(y) = R(y) - C(y).$$

---

* After W. J. Baumol, *Calculus and Economics Concepts.* Princeton notes.

233

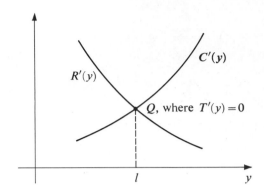

FIGURE 6.1   Marginal cost and marginal revenue.

We are assuming that the revenue function $R(y)$ has a derivative $R'(y)$ called *marginal revenue* in economics and $C(y)$ has a derivative $C'(y)$ called *marginal cost*. Therefore $T(y)$ has a derivative $T'(y)$ (marginal profit) given by

$$T'(y) = R'(y) - C'(y),$$

defined for every production level y. We observe that the theorem on critical points (§13.1) implies that if there is a maximum value of $T(y)$, it must occur at a point where $T'(y) = 0$, therefore at a point where $R'(y) = C'(y)$.

We now plot (Figure 6.1) on the same coordinate axes the graphs of marginal revenue $R'(y)$ and marginal cost $C'(y)$. The data for constructing these graphs will ordinarily be obtained from the records of the firm rather than from some formula. For our firm, the graphs of $R'(y)$ and $C'(y)$ intersect at a point Q, where $y = l$. Therefore at this point $R'(y) = C'(y)$ and hence l is a critical value of y.

But since not all critical values determine maximum points, how do we know that $(l, T(l))$ is the maximum point of the profit function $T(y)$? We can also answer this question from the graphs of the marginal revenue and cost. We observe that when $y < l$, the graph of $R'(y)$ is above the graph of $C'(y)$, and hence $R'(y) - C'(y)$ is positive. Similarly, if $y > l$, the value of $R'(y) - C'(y)$ is negative. We tabulate these results and their implications for the profit function $T(y)$ (see accompanying table).

Interval	$T'y = R'(y) - C'(y)$	$T(y)$
$y < l$	$T'(y) = +$	Increasing
$y = l$	$T'(y) = 0$	Critical
$y > l$	$T'(y) = -$	Decreasing

This tells us that the graph of $T(y)$ increases with y until the level l is reached and thereafter decreases. Hence $(l, T(l))$ is actually a point of maximum profit within the set of production levels y for which these graphs represent the experience of the firm. This completes the solution of Example A.

35.2 The Monopoly Case.

The method of solution of Example A depended only on very general considerations: namely, that revenue and cost $R(y)$ and $C(y)$ are functions of the production level y and that the profit $T = R - C$. In particular, the method is applicable both to firms engaged in competitive operations and to firms that have a monopoly. Let us now specialize the analysis by introducing a "demand function" $P(y)$, where $P(y)$ is the highest price at which y units of the product can be sold. Ordinarily under monopoly conditions, the "demand curve," which is the graph of $P(y)$, will be a down-sloping curve since under monopoly conditions, to increase the number of units sold the firm must decrease the price. This is true in the following example.

Example B.   A firm produces ball-point pens. The relation between the price $P(y)$ in cents and the number of pens y the market will absorb each week is given by

$$P(y) = 100 - 0.02y$$

for $0 \leq y < 5000$. The total cost of producing y pens is given by

$$C(y) = 40y + 25{,}000.$$

Find the number of pens that will give the maximum profit, and find the maximum profit.

Solution.   The revenue from the sale of y pens at the price $P(y)$ per pen is $yP(y)$. Thus, $R(y) = yP(y)$. The profit function $T(y)$ is therefore given by

$$
\begin{aligned}
T(y) &= yP(y) - C(y) \\
&= y(100 - 0.02y) - (40y + 25{,}000) \\
&= 100y - 0.02y^2 - 40y - 25{,}000 \\
&= -0.02y^2 + 60y - 25{,}000.
\end{aligned}
$$

We compute $T'(y)$,

$$T'(y) = -0.04y + 60,$$

and we set $T'(y) = 0$. This gives us $-0.04y + 60 = 0$, or $y = 1500$. We shall test the critical points by the sign of the first derivative $T'(y)$, tabulating the work as usual (see accompanying table). This shows that the profit $T(y)$ increases from the production

Interval	$T'(y) = -0.4(y - 1500)$	$T(y)$
$0 \leq y < 1500$	$T'(y) = (-)(-) = +$	Increasing
$y = 1500$	$T'(y) = (-)(0) = 0$	Critical
$1500 < y < 5000$	$T'(y) = (-)(+) = -$	Decreasing

level 0 to 1500 pens and then decreases from 1500 to 5000. Therefore a production level of 1500 pens per week gives the maximum profit and the price at which these pens are sold is found from the demand function

$$P(1500) = 100 - 0.02(1500) = 100 - 30 = 70 \text{ cents.}$$

The maximum total profit is

$$
\begin{aligned}
T(1500) &= 1500P(1500) - C(1500) \\
&= (1500)(70) - [40(1500) + 25{,}000] \\
&= \$200 \text{ per week.}
\end{aligned}
$$

This completes the solution of the problem.

35.3   The Effect of Excise Taxes

Example C.   The government decides to levy a tax of 10 cents per pen. Find the production level at which the firm of Example B should operate for maximum profit under this new condition. Find the new selling price and weekly profits.

Solution.   The demand function $P(y)$ is the same but the total weekly cost is now given by

$$C(y) = 40y + 10y + 25,000 = 50y + 25,000,$$

and the total profit is

$$T(y) = yP(y) - C(y) = -0.02y^2 + 50y - 25,000,$$

We compute $T'(y)$ and set $T'(y) = 0$. This gives us

$$T'(y) = -0.04y + 50 = 0,$$

which has the solution $y = 1250$. Testing this critical value by use of the sign of $T'(y)$ as before, we find that the profit increases with increased production until the level $y = 1250$ pens is reached, and then decreases. So weekly production of 1250 pens gives maximum profit. The new selling price is $P(1250) = 100 - 0.02(1250) = 75$ cents. Furthermore, the total weekly profit is now given by

$$T(1250) = 1250(75) - [50(1250) + 25,000] = \$62.50 \text{ per week.}$$

This completes the solution of the problem. We observe that it did not pay to pass on all of the tax to the consumer.

35.4   The Effect of Overhead Costs

Example D.   "Overhead costs," or "fixed costs," are defined as costs that do not vary with the level of production. Suppose that in Example B the overhead costs increase by $100 per week. What change should be made in the level of production to maximize profit under this new condition?

EXPLORATION OF THE PROBLEM. We might argue that we should increase sales to spread overhead. We might increase production to a level where the overhead cost per pen is back to its original level in our solution of Example B. But under monopoly conditions, to do this we must accept a lower price to sell the increased output. This will cut our profit. On the other hand, we might argue that we should increase the price in order to help cover the increase in costs. But according to the demand function we will then accept a lower production level. Which of these arguments is right?

Solution.   The total cost function in Example B is now replaced by one in which

$$C(y) = 40y + 25,000 + 10,000 \text{ cents,}$$
$$= 40y + 35,000.$$

As a result, the total profit becomes

$$T(y) = -0.02y^2 + 60y - 35,000,$$

and the marginal profit,

$$T'(y) = -0.04y + 60,$$

which is the same as it was in Example B. The remainder of the analysis of critical values now proceeds exactly as in Example B to give the conclusion that a production level of 1500 pens again gives maximum profit. The price is 70 cents as before and the maximum profit is now $T(1500) = \$100$ per week.

This is our solution to the problem. Neither of the exploratory arguments was correct. It turned out that there is nothing the firm can do to offset the increased overhead by varying its price or production level. The increase in overhead is simply subtracted from the total revenue at the same optimal level of production as before.

35.5   A Minimum Cost Problem.   We now turn to an economic problem that calls for finding a minimum point.

Example E.   The cost C of transmitting a constant electric current i over a wire is the sum of the two kinds of costs. There is a heat loss proportional to Ri^2 (Ohm's law), where R is the electrical resistance of the conductor. This cost could be reduced by reducing R but this calls for a larger wire, heavier supports, and therefore higher capital costs inversely proportional to R. The total cost function for any resistance R is of the form

$$C(R) = B\left(Ri^2 + \frac{k}{R}\right),$$

where B and k are known positive constants. Find the resistance R that will give the minimum cost.

Solution.   The domain of $C(R)$ is all positive numbers R. Since there are no end-points and no points where $C'(R)$ fails to exist, the only critical values are the positive numbers R, where $C'(R) = 0$. We compute $C'(R)$ and set $C'(R) = 0$. This gives us

$$B\left(i^2 - \frac{k}{R^2}\right) = 0.$$

This equation has only one positive root $R = \sqrt{k}/i$. We shall test this critical value to see if it gives us a minimum point. We can write $C'(R)$ in the form

$$C'(R) = \frac{Bi^2}{R^2}\left(R^2 - \frac{k}{i^2}\right).$$

From this we see that if $R^2 < k/i^2$, the derivative $C'(R)$ is negative; and if $R^2 > k/i^2$, the derivative is positive. Therefore $\sqrt{k}/i$ is the resistance that gives minimum cost. This completes the solution of the problem.

35.6   Exercises

1. Compute the marginal revenue function in Example B.
2. In Example C, did it pay the firm to pass on the new tax to the consumer?
3. Generalize Example D to prove that a change in overhead costs never changes the level of production or the price to be charged for maximum profit.
4. As in Figure 6.1, plot the marginal revenue and marginal cost curves and find the y-coordinate of the point of intersection: (a) For Example B; (b) for Example C; (c) for Example D.
5. The price P at which y units of an article can be marketed is $36 - 0.02y$ in dollars. The total cost of producing y units is $600 + 4y$ in dollars. Find the production y that gives maximum profit to the producer.

6. The demand function for a product is $P(y) = 150 - 0.01y$ and the total cost function is $C(y) = 360 + 50y$. Find the production level for maximum profit.

7. A taxi company finds that at a charge of 20 cents a mile it does 800 miles of business per day. The number of passenger miles is 40 less for each cent increase in fare. What rate yields the greatest total revenue?

8. In Exercise 7, the operating costs are 15 cents per mile. What rate yields the greatest total profit?

9. If the total cost of producing and selling y units is

$$C(y) = 120y - 0.4y^2 + 0.0005y^3$$

and the demand function is $P(y) = 100 - 0.1y$:
(a) What level of production yields minimum cost per unit?
(b) Does this level of production yield maximum profit?
(c) At how many levels of production does marginal cost equal marginal revenue?
(d) Find the level of production for maximum profit.

10. A manufacturer contracts to make 50,000 articles at $80 per hundred. If the number exceeds 50,000, the price is 10 cents less for each hundred in excess of 50,000. How large an order will maximize revenues?

11. In a factory, the monthly cost function C for y units produced per month is $C = 2000 + 18y - 0.01y^2$. The manufacturer finds that at price P he can sell $y = 2000 - 20P$ units. What price gives the greatest profit? What is the maximum profit?

FIGURE 6.2   Exercise 11.

12. A strip of sheet metal 12 in. wide is to be bent to form a rain gutter of rectangular cross section as indicated in Figure 6.2. How deep should the trough be to give maximum water capacity?

13. Economic functions are often simple polynomials of degree 1. The function $P = 2x + 3y - 7$ is defined on the triangle with vertices $A:(2, 1)$; $B:(3, 4)$; $C:(8, 2)$. Find the maximum and minimum points of P.

14. Show that the maximum and minimum points of the function $P = ax + by + c$ defined on the triangle with vertices $A:(x_1, y_1)$; $B:(x_2, y_2)$; $C:(x_3, y_3)$ must occur at a vertex.

36 Scientific Measurement

36.1 General Principles in Scientific Measurement. When we establish a mathematical scale for the states of some observable physical quantity such as distance, time, mass, temperature, entropy, concentration, density, pressure, or perhaps intelligence economic utility, demand, social attitude, or intensity of emotion, we are in every case using the real numbers as a model representing the magnitude of the physical quantity. To establish the relationship between states, or magnitudes, of the physical quantity and the real numbers in the number line, it is not enough to identify experimentally two states, label one 0 and the other 100, and then claim that the numerical values of all the remaining states, are determined. There is also an operational principle* that

---

* P. W. Bridgeman, *Dimensional Analysis*. Article in *Encyclopedia Brittanica*.

asserts that it is necessary to exhibit experimentally a composition of the physical states under which any two states combine to form uniquely a third state. Then it is required that the measurement shall map these physical states into the real numbers in such a way that the composition of states corresponds to addition (or multiplication) of the numerical values. When we have such a mapping, we call it a *measurement* and the numerical image of the array of physical states is called a *scale* for the physical quantity.

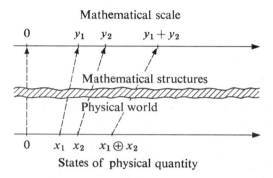

FIGURE 6.3   Measurement of a physical quantity.

We may picture this by a diagram (Figure 6.3), which has one foot in the physical world and the other in the mathematical domain of real numbers. To discuss the mathematical nature of measurement, this half-physical, half-mathematical picture is unsatisfactory. We replace the array of states of the physical quantity by a real-number line labeled by the name of the physical quantity. Thus we have two copies of the real-number line, which we represent as perpendicular axes. Our mathematical model of a measurement is then a function on the labeled real line into the real line that preserves one of the basic operations. For example, an addition-preserving measurement f has the property that

$$f(x_1 + x_2) = f(x_1) + f(x_2).$$

One such function is $y = mx$, where x is distance (Figure 6.4). In fact, if (x_1, y_1) and (x_2, y_2) are two points in the measurement, $(x_1 + x_2, y_1 + y_2)$ is also in it; for $y_1 = mx_1$ and $y_2 = mx_2$ implies that

$$y_1 + y_2 = m(x_1 + x_2).$$

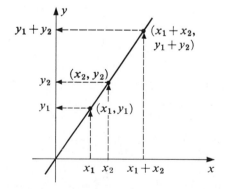

FIGURE 6.4   Addition preserving measurement.

In view of the requirement that a measurement preserve a composition of states, it is no mere accident that certain special functions are prevalent in measurement; and that fitting just any convenient curve, such as a polynomial, to a set of points found by experimental measurement ordinarily has no physical significance. We find that scientific measurements $y = f(x)$ of real-number-like quantities appear over and over as linear laws $y = mx$ or $y = mx + b$, power laws $y = Cx^p$, exponential laws $y = Ce^{kx}$, or logarithmic laws $y = k \ln Cx$. Let us examine conditions that prescribe that this shall be true, and find techniques for relating experimental data to these mathematical models and evaluating the constants involved.

36.2 Additive Measurements. For a physical quantity that has a composition of states we regard as addition, the measurement model is a continuous function on these states that preserves the addition of states. Mathematically, we convert this idea into the following definition.

DEFINITION. An *additive measurement* of the real line is a continuous function $f(x)$ on the real line such that for every pair of numbers x and h,

$$f(x + h) = f(x) + f(h).$$

We have seen that $f(x) = mx$ for every constant m is such a function, and we now give a calculus proof that these are the only ones.

THEOREM (On additive measurements). The only additive measurements of the real line are of the form $f(x) = mx$, for some number m.

Proof. Since $f(x)$ is additive,

$$f(x) = f(x + 0) = f(x) + f(0),$$

which implies that $f(0) = 0$. Also

$$0 = f(0) = f(x - x) = f(x) + f(-x)$$

implies that $f(-x) = -f(x)$ and hence that

$$f(x - h) = f(x) - f(h).$$

The hypothesis that $f(x)$ is continuous gives us two pieces of information, that $f(x)$ is integrable (§19.2) and that the indefinite integral $g(x)$, defined by

$$g(x) = \int_0^x f(t)\, dt,$$

is differentiable and $g'(x) = f(x)$ (§21.7).

On the other hand, we can calculate $g'(x)$ from the limit, as $h \to 0$, of the divided difference

(1)
$$\frac{g(x + h) - g(x)}{h} = \frac{1}{h} \int_0^{x+h} f(t)\, dt - \frac{1}{h} \int_0^x f(t)\, dt$$

in the last integral we make the substitution $t = u - h$, which implies that $dt = du$, $u = h$ when $t = 0$ and $u = x + h$ when $t = x$. We then find, remembering additivity,

$$-\frac{1}{h}\int_0^x f(t)\,dt = -\frac{1}{h}\int_h^{x+h} f(u-h)\,du = -\frac{1}{h}\int_h^{x+h}[f(u)-f(h)]\,du$$

$$= -\frac{1}{h}\int_h^{x+h} f(u)\,du + \frac{f(h)}{h}\int_h^{x+h} du$$

$$= \frac{1}{h}\int_{x+h}^h f(u)\,du + \left[\frac{f(h)}{h}\right]x.$$

We substitute this expression into (1) and find that we can combine two integrals, $\int_0^{x+h} + \int_{x+h}^h = \int_0^h$. When we do this we obtain

$$(2) \qquad \frac{g(x+h)-g(x)}{h} = \frac{1}{h}\int_0^h f(t)\,dt + \left[\frac{f(h)}{h}\right]x$$

$$= \frac{g(h)}{h} + \left[\frac{f(h)}{h}\right]x.$$

Now we proceed to the limit as $h \to 0$. We have seen that the left member approaches the limit $g'(x) = f(x)$. This includes, as a special case, $\lim_{h \to 0}$ of the first term on the right. In fact, since $g(0) = 0$,

$$\lim_{h \to 0}\frac{g(h)}{h} = \lim_{h \to 0}\frac{g(h)-g(0)}{h} = g'(0) = f(0) = 0.$$

Hence, for the last term in (2),

$$\lim_{h \to 0}\frac{f(h)}{h}$$

exists as some number, which we call m. Therefore when we pass to the limit as $h \to 0$ in (2), we find that for every x, $f(x) = 0 + mx$, which is what we set out to prove.

COROLLARY.   If $y = f(x)$ is a continuous mapping of the real line that preserves the addition of differences, then $f(x)$ has the form

$$f(x) = mx + b$$

for some m and some b.

For, by the theorem, $f(x)$ has such a form that $y - y_1 = m(x - x_1)$, which implies that $y = mx + b$, where $b = mx_1 + y_1$.

Time and temperature measurements have this form, since it is the intervals of time $t_2 - t_1$ and the differences of temperature $x_2 - x_1$ that we "add" experimentally.

Example.   A pendulum that is assumed to have a constant period in physical time is set in motion. Every time it passes through the vertical (equilibrium) position going from left to right the observer reads a clock and records the clock time t in seconds required to complete s oscillations. These observations are given by the accompanying table. The questions are, Is this a valid time measurement, and if so, what is the period of the pendulum in clock seconds?

s natural units	0	1	2	3	4	5	6	7
t clock seconds	1.3	3.8	6.2	8.7	11.0	13.5	15.9	18.3

Solution.   The experiment meets the requirements that there be a physical composition of time states. This composition of physical time units consists of the accumulation of times of oscillation as the pendulum swings, whose "sum" at whole intervals is given

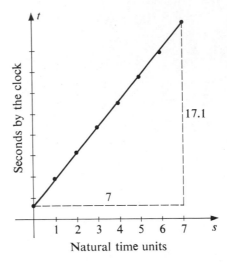

FIGURE 6.5   Measurement of time.

simply by the observer's counting. We must then determine whether the mapping onto the clock scale of seconds is additive. To do this we plot the data in Cartesian coordinates and find that a straight-line graph fits the data with acceptable accuracy (Figure 6.5). By graphical calculation, the slope of this line $m = 17.1/1 = 2.44$ seconds approximately. The fact that the graph is a straight line verifies that the experimental mapping is of the form $t = ms + b$. Hence it preserves the addition of natural time intervals and is therefore a measurement. This measurement gives 2.44 clock seconds as the period of the pendulum.

DEFINITION (Linear measurement).   A mapping $y = f(x)$ of the real numbers is *linear* if and only if it is continuous and preserves the addition of differences $x_2 - x_1$.

Clearly, additive measurements are a special case of linear measurements.

We find that the same mathematics applies to the experimental comparison of one previously measured scale with another. For example, using two thermometers for Fahrenheit and Centigrade temperature measurements F and C, we find that these measurements are related by $F = \frac{9}{5}C + 32$. This transformation from one scale of temperature measurement to another preserves the addition of temperature differences. Still more generally we may find that the mathematical representations of two *different* physical observables are related linearly, and from this we can conclude that differences and sums of differences are preserved, and so if we have a valid measurement of one quantity, we have one of the other. For example, the demand q for a certain good might be expressed experimentally in terms of the price p by a formula like $q = -0.05p + 65$, whose form assures us that we have a valid measurement of demand differences if we have a valid measurement of price differences.

We summarize by stating some easy consequences of the theorem on additive measurements.

COROLLARY (Tests of empirical data).   For a continuous mapping $f(x)$ of the real numbers the following are equivalent statements.

(a) $f(x)$ is a linear measurement.
(b) Its graph is a straight line that is not vertical.

(c)   For some numbers m and b, $f(x) = mx + b$.

(d)   Every arithmetic progression of states $x_1, x_2, \ldots, x_n \cdots$ maps into an arithmetic progression $f(x_1), f(x_2), \ldots, f(x_n) \ldots$.

(e)   There is a constant ratio m of differences; that is, for every pair of distinct states $x_1, x_2,$

$$m = \frac{f(x_2) - f(x_1)}{x_2 - x_1}.$$

(f) $f(x)$ satisfies for some number m a differential equation $Df(x) = m$.

These tests do not take account of the deviations from an exact measurement due to random errors. For such a consideration, we should turn to probability analysis, which is outside the subject of calculus.

A consequence of the theorem is that for linear measurements when we know the mappings of two points, the complete measurement $f(x)$ is determined. That is, if we know that we have a linear measurement $y = f(x)$, and we know two points (x_1, y_1) and (x_2, y_2), then the measurement has the form $f(x) = mx + b$, and m and b are determined real numbers. However, this is not entirely elementary, for if we examine the proof of the theorem, we find that it depends not only on the additive property and continuity of $f(x)$ but also essentially, through the fundamental theorem of calculus, on the completeness property of the real numbers (§7.5), which says that every nonempty, bounded set of real numbers has an infimum and a supremum.

A classical example that illustrates the essential role of the completeness property in this analysis is the continuous, additive mapping $c = f(x)$ of the sides x of isosceles right triangles on their hypotenuses c (Figure 6.6). It was known to the ancient Greeks (in somewhat different mathematical formulation) that the rational numbers, which lack the completeness property, do not suffice to express this continuous additive measurement. In fact, the measurement is given by $c = \sqrt{2}x$, where $\sqrt{2}$ is not a rational number but requires the completeness property to establish its existence.

36.3 Exercises

In Exercises 1–4, find which tables of data are measurements that preserve addition of intervals. Which ones are additive? For those that preserve addition of intervals, find the mapping function.

1.

x	3.2	6.7	10.3	13.8	17.3	20.8
y	2.3	9.3	16.3	23.3	30.3	37.3

2.

p	1.6	4.0	6.4	8.8	11.2	13.6
s	8.5	3.7	-1.1	-5.9	-10.1	-15.5

3.

x	3.1	7.6	12.1	16.6	21.1	25.6
t	8.37	20.52	32.67	44.82	56.97	69.12

4.

x	0.31	0.76	1.21	1.66	2.11	2.56
y	2.20	10.89	18.68	25.57	31.56	36.65

FIGURE 6.6   Irrational measurement.

5. A linear measurement of x onto y contains the points (2, 3) and (5, 7). Find the function that defines it.

6. Prove that a linear measurement $f(x) = mx + b$ of x is an additive measurement of differences $x_2 - x_1$.

7. Prove that the statement "y is directly proportional to x" (§3.2) describes an additive measurement of the numbers x onto numbers y. Is "y is inversely proportional to x" a measurement?

8. Prove that under an additive measurement $f(x)$, a set of n states with sum $s = x_1 + x_2 + \cdots + x_n$ maps into a set of n numbers with sum ms.

9. A set of n states, $x_1, x_2 \ldots, x_n$, has the average $\bar{x}$. Find the average of the values of y under the linear measurement $y = mx + b$.

The following problems are concerned with the extension of additive measurements of numbers to additive measurements of pairs of numbers $w = f(x, y)$. We define the sum of two pairs

$$(x_1, y_1) + (x_2, y_2) = (x_1 + x_2, y_1 + y_2)$$

and define $f(x, y)$ to be an additive measurements of pairs (x, y) if f is continuous in each variable and

$$f[(x_1, y_1) + (x_2, y_2)] = f(x_1, y_1) + f(x_2, y_2).$$

10. Prove that if $f(x, y)$ is an additive measurement of pairs (x, y), then for some numbers a and b,

$$f(x, y) = ax + by.$$

11. In Exercise 10, show that the additive measurement $f(x, y)$ is completely determined if we know what f does to the pairs (1, 0) and (0, 1).

12. A black box (Figure 6.7) has the numerical inputs x and y, which are physical quantities such as electric current, or voltage. The output quantity w is known to be a linear measurement of the pairs (x, y), but it is not known what is inside the black box. By experiment it is found that (1, 0) produces the output $w = 5$ and (0, 1) produces the output $w = 9$. Find a function that represents the action of the black box.

13. In a black box like that of Exercise 12, the input $(3, -2)$ produces the output $w = 4$ and the input (2, 5) produces the output $w = 11$. Find a function that represents the action of the box.

FIGURE 6.7   Exercise 12.

14. An additive black box has three inputs (x, y, z) and two outputs (v, w). Show that if its action is continuous, it is represented by the pairs of functions

$$v = a_1 x + b_1 y + c_1 z,$$
$$w = a_2 x + b_2 y + c_2 z.$$

15. In Exercise 14 the input $(x, y, z) = (1, 0, 0)$ produces the output $(1, -3)$; $(0, 1, 0)$ produces $(-2, 4)$; and $(0, 0, 1)$ produces $(5, 6)$. (a) Find the function that represents the action of the linear black box on any input (x, y, z). (b) Describe the pattern that enables one to write down the output function immediately for all such problems.

37   Multiplicative and Exponential Measurements

37.1   Multiplicative Measurements.   It is natural to represent the states of certain physical quantities by real numbers but to express their compositions of states by multiplication rather than addition (§36). For example, if $s = rt + s_0$ gives a distance s in uniform motion as a function of time t, the coefficient r is a physical quantity called the *velocity*. Similarly, if the time measurements in two time scales are related by $t = ku + t_0$, the coefficient k can be identified as a physical observable also. Then, when we combine the two observables r and k by substitution, we find that

$$s = r(ku + t_0) + s_0 = (rk)u + (rt_0 + s_0).$$

This is an additive measurement from the scale of u to that of s. The coefficient rk combined the quantities r and k by multiplication rather than addition. A number of common physical observables admit multiplicative composition of states, or in some cases both multiplicative and additive compositions. Some of these are rates, densities, "coefficients" like coefficient of friction, modulus of elasticity, and chemical concentrations.

Thus we are led to define a *multiplicative measurement* as a continuous function $f(x)$ such that $f(ax) = f(a)f(x)$. In this case, we find it appropriate to restrict the set of real numbers represented by a, x, and $f(a), f(x)$, to be *positive* real numbers.

Our problem is, What is the form of the functions $y = f(x)$, which represent multiplicative measurements of the positive real numbers x into the positive real numbers y?

Solution.   We can reduce the problem to that of additive measurements if we replace f by a mapping of $\ln x$ into $\ln y$. Then

$$\ln ax = \ln a + \ln x, \quad \text{and} \quad \ln [f(a)f(x)] = \ln f(a) + \ln f(x).$$

Let $X = \ln x$, $Y = \ln y$, $A = \ln a$, and let $Y = F(X)$ if and only if $y = f(x)$. Hence f is continuous and multiplicative if and only if F is continuous and additive, $F(A + X) = F(A) + F(X)$. The theorem on additive measurements then tells us that $Y = F(X)$ has the form $Y = pX$, for some number p. But this means that

$$\ln y = p \ln x = \ln x^p,$$

which implies that $y = x^p$. We state this result as a theorem.

THEOREM (On multiplicative measurements).   Every multiplicative measurement on the positive real numbers has the form $f(x) = x^p$, for some real exponent p.

We regard the power functions x^p as the multiplicative analogues of the additive functions mx. Similarly the multiplicative analogues of differences $x_2 - x_1$ are ratios x_2/x_1. The multiplicative analogue of our corollary on linear measurements is the following.

COROLLARY.   Every continuous mapping $f(x)$ of the positive real numbers that is a multiplicative measurement of the ratios has the form $y = cx^p$ for some real numbers c and p.

DEFINITION.   A continuous mapping $y = f(x)$ of the positive real numbers is a *power law* if and only if it is a multiplicative measurement of the ratios x_2/x_1.

It is easily seen that multiplicative measurements are a special case of power laws.

37.2   Log-Log Plots.   Suppose that we have some empirical data giving a set of pairs (x, y) of positive real numbers whose law of measurement $f(x)$ is unknown, but we think it may be a multiplicative measurement. We may test this hypothesis by the same device that we used for the proof. That is, we plot the pairs $(\ln x, \ln y)$ and see if a straight-line plot approximates the data acceptably. If so, the data support the hypothesis that the law of measurement is of the form $f(x) = cx^p$, where p is the slope of the line and $\ln c$ is the intercept on the $\ln y$ axis.

Example.   A psychologist conducted an experiment to see if a subject's perception of lengths ψ laid out on the ground constituted a valid measurement in comparison with measurements s obtained from a graduated steel tape. If so, he wanted to find the "psychophysical law" $\psi = f(s)$. In one run his data were as in the accompanying table. Test for the power law.

Physically measured length, s	5	6	7	8	9	10
Perceived length in meters, ψ.	6.10	7.59	9.39	11.08	13.10	15.34

Solution.   We plot pairs $(\ln s, \ln \psi)$. This can be done most conveniently by using log-log paper (Figure 6.8). The points may be well fitted by a straight line, which supports the hypothesis that the perceived measurement is valid as preserving multiplication of ratios. That is, the psychophysical law has the form $\psi = cs^p$. We find that on the line $\psi = 0.811$ when $s = 1$ and that the slope of the line measured in a *uniform* scale is 1.27. This completes the evaluation of the constants c and p and gives the power function $\psi = 0.811s^{1.27}$.

37.3   Exponential Measurements.   We have found the form of measurement functions that map an addition scale onto an addition scale (§36.2) and those that map a multiplication scale onto a multiplication scale (§37.1). Among the remaining possibilities for continuous measurements that preserve compositions of states of observable quantities

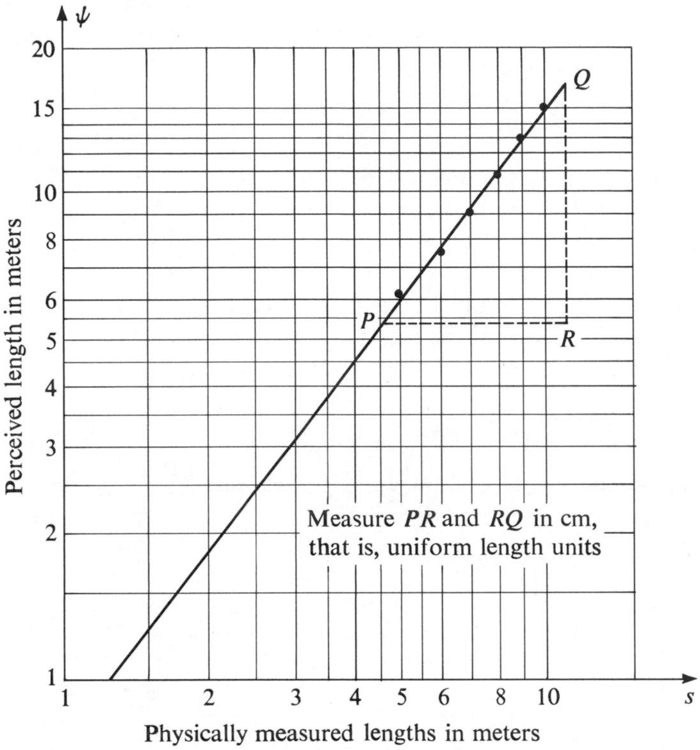

FIGURE 6.8   Log-log plot of psychological data.

are those that transform addition into multiplication. Thus we seek the continuous mappings $f(x)$ of the real numbers into the positive numbers for which

$$f(x + a) = f(x)f(a).$$

Again we can reduce this to additive mappings. Indeed, if instead of $f(x)$ we seek the function $F(x)$ for which $F(x) = \ln y = \ln f(x)$, we find that

$$\ln f(x + a) = \ln f(x) + \ln f(a)$$

can be written

$$F(x + a) = F(x) + F(a).$$

Now the theorem on additive measurements (§36.2) says that since $F(x)$ is continuous and $F(x)$ is additive, for some number k, $F(x) = kx$. That is, $\ln f(x) = kx$ or $f(x) = \exp kx$. We have thus established the following.

THEOREM (On exponential measurements).   The only continuous measurements $y = f(x)$ of the real line into the positive real numbers that represent addition in the x-scale by multiplication in the y-scale are of the form $y = \exp kx$ for some k.

COROLLARY.   The only continuous measurements of the real line into the positive real numbers that represent the addition of differences as the product of ratios are of the form

$$y = C \exp kx$$

for some C and some k.

Example. The gross national product of the U.S. economy in billions of dollars was given by the accompanying table over a 10-year period. Show that these data support the conclusion that this mapping of time into GNP is exponential, and find the growth rate.

Years	1957	1958	1959	1960	1961	1962	1963	1964	1965	1966	1967
GNP	441	447	484	504	520	560	591	632	681	740	777

Solution. We let t be time in years after 1957, and y the gross national product. Then we plot the pairs $(t, \ln y)$ on semilog paper. The paper can be fitted by a straight line that has slope 0.0275 (Figure 6.9). Hence the observations support the conclusion that

$$y = 420 \exp [0.0275(t - 1957)],$$

which is an exponential growth (§29) with growth rate 2.75 percent per year. This completes the problem.

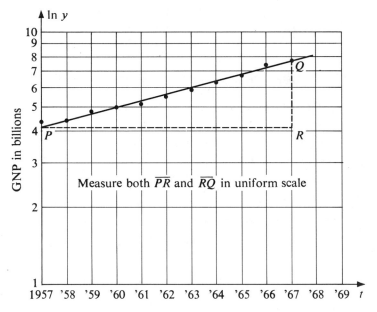

FIGURE 6.9   Measurement of U.S. gross national product.

37.4 Logarithmic Laws.   We state the result in a theorem.

THEOREM (On logarithmic measurements).   The only continuous measurements $y = f(x)$ of the positive real line into the real numbers that represent multiplication in the x-scale as addition in the y-scale are of the form $y = C \ln x$ for some constant C.

COROLLARY.   The only continuous measurements of the positive real line into the real line that represent multiplication of ratios as addition of differences are of the form $y = C \ln x + B$ for some numbers C and B.

37.5   Exercises

In Exercises 1–9, the tables of data may represent linear laws, power laws, exponential laws, or logarithmic laws, or none of these. Determine in each case whether the data represent one of the measurement laws, and if so, find the mapping formula with approximate values of the constants.

1.

t	1.2	2.4	4.8	9.6	19.2
s	1.44	5.76	23.04	92.16	368.64

2.

u	1	2	4	8	16
v	2.7	3.84	5.40	7.65	10.80

3.

x	3	5	7	9	11
y	0.48	0.70	0.85	0.95	1.04

4.

x	1.5	2.7	3.9	4.8	5.6
y	6.30	20.31	42.59	64.51	87.81

5.

r	1	1.5	2	2.5	3.0
s	0.368	0.233	0.135	0.082	0.050

6.

t	0	2	4	6	8
A	100	108.2	117.0	126.5	136.9

7.

x	0.7	0.9	1.1	1.3	1.5
y	-0.155	-0.046	0.041	0.114	0.176

8.

t	1	2	4	8	16
x	-3.30	-1.60	1.80	8.60	22.2

9.

x	1	2	4	8	16
$10^3 y$	6.2	20.01	65.09	211.50	686.70

10. In a psychological experiment to relate the perceived loudness ψ of a sound to the physical loudness in decibels s, the following table of data was recorded.

s	2	4	6	8	10
ψ	1.3	1.5	1.8	1.8	2.1

Does this experiment support the hypothesis that the psychophysical law involved is a power law? If so, with what exponent?

11. In a psychological experiment to relate the perceived lengths ψ of distances in meters to the physically measured lengths s in meters, the following table of data was recorded. The perceived lengths were averages from the estimates of several subjects.

s	150	250	500	1000	2000	3000	5000	10,000
ψ	85	111	206	373	633	1069	1437	2064

Does the experiment support the hypothesis that the power law applies? If so, fit a power law to the data. This shows that perception of length is accurate in what sense?

12. A chemist found that 5.11 g of a substance decomposed in a hot-water solution leaving the following amounts of A after t minutes.

t (min)	0	10	20	30	40	50	60
A (g)	5.11	3.77	2.74	2.02	1.48	1.08	0.80

Is there a physical law relating t and A, and if so, what is the form of it?

37.6 Problems

T1. Prove that a measurement is a power law if and only if every geometric progression of states $x_1, x_2, x_3, \ldots$ maps into a geometric progression of states $y_1, y_2, y_3, \ldots$.

T2. Show that a measurement $y = f(x)$ is a power law if and only if it satisfies a differential equation

$$Dy = p\left(\frac{y}{x}\right)$$

for some number p.

T3. Prove that a measurement is an exponential law if and only if every arithmetic progression states $x_1, x_2, x_3, \ldots$ maps into a geometric progression of states $y_1, y_2, y_3, \ldots$.

T4. Prove the theorem on logarithmic measurement (§37.4).

T5. Consider a continuous mapping of the real line onto the unit circle in which the addition of real numbers x corresponds to addition of rotations of the circle. What is the form of any such mapping?

T6. In the problem of determining the multiplicative measurements (§37.1), we let $y = f(x)$ induce a function $Y = F(X)$, where $X = \ln x$ and $Y = \ln y$. The various mappings are indicated by arrows in the accompanying diagram (Figure 6.10). We can determine F explicitly by starting from X and going down to x via the antilogarithm, then to y via f, then to Y via the logarithm. Show that

$$F(X) = \ln f(\exp X).$$

T7. In Problem T6, show explicitly that if $A = \ln a$, then f is multiplicative if and only if, for every A and X,

$$F(A + X) = F(A) + F(X).$$

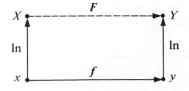

FIGURE 6.10   Problem T6.

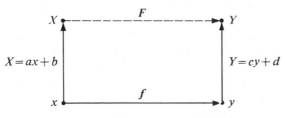

FIGURE 6.11   Problem T8.

T8.* A physically significant function $y = f(x)$ on physical measurements x to physical measurements y must have the same form if we change the scales of x and y independently. Nature does not recognize the particular scales in which physical states are measured. If the requirement for physical significance is

$$F(X_2 - X_1) = kf(x_2 - x_1) \qquad \text{for some constant } k \text{ and all } x_1, x_2$$

(Figure 6.11), show that a continuous f is physically significant if and only if f is itself a linear measurement.

T9.* In Problem T8, if the requirement for f to be physically significant is that it preserves ratios under the transformations of scale $X = ax$ and $Y = cy$, that is, if

$$F\left(\frac{X_2}{X_1}\right) = kf\left(\frac{x_2}{x_1}\right),$$

show that f must be a power law, $f(x) = mx^p$.

T10.* In Problem T9, what further restriction of f occurs if we require that f preserve the absolute significance of ratios,

$$F\left(\frac{X_2}{X_1}\right) = f\left(\frac{x_2}{x_1}\right) ?$$

38   Dynamic Processes in Biology*

38.1   Calculus Models for Life Processes. Calculus is an appropriate tool to describe biological processes that involve time variation. Typically some measurable substance is formed, or transported, or absorbed, or excreted, or diffused, or transferred through a permeable membrane in the living system. Regarding the amounts or concentrations of such substances as functions of time, we may often describe these physiological processes by relations between derivatives with respect to time. Here, calculus is a prime tool.

Other types of mathematical models apply to the static structural aspects of biology and to the chance phenomena involved in the encounters between populations of cells or organisms and their environment. The mathematics for these aspects of biology tends to be more algebraic, structural, and combinatorial in character and hence to involve less calculus. Here let us study a few examples of dynamic processes where calculus is a suitable means of expression. We use differential equations, like those of exponential growth (§29).

---

* This section is based in part on Robert M. Thrall et al. (eds.), *Some Mathematical Models in Biology*, 2nd ed. (University of Michigan) 1967. Particular use is made of articles attributed to Richard F. Baum, James O. Brooks, and "Hebrew University Report."

38.2   Diffusion Through a Cell Membrane

Example.   A solute is present in constant concentration c outside a cell. The same substance is present inside the cell initially at a lower concentration y_0. Assuming Fick's law, which says that the time rate of movement of a solute across a thin membrane is proportional to the area of the membrane and to the difference in concentration of the solute on the two sides of the membrane, find the function representing the concentration of the solute inside the cell (Figure 6.12).

Solution.   Let y be the function of t that gives the concentration inside the cell and let s give the amount of the solute inside the cell at time t. Then

$$D_t y = \frac{1}{V} D_t s,$$

where V is the volume of the cell. Then Fick's law says that

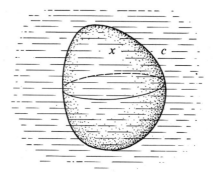

FIGURE 6.12   Cell in liquid solute.

where k is a constant, the permeability of the membrane. Combining these two equations, we have

$$Dy = \frac{kA}{V}(c - y),$$

subject to the initial condition $y(0) = y_0$. Since $Dy = -D(c - y)$, this can be written

$$D(c - y) = -\frac{kA}{V}(c - y),$$

which has the general solution (§29)

$$c - y = \exp\left(-\frac{kA}{V}t\right).$$

Since $\exp 0 = 1$, the initial condition implies that $C = c - y_0$. Hence the final solution is

$$y = c - (c - y_0)\exp\left(-\frac{kA}{V}t\right),$$

whose graph is a function increasing from y_0 to an asymptotic limit at $y = c$ (Figure 6.13).

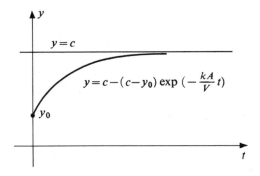

FIGURE 6.13   Concentration build-up in a cell.

38.3   Diffusion of a Solute Across the Wall of a Blood Vessel

Example.   A solution moves at constant velocity v in a tube with a thin membrane for its wall, diffusing to lower constant concentration c outside the tube. Find the function $y(x)$ that represents the concentration at a point x centimeters from the point at which the solution entered the tube at concentration y_0.

Solution.   We consider a small chamber, of length h, from x to $x + h$ (Figure 6.14). Its volume is $\pi r^2 h$. During a time interval $[t, s]$ the solution enters at x with concentration $y(x)$ and leaves at h with concentration $y(x + h)$. Hence the amount left behind by diffusion across the tube wall is the difference $[y(x) - y(x + h)]\pi r^2 h$. Assuming that Fick's law holds locally, we calculate the amount that diffuses through the membrane in the time from t to s as

$$2\pi rh[y(x) - c]k(s - t).$$

But since $h = v(s - t)$, the equality can be written

$$[y(x) - y(x + h)]\pi r^2 h = 2\pi rh[y(x) - c]\frac{k}{v} h.$$

Dividing both sides by $\pi r^2 h^2$, we find that

$$\frac{y(x + h) - y(x)}{h} = -\frac{2k}{vr} [y(x) - c].$$

Proceeding to the limit as $h \to 0$, we find

$$D_x y = -\frac{2k}{vr} [y - c].$$

Since $Dy = D(y - c)$, this differential equation has the general solution (§29)

$$y - c = C \exp \left(-\frac{2k}{vr} x \right).$$

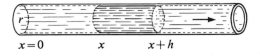

$$x = 0 \qquad x \qquad x + h$$

FIGURE 6.14   Flow through tube.

Evaluating the constant C by means of $y(0) = y_0$, we find

$$y(x) = c + (y_0 - c) \exp\left(-\frac{2k}{vr}x\right).$$

The graph of this function shows the concentration decreasing along the tube and approaching the outside level c asymptotically (Figure 6.15).

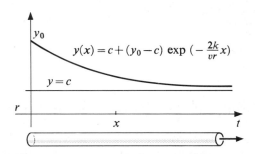

FIGURE 6.15   Loss of solute flowing through a permeable tube.

38.4   An Epidemic

Example.   The rate of increase of the fraction of a population that becomes infected with a contagious disease is directly proportional to both the fraction that has been infected and the fraction that has not been infected. Find the function giving the fraction of the population infected, starting from the fraction p_0 at time 0.

COMMENT ON THE MODEL.   We approximate the fraction infected by a differentiable $p(t)$, though the actual function in a finite population will vary by discrete jumps as individual members become infected one at a time. The approximation is a good one if the population is large. The assumed transmission law is that the rate of infection is proportional to the likelihood of an infected coming in contact with a healthy. This turns out to be realized in nature if the disease is one to which everybody is susceptible and one in which infected people are not removed, by either their death or getting well.

Solution.   If $p(t)$ is the fraction infected, then $1 - p(t)$ is the fraction of the population that is healthy. To say that the rate of infection is directly proportional to both $p(t)$ and $1 - p(t)$ gives us the differential equation

$$Dp = kp(1 - p),$$

where k is a constant. We have the initial condition $p(0) = p_0$. We can rewrite this

$$\frac{Dp}{p(1 - p)} = k;$$

or, since we find by inspection that

$$\frac{1}{p(1 - p)} = \frac{1}{p} + \frac{1}{1 - p}.$$

this becomes, on the signs being changed,

$$-\frac{Dp}{p} - \frac{Dp}{1 - p} = -k.$$

Integrating both sides with respect to t, we find that

$$-\ln p + \ln (1 - p) = -kt + C.$$

If we set $t = 0$ to evaluate C, we find that

$$C = \ln \frac{1 - p_0}{p_0}.$$

Since $-kt = \ln \exp(-kt)$, this can be written as

$$\frac{1 - p}{p} = \frac{1 - p_0}{p_0} \exp(-kt).$$

This is our solution of the problem, where we express the ratio of the healthy to the infected fraction as a function of t. Since $\lim_{t \to \infty} \exp(-kt) = 0$, this implies that $\lim_{t \to \infty} p(t) = 1$. Everybody eventually becomes infected. We sketch the graph of $p(t)$ (Figure 6.16).

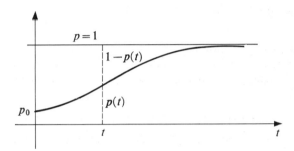

FIGURE 6.16   Spread of an epidemic.

38.5   Exercises

1. In Example 38.2, at what time is the diffusion rate a maximum?
2. In Example 38.2, find the permeability coefficient k if $y = y_1$ when $t = t_1$.
3. In Example 38.2, if $kA/V = 0.1$, what is the time required for the difference between the inside and outside concentrations to reach half its initial value. Show that this does not depend on c.
4. Consider the effect of the size of the cell in 38.2. If the cell is a sphere of radius r, what is the ratio A/V? If the cell radius is doubled, what is the effect on the rate of increase of the concentration inside the cell?
5. In Example 38.3, find the average concentration of the solute in an infinitely long tube.
6. In Example 38.3, find the total rate of diffusion of solute from a length L of the tube.
7. In the example of the epidemic, what is the value of p where the rate of increase of $p(t)$ is a maximum? What is the maximum rate, $p'(t)$?
8. For the epidemic, find $p(t)$ if $k = 0.5$ and $p_0 = \frac{1}{11}$.
9. For Exercise 8, how long will it take for half the population to be infected?
10. Describe an epidemic whose mathematical model is $Dp = k(1 - p)$. Does this model require that a healthy person come in contact with an infected person? Solve for $p(t)$ with $p(0) = 0$.
11. Glucose is infused into veins of fluid volume V at the rate of A mg/min and it is diffused at a rate $-ky$, where y is the concentration in the veins. Write a differential equation for the concentration at time t, starting with concentrations y_0 at time $t = 0$. Show that the

solution is

$$y(t) = \frac{A}{kV} [1 - \exp(-kt)] + y_0 \exp(-kt).$$

12. In Exercise 11, what is the steady-state concentration approached as $t \to \infty$?

13. In Exercise 11, show that for an initial infusion rate A/V greater than an initial diffusion rate ky_0, $y'(t) > 0$ for every value of t. Use this information to sketch the curve.

14. It was found by experiment that the subjective brightness ϕ of a light source judged by viewers is related to the physical brightness y by power law, $\phi = ky^p$. In this relation, for viewers who have previously been dark-adapted, $k = 10 \times 10^{-3}$ and $p = 0.35$. For viewers who have previously been light-adapted, $k = 0.5 \times 10^{-3}$ and $p = 0.60$. For what range of physical brightness y will the light-adapted people report the light to be brighter than the dark-adapted people do?

15. An agent that kills bacteria has been added to a culture initially containing 10^6 viable bacteria per milliliter. The number x of bacteria at each of a series of times after the bactericidal agent was added were found to be approximately

t (min)	0	10	20	30	40	50	60
x (no. bacteria)	10^6	10^5	10^4	10^3	10^2	10	1

Make a suitable plot to discover the form of the function $x(t)$.

16. A subject of finite brain capacity learns at a rate proportional to the number of his uncommitted brain cells. Show that the percentage y of committed brain cells satisfies the differential equation

$$D(1 - y) = -k(1 - y), \quad \text{where } k > 0.$$

17. Find a formula for the solution y of the differential equation in Exercise 16 where the subject starts with an empty mind. Sketch the graph of y. Show that at the start $y' = k$.

18. In Exercises 16 and 17, after 3 hours the rate y' has slowed to half its initial value. (a) Find k. (b) Does this mean that after 3 hours the subject has used up half his brain capacity? Show calculations or reasoning to justify the answer.

19.* A learning computer with finite memory acquires information at a rate proportional to the number of uncommitted memory elements, and simultaneously forgets at a rate proportional to the number of committed memory elements. The net retention $y(t)$ at time t, expressed as a percentage of the total memory, satisfies the differential equation $y' = \lambda(1 - y) - \phi y$, where the learning coefficient λ and the forgetting coefficient ϕ are constants.
(a) Show that this differential equation is equivalent to

$$\left(\frac{\lambda}{\lambda + \phi} - y \right)' = -(\lambda + \phi) \left(\frac{\lambda}{\lambda + \phi} - y \right).$$

Solve it for the case in which the computer starts with cleared memory. (b) Show that the net retention, $y(t)$, approaches the saturation level, $\frac{\lambda}{\lambda + \phi}$, as $t \to \infty$. (c) Thinking of λ as fixed and ϕ as variable, show that the computer's net retention will reach a prescribed, below-saturation, level m faster if there is some forgetting, $\phi > 0$, than if $\phi = 0$. This suggests the very difficult problem: For prescribed λ and m find ϕ so that $y(t)$ approaches the level m in minimum time.

(*Note.* Fast learners are fast forgetters.)

Part Two
Multivariable Calculus

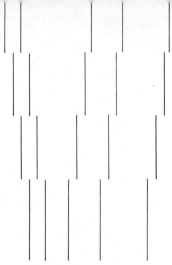

Techniques of Differentiation and Integration

CHAPTER 7

ACADEMIC NOTE. This first chapter of Part II serves three purposes. (1) It reviews differentiation and integration for those who have previously studied the calculus of functions of one variable, whether in Part I or other standard text. (2) It extends and elaborates these techniques, making full use of the chain rule. (3) It initiates a change of notation into the differential style of calculus, which is more appropriate for functions of several variables, where there is no requirement to build up methods of integration. This chapter is not prerequisite to the remainder of Part II but is needed in Part III.

1 Best Linear Approximation

1.1 Historical and Critical Preface. We recall that in the calculus of functions of one real variable we studied the relationship between functions $f(x)$ and their derivative functions $f'(x)$, expressing this relationship by means of a differentiation operator D and its inverse, writing $Df(x) = f'(x)$. This style of calculus was invented by Isaac Newton (1642–1727) and was later developed, in these notations especially, by the French mathematician A. L. Cauchy (1789–1857). Although it is somewhat restricted, this operator style is perhaps the simplest, and it conforms well to modern mathematics. Using functions in the sense of mappings, it lends itself readily to the precise statement and proof of the theorems of calculus. Moreover, it clearly reveals the relation of the algebraic properties of the differentiation operator D to the algebra of functions.

On the other hand, the German philosopher-mathematician Gottfried Leibniz (1646–1716), independently of Newton, invented a version of calculus that is still in wide use. We shall call it the *differential style*. In the Leibniz notation, we study relations $y = f(x)$, and the derivative is represented by the symbol dy/dx. The slope of a secant line joining the point $P:(x, y)$ in the graph of $y = f(x)$ to a neighboring point $Q:(x = \Delta x, y + \Delta y)$ is $\Delta y/\Delta x$ (Figure 7.1), so that

$$\frac{dy}{dx} = \lim_{\Delta x \to 0} \frac{\Delta y}{\Delta x}.$$

The differential style is well adapted to geometry and to scientific writing in which it is advantageous to express relationships between physical observables by means of relations that keep in view both independent and dependent variables and their physical interpretations. For example, Boyle's law, $pv = RT$, is a relation between the pressure of a gas p, its volume v, the gas constant R and the absolute temperature T of the gas. The differential style handles this better than the operator style; functions in the strict mathematical sense tend to suppress the variables or even to eliminate them altogether. Moreover, there is great freedom, somewhat greater generality, and almost magic maneuverability in the differential style. Its simple maneuvers predict many theorems, which turn out to be provable when one gives the proper interpretation to the mathematical forms that appear, though in some cases the proofs are difficult. Most important, the differential style is better suited for multivariable calculus.

Since so much existing scientific writing employs the Leibniz notation, and since it has advantages of its own, let us review the basic ideas of calculus in the differential style.

1.2   Local Coordinates.   We think of a function as a relation $y = f(x)$ consisting of a set of ordered pairs (x, y). We fix a point $P:(x, y)$ in the graph of $y = f(x)$ and set up in the xy-plane a second system of coordinate axes with origin at P (Figure 7.1), with axes parallel to the xy-axes, using the same units. Coordinates referred to in this system of coordinates with origin at P are called local coordinates. We denote the local coordinates of a point by (dx, dy) or $(\Delta x, \Delta y)$ or (h, k). The symbols d and Δ refer to differences. For if Q is another point in the graph with xy-coordinates (x_1, y_1) we see that the local coordinates $(\Delta x, \Delta y)$ of Q are given by the differences

$$\Delta x = x_1 - x, \qquad \Delta y = y_1 - y.$$

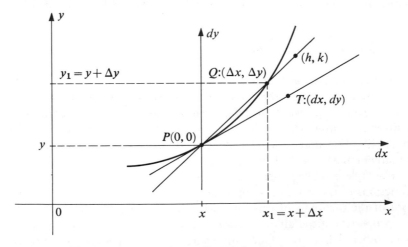

FIGURE 7.1   Local coordinates in the xy-plane.

Or, given the local coordinates $(\Delta x, \Delta y)$ we can solve these equations for (x_1, y_1) to find that the xy-coordinates of Q are $(x + \Delta x, y + \Delta y)$. We observe that the local coordinates of P are $(0, 0)$ and that in local coordinates the equation $y = f(x)$ becomes $\Delta y = f(x + \Delta x) - f(x)$. Our problem is to find the tangent line as a relation in the local coordinates (dx, dy). Traditionally, the local coordinates of a point in the tangent line are denoted by (dx, dy), while the local coordinates of a point in the graph of $y = f(x)$ are $(\Delta x, \Delta y)$. Local coordinates are also called *increments*.

In local coordinates, the equation of the straight line through P with slope m is

$$dy = m\, dx.$$

In the differential style of calculus, we first find the line $dy = m\, dx$ through P that is the "best linear approximation" to the curve at P in a sense we shall shortly make precise. Then we define this best linear approximation, or "best-fitting line," to be the tangent line at P. The idea becomes consistent with the operator style when we show that the slope of the best linear approximation is $f'(x)$.

1.3   Best Linear Approximation.   Let $P:(x, y)$ be a fixed point in the graph of the continuous relation $y = f(x)$ and let $Q:(x + \Delta x, y + \Delta y)$ be a neighboring point on the curve, where $\Delta x \neq 0$. We set up a local coordinate system with origin at P, giving local coordinates (dx, dy) or $(\Delta x, \Delta y)$. The local coordinates of P are $(0, 0)$ and those of Q are $(\Delta x, \Delta y)$. We consider a line $dy = m\, dx$ through P, with m as yet undetermined. We take a point $R:(\Delta x, m\, \Delta x)$ on this line having the same first coordinate as Q (Figure 7.2). The vertical distance $|RQ| = |\Delta y - m\, \Delta x|$ represents the error at Δx of approximating the function $\Delta y = f(x + \Delta x) - f(x)$ by the linear function $dy = m\, dx$. This error approaches zero with $|\Delta x|$ for every line $dy = m\, dx$. We consider the *relative* error of the linear approximation, expressed as a percentage of $|\Delta x|$.

Relative error of the linear approximation equals $|\Delta y - m\, \Delta x|/|\Delta x|$. We prove the following preliminary result.

LEMMA.   Only one line $dy = m\, dx$ can approximate $\Delta y = f(x + \Delta x) - f(x)$ so that

$$\lim_{\Delta x \to 0} \frac{|\Delta y - m\, \Delta x|}{|\Delta x|} = 0.$$

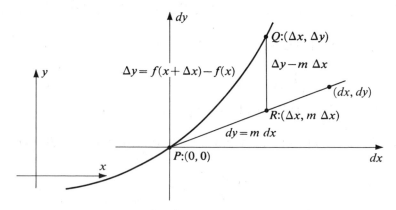

FIGURE 7.2   Curve and approximating line in local coordinates.

Proof.   Suppose there is another line $dy = p\,dx$, $p \neq m$, such that also

$$\lim_{\Delta x \to 0} \frac{|\Delta y - p\,\Delta x|}{|\Delta x|} = 0.$$

Since

$$|m - p| = \left|\frac{(\Delta y - m\,\Delta x) - (\Delta y - p\,\Delta x)}{\Delta x}\right| \leqq \left|\frac{\Delta y - m\,\Delta x}{\Delta x}\right| + \left|\frac{\Delta y - p\,\Delta x}{\Delta x}\right|.$$

and by hypothesis the limits of the last two terms are both zero, this implies that $\lim_{\Delta x \to 0} |m - p| = 0$. Therefore $m = p$, and the lemma is proved. It justifies the following definition.

DEFINITION.   Let $y = f(x)$ be a function and let $\Delta y = f(x + \Delta x) - f(x)$ describe the function in local $(\Delta x, \Delta y)$ coordinates, referred to as point $P:(x, y)$ in its graph. Then the line $dy = m\,dx$ is the *best linear approximation* to the function at P if and only if the relative error of the approximation approaches zero with $|\Delta x|$. That is,

$$\lim_{|\Delta x| \to 0} \frac{|\Delta y - m\,\Delta x|}{|\Delta x|} = 0.$$

DEFINITION.   The *tangent line* to the graph of $y = f(x)$ at $P:(x, y)$ is the line whose equation is the best linear approximation at P.

THEOREM.   If $y = f(x)$ has a derivative

$$f'(x) = \lim_{\Delta x \to 0} \frac{f(x + \Delta x) - f(x)}{\Delta x}$$

at $P:(x, y)$, then the best linear approximation exists, and in local coordinates (dx, dy) is given by

$$dy = f'(x)\,dx.$$

Proof.   In local coordinates $(\Delta x, \Delta y)$, the relation defining the function can be written

$$\Delta y = f(x + \Delta x) - f(x).$$

The relative error at Δx of the linear approximation $dy = m\,dx$ is

$$\left|\frac{\Delta y - m\,\Delta x}{\Delta x}\right| = \left|\frac{f(x + \Delta x) - f(x) - m\,\Delta x}{\Delta x}\right| = \left|\frac{f(x + \Delta x) - f(x)}{\Delta x} - m\right|.$$

Since $y = f(x)$ has the derivative $f'(x)$, it follows that

$$\lim_{|\Delta x| \to 0} \left|\frac{y - m\,\Delta x}{\Delta x}\right| = f'(x) - m = 0$$

if and only if $m = f'(x)$. This completes the proof.

COROLLARY 1.   The slope of the tangent line to the graph of $y = f(x)$ at $P:(x, y)$ is given by

$$\frac{dy}{dx} = f'(x).$$

Here dy/dx means what it says, the ratio of the local coordinate dy to dx.

COROLLARY 2.   If $(\Delta x, \Delta y)$ are local coordinates of a point Q in the relation $y = f(x)$, referred to a point $P:(x, y)$ where $f(x)$ is differentiable, then

$$\lim_{\Delta x \to 0} \frac{\Delta y}{\Delta x} = f'(x).$$

Calculating a derivative of $y = f(x)$ from the definition, in the Leibniz notation, that is, by

$$\frac{dy}{dx} = \lim_{\Delta x \to 0} \frac{\Delta y}{\Delta x} = \lim_{\Delta x \to 0} \frac{f(x + \Delta x) - f(x)}{\Delta x}.$$

is called "finding the derivative by the Δx-process."

Example.   Compute dy/dx by the Δx-process for the relation $y = x^2 - 3x + 4$. Also find the equation of the tangent line at (x, y) in local coordinates.

Solution.   We take two points $P:(x, y)$ and $Q:(x + \Delta x, y + \Delta y)$ in the relation. Since Q is in the relation, it is true that

$$y + \Delta y = (x + \Delta x)^2 - 3(x + \Delta x) + 4$$
$$= x^2 + 2x(\Delta x) + (\Delta x)^2 - 3x - 3(\Delta x) + 4.$$

Since P is in the relation, it is true that

$$y \qquad = x^2 \qquad\qquad\qquad - 3x \qquad\qquad + 4.$$

Subtracting, we find that

$$\Delta y = 2x(\Delta x) + (\Delta x) - 3(\Delta x).$$

We divide both sides by Δx, $\Delta x \neq 0$, and find that

$$\frac{\Delta y}{\Delta x} = 2x + (\Delta x) - 3.$$

Then proceeding to the limit as $\Delta x \to 0$, we obtain

$$\frac{dy}{dx} = \lim_{\Delta x \to 0} \frac{\Delta y}{\Delta x} = \lim_{\Delta x \to 0} [2x + (\Delta x) - 3] = 2x - 3.$$

Hence the derivative is the ratio given by $dy/dx = 2x - 3$. The equation of the tangent line at $P:(x, y)$ in local (dx, dy) coordinates is

$$dy = (2x - 3)\, dx.$$

This completes the problem.

1.4   Reconciliation of Differential Style with Operator Style.   In the calculation of derivatives, the differential style of calculus is equivalent to the operator style we have developed. All of the rules of differentiation (I, §12.3) apply. We may bridge the gap somewhat between the two styles by introducing a derivative operator d/dx into the differential style. Since $y = f(x)$, we can get the symbol dy/dx on the left by operating on it with d/dx. This gives us

$$\frac{dy}{dx} = \frac{d}{dx} f(x).$$

Hence, if we understand $d/dx\,f(x)$ to be $f'(x)$, we have a derivative operator d/dx in the differential style, which means the same as the operator D_x in operator style. That is,

$$\frac{d}{dx} f(x) = f'(x) \qquad \text{and} \qquad D_x f(x) = f'(x).$$

Similarly, we may use

$$\left(\frac{d}{dx}\right)\left(\frac{d}{dx}\right) = \frac{d^2}{dx^2}$$

as a second-derivative operator meaning the same as D_x^2.

It is more consistent with differential-style calculus to say the same thing in a slightly different way. We change the meaning of the operator D to mean $Df(x) = f'(x)\,dx$. This is consistent with the idea that the process of differentiation is changed from finding the numerical slope $f'(x)$ to that of finding the best linear approximation $f'(x)\,dx$ of $f(x)$ at the point $(x, f(x))$. With this change in the meaning of the operator D no changes in the formulas for differentiation are necessary. We observe that when the function is the identity function x, we get $Dx = 1\,dx = dx$. We treat variables x and y as identity functions, so if y is a variable, $Dy = 1\,dy = dy$. This enables us to apply the differentiation operator D to both sides of a relation $y = f(x)$ and interpret $Dy = Df(x)$ to mean the equation of the tangent line, $dy = f'(x)\,dx$, where (dx, dy) are local coordinates. We do not use small d as an operator.

Example. Differentiate

$$y = (2x - 1)^5 (7x + 4)^3.$$

Solution. We treat D as an operator and apply the rules of differentiation (I, §12.3) or Table I (Appendix).

$$Dy = (2x - 1)^5\,D(7x + 4)^3 + (7x + 4)^3\,D(2x - 1)^5,$$
$$= (2x - 1)^5 3(7x + 4)^2(7)\,dx + (7x + 4)^3(5)(2x - 1)^4(2)\,dx,$$

or

$$dy = (2x - 1)^4(7x + 4)^2(112x + 19)\,dx.$$

This is the equation of the tangent line in local coordinates. The coefficient of dx is dy/dx, the slope of the tangent line at P.

To calculate second derivative, we proceed as follows. If $y = 3x^2 - 1$,

$$Dy = 6x\,dx \qquad \text{and} \qquad D^2y = D(Dy) = [D(6x)]\,dx = [6\,dx]\,dx = 6\,dx^2.$$

We interpret D^2y/dx^2 as the second derivative d^2y/dx^2. Hence $d^2y/dx^2 = 6$. Notice that in the second application of the operator D, we are still differentiating with respect to x. The factor dx already present from the first differentiation acts like a constant factor.

1.5 *Exercises*

In Exercises 1–6, compute dy/dx by the Δx-process. In each case write the equation of the tangent line at $P:(x, y)$ in local coordinates (dx, dy).

1. $y = x^2$.
2. $y = x^3$.
3. $y = mx + b$.
4. $y = 3x^2 - 7x + 4$.
5. $y = 1$.
6. $y = 1 - x - x^2$.

In Exercises 7–17, use the operator D to find the local equation of the tangent line and the indicated derivative.

7. $\dfrac{dy}{dx}$ for $y = (3x - 1)^5$.

8. $\dfrac{dy}{dx}$ for $y = (4 - 2x)^7$.

9. $\dfrac{dy}{dt}$ for $y = (2t - 1)(3t + 1)$.

10. $\dfrac{ds}{dt}$ for $s = (1 - 2t)^4(2 - 3t)^5$.

11. $\dfrac{dx}{dt}$ for $x = \{1 - [1 - (2t - 1)^4]^3\}$.

12. $\dfrac{dp}{dv}$ for $p = \dfrac{(2 - v)^3}{6}$.

13. $\dfrac{dr}{d\theta}$ for $r = \tfrac{1}{2}\theta(1 - \theta)$.

14. $\dfrac{d^2y}{dx^2}$ for $(x^4 - 1)^2$.

15. $\dfrac{d^2x}{dt^2}$ for $x = -2t + 1$.

16. $\dfrac{d^2y}{dr^2}$ for $y = 10(1 - r^2)^3$.

17. $\dfrac{d^2z}{du^2}$ for $z^2 - u^2 + 2u - 1 = 0$.

18. Find the numerical value of y, dy/dx, d^2y/dx^2, and sketch the graph of $y = (3x - 4)^7$ near the point where $x = 1$.

19. Find maximum and minimum values of y for $y = x^3 - 7x + 1$ when $-1 \leqq x \leqq 2$.

1.6 Problems

T1. Apply the binomial theorem,

$$(x + \Delta x)^n = x^n + nx^{n-1}\,\Delta x + \frac{n(n - 1)}{2!}\,x^{n-2}\,\Delta x^2 + \cdots,$$

to calculate by the Δx-process the derivative dy/dx of $y = x^n$, where n is any positive integer.

T2. Apply the Δx-process to compute the derivative dy/dx of the quotient $y = u(x)/v(x)$, where $u(x)$ and $v(x)$ are differentiable functions, and $v(x) \neq 0$.

2 Applications of the General Chain Rule

2.1 Examples

Example A.   Differentiate $\ln (1 + x^2)$.

Solution.   This is in the form $\ln u$, with $u = 1 + x^2$. The function $u = 1 + x^2$ is everywhere positive and hence $\ln u$ is defined for every value of x. Since both u and $\ln u$ are differentiable, the chain rule (I, §20.1),

$$Df(u) = f'(u)\,Du,$$

gives the derivative

$$D \ln u = \frac{Du}{u} = \frac{D(1 + x^2)}{1 + x^2} = \frac{2x}{1 + x^2}\,dx.$$

This completes the problem.

Example B.   Find the slope of the tangent line to the circle $x^2 + y^2 = 25$ at the point $(3, -4)$ and plot.

Solution.   We solve the equation of the circle for y as a function of x. This gives us either

$$y = \sqrt{25 - x^2} \qquad \text{or} \qquad y = -\sqrt{25 - x^2},$$

both on $[-5, 5]$. The first of these solution functions has as its graph the upper half of the circle (Figure 7.3(b)) and the second solution has as its graph the lower half of the circle (Figure 7.3(c)). The solution $y = -\sqrt{25 - x^2}$ is the only one that includes the given point $(3, -4)$ in its graph. We differentiate it, using the chain rule formula $V_{1/2}$, which says that $D\sqrt{u} = Du/2\sqrt{u}$. In this case $u = 25 - x^2$. This gives us

$$dy = -\sqrt{25 - x^2} = -\frac{D(25 - x^2)}{2\sqrt{25 - x^2}} = \frac{x\,dx}{\sqrt{25 - x^2}}.$$

When $x = 3$, this gives us $dy/dx|^{x=3} = \frac{3}{4}$, the slope of the tangent line, which we plot (Figure 7.3(c)). This completes the problem.

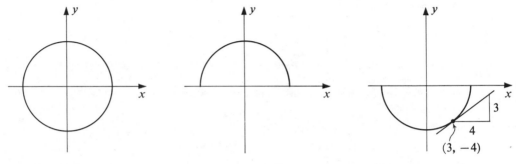

(a) The complete circle   (b) The solution   (c) The solution
$y = \sqrt{25 - x^2}$, $y = \sqrt{25 - x^2}$

FIGURE 7.3   The circle $x^2 + y^2 = 25$ and two solutions.

Example C.   Differentiate $y = \arccos \sqrt{1 - x^2}$ on $[-1, 1]$.

Solution.   This is a repeated composition. If we start with $u = 1 - x^2$, it is in the form $\arccos \sqrt{u}$. Then if $v = \sqrt{u}$, it is $\arccos v$, whose differential is

$$dy = D \arccos v = -\frac{Dv}{\sqrt{1 - v^2}} = -\frac{D\sqrt{u}}{\sqrt{1 - u}} = -\frac{Du/2\sqrt{u}}{\sqrt{1 - u}}.$$

Then, since $Du = -2x\,dx$, this gives us

$$D \arccos \sqrt{1 - x^2} = \frac{x\,dx}{\sqrt{1 - (1 - x^2)}\sqrt{1 - x^2}} = \frac{x\,dx}{|x|\sqrt{1 - x^2}}$$

$$= \begin{cases} +\dfrac{dx}{\sqrt{1 - x^2}} & \text{when } 0 < x < 1, \\[2mm] -\dfrac{dx}{\sqrt{1 - x^2}} & \text{when } -1 < x < 0, \end{cases}$$

and the derivative does not exist when $x = -1$, $x = 0$, and $x = 1$.

2.2 A Maximum-Minimum Problem

Example.   Investigate for maximum and minimum points the function $f(x) = xe^{-x^2}$ on the real line.

Solution.   By the product formula, and then the chain rule applied to e^{-x^2}, we differentiate and find

$$f'(x) = x \frac{d}{dx} e^{-x^2} + \frac{dx}{dx} e^{-x^2} = xe^{-x^2} \frac{d}{dx}(-x^2) + (1)e^{-x^2}$$

$$= xe^{-x^2}(-2x) + e^{-x^2} = e^{-x^2}(-2x^2 + 1).$$

Since $f(x)$ is differentiable for every x, and all points are interior points, the only critical points are those where

$$f'(x) = e^{-x^2}(-2x^2 + 1) = 0.$$

Since e^{-x^2} is everywhere positive, the only critical points are those where $-2x^2 + 1 = 0$, or $x = -1/\sqrt{2}$ and $x = +1/\sqrt{2}$. We can factor $f'(x)$ in the form

$$f'(x) = -2e^{-x^2}\left(x^2 - \frac{1}{2}\right) = -2e^{-x^2}\left(x + \frac{1}{\sqrt{2}}\right)\left(x - \frac{1}{\sqrt{2}}\right)$$

and tabulate the signs of $f'(x)$ as in the accompanying table (I, §§1.6 and 16.3).

	$e-2^{-x^2}\left(x + \frac{1}{\sqrt{2}}\right)\left(x - \frac{1}{\sqrt{2}}\right)$	$f'(x)$
$x < -\dfrac{1}{\sqrt{2}}$	$(-1)\,(+)\,(-)\,(-1)$	$-$
$x = -\dfrac{1}{\sqrt{2}}$	$(-)\,(+)\,(0)\,(-)$	0 (min)
$-\dfrac{1}{\sqrt{2}} < x < \dfrac{1}{\sqrt{2}}$	$(-)\,(+)\,(+)\,(-1)$	$+$
$x = \dfrac{1}{\sqrt{2}}$	$(-)\,(+)\,(+)\,(0)$	0 (max)
$\dfrac{1}{\sqrt{2}} < x$	$(-)\,(+)\,(+)\,(+)$	$-$

The signs of $f'(x)$ tells us that $x = -1/\sqrt{2}$ gives a minimum and $x = 1/\sqrt{2}$ a maximum. We plot the graph of $f(x)$, observing that

$$f\left(-\frac{1}{\sqrt{2}}\right) = -\frac{1}{\sqrt{2}} e^{-1/2} = -0.429 \text{ approx.}$$

and

$$f\left(+\frac{1}{\sqrt{2}}\right) = +\frac{1}{\sqrt{2}} e^{-1/2} = +0.429,$$

(Figure 7.4), and using the information about the signs of $f'(x)$. This completes the problem.

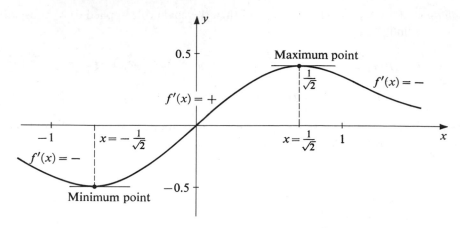

FIGURE 7.4   Graph of $f(x) = xe^{-x^2}$.

2.3   Length of a Chain

Example.   Calculate $D \cos^n ax$, where a is constant.

Solution.   This is in the power form $(\cos ax)^n$. We apply the chain rule twice to find that

$$D \cos^n ax = n \cos^{n-1} ax(D \cos ax) = n \cos^{n-1} ax(-\sin ax) D ax$$
$$= -na \cos^{n-1} ax \sin ax \, dx.$$

In the first differentiation, the factor $D \cos ax$ was the chain rule factor for the form $Du^n = nu^{n-1} Du$. In the next expression, we differentiated $D \cos ax = -\sin ax \, D ax$, introducing $D ax$ as a chain rule factor. That chain ended when we found that

$$D ax = a \, Dx = a(1) \, dx = a \, dx.$$

In every calculation of the derivative of a composite function that starts with a function of the independent variable x, the chain rule factors continue through each composition until we finally come to $Dx = 1 \, dx$. This tells us where the chain stops. It is useful to remember this and to pursue every chain rule differentiation until it stops naturally with $Dx = 1 \, dx$. If we do not use this criterion, it is easy to make the mistake of omitting a chain rule factor.

2.4   Derivative of Indefinite Integral.

We recall that if $f(x)$ is a continuous function on the interval $[a, b]$ and we integrate it from a to a variable point x in the interval, we get the indefinite integral (I, §21.7),

$$F(x) = \int_a^x f(t) \, dt.$$

The derivative of this indefinite integral at x is

$$\frac{d}{dx} F(x) = f(x).$$

When the upper limit of the integral is replaced by a function $u(x)$, the definite integral becomes a composite function whose value at x is

$$F[u(x)] = \int_a^{u(x)} f(t)\, dt.$$

The chain rule says that

$$\frac{d}{dx} \int_a^{u(x)} f(t)\, dt = f[u(x)] \frac{d}{dx} u.$$

A similar formula holds when the lower limit is a differentiable function of x. In fact,

$$\frac{d}{dx} \int_{u(x)}^{v(x)} f(t)\, dt = f[v(x)] \frac{d}{dx} v - f[u(x)] \frac{d}{dx} u.$$

2.5   Logarithmic Differentiation.

We return to the theorem on logarithmic differentiation (I, §26.4) to restate it in a slightly more general form.

THEOREM (Logarithmic differentiation).   If $f(x)$ is a function, and if $\ln |f(x)|$ is differentiable, then $f(x)$ is differentiable and

$$f'(x) = f(x) \frac{d}{dx} \ln |f(x)|.$$

Moreover, if $f(a) = 0$, it is still true that

$$f'(a) = \lim_{x \to a} f(x) \frac{d}{dx} \ln |f(x)|,$$

provided that this limit exists.

   Proof.   Theoretical (§2.7).

Example.   Differentiate

$$y = \frac{(x - 1)^{2/3}(2x + 3)^5}{(x + 1)^{3/5}}.$$

   Solution.   We may expect that to differentiate y, it is easier to take the natural logarithm $\ln |y|$ and differentiate it. In fact, we have

$$\ln |y| = \tfrac{2}{3} \ln |x - 1| + 5 \ln |2x + 3| - \tfrac{3}{5} \ln |x + 1|.$$

We now differentiate setting $dy/dx = y'$ and find

$$\frac{y'}{y} = \frac{2}{3}\left(\frac{1}{x - 1}\right) + 5\left(\frac{2}{2x + 3}\right) - \frac{3}{5}\left(\frac{1}{x + 1}\right).$$

provided that $x \neq 1$ and $x \neq -\tfrac{3}{2}$ and $x \neq -1$. Finally, we multiply both sides by the function y and find

$$y' = \frac{2}{3} \frac{(2x + 3)^5}{(x - 1)^{1/3}(x + 1)^{3/5}} + 10 \frac{(x - 1)^{2/3}(2x + 3)^4}{(x + 1)^{3/5}} - \frac{3}{5} \frac{(x - 1)^{2/3}(2x + 3)^5}{(x + 1)^{8/5}}$$

We found this complicated derivative much more simply than with direct differentiation. Direct differentiation would imply that y is differentiable except when $x = +1$ or $x = -1$. However, logarithmic differentiation also rules out $x = -\tfrac{3}{2}$, where $\ln |2x + 3|$

is not defined. The extended theorem above takes care of this since the limit, with $-\frac{3}{2}$ deleted, $\lim_{x \to -3/2} y'$ exists. Hence, this formula is valid even where $x = -\frac{3}{2}$. Thus by the devices of differentiating the logarithm of the *absolute value* of $f(x)$, and then of extending the validity of the resulting formula by use of the limit (continuity of $f'(x)$), we make logarithmic differentiation as general as direct differentiation.

2.6 Exercises

In Exercises 1–18, differentiate with respect to x, using the chain rule.

1. $\ln(1 + \sqrt{x})$.

2. $\sqrt{1 - x^2}$.

3. $(x^2 + 1)^{-3/2}$.

4. $\ln|x + \sqrt{1 + x^2}|$.

5. $\arccos \dfrac{t}{\sqrt{t^2 + 1}}$.

6. $e^{\cos x}$.

7. $e^{\ln x}$.

8. $\ln(e^x)$.

9. $(a^2 - x^2)^{-1/2}$.

10. $\ln \sqrt{1 - x^2}$.

11. $\sin^3 5x$.

12. $x^2 \sin^3 5x$.

13. $xe^{\sin 2x}$.

14. $a^{x^2}, a > 0$.

15. $\arctan \dfrac{x}{\sqrt{1 - x^2}}$.

16. $e^{ax} \sin bx$.

17. $\displaystyle\int_a^{x^2} f(u)\, du$, where $f(x)$ is continuous.

18. $e^{-\cos x^2}$.

In Exercises 19–27, differentiate with respect to x, where y is a differentiable function of x whose derivative is $y'\,dx$.

19. $x \cos y$. $\left(\text{Solution}: \dfrac{d}{dx} x \cos y = x \dfrac{d}{dx} \cos y + \left(\dfrac{dx}{dx}\right) \cos y = -xy' \sin y + \cos y.\right)$

20. xy^2.

21. $\sqrt{x^2 + y^2}$.

22. $\dfrac{y}{x}$.

23. $\dfrac{x}{y}$.

24. $\dfrac{1}{\sqrt{x^2 + y^2}}$.

25. e^{axy}.

26. $e^{-y} \sin ax$.

27. $\arctan \dfrac{y}{x}$.

In Exercises 28–29, differentiate using logarithmic differentiation. State where the differentiation is not valid.

28. $y = \dfrac{\sqrt[3]{(x + 1)^2}(2x - 1)^5}{(x + 4)^2}$.

29. $y = \sqrt[3]{\dfrac{x^2 + 4}{(2x - 1)^5}}$.

30. Establish mathematically the maximum value of te^{-t} for $t \geq 0$.

31. Among all circular sectors with a given perimeter, find the one that has maximum area.

32. What is the longest pole that can be carried horizontally around a right-angle corner from a corridor $13\frac{1}{2}$ ft wide to one 4 ft wide?

33. An open gutter with sloping sides (Figure 7.5) is to be made from sheet metal 18 in. wide by bending up one-third of the width on each side. For what bending angle θ is the maximum water capacity attained?

34. A man standing at A has eyes 6 ft above the ground and looks at a vertical sign at B that is 12 ft high and raised 10 ft above the ground (Figure 7.6). At what distance $|AB|$ will his viewing angle ϕ be a maximum?

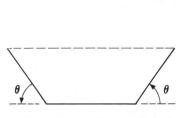

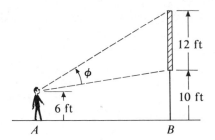

FIGURE 7.5   Exercise 33.   FIGURE 7.6   Exercise 34.

35. Establish the minimum point on the real line of the function given by

$$F(x) = \int_0^{(x-1)^2} e^{-t^2}\, dt.$$

Note. One cannot integrate e^{-t^2} in terms of elementary functions.

36.* (The wineglass problem.)
A full conical wineglass has depth h and generating angle α (Figure 7.7). We ease into it a ball of radius r. Show that the ball of radius

$$r = \frac{h \sin \alpha}{\sin \alpha + \sin 2\alpha}$$

produces maximum overflow.

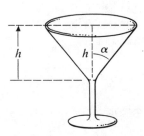

FIGURE 7.7   Exercise 36.

2.7 (Theory section) Theorem on the Limit of the Derivative.

From the mean value theorem (I, §13.2) we know that if f is continuous on the interval $[x, w]$ and differentiable at all points between x and w, then

$$\frac{f(w) - f(x)}{w - x} = f'(c),$$

where c is some number between x and w. Suppose that f' has a limit at x, that is, $\lim_{w \to x} f'(w) = m$. The $\lim_{w \to x} f'(c) = m$ also, because c is always a number between x and w. Hence in this case,

$$\lim_{w \to x} \frac{f(w) - f(x)}{w - x} = \lim_{w \to x} f'(c).$$

But the limit on the left is $f'(x)$, by definition. This observation completes the proof of the following theorem.

THEOREM.   If $f(x)$ is continuous at x and differentiable in a neighborhood of x, with x itself deleted, and $\lim_{w \to x} f'(w)$ exists, then $f(x)$ is differentiable at x also and $f'(x) = \lim_{w \to x} f'(w)$.

3   Antiderivatives of Chains

3.1   Recognizing Chains.   The art of skillful integration in closed form is one aspect of the high algebraic art of recognition of form. To practice it, we must be able to recognize chains. Theoretically there is nothing more to it than the rule of "integration by substitution"

$$\int \text{VI.} \qquad\qquad \int f'(u) \, Du = f(u) + C,$$

together with formulas for derivatives of a standard collection of "elementary" functions. In practice, it is difficult to recognize the chain $f'(u) \, Du$. We observe that in VI the chain rule factor Du must be present in exact form before we can get the antiderivative $f(u) + C$, but in the antiderivative itself the chain rule factor has disappeared. One proceeds by trial and error, trying to factor the expression to be integrated, or *integrand*, into two factors, one of which is the derivative $f'(u)$ and the other, the derivative Du, called the chain rule factor, so that the product $f'(u) \, Du$ is a perfect chain. This cannot always be done, and it is not only difficult, but even beyond our mathematical scope to decide when it is not possible. We recall that $Du = u' \, dx$.

It is a fact, which we cannot prove at this stage, that there exist no integrals in terms of finite combinations of elementary functions for

$$\int e^{x^2} \, dx, \qquad \int \sin(t^2) \, dx, \qquad \int \sqrt{x^3 + x + 1} \, dx, \qquad \int \ln \cos x \, dx,$$

even though the integrands are integrable. For example, e^{x^2} has an antiderivative given by the indefinite integral $\int_0^x e^{u^2} \, du$.

On the other hand,

$$\int e^{x^2} x \, dx, \qquad\qquad \int t \sin(t^2) \, dt,$$

$$\int \sqrt{x^3 + x + 1} \, (3x^2 + 1) \, dx, \qquad \int \sin x \ln \cos x \, dx$$

are all easily integrated. We proceed by examples.

3.2   Integrating Power Forms

Example A.   Calculate $\int (x^2 + 1)^3 \, x \, dx$.

Solution.   The integrand looks like $u^3 \, Du$, where $u = x^2 + 1$. Trying this, we calculate $Du = 2x \, dx$. Looking at the integrand, we find that we have $x \, dx$, but not $2x \, dx$, as a factor. We can correct this defect when only a missing constant factor is involved. In fact, we multiply $x\,dx$ in the integrand by 2 and compensate by multiplying outside the integral sign by $\frac{1}{2}$. Thus

$$\int (x^2 + 1)^3 \, x \, dx = \frac{1}{2} \int (x^2 + 1)^3 (2x \, dx) = \frac{1}{2} \int u^3 \, Du$$

$$= \frac{1}{2} \frac{u^4}{4} + C = \frac{1}{8} (x^2 + 1)^4 + C.$$

A slight variation of the procedure puts less strain on the imagination by using direct substitution. When we try $u = x^2 + 1$, $Du = 2x \, dx$, we solve for $dx = Du/2x$ and substitute into the integrand as follows,

$$\int (x^2 + 1)^3 \, x \, dx = \int u^3 \, x \left(\frac{Du}{2x} \right) = \frac{1}{2} \int u^3 \, Du$$

completing as before.

Example B.   Calculate $\int (x^2 + 1)^3 \, dx$.

Solution.   This looks like $u^3 \, Du$ with $u = x^2 + 1$. This will require $Du = 2x \, dx$ as a chain factor. We can introduce the constant factor 2, as in Example A, but not the function x. Thus the trial fails. The variant procedure of direct substitution leads to the same difficulty.

Hence, we abandon the attempt to make the power-formula chain with $u = x^2 + 1$, and we start over. This time we apply the binomial theorem and reduce the integrand to a sum of powers of x for which dx is the appropriate chain factor.

$$\int (x^2 + 1)^3 \, dx = \int (x^6 + 3x^4 + 3x^2 + 1) \, dx$$
$$= \tfrac{1}{7} x^7 + \tfrac{3}{4} x^5 + x^3 + x + C.$$

This completes the problem.

Example C.   Calculate $\displaystyle \int \frac{x \, dx}{x^2 + 1}$.

Solution.   This looks like $\int (Du/u)$ with $u = x^2 + 1$, $Du = 2x \, dx$. Introducing the missing factor 2 as in Example A, we find

$$\int \frac{x \, dx}{x^2 + 1} = \frac{1}{2} \int \frac{2x \, dx}{x^2 + 1} = \frac{1}{2} \int \frac{Du}{u} = \frac{1}{2} \ln |u| + C$$

$$= \frac{1}{2} \ln (x^2 + 1) + C = \ln \sqrt{x^2 + 1} + C.$$

This completes the problem.

Example D.   Calculate $\displaystyle \int \frac{dx}{x^2 + x + 1}$.

Solution.   This looks like $\int (Du/u)$ where $u = x^2 + x + 1$, $Du = (2x + 1) \, dx$. As in Example B, there is no way to introduce the missing factor $(2x + 1)$ into the chain factor. Hence we abandon the trial to make the chain Du/u and start over.

This time we cannot use the binomial theorem, as in Example C. Perhaps after other vain attempts, we finally decide to complete the square of the quadratic $x^2 + x + 1$ in the denominator. The result is

$$x^2 + x + 1 = \left(x^2 + x + \frac{1}{4}\right) + \left(1 - \frac{1}{4}\right) = \left(x + \frac{1}{2}\right)^2 + \left(\frac{\sqrt{3}}{2}\right)^2.$$

We set $u = x + \frac{1}{2}$, $Du = 1\ dx$, and find with this transformation the integral

$$\int \frac{dx}{x^2 + x + 1} = \int \frac{dx}{(x + \frac{1}{2})^2 + (\sqrt{3}/2)^2} = \int \frac{Du}{u^2 + (\sqrt{3}/2)^2}$$

$$= \frac{2}{\sqrt{3}} \arctan \frac{2}{\sqrt{3}} u + C = \frac{2}{\sqrt{3}} \arctan \frac{2}{\sqrt{3}} (x + \frac{1}{2}) + C,$$

where the integration was completed by $\int$ XVI$_a$. This completes the problem and illustrates the difficulty of recognizing a chain that was perfect as it occurred originally. We can check by differentiating the last expression and verify that its derivative is $1/(x^2 + x + 1)$.

3.3   Simple Trigonometric Chains

Example E.   Integrate $\int \cos^6 2x \sin^3 2x\ dx$.

Solution.   We take advantage of the odd power of the sine factor to find a chain. In fact, we factor

$$\sin^3 2x = \sin^2 2x \sin 2x = (1\ \cos^2 2x) \sin 2x,$$

so that the integral becomes

$$\int \cos^6 2x \sin^3 2x\ dx = \int \cos^6 2x(1 - \cos^2 2x) \sin 2x\ dx$$

$$= \int \cos^6 2x(\sin 2x\ dx) - \int \cos^8 2x \sin 2x\ dx.$$

Each of these two integrals now looks like a power formula $\int u^n\ Du$ with $u = \cos 2x$, $Du = -\sin 2x\ (2\ dx)$. We must introduce the factor -2 into the integrand to complete the chains, thus

$$\int \cos^6 2x \sin^3 2x\ dx = -\frac{1}{2} \int \cos^6 2x(-\sin 2x)\, 2dx + \frac{1}{2} \int \cos^8 2x(-\sin 2x)\, 2dx$$

$$= -\frac{1}{2} \frac{\cos^7 2x}{7} + \frac{1}{2} \frac{\cos^9 2x}{9} + C.$$

This completes the problem.

We observe that this technique will calculate any integral $\int \cos^m ax \sin^n ax\ dx$, provided that one of the exponents m or n is an odd integer so that we can factor out $Du = -\sin ax(a\ dx)$ or $Du = \cos ax(a\ dx)$.

Example F.   Calculate $\int \tan u\ Du$.

Solution. Since $\tan u = \sin u/\cos u$, the integral becomes

$$\int \tan u \, Du = \int \frac{\sin u \, Du}{\cos u} = -\int -\frac{\sin u \, Du}{\cos u}$$

$$= -\ln |\cos u| + C, \qquad \text{or} \qquad \ln |\sec u| + C.$$

3.4 Some Chains with Exponentials

Example G. Calculate

$$\int_0^1 \frac{e^x}{1 + e^{-x}} \, dx.$$

Solution. We first try to find an antiderivative

$$\int \frac{e^x}{1 + e^{-x}} \, dx.$$

If we try to make a chain in the form Du/u with $u = 1 + e^{-x}$ and $Du = -e^{-x} \, dx$, we find that we can introduce the minus sign to put $-e^x \, Dx$ in the numerator. This does not give us the Du required to complete the chain. Hence we try another way.

We multiply numerator and denominator by e^x to eliminate the negative exponent. Then we carry out one step of division and thus we find by algebra that

$$\frac{e^x}{1 + e^{-x}} = \frac{(e^x)^2}{e^x + 1} = e^x - \frac{e^x}{e^x + 1}.$$

This gives us

$$\int \frac{e^x}{1 + e^{-x}} \, dx = \int \left(e^x - \frac{e^x}{e^x + 1} \right) dx = e^x - \ln (e^x + 1) + C.$$

Returning to the definite integral in the problem, we find by the fundamental theorem of integral calculus

$$\int_0^1 \frac{e^x}{1 + e^{-x}} \, dx = e^x - \ln (e^x + 1) \Big|_0^1 = e - \ln (e + 1) - 1 + \ln 2.$$

A numerical evaluation, if desired, will now complete the problem.

Example H. Find the curve $y = f(x)$ through the origin $(0, 0)$ having slope $f'(x) = 1/e^x + e^{-x}$ at every point.

Solution. For some value of the constant of integration

$$y = \int \frac{1}{e^x + e^{-x}} \, dx = \int \frac{e^x \, dx}{(e^x)^2 + 1} = \arctan e^x + C.$$

Now, since $y = 0$ when $x = 0$, we find that $0 = \arctan e^0 + C$, or $0 = \pi/4 + C$. Hence $C = -\pi/4$, and the required curve is given by

$$y = \arctan e^x - \frac{\pi}{4}.$$

3.5 Exercises

In Exercises 1–42, calculate the antiderivatives and definite integrals as indicated. Exercises 1–10 are like Examples A and C.

1. $\int (x^3 + 4)x^2 \, dx$.

2. $\int \sqrt{1 + x^2}\, x \, dx$.

3. $\int_0^{1/2} \dfrac{x}{\sqrt{1 - x^2}} \, dx$.

4. $\int_0^1 \dfrac{x \, dx}{1 + x^2}$.

5. $\int \dfrac{x}{(1 + x^2)^2} \, dx$.

6. $\int \dfrac{x^3}{x^4 + 1} \, dx$.

7. $\int \sqrt{2 + 4x + 4x^2}\,(2x + 1) \, dx$.

8. $\int \left(\dfrac{y}{x}\right)^2 \dfrac{(xy' - y)\, dx}{x^2}$.

9. $\int \dfrac{yy'\, dx}{1 + y^2}$.

10. $\int \dfrac{Du}{\sqrt{u}(1 + \sqrt{u})}$.

Exercises 11–16 are like Example B.

11. $\int (x^2 - 3)^2 \, dx$.

12. $\int \left(\dfrac{x^2 + x + 1}{x + 1}\right)^2 \, dx$.

13. $\int_0^2 e^{\ln x} \, dx$.

14. $\int x \cos (\arcsin x) \, dx$.

15. $\int \sqrt{1 + \sin 2x} \, dx$.

16. $\int_0^T e^{\sin^2 \theta + \sin^2 \theta} \, d\theta$.

Exercises 17–22 are like Example D. Some involve improper integrals that require an investigation of convergence (I, §30).

17. $\int \dfrac{dx}{x^2 - x + 1}$.

18. $\int_0^\infty \dfrac{dx}{x^2 + 2x + 2}$.

19. $\int_0^1 \dfrac{dx}{\sqrt{x - x^2}}$.

20. $\int_{-1/2}^0 \dfrac{dx}{4x^2 + 4x + 2}$.

21. $\int \dfrac{x \, dx}{\sqrt{(1 + x)(2 - x)}}$.

22. $\int_0^a \sqrt{2ax - x^2}\,(x - a) \, da$.

Exercises 23–26 are like Examples E and F.

23. $\int \sin^2 x \cos^3 x \, dx$.

24. $\int_0^a \cos^3 \left(\dfrac{x}{a}\right) \, dx$.

25. $\int_0^{\pi/a} \sin^7 ax \, dx$.

26. $\int 2 \sin ax \cos ax \, dx$.

27. $\int \dfrac{\sin 2x}{\cos^2 2x} \, dx$.

28. $\int_0^\pi \left[1 - \sin^3 \left(\dfrac{\theta}{2}\right)\right] d\theta$.

29. $\int_{\pi/4}^\theta \dfrac{\sec^2 x}{\tan x} \, dx$.

30. $\int \dfrac{\sin ax}{\sqrt{2 - \cos ax}} \, dx$.

31. $\int \sqrt{1 + \sin^2 \theta}\, \sin \theta \cos \theta \, d\theta$.

32. $\int \dfrac{\arcsin x}{\sqrt{1 - x^2}} \, dx$.

33. $\int (1 + 2x \cos x)^3 (\cos x - x \sin x) \, dx$.

34. $\int_0^{\pi/4} \dfrac{dx}{\sec x}$.

35. $\int \dfrac{dx}{\sqrt{1 - x^2} \, \arcsin x}$.

36. $\int \dfrac{\sin x}{1 - \cos x} \, dx$.

Exercises 37–42 are like Example G.

37. $\int \sqrt{1 + e^{2x}}\, e^{2x}\, dx.$ 38. $\int \dfrac{e^x}{1 + e^{2x}}\, dx.$

39. $\int \dfrac{e^x}{e^{2x} + e^x + 1}\, dx.$ 40. $\int_{-1}^{1} (e^x + e^{-x})^2\, dx.$

41. $\int_{0}^{1} \dfrac{e^x - e^{-x}}{e^x}\, dx.$ 42. $\int_{0}^{\infty} e^{-x^2} x\, dx.$

In Exercises 43–46, find the unknown function that satisfies the differential equation and initial condition. They are like Example D.

43. $x \dfrac{du}{dx} = \ln x,\ u(1) = 4.$ 44. $\dfrac{d}{dx} f(x) = x\sqrt{1 + x^2},\ f(1) = 0.$

45. $f(x)f'(x) = x^2,\ f(1) = 0.$ 46. $f'(x) = \dfrac{\arctan x}{1 + x^2},\ f(1) = \pi^2/4.$

In Exercises 47–64, calculate the antiderivatives and integrals. They are of mixed types.

47. $\int \sqrt{e^{3x} + e^{2x}}\, dx.$ 48. $\int_{0}^{\pi a} \sin^3 \left(\dfrac{x}{a} \right) dx.$

49. $\int \dfrac{dx}{e^x + e^{-x}}.$ 50. $\int_{0}^{\pi/2} \dfrac{\sin x\, dx}{\cos^2 x + 1}.$

51. $\int \sin x \sin 2x\, dx.$ 52. $\int \sin x \cos^2 2x\, dx.$

53. $\int \cos \dfrac{\theta}{2} \sin \theta\, d\theta.$ 54. $\int \tan^3 x \cos^4 x\, dx.$

55. $\int \dfrac{x}{x^2 + x + 1}\, dx.$ 56. $\int \dfrac{e^{3x} + 1}{e^x}\, dx.$

57. $\int \dfrac{\sin \sqrt{x}}{\sqrt{x}}\, dx.$ 58. $\int \left(\dfrac{x}{x^3 + 1} \right)^2 dx.$

59. $\int_{1}^{e} \dfrac{\ln x}{x}\, dx.$ 60. $\int_{e}^{e^2} \dfrac{dx}{x \ln x}.$

61. $\int_{0}^{a} (\sqrt{a} - \sqrt{x})^2\, dx.$ 62. $\int \dfrac{1 + \sin x}{\cos x}\, dx.$

63. $\int \dfrac{2x + 3}{x^2 + 2x - 5}\, dx.$ 64. $\int \dfrac{x}{1 + x^4}\, dx.$

In Exercises 65–74, calculate the definite integrals. Some are improper integrals (I, §30), requiring an investigation of convergence.

65. $\int_{0}^{\pi} \sin^3 x \cos x\, dx.$ 66. $\int_{0}^{\pi} \cos^5 x\, dx.$

67. $\int_{0}^{1} \dfrac{dx}{\sqrt{2x - x^2}}$ (Improper).   68. $\int_{0}^{\infty} \dfrac{dx}{2 + 2x + x^2}$ (Improper).

69. $\int_{1}^{\infty} \dfrac{dx}{x \ln x}.$ 70. $\int_{1}^{\infty} \dfrac{dx}{x^{3/2}}.$

71. $\int_{0}^{a} \dfrac{x\, dx}{(a^2 - x^2)^{3/2}}$ 72. $\int_{0}^{\infty} e^{-3x^2} x\, dx.$

73. $\int_{-\pi}^{\pi} (\cos x + \sin x)^2\, dx.$ 74. $\int_{-\infty}^{\infty} \dfrac{dx}{4 + x^2}$

75. Sketch the graph of the integrand function in Exercise 74 and represent the integral as the area of a region related to this curve.

In Exercises 76–78, m and n are positive integers.

76. $\displaystyle\int_0^{2\pi} \sin mx \, \sin nx \, dx = \begin{cases} 0 \text{ if } m \neq n, \\ \pi \text{ if } m = n. \end{cases}$

77. $\displaystyle\int_0^{2\pi} \cos mx \, \cos nx \, dx = \begin{cases} 0 \text{ if } m \neq n, \\ \pi \text{ if } m = n. \end{cases}$

78. $\displaystyle\int_0^{2\pi} \sin mx \, \cos nx \, dx = 0.$

4 Trigonometric Substitution

4.1 Squares of Sine and Cosine.

The antiderivatives $\int \sin^2 x \, dx$ and $\int \cos^2 x \, dx$ occur frequently and do not form recognizable chains directly. We may use integration by parts,

$$\int u \, Dv = uv - \int v \, Du.$$

In the integral

$$I = \int \cos^2 x \, dx$$

let $u = \cos x$, $Dv = \cos x \, dx$; then

$$Du = -\sin x \, dx, \qquad v = \sin x.$$

Integration by parts then gives

$$I = \cos x \sin x + \int \sin^2 x \, dx$$

$$= \cos x \sin x + \int (1 - \cos^2 x) \, dx$$

$$= \cos x \sin x + x - \int \cos^2 x \, dx.$$

The last integral is $-I$. Transposing this term, we find

$$2I = \cos x \sin x + x,$$

or

$$\int \cos^2 x \, dx = \frac{1}{2} x + \frac{1}{2} \cos x \sin x + C = \frac{1}{2} x + \frac{1}{4} \sin 2x + C.$$

A similar procedure gives

$$\int \sin^2 x \, dx = \frac{1}{2} x - \frac{1}{2} \cos x \sin x + C = \frac{1}{2} x - \frac{1}{4} \sin 2x + C.$$

4.2 Trigonometric Substitution.

The antiderivative

$$\int \sqrt{a^2 - x^2} \, dx, \qquad |x| \leq |a|,$$

does not convert into a power formula with $u = a^2 - x^2$, for we cannot complete the chain $u^{1/2} \, Du$ with the required $Du = -2x \, dx$. However, we can take advantage of the Pythagorean identity $\sin^2 \theta + \cos^2 \theta = 1$ to transform it into an integral with the radical removed. We introduce a new independent variable θ by means of the inverse substitution

$$x = a \sin \theta, \qquad dx = a \cos \theta \, d\theta.$$

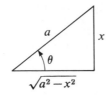

FIGURE 7.8 $\theta = \arcsin x/a.$

Postponing the theoretical justification of the procedure, we write

$$\int \sqrt{a^2 - x^2}\, dx = \int \sqrt{a^2 - a^2 \sin^2 \theta}\, (a \cos \theta\, d\theta),$$

where the equals sign means that when the left member is regarded as a composite function of θ by substituting $x = a \sin \theta$, the two members are equal as functions of θ. We simplify the right member and use the antiderivative of $\cos^2 \theta$ and find that

$$\int \sqrt{a^2 - x^2}\, dx\Big|^{x = a \sin \theta} = a^2 \int \cos^2 \theta\, d\theta = a^2 \left(\frac{\theta}{2} + \frac{1}{4} \sin 2\theta\right) + C$$

$$= \frac{a^2}{a} (\theta + \sin \theta \cos \theta) + C.$$

The inverse of the substitution $x = a \sin \theta$ is $\theta = \arcsin x/a$, $-1 \leq x/a \leq 1$, which we now apply to restore the variable x into the antiderivative formula. We obtain the antiderivative

$$\int \sqrt{a^2 - x^2}\, dx = \frac{a^2}{a} (\theta + \sin \theta \cos \theta)\Big|^{\theta = \arcsin x/a}$$

$$= \frac{a^2}{a} \left(\arcsin \frac{x}{a} + \frac{x}{a} \frac{\sqrt{a^2 - x^2}}{a}\right) + C.$$

In the last term we computed $\cos (\arcsin (x/a))$ by the symbolic triangle technique (I, §33.1) (Figure 7.8). This completes the integration of $\sqrt{a^2 - x^2}$.

4.3   Area of Ellipse

Example.   Compute the area of the ellipse (Figure 7.9)

$$\frac{x^2}{x^2} + \frac{y^2}{b^2} = 1, \qquad a > 0, b > 0.$$

Solution.   We solve the equation of the ellipse for y as a function of x. The solution function representing the upper half of the ellipse is

$$y = \frac{b}{a} \sqrt{a^2 - x^2}, \qquad |x| \leq a.$$

The area $A/2$ of the upper half of the region inside the ellipse is then given by the definite integral

$$\frac{A}{2} = \int_{-a}^{a} \frac{b}{a} \sqrt{a^2 - x^2}\, dx.$$

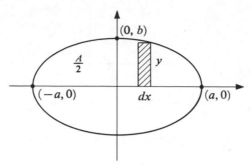

FIGURE 7.9   Ellipse $\dfrac{x^2}{a^2} + \dfrac{y^2}{b^2} = 1$.

To evaluate the integral, we apply the fundamental theorem of integral calculus (I, §19.3) using the antiderivative of $\sqrt{a^2 - x^2}$. We find that

$$\frac{A}{2} = \frac{b}{a} \int_{-a}^{a} \sqrt{a^2 - x^2}\, dx = \frac{1}{2}\frac{b}{a}\left(a^2 \arcsin\frac{x}{a} + x\sqrt{a^2 - x^2}\right)\Big|_{-a}^{a}$$

$$= \frac{1}{2}\frac{b}{a}\left(a^2\frac{\pi}{2} + 0 - a^2\left(-\frac{\pi}{2}\right) - 0\right) = \frac{\pi ab}{2}.$$

Hence $A = \pi ab$. We observe that in evaluating **arcsin** (x/a) at points where $x = a$ and $x = -a$ it is important to remember the restricted values of the arc sine function in the range $[-\pi/2, \pi/2]$ (I, §33.3), that is, arcsin $1 = \pi/2$ and arcsin $(-1) = -\pi/2$.

Alternative shorter procedure.   We need not transform the variable back from θ to x, but if we transform the limits from x-limits $x = -a$ and $x = b$ to θ-limits we can evaluate the area integral directly in terms of the transformed variables. The substitution $x = a \sin \theta$ replaces the chain $\sqrt{a^2 - x^2}\, dx$ by $a^2 \cos^2 \theta\, d\theta$, which we found (§4.1) has the antiderivative $(a^2/2)(\theta + \sin \theta \cos \theta)$. Now, instead of going back through the inverse substitution to a function of x, let us transform the limits of the definite integral as follows. In $x = a \sin \theta$, when $x = -a$, $\sin \theta = -1$ and $\theta = -\pi/2$, and when $x = a$, $\sin \theta = +1$ and $\theta = \pi/2$. With these transformed limits, we can write

$$\frac{A}{2} = \frac{b}{a}\int_{-a}^{a}\sqrt{a^2 - x^2}\, dx = \frac{b}{a}\int_{-\pi/2}^{\pi/2} a^2 \cos^2 \theta\, d\theta = ab\left(\frac{1}{2}\right)(\theta + \sin \theta \cos \theta)\Big|_{-\pi/2}^{\pi/2}$$

$$= \frac{ab}{2}\left[\frac{\pi}{2} + 0 - \left(-\frac{\pi}{2}\right) - 0\right] = \frac{\pi ab}{2}.$$

We must be careful to choose the values of θ from the principal values of **arcsin** (x/a). Hence $A = \pi ab$ as before. We justify this later (§4.7).

4.4   Powers of Sine and Cosine

Example.   Integrate

$$J = \int \cos^n ax\, dx,$$

where n is a positive integer.

Solution.   We apply integration by parts choosing $u = \cos^{n-1} ax$, $Dv = \cos ax\, dx$. Then

$$Du = (n-1)\cos^{n-2} ax(-a\sin ax)\, dx, \qquad v = \frac{1}{a}\sin ax.$$

Integration by parts then says that

$$J = \frac{1}{a}\cos^{n-1} ax \sin ax + (n-1)\int \cos^{n-2} ax \sin^2 ax\, dx.$$

We use the Pythagorean relation $\sin^2 ax = 1 - \cos^2 ax$ in the new integrand and find that it brings back the original integral J,

$$J = \frac{1}{a}\cos^{n-1} ax \sin ax + (n-1)\int \cos^{n-2} ax\, dx - (n-1)J.$$

This is not taking us in circles as it first appears; for we can transpose the term $-(n-1)J$, combine it with the original J to form nJ on the left. Then, dividing both sides by n, we get the *reduction formula*,

$$\int \cos^n ax\, dx = \cos^{n-1}\frac{ax}{na}\sin ax + \frac{n-1}{n}\int \cos^{n-2} ax\, dx.$$

This essentially solves the problem, because it replaces the problem of integrating $\int \cos^n ax\ dx$ by another integration of the same form with the exponent of $\cos ax$ reduced by 2. Without repeating the integration by parts itself, we can repeatedly apply the reduction formula until the exponent goes down to either $n = 0$ or $n = 1$. If $n = 0$, the last integration is just $\int dx$, and if $n = 1$, the last integration is $\int \cos ax\ dx$.

Example.   Integrate

$$J = \int \cos^4 ax \sin^2 ax\ dx.$$

Solution.   We first use $\sin^2 \theta + \cos^2 \theta = 1$ to transform J into

$$J = \int \cos^4 ax\ dx - \int \cos^6 ax\ dx.$$

Then we apply the reduction formula to the integral of $\cos^6 ax$ with $n = 6$. We get

$$J = \int \cos^4 ax\ dx - \left[\frac{\cos^5 ax \sin ax}{6a} + \frac{5}{6}\int \cos^4 ax\ dx\right]$$

$$= \frac{1}{6}\int \cos^4 ax\ dx - \frac{\cos^5 ax \sin ax}{6a}.$$

Now we apply the reduction formula again, with $n = 4$.

$$J = \frac{1}{6}\left[\frac{\cos^3 ax \sin ax}{4a} + \frac{3}{4}\int \cos^2 ax\ dx\right] - \frac{\cos^5 ax \sin ax}{6a}.$$

Applying the reduction formula again, with $n = 2$, we get for the integral

$$\int \cos^2 ax\ dx = \frac{\cos ax \sin ax}{2a} + \frac{1}{2}\int dx = \frac{\cos ax \sin ax}{2a} + \frac{x}{2} + C$$

Inserting this into the preceding expression for J and simplifying, we find

$$J = \frac{x}{16} + \frac{\sin ax}{2a} \left[\frac{\cos ax}{8} + \frac{\cos^3 ax}{12} - \frac{\cos^5 ax}{3} \right] + C.$$

This completes the problem.

The similar reduction formula for $\int \sin^n ax\, dx$ when n is a positive integer is

$$\int \sin^n ax\, dx = -\frac{\sin^{n-1} ax \cos ax}{na} + \frac{n-1}{n} \int \sin^{n-2} ax\, dx.$$

Proof.   Exercise 23, page 284.

4.5   The Substitution $x = a \tan \theta$.   We consider integrals of the form

$$\int (x^2 + a^2)^p\, dx.$$

There is no way of introducing the necessary Du to treat this as a power form $\int u^n\, Du$

Example.   Integrate

$$J = \int \frac{Du}{(a^2 + u^2)^{n+1}}.$$

where n is a positive integer.

Solution.   We use the substitution $x = a \tan \theta$ in the form

$$u = a\,\frac{\sin \theta}{\cos \theta}, \qquad Du = a\,\frac{d\theta}{\cos^2 \theta}.$$

Under this substitution,

$$a^2 + x^2 = a^2 + a^2\,\frac{\sin^2 \theta}{\cos^2 \theta} = \frac{a^2(\cos^2 \theta + \sin^2 \theta)}{\cos^2 \theta} = \frac{a^2}{\cos^2 \theta}.$$

The integral then transforms into

$$J \to \frac{1}{a^{2n+1}} \int \cos^{2n} \theta\, d\theta.$$

We perform one stage of reduction of the exponent of power of $\cos \theta$ using the reduction formula for $\int \cos^n \theta\, d\theta$. Then we transform back to functions of u, using the symbolic triangle for $u/a = \tan \theta$ to evaluate the terms that appear in the reduction formula for $\int \cos^n \theta\, d\theta$ (Figure 7.10). The result is the reduction formula

$$J = \frac{1}{2a^2} \left[\frac{u}{n(a^2 + u^2)^n} + \frac{2n-1}{n} \int \frac{Du}{(a^2 + u^2)^n} \right],$$

which essentially solves the problem.

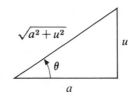

FIGURE 7.10 $\theta = \arctan u/a$.

Proof.   Exercises.

Example.   Integrate

$$\int \frac{dx}{(x^2 + 2x + 2)^2}.$$

Solution.   We complete the square of the quadratic in the denominator,

$$x^2 + 2x + 2 = (x + 1)^2 + 1.$$

The integral then becomes

$$J = \int \frac{dx}{[1 + (x + 1)^2]^2}.$$

which is in the form $\int \boldsymbol{Du}/(a^2 + \boldsymbol{u}^2)^{n+1}$, with $\boldsymbol{u} = x + 1$, $a = 1$, $n = 1$. We apply the reduction formula for this integral and find

$$J = \frac{1}{2}\left[\frac{x + 1}{1[1 + (x + 1)^2]^1} + \frac{1}{1}\int \frac{dx}{1 + (x + 1)^2} \right].$$

The last term is **arctan** $(x + 1)$. Hence we have the antiderivative

$$J = \frac{1}{2} \frac{x + 1}{x^2 + 2x + 2} + \frac{1}{2} \text{arctan } (x + 1) + \boldsymbol{C}.$$

This completes the problem.

4.6   Exercises

In Exercises 1–22, calculate the antiderivatives and definite integrals as indicated. Exercises 1–8 are like the example in Part I, Sections 37.1 and 37.2.

1. $\displaystyle\int_0^\pi \sin^2\frac{t}{2}\, dt.$ 2. $\displaystyle\int \cos^2 3x\, dx.$

3. $\int \sqrt{a^2 - x^2}\, dx$, using $x = a \cos \theta$. Explain the apparent difference between the result and that of Section 4.2.

4. $\int dx/\sqrt{a^2 - x^2}$, using the substitution $x = a \sin \theta$.

5. $\int x\sqrt{a^2 - x^2}\, dx$ in two ways: using the substitution $x = a \sin \theta$, and also by direct integration as a power form (§4.2).

6. $\int x(a^2 - x^2)^{3/2}\, dx$ in two ways, as in Exercise 5.

7. $\displaystyle\int_0^a x^3\sqrt{a^2 - x^2}\, dx.$ 8. $\displaystyle\int_{-1}^1 \sqrt{1 - x^2}\, dx.$

In Exercises 9–12, use the alternative shorter procedure, transforming the limits of integration (§4.3).

9. $\int_0^{\sqrt{2}} \sqrt{4 - x^2}\, dx.$ 10. $\int_{-1/2}^{1/2} \sqrt{1 - x^2}\, dx.$

11. $\int_0^{1/2} \sqrt{1 - 4x^2}\, dx.$ 12. $\int_0^a \sqrt{1 - (x/a)^2}\, dx.$

Exercises 13–16 involve even powers of sine and cosine (§4.4).

13. $\int \sin^4 ax\, dx.$ 14. $\int_0^\pi \sin^2 x \cos^2 x\, dx.$

15. $\int_0^\pi \sin^4 2x \cos^2 2x\, dx.$ 16. $\int_0^1 x^2\sqrt{1 - x^2}\, dx.$

Exercises 17–22 involve the substitution $x = a \tan \theta$. Recall the antiderivative (I, §31.7).

$$\int \sec u\, Du = \int \frac{Du}{\cos u} = \ln \left| \frac{1 + \sin u}{\cos u} \right| + C.$$

17. $\int \frac{dx}{(a^2 + x^2)^2}.$ 18. $\int \frac{dx}{(4 + x^2)^{3/2}}.$

19.* $\int_0^1 \frac{dt}{(1 + t + t^2)^3}.$ 20. $\int \frac{dx}{\sqrt{x^2 - 2x + 2}}.$

21. $\int \frac{x\, dx}{x^2 + x + 1}.$ 22.* $\int_{-a}^a \frac{dx}{(a^2 + x^2)^4}.$

23. Derive the reduction formula for the antiderivative $\int \sin^n ax\, dx$ (§4.4).

24. Complete the details to derive the reduction formula for the antiderivative (§4.5).

$$J = \int \frac{Du}{(a^2 + u^2)^{n+1}}.$$

25. Apply the substitution $x = a \tan \theta$ to integrate $\int dx/(x^2 + a^2).$

26. Does the reduction formula of Exercise 24 apply to the integral of Exercise 25? Explain.

Exercises 27–38 are of mixed types. Integrate.

27. $\int_0^a \frac{dx}{\sqrt{a^2 + x^2}}.$ 28. $\int_0^{\pi/12} \sin^2 3\theta\, d\theta.$

29. $\int \sin^4 \theta\, d\theta.$ 30. $\int \sin x \cos 2x\, dx.$

31. $\int \cos^3 x\sqrt{\sin x}\, dx.$ 32. $\int \frac{dx}{(x^2 + 4x + 5)^3}.$

33. Apply the reduction formula for $\int \cos^n ax\, dx$ in the case $n = -1$. Is it valid? Does it solve the problem?

34. Derive the reduction formula

$$\int \frac{dx}{\cos^n ax} = \frac{1}{n - 1} \left[\frac{1}{a} \frac{\sin ax}{\cos^{n-1} ax} + (n - 2) \int \frac{dx}{\cos^{n-2} ax} \right]$$

when the integer $n > 1$.

35. Use the reduction formula of Exercise 34 with the antiderivative $\int Du/\cos u$ (I, §31.7) to find the antiderivative

$$\int \sec^3 \theta\, d\theta = \int \frac{d\theta}{\cos^3 \theta} = \frac{1}{2} \left[\frac{\sin \theta}{\cos^2 \theta} + \ln \left| \frac{1 + \sin \theta}{\cos \theta} \right| \right] + C.$$

36. Using the result of Exercise 35, compute $\int_0^a \sqrt{a^2 + x^2}\, dx.$

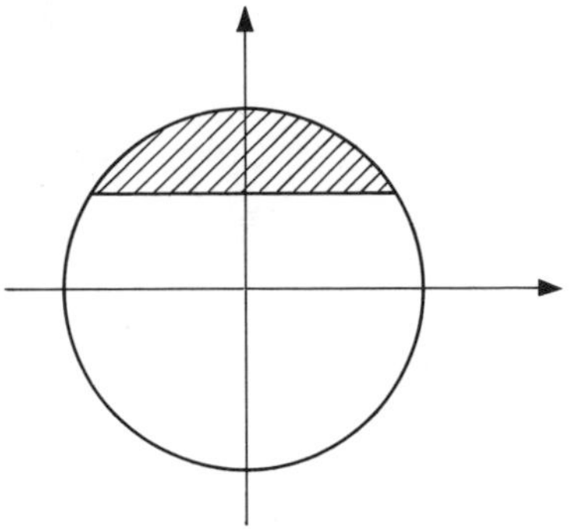

FIGURE 7.11   Exercise 38.

37.   Find the area of the region enclosed between the hyperbola $b^2x^2 - a^2y^2 = a^2b^2$ and the line $x = h$, where $a < h$.
38.   Find the area of the segment of the circle $x^2 + y^2 = r^2$ cut off by the line $y = k$, where $0 < k < r$ (Figure 7.11).

4.7   Theory of Inverse Substitution.   We state the general rule of inverse substitution in the following theorem.

THEOREM (Inverse substitution).   Let the chain $f(x)\ Dx$ be transformed by the invertible differentiable substitution $x = h(t)$ into the chain $f[h(t)]h'(t)\ dt$, which has the anti-derivative $F(t)$. Then, if $t = g(x)$ is the inverse of $x = h(t)$, the composite function $F[g(x)]$ is an antiderivative of the chain $f(x)\ Dx$.

Proof.   The general chain rule (I, §20.1) implies that $f(x)\ Dx = f[h(t)]h'(t)\ Dt$ when x in the left member is replaced by $h(t)$.

By hypothesis, this chain has an antiderivative $F(t)$, so that

$$DF(t) = f[h(t)]h'(t)\ Dt = f[h(t)]\ Dh(t).$$

We now transform this equation back to a function of x by the substitution $t = g(x)$, whose inverse is $h(t)$. That is, $h[g(x)] = x$. This gives us, by the chain rule again,

$$DF[g(x)] = f[h(g(x))]\ Dh[g(x)] = f(x)\ Dx.$$

Thus $F[g(x)]$ is an antiderivative of the chain $f(x)\ Dx$, and this completes the proof of the theorem.

This statement of the substitution rule as a theorem of differential calculus emphasizes the fact that we did not transform the definite integral $\int_a^b f(x)\ dx$. However, we can use it to evaluate that integral under conditions described in the following corollary.

COROLLARY.   Under the hypotheses of the theorem, if $f(x)$ is integrable, then

$$\int_a^b f(x)\ dx = F[g(a)] - F[g(b)] = \int_{g(a)}^{g(b)} f[h(t)]h'(t)\ dt.$$

Proof.   The antiderivative $F[g(x)]$ of $f(x)$ satisfies requirements of the fundamental theorem of calculus. This gives us the evaluation of the first integral. On the other hand, $F(t)$ is an antiderivative of the chain $f[h(t)]h'(t)\,dt$, and hence the same number $F[g(b)] - F[g(a)]$ evaluates the second integral as well, and the proof is complete.

The corollary justifies the alternative method of evaluating a definite integral by transforming both the integrand chain and the limits of integration (§4.3).

5 Integration of Rational Functions by Partial Fractions

5.1 A Reverse Problem.

A rational function is one whose values are determined as the quotient of two polynomials $P(x)/Q(x)$ at points x where $Q(x) \neq 0$. Our objective is to find a technique for computing antiderivatives

$$\int \frac{P(x)}{Q(x)}\,dx.$$

Let us begin by calculating an antiderivative for a rational function $f(x)$, which is given as the sum of a polynomial plus some rational fractions with powers of linear and quadratic terms as denominators,

$$(1) \qquad f(x) = x - 4 + \frac{1}{x} - \frac{1}{x^2} + \frac{5}{x^3} + \frac{2x - 1}{x^2 + x + 1} + \frac{x + 3}{(x^2 + x + 1)^2},$$

defined wherever $x \neq 0$. We can clearly convert this sum into the form of a rational fraction as given by the definition by adding these terms into one fraction with denominator $x^3(x^2 + x + 1)^2$. In fact, the result is

$$(2) \qquad f(x) = \frac{x^8 - 2x^7 - 2x^6 - 8x^5 + x^4 + 7x^3 + 14x^2 + 9x + 5}{x^7 + 2x^6 + 3x^5 + 2x^4 + x^3}.$$

We can readily compute an antiderivative of the sum (1) but it is by no means obvious in the form (2) of a single rational fraction.

Let us compute the antiderivative $\int f(x)\,dx$ in the form (1). We can write down the first five terms by inspection,

$$\int f(x)\,dx = \frac{x^2}{2} - 4x + \ln|x| + \frac{1}{x} - \frac{5}{2x^2} + \int \frac{2x - 1}{x^2 + x + 1}\,dx$$

$$+ \int \frac{x + 3}{(x^2 + x + 1)^2}\,dx.$$

For the term preceding the last one, we can complete a chain in the form $\int (Du/u)$. We let $u = x^2 + x + 1$ and $Du = (2x + 1)\,dx$, so that by adding and subtracting $+1$ in the numerator we find the antiderivative

$$\int \frac{2x - 1}{x^2 + x + 1}\,dx = \int \frac{(2x + 1)}{x^2 + x + 1}\,dx - 2\int \frac{dx}{x^2 + x + 1}$$

$$= \ln|x^2 + x + 1| - 2\int \frac{dx}{[x + (1/2)]^2 + (\sqrt{3}/2)^2},$$

where we have completed the square of the denominator quadratic in the last integral (§3.2, Example D). Thus for this term we finally get

$$\int \frac{(2x - 1)\,dx}{x^2 + x + 1} = \ln|x^2 + x + 1| - \frac{4}{\sqrt{3}}\arctan\left(\frac{2x + 1}{\sqrt{3}}\right) + C.$$

Similarly, in the last term of $\int f(x)\, dx$ we complete a chain of the form $\int u^{-2}$ with $u = x^2 + x + 1$ and $Du = (2x + 1)\, dx$, obtaining

$$\int \frac{x+3}{(x^2+x+1)^2}\, dx = \frac{1}{2}\int \frac{2x+1}{(x^2+x+1)^2}\, dx + \frac{5}{2}\int \frac{dx}{(x^2+x+1)^2}$$

$$= -\frac{1}{2(x^2+x+1)} + \frac{5}{2}\int \frac{dx}{[(x+(1/2))^2 + (\sqrt{3}/2)^2]^2},$$

where in the denominator of the last integral we have once more completed the square of $x^2 + x$. This last term is a form in which the substitution $x = a\tan\theta$ produces a reduction formula (§4.5),

$$\int \frac{Du}{(u^2+a^2)^{n+1}} = \frac{1}{2a^2}\left[\frac{u}{n(u^2+a^2)^n} + \frac{2n-1}{n}\int \frac{Du}{(u^2+a^2)^n}\right],$$

In this case $n = 1$, $u = x + \text{M}$, $a = \sqrt{3/2}$. Completing the integration by this formula, we have the final result

$$\int f(x)\, dx = \frac{x^2}{2} - 4x + \ln|x| + \frac{1}{x} - \frac{5}{2x^2} + \ln|x^2 + x + 1|$$

$$+ \frac{1}{3}\frac{5x+1}{x^2+x+1} - \frac{2\sqrt{3}}{9}\arctan\frac{2x+1}{\sqrt{3}} + C.$$

Thus we could integrate the rational function $f(x) = P(x)/Q(x)$ when it was expressed as a polynomial plus a sum of elementary fractions of the form $r(x)/[q(x)]^n$, where $q(x)$ is an irreducible linear or quadratic factor of the denominator, $Q(x)$, and $r(x)$ is of lower degree than $q(x)$. When $f(x)$ was represented as a sum of such "partial" fractions, we found that its integral could be completed as a sum of polynomials, powers, and logarithms of the denominator factors, and arc tangents. This completes the problem. We will see that this is true of any rational function.

5.2   Partial Fractions.

We now undertake to reverse this procedure. Starting with a given rational function with real-number coefficients

$$f(x) = \frac{P(x)}{Q(x)},$$

we seek to decompose it into a sum of partial fractions whose antiderivatives we can compute. Our procedure is as follows.

(1)   If $P(x)$ is not of lower degree than $Q(x)$, we divide it by $Q(x)$ to obtain a quotient polynomial plus a remainder fraction in which the degree of the numerator is less than the degree of $Q(x)$.

(2)   Factor $Q(x)$ into its irreducible factors with real coefficients. It follows from the fundamental theorem of algebra that this factorization will be a product of powers of linear and quadratic factors.

(3)   Write a sum of partial fractions

$$\frac{A_1}{x - r_1} + \frac{A_2}{(x - r_1)^2} + \cdots + \frac{A_p}{(x - r)^p},$$

corresponding to each power of a linear factor, $(x - r_1)^p$, with arbitrary unknown coefficients $A_1, A_2, \ldots, A_p$.

(4)   Write a sum of partial fractions

$$\frac{B_1 x + C_1}{ax^2 + bx + c} + \frac{B_2 x + C_2}{(ax^2 + bx + c)^2} + \cdots + \frac{B_q x + C_q}{(ax^2 + bx + c)^q},$$

corresponding to each power of a quadratic factor $(ax^2 + bx + c)^q$ with unknown coefficients $B_1 \cdots B_q$, $C_1 \cdots C_q$.

(5)   Combine all these partial fractions into one fraction with $Q(x)$ as its denominator. The numerator of this fraction must then be identically equal to $P(x)$.

(6)   Obtain a system of simultaneous linear equations for the unknown coefficients by equating like powers of x in $P(x)$ and the numerator of the sum of all the partial fractions. Alternatively assign x chosen numerical values to get these equations.

(7)   Solve for the unknown coefficients. Thus we complete the decomposition of $P(x)/Q(x)$ into partial fractions.

(8)   We now calculate $\int (P(x)/Q(x) \, Dx$ by finding the antiderivative of each partial fraction.

Let us see how these steps apply in the problem we worked above, given the fraction $f(x)$ in the form (2) (§5.1).

(1)   We observe that the degree of the numerator exceeds that of the denominator so we perform two steps of polynomial division.

(2)   Then we factor the denominator. The result of these two steps is

$$f(x) = x - 4 + \frac{3x^6 + 5x^5 + 8x^4 + 11x^3 + 14x^2 + 9x + 5}{x^3(x^2 + x + 1)^2},$$

(3 and 4).   We write

$$\frac{A_1}{x} + \frac{A_2}{x^2} + \frac{A_3}{x^3}$$

corresponding to the factor x^3, and

$$\frac{B_1 x + C_1}{x^2 + x + 1} + \frac{B_2 x + C_2}{(x^2 + x + 1)^2},$$

corresponding to $(x^2 + x + 1)^2$.

(5)   Combining all these terms into one fraction and equating numerators, we find in part,

$$3x^6 + 5x^5 + 8x^4 + 11x^3 + 14x^2 + 9x + 5$$
$$= (A_1 + B_1)x^6 + (2A_1 + A_2 + B_1 + C_1)x^5 + \cdots$$
$$+ (A_1 + 2A_2 + 3A_3)x^2 + (A_2 + 2A_3)x + A_3.$$

(6)   This gives equations

$$A_1 + B_1 = 3, \ldots, A_1 + 2A_2 + 3A_3 = 14, \quad A_2 + 2A_3 = 9, \quad A_3 = 5.$$

(7)   Solving, we find, in part, $A_3 = 5$, $A_2 = -1$, $A_1 + 1, \ldots$. This gives us the numerators of the partial fractions.

(8)   We then find antiderivatives as in the example to complete the problem.

5.3   *Example.* Calculate the antiderivative $\int dx/(x^2 - a^2)$, where $a \neq 0$.

Solution.   We decompose the integrand into partial fractions as follows:

$$\frac{1}{x^2 - a^2} = \frac{1}{(x - a)(x + a)} = \frac{A}{x - a} + \frac{B}{x + a}.$$

which gives us the identity $1 = A(x + a) + B(x - a)$. Setting $x = a$, we find $A = \frac{1}{2}a$. Setting $x = -a$, we find $B = -\frac{1}{2}a$. Hence

$$\int \frac{dx}{x^2 - a^2} = \frac{1}{2a} \int \frac{dx}{x - a} - \frac{1}{2a} \int \frac{dx}{x + a} = \frac{1}{2a} (\ln |x - a| - \ln |x + a|) + C$$

$$= \frac{1}{2a} \ln \left| \frac{x - a}{x + a} \right| + C.$$

5.4   *Example.*   Find

$$\int_1^2 \frac{3x^4 + x^3 + 2x^2 + x - 2}{x^4 + x^2} \, dx.$$

Solution.   We carry out one step of division to reduce the degree of the numerator, and factor the denominator. This gives us

$$\frac{3x^4 + x^3 + 2x^2 + x - 2}{x^4 + x^2} = 3 + \frac{x^3 - x^2 + x - 2}{x^2(x^2 + 1)} = 3 + \frac{A_1}{x} + \frac{A_2}{x^2} + \frac{Bx + C}{x^2 + 1}.$$

Combining the partial fractions into a single fraction and equating numerators, we find

$$x^3 - x^2 + x - 2 = (A_1 + B)x^3 + (A_2 + C)x^2 + A_1 x + A_2,$$

which, by the equating of coefficients, implies that

$$A_1 + B = 1, \qquad A_2 + C = -1, \qquad A_1 = 1, \qquad A_2 = -2.$$

Solving, we find $A_1 = 1$, $A_2 = -2$, $B = 0$, $C = 1$. Hence, the antiderivative needed to evaluate the definite integral becomes

$$\int \left(3 + \frac{1}{x} - \frac{2}{x^2} + \frac{1}{x^2 + 1} \right) dx = 3x + \ln |x| + \frac{2}{x} + \arctan x.$$

We observe that the integrand is continuous between the limits of integration so that the fundamental theorem may be applied. It gives us

$$\int_1^2 \frac{3x^4 + x^3 + 2x^2 + x - 2}{x^4 + x^2} \, dx = \left[3x + \ln |x| + \frac{2}{x} + \arctan x \right]_1^2$$

$$= (6 + \ln 2 + 1 + \arctan 2) - \left(3 + 0 + 2 + \frac{\pi}{4} \right)$$

$$= 2 + \ln 2 + \arctan 2 + \frac{\pi}{4}.$$

This completes the problem.

5.5   Remarks on the Theory.   From the theoretical standpoint our argument is incomplete. We did not prove that every rational fraction can be represented, in some sense uniquely, as a sum of partial fractions with denominators that are powers of linear factors and irreducible quadratic factors. To establish this it would be desirable to use the theory of division in polynomial algebra and the fundamental theorem of algebra.*

* *See* Birkhoff-MacLane, *Survey of Modern Algebra*, 3rd ed. (New York: Macmillan, 1966).

For our purposes the theory of partial fractions is unnecessary, provided that in each problem we succeed in finding a decomposition into elementary partial fractions. It is a fact that the rules will produce such a decomposition in every case, and once we have it for the given rational function, our argument is rigorous for that function. We do not even need uniqueness.

5.6 Exercises

In Exercises 1–20, use expansion into partial fractions to find the indicated antiderivatives and integrals.

1. $\displaystyle\int \frac{dx}{x^2 - x - 6}$.

2. $\displaystyle\int \frac{dx}{x^2 - 2x + 1}$.

3. $\displaystyle\int \frac{x\,dx}{x^2 - 4x + 4}$.

4. $\displaystyle\int \frac{x - 1}{x^2 + 1}\,dx$.

5. $\displaystyle\int \frac{3x + 5}{x^2 + x + 1}\,dx$.

6. $\displaystyle\int \frac{x^2 + 3x + 2}{x^2 - x - 6}\,dx$.

7. $\displaystyle\int \frac{x^3}{x^2 - x + 1}\,dx$.

8. $\displaystyle\int \frac{dx}{x^4 - 16}$.

9. $\displaystyle\int \frac{x^3 - x}{x^4 + 2x^2 + 1}\,dx$.

10. $\displaystyle\int \frac{z^2\,dz}{(z - 1)^3}$.

11. $\displaystyle\int_0^1 \frac{5z\,dz}{(z + 2)(z^2 + 1)}$.

12. $\int_{-2}^{2} x/(x^2 - 1)\,dx$ does not exist. The improper integral does not converge.

13. $\displaystyle\int \frac{x\,dx}{x^3 + 1} = -\frac{1}{3}\ln(x + 1) + \frac{1}{6}\ln(x^2 - x + 1) + \frac{1}{\sqrt{3}}\arctan\frac{2x - 1}{\sqrt{3}} + C$.

14. Using Exercise 13, we find

$$\int_0^1 \frac{x\,dx}{x^3 + 1} = \frac{\sqrt{3}}{9}\pi - \ln\sqrt[3]{2}$$

but $\int_{-1}^{1} x\,dx/(x^3 + 1)$ does not exist.

15. $\displaystyle\int_1^\infty \frac{x^2\,dx}{(x^2 + 1)^2}$.

16. $\displaystyle\int_1^\infty \frac{dx}{x^2 + x^4}$.

17. If $a \neq 0$ and $b^2 - 4ac > 0$, the quadratic equation $ax^2 + bx + c = 0$ has two distinct real roots, r_1 and r_2. In this case, show that

$$\int \frac{dx}{ax^2 + bx + c} = \frac{1}{a}\frac{1}{r_2 - r_1}\ln\left|\frac{x - r_2}{x - r_1}\right| + C, \qquad x \neq r_1, \quad x \neq r_2.$$

18. If $a \neq 0$ and $b^2 - 4ac < 0$, show that

$$\int \frac{dx}{ax^2 + bx + C} = \frac{2}{\sqrt{4ac + b^2}}\arctan\frac{2ax + b}{\sqrt{4ac - b^2}} + C.$$

19. If $a \neq 0$ and $b^2 - 4ax = 0$, show that

$$\int \frac{dx}{ax^2 + bx + c} = -\frac{2}{2ax + b} + C, \qquad x \neq -\frac{b}{2a}.$$

20. If $a \neq 0$ with any value of $b^2 - 4ac$, show that

$$\int \frac{x\,dx}{ax^2 + bx + c} = \frac{1}{2a}\ln|ax^2 + bx + c| - \frac{b}{2a}\int \frac{dx}{ax^2 + bx + c}.$$

where the last integral is to be calculated as in Exercises 17, 18, or 19.

6 Extended Tables of Integrals

6.1 Resumé of Techniques of Integral Calculus. We have extended somewhat the techniques of the analytic program for integral calculus (§2.5). We recall (I, §34R) that in this analytic program we evaluate the definite integral $\int_a^b f(x)\,dx$ by solving the differential equation $DF(x) = f(x)\,dx$ on $[a, b]$ for the unknown function $F(x)$; and then, using the fundamental theorem of integral calculus (I, §19.3), we evaluate $\int_a^b f(x)\,dx = F(b) - F(a)$. To implement this analytic program, we developed families of functions that could be described by explicit algebraic formulas in terms of a finite generating set of functions. So far, our set of generating functions includes: $\mathbf{1}$, x, $\ln |x|$, $\exp x$, x^p for real numbers p, and the trigonometric functions with their inverses. The family of functions that we can differentiate in closed form under the analytic program (Formulas I–XXII) consists of all the generator functions together with all functions that can be formed from them by any finite sequence of the operations of addition, subtraction, multiplication by real numbers; multiplication and division of functions; and composition of functions (chain rule). Integration, however, has typically forced us to go out of this family of "known" functions and bring in additional generators (I, §34R). Moreover each new generator we affix to our list imposes a heavier load on our memory. Each new generator function brings along its set of identities, like the trigonometric identities, and to make use of them we have been forced to introduce more elaborate techniques such as repeated integrations by parts, ingenious substitutions, and expansion in partial fractions. In the practice of the analytic program for integral calculus, there appears to be no end to this proliferation of new generator functions and associated techniques.

Seeking relief from this proliferation of technique in the analytic program for calculus, we may attempt to implement the program with a really comprehensive table of antiderivatives, covering almost anything we should encounter in practice. We shall use it as an external memory to supplement our own memory and relieve us of applying laborious techniques. Let us explore this possibility to get some idea of what it can do for us and what it cannot do.

6.2 Use of the Table of Integrals to Integrate Rational Functions. We understand at once that it is impossible to tabulate all of the infinitude of functions and their antiderivatives. Even a very large table of integrals will have only a finite set of generator functions. Hence we still must use some technique to reduce a given integral to such a form that we can adapt the tabulated antiderivatives to it. We cannot eliminate the partial fraction procedure for rational fractions, and after we have the expansion in partial fractions (§5.2), we integrate by inspection the polynomial terms and terms with denominators that are powers of a first-degree expression in x. That leaves the technically more difficult integration of terms of the form

$$\frac{Ax + B}{(ax^2 + bx + c)^n}, \qquad \text{where} \quad b^2 - 4ac < 0.$$

Our extended table (Appendix 1) includes several integrals for this purpose. We refer to them by such symbols as $\int 24$, meaning 24 in the extended table of integrals.

Example. Using the table of integrals, calculate (§5.1)

$$\int \frac{2x - 1}{x^2 + x + 1}\,dx + \int \frac{x + 3}{(x^2 + x + 1)^2}\,dx.$$

Solution.   These integrals involve the quadratic $ax^2 + bx + c$, with $a = b = c = 1$ and $b^2 - 4ac = -3 < 0$. Hence the quadratic is not factorable into real linear factors. Looking at $\int 23$ to $\int 26$ in the table, we decompose these two integrals into

$$2\int \frac{x\,dx}{x^2 + x + 1} - \int \frac{dx}{x^2 + x + 1} + \int \frac{x\,dx}{(x^2 + x + 1)^2} + 3\int \frac{dx}{(x^2 + x + 1)^2}.$$

By $\int 24$ and then $\int 23b$, the first term gives

$$2\int \frac{x\,dx}{x^2 + x + 1} = 2\left[\frac{1}{2(1)} \ln |x^2 + x + 1| = \frac{1}{2(1)} \int \frac{dx}{x^2 + x + 1}\right]$$

$$= \ln |x^2 + x + 1| - \frac{2}{\sqrt{3}} \arctan \frac{2x + 1}{\sqrt{3}}.$$

The second integral, by $\int 23b$, is

$$-\int \frac{dx}{x^2 + x + 1} = -\frac{2}{\sqrt{3}} \arctan \frac{2x + 1}{\sqrt{3}}.$$

In the third integral, by $\int 26$, $n = 1$, and it gives us

$$\int \frac{x\,dx}{(x^2 + x + 1)^2} = \frac{-(x + 2)}{(1)(3)(ax^2 + bx + c)} - \frac{(1)(1)}{(1)(3)} \int \frac{dx}{x^2 + x + 1}.$$

$$= \frac{-x - 2}{3(x^2 + x + 1)} - \frac{1}{3}\left(\frac{2}{\sqrt{3}} \arctan \frac{2x + 1}{\sqrt{5}}\right).$$

Finally, the fourth integral, by $\int 25$, with $n = 1$, gives

$$3\int \frac{dx}{(x^2 + x + 1)^2} = 3\left[\frac{2x + 1}{(1)(3)(x^2 + x + 1)} + \frac{2(1)(1)}{(1)(3)} \int \frac{dx}{x^2 + x + 1}\right]$$

$$= \frac{2x + 1}{x^2 + x + 1} + 2\left(\frac{2}{\sqrt{3}} \arctan \frac{2x + 1}{\sqrt{3}}\right).$$

Adding these four antiderivatives, we find that

$$\int \frac{2x - 1}{x^2 + x + 1}\,dx + \int \frac{(x + 3)\,dx}{(x^2 + x + 1)^2}$$

$$= \ln |x^2 + x + 1| + \frac{5x + 1}{3(x^2 + x + 1)} - \frac{2}{3\sqrt{3}} \arctan \frac{2x + 1}{\sqrt{3}}.$$

This completes the problem.

The answer agrees with the result we obtained (§4.5) by adjusting the numerators to complete chains of the form Du/u and Du/u^2, completing the square, recognizing the form as an arc tangent form, carrying through a trigonometric substitution, integrating $\cos^2 \theta$, and transforming back to the original variables by means of inverse trigonometric functions. None of this technique was required here when we used the table, but the actual work was not much shorter, or simpler in algebra.

In this connection, we look briefly at $\int 23a$. When $b^2 - 4ac > 0$, the roots of $ax^2 + bx + c = 0$ are real and hence the quadratic denominator in $(ax^2 + bx + c)^{-1}$ is factorable into linear factors, and the rational fraction $(ax^2 + bx + c)^{-1}$ will decompose into partial fractions, which can be integrated as logarithms. These, in turn, will recombine to give the result of $\int 23a$. Using the table, we need not do all this; in fact, it is not necessary that we know it. Even if, failing to check the sign of $b^2 - 4ac$, we attempt to use a wrong one of the three $\int 23$ formulas, we shall find that either we have an imaginary number or we have to divide by zero.

6.3 A Power of cos 2x

Example. Evaluate

$$\int_0^{\pi/6} \frac{\sin^4 2x}{\cos^6 2x}\,dx.$$

Solution. We must use a little trigonometry to get the required antiderivative into the form $\int 34$. Since $\sin^2 2x = 1 - \cos^2 2x$, the integrand can be expressed in powers of cosine alone as follows.

$$\int \frac{\sin^4 2x}{\cos^6 2x}\,dx = \int \frac{(1 - \cos^2 2x)^2}{\cos^6 2x}\,dx$$

$$= \int \frac{1}{\cos^6 2x}\,dx - 2\int \frac{1}{\cos^4 2x}\,dx + \int \frac{1}{\cos^2 2x}\,dx.$$

We now apply the second form of $\int 34$ to $\int \cos^{-6} 2x\,dx$ in order to reduce the negative exponent, $n = -6$. We find, since $a = 2$, that

$$\int \cos^{-6} 2x\,dx = -\frac{\cos^{-5} 2x \sin 2x}{(-5)(2)} + \frac{-4}{-5}\int \cos^{-4} 2x\,dx.$$

We simplify and combine the last integral with the middle term above and this gives us

$$\int \frac{\sin^4 2x}{\cos^6 2x}\,dx = \frac{\sin 2x}{10 \cos^5 2x} - \frac{6}{5}\int \cos^{-4} 2x\,dx + \int \cos^{-2} 2x\,dx.$$

Now we apply $\int 34$ again to reduce the exponent -4. We find

$$-\frac{6}{5}\int \cos^{-4} 2x\,dx = -\frac{6}{5}\left[-\frac{\cos^{-3} 2x \sin 2x}{(-3)(2)} + \frac{-2}{-3}\int \cos^{-2} 2x\,dx \right]$$

$$= -\frac{1}{5}\frac{\sin 2x}{\cos^3 2x} - \frac{4}{5}\int \cos^{-2} 2x\,dx.$$

The integrands $\cos^{-2} 2x$ integrate directly by $\int 33$. Then, combining terms and evaluating at the limits, we obtain the final result from the fundamental theorem.

$$\int_0^{\pi/6} \frac{\sin^4 2x}{\cos^6 2x}\,dx = \left(\frac{\sin 2x}{10 \cos^5 2x} - \frac{1}{5}\frac{\sin 2x}{\cos^3 2x} + \frac{1}{10}\tan 2x \right)\Big|_0^{\pi/6} = \frac{3}{2}\sqrt{3}.$$

This completes the problem.

6.4 Comments on the Value of a Table of Integrals.

These examples give us some idea of what help we can expect from an extended table of integrals. It can enlarge our generating set of functions and reduce the technical difficulty of using them. On the other hand, when we have a function to integrate, we cannot ordinarily expect to find it tabulated in exactly the same form. We must still be able to scan the table and recognize a tabulated integral into which we can transform our given integral. Moreover, we must still expect to have to perform such algebraic maneuvers as expansion in partial fractions in order to resolve our given integrand into components whose antiderivatives are tabulated. All methods of integration "in closed form" reduce the given integral to a known integral. The table of integrals merely expands the list of known integrals.

Finally there can be no assurance that integration will not force us to go outside the known family of functions, and hence out of the table again, to obtain an antiderivative.

Example. Integrate

$$\int \frac{\cos x}{x^2}\, dx.$$

Solution. By $\int 36$, second form,

$$\int \frac{\cos x}{x^2}\, dx = \frac{1}{(-1)}\, x^{-1} \cos x + \frac{1}{(-1)}\int \frac{\sin x}{x}\, dx.$$

Now if we turn to $\int 35$ to find $\int (\sin x/x)\, dx$, the formula fails because of division by zero when $n = -1$. This table does not give an antiderivative in the family of accepted elementary functions, nor do others. We later discover that it would require the introduction of a new transcendental function if we are to have an explicit antiderivative for it. In a more extensive table, we find the definite integral

$$\int_0^\infty \frac{\sin x}{x}\, dx = \frac{\pi}{2},$$

a result which might serve our purposes in some applications. Otherwise we would resort to numerical computation. The more advanced our technical application of calculus becomes, the more we expect to turn away from the analytic program to the numerical one, even with a big table of integrals available.

We observe that many formulas in the extended table of integrals are reduction formulas, saving us the work of repeated integration by parts.

6.5   Exercises on Tables of Integrals

Calculate the indicated antiderivatives on definite integrals by use of the extended table of integrals.

1. $\displaystyle\int \frac{dx}{x^2 - x - 6}.$

2. $\displaystyle\int \frac{dx}{x^2 + x + 1}.$

3. $\displaystyle\int \frac{(3x + 2)\, dx}{x^2 + x + 1}.$

4. $\displaystyle\int \frac{x^2 - x + 1}{x^2 + x + 1}\, dx.$

5. $\displaystyle\int \frac{dx}{(x^2 - x + 1)^3}.$

6. $\displaystyle\int \frac{dx}{x^2 - 6x + 9}.$

7. $\displaystyle\int \frac{x\, dx}{(x^2 - x - 6)^3}.$

8. $\displaystyle\int \frac{5x - 4}{(x^2 + x + 1)^3}\, dx.$

9. $\displaystyle\int \frac{3x^3 + 4x^2 + x + 5}{(x^2 + 1)^2}\, dx.$

10. $\displaystyle\int \frac{x^4 + 2x^3 + 3x^2 - x}{x^4 + x^2}\, dx.$

11. $\displaystyle\int_0^3 \frac{x\, dx}{(x^2 - x - 6)^3}.$

12. $\displaystyle\int_0^1 \frac{dx}{(x^2 + 1)^3}.$

13. $\displaystyle\int_0^\pi \sin^4 3x\, dx.$

14. $\displaystyle\int \sin^3 \pi x\, dx.$

15. $\int_{1/4}^{1/2} \dfrac{1}{\sin^3 \pi x}\, dx.$

16. $\int \dfrac{\sin \pi x}{x^2}\, dx.$

17. $\int_0^\infty \dfrac{\cos \pi x}{x^2}\, dx.$

18. $\int x^3 \sin \pi x\, dx.$

19. $\int_0^\pi x^3 \cos 2x\, dx.$

20. $\int_0^1 (1 + x^2) \cos \pi x\, dx.$

21. $\int_0^1 \tan^3 \dfrac{\pi x}{4}\, dx.$

22. $\int_0^a \arccos \dfrac{x}{a}\, dx.$

23. $\int \sec^4 2x\, dx.$

24. $\int \sec^5 ax\, dx.$

25. $\int_0^a \arctan \dfrac{x}{a}\, dx.$

26. $\int_0^\infty x^3 e^{-x}\, dx.$

27. $\int_0^a \sqrt{x^2 + a^2}\, dx.$

28. $\int_0^a \sqrt{a^2 - x^2}\, dx.$

29. $\int_0^4 e^{-x} \sin \pi x\, dx.$

30. $\int_1^2 \dfrac{dx}{x^3 \sqrt{4 - x^2}}.$

31. $\int_0^2 e^{-\mu t} \sin^3 \pi x\, dx.$

32. $\int_0^{3/2} \dfrac{x^3\, dx}{\sqrt{2x - 3}}.$

33. $\int_0^e \dfrac{dx}{\sqrt{x^2 - e^2}}.$

34. $\int_e^{2e} \dfrac{dx}{\sqrt{x^2 - e^2}}.$

35. $\int x^3 (1 + \ln x)^2\, dx = \dfrac{x^4}{4} \left[(\ln ex)^2 - \dfrac{1}{2} \ln ex + \dfrac{1}{4} \right] + C.$

36. $\int_{-\infty}^0 e^{2x} \sin 3x\, dx = -\dfrac{3}{13}.$

7 Use of Integration by Parts

7.1   Integration by Parts.

We recall integration by parts (I, §20.5), which says that if u and v are differentiable functions,

$\int$ III. $\int u\, Dv = uv - \int v\, Du.$

We apply it first to find the antiderivative of **arcsin** $x\, dx$. Let

$$u = \arcsin x \qquad \text{and} \qquad Dv = dx,$$

so that

$$Du = \frac{dx}{\sqrt{1 - x^2}} \qquad \text{and} \qquad v = x.$$

This gives us

$$\int \arcsin x\, dx = x \arcsin x - \int \frac{x\, dx}{\sqrt{1 - x^2}}.$$

The last term can now be converted into the power formula $\int z^{-1/2}\, Dz$, if we substitute $z = 1 - x^2$, $D_z = -2x\, dx$. Introducing the factor -2, we have

$$\int \arcsin x\, dx = x \arcsin x + \tfrac{1}{2} \int (1 - x^2)^{-1/2}(-2x\, dx)$$

$$= x \arcsin x + \sqrt{1 - x^2} + C.$$

Similar calculations will work for $\int \arccos x\, dx$ and $\int \arctan x\, dx$.

7.2   A Cyclic Device in Integration by Parts

Example A.   Compute $\int e^{ax} \cos bx \, dx$.

Solution.   We apply integration by parts with

$$u = e^{ax} \qquad \text{and} \qquad Dv = \cos bx \, dx,$$

so that

$$Du = ae^{ax} \, dx \qquad \text{and} \qquad v = -\frac{1}{b} \cos bx.$$

This now gives us

$$\int e^{ax} \cos bx \, dx = \frac{1}{b} e^{ax} \sin bx - \frac{a}{b} \int e^{ax} \sin bx \, dx.$$

We now apply integration by parts to the last integral, this time with

$$u = e^{ax} \qquad \text{and} \qquad Dv = \sin bx \, dx,$$

so that

$$Du = ae^{ax} \, dx \qquad \text{and} \qquad v = -\frac{1}{b} \cos bx.$$

This now gives us

$$\int e^{ax} \cos bx \, dx = \frac{1}{b} e^{ax} \sin bx - \frac{a}{b} \left[-\frac{1}{b} e^{ax} \cos bx + \frac{a}{b} \int e^{ax} \cos bx \, dx \right]$$

$$= \frac{1}{b} e^{ax} \sin bx + \frac{a}{b^2} e^{ax} \cos bx - \frac{a^2}{b^2} \int e^{ax} \cos bx \, dx.$$

At first glance it appears that we have gotten nowhere, since the last integral is exactly the same as the original, with no reduction. On closer inspection, however, we see that it has a different coefficient from the original integral on the left. So we transpose the last term and combine it with the original integral. We add a constant of integration and we have finished! Thus

$$\left(1 + \frac{a^2}{b^2} \right) \int e^{ax} \cos bx \, dx = \frac{1}{b^2} e^{ax} (b \sin ax + a \cos bx) + C,$$

or

$$(1) \qquad \int e^{ax} \cos bx \, dx = \frac{1}{a^2 + b^2} e^{ax} (b \sin ax + a \cos bx) + C_1,$$

where the arbitrary constant

$$C_1 = \frac{b^2}{a^2 + b^2} C.$$

Example B.   Compute the definite integral $\int_0^{\pi/2} \sin^n x \, dx$, where n is a positive integer.

Solution.   Using integration by parts with

$$u = \sin^{n-1} x \qquad \text{and} \qquad Dv = \sin x \, dx,$$

so that

$$Du = (n-1) \sin^{n-2} x \cos x \, dx \qquad \text{and} \qquad v = -\cos x,$$

gives us

$$\int_0^{\pi/2} \sin^n x \, dx = -\sin^{n-1} x \cos x \Big|_0^{\pi/2} + (n-1) \int_0^{\pi/2} \sin^{n-2} x \cos^2 x \, dx.$$

The first term on the right is zero. In the second term, we replace $\cos^2 x = 1 - \sin^2 x$ and break it into two integrals. This gives us

$$\int_0^{\pi/2} \sin^n x \, dx = (n-1) \int_0^{\pi/2} \sin^{n-2} x \, dx - (n-1) \int_0^{\pi/2} \sin^n x \, dx.$$

Again we use the cyclic device and transpose the last integral to combine it with the original integral on the left, which then becomes $n \int_0^{\pi/2} \sin^n x \, dx$. Dividing by n, we find

(2)
$$\int_0^{\pi/2} \sin^n x \, dx = \frac{n-1}{n} \int_0^{\pi/2} \sin^{n-2} x \, dx.$$

This result is sufficient to calculate the integral; for it replaces the integral of $\sin^n x$ by one of the same form involving the lower power $\sin^{n-2} x$. We can apply the formula (2) to the last term iteratively until it comes down to either $\int_0^{\pi/2} \sin x \, dx$ or $\int_0^{\pi/2} 1 \, dx$. To be explicit, let us calculate it for $\int_0^{\pi/2} \sin^7 x \, dx$ and $\int_0^{\pi/2} \sin^8 x \, dx$. Using (2) iteratively, we obtain

$$\int_0^{\pi/2} \sin^7 x \, dx = \frac{6}{7} \int_0^{\pi/2} \sin^5 dx = \left(\frac{6}{7}\right)\left(\frac{4}{5}\right) \int_0^{\pi/2} \sin^3 dx$$

$$= \left(\frac{6}{7}\right)\left(\frac{4}{5}\right)\left(\frac{2}{3}\right) \int_0^{\pi/2} \sin x \, dx = \left(\frac{6}{7}\right)\left(\frac{4}{5}\right)\left(\frac{2}{3}\right) (1).$$

With the even exponent 8 at the end is different. Again using (2) iteratively, we find that

$$\int_0^{\pi/2} \sin^8 x \, dx = \frac{7}{8} \int_0^{\pi/2} \sin^6 x \, dx = \left(\frac{7}{8}\right)\left(\frac{5}{6}\right) \int_0^{\pi/2} \sin^4 x \, dx$$

$$= \left(\frac{7}{8}\right)\left(\frac{5}{6}\right)\left(\frac{3}{4}\right) \int_0^{\pi/2} \sin x \, dx = \left(\frac{7}{8}\right)\left(\frac{5}{6}\right)\left(\frac{3}{4}\right)\left(\frac{1}{2}\right) \int_0^{\pi/2} 1 \, dx$$

$$= \frac{7 \cdot 5 \cdot 3 \cdot 1}{8 \cdot 6 \cdot 4 \cdot 2} \left(\frac{\pi}{2}\right).$$

7.3   Reducing a Power Factor

Example.   Find the antiderivative $\int x^2 \sin ax \, dx$.

Solution.   In integration by parts we let

$$\boldsymbol{u} = \boldsymbol{x^2} \qquad \text{and} \qquad \boldsymbol{Dv} = \boldsymbol{\sin ax \, dx},$$

so that

$$\boldsymbol{Du} = \boldsymbol{2x \, dx} \qquad \text{and} \qquad \boldsymbol{v} = -\frac{1}{a} \cos ax.$$

This gives us

$$\int x^2 \sin ax \, dx = -\frac{x}{a^2} \cos ax + \frac{2}{a} \int x \cos ax \, dx.$$

In a second round of integration by parts in the last integral, we let

$$u = x \qquad \text{and} \qquad Dv = \cos ax \, dx,$$

so that

$$Du = dx \qquad \text{and} \qquad v = \frac{1}{a} \sin ax.$$

This gives us

$$\int x^2 \sin ax \, dx = -\frac{x}{a^2} \cos ax + \frac{2}{a} \left[\frac{x}{a} \sin ax - \frac{1}{a} \int \sin ax \, dx \right]$$

$$= -\frac{x}{a^2} \cos ax + \frac{2x}{a^2} \sin ax + \frac{2}{a^3} \cos ax + C.$$

Thus to calculate the antiderivative $\int x^2 \sin ax \, dx$, we used two rounds of integration by parts to reduce the factor x^2 of $\sin ax$ by differentiation.

7.4   A Pair of Reduction Formulas.   The integration of $\int x^2 \sin ax \, dx$ above makes it clear that we could complete the integration if we had a pair of reduction formulas for

$$\int x^n \sin ax \, dx \qquad \text{and} \qquad \int x^n \cos ax \, dx.$$

To derive the first of these, we apply integration by parts with

$$u = x^n, \qquad\qquad Dv = \sin ax \, dx,$$

$$Du = nx^{n-1} \, dx, \qquad v = -\frac{1}{a} \cos ax.$$

Then integration by parts gives us immediately the reduction formula $\int 35$,

$$\int x^n \sin ax \, dx = -\frac{x^n}{a} \cos ax + \frac{n}{a} \int x^{n-1} \cos ax \, dx.$$

Similarly, one round of integration by parts gives $\int 36$,

$$\int x^n \cos ax \, dx = \frac{x^n}{a} \sin ax - \frac{n}{a} \int x^{n-1} \sin ax \, dx.$$

We may now use these to integrate $\int x^3 \sin ax \, Dx$. Applying $\int 35$ with $n = 3$, we reduce the exponent by 1,

$$\int x^3 \sin ax \, dx = -\frac{x^3}{a} \cos ax + \frac{3}{a} \int x^2 \cos ax \, dx.$$

Then $\int 36$ with $n = 2$ reduces the exponent by 1 again,

$$\int x^3 \sin ax \, dx = -\frac{x^3}{a} \cos ax + \frac{3}{a} \left[\frac{x^2}{a} \sin ax - \frac{2}{a} \int x \sin ax \, dx \right].$$

Then back to $\int 35$ with $n = 1$ gives

$$\int x^3 \sin ax \, dx = -\frac{x^3}{a} \cos ax + 3\frac{x^2}{a^2} \sin ax - \frac{6}{a^2} \left[-\frac{x}{a} \cos ax + \frac{1}{a} \int \cos ax \, dx \right]$$

$$= -\frac{x^3}{a} \cos ax + 3\frac{x^2}{a^2} \sin ax + 6\frac{x}{a^3} \cos ax - \frac{6}{a^4} \sin ax + C.$$

7.5   Stepping up a Negative Power Factor

Example.   Integrate

$$\int \frac{e^{kx}}{x^n} \, dx, \, n > 1.$$

Solution.   We use integration by parts in successive rounds so as to step up the power of x^{-n} by 1 each time until it is eliminated. Let

$$u = e^{kx}, \qquad Dv = x^{-n} \, dx,$$

$$Du = ke^{kx} \, dx, \qquad v = \frac{x^{-n+1}}{-n+1} = -\frac{1}{n-1} \frac{1}{x^{n-1}}.$$

Thus we find the reduction formula

$$\int \frac{e^{kx}}{x^n} \, dx = \frac{1}{n-1} \left[-\frac{e^{kx}}{x^{n-1}} + k \int \frac{e^{kx}}{x^{n-1}} \, dx \right].$$

This does not solve the problem completely because the reduction formula fails when we come to $\int (e^{kx}/x) \, dx$. For in this case integration by parts with $Dv = dx/x$ introduces **ln** x, which is not in the formula. In fact, all of our devices fail to produce an elementary function that is an antiderivative of e^{kx}/x.

7.6   Use of Taylor's Theorem.

In such cases as $\int (e^x/x) \, dx$, we may resort to the Taylor's expansion (I, §14.4), with remainder. We expand e^x in powers of $x - 0$ and find that

$$e^x = 1 + x + \frac{x}{2!} + \cdots + \frac{x^n}{n!} + R_{n+1},$$

where the remainder term

$$R_{n+1} = e^z \frac{x^{n+1}}{(n+1)!},$$

and z is some unspecified number between 0 and x. (See also I, §29.5, where we used this to compute e).

We know (I, §21.7) that an antiderivative $F(x)$ is given for each x by the indefinite integral.

$$F(x) = \int_1^x \frac{e^u}{u} \, du.$$

Using Taylor's expansion for e^u and dividing by the u in the denominator of the integrand gives

$$F(x) = \int_1^x \left[\frac{1}{u} + 1 + \frac{u}{2!} + \frac{u^2}{3!} + \cdots + \frac{u^{n-1}}{n!} + \frac{R_{n+1}}{u} \right] du.$$

Integrating gives us

$$F(x) = \ln x + x + \frac{x^2}{2 \cdot 2!} + \frac{x^3}{3 \cdot 3!} + \cdots + \frac{x^n}{n \cdot n!} + S_{n+1} + C,$$

where $-F(1)$ has been absorbed into the constant C. For the remainder term, S_{n+1}, we have

$$|S_{n+1}| = \left| \int_0^x e^z \frac{u^n}{(n+1)!} \, du \right| \leq M \frac{x^{n+1}}{(n+1)(n+1)!},$$

where M is the maximum value of e^u on the interval $[0, x]$. For a fixed number x, this remainder approaches zero as n increases. Hence the cut-off series of terms

$$F(x) = \ln x + x + \frac{x^2}{2 \cdot 2!} + \cdots + \frac{x^n}{n \cdot n!}$$

gives an antiderivative of e^x/x to any prescribed accuracy when we choose n large enough.

In an extended table of integrals, we encounter the use of this expansion method to represent such integrals as

$$\int \frac{\sin ax}{x} \, dx, \qquad \int \frac{\arcsin ax}{x} \, dx,$$

and their cosine analogues.

7.7 Exercises

In Exercises 1–38, calculate the antiderivatives and integrals as indicated.

Exercises 1–6 involve direct integration by parts.

1. $\int \arccos x \, dx.$ 2. $\int \arctan x \, dx.$

3. $\int \operatorname{arcsec} x \, dx.$ 4. $\int \frac{x}{\cos^2 x} \, dx.$

5. $\int x \arctan \frac{x}{a} \, dx.$ 6. $\int x \ln x \, dx.$

Exercises 7–12 involve the cyclical device (§7.2).

7. $\int e^{-x} \sin x \, dx.$ 8. $\int e^{ax} \sin bx \, dx.$

9. $\int \sin^2 x \, dx$ with $u = \sin x.$ 10. $\int \sin (\ln ax) \, dx.$

11. $\int_0^\infty e^{-x} \sin x \, dx.$ 12. $\int \sqrt{1 - x^2} \, dx$ with $u = \sqrt{1 - x^2}.$

Exercises 13–16 involve reducing a power factor.

13. $\int x^3 \sin ax \, dx.$ 14. $\int t^2 e^{-kt} \, dt.$

15. $\int x^2 \sin^2 ax \, dx.$ 16. $\int \sin^3 ax \, e^{-x} \, dx.$

Exercises 17–24 involve the general reduction formula for $\int x^n \cos ax \, dx$ (§7.4) for positive integers n.

17. Use the reduction formula to calculate $\int x^3 \cos 5x \, dx.$
18. Use the reduction formula to calculate $\int_0^{2\pi} x^4 \cos ax \, dx.$
19. Use the reduction formula to calculate $\int_0^{2\pi} x^n \cos x \, dx.$ Observe the two cases when n is odd and when n is even.
20. Use the reduction formula to calculate $\int_{-\pi/2}^{\pi/2} x^n \cos x \, dx.$

21. $\int_1^x \frac{\sin u}{u} \, du.$

22. $\int \frac{\sin x}{x^2} \, dx.$

23. $\int \dfrac{\ln x}{(x-1)^2}\, dx$. Expand $\ln x$ in powers of $x - 1$.

24. $\int_1^t \dfrac{e^{-x}}{x^2} \cdot$ Expand e^{-x} in powers of $x - 1$.

Exercises 25–38 are of mixed types. Integrate.

25. $\int \ln ax\, dx.$

26. $\int_0^\infty e^{-x} \sin ax\, dx.$

27. $\int \ln^2 ax\, dx.$

28. $\int t^2 \ln^3 t\, dt.$

29. $\int_0^1 x \arcsin x\, dx.$

30. $\int_0^\infty x^3 e^{-x}\, dx = 3!.$

31. $\int_0^{\pi/2} \cos^3 x\, dx.$

32. $\int_0^{\pi/2} \cos^4 x\, dx.$

33. $\int_0^{\pi/2} \cos^n x\, dx = \dfrac{2.4 \cdots (n-1)}{1.3 \cdots n}$ when n is an odd integer.

34. $\int_0^{\pi/2} \cos^n dx = \dfrac{1.3 \cdots (n-1)}{2.4 \cdots n}$ when n is an even integer.

35. $\int_0^e x \ln x\, dx.$

36. $\int_0^{2\pi} \sin^2 \theta \cos^4 \theta\, d\theta.$

37. $\int x \cos^2 ax\, dx = \dfrac{x^2}{4} + \dfrac{x}{4a} \sin 2ax + \dfrac{1}{8a^2} \cos 2ax + C.$

38. $\int \cos \sqrt{x}\, dx$ using the substitution $z = \sqrt{x}.$

In Exercises 39–44, use integration by parts to compute general reduction formulas that reduce the integer exponent n, where $n > 1$.

39. $\int x^n \sin ax\, dx.$

40. $\int x^n e^{ax}\, dx.$

41. $\int \cos^n ax\, dx.$

42. $\int \tan^n ax\, dx.$

43. $\int \sin^m x \cos^n x\, dx$, m fixed.

44. $\int \sin^n x \cos^m x\, dx$, m fixed.

45. Use Exercise 39 to compute $\int_0^{\pi/2} x^5 \sin x\, dx.$

46. Use Exercise 39 to calculate $\int x^4 \sin ax\, dx.$

47. Use Exercises 43 and 44 to calculate $\int \sin^6 x \cos^4 x\, dx$ by first reducing the exponent of $\cos x$, then that of $\sin x$.

48. Use Exercise 40 to compute $\int_0^\infty x^5 e^{-x}\, dx.$

7R Review of Methods of Integration

7.1R Summary. In extending the methods of differentiation and integration we first recognize the effect of the general chain rule on composite functions (§2). One useful application of the chain rule is in logarithmic differentiation, where we differentiate $\ln |u|$ in order to differentiate u (§2.5). When we undertake to use the general chain rule in practice it becomes particularly difficult to recognize exact chains for the purpose of finding antiderivatives (§3). Thus the recognition of form, one of the great arts of the mathematician, becomes essential in the practice of integration. Certain forms such as $\sqrt{a^2 - x^2}\, dx$ and $(a^2 + x^2)^n\, dx$ transform into recognizable chains under trigonometric substitutions $x = a \sin \theta$, or $x = a \tan \theta$ (§4). We found that any rational function, the quotient of two polynomials, can be integrated by partial fraction expansions (§5). We found that an extended table of integrals (§6) is of some assistance for completing the final stages of an integration by partial fraction expansions, and for reduction of exponents on integrals of the type $\int u^m v^n \, dx$ step by step until an elementary integral is reached. The use of integration by parts to derive reduction formulas and for direct integration is a powerful method of integration (§7).

7.2R   Self Quiz

1. State the chain rule for differentiating the composite $f[u]$, where u and f are differentiable functions (I, §20.1).
2. State the chain rule for $f[g(u)]$, where f, g, u are differentiable.
3. What determines where the chain stops in differentiating a composite function (§2.3)?
4. How can we differentiate u by differentiating $\ln |u|$? What are the advantages (§2.5)?
5. Does the power formula apply to integrating $\int u^n \, dx$ if u is a function of x? If not, what adjustments can be made to make it a power form where possible (§3.2)?
6. Explain the use of a substitution to convert an integral into one of the forms whose antiderivatives are known (§3.1).
7. How can we integrate powers of sine and cosine in integrals $\int \sin^m x \cos^n x \, dx$ (§§3.3, 4.1, 4.4)?
8. What algebraic device will make a recognizable chain out of $dx/(ax^2 + bx + c)$ (§§3.2)?
9. What substitutions will remove the radical for integrating functions involving $\sqrt{a^2 - x^2}$, $\sqrt{a^2 + x^2}$, $\sqrt{x^2 - a^2}$ (§4.2)?
10. What must be done first to integrate $(x^3 + x + 1)/(x^2 - 1) \, dx$ (§5.2)?
11. Describe the procedure for expanding $(x^3 - 1)/[x^2(x^2 + x + 1)]$ into partial fractions (§5).
12. Can an exhaustive table of integrals replace the need for skill in integration? for approximate numerical integration or Taylor's expansions (§6.4)?
13. What device will integrate $\int \ln x \, dx$ and $\int \arcsin x \, dx$ (§7.1)?
14. If it is found after two stages of integration by parts that

$$\int e^x \cos x \, Dx = e^x \cos x + e^x \sin x - \int e^x \cos x \, dx,$$

does this imply that the method fails because it comes back to the integral with which it started (§7.2)?

7.3R   A Miscellaneous List of Integrals

The following mixed list of integrals will serve to review the methods of integration at a level of standard competence. Some require use of the table of integrals.

1. $\int \dfrac{dx}{x^2 - 4}.$

2. $\int x \ln (x - 1) \, dx.$

3. $\int_0^a \sqrt{a^2 - x^2} \, dx.$

4. $\int e^{-2x} \sin 3x \, dx.$

5. $\int_0^a \sqrt{a^2 + x^2} \, dx.$

6. $\int_0^1 \dfrac{dx}{(1 + x^2)^2}.$

7. $\int \dfrac{x \, dx}{x^2 - x + 1}.$

8. $\int \dfrac{x \, dx}{x^2 - 2x - 5}.$

9. $\int \dfrac{1 + \cos x}{\sin x} \, dx.$

10. $\int \cos^2 ax \, dx.$ (Two methods).

11. $\int \dfrac{\cos x}{1 + \sin x}\, dx.$

12. $\int \dfrac{\cos^2 x}{\sin 2x}\, dx.$

13. $\int_0^{\pi/4} \sec^3 \theta\, d\theta.$

14. $\int x \arcsin x\, dx.$

15. $\int x^2 e^{-2x}\, dx.$

16. $\int_1^e \dfrac{\ln 2x}{x}\, dx.$

17. $\int x^3 \ln x\, dx.$

18. $\int \dfrac{\sin 2\sqrt{x}}{\sqrt{x}}\, dx.$

19. $\int \dfrac{dy}{e^y - e^{-y}}.$

20. $\int x \cos (\arcsin x)\, dx.$

21. $\int \dfrac{x^3 + 4x^2 + 5x + 7}{x^3 + x^2 + 4x + 4}\, dx.$

22. $\int \sin^5 ax\, dx.$

23. $\int \sinh x\, dx.$

24. $\int_0^1 \dfrac{1 - x}{1 + x}\, dx.$

25. $\int_0^{\pi/4} \cos^6 2x\, dx.$

26. $\int \dfrac{dx}{x\sqrt{x^2 - x^2}}.$

27. $\int e^{\ln \sin x} \cos x\, dx.$

28. $\int \sqrt{\dfrac{1 - x}{1 + x}}\, dx.$

29. $\int (\cos \theta + \sin \theta)^2\, d\theta.$

30. $\int \dfrac{e^x\, dx}{1 + e^{2x}}.$

7.4R   Miscellaneous Exercises

The following list includes some more difficult exercises and ones of more specialized interest than those of the preceding list.

1. Find the maximum and minimum values of $f(x) = 1 + \cos x + \frac{1}{2}\cos 2x$ on $[-\pi, \pi]$.
2. Differentiate x^x on $[0 < x < \infty]$.
3. Differentiate $\log_{10} x$.
4. Show by Taylor's theorem that for positive numbers x

$$\ln x = (x - 1) - \frac{(x - 1)^2}{2} + \frac{(x - 1)^3}{3} - \cdots + (-1)^{n-1}\frac{(x - 1)^n}{n} + R_n,$$

where

$$|R_n| = \frac{1}{z^{n+1}} \frac{(x - 1)^{n+1}}{n + 1} \qquad \text{for some number } z \text{ between } 1 \text{ and } x.$$

5. Can we apply Taylor's theorem to expand $\ln x$ in powers of x?
6. Find the area of the region bounded by the curve $y = \sqrt{x}$, the x-axis, and the line $ay = 2b(x - a)$.
7. Find the area of the region bounded by the curve

$$\frac{x^2}{a^2} - \frac{y^2}{b^2} = 1$$

and the line $x = 2a$.

8. Find the area of the region bounded by the curve $x^2 - y^2 + 4 = 0$ and the lines $x = 2$ and $x = -2$.
9. Find the mean value of $\sqrt{x}$ on the interval $[0, 1]$.
10. Find the work done when a point is moved along the x-axis from 1 to 4 against a force F, which varies as the 1.6 power of x, if $F = 10$ when $x = 1$.
11. Show that if $y = Cx^p$, then $Dy = p(y/x)$.
12. Find the xy-equation of a curve whose graph in XY coordinates, where $X = \ln x$ and $Y = \ln y$, is a line through points $X = 1$, $Y = 2$ and $X = 3$, $Y = 6$.

13. Differentiate to find $(d/dx) \int_0^{\ln x^2} e^u \, du$.

14. Show that $\int [dx/(x^2 + c)]$ has different antiderivative formulas according to the sign of c, and find them.

15. For what values of x does the integral $\int_x^1 \sqrt{1 - u^2} \, du$ have a value? For these values compute it.

16. We may define the inverse $f^{-1}(x)$ of a function $f(x)$ with derivative $f'(x) \neq 0$ by $f[f^{-1}(x)] = x$, for every x. Show that

$$\frac{d}{dx} f^{-1}(x) = \frac{1}{f'[f^{-1}(x)]}.$$

17. Without integrating prove that $\int_{-1}^1 xe^{-x^4} \, dx = 0$.

18. Without integrating prove that $\int_{-\pi}^\pi \sin^3 x \cos^n x \, dx = 0$.

19. Evaluate $\int_1^3 4z^{1.3} \, dz$.

20. Evaluate $\int_4^6 (2t - 7)^{2.7} \, dt$.

21. Integrate $\int_0^1 e^{-xt} \, dt$.

22. Integrate $\int_0^x e^{-xt} \, dt$.

23. Differentiate $\frac{1}{2} \ln |\tan \theta|$. Does it have a maximum point?

24. If $f(x) = x \sin (1/x)$ when $x \neq 0$ and $f(0) = 0$, find $f'(x)$ and $f'(0)$. Sketch the graph.

25. Show that if $y = A \cos kt + B \sin kt$, $(d^2y/dx^2) + k^2y = 0$.

26. Show that if $y = Ae^{kt} + Be^{-kt}$, $D^2y - k^2y = 0$.

27. Explain why we may calculate antiderivatives of $\int \sin x \cos x \, dx$ in two different ways and get different answers:

(a) $\int \sin x(\cos x \, dx) = \int u \, Du = \dfrac{\sin^2 x}{2}$.

(b) $\int \sin x \cos x \, dx = - \int \cos x(-\sin x \, dx) = - \int u \, Du = -\dfrac{\cos^2 x}{2}$.

Show how to get one answer from the other.

28. Find formulas for $(d^n/dx^n) \cos x$, for all n.

29. Find formulas for $(d^n/dx^n) \sin x$, for all n.

30. Find the maximum and minimum of $\sin x - \cos x$ on $[-\pi, \pi]$, if it exists.

31. Find the maximum and minimum of $\sin 2x + 2 \sin x$ on $[-\pi, \pi]$, if it exists.

32. Find the maximum and minimum of $\sin x/\cos x$ on the real numbers, if it exists.

33. An unseen flying saucer, being tracked by a rotating radar antenna, is moving on a straight-line path whose nearest approach to the tracking station is 4 miles due north of it. At the instant that the antenna is pointed at N 60° E, the rate of rotation of the antenna is $-\frac{1}{2}$ rad/sec. How fast is the flying saucer moving?

34. The functions x and y are related by

$$x \cos y + y \cos x = 1.$$

Find x' when $x = \pi/2$, $y = 0$, and $y' = -1$.

35.* An open gutter with sides at angle θ to the horizontal is to be made from a long piece of sheet metal 12 inches wide by bending it into a circular arc that subtends an angle θ at the center (Figure 7.12). Find the angle θ for maximum capacity.

36. A freely falling particle (I, §18.4) is thrown at a velocity of V_0 in a direction that makes an angle α with the horizontal. Show that its height at any time t thereafter is given by

$$y = -g \frac{t^2}{2} + (V_0 \sin \alpha)t,$$

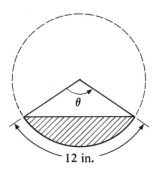

FIGURE 7.12   Exercise 35.

and its horizontal distance from the starting point is given by $x = (V_0 \cos \alpha)t$. Show that the range h, or horizontal distance traveled before it strikes the level ground again, is given by

$$h = \frac{2V_0^2 \sin \alpha \cos \alpha}{g}.$$

37. (a) Show by calculus that the maximum range of the particle (Exercise 36) for a fixed muzzle velocity V_0 is attained when $\alpha = \pi/4$.
    (b) Use only trigonometry to get the same result.
38. A right circular cone is inscribed in a sphere of radius r. Find the angle at the vertex that gives the cone of maximum lateral area.
39. A right circular cone is circumscribed about a sphere of radius r. Show that the angle θ at the vertex that gives the cone of minimum volume is given by $\sin \theta/2 = \frac{1}{3}$.
40. Show that $\int_{-\infty}^{\infty} \arctan x \, dx = \pi$.

In Exercises 41–53, integrate. These include some more difficult problems.

41. $\displaystyle\int_{-\pi}^{\pi} \sin mx \cos nx \, dx.$

42. $\displaystyle\int \frac{x^4 + 1}{x(x^2 + 1)^2} \, Dx.$

43. $\displaystyle\int_0^{\infty} e^{-st} \sin \omega t \, dt.$

44. $\displaystyle\int \frac{e^x \, dx}{e^{2x} + 2e^x - 3}.$

45. $\displaystyle\int \frac{x \, dx}{\sqrt{x + 1}}.$

46. $\displaystyle\int_0^1 \frac{1 + \sqrt{x}}{1 - \sqrt{x}} \, dx.$

47. $\displaystyle\int_0^a \frac{x \, dx}{\sqrt{a^4 - x^4}}$ (Improper).

48. $\displaystyle\int \frac{dx}{(a^2 - x^2)^{3/2}}.$

49. $\displaystyle\int \frac{dx}{\sqrt{ax^2 + c}}$ for all signs of a and c.

50. $\displaystyle\int_0^1 \frac{dx}{x^{2/3}}.$

51. $\displaystyle\int \frac{dt}{\sqrt{1 - e^{-2t}}}.$

52. $\displaystyle\int_1^{\infty} \frac{dx}{x \ln x}.$

53. $\displaystyle\int_0^1 \frac{dx}{x^k}.$

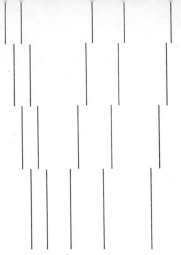

Vectors in Analytic Geometry

INTRODUCTION.   Before we undertake the study of calculus for functions of several variables, $w = f(x, y, z)$, we need both algebraic and graphical tools for representing such functions.   We soon find that the common graphing scheme for functions, $y = f(x)$, does not apply.   Many other devices used for $y = f(x)$, when x and y are variables in the real number line, do not serve when we go to functions of several variables.

We introduce vectors as a new tool and replace the geometry of graphs with strict Euclidean geometry.

HISTORICAL NOTE

The nature of Euclidean geometry. (To be read rather than studied). Up to now we have set the calculus in a form of geometry, which we may think of as the geometry of graphs. That is, the plane has been composed of points located by coordinates in a Cartesian xy-coordinate system, in which we admit transformations of scales on each axis independently. This type of analytic geometry, technically a special case of affine geometry, permitted us to consider lines, line segments, a property of betweenness among three points on a line, midpoints, slopes, parallel lines, directed line segments along lines parallel to the axes, tangent lines to curved graphs, velocities, accelerations, and something we called area. However, many other familiar geometric entities were not invariant under the independent transformations of scales, and hence could not be described in our geometry of graphs.

We now introduce the strict Euclidean* geometry, in which all of the properties of the

---

* The Alexandrian Greek mathematician Euclid lived about 300 B.C. He wrote a textbook on geometry, called "Elements," that became one of the great classics of Western literature. Except for the Bible, Euclid's *Elements* has probably been more widely disseminated than any other book. School texts on geometry still consist almost entirely of diluted Euclid. His geometry was not universally applicable but it was so dominant in literature that it took an intellectual revolution in the nineteenth century to gain any recognition at all for non-Euclidean geometry.

geometry of graphs are still valid, so that nothing has to be changed, and in addition we have some new properties. These new properties belonging to Euclidean geometry include, as the fundamental idea, that of the distance between any two points, whether the line segment joining them is parallel to an axis or not. From this follow the ideas of perpendicular lines, rotations of the plane, angle between two lines, the length of arc, and curvature of a curve.

In Euclidean geometry we assume (Figure 8.1) that any line segment $\overrightarrow{AB}$ in the plane has a numerical length $|\overrightarrow{AB}|$ such that $|\overrightarrow{AB}| \geq 0$, $|\overrightarrow{AB}| = |\overrightarrow{BA}|$, $|\overrightarrow{AB}| = 0$ if and only if points A and B coincide and in a triangle ABC, $|\overrightarrow{AC}| \leq |\overrightarrow{AB}| + |\overrightarrow{BC}|$. We assume that this length is invariant under translations, and rotations of the plane about any point. Thus we are able to compare the lengths of any two line segments, parallel or not.

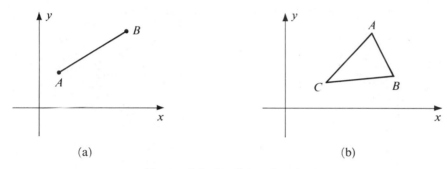

(a)   (b)

FIGURE 8.1   Euclidean length.

We observe that the geometry of graphs is not Euclidean geometry. For example, if one coordinate axis is a scale for distances and the other is a scale for time it makes no sense to say that a distance $|\overrightarrow{AB}|$ on the distance axis is equal to a time interval $|\overrightarrow{CD}|$ on the time axis. Whether $|\overrightarrow{AB}| = |\overrightarrow{CD}|$ depends entirely on the arbitrary choice of scales on the two axes.

We may use the ideas of Euclidean geometry to define perpendicular lines, not possible with the geometry of graphs. Two lines that intersect at O are perpendicular if and only if for every point P (Figure 8.2) on one line and two points A and B on the other, such that $|\overrightarrow{AP}| = |\overrightarrow{BP}|$, it is true that $|\overrightarrow{OA}| = |\overrightarrow{OB}|$. The definition of right triangles and squares

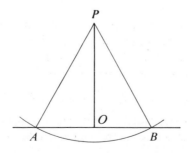

FIGURE 8.2   Perpendicular lines.

follows from this. Moreover, the congruence of triangles depends on the assumption of Euclidean geometry that we can translate and rotate one triangle so that any side coincides with a side of equal length in another triangle.

The theorem of Pythagoras is a theorem of Euclidean geometry of such central importance that we should recall a proof of it, at least in outline.

THEOREM (Pythagoras). A triangle ACB with sides a, b, c in which $a \neq 0$ and $b \neq 0$ is a right triangle, with c as hypotenuse, if and only if

$$a^2 + b^2 = c^2.$$

Sketch of the proof. We will first show that if C is a right angle, then $a^2 + b^2 = c^2$. To do this, we construct two congruent copies of a square with side $a + b$. On one, we locate four points on the perimeter by alternating the lengths, b, a, b, a, b, a, b, a, around the square (Figure 8.3(a)) so as to form a square in the center whose area is

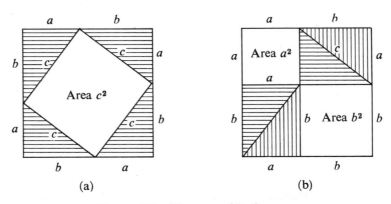

FIGURE 8.3   Theorem of Pythagoras.

c^2. On the other copy of the big square, we locate four points on the perimeter by marking off sides a and b in the sequence a, b, a, b, b, a, b, a, around the square and connecting the partition points to form two squares of area a^2 and b^2 (Figure 8.3(b)). We partition the remaining figures (shaded) into four right triangles, each of which is congruent to the original triangle ACB. The total area of these four triangles is $4(\frac{1}{2}ab) = 2ab$. Since the areas of the big squares are the same, we have

$$c^2 + 2ab = a^2 + b^2 + 2ab.$$

Therefore

$$c^2 = a^2 + b^2.$$

Conversely, if $c^2 = a^2 + b^2$, we can prove that the angle at C in the triangle ACB is a right angle. We construct a perpendicular AD from vertex A on the line containing c (Figure 8.4). Let $|CD| = d$ and $|AD| = h$. This forms two right triangles CDA and ADB. There are two cases. If D is between C and B, we have

$$c^2 = h^2 + (a - d)^2 \qquad \text{and} \qquad b^2 = h^2 + d^2.$$

This, with $c^2 = a^2 + b^2$, implies that $2ad = 0$. Hence, since $a \neq 0$, $d = 0$, which implies that C is a right angle. If D is outside the segment CB, we have

$$c^2 = h^2 + (a + d)^2 \qquad \text{and} \qquad b^2 = h^2 + d^2,$$

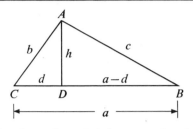
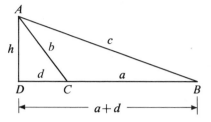

(a)   Case where D is between C and B (b)   Case where D is outside CB

FIGURE 8.4   Converse of theorem of Pythagoras.

which, with $c^2 = a^2 + b^2$, implies that $2ad = 0$. Hence $d = 0$. In either case, the point D coincides with C, and the triangle ACB is congruent to the right triangle ADB. When this is proved, the proof of the theorem is complete.

THEOREM (Law of cosines).   If ABC is any triangle with sides of lengths a, b, c and angles α, β, γ, then

$$c^2 = a^2 + b^2 - 2ab \cos \gamma.$$

Proof.   We use the figure for the converse of the theorem of Pythagoras (Figure 8.5). Then in the two right triangles ADB and ADC, the theorem of Pythagoras gives us

$$y^2 + (x + a)^2 = c^2 \qquad \text{and} \qquad y^2 + x^2 = b^2.$$

Equating the two values of y^2, we find $c^2 - (x + a)^2 = b^2 - x^2$. Expanding and collecting terms, and making use of the fact that $x = b \cos \gamma$, we establish the law of cosines.

We observe that if $\gamma = \pi/2$, then $\cos \gamma = 0$, and the law of cosines reduces to the theorem of Pythagoras. Conversely, if $c^2 = a^2 + b^2$, then $-2ab \cos \gamma = 0$, and if $a \neq 0$ and $b \neq 0$, then $\cos \gamma = 0$, which implies that $\gamma = \pi/2$, as stated in the theorem of Pythagoras.

THEOREM (Law of sines).   In any triangle with no zero side,

$$\frac{\sin \alpha}{a} = \frac{\sin \beta}{b} = \frac{\sin \gamma}{c}.$$

Proof.   In the figure for the law of cosines, we observe that

$$\frac{y}{b} = \sin \gamma \qquad \text{and} \qquad \frac{y}{c} = \sin \beta.$$

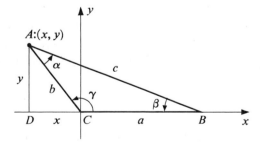

FIGURE 8.5   Law of cosines.

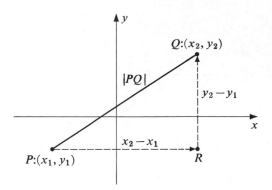

FIGURE 8.6   The distance formula.

Equating the values of y and dividing by bc, we find

$$\frac{\sin \beta}{b} = \frac{\sin \gamma}{c}.$$

The other statements are proved similarly.

8. Euclidean Analytic Geometry

8.1 Distance Formula.   In the Euclidean plane we take any coordinate system with axes perpendicular, and equal units on the two axes. Let $P:(x_1, y_1)$ and $Q:(x_2, y_2)$ be two points* in the plane. Then the distance $|\overrightarrow{PQ}|$ is given for every choice of an admissible coordinate system by the formula

(1) $$|\overrightarrow{PQ}| = \sqrt{(x_2 - x_1)^2 + (y_2 - y_1)^2}.$$

Proof.   We construct $\overrightarrow{PR}$ parallel to the x-axis and $\overrightarrow{RQ}$ parallel to the y-axis (Figure 8.6). Then the triangle PRQ is a right triangle, and by the theorem of Pythagoras

$$|\overrightarrow{PR}|^2 + |\overrightarrow{RQ}|^2 = |\overrightarrow{PQ}|^2.$$

Since $|\overrightarrow{PR}| = |x_2 - x_1|$ and $|\overrightarrow{RQ}| = |y_2 - y_1|$, this gives

$$|\overrightarrow{PQ}|^2 = (x_2 - x_1)^2 + (y_2 - y_1)^2,$$

which implies the distance formula (1) as stated.

8.2 Perpendicular Line Segments.   If $\overrightarrow{PA}$ and $\overrightarrow{QB}$ are two directed line segments in the plane, having projections on the coordinate axes given by $\overrightarrow{PA}:(a_1, a_2)$ and $\overrightarrow{PB}:(b_1, b_2)$, then $\overrightarrow{PA}$ and $\overrightarrow{PB}$ are perpendicular if and only if the "inner product" vanishes, (Figure 8.7)—

$$a_1b_1 + a_2b_2 = 0.$$

---

* We use two names for a point. Its geometric name is P. After the colon, the row of coordinates may be regarded as another name for the point. So we read $P:(x, y)$ to say, "point P, with coordinates (x, y)." We often shorten this by saying "the point P" or "the point (x, y)."

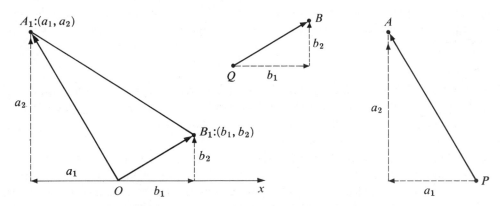

FIGURE 8.7   Perpendicular line segments.

Proof. Through the origin we construct a line segment $\overrightarrow{OA_1}$ parallel to $\overrightarrow{PA}$, where A_1 has the coordinates (a_1, a_2). Similarly, we construct $\overrightarrow{OB_1}$ parallel to $\overrightarrow{QB}$, where B_1 has coordinates (b_1, b_2). Then $\overrightarrow{PA}$ and $\overrightarrow{QB}$ are perpendicular if and only if $\overrightarrow{OA_1}$ and $\overrightarrow{OB_1}$ are perpendicular. The theorem of Pythagoras implies that $\overrightarrow{OA_1}$ and $\overrightarrow{OB_1}$ are perpendicular if and only if

$$|\overrightarrow{A_1B_1}|^2 = |\overrightarrow{OA_1}|^2 + |\overrightarrow{OB_1}|^2,$$

or

$$(b_1 - a_1)^2 + (b_2 - a_2)^2 = (a_1^2 + a_2^2) + (b_1^2 + b_2^2).$$

Expanding and simplifying, we find that this is true if and only if $a_1b_1 + a_2b_2 = 0$, which completes the proof.

We can express the same criterion in terms of slopes with some loss in generality and simplicity. If neither a_1 nor b_1 is zero, the inner product condition is equivalent to (I, §4)

$$\left(\frac{a_2}{a_1}\right)\left(\frac{b_2}{b_1}\right) = -1.$$

The ratio a_2/a_1 is the slope of $\overrightarrow{PA}$ and b_2/b_1 is the slope of $\overrightarrow{QB}$. Hence when $a_1b_1 \neq 0$, the inner product condition is equivalent to the statement that the slope m_B of $\overrightarrow{QB}$ is the negative reciprocal of the slope m_A of $\overrightarrow{PA}$. That is

$$m_B = -\frac{1}{m_A}.$$

Example A. Two directed line segments $\overrightarrow{PA}$ and $\overrightarrow{QB}$ have components $(-1, 3)$ and $(6, 2)$. Are they perpendicular?

Solution. By the inner product condition, $(-1)(6) + (3)(-2) = 0$. Hence they are perpendicular. In terms of slopes, the slope $m_A = -\frac{3}{1}$ and the slope $m_B = \frac{2}{6}$, which are negative reciprocals.

Example B. Two directed line segments $\overrightarrow{PA}$ and $\overrightarrow{QB}$ have components $(1, 0)$ and (01), respectively. By the inner product condition, $(1)(0) + (0)(1) = 0$. Hence the line segments are perpendicular. The slope condition fails since the slope $m_B = \frac{1}{0}$, which is meaningless.

8.3 Circles. For each fixed point C in the Euclidean plane and each distance r, $C \leqq r$, we define the *circle* with center at C and radius r to be the set of points P in the plane such that $|\overrightarrow{CP}| = r$. The *circular disk* with center at C and radius r is the set of points P such that $|CP| \leqq r$, (Figure 8.8).

In order to find an equation of the circle with center at C and radius r, we consider a "locus problem." We choose a coordinate system for the plane and this determines coordinates for the fixed center C. Then we assign *variable* coordinates (x, y) to the point P and write an equation in the variables (x, y) that is satisfied if and only if the defining condition $|\overrightarrow{CP}| = r$ is true. The graph of this equation is then the required circle.

After a coordinate system has been chosen let (x_0, y_0) be the coordinates of C and let the variable coordinates (x, y) represent P. Then $CP = r$ is true if and only if

$$\sqrt{(x - x_0)^2 + (y - y_0)^2} = r.$$

Since $0 \leqq r$, this condition is satisfied if and only if

$$(x - x_0)^2 + (y - y_0)^2 = r^2,$$

which is called the standard equation of the circle with center at (x_0, y_0) and radius r.

We can expand the squares in the standard equation of the circle and multiply both members by the nonzero constant a and find that the standard equation is equivalent to one of the form

(2) $$ax^2 + ay^2 + dx + ey + f = 0, \qquad a \neq 0.$$

This is a second-degree equation in the Cartesian coordinates (x, y), in which the coefficient of x^2 and y^2 are equal and the coefficient of xy is zero. Do all equations of this form represent circles?

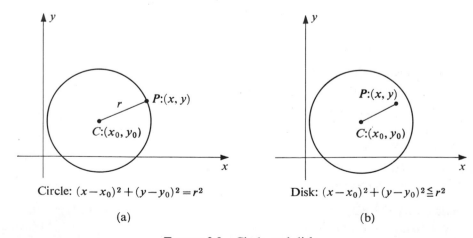

Circle: $(x-x_0)^2+(y-y_0)^2=r^2$ Disk: $(x-x_0)^2+(y-y_0)^2\leqq r^2$

(a)   (b)

FIGURE 8.8   Circle and disk.

Example A.   Find the graph of the equation

$$2x^2 + 2y^2 - 3x + 7y - 8 = 0.$$

Solution.   We divide both members by 2 and write the equivalent equation

$$(x^2 - \tfrac{3}{2}x) + (y^2 + \tfrac{7}{2}y) = 4.$$

Then we complete the two squares on the left by adding $\tfrac{9}{16}$ and $\tfrac{49}{16}$ to both members to obtain

$$\left(x - \frac{3}{4}\right)^2 + \left(y + \frac{7}{4}\right)^2 = 4 + \frac{9}{16} + \frac{49}{16} = \frac{122}{16}.$$

Comparing this with the standard equation, we see that this is a circle with center at $(\tfrac{3}{4}, -\tfrac{7}{4})$ and radius $\sqrt{122}/4$. Using the compass, with this information we construct the graph of the circle $2x^2 + 2y^2 - 3x + 7y - 8 = 0$ without plotting points.

Example B.   Find the graph of $2x^2 + 2y^2 - 3x + 7y + 8 = 0$.

Solution.   This is a second-degree equation of apparently the right type for a circle. However, completing the squares as in Example A, we find the equivalent equation

$$\left(x - \frac{3}{4}\right)^2 + \left(y + \frac{7}{4}\right)^2 = -4 + \frac{9}{16} + \frac{49}{16} = \frac{-6}{16}.$$

Since the left member is positive or zero for every (x, y) and the right member is negative, there are no points (x, y) that satisfy the equation. Its graph is the empty set, not a circle. These two examples indicate how, in every case, completing the squares will determine whether a second-degree equation of the form (2) is a circle or not.

8.4   Slope as Tangent of Inclination.   In Euclidean geometry each line has an angle of inclination θ measured by rotation from the positive ray of the x-axis to the line (Figure 8.9). The slope of the line $m = \tan \theta$. In the geometry of graphs, in which the two coordinates are subject to independent change of scale, there could be no angle of inclination. While the slope m could be expressed by the ratio

$$m = \frac{y_2 - y_1}{x_2 - x_1},$$

it could not be represented by the Euclidean geometry statement $m = \tan \theta$.

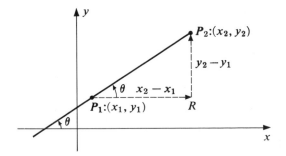

FIGURE 8.9   Tangent of angle of inclination.

8.5   Distance Formula in Euclidean 3-Space.   Euclidean 3-dimensional space, $\mathscr{E}_3$, is also distinguished by existence of a distance, which is invariant under rotation of space. The properties of $\mathscr{E}_3$ make up what is called *solid geometry*. Every plane in 3-space is a Euclidean plane and in it the theorem of Pythagoras is valid. We introduce a coordinate system made up of three mutually perpendicular axes with equal units on them. Then points P have three coordinates (x, y, z) determined by projection of P onto the three axes (Figure 8.10). If $P:(x_1, y_1, z_1)$ and $Q:(x_2, y_2, z_2)$ are two points, then line segment $\overrightarrow{PQ}$ may be represented as the diagonal of a box with edges $x_2 - x_1$, $y_2 - y_1$, $z_2 - z_1$ parallel to the coordinate axes. Then it follows from the plane theorem of Pythagoras that the distance $|PQ|$ is given by

$$|PQ| = \sqrt{(x_2 - x_1)^2 + (y_2 - y_1)^2 + (z_2 - z_1)^2}.$$

Proof.   Exercises.

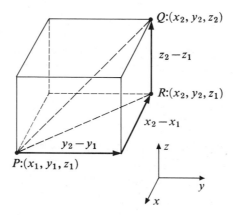

FIGURE 8.10   Distance formula in 3-space.

Moreover if $\overrightarrow{PA}$ and $\overrightarrow{QB}$ are two directed line segments having projections on the three coordinate axes given by $\overrightarrow{PA}:(a_1, a_2, a_3)$ and $\overrightarrow{QB}:(b_1, b_2, b_3)$, then by an argument similar to that of the plane (§8.2) we find the following:

CONDITION FOR PERPENDICULARITY.   The line segments $\overrightarrow{PA}$ and $\overrightarrow{QB}$ are perpendicular if and only if the inner product

$$a_1b_1 + a_2b_2 + a_3b_3 = 0.$$

In 3-space we cannot determine a direction by one angle of inclination and hence the condition of negative reciprocal slopes cannot apply. The inner product condition, however, extends naturally to 3-space.

8.6   Sphere and Ball.   The set of all points $P:(x, y, z)$ at a constant distance r from a fixed center $C:(x_0, y_0, z_0)$ forms a sphere (Figure 8.11(a)). It is a surface. The equation of the sphere is derived from the fact the point P is on the sphere if and only if the distance $|\overrightarrow{CP}| = r$; it is

$$(x - x_0)^2 + (y - y_0)^2 + (z - z_0)^2 = r^2.$$

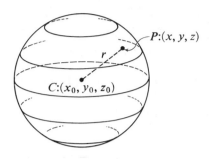

(a) Sphere, a surface
$(x - x_0)^2 + (y - y_0)^2 + (z - z_0)^2 = r^2$

(b) Ball, a solid
$(x - x_0)^2 + (y - y_0)^2 + (z - z_0)^2 \leqq r^2$

FIGURE 8.11   Sphere and ball.

The set of all points $P:(x, y, z)$ for which $|\overrightarrow{CP}| \leqq r$ is the solid ball consisting of all points on the sphere and inside the sphere with center at C and radius r. In coordinates, the ball is represented by the relation (Figure 8.11(b))

$$(x - x_0)^2 + (y - y_0)^2 + (z - z_0)^2 \leqq r^2.$$

The "open ball," consisting of points inside but not on the sphere, is given by the inequality

$$(x - x_0)^2 + (y - y_0)^2 + (z - z_0)^2 < r^2.$$

8.7   Exercises

In Exercises 1–8, find the distance $|\overrightarrow{PQ}|$ between the points.

1. $P:(3, -1)$ and $Q:(2, 5)$.
2. $P:(-3, 10)$ and $Q:(0, -9)$.
3. $P:(-1, 8)$ and $Q:(4, 8)$.
4. $P:(\sqrt{2}, -1)$ and $Q:(x, y)$.
5. $P:(-1, 0, 4)$ and $Q:(3, -2, 5)$.
6. $P:(1, 1, -1)$ and $Q:(x, 1, -1)$.
7. $P:(a, b, c)$ and $Q:(0, 0, 0)$.
8. $P:(x, y, z)$ and $Q:(a, b, c)$.

9.   Show that $\overrightarrow{PQ}$ and $\overrightarrow{PR}$ are perpendicular line segments: $P:(1, 3, -4)$, $Q:(2, 5, 1)$, $R:(0, 1, -3)$.

10.   Find the tangent of the inclination to the x-axis of the line $\overrightarrow{PQ}$, $P:(3, -1)$, $Q:(2, 5)$.

11.   In Exercise 10, find the cosines of the angles $\overrightarrow{PQ}$ makes with x-axis and y-axis.

12.   In Exercise 9, find the cosines of the angles $\overrightarrow{PQ}$ makes with x-axis, y-axis, and z-axis.

13.   Find the equation of the circle with center at $C:(-1, 3)$ and radius 4.

14.   Find the equation of the circle with center at $C:(-1, 3)$ which passes through the origin.

15.   Plot the graph of the relation

$$(x - 1)^2 + (y + 2)^2 \leqq 4.$$

16.   Find the equation of the sphere of radius 5 with center at $C:(-3, 0, 1)$.

17.   For the points $P:(x_1, y_1, z_1)$ and $Q:(x_2, y_2, z_2)$ show that the midpoint of the line segment $\overrightarrow{PQ}$ is

$$M: \left(\frac{x_1 + x_2}{2}, \frac{y_1 + y_2}{2}, \frac{z_1 + z_2}{2} \right).$$

In Exercises 18–21, find the equation of the line.

18. The line through $(-2, 3)$ perpendicular to $3x + 2y - 7 = 0$.
19. The line through $(-1, 4)$ perpendicular to the line segment joining $A:(-1, 6)$ and $B:(5, -7)$.
20. The line which is the perpendicular bisector of the line segment joining $A:(-5, 7)$ and $B:(\frac{3}{2}, -2)$.
21. The medians of the triangle $A:(2, -3)$, $B:(4, -1)$, $C:(-5, 2)$.
22. Plot the graph in $\mathscr{E}_2$ of $x^2 + y^2 - 3x - 6y - 1 = 0$.
23. Plot the graph in $\mathscr{E}_3$ of

$$x^2 + y^2 + z^2 - 2x + 4y + 6z = 0.$$

24. Find the center and radius of the figure in $\mathscr{E}_2$,

$$2x^2 + 2y^2 - 3x + 5y - 1 \leqq 0.$$

What is the figure?
25. Find the equation whose graph consists of all points in the xy-plane that are at distance 4 from the point $C:(2, 1, 1)$.
26. Plot the graph in $\mathscr{E}_3$ of $x = 4$.
27. Find the inequality describing the ball with center at $C:(-1, 2, 3)$ and radius 5.
28. Find the equation of the graph in $\mathscr{E}_3$ traced by a point P such that $\overrightarrow{AP}$ is perpendicular to $\overrightarrow{BP}$ where $A:(1, -1, 2)$, $B:(2, 3, -2)$. What is the graph?

9 Multicoordinate Spaces of Physical Quantities

9.1 Functions of Several Variables in Nature. To start with a simple example, the volume V of a box whose edges have length x, width y, and height z (Figure 8.12) is determined when the numbers x, y, z are assigned independently by $V = xyz$. This is what we mean when we say that V is a function of x, y, z.

The surface area of S of the box is another function of the three variables x, y, z given by $S = 2xy + 2yz + 2zx$.

In physics, Coulomb's law gives the magnitude F of attraction between two electric charges as a function of their charges, q_1 and q_2, and the distance r between them by the formula

$$F = k\frac{q_1q_2}{r^2},$$

where k is a constant.

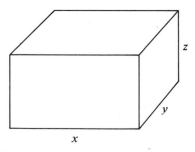

FIGURE 8.12　Rectangular box.

The tidal lifting force T at any point P in the ocean is a function

$$F(\theta, \phi, \theta_s, \phi_s, r_s, \theta_m, \phi_m, r_m),$$

where θ, ϕ are the latitude and longitude of the point P; θ_s, ϕ_s, r_s are the latitude, longitude and distance of the sun; and θ_m, ϕ_m, r_m are the latitude, longitude, and distance of the moon.

In a living cell, the concentration C of a substance can be expressed as a function of the synthesis rate S of the substance, the surface area A of the cell, and the diffusion coefficient D by

$$C = k\frac{SA}{D}.$$

A simple example of a function of several variables occurs when we express the profit of a corporation in terms of the dollar value S of sales, the cost M of materials purchased, the dollar value L of labor and capital hired, and the corporate income tax c by $P = (1 - c)(S - M - L)$.

A function that requires several variables to express its values occurs when we express the position $\vec{r}$ of a space vehicle launched from Cape Kennedy as a function of time t after the launch. For the position $\vec{r}$ consists of its latitude ϕ, longitude θ, and distance r from the earth.

Finally, we encounter functions in which both the points in the domain and the values of the function require several variables to determine them. For example, at any one time t the air at a point $P:(\phi, \theta, r)$ in the earth's atmosphere has a velocity that requires three numbers (v_1, v_2, v_3) to describe it. Thus the velocity of air in the earth's atmosphere is a function of the four variables (t, ϕ, θ, r) and has values (v_1, v_2, v_3).

9.2   Row Space.* The word *space* in mathematics is a conveniently vague one and means simply a set of elements with structure. We consider the set $\mathcal{R}_n$ of all rows $(x_1, x_2 \ldots, x_n)$ of real numbers and make a space out of it by defining an algebraic structure in it. For brevity, let $\vec{x}$ denote the row* $(x_1, x_2, \ldots, x_n)$, $\vec{y}$ the row $(y_1, y_2 \ldots, y_n)$, and $\vec{z}$ the row $(z_1, z_2, \ldots, z_n)$. Then we define the sum $\vec{x} + \vec{y}$ by

$$\vec{x} + \vec{y} = (x_1 + y_1, x_2 + y_2, \ldots, x_n + y_n).$$

Also the *scalar product* $c\vec{x}$ of the row $\vec{x}$ by the number c is defined by

$$c\vec{x} = (cx_1, cx_2, \ldots, cx_n).$$

We can prove (Exercises) that with these two operations the space $\mathcal{R}_n$ is a linear space as described by the following properties. We list the properties as "axioms" since many other systems have algebraic operations that have the same properties.

AXIOMS FOR REAL LINEAR SPACE.   A real linear space is a set of elements $\mathcal{L} = \{\vec{x}, \vec{y}, \vec{z}, \ldots\}$ and a field of scalars consisting of the real numbers, with two operations.

A1.   There is an addition such that to each pair of elements $\vec{x}$ and $\vec{y}$ there corresponds a unique sum $\vec{x} + \vec{y}$ in $\mathcal{L}$.

A2.   This sum is associative, $(\vec{x} + \vec{y}) + \vec{z} = \vec{x} + (\vec{y} + \vec{z})$; and

---

* Here *row* means only the ordered array of numbers $(x_1, x_2, \ldots, x_n)$, usually called an *n-tuple*. We do not indicate that our row is a horizontal row taken from a rectangular matrix of elements.

A3.   commutative, $\vec{x} + \vec{y} = \vec{y} + \vec{x}$.
A4.   There is a zero element $\vec{0}$ in $\mathscr{L}$ such that for every $\vec{x}$, $\vec{x} + \vec{0} = \vec{x} = \vec{0} + \vec{x}$.
A5.   For every $\vec{x}$ in $\mathscr{L}$, there is an inverse element $-\vec{x}$ such that $\vec{x} + (-\vec{x}) = \vec{0}$.
S1.   There is also a multiplication by scalars k such that to every $\vec{x}$ in $\mathscr{L}$ and every real number k there is defined a unique element $k\vec{x}$ in $\mathscr{L}$.
S2.   For every pair of elements $\vec{x}$ and $\vec{y}$ and every scalar k, $k(\vec{x} + \vec{y}) = k\vec{x} + k\vec{y}$.
      For any two scalars and any element $\vec{x}$,
S3. $(k_1 + k_2)\vec{x} = k_1\vec{x} + k_2\vec{x}$, and
S4. $(k_1 k_2)\vec{x} + k_1(k_2\vec{x})$.
S5.   The scalar 0 and the element $\vec{0}$ of the linear space are related by $0\vec{x} = \vec{0}$. Also,
S6.   for every real number k, $k\vec{0} = \vec{0}$. For every linear space element $\vec{x}$, the scalar 1 multiplies to give $1\vec{x} = \vec{x}$.

We observe that the 1-rows are just real numbers, so the real number system itself is a row-space in which the rows have only one element.

9.3   Function of Several Variables Abstractly Defined.

We consider a set $\mathscr{D}$ in a row-space $\mathscr{X}$ having elements $\vec{x}:(x_1, x_2, \ldots, x_n)$ and row-space $\mathscr{Y}$ having elements $\vec{y}:(y_1, y_2, \ldots, y_m)$. A system $f:\mathscr{D} \to \mathscr{Y}$ is a function on the domain $\mathscr{D}$ to $\mathscr{Y}$ if and only if for every row $\vec{x}$ in $\mathscr{D}$, f selects one and only row $\vec{y}$ in $\mathscr{Y}$.

We may think of the function $f:\mathscr{D} \to \mathscr{Y}$ as a mapping (Figure 8.13).

9.4   Difficulties with Graphs.

The definition of a function of several variables is a simple generalization of the definition of a function f of one variable that selects for each number x one and only one number y, written $y = f(x)$. We have only generalized the idea of replacing the number x by the row of numbers $\vec{x}$, and the number y by the row $\vec{y}$. Yet, even with so simple a function as the volume of the box (§8.12)

$$V = xyz,$$

when we attempt to graph it we find that we need four dimensions to plot the "point" (x, y, z, V). The human limitation of perception to spaces of dimension 3 or less precludes the possibility of picturing the graph of the volume function in the same way that we graph

$$y = 2x^2 + 1.$$

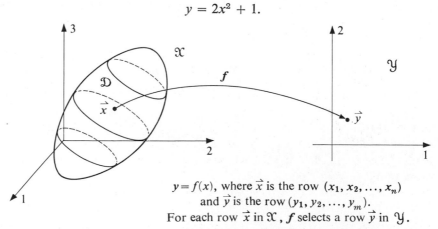

$y = f(x)$, where $\vec{x}$ is the row $(x_1, x_2, \ldots, x_n)$
and $\vec{y}$ is the row $(y_1, y_2, \ldots, y_m)$.
For each row $\vec{x}$ in $\mathscr{X}$, f selects a row $\vec{y}$ in $\mathscr{Y}$.

FIGURE 8.13   Function of several variables as a mapping.

Hence we must do without the physical picture of the graph when too many variables are involved; and proceed with algebraic tools, which impose no limitation on the allowable number of elements in a row.

9.5 Algebraic Difficulties. When we have a function of several different physical variables, such as time, mass, velocity, force, length, or area, we *may* distinguish them by their names or physical properties. In the abstract formulation

$$f(x_1, x_2, x_3, \ldots, x_n),$$

we preserve the distinction by writing them down in order: x_1 is the first variable; x_2, the second; and so on. In either case we admit an arbitrary change of scale in each of the variables according to a substitution of the form

$$\vec{x} = a\vec{x}' + \vec{b}.$$

Here in row notation, $\vec{x} = a\vec{x}' + \vec{b}$ denotes the entire set

$$\{x_1 = a_1x' + b_1, x_2 = a_2x_2' + b_2, \ldots, x_n = a_nx_n' + b_n\}$$

of independent transformations of scale. We require that each $a_i > 0$, $i = 1, 2, \ldots, n$. We recognize two special behaviors of a function $f(\vec{x})$ of several variables when we transform the variables $\vec{x}:(x_1, x_2, \ldots, x_n)$ by independent changes of scale.

DEFINITION. A function $f(\vec{x})$ of several variables $\vec{x}:(x_1, x_2, \ldots, x_n)$ is an *absolute invariant* (*a dimensionless quantity* or a *unit-free* quantity) if and only if under every change of scale $\vec{x} = a\vec{x}' + \vec{b}$,

$$f(a\vec{x}' + \vec{b}) = f(\vec{x}').$$

DEFINITION. A function $f(\vec{x})$ is *homogeneous* (a *homogeneous quantity* or a *relative invariant*) if and only if for every change of scale $\vec{x} = a\vec{x}' + \vec{b}$, there is a number k such that

$$f(a\vec{x}' + \vec{b}) = k f(\vec{x}'), \qquad k \neq 0.$$

We assert as a physico-mathematical principle that a function of physical variables is itself a physical (or geometric) quantity only if it is homogeneous. This includes absolute invariant as a special case. For the choice of scale is made arbitrarily for convenience in computation by the mathematician. Therefore any mathematical function that changes form under transformation of scale cannot have physical or geometric significance. By custom, homogeneous quantities are described by number symbols, after which the physical units are designated by phrases such as "feet per second," ft/sec; "square centimeters" cm^2; or "pounds per square foot," lb/ft^2.

Example A. Does run $\overrightarrow{P_1P_2} = x_2 - x_1$ represent a physical quantity?

Solution. We subject the variable x to the change of scale $x = ax' + b$, so that

$$x_2 = ax_2' + b, \qquad x_1 = ax_1' + b, \qquad x_2 - x_1 = a(x_2' - x_1').$$

In the new scale, we have run $\overrightarrow{P_1P_2} = a(x_2' - x_1')$, which is of the same form as the formula in the original scale except that it is multiplied by the constant a. It is a homogeneous quantity and therefore a physical quantity.

Example B. Does the velocity $v = (x_2 - x_1)/(t_2 - t_1)$ in ft/sec represent a physical quantity?

Solution. We subject the distance x to the change of scale $x = ax' + b$, where $a > 0$, and the time to a change of scale $t = ct' + d$, where $c > 0$. Then the velocity v in the new scales is given by

$$v = \left(\frac{a}{c}\right)\frac{x_2' - x_1'}{t_2' - t_1'}.$$

Since the velocity v' in the new units is given by the ratio

$$v' = \frac{x_2' - x_1'}{t_2' - t_1'}.$$

this shows that $v' = (c/a)v$. That is, velocity is a homogeneous physical quantity.

Example C. Is the ratio $m = (y_2 - y_1)/(x_2 - x_1)$ of two line segments measured in the same scale a physical quantity?

Solution. Since they are measured in the same scale we subject both x and y to the same change of scale $x = ax' + b$, $y = ay' + b$, $a > 0$, and find that

$$m = \left(\frac{a}{a}\right)\frac{y_2' - y_1'}{x_2' - x_1'} = m'.$$

Hence the ratio m is an absolute invariant (dimensionless quantity).

Example D. Does the function given by $s = x + t$, where x is a length and t is a time, represent a physical quantity?

Solution. Introducing the changes of scale $x = ax' + b$, and $t = ct' + d$, $a > 0$, $c > 0$, we find that in the new units

$$s = ax' + ct',$$

which is not any constant multiple of $x' + t'$. Hence the sum $x + t$ is not a physical quantity.

Similar arguments apply to polynomials like $x^2y - 3xy + 6y^2 - 2x + y + 7$. If x and y are nonzero physical variables, this polynomial does not represent a physical quantity.

9.6 Noninvariance of Euclidean Distance.

We also find that the Euclidean distance formula

$$|\overrightarrow{PQ}| = \sqrt{(x_2 - x_1)^2 + (y_2 - y_1)^2}$$

is not even homogeneous when the x- and y-coordinates are subject to independent change of scale. Hence, the Euclidean distance formula has no geometric significance unless x- and y-coordinates are restricted to having the same scale. Therefore we cannot assume that the geometry of row space is a Euclidean geometry.

9.7   Exercises

1. Verify that a row space satisfies the properties of a real linear space.

In Exercises 2–5, perform the indicated operations on the rows

$$\vec{x}:(-1, 0, 1, 4), \quad \vec{y}:(3, -1, 2, 5), \quad \vec{z}:(0, 1, 0, -1).$$

2. $(\vec{x} + \vec{y}) - \vec{z}$.
3. $2\vec{x} - \vec{y} + \vec{z}$.
4. $3(\vec{x} - \vec{y}) - 3(\vec{x} + \vec{y})$.
5. $\vec{x} - 7\vec{y} - \vec{z}$.

In Exercises 6–9, draw a system of coordinates in 3-space and plot the given points in it.

6. Plot $(-3, 4, -1)$.
7. Plot $(0, 0, 4)$.
8. Plot $(-1, 0, 0)$.
9. Plot $(1, -5, 1)$.

10. Express the distance traveled s as a function of constant velocity $\vec{v} = (a, b, c)$ and time traveled t. What are the units of s?

11. Show that the surface area of the box (§9.1) is a homogeneous quantity if the edges are given on the same scale but is not a geometric quantity if the edges are subject to independent change of scale.

12. Show that radian measure of an angle is an absolute invariant, that is, is dimensionless or unit-free.

13. For motion in a fluid, the Mach number is the ratio of the local rate of motion to the speed of sound. What are its dimensions?

14. Show that the values of $\sin \theta$, $\cos \theta$, $\tan \theta$ are unit-free.

15. Show that $\ln x$ defined for each positive x by

$$\ln x = \int_1^x \frac{du}{u}$$

is dimensionless even if x is in physical units.

16. Show that the statement, "Mass times acceleration equals force," does not have physical significance, whereas "Mass times acceleration is proportional to the force," does.

17. If x and $f(x)$ are measured in the same units, what are the units of the derivative, $f'(x)$?

18. If x has units of length and $f(x)$ has units of force, what are the units of the integral $\int_a^b f(x)\, dx$?

19. In the graphs geometry of the plane, show that under independent transformations of scale straight lines go into straight lines; points of intersection into points of intersection; and parallel lines into parallel lines.

20. Show that the power function, described for all positive numbers x and real powers p by $f(x) = x^p$, preserves multiplication $f(uv) = f(u)f(v)$ and preserves the dimensionless property of ratios. That is, $(x_2/x_1)^p$ is dimensionless.

21. What are the units of velocity dx/dt and acceleration d^2x/dt^2?

22. In the motion of a particle of mass m at an instant when the velocity $dx/dt = v$ and the acceleration is a, the kinetic energy $T = \frac{1}{2}mv^2$ and the force acting on the particle is given by $F = ma$. What are the units of T and F? Show that work, Fx, and T have the same units.

23. Show that in the graphs plane where slant distances $|PQ|$ have no significance, it is still possible to define a midpoint of the line segment $\overrightarrow{PQ}$ joining $P:(x_1, y_1)$ to $Q:(x_2, y_2)$. Show that under independent changes of scale midpoints go into midpoints.

24. In the graphs plane there are fixed points A on the x-axis and B on the y-axis (Figure 8.14). With respect to any chosen scales their coordinates are $A:(a, 0)$, $B:(0, b)$. Show that the graph of the ellipse

$$\frac{x^2}{a^2} + \frac{y^2}{b^2} = 1$$

is not changed by independent transformations of scale on the two axes.

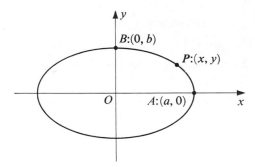

FIGURE 8.14   Exercise 24.

25. The equation of parabolas with respect to coordinates measured in the same scale is $y^2 = 4px$. Show that all parabolas have the same shape, that every parabola is obtainable by magnification of a standard one.

26.* In a flowing fluid we have the following physical quantities with dimensions given in terms of length L, mass M, and time T.

Quantity: velocity v, density ρ, diameter D, viscosity μ.

Dimensions: LT^{-1}, ML^{-3}, L, $ML^{-1}T^{-1}$.

Show that all dimensionless products $v^a \rho^b D^c \mu^d$ have the form

$$\left(\frac{v\rho D}{\mu}\right)^p$$

for some power p. The dimensionless ratio is called the *Reynolds number*.

27. In the rows (t, x, y, z) of time-space, t has units of time and x, y, z all have the same length units. It was an important milestone in the history of science when, for his special theory of relativity, Einstein introduced in this row space the distance

$$\sqrt{c^2(t_2 - t_1)^2 - (x_2 - x_1)^2 - (y_2 - y_1)^2 - (z_2 - z_1)^2},$$

where c is the speed of light, instead of the Euclidean distance formula

$$\sqrt{(t_2 - t_1)^2 + (x_2 - x_1)^2 + (y_2 - y_1)^2 + (z_2 - z_1)^2}.$$

Show that the distance formula of relativity has physical significance, while the Euclidean distance does not.

10   Vectors

We undertake the study of vectors as an escape from the complications of dimensional scales in physical spaces. By means of vectors we can set up coordinates that are dimensionless and thereby regain the algebraic freedom and simplicity of Euclidean geometry for the representation of physical quantities (§10.8).

10.1   Vectors of the Cartesian Plane.   We consider the familiar Cartesian plane constructed from a pair of perpendicular number lines called coordinate axes, with equal units in the scale on each axis. The points are pairs of numbers (x, y), where x is a number on the x-axis and y is a number on the y-axis. It becomes a Euclidean plane if we measure the distance between any two points (x_1, y_1) and (x_2, y_2) by the distance formula $\sqrt{(x_2 - x_1)^2 + (y_2 - y_1)^2}$ (Figure 8.15(a)).

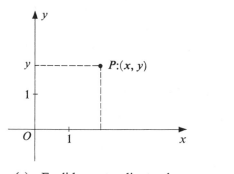

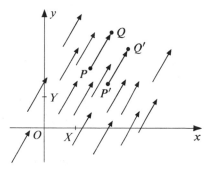

(a)   Euclidean coordinate plane   (b)   Parallel displacement

FIGURE 8.15   Vectors of the plane.

We define vectors as parallel displacements of the plane. We may picture a parallel displacement by placing a sheet of paper on a tabletop with edges parallel to the table's edges. The tabletop and paper represent two copies of the plane. At the start, a point P in the paper is directly over the same point P in the tabletop. Then we slide the paper, keeping the edges parallel to the table edges. Every point P moves to a new position Q along a straight-line track $\overrightarrow{PQ}$ (Figure 8.15(b)). This track $\overrightarrow{PQ}$ is called the *directed line segment from P to Q*. Thus a parallel displacement produces infinitely many directed line segments $\overrightarrow{PQ}$, $\overrightarrow{P'Q'}$, ... in the plane. They are parallel and equal in length.

In general, we postulate that the plane admits parallel displacements. These are mappings of the plane onto itself that either send every point P to a different point Q, or leave every point fixed (the zero displacement). Given two points P and Q in the plane, we have one and only one parallel displacement that sends P to Q.

In the case of the Cartesian plane, this is easy to realize. If a parallel displacement sends $P:(x, y)$ to $Q:(x + a, y + b)$, then it sends every point $P':(x', y')$ to $Q':(x' + a, y' + b)$. Thus to carry out the action of the vector $\vec{v}:(a, b)$ on the plane, we add the same pair (a, b) to every point (x, y).

10.2   Position Vectors.   As we have seen, a vector $\vec{v}$ acts on the plane to displace every point P to a point Q. Conversely, any one of these directed line segments $\overrightarrow{PQ}$, $\overrightarrow{P'Q'}$ determines $\vec{v}$. We can establish a one-to-one correspondence between vectors and directed line segments by considering the action of the vector only on the origin $(0, 0)$. A vector $\vec{v}$ sends O to just one point R and $\vec{v}$ is represented by just one directed line segment $\overrightarrow{OR}$. On the other hand, the point R determines the directed line segment $\overrightarrow{OR}$ and hence $\vec{v}$. The directed line segment $\overrightarrow{OR}$ is called the *position vector* of $\vec{v}$, and also the *position vector of the point R*.* Thus we have the one-to-one correspondences indicated by the two-way arrow $\leftrightarrow$ (Figure 8.16):

$$\text{vector} \leftrightarrow \text{position vector} \leftrightarrow \text{point,}$$

$$\vec{v} \quad \leftrightarrow \quad \overrightarrow{OR} \quad \leftrightarrow \quad R.$$

---

* In the literature, position vectors are sometimes called *bound vectors*, in contrast to vectors as parallel displacements, which are called *free vectors*.

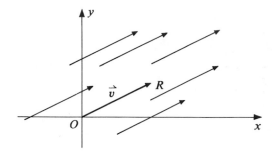

FIGURE 8.16 Position vector.

10.3 Vector Operations. We define the sum, $\vec{u} + \vec{v}$, of two vectors as the vector which results from applying first $\vec{u}$ and then $\vec{v}$ to the plane (Figure 8.17(a)).

Let us choose arbitrarily an origin O in the plane. Then every nonzero vector $\vec{v}$ determines a position vector $\overrightarrow{OR}$, and this determines a line OR, called the line of $\vec{v}$. In order to define *scalar multiplication* of a vector $\vec{v}$ by a number c, we describe the action of $c\vec{v}$ on O. First we define $c\vec{0} = \vec{0}$. Then if $\vec{v} \neq \vec{0}$, we use a geometric construction (Figure 8.17(b)) to determine $c\vec{v}$ by means of similar triangles in the plane.* Thus the action of $c\vec{v}$ on O is to send O to a point S such that $\overrightarrow{OS} = c\overrightarrow{OR}$. We readily see that the definition is independent of the choice of origin O. We observe that the similar triangle construction of $c\overrightarrow{OR}$ makes no use of the distance scale in the line OS.

Example. With vectors $\vec{u}$ and $\vec{v}$ given by the position vectors $\overrightarrow{OQ}$ and $\overrightarrow{OR}$ (Figure 8.18), find the position vectors of $\vec{u} + \vec{v}, \vec{u} - \vec{v}, 3\vec{v}$.

Solution. We complete the parallelogram $ORSQ$ having sides $\overrightarrow{OR}$ and $\overrightarrow{OQ}$. Then $\overrightarrow{OS}$ represents the $\vec{u} + \vec{v}$. To represent $\vec{u} - \vec{v}$ we add $\vec{u} + (-\vec{v})$. We represent $-\vec{v}$ in the line of $\vec{v}$ as $-1(\vec{v})$, shown by the line segment $\overrightarrow{OR'}$ (Figure 8.18). Then we complete the parallelogram $OQTR'$. The position vector $\overrightarrow{OT}$ represents $\vec{u} - \vec{v}$. To plot a position vector representing $3\vec{v}$, we install a number line in the plane with origin at O. Then we draw $\overrightarrow{1R}$ and draw $\overrightarrow{3P}$ parallel to $\overrightarrow{1R}$. Now $\overrightarrow{OP}$ represents $3\vec{v}$.

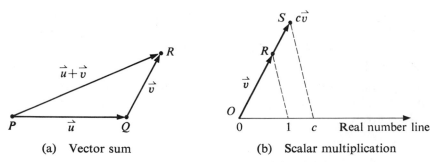

(a) Vector sum

(b) Scalar multiplication

FIGURE 8.17 Vector operations of the plane.

---

* A rigorous arithmetic definition of $c\vec{v}$ is essentially equivalent to retracing the construction of the real-number system in the line of $\vec{v}$, making use of the existence of least upper bounds.

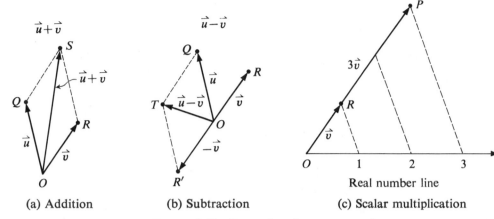

(a) Addition   (b) Subtraction   (c) Scalar multiplication

FIGURE 8.18   Examples of vector operations.

These operations can be carried out numerically for vectors of the Cartesian coordinate plane. Since a displacement that sends (x, y) to $(x + a, y + b)$ also sends any other (x', y') to $(x' + a, y' + b)$, we denote the vector by $\vec{v}:(a, b)$. .It displaces the origin to the point with coordinates (a, b). Let $\vec{u}:(a_1, b_1)$ be another vector. Then the sum $\vec{u} + \vec{v}$ is given by $(a_1 + a, b_1 + b)$ and the scalar product $c\vec{v}$ is given by (ca, cb). Hence for vectors of the coordinate plane, the operations coincide with those of the linear space of rows (§9.2). For this reason the set of vectors of the plane, with the operations of addition and scalar multiplication among themselves, form a linear space. We call it a *vector space*.

10.4   Basis. We may choose arbitrarily two points X and Y, so that X, O, Y are not in the same line. Let $\vec{i}$ be the vector that sends O to X and let $\vec{j}$ be the vector that sends O to Y. Then $\vec{i}$ and $\vec{j}$ are *linearly independent*. This means that if for some numbers a and $b, a\vec{i} + b\vec{j} = \vec{0}$, then $a = b = 0$. We wish to show that the set $\{\vec{i}, \vec{j}\}$ is a *basis* for the vector space. That is, every vector $\vec{v}$ in the space is uniquely represented by a linear combination,

$$\vec{v} = x\vec{i} + y\vec{j}.$$

To show this, we consider the action of $\vec{v}$ on the origin, producing the position vector $\overrightarrow{OR}$ (Figure 8.19). By parallel projection of R on the line of $\vec{i}$, we locate the point S. Then the position vector $\overrightarrow{OS}$ determines uniquely the scalar product $x\vec{i}$. Similarly, we project R on the line of $\vec{j}$ to locate T, and the position vector $\overrightarrow{OT}$ determines the scalar product $y\vec{j}$. By the parallelogram construction of addition, $\vec{v} = x\vec{i} + y\vec{j}$.

The chosen basis $\{\vec{i}, \vec{j}\}$ establishes a 1–1 correspondence between the vectors of the plane and rows (x, y) of scalars:

$$\vec{v} = x\vec{i} + y\vec{j} \leftrightarrow (x, y), \qquad \vec{u} = x'\vec{i} + y'\vec{j} \leftrightarrow (x', y').$$

Moreover,

$$\vec{v} + \vec{u} = (x + x')i + (y + y')j \leftrightarrow (x + x', y + y'),$$

and

$$c\vec{v} = cx\vec{i} + cy\vec{j} \leftrightarrow (cx, cy).$$

Also we observe that $\vec{i} \leftrightarrow (1, 0)$ and $\vec{j} \leftrightarrow (0, 1)$.

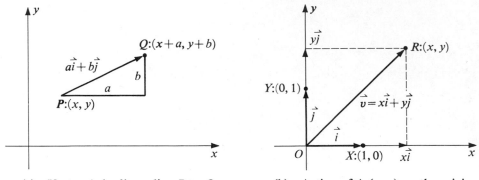

(a)   Vector $\vec{v}:(a, b)$ sending P to Q (b)   Action of $\vec{v}:(x, y)$ on the origin

FIGURE 8.19   Vectors of the coordinate plane.

Example.   With respect to a basis $\{\vec{i}, \vec{j}\}$, $\vec{u} = 2\vec{i} - 3\vec{j}$ and $\vec{u} = \vec{i} + 5\vec{j}$. Find $\vec{u} + \vec{v}$, $\vec{u} - \vec{v}$, $3\vec{v}$.

Solution.   We simply perform the indicated operations algebraically.

$$\vec{u} + \vec{v} = (2 + 1)\vec{i} + (-3 + 5)\vec{j} = 3\vec{i} + 2\vec{j}.$$
$$\vec{u} - \vec{v} = (2 - 1)\vec{i} + (-3 - 5)\vec{j} = \vec{i} - 8\vec{j}.$$
$$3\vec{v} = 3(\vec{i} + 5\vec{j}) = 3\vec{i} + 15\vec{j}.$$

The *standard basis* for vectors of the coordinate plane consists of the vector $\vec{i}$, which sends $(0, 0)$ to $(1, 0)$, and $\vec{j}$, which sends $(0, 0)$ to $(0, 1)$ (Figure 8.20). This is equivalent to choosing as a coordinate system a pair of perpendicular axes with equal units on the two axes.

10.5   Inner Product and Norm.   In the Euclidean coordinate plane, with standard basis $\{\vec{i}, \vec{j}\}$, for any two vectors $\vec{u} = x'\vec{i} + y'\vec{j}$ and $\vec{v} = x\vec{i} + y\vec{j}$ we define an *inner product*, $\vec{u} \cdot \vec{v}$, also called a "dot product," by

$$\vec{u} \cdot \vec{v} = x'x + y'y.$$

We observe that $\vec{u} \cdot \vec{v}$ is a scalar and that it occurs in connection with the theorem of Pythagoras (§8.1). For example, if $\vec{u} = 2\vec{i} - 3\vec{j}$ and $\vec{v} = -\vec{i} + 5\vec{j}$,

$$\vec{u} \cdot \vec{v} = -2 - 15 = -17.$$

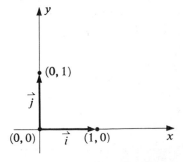

FIGURE 8.20   Standard basis.

Also if $\overrightarrow{OR}$ is the position vector of $\vec{v}$, then the length $|OR|$ is given by

$$|OR| = \sqrt{x^2 + y^2} = \sqrt{\vec{v}\cdot\vec{v}}.$$

We define the *norm* of $\vec{v}$, written $\|\vec{v}\|$, in terms of the inner product to be the length of the position vector $\overrightarrow{OR}$,

$$\|\vec{v}\| = \sqrt{\vec{v}\cdot\vec{v}}.$$

A vector $\vec{v}$ is called a *unit vector* if $\|\vec{v}\| = 1$.

Finally, let $P_1:(x_1, y_1)$ and $P_2:(x_2, y_2)$ be any two points in the plane. Let $\vec{v}_1 = \overrightarrow{OP_1}$, $\vec{v}_2 = \overrightarrow{OP_2}$. Then the distance relationship is

$$|P_1P_2| = \|\vec{v}_2 - \vec{v}_1\| = \sqrt{(x_2 - x_1)^2 + (y_2 - y_1)^2}$$

(Exercises).

10.6   Exercises

1. For vectors $\vec{u}$ and $\vec{v}$ given by position vectors $\overrightarrow{OQ}$ and $\overrightarrow{OP}$ (Figure 8.21), construct position vectors of $\vec{u} + \vec{v}$, $-\vec{v}$, $\vec{u} - \vec{v}$, $2u$, $-3v$.
2. For vectors $\vec{u} = 3\vec{i} - 4\vec{j}$ and $v = \vec{i} - \vec{j}$, find: $\vec{u} + \vec{v}$, $\vec{u} - \vec{v}$, $\|\vec{u}\|$, $\|\vec{v}\|$, $\|\vec{u} + \vec{v}\|$, $\|\vec{u} - \vec{v}\|$, $\vec{u}\cdot\vec{v}$, $3\vec{u} - 4\vec{v}$, $\|-4v\|$, $2\vec{u} - 5\vec{v}$.
3. If $\overrightarrow{OP}$ is the position vector of $\vec{v}$, and $\overrightarrow{OQ}$ is the position vector of $\vec{u}$, show that distance $|PQ| = \|\vec{u} - \vec{v}\|$.
4. Prove the theorem: For any three vectors $\vec{u}$, $\vec{v}$, $\vec{w}$ of the plane,
    (a) $\vec{u}\cdot\vec{v} = \vec{v}\cdot\vec{u}$;
    (b) $(c\vec{u})\cdot\vec{v} = c(\vec{u}\cdot\vec{v}) = \vec{u}\cdot c\vec{v}$;
    (c) $\vec{u}\cdot(\vec{v} + \vec{w}) = (\vec{u}\cdot\vec{v}) + (\vec{u}\cdot\vec{w})$;
    (d) $\vec{v}\cdot\vec{v} \geqq 0$ and $\vec{v}\cdot\vec{v} = 0$ if and only if $\vec{v} = \vec{0}$.
5. Prove that the vector space $\mathscr{V}$ of the vectors of the plane is a real linear space (§10.3 and §9.2).
6. Prove that for any two vectors $\vec{u}$ and $\vec{v}$,
    $$\|\vec{u} + \vec{v}\|^2 + \|\vec{u} - \vec{v}\|^2 = 2\|\vec{u}\|^2 + 2\|\vec{v}\|^2.$$
7. In Exercise 6, draw the position vectors of $\vec{u}$ and $\vec{v}$ and complete the parallelogram with its diagonals. State geometrically what Exercise 5 says about this parallelogram.
8. Show that the law of cosines (§8.7) says in vector language (Figure 8.22)
    $$\|\vec{v} - \vec{u}\|^2 = \|\vec{v}\|^2 + \|\vec{u}\|^2 - 2\|\vec{u}\| \, \|\vec{v}\| \cos \theta,$$

where θ is an angle between the position vectors of $\vec{u}$ and $\vec{v}$.

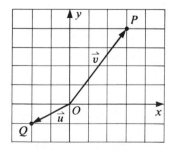

FIGURE 8.21   Exercise 1.

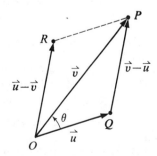

FIGURE 8.22   Exercise 8.

9. In Exercise 8, expand the left member and show that the law of cosines can be written

$$\cos \theta = \frac{\vec{u} \cdot \vec{v}}{\|\vec{u}\| \, \|\vec{v}\|}.$$

10. Use Exercise 9 to prove that for every two vectors $\vec{u}$ and $\vec{v}$

$$\vec{u} \cdot \vec{v} \leqq \|\vec{u}\| \cdot \|\vec{v}\|.$$

Be sure to include the cases where $\vec{u} = \vec{0}$ or $\vec{v} = \vec{0}$.
11. Find the angle between the position vectors of $\vec{i} + \vec{j}$ and $-\vec{i} - 3\vec{j}$; of $4\vec{j}$ and $4\vec{i} - 3\vec{j}$.
12. Show that the position vectors of $2\vec{i} + 4\vec{j}$ and $8\vec{i} - 4\vec{j}$ are perpendicular.
13. Show that if $\vec{u} = a\vec{i} + b\vec{j}$ is a unit vector, then $a = \cos \alpha$ and $b = \cos \beta$ (Figure 8.23).
14. In Exercise 13, find $\cos \alpha$, $\cos \beta$ if $\vec{u} = a\vec{i} + b\vec{j}$ is not a unit vector but $\vec{u} \neq \vec{0}$.
15. As in Exercise 14, show that every vector $\vec{v}$ can be written in the form

$$\vec{v} = \|\vec{v}\|(\vec{i} \cos \alpha + \vec{j} \sin \alpha).$$

16. Use Exercise 9 to show (Figure 8.24) that if $\vec{u}$ is a unit vector and $\vec{v}$ is any vector, then the length $|OM|$ of the projection of the position vector of $\vec{v}$ on the line of $\vec{u}$ is given by $\vec{v} \cdot \vec{u}$.

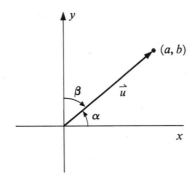

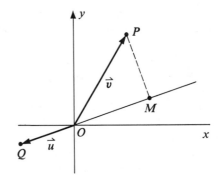

FIGURE 8.23   Exercise 13.   **FIGURE 8.24**   Exercise 16.

17. If $\vec{v} = x\vec{i} + y\vec{j}$, what is the set of points $R:(x, y)$ such that $\|\vec{v}\| = 4$?
18. If $\vec{v} = x\vec{i} + y\vec{j}$ and $\vec{c} = a\vec{i} + b\vec{j}$, what is the set of points $P:(x, y)$ such that $\|\vec{v} - \vec{c}\| = r$. where $r \geqq 0$ and is a constant?
19. If $\vec{v} = x\vec{i} + y\vec{j}$ and $\vec{c} = a\vec{i} + b\vec{j}$, what is the set of points such that $\vec{v} \cdot \vec{c} = 4$?
20. Use Exercise 9 to prove that the position vectors of $\vec{u}$ and $\vec{v}$ are perpendicular if and only if $\vec{u} \cdot \vec{v} = 0$.
21. Use Exercise 9 to prove that the position vectors of $\vec{u}$ and $\vec{v}$ are parallel if and only if $|\vec{u} \cdot \vec{v}| = \|\vec{u}\| \cdot \|\vec{v}\|$.
22. Prove geometrically the triangle inequality.

$$\|\vec{u} + \vec{v}\| \leqq \|\vec{u}\| + \|\vec{v}\|.$$

23. In Exercise 22, prove that if $\|\vec{u} + \vec{v}\| = \|\vec{u}\| + \|\vec{v}\|$, then the position vectors of $\vec{u}$ and $\vec{v}$ are parallel.
24. In Exercise 23, if the position vectors of $\vec{u}$ and $\vec{v}$ are parallel, does this imply that $\|\vec{u} + \vec{v}\| = \|\vec{u}\| + \|\vec{v}\|$? Consider vectors $\vec{u}$ and $-\vec{v}$.
25. Prove that if $\vec{u} = \vec{0}$ or $v = \vec{0}$, then $\vec{u} \cdot \vec{v} = 0$. If $\vec{u} \cdot \vec{v} = 0$ does this imply that either $\vec{u} = \vec{0}$ or $\vec{v} = \vec{0}$? If the answer is *yes*, prove it. If the answer is *no*, give a counterexample.
26. Let $\vec{u}$ be a fixed vector. If $\vec{u} \cdot \vec{v} = 0$ for every vector $\vec{v}$ of the plane, what can be said about $\vec{u}$? Proof.

10.7   Problems

T1. Define the vectors of a line and define addition and scalar multiplication for them.
T2. Prove that addition of vectors of the line is commutative.
T3. Define the order relation $<$ for vectors of the line.
T4. Prove for vectors of the line that if $\vec{u} < \vec{v}$ and $\vec{0} < \vec{w}$, then $\vec{u} + \vec{w} < \vec{v} + \vec{w}$.

10.8   Physical Plane and Coordinate Plane.

(A theoretical discussion on the relation of mathematical and physical quantities.) We consider two physical quantities, like time and temperature, each of whose array of states can be arrayed in a line. We form a geometric model for the states of these two physical quantities by imbedding their lines of states, $\mathcal{S}$ and $\mathcal{T}$, in a plane, which we shall call the *physical plane*. We are assuming that we can locate points S in $\mathcal{S}$ and T in $\mathcal{T}$ without reference to any number scales in these lines. Then points R in the plane are located by parallel projections on the lines $\mathcal{S}$ and $\mathcal{T}$. In this way every point R in the plane determines a pair of points (S, T) in the reference lines, and each point S in $\mathcal{S}$ and T in $\mathcal{T}$ together determine a point R in the plane (Figure 8.25).

A parallel displacement of the plane is a mapping of the plane onto itself that either sends every point P to a different point Q or leaves every point fixed (the zero displacement). By saying that it is a parallel displacement, we mean that the displacement sends the points of $\mathcal{S}$ into a line parallel to $\mathcal{S}$ and the points of $\mathcal{T}$ into a line parallel to $\mathcal{T}$. This implies that every line in the plane goes into a line parallel to itself. Given two points P and Q in the plane, we find one and only one parallel displacement that sends P to Q.

In the same way that we defined vectors $\vec{u}$, $\vec{v}$, $\vec{w}$ $\cdots$ of the coordinate plane we can define vectors of the physical plane. The sum $\vec{u} + \vec{v}$ and scalar product $c\vec{v}$ are defined by the same parallel displacement construction (§10.3) that we used for the coordinate plane.

Now we make use of two fixed states in each of the two physical lines $\mathcal{S}$ and $\mathcal{T}$. These must be locatable by experiment. For example, in the temperature line we locate the freezing point of water and the boiling point. Identifying one of these on each line as the zero point O, we may take their zero points as the point of intersection of the lines $\mathcal{S}$ and $\mathcal{T}$. Then we still have on $\mathcal{S}$ the fixed state I, and on $\mathcal{T}$ the fixed state J (Figure 8.25). We choose as a basis for vectors of the physical plane, the vector $\vec{i}$, which sends O to I, and the vector $\vec{j}$, which sends O to S. Then every vector $\vec{v}$ of the physical plane can be represented (Figure 8.26) as a linear combination

$$\vec{v} = x\vec{i} + y\vec{j}.$$

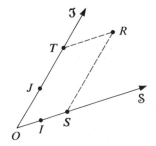

(a)   Physical plane with fixed axes   (b)   Vectors of the physical plane

FIGURE 8.25

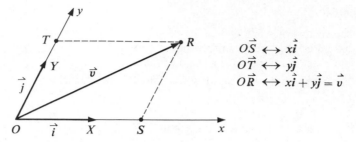

FIGURE 8.26   Basis for plane vectors.

The determination of x and y is made by parallel projection in the same way that we did it for the coordinate plane (§10.4). It is important to observe that the number x is dimensionless, independent of the scale in the line $\mathcal{S}$, for it is the ratio of two directed line segments in $\mathcal{S}$. Similarly y is dimensionless, independent of the scale chosen in $\mathcal{T}$. Hence we may freely perform any algebraic operations of the real numbers on x and y without the restrictions imposed by the requirements of invariance (§9).

In summary, we have replaced the physical plane by the coordinate plane in the following way. We represent the zero point O in the physical plane by $(0, 0)$ in the coordinate plane. We represent the scale point I by $(1, 0)$ and the scale point J by $(0, 1)$. Then we have made the basis vectors $\{\vec{i}, \vec{j}\}$, for the two models correspond. Hence all positions determined by

$$\vec{v} = x\vec{i} + y\vec{j}$$

in the physical plane correspond uniquely to points with coordinates (x, y) in the Euclidean coordinate plane. The vectors of the two planes were the essential link.

Not all of the mathematical results that can be derived by operations in the Euclidean coordinate plane can be interpreted in the physical plane. For example, the Euclidean distance has no meaning in the physical plane if $\mathcal{S}$ and $\mathcal{T}$ are lines of states of two dissimilar quantities, like distance and time. Hence it makes no sense in the physical plane to speak of the lines $\mathcal{S}$ and $\mathcal{T}$ as perpendicular, or to have the same units on $\mathcal{S}$ as on $\mathcal{T}$. But, in general, the properties of affine geometry, derived in the coordinate plane, have meaning in the physical plane. For the properties of the plane, which can be described by the action of vectors of the plane, are affine properties. More specifically, they are properties of the affine plane with two distinguished lines in it. But, even if we perform an operation on the coordinates (x, y) that has no physical interpretation, it still does no violence to the restrictions on scaled physical quantities s and t in the physical plane. Moreover, any statement that does have physical meaning is automatically stated in invariant form.

From the standpoint of elementary applied mathematics, one simply ignores the physical meanings of the coordinates (x, y). Our vector-linked model justifies this, and provides us with a basis for doing dimensional analysis with scaled values of physical quantities when that becomes necessary. For more advanced mechanics, such as relativity and quantum mechanics, this program, or some similar one, becomes essential. In particular, quantum mechanics says, "To every physical quantity there corresponds a linear operator." Our coordinates x, y are linear operators over the vectors of the plane in the sense that

$$x(\vec{u} + \vec{v}) = x\vec{u} + x\vec{v}, \quad \text{and} \quad x(c\vec{v}) = c(x\vec{v}).$$

Hence the model vector-linked to the Euclidean coordinate plane establishes the linear operator representation of physical quantities in elementary analysis, though we need not think of it in these terms.

11   Line and Plane

11.1   Vectors of Higher-Dimensional Spaces. We found (§10.5) that we can represent the vectors of the plane as displacements of a plane in which the "points" are rows of numbers (x, y). We use this idea to extend to a space $\mathscr{P}$ in which the points are rows of numbers $P:(x_1, x_2, \ldots, x_n)$. Vectors are also represented by rows of numbers $\vec{v}:(a_1, a_2, \ldots, a_n)$. The vector $\vec{v}$ is the displacement that sends every point $P:(x_1, x_2, \ldots, x_n)$ to the point $Q:(x_1 + a_1, x_2 + a_2, \ldots, x_n + a_n)$. We can picture it in 3-space, where a row of three numbers (x_1, x_2, x_3) determines a point (Figure 8.27).

As we did for vectors of the plane, we define the sum $\vec{v} + \vec{w}$ of the vector $\vec{v}$ and the vector $\vec{w}:(b_1, b_2, \ldots, b_n)$ as the displacement consisting of $\vec{v}$ and then $\vec{w}$. That is, $\vec{v} + \vec{w}$ is given by adding $(a_1 + b_1, a_2 + b_2, \ldots, a_n + b_n)$ to any row $P:(x_1, x_2, \ldots, x_n)$ representing a point. We define $c\vec{v}$ as the displacement that results from adding $(ca_1, ca_2, \ldots, ca_n)$ to every point $P:(x_1, x_2, \ldots, x_n)$. The set $\mathscr{V}$ of all vectors of the space of n-rows generated in this way is thus a real linear space (§9.2).

11.2   Rows, Vectors, and Position Vectors. For 3-space we may choose the set of three vectors $\mathscr{B} = \{\vec{i}, \vec{j}, \vec{k}\}$ given by

$$\vec{i}:(1, 0, 0), \qquad \vec{j}:(0, 1, 0), \qquad \vec{k}:(0, 0, 1)$$

as a basis. Here, for example, $\vec{i}$ means the displacement of 1 parallel to the 1-axis that sends every point $P:(x_1, x_2, x_3)$ to $Q:(x_1 + 1, x_2, x_3)$. We can then express every vector $\vec{v}$ as a linear combination of these three,

$$\vec{v} = x_1\vec{i} + x_2\vec{j} + x_3\vec{k}.$$

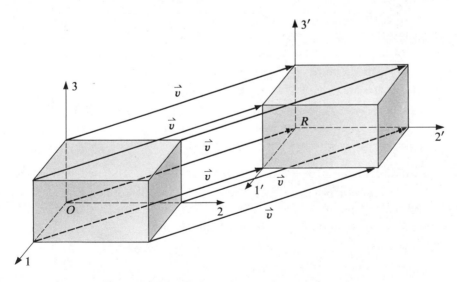

FIGURE 8.27   Action of a vector $\vec{v}$ on 3-space.

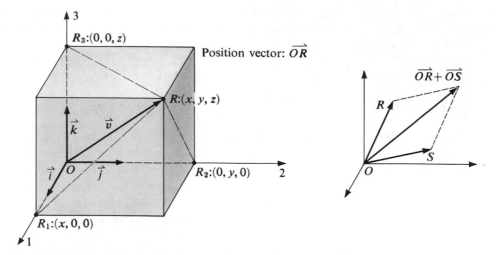

FIGURE 8.28   Position vectors of 3-space, $\vec{v} = x\vec{i} + y\vec{j} + z\vec{k}$.

The scalar multipliers (x_1, x_2, x_3) are called the coordinates of the vector $\vec{v}$ with respect to the basis $\mathscr{B} = \{\vec{i}, \vec{j}, \vec{k}\}$.

The vector $\vec{v}$ sends the origin $O:(0, 0, 0)$ to the point R, and the line segment $\overrightarrow{OR}$ is called the *position vector* of $\vec{v}$ (Figure 8.28). These position vectors, or directed line segments, have an addition $\overrightarrow{OR} + \overrightarrow{OS}$ in which the sum is the diagonal of the parallelogram having sides $\overrightarrow{OR}$ and $\overrightarrow{OS}$. Also they have a scalar multiplication $c\overrightarrow{OR}$ in the line of $\overrightarrow{OR}$. They also form a real linear space with these operations. Thus we have 1–1 correspondences between vectors $\vec{v}$ and their coordinate rows on the one hand, and their position vectors on the other. These correspondences preserve the linear operations. We express this idea in the accompanying table, using the symbol $\leftrightarrow$ for the pairing of the 1–1 correspondence.

Rows		Vectors		Position Vectors
(x_1, x_2, x_3)	$\leftrightarrow$	$\vec{v} = x_1\vec{i} + x_2\vec{j} + x_3\vec{k}$	$\leftrightarrow$	$\overrightarrow{OR}$
(y_1, y_2, y_3)	$\leftrightarrow$	$\vec{w} = y_1\vec{i} + y_2\vec{j} + y_3\vec{k}$	$\leftrightarrow$	$\overrightarrow{OS}$
$(x_1 + y_1, x_2 + y_2, x_3 + y_3)$	$\leftrightarrow$	$\vec{v} + \vec{w}$	$\leftrightarrow$	$\overrightarrow{OR} + \overrightarrow{OS}$
(cx_1, cx_2, cx_3)	$\leftrightarrow$	$c\vec{v}$	$\leftrightarrow$	$\overrightarrow{OR}$

Finally, without the linear operations, there is a 1–1 correspondence between position vectors $\overrightarrow{OR}$ and their endpoints R. We shall emphasize the representation of vectors, position vectors, and points by rows of coordinates, since these are most explicit and convenient. We shall abbreviate the language and speak of:

(1)   The vector $\vec{v}:(x_1, x_2, x_3)$, meaning $\vec{v} = x_1\vec{i} + x_2\vec{j} + x_3\vec{k}$.
(2)   The position vector $\overrightarrow{OR}:(x_1, x_2, x_3)$.
(3)   The point $R:(x_1, x_2, x_3)$.

When it is needed we can always return to writing out the vector $\vec{v}$ in terms of the basis $\{\vec{i}, \vec{j}, \vec{k}\}$.

11.3 Inner Product and Norm. We define the *inner product* (or dot product) $\vec{v} \cdot \vec{w}$ of the vectors $\vec{v}: (x_1, x_2, \ldots, x_n)$ and $\vec{w}: (y_1, y_2, \ldots, y_n)$ to be the scalar

$$\vec{v} \cdot \vec{w} = x_1 y_1 + x_2 y_2 + \cdots + x_n y_n.$$

The algebraic properties of rows enable us to prove the following theorem (Exercises).

THEOREM (Properties of inner product).
(a) $\vec{v} \cdot (\vec{w} + \vec{u}) = \vec{v} \cdot \vec{w} + \vec{v} \cdot \vec{u}.$
(b) $\vec{v} \cdot \vec{w} = \vec{w} \cdot \vec{v}.$
(c) $(c\vec{v}) \cdot \vec{w} = c(\vec{v} \cdot \vec{w}) = \vec{v} \cdot (c\vec{w}).$
(d) $\vec{v} \cdot \vec{v} \geq 0$ and $\vec{v} \cdot \vec{v} = 0$ if and only if $\vec{v} = \vec{0}.$

DEFINITION.   The *norm* of the vector $\vec{v}$, written $\|\vec{v}\| = \sqrt{\vec{v} \cdot \vec{v}}$.

In the plane and in 3-space, when the units on the axes are equal and the axes are perpendicular the $\|\vec{v}\|$ is the length of its position vector $\overrightarrow{OR}$. In general, for $\vec{v}: (x_1, x_2, \ldots, x_n)$,

$$\|\vec{v}\| = \sqrt{x_1^2 + x_2^2 + \cdots + x_n^2}.$$

THEOREM (Schwarz inequality). $|\vec{u} \cdot \vec{v}| \leq \|\vec{u}\| \, \|\vec{v}\|.$

Proof.   (Exercises).

11.4 Located Vector and Distance Formula. In the triangle ORS (Figure 8.29) we add directed line segments to find that $\overrightarrow{OR} + \overrightarrow{RS} = \overrightarrow{OS}$, or

$$\overrightarrow{RS} = \overrightarrow{OS} - \overrightarrow{OR}.$$

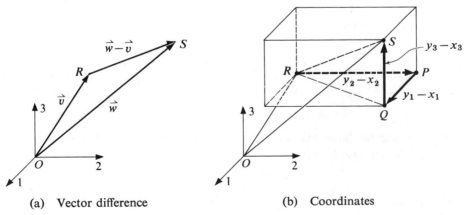

(a) Vector difference   (b) Coordinates

FIGURE 8.29   Located vector.

The directed line segment $\overrightarrow{RS}$ is called a *located vector*. If $\overrightarrow{OR}$ is the position vector of $\vec{v}$ and $\overrightarrow{OS}$ the position vector of $\vec{w}$, then RS represents $\vec{w} - \vec{v}$. We assign to $\overrightarrow{RS}$ the distance

$$|RS| = \|\vec{w} - \vec{v}\|.$$

For points $R:(x_1, x_2, \ldots, x_n)$ and $S:(y_1, y_2, \ldots, y_n)$ the located vector $\overrightarrow{RS}$ is produced by the vector

$$(y_1 - x_1, y_2 - x_2, \ldots, y_n - x_n)$$

and the distance formula becomes

$$|RS| = \sqrt{(y_1 - x_1)^2 + (y_2 - x_2)^2 + \cdots + (y_n - x_n)^2}.$$

We can prove that this distance formula gives the distance $|RS|$ in 3-space (Figure 8.29(b)) when the basis vectors are perpendicular and of the same length (Exercises).

Example.   Find the distance between the points $R:(-3, 1, 0)$ and $S:(4, -2, 5)$.

Solution.   By the distance formula

$$|RS| = \sqrt{(4 + 3)^2 + (-2 - 1)^2 + (5 - 0)^2} = \sqrt{83}.$$

11.5   Parallel and Orthogonal Vectors.

DEFINITION.   Vector $\vec{w}$ is *parallel* to vector $\vec{v}$ if for some number c, $\vec{w} = c\vec{v}$.

DEFINITION.   Two sectors, $\vec{v}$ and $\vec{w}$, are *orthogonal* if and only if $\vec{u} \cdot \vec{v} = 0$.

These conditions replace the slope conditions, which are useful only for lines in the plane.

We now prove that in the plane or 3-space, when the basis vectors $\{\vec{i}, \vec{j}, \vec{k}\}$ determine position vectors that are mutually perpendicular and of the same length, two nonzero vectors are orthogonal if and only if their position vectors are perpendicular. To so prove, we first use vector algebra to show that

$$\|\vec{v} - \vec{w}\|^2 = (\vec{v} - \vec{w}) \cdot (\vec{v} - \vec{w}) = \vec{v} \cdot \vec{v} - 2(\vec{v} \cdot \vec{w}) + (\vec{w} \cdot \vec{w}).$$

This can be written

$$\|\vec{v} - \vec{w}\|^2 = \|\vec{v}\|^2 + \|\vec{w}\|^2 - 2(\vec{v} \cdot \vec{w}).$$

We look again at the figure (Figure 8.29(a)). Remembering the law of cosines (§8.7), we see that the law of cosines can be written

$$\|\vec{v} - \vec{w}\|^2 = \|\vec{v}\|^2 + \|\vec{w}\|^2 - 2\|\vec{v}\| \, \|\vec{w}\| \cos \theta,$$

where θ is an angle between the position vectors of $\vec{v}$ and $\vec{w}$. Comparing this with the vector relation above, we find

$$\vec{v} \cdot \vec{w} = \|\vec{v}\| \, \|\vec{w}\| \cos \theta.$$

This implies that if $\vec{u}$ and $\vec{v}$ are nonzero orthogonal vectors, then $\cos \theta = 0$ and hence the position vectors of $\vec{v}$ and $\vec{w}$ are perpendicular.

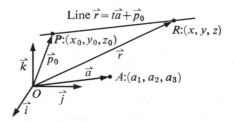

Line $\vec{r} = t\vec{a} + \vec{p}_0$

$P:(x_0, y_0, z_0)$

$R:(x, y, z)$

$\vec{k}$ $\vec{p}_0$ $\vec{r}$

$\vec{a}$ $A:(a_1, a_2, a_3)$

O $\vec{j}$

$\vec{i}$

FIGURE 8.30   Line in 3-space.

We use the same formula to define $\cos \theta$ for higher-dimensional spaces. If $\|\vec{v}\| \neq 0$ and $\|\vec{w}\| \neq 0$,

$$\cos \theta = \frac{(\vec{v} \cdot \vec{w})}{\|\vec{v}\| \, \|\vec{w}\|}.$$

11.6   Equations of a Line. It is customary in 3-space to represent coordinates by rows (x, y, z) instead of (x_1, x_2, x_3). We represent a fixed point by coordinates (x_0, y_0, z_0).

Example. Find the equation of the line in 3-space through the point $P:(x_0, y_0, z_0)$ parallel to vector $\vec{a}:(a, b, c)$.

Solution. Let a variable point in the line be $R:(x, y, z)$. The point R is on the line if and only if the located vector $\overrightarrow{PR}$ is parallel to the position vector $\overrightarrow{OA}$ of $\vec{a}$ (Figure 8.30). That is, for some number t,

$$\overrightarrow{PR} = t\overrightarrow{OA}.$$

Since $\overrightarrow{PR} = \overrightarrow{OR} - \overrightarrow{OP}$, this says in terms of rows that

$$(x - x_0, y - y_0, z - z_0) = t(a, b, c).$$

Equating the three components of these rows tells us that the point $R:(x, y, z)$ is on the line if and only if for some number t

$$x = ta + x_0, \quad \text{and}$$
$$y = tb + y_0, \quad \text{and}$$
$$z = tc + z_0.$$

These are the scalar parametric equations of the line in which the variable t is called the *parameter*. The same equation may be expressed more compactly in vector notation for vectors $\vec{r}:(x, y, z)$ and $\vec{p}_0:(x_0, y_0, z_0)$ as

$$\vec{r} = t\vec{a} + \vec{p}_0.$$

11.7   Equation of a Plane in 3-Space

Example. Find an equation representing the plane containing the point $P_0:(x_0, y_0, z_0)$ and orthogonal to vector $\vec{n}:(a, b, c)$.

Solution. Let a variable point in the plane be $R:(x, y, z)$. Then the point R is in the plane if and only if the located vector $\overrightarrow{P_0R}$ is orthogonal to $\vec{n}$ (Figure 8.31). This is true if and only if

$$a(x - x_0) + b(y - y_0) + c(z - z_0) = 0.$$

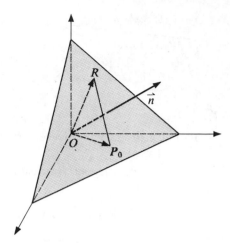

FIGURE 8.31   Plane and normal.

This is the required equation of the plane. The vector $\vec{n}:(a, b, c)$ is called its *normal*.

Example. Show that $2x + 3y - 7z + 5 = 0$ is the equation of a plane and find a normal vector for it.

Solution. By substitution we find that $P_0:(1, 0, 1)$ is a point on the plane, since $2(1) + 3(0) - 7(1) + 5 = 0$. Subtracting, we find that $R:(x, y, z)$ is in the graph if and only if

$$2(x - 1) + 3(y - 0) - 7(z - 1) = 0.$$

This is in the form of the equation of the plane through $P_0:(1, 0, 1)$ with normal $\vec{n}:(2, 3, -7)$.

11.8 Exercises

1. Plot the vectors $\vec{v}:(-1, 2, 0)$ and $\vec{w}:(-1, 3, 4)$ in 3-space. Find $\vec{v} + \vec{w}, 2\vec{v}, \vec{v} - \vec{w}$, $\|\vec{v} - \vec{w}\|$.

2. Without a picture, for vectors $\vec{v}:(3, -2, -2, 4)$ and $\vec{w}:(1, 0, 3, 1)$, find $\vec{v} + \vec{w}, 3\vec{w}$, $\|\vec{w}\|$.

3. For $\vec{v}:(1, 0, 3, 1)$, find c so that $\|c\vec{v}\| = 1$.

4. For $\vec{v}:(x, y, z)$, show that a unit vector in the direction of $\vec{v}$ is $(1/\|\vec{v}\|)\vec{v}$.

5. For $\vec{v}:(2, 1, -1)$ and $\vec{w}:(-1, 3, 4)$, find $\vec{v}\cdot\vec{w}$ and $\cos\theta$, where θ is an angle between the position vectors of $\vec{v}$ and $\vec{w}$.

6. Show that if $\vec{v}$ has the position vector $\overrightarrow{OR}$, and $\|\vec{u}\| = 1$, then $\vec{v}\cdot\vec{u}$ gives the scalar projection of $\overrightarrow{OR}$ on the line of $\vec{u}$.

7. Show from the inner product

$$\left(\frac{\vec{u}}{\|\vec{u}\|} - \frac{\vec{v}}{\|\vec{v}\|}\right)\cdot\left(\frac{\vec{u}}{\|\vec{u}\|} - \frac{\vec{v}}{\|\vec{v}\|}\right)$$

that if $\vec{u} \neq \vec{0}$ and $\vec{v} \neq \vec{0}$, then

$$\vec{u}\cdot\vec{v} \leqq \|\vec{u}\| \, \|\vec{v}\| \text{(Schwarz inequality)}.$$

8. Show that the Schwarz inequality, Exercise 7, holds also when $\|\vec{u}\| = 0$ or $\|\vec{v}\| = 0$.

9. Find the equations of the line through point $P_1:(-1, 3, 4)$ parallel to $\vec{a}:(1, -1, 2)$.
10. Find the equation of the plane through $P_1:(-1, 0, 3)$ with normal $\vec{n}:(2, 3, -2)$.
11. Show that $ax + by + cz = d$ is the equation of a plane with normal $\vec{n}:(a, b, c)$.
12. Find the equations of a line through the point $P_1:(4, -3, 0)$ perpendicular to the plane $2x - 6y + 11z = 8$.
13. Find the equations of the line through the two points $P_1:(3, 1, -1)$ and $P_2:(1, -2, 0)$.
14. Find the equations of the line through the points $P_1:(x_1, y_1, z_1)$ and $P_2:(x_2, y_2, z_2)$.
15. Express the equation of the line in Exercise 14 in vector notation.
16. Find the equations of the line of intersection of the planes, $2x + 3y - 5z = 1$ and $x + y - z = 8$. What vector gives the direction of this line?
17. A vector $\vec{u}$ is called a *unit vector* if $\|\vec{u}\| = 1$. Figure 8.32 shows a unit vector and the angles (α, β, γ) that it makes with the basis vectors. Show that if $\vec{u}:(x, y, z)$ is a unit vector, then $x = \cos\alpha$, $y = \cos\beta$, $z = \cos\gamma$. These three numbers are called direction cosines.

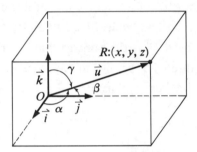

FIGURE 8.32   Exercise 17.

18. Derive the properties of inner products (§11.3, Theorem).
19. Prove that in Euclidean 3-space the distance $|RS|$ (Figure 8.29) between $R:(x_1, x_2, x_3)$ and $S:(y_1, y_2, y_3)$ is given by

$$|RS| = \sqrt{(y_1 - x_1)^2 + (y_2 - x_2)^2 + (y_3 - x_3)^2}.$$

20. Find the parametric equations

$$x = at + x_0, \qquad y = bt + y_0$$

of the line in 2-space given by

$$2x - 3y - 5 = 0.$$

21. In 2-space, the idea of the line and plane coincide. Regarding $ax + by + c = 0$ as a plane, find the normal vector to it. Regarding it as a line, find the vector giving its direction, and a vector equation for it.

12 Orthogonal Projection

12.1 Projection of a Line Segment on a Line

Example. Find the scalar projection of the located vector $\overrightarrow{PQ}$ of $\vec{v}$ on the line of the unit vector $\vec{u}$ (Figure 8.33(a)).

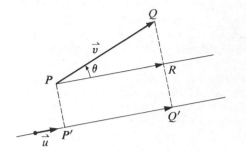

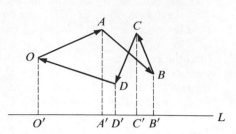

(a) Projection of a line segment (b) Projection of a closed polygon

FIGURE 8.33 Projection.

Solution. The projection $\overrightarrow{P'Q'}$ of $\overrightarrow{PQ}$ on the line of $\vec{u}$ is determined geometrically by the points P' and Q', which we find by drawing $\overrightarrow{PP'}$ and $\overrightarrow{QQ'}$ orthogonal to the line of $\vec{u}$. The vector determined by $\overrightarrow{P'Q'}$ is the same as that determined by $\overrightarrow{PR}$, which is the projection of $\overrightarrow{PQ}$ on a line through P parallel to $\vec{u}$. Hence we treat both $\vec{u}$ and $\vec{v}$ as acting on P, and find run $\overrightarrow{PR} = |PR| \cos \theta$, where run $\overrightarrow{PR}$ is the distance from P to R with appropriate sign.

We recall that $\vec{v} \cdot \vec{u} = \|\vec{v}\| \, \|\vec{u}\| \cos \theta$, and since $\|\vec{u}\| = 1$ and $\|\vec{v}\| = |PR|$, we have (§12.5)

$$\text{Proj } \overrightarrow{PQ} = \vec{v} \cdot \vec{u}.$$

Here Proj $\overrightarrow{PQ}$ = run $\overrightarrow{PR}$, and is the scalar length of the projection of $\overrightarrow{PQ}$ on the line of u, positive if it has the same orientation as u, negative if it has the opposite orientation. This solves the problem.

Example. Find the projection of the position vector $\overrightarrow{OR}$ of $\vec{v}:(2, -3, 1)$ on the line of $\vec{w}:(-1, 1, -4)$.

Solution. We find the unit vector

$$\frac{\vec{w}}{\|\vec{w}\|} : \left(-\frac{1}{\sqrt{18}}, \frac{1}{\sqrt{18}}, -\frac{4}{\sqrt{18}} \right)$$

parallel to $\vec{w}$. Then the projection of $\overrightarrow{OR}$ on the line of $\vec{w}$ is

$$\text{Proj } \vec{v} = \vec{v} \cdot \frac{\vec{w}}{\|\vec{w}\|} = \frac{2(-1) - 3(1) + 1(-4)}{\sqrt{18}} = -\frac{3}{\sqrt{2}}.$$

The minus sign indicates that Proj $\vec{v}$ runs in the opposite orientation to $\vec{w}$.

12.2 Projection of a Sum of Vectors. The projection on the line of unit vector $\vec{u}$ of the sum $\vec{v} + \vec{w}$ is given by

$$\text{Proj } (\vec{v} + \vec{w}) = (\vec{v} + \vec{w}) \cdot \vec{u} = \vec{v} \cdot \vec{u} + \vec{w} \cdot \vec{u} = \text{Proj } \vec{v} + \text{Proj } \vec{w}.$$

That is, the projection of the sum of two vectors is the sum of their projections.

We can apply this to any directed closed polygon (Figure 8.33(b)). Since the final point coincides with the initial point, the sum of the vectors themselves is zero. Hence the sum of their projections on any line L is also zero.

12.3   Vector Projections.   Given a vector $\vec{v}$ and a unit vector $\vec{u}$, we seek to find two orthogonal vectors $\vec{w}_1$ and $\vec{w}_2$ such that $\vec{v} = \vec{w}_1 + \vec{w}_2$ and $\vec{w}_1$ is parallel to $\vec{u}$ (Figure 8.34). The scalar projection of $\vec{v}$ on the line of $\vec{u}$ is $\vec{v} \cdot \vec{u}$. Hence a vector in the line of $\vec{u}$ having this length and direction is the scalar product $(\vec{v} \cdot \vec{u})\vec{u}$. This is $\vec{w}_1$. To find $\vec{w}_2$, we simply subtract $\vec{w}_1$ from $\vec{v}$. Hence, we find two orthogonal components of $\vec{v}$ with $\vec{w}_1$ in the chosen direction $\vec{u}$ to be

$$\vec{w}_1 = (\vec{v} \cdot \vec{u})\vec{u} \qquad \text{and} \qquad \vec{w}_2 = \vec{v} - (\vec{v} \cdot \vec{u})\vec{u}.$$

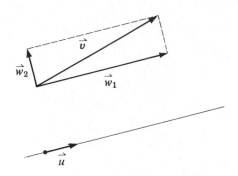

FIGURE 8.34   Orthogonal components.

Example A.   Resolve the vector $\vec{v}:(3, -1, 4)$ into two orthogonal components, one in the plane $2x - 3y + 4z = 7$ and one orthogonal to this plane.

Solution.   This is a variation of the procedure above. The coefficient vector $(2, -3, 4)$ is orthogonal to the plane. It has the unit normal

$$n: \left(\frac{2}{\sqrt{29}}, \frac{3}{\sqrt{29}}, \frac{4}{\sqrt{29}} \right).$$

The scalar projection of $\vec{v}$ on the unit normal is

$$\vec{v} \cdot \vec{n} = \frac{(3)(2) - 1(-3) + 4(4)}{\sqrt{29}} = \frac{25}{\sqrt{29}}.$$

Then the vector projection of $\vec{v}$ orthogonal to the plane is

$$\vec{w}_1 = \frac{25}{\sqrt{29}} \, \vec{n}: \left(\frac{50}{29}, \frac{75}{29}, \frac{100}{29} \right).$$

To find the component $\vec{w}_2$ parallel to the plane we subtract $\vec{w}_1$ from $\vec{v}$. We find $\vec{w}_2:\left(\frac{37}{29}, \frac{21}{29}, \frac{16}{29}\right)$.

Example B.   Find the perpendicular distance d from the plane $ax + by + cz - k = 0$ to the point $P:(x_0, y_0, z_0)$.

Solution. The plane has the normal (a, b, c) and the unit normal $\bar{n}:(a/r, b/r, c/r)$, where $r = \pm \sqrt{a^2 + b^2 + c^2}$. Let $\bar{v}:(x, y, z)$ give a position vector $\overrightarrow{OR}$ to the point $R:(x, y, z)$ in the plane (Figure 8.35). The projection of the located vector $\overrightarrow{RP}:(x_0 - x, y_0 - y, z_0 - z)$ on the normal gives the required distance d. In fact,

$$d = \frac{a}{r}(x_0 - x) + \frac{b}{r}(y_0 - y) + \frac{c}{r}(z_0 - z).$$

But, since $R:(x, y, z)$ is in the plane, $ax + by + cz = k$. Hence the expression for d reduces to

$$d = \frac{ax_0 + by_0 + cz_0 - k}{r}.$$

Arbitrarily we choose the sign of r so that $k/r \geqq 0$. This has the effect of pointing the normal from the origin to the plane.

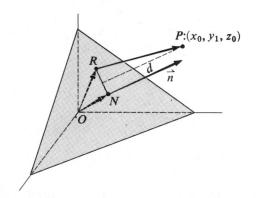

FIGURE 8.35   Distance of point from plane.

12.4   Normal Form of Equation of a Line.   In the xy-plane the equation of a line is the plane equation $ax + by = c$. The unit normal to this line is the vector $\bar{n}:(a/r, b/r)$, where $r = \pm \sqrt{a^2 + b^2}$ (Figure 8.36). In this case, we recognize that the components of the unit normal are $(\cos \alpha, \sin \alpha)$, where α is the angle of inclination of the normal. We have seen (§12.3, Example B) that the perpendicular distance of a point $R:(x, y)$ from this "plane" is given by

$$d = \frac{ax + by - c}{r},$$

where we choose the sign of r so that we obtain the normal intercept $p = c/r \geqq 0$. The point $R:(x, y, z)$ is in the line if and only if the distance $d = 0$, that is, if

$$x \cos \alpha + y \sin \alpha - p = 0.$$

This is known as the *normal form* of the equation of the line, since it is expressed in terms of the normal intercept p and the inclination α of the normal. The distance d of any point $P:(x_0, y_0)$ from the line is then given by

$$d = x_0 \cos \alpha + y_0 \sin \alpha - p.$$

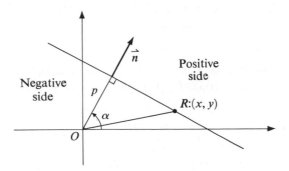

FIGURE 8.36   Normal form of a line.

Example.   Find the distance from the line $4x + 3y + 7 = 0$ to the point $P:(-5, 6)$.

Solution.   To transform the equation to normal form, we divide by $-\sqrt{4^2 + 3^2} = -5$, choosing the sign to be opposite that of the term 7. Then the normal form is

$$-\tfrac{4}{5}x - \tfrac{3}{5}y - \tfrac{7}{5} = 0.$$

The distance from the line to the point $P:(-5, 6)$ is

$$d = \frac{-4(-5) - 3(6) - 7}{5} = -1.$$

The minus sign means that P is on *the same side of the line as* the origin.

12.5   **Line of Intersection of Two Planes.**   We review here a special case of solving simultaneously two linear equations.

Example.   Find the equation of the line of intersection in 3-space of the two planes

$$2x_1 - 4x_2 + x_3 = 12 \quad \text{and} \quad x_1 - x_3 = 4.$$

Solution.   We use the method of elimination to solve simultaneously the system

$$x_1 + 0x_2 - x_3 = 4, \quad \text{and}$$
$$2x_1 - 4x_2 + x_3 = 12.$$

The method of elimination makes use of the following elementary row operations. The system is logically equivalent to any one obtainable from it by (1) interchanging two rows; (2) multiplying one row by a nonzero number; (3) replacing any row by one obtained by adding to it a linear combination of any of the others.

Here we eliminate x_1 from the second equation by adding to it -2 times the first. We have

$$- x_1 + 0x_2 - x_3 = 4, \quad \text{and}$$
$$0x_1 - 4x_2 + 3x_3 = 4.$$

To make the coefficient of x_2 into 1, we multiply the second row by $-\tfrac{1}{4}$. The system then reads

$$x_1 + 0x_2 - x_3 = 4, \quad \text{and}$$
$$0x_1 + x_2 - \tfrac{3}{4}x_3 = -1.$$

The unknowns x_1 and x_2 now appear with their coefficients in the diagonal array

$$\begin{pmatrix} 1 & 0 \\ 0 & 1 \end{pmatrix}.$$

Taking advantage of this "reduced echelon form," we find that we can set $x_3 = t$ and read off values of x_1 and x_2 in terms of t. This gives us all solutions of the original system

$$\begin{aligned} x_1 &= & t + 4, & \quad \text{and} \\ x_2 &= +\tfrac{3}{4}t - 1, & & \quad \text{and} \\ x_3 &= & t, & \end{aligned}$$

for all values of t. These are the parametric equations of a line (§11.6). They represent a line through the point $P_0:(4, -1, 0)$ in the direction given by $\vec{v}:(1, \tfrac{3}{4}, 1)$.

12.6 Exercises

1. Find the scalar projection of the position vector $\overrightarrow{OR}:(-1, 3, 5)$ on the line of $\vec{w}:(-2, 1, 2)$.
2. Resolve the vector $\vec{v}:(-2, 0, 1)$ into two orthogonal components, with one parallel to $\vec{w}:(-2, 1, 2)$.
3. Resolve the vector $\vec{v}:(-2, 0, 1)$ into two components, one orthogonal to the plane $2x - y - 2z = 6$, and one parallel to it.
4. Find the distance from the line $2x - 3y + 6 = 0$ to the point $(-1, 4)$.
5. Find the distance between the parallel lines $2x + 3y - 8 = 0$ and $4x + 6y + 11 = 0$.
6. Find the area of the triangle $(2, -3), (1, 4), (-3, 5)$.
7. Find the normal equation, inclination of normal, and normal intercept of the line joining the points $(2, -1)$ and $(3, 0)$.
8. Find all vectors orthogonal to both $\vec{u}:(-1, 2, 2)$ and $\vec{v}:(4, -1, 0)$.
9. Find an equation of the plane containing the three points $P:(1, 0, 0)$, $Q:(-1, 1, 1)$ and $R:(0, 2, -1)$.
10. Find the equation of the line in 2-space with normal vector $\vec{p}:(m, -1)$ and containing the point $(0, b)$.
11. Find *some* plane in 3-space containing the three points $(2, -1, 3)$, $(4, 2, -1)$ and $(8, 8, -9)$. What is the source of the difficulty?
12. Find the area of the triangle joining the points $A:(2, 0, 0)$, $B:(0, 3, 0)$, and $C:(0, 0, 5)$.
13. For the three points in Exercise 12, find the projection of the polygon ABC on the line of the vector $\vec{v}:(1, 1, -1)$.
14. Find the coordinates of the point where the line $x = 2t - 1, y = -6t + 4, z = -2t + 4$ intersects the plane $3x - y + 4z - 5 = 0$.
15. Find the angle between the planes $2x - 3y + 6z = 0$ and $x + 2y - 8z = 4$.
16. Find the vector projection on the plane $2x - 3y + 6z = 0$ of the line segment $\overrightarrow{PQ}$, $P:(1, -2, 3)$, $Q:(4, 1, 5)$.
17. A line has the vector equation $\vec{r} = \vec{a}s + \vec{b}$, where $\vec{a}$ is a unit vector. Prove that s is the distance from the point $\vec{b}$ to the point $\vec{r}$.
18. A triangle in the plane $x - 2y + 2z = 8$ has area 5. What are the areas of the projections of this triangle on the xy-plane? on the yz-plane? on the zx-plane?
19. A triangle in 3-space has projections on the coordinate planes whose areas are a_1, a_2, a_3. What is the area of the triangle?
20. What is the graph in xyz-space of the following set of parametric equations with two parameters, s and t?

$$\begin{aligned} x &= 2s - 3t + 4, & \quad \text{and} \\ y &= s + t - 1, & \quad \text{and} \\ z &= -s + 2t + 5. \end{aligned}$$

21.   Show that the system of equations

$$2x_1 + 3x_2 - 6x_3 + 8x_4 = 8 \quad \text{and}$$
$$x_1 - x_2 + x_3 - x_4 = 1$$

intersect in a plane in $x_1x_2x_3$-space.

22.   Find all solutions of the equations

$$x + y - z = 5, \quad \text{and}$$
$$3x - y + 4z = 2, \quad \text{and}$$
$$9x + y + 5z = 11.$$

What does the result mean about the intersections of the planes represented by these equations?

23.   Find all vectors $\vec{n}:(x, y, z)$ that are orthogonal to both $\vec{v}_1:(a_1, b_1, c_1)$ and $\vec{v}_2:(a_2, b_2 c_2)$. Show that all of the vectors $\vec{n}$ have the form $\vec{n}:s[(b_1c_2 - b_2c_1), -(a_1c_2 - a_2c_1), (a_1b_2 - a_2b_1)]$, where s is any scalar.

24.   Show that the three planes,

$$2x + 3y - 6z = 0, \quad \text{and}$$
$$x + 12y + 3z = 0, \quad \text{and}$$
$$x - 2y - 5z = 0, \quad \cdot$$

intersect in a common line. Find the direction and equations of this line of intersection.

25.* Find the minimum distances between the lines $x = 2t + 3$, $y = -2t + 1$, $z = 5t + 5$ and $x = -t + 1$, $y = 3t - 5$, $z = -t + 6$.

26.   Draw a graph and represent by a shaded area all points $P:(x, y)$ in the plane where $2x - 3y \geq 1$ and $x + 5y \geq 3$ and $3x + 4y - 6 \geq 0$.

13   Graphs of Relations

ACADEMIC NOTE ON SCALAR AND VECTOR METHODS.   Having set up the isomorphisms between spaces of points, position vectors, vectors and coordinate rows, we can choose either methods that emphasize vectors or methods that emphasize points and their coordinates. The latter, scalar, approach is a direct continuation of the methods of coordinate geometry with which the student is familiar. In the following sections we shall emphasize scalar methods but we shall often translate results into vector language. We shall also use vector methods themselves when they are advantageous, as in the case of directions, velocities, acceleration, and forces. Our general purpose is to make a gradual transition to vector methods.

13.1   Relations and Functions in the Plane.

Denoting by $\mathscr{R}^2$ the set of all rows of two real numbers (x, y), we define a *relation* in $\mathscr{R}^2$ as any subset of $\mathscr{R}^2$. We shall be concerned with a special kind of relation, one that is determined by a function $F(x, y)$ of the two variables x and y. For any constant C, the equation

$$F(x, y) = C$$

determines a relation; it selects a set of pairs (x, y) for which the equation is true and rejects the complementary set of pairs for which the equation is false. We regard this

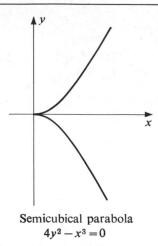

Semicubical parabola
$4y^2 - x^3 = 0$

FIGURE 8.37   Semicubical parabola, $4y^2 - x^3 = 0$.

equation as an open statement involving the values $F(x, y)$ of the function $F(x, y)$. The *graph of the relation* $F(x, y) = C$ is the set of all points $P:(x, y)$ whose coordinates satisfy the equation.

For example,

$$4y^2 - x^3 = 0$$

is a relation whose graph is a curve in the plane (Figure 8.37).

All functions of one variable $f(x)$ can be expressed as relations

$$y - f(x) = 0,$$

so if we study a calculus of relations we are including the calculus of functions of one variable (Part I, *Calculus of Elementary Functions*). On the other hand, some relations in the plane are not functions of x. For example, from the graph of the relation $4y^2 - x^3 = 0$, we see that for a positive value of x, the relation does not select a single value of y, as required for a function of x. Instead, it selects two values of y for each positive x,

$$y = +\tfrac{1}{2}x^{3/2} \quad \text{and} \quad y = -\tfrac{1}{2}x^{3/2}.$$

The graph of the relation consists of the union of the graphs of the solution functions, $\tfrac{1}{2}x^{3/2}$ and $-\tfrac{1}{2}x^{3/2}$.

Since the solution functions are difficult to obtain, both in theory and in practice, and are often more complicated in algebraic form than the function $F(x, y)$ that determines them, it is desirable that we apply calculus methods directly to the relation $F(x, y) = C$ without finding the solution functions.

We exhibit a catalogue of graphs for standard relations in the plane (Figure 8.38).

We use the same type of graph to represent relations $F(x, y, z) = C$ in three variables. The graph is a surface in $\mathscr{R}^3$. Such 3-dimensional graphs are difficult to plot and portray on plane paper. So analytic geometry in 3-space must devise methods for representing such graphs. One way is to use a parallel perspective projection, a simplified form of the method used by artists. We list some basic principles in this technique.

$(x-a)^2 + (y-b)^2 = r^2$

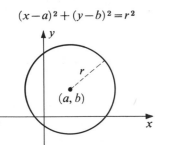

$y^2 - 4px = 0$

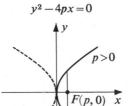

$x^2 - 4py = 0$

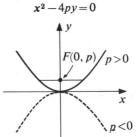

(a) Circle

(b) Parabola

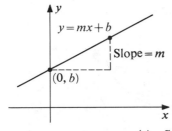

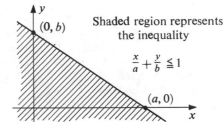

Shaded region represents the inequality

$$\frac{x}{a} + \frac{y}{b} \leqq 1$$

(c) Straight lines

$$\frac{x^2}{a^2} + \frac{y^2}{b^2} = 1 \qquad \begin{array}{c} a > b \\ c = \sqrt{a^2 - b^2} \end{array} \qquad \frac{y^2}{a^2} + \frac{x^2}{b^2} = 1$$

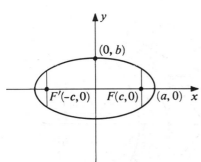

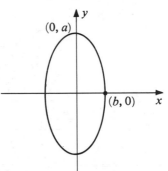

(d) Ellipse

$$\frac{x^2}{a^2} - \frac{y^2}{b^2} = 1 \qquad c = \sqrt{a^2 + b^2} \qquad \frac{y^2}{a} - \frac{x^2}{b^2} = 1$$

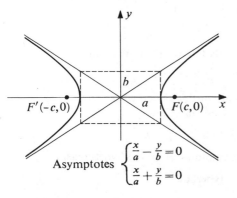

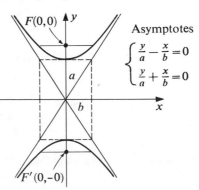

Asymptotes $\begin{cases} \dfrac{y}{a} - \dfrac{x}{b} = 0 \\ \dfrac{y}{a} + \dfrac{x}{b} = 0 \end{cases}$

(e) Hyperbola

(*continued*)

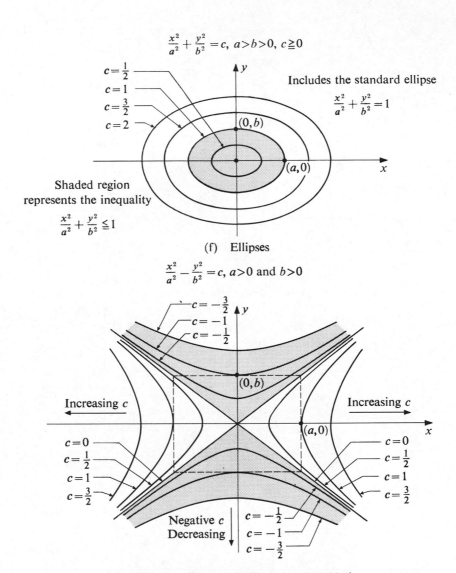

$$\frac{x^2}{a^2} + \frac{y^2}{b^2} = c, \ a>b>0, \ c \geqq 0$$

Includes the standard ellipse

$$\frac{x^2}{a^2} + \frac{y^2}{b^2} = 1$$

$c = \frac{1}{2}$

$c = 1$

$c = \frac{3}{2}$

$c = 2$

$(0,b)$

$(a,0)$

Shaded region represents the inequality

$$\frac{x^2}{a^2} + \frac{y^2}{b^2} \leqq 1$$

(f)   Ellipses

$$\frac{x^2}{a^2} - \frac{y^2}{b^2} = c, \ a>0 \text{ and } b>0$$

$c = -\frac{3}{2}$

$c = -1$

$c = -\frac{1}{2}$

$(0,b)$

Increasing c Increasing c

$(a,0)$

$c = 0$ $c = 0$

$c = \frac{1}{2}$ $c = \frac{1}{2}$

$c = 1$ $c = 1$

$c = \frac{3}{2}$ $c = \frac{3}{2}$

Negative c $c = -\frac{1}{2}$
Decreasing $c = -1$
              $c = -\frac{3}{2}$

Shaded region represents the inequality relation

$$\frac{x^2}{a^2} - \frac{y^2}{b^2} \leqq 0$$

Observe that the boundary of this region where $c=0$ consists of two straight lines

$$\frac{x}{a} - \frac{y}{b} = 0 \text{ and } \frac{x}{a} + \frac{y}{b} = 0.$$

These are the common asymptotes of all the hyperbolas.

(g)   Hyperbolas

FIGURE 8.38   Standard relations of the plane.

13.2   Perspective Representation of Coordinate Systems

(1)   Place the z-axis vertical and the y-axis horizontal. The x-axis is drawn at about 60 degrees to the negative end of the y-axis, representing a perspective view of the x-axis perpendicular to the yz-plane with its positive direction toward the viewer. The viewer is then in the first octant where all three coordinates (x, y, z) are positive (Figure 8.39(a)). It is as if the viewer is in a room where the lower front left corner is the origin, the z-axis is the vertical through it, the front wall is the yz-plane, the floor is the xy-plane, and the left wall is the xz-plane.

Lines and curves visible to the viewer in this position are drawn continuously black. Lines behind the coordinate planes or behind other parts of the figure are dotted.

(2)   Choose equal units on the y-axis and z-axis but represent the same unit as half as long on the x-axis, due to the perspective.

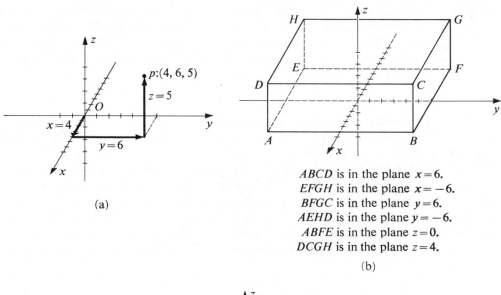

(a)

$ABCD$ is in the plane $x=6$.
$EFGH$ is in the plane $x=-6$.
$BFGC$ is in the plane $y=6$.
$AEHD$ is in the plane $y=-6$.
$ABFE$ is in the plane $z=0$.
$DCGH$ is in the plane $z=4$.

(b)

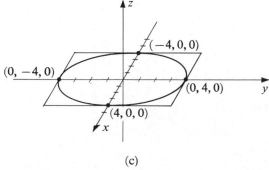

(c)

FIGURE 8.39   Perspective view of coordinate system in 3-space. (a) Coordinate axes and point $(4, 6, 5)$, (b) Planes parallel to coordinate planes, (c) Perspective view of circle $x^2 + y^2 = 16$ in plane $z = 0$.

(3) Straight lines appear in perspective as straight lines and parallel lines remain parallel.

(4) Tangents appear in perspective as tangents.

(5) Equal distances along parallel lines appear in perspective as equal distances. But line segments of equal length on divergent lines do not ordinarily appear equal in perspective. We observe this in the laying of the measurements to locate the point (4, 6, 5) (Figure 8.39(a)).

(6) Equal angles do not ordinarily appear equal in perspective. This occurs in the right angle between the positive rays of the x-axis and y-axis, which is depicted as 120 degrees. Similarly the right angle between the negative rays of the x- and y-axes appears as 60 degrees (Figure 8.39(a)).

The set of all points satisfying the equation $x = 6$ is a plane parallel to the yz-plane. For every constant C, the graph of $x = C$ is a plane parallel to the yz-plane, $x = 0$. Similarly $y = 6$ and $y = -6$ are planes parallel to the xz-plane, $y = 0$. Similarly $z = 4$ is a plane parallel to the xy-plane, $z = 0$. We picture these planes as faces of a rectangular box (Figure 8.39(b)). We use Principle (5) to plot the vertices of the box.

We use the principle (4), that tangents appear as tangents, to sketch the perspective view of the circle $x^2 + y^2 = 16$ in the plane $z = 0$. The x-axis and the y-axis contain perpendicular diameters of this circle. The tangents at $(0, 4, 0)$ and $(0, -4, 0)$ on this circle must be parallel to the x-axis. Similarly, the tangents at $(4, 0, 0)$ and $(-4, 0, 0)$ must be parallel to the y-axis. We use these lines as guides and sketch the circle as tangent to them at the four plotted points (Figure 8.39(c)).

13.3 Cylinders with Axes Parallel to the Coordinate Axes. A cylinder in n-space is determined by a directrix curve C and a generator line G that intersects C. The cylinder having directrix C and generators parallel to G is the set of all points that are on lines that intersect C and are parallel to G. We illustrate by drawing a perspective figure of the familiar right circular cylinder with directrix $x^2 + y^2 = 1$ in the plane $z = 0$ and generators that are parallel to the z-axis (Figure 8.40). Since the surface is infinite in extent along the vertical generators, we must cut off the picture somewhere and we choose to do this between two planes parallel to the xy-plane. The picture (Figure 8.40) is therefore only a partial graph, suggestive of the infinitely long cylindrical surface.

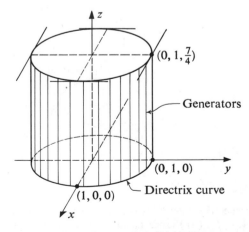

FIGURE 8.40 Cylinder $x^2 + y^2 = 1$ cut off by planes, $z = 0$ and $z = \frac{7}{4}$.

We observe that when the cylinder is cut by any plane $z = C$ parallel to the plane of the directrix, the curve of intersection is simply a translation of the directrix circle $x^2 + y^2 = 1$, $z = 0$. The equation of the intersection is the simultaneous system $x^2 + y^2 = 1$, and $z = C$. If (x, y) satisfies $x^2 + y^2 = 1$, then for every number z, the point (x, y, z) in $\mathscr{R}^3$ is in the cylindrical surface. That is, if (x, y) is fixed in the relation $x^2 + y^2 = 1$ and z varies over all real numbers, the points (x, y, z) fill up the generator line through $(x, y, 0)$. Thus we regard the single equation

$$x^2 + y^2 = 1$$

as the equation of a cylinder with generators parallel to the z-axis, and with the directrix the circle determined by the simultaneous system $x^2 + y^2 = 1$ and $z = 0$. We can express this idea in set-builder notation by saying that the cylinder $x^2 + y^2 = 1$ in $\mathscr{R}^3$ is the set $\{(x, y, z) \in \mathscr{R}^3 : x^2 + y^2 = 1\}$.

In general, $F(x, y) = C$ in $\mathscr{R}^3$ is an equation of a cylinder with directrix $F(x, y) = C$, $z = 0$, and generators parallel to the z-axis. Similarly, $F(y, z) = C$ in $\mathscr{R}^3$ is an equation of a cylinder with directrix $F(y, z) = C$, $x = 0$, and generators parallel to the x-axis.

13.4   Vertical Sections of $z = f(x, y)$

Example.   Sketch the graph in $\mathscr{R}^3$ of the surface $z = 12 - (x^2/4) - (y^2/3)$ on an xy-domain, which is the rectangular region defined by $|x| \leq 4$ and $|y| \leq 3$.

Solution.   We reduce the problem to one of plotting functions of one variable by choosing successively several fixed values of x and plotting the resulting curves in parallel vertical planes. A typical one is the vertical plane section with $x = 0$ and $z = 12 - (y^2/3)$, which is a parabola, and we plot this parabola on the interval $|y| \leq 3$ (Figure 8.41, Parabola in the plane ABC). Similarly, we plot parabolas in planes $x = -4$, $x = -3, \ldots, x = 4$. The last one is $x = 4$ and $z = 8 - (y^2/3)$ (Figure 8.41, Parabola in plane HJK).

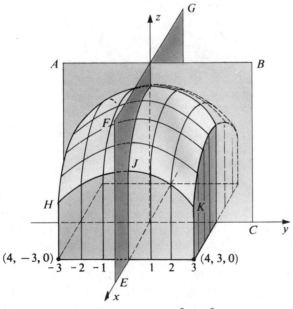

FIGURE 8.41   Graph of $z = 12 - \dfrac{x^2}{4} - \dfrac{y^2}{3}$, $|x| \leq 4$ and $|y| \leq 3$.

Then we fix values of y, giving us a series of vertical planes $y = -3$, $y = -2$, ..., $y = 3$, which are at right angles to the planes with $x = c$. A typical one is the parabola $y = 0$ and $z = 12 - (x^2/4)$ for $|x| \leq 3$ (Figure 8.41, Parabola in plane *EFG*). When these two series of vertical sections are plotted, we have a net of curves in the surface that creates a perspective view of the whole surface. This completes the problem.

13.5 Graphs of Relations $F(x, y, z,) = C$ in $\mathcal{R}^3$

Example. Sketch the graph of $x^2 + y^2 - z^2 = 1$ in $\mathcal{R}^3$.

Solution. We plot (Figure 8.42) plane sections on each of the three coordinate planes. These are called *traces*. The xy-trace is the circle $z = 0$ and $x^2 + y^2 = 1$, which we sketch. The yz-trace is the hyperbola $x = 0$ and $y^2 = z^2 = 1$. The xz-trace is the hyperbola $y = 0$ and $x^2 - z^2 = 1$.

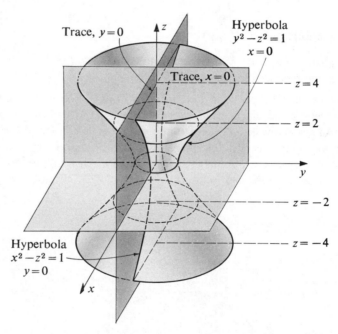

FIGURE 8.42   Hyperboloid of one sheet, $x^2 + y^2 - z^2 = 1$.

Then we plot a series of plane sections for different values of z. These are called level curves. We find that they are the circles $z = -4$ and $x^2 + y^2 = 17$; $z = -2$ and $x^2 + y^2 = 5$; $z = 2$ and $x^2 + y^2 = 5$; and $z = 4$ and $x^2 + y^2 = 17$. When these level curves in planes $z = c$ are sketched along with the two vertical traces for $x = 0$ and $y = 0$, we find that we have constructed a perspective view of the surface. The surface is called a hyperboloid of one sheet. This completes the problem.

13.6 Standard Surfaces Given by Relations in Cartesian Coordinates.
We exhibit a collection of surfaces (Figure 8.43) given by relations of the form $F(x, y, z) = C$, particularly the so-called "quadric surface" determined by certain standard second-degree polynomial forms $F(x, y, z)$.

$$(x_1 - a_1)^2 + (x_2 - a_2)^2 + (x_3 - a_3)^2 = r^2$$

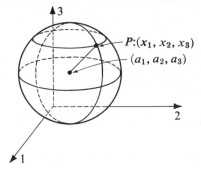

(a)   Sphere

$$\frac{x^2}{a^2} + \frac{y^2}{b^2} + \frac{z^2}{c^2} = 1$$

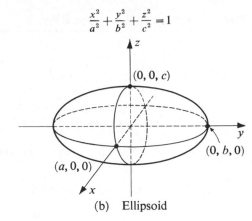

(b)   Ellipsoid

$$\frac{x^2}{a^2} + \frac{y^2}{b^2} - 4pz = 0$$

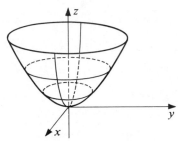

(c)   Elliptic paraboloid
The curves $z = $ const are ellipses.

$$\frac{x^2}{a^2} - \frac{y^2}{b^2} + \frac{z^2}{c^2} = 1$$

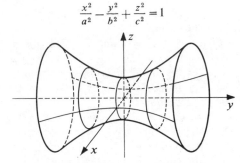

(d)   Hyperboloid (See also Fig. 2.42)
(One negative sign in quadratic form)

$$\frac{x^2}{a^2} + \frac{y^2}{b^2} - \frac{z^2}{c^2} = 1$$

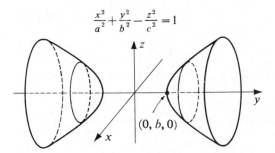

(e)   Hyperboloid
(Two negative signs in quadratic form)
Sections $x = $ const and $z = $ const are hyperbolas.
Sections $y = k$ $(k \geq b)$ are ellipses.

$$\frac{x^2}{a^2} - \frac{y^2}{b^2} + 4pz = 0$$

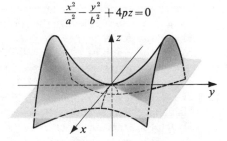

(f)   Hyperbolic paraboloid
A saddle surface made up
of straight lines.

$$\frac{y^2}{a^2} + \frac{z^2}{c^2} = 1$$

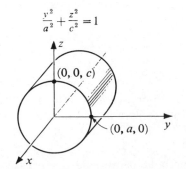

(g)   Cylinder (See also Fig. 2.40)

$$\frac{x^2}{a^2} - \frac{y^2}{b^2} + \frac{z^2}{c^2} = 0$$

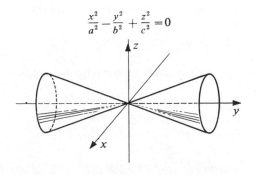

(h)   Cone

FIGURE 8.43   Standard surfaces in $\mathscr{R}^3$.

13.7 Exercises

1. Draw a perspective view of a Cartesian 3-space coordinate system and plot the points:
   (a) $(5, 4, 6)$;   (b) $(5, -4, 6)$;   (c) $(5, 4, -6)$;   (d) $(-5, 4, 6)$.
2. In a Cartesian coordinate system for $\mathscr{R}^3$, plot the points $(4, 2, 0)$, $(4, 2, 1)$, $(4, 2, 3)$, $(4, 2, 10)$, $(4, 2, -1)$, $(4, 2, -5)$. In general, what is the location of $(4, 2, z)$ for any number z?
3. Sketch a box whose faces lie in one of the planes $x = 6$, $x = -6$, $y = 0$, $y = 4$, $z = -1$, $z = 3$.

In Exercises 4–14, sketch in Cartesian 3-space coordinates the graphs of the relations given.

4. The relations $y = -1$, $y = 0$, $y = 1$, $y = 2$ as separate graphs in the same coordinate system. In words, what do these equations represent?
5. The plane $y = x$ in $\mathscr{R}^3$.
6. The planes $y = mx$ in $\mathscr{R}^3$ for $m = 0, 1, 2, 5, 10$.
7. The curves:
   (a) $x^2 + z^2 = 4$ and $y = 0$;   (b) $x^2 + z^2 = 4$ and $y = 3$;   (c) $x^2 + z^2 = 4$ and $y = -3$.
8. The line in $\mathscr{R}^3$ given by the system $y = x$ and $x + y = 2$.
9. The parabola given by the system $z = x^2$ and $y = 4$.
10. The cylinders:
    (a) $y^2 + z^2 = 1$;   (b) $x^2 + z^2 = 1$, as separate graphs.
11. The graph of $x + y = 2$ in $\mathscr{R}^3$.
12. The parabolic cylinder $y = x^2$.
13. The parabolic cylinder $z = y^2$.
14. The cylinder $z = y^3$.

15. Plot the line through $(4, 6, -2)$ that is parallel to the direction $\vec{a}:(3, 3, 3)$.
16. Sketch the graph of the relation $x^2 + y^2 - z^2 = 0$, using level curves $z = c$ and vertical sections $y = mx$. Identify the surface by its geometric name.
17. Sketch the graph of $y^2 - 4x = z$, using level curves $z = c$ and vertical sections $y = mx$.
18. Sketch the graph of $y^2 + z^2 = 4$. Identify by geometric name.
19. Sketch the graph of $x^2 + y^2 = e^z$ in perspective, using level curves. What happens as z approaches $-\infty$?
20. Sketch in perspective the graph of $x^2 - y^2 + z^2 = 1$.
21. Using level curves and vertical sections, sketch the graph of $z = x^2 + y^2 + 4$.
22. Sketch the graph in $\mathscr{R}^3$ of $x^2 - 2y^2 = 0$.
23. Sketch the graph in $\mathscr{R}^3$ of $(x^2 + y^2 - 1)(x^2 + z^2 - 1) = 0$.
24. Show by a graph in $\mathscr{R}^3$ all simultaneous solutions of

$$x + y + z = 4, \quad \text{and}$$
$$x - y \quad\;\; = 0.$$

25. Sketch in $\mathscr{R}^3$ the plane containing the position vectors of $\vec{u}:(4, 2, 1)$ and $\vec{v}:(1, 0, 0)$.

14 Contour Representation of Surfaces

14.1 Perspective Graphs and Contour Lines. As a means of avoiding the difficulties of perspective projections of 3-dimensional figures we now focus our attention on the level contours of the graph.

Example. Sketch a perspective graph of the hyperbolic paraboloid $z = -xy$.

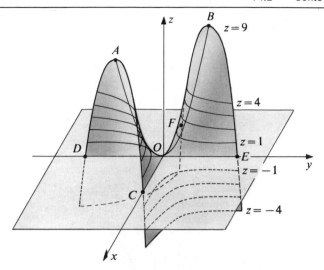

FIGURE 8.44   Saddle point of hyperbolic paraboloid. (Compare with Figure 8.43(f).)

Solution.   The horizontal sections are the hyperbolas $-xy = c$ with asymptotes parallel to the x- and y-axes in the planes $z = c$. The xy-trace consists of the x- and y-axes themselves; for when we set $z = 0$, we get $-xy = 0$, which is equivalent to $x = 0$ or $y = 0$ (Figure 8.44).

The trace in the plane $x = 0$ is the y-axis; for when we set $x = 0$, we find that $z = 0$, and the two planes $x = 0$ and $z = 0$ define the y-axis. Similarly, the trace in the plane $y = 0$ is the x-axis.

These two vertical sections added no new information, so we find another vertical section. We choose $y = -x$ and get the parabola $z = x^2$ in the plane $y = -x$ (Figure 8.44). At right angles to this vertical plane section, we find other plane sections $x - y = c$. These are all parabolas $x - y = c$ and $z = -x^2 + cx$. Typical sections in this series are $x - y = 6$, $z = -x^2 + 6x$ (Figure 8.44, Parabola in plane CAD) and also $x - y = -6$, $z^2 = -x^2 - 6x$ (Figure 8.44, Parabola in plane EBF). When we plot these vertical sections together with the level lines $z = c$ using the principles of perspective plotting (§13.2), they give us a perspective view of the graph of this surface showing its saddle shape in the neighborhood of the origin $(0, 0, 0)$. It is called a *hyperbolic paraboloid*. This completes the problem.

14.2   Contour Lines.

In the perspective graph of the relation $z = -xy$ (Figure 8.44), we observe the relationship of the curves $-xy = c$ in the xy-plane and the level lines $z = c$ of the surface. If we project these level lines down onto the xy-plane, they coincide with the series of hyperbolas $-xy = c$ in the xy-plane. These projections of level lines onto the horizontal plane are called *contour lines* (Figure 8.45). We observe that we can interpret the graph of the surface $z = -xy$ in $\mathcal{R}^3$ from its plane contour map. Indeed, the contour map is a suitable substitute for the perspective sketch of the graph in $\mathcal{R}^3$ and much easier to construct. We should become skillful in reading contour maps as surfaces in a space of dimension one higher. Examples of the use of contour maps to indicate surfaces include topographic maps of the earth's surface and weather maps, showing isobars as contour lines of constant air pressure. Heat flow and steady state fluid flow are represented in the same way.

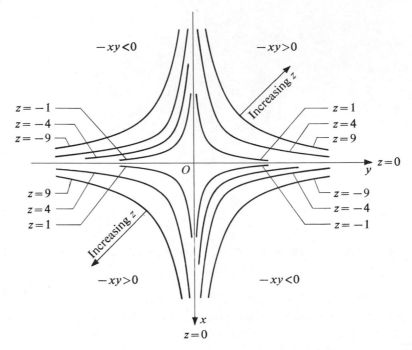

$-xy<0$

$-xy>0$

Increasing z

$z=-1$

$z=-4$

$z=-9$

$z=1$

$z=4$

$z=9$

$z=0$

y

$z=9$

$z=4$

$z=1$

$z=-9$

$z=-4$

$z=-1$

Increasing z

$-xy>0$

$-xy<0$

O

x

$z=0$

FIGURE 8.45   Contour map of $z = -xy$. The value of $-xy$ is positive in the first and third quadrants, negative in the second and fourth. The point O is a saddle point.

14.3   Inequalities.   We may use contour maps to give a graphical representation of inequalities of the form $f(x, y) \leqq c$. We think of the relation $w = f(x, y)$ and its contour map in the xy-plane. Then the graph of the inequality $f(x, y) \leqq c$ consists of all points $P:(x, y)$ on some contour for a fixed level w for which $w \leqq c$. We plot the contour $f(x, y) = c$ and indicate by shading the region on one side of it where $f(x, y) \leqq c$.

Example.   Plot the graph of the inequality $4x^2 + y^2 \leqq 4$.

Solution.   For the relation $w = 4x^2 + y^2$, we plot the level contour $w = 4$ (Figure 8.46). It is the ellipse

$$\frac{x^2}{1} + \frac{y^2}{4} = 1.$$

For other values of w, the graph of the level contour is the ellipse

$$\frac{4x^2}{w} + \frac{y^2}{w} = 1,$$

which is inside the ellipse $w = 4$ when $0 \leqq w < 4$. Hence, the graph of the inequality $4x^2 + y^2 \leqq 4$ consists of points on and inside the ellipse $4x^2 + y^2 = 4$.

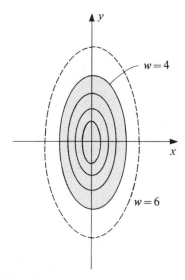

FIGURE 8.46   Inequality $4x^2 + y^2 \leq 4$. Contour map of $w = 4x^2 + y^2$. Shaded region consists of points where $4x^2 + y^2 \leq 4$.

14.4   Examples of Contour Maps and Inequalities. We plot some contour maps of functions with related inequalities (Figure 8.47). We observe that for $w = x^2 + y^2$ (Figure 8.47(b)) and for the cone $w^2 = x^2 + y^2$ (Figure 8.47(d)) the contour lines are concentric circles. For the cone, the level contours $w = 0, 1, 2, 3, \ldots$ are equally spaced, indicating that w increases linearly with x and with y. On the other hand, for the paraboloid the level contours $w = 0, 1, 2, 3, 4$, become more closely spaced for larger values of w. This closer spacing of the level contours indicates a faster rate of increase (or decrease) of w. We plot vertical sections, $x = 0$, of the paraboloid $w = x^2 + y^2$ and the cone $w^2 = x^2 + y^2$ (Figure 8.47(f)), showing their projections $P(3)$ for the paraboloid and $C(3)$ for the cone, at the level $w = 3$.

14.5   Level Surfaces to Represent Relations of Four Variables. The idea that the contours can represent a relation $z = f(x, y)$ in $\mathscr{R}^3$ by a picture in its domain in $\mathscr{R}^2$ enables us to represent a relation $w = f(x, y, z)$ in $\mathscr{R}^4$ by a picture in $\mathscr{R}^3$. The level contours $w = c$ are the surfaces $f(x, y, z) = c$, which we can picture in $\mathscr{R}^3$ (Figure 8.48). In this graph the value of f increases from level surface to level surface. At a point R we see intuitively that there is a rate of change of f in every direction and guess that the maximum rate of change of f occurs in a direction perpendicular to the level surface at R. We picture this by a located vector $\vec{g}$ at R, which gives both the direction and magnitude, $\|\vec{g}\|$, of the maximum rate of change of f. This idea is important in 3-dimensional fluid flow and in the representation of fields of force by scalar potentials.

14.6   Relations in n-Space. We cannot draw graphs in the space $\mathscr{R}^n$ whose points are rows $(x_1, x_2, \ldots, x_n)$ when $n > 3$, but there is no difficulty in extending the idea to relations $F(x_1, x_2, \ldots, x_n) = C$ determined by a function of several variables $F(\pmb{x}_1, \pmb{x}_2,$

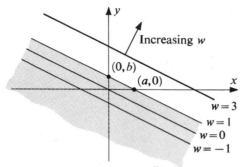

(a)   Plane $w = \frac{x}{a} + \frac{y}{b}$
and (shaded) inequality $\frac{x}{a} + \frac{y}{b} \leqq 1$

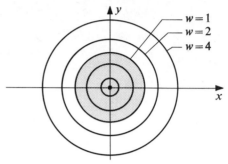

(b)   Paraboloid $w = x^2 + y^2$
and (shaded) inequality $x^2 + y^2 \leqq 1$

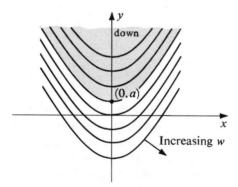

(c)   Parabolic cylinder $w = x^2 - 4ay$, $a > 0$,
and (shaded) inequality $x^2 - 4ay \leqq 0$

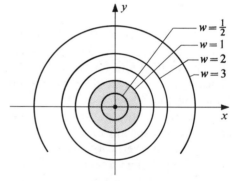

(d)   Cone $w^2 = x^2 + y^2$
and (shaded) inequality $x^2 + y^2 \leqq 1$

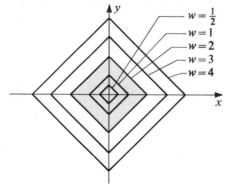

(e) $w = |x| + |y|$
and (shaded) inequality $|x| + |y| \leqq 2$

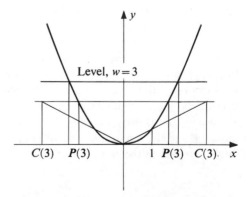

(f)   Vertical section, $x = 0$, of $w = x^2 + y^2$
and $w^2 = x^2 + y^2$ and level $w = 3$

FIGURE 8.47   Contour maps and inequalities.

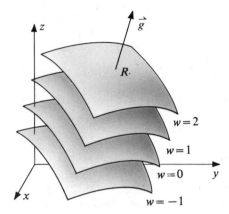

FIGURE 8.48   Contour graph of relation in $\mathcal{R}^4$.

$\ldots, x_n)$ defined in $\mathcal{R}^n$. We even use the same geometric language. For example, we speak of the "plane"

$$a_1 x + a_2 x_2 + \cdots + a_n x_n = C.$$

We can also use the vector notation $\vec{x}$ as a shorter symbol for the row

$$\vec{x}:(x_1, x_2, \ldots, x_n).$$

Moreover, the norm $\|\vec{x}\| = \sqrt{x_1^2 + x_2^2 + \cdots + x_n^2}$. In this way, we can speak of the sphere $\|\vec{x}\| = r$ in $\mathcal{R}^n$ and the ball $\|\vec{x}\| \leq r$.

We observe that when we study relations $F(x_1, x_2, \ldots, x_n) = C$, we are using the idea of the function $F(x_1, x_2, \ldots, x_n)$ as fundamental and we are not discarding functions in favor of relations.

14.7 Exercises

In Exercises 1–6, sketch both perspective and contour representations, using vertical sections in the perspective representations of the functional relations.

1. $w = x - y$.   2. $w = y - x^2$.

3. $z = \dfrac{x^2}{a^2} + \dfrac{y^2}{b^2}$.   4. $w = x^2 y$.

5. $w = -2x + 3y + 12$.   6. $(x^2 + y^2)z = 1,\ (x, y) < (0, 0)$.

In Exercises 7–14, plot the inequalities by contours.

7. $y < x$.   8. $x + 2y + 6 \leq 0$.
9. $x^2 + y^2 \leq 2x$.   10. $x^2 + y^2 - 1 \geq 0$.

11. $y \leq x^2$.   12. $\dfrac{1}{x^2 + y^2} \leq 2$.

13. $x^2 - y^2 \leq 0$.   14. $z \leq x$.

In Exercises 15–21, construct graphical representations of the relations, using level contours, and also perspective projections where possible.

15. The plane $z = 2x + 3y$ on the triangular region with vertices $(6, 0, 0)$, $(0, 4, 0)$, and $(-6, 0, 0)$.
16. The plane $z = -2x + 3y + 12$ on the circular disk $\{x^2 + y^2 \leq 4\}$.
17. The functional relation $z = \sqrt{4 - x^2 - y^2}$ on the disk, $\{x^2 + y^2 \leq 4\}$.

18.   The paraboloid $z = -(x^2 + y^2)/4$ on the square region $\mathscr{D}:\{|x| \leq 4$ and $|y| \leq 4\}$.
19.   The hyperbolic paraboloid $z = (x^2/a^2) - (y^2/b^2)$ (similar to Figures 8.44 and 8.45).
20.   The waveform surface $z = \sin(x + y)$ on the xy-plane.
21.   The relation $x^2 + 2z + y^2 = 0$ in xyz-space.

22.   Sketch the graph of the relation determined by the simultaneous inequalities $2x + 3y - 7 \leq 0$ and $x + y - 1 \geq 0$.
23.   Sketch the graph of the relation determined by the simultaneous inequalities $y \geq x^2$ and $y \leq 2x$.
24.   Sketch the graph of the relation given by the simultaneous inequalities $x^2 + y^2 \leq 1$ and $x^2 + y^2 \leq 2x$.
25.   Draw in perspective the solid figure that is in both $x^2 + y^2 \leq 1$ and $y^2 + z^2 \leq 1$. Show that sections of this figure parallel to the xz-plane are square disks.
26.   Use level contours to represent the function $f(x, y, z) = x^2 + y^2 + z^2$.
27.   Draw level contours to represent the function $f(x, y, z) = 2x - 3y + 4z + 1$. Show that the level contours are planes in $\mathscr{R}^3$ with common normal $\vec{n}:(2, -3, 4)$.
28.   Draw contour lines to represent the function $f(x, y) = (x^2 + y^2)/x$ in the xy-plane with $(0, 0)$ excluded. What happens at $(0, 0)$?
29.   Sketch the figure in $\mathscr{R}^3$ determined by all points R such that $\|\vec{r}\| \leq 1$.
30.   Draw contour lines to represent the function

$$f(x, y) = 4(x - 1)^2 + (y + 2)^2 + 16.$$

From the contour lines locate the minimum point of the function.
31.   Show that the surface $z = x^2 - 4py$ is a parabolic cylinder (Figure 8.47(c)). Find the direction $\vec{v}:(a, b, c)$ of the generator lines.
32.   The xy-contour lines of the relation $x^2 + y^2 + 2zx = 0$ have a common tangent line at the origin. What does this imply about the graph of the relation in $\mathscr{R}^3$?

14.8   The Hyperbolic Paraboloid as a Ruled Surface

Example.   Sketch in perspective the graph of the hyperbolic paraboloid $z = -\frac{1}{4}xy$ using the vertical sections $x = c$ and $y = k$.

Solution.   We know (Figure 8.44) that the graph is an everywhere-curved surface with parabolic vertical sections in planes $x + y = a$ and $x - y = b$. Also it has hyperbolic level curves in the planes $z = h$.

When we determine the curves of intersection of this surface with the planes $x = c$ and $y = k$, we discover that they are straight lines! In fact, when $x = c$, $z = -(c/4)y$. This is a straight line with slope $-(c/4)$ in the plane $x = c$. Also, when $y = k$, $z = -(k/4)y$. This is a straight line with slope $-(k/4)$ in the plane $y = k$. If we plot a series of these lines from $x = -6$ to $x = 6$ and $y = -6$ to $y = 6$, using the principles of perspective plotting (§13.2), we find with the aid of the parabola $y = -x$, $z = \frac{1}{4}x^2$ that we have a picture of a portion of the surface (Figure 8.49).

The straight lines $x = c$, $z = -\frac{1}{4}xy$ and $y = k$, $z = \frac{1}{4}xy$ are called *rulings* in the surface and any surface that can be constructed entirely of rulings is a *ruled surface*. We observe that the rulings in the planes $x = 6$ and $y = 6$ intersect at $B:(6, 6, -9)$ and the rulings in the planes $x = -6$ and $y = -6$ intersect at $D:(-6, -6, -9)$. Different pairs of rulings intersect at the top in points $P:(6, -6, 9)$ and $Q:(-6, 6, 9)$. Vertical projections from these top corners down into the plane $z = -9$ therefore determine points $A:(6, -6, -9)$ and $C:(-6, 6, -9)$. The portion of the surface $z = -\frac{1}{4}xy$ that we have plotted therefore has a square domain in the xy-plane given by $|x| \leq 6$ and $|y| \leq 6$.

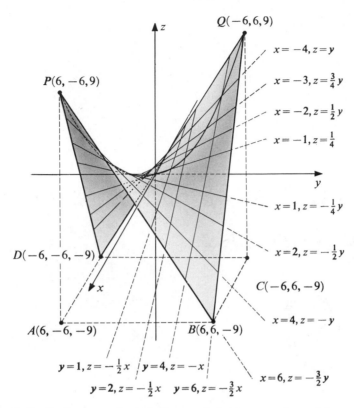

$Q(-6,6,9)$

$x=-4, z=y$

$x=-3, z=\frac{3}{4}y$

$x=-2, z=\frac{1}{2}y$

$x=-1, z=\frac{1}{4}$

$P(6,-6,9)$

$x=1, z=-\frac{1}{4}y$

$x=2, z=-\frac{1}{2}y$

$D(-6,-6,-9)$

$C(-6,6,-9)$

$x=4, z=-y$

$A(6,-6,-9)$

$B(6,6,-9)$

$y=1, z=-\frac{1}{2}x$ $y=4, z=-x$

$x=6, z=-\frac{3}{2}y$

$y=2, z=-\frac{1}{2}x$ $y=6, z=-\frac{3}{2}x$

FIGURE 8.49 Rulings on hyperbolic paraboloid $z = -\frac{1}{4}xy$. (Compare with Figure 8.44.)

14.9 An Application to Architecture. Now let the units in the graph of the hyperbolic paraboloid (Figure 8.49) be feet. The mathematical properties of the hyperbolic paraboloid, which we have discovered, show that we can construct a roof that is in the shape of a hyperbolic paraboloid and support this roof by straight rafters along the rulings. If we take $z = -9$ as the ground level, we can support this roof at the points B and D with only light support at A and C to keep it balanced. If we put in light vertical posts AP and CQ, our roof covers a 12 ft $\times$ 12 ft square and has a height of 9 ft in the center, a maximum height of 18 ft at A and C, and a zero minimum height at B and D.

This roof-form has been used often in recent architecture. The vertical scale can be adjusted by means of a parameter m in the equation $z = mxy$, which is, for every number m, a hyperbolic paraboloid.

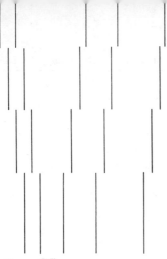

Differentiation of Functions of Several Variables

CHAPTER 9

15 Partial Derivatives

15.1 Idea of Partial Differentiation.   We recall (I, §10.4) that for functions $f(x)$ of one real variable, that is, functions defined on some interval of real numbers, the numerical derivative $f'(x)$ at x was defined to be the limit of the divided difference

$$f'(x) = \lim_{w \to x} \frac{f(w) - f(x)}{w - x}.$$

It is possible to calculate this divided difference when w and x are real numbers in the domain of f and $w \neq x$; for there is a division operation in the real numbers. Now if $f(\vec{x})$ becomes a function of several variables, where $\vec{x}$ is a row $(x_1, x_2, \ldots, x_n)$ and $\vec{w}$ another row in the domain of f, the analogous divided difference makes no sense; for in the algebra of row spaces there is no division operation that would permit us to divide by the row $\vec{w} - \vec{x}$. Therefore, the idea of a derivative does not extend without modification to functions of several real variables. Accordingly, we make the simplest modification that will permit us to use the differentiation of functions of one real variable with all of its formulas and techniques. This modification is letting the elements of the row $\vec{x} : (x_1, x_2, \ldots, x_n)$ vary only one at a time, keeping the others fixed, a procedure called partial variation.

We consider first a function $f(x_1, x_2)$ defined on some rectangular domain in the 2-space of rows (x_1, x_2) and think of $\vec{x} : (x_1, x_2)$ as a fixed point in the domain of f. We keep the second element fixed at x_2 and consider other rows (w_1, w_2) in the domain of f such that

$w_1 \neq x_1$. With this restricted variation of the row, we can write a meaningful divided difference and take its limit as the single real variable as w_1 approaches x_1. This defines an ordinary numerical derivative $f_1(x_1, x_2)$ with x_2 held constant:

$$f_1(x_1, x_2) = \lim_{w_1 \to x_1} \frac{f(w_1, x_2) - f(x_1, x_2)}{w_1 - x_1}.$$

Similarly, at the same point (x_1, x_2) we can think of the first element x_1 as fixed, consider other rows (x_1, w_2) where $w_2 \neq x_2$, and calculate the derivative

$$f_2(x_1, x_2) = \lim_{w_2 \to x_2} \frac{f(x_1, w_2) - f(x_1, x_2)}{w_2 - x_2}.$$

The derivative $f_1(x_1, x_2)$ is called *the partial derivative of $f(x_1, x_2)$ with respect to the first variable at (x_1, x_2)*. The derivative $f_2(x_1, x_2)$ is called *the partial derivative with respect to the second variable at (x_1, x_2)*.

It is clear that the idea extends to functions $f(x_1, x_2, \ldots, x_n)$ on domains in $\mathscr{R}^n$ and yields n partial derivatives

$$f_1(x_1, \ldots, x_n), f_2(x_1, \ldots, x_n), \ldots, f_n(x_1, \ldots, x_n).$$

15.2   Other Notations for Partial Derivatives.   Several different notations for partial derivatives are used.

For a function $f(x, y, z)$, the following symbols denote partial derivatives:

$$f_1(x, y, z) = f_x(x, y, z) = \frac{\partial}{\partial x} f(x, y, z),$$

$$f_2(x, y, z) = f_y(x, y, z) = \frac{\partial}{\partial y} f(x, y, z),$$

and

$$f_3(x, y, z) = f_z(x, y, z) = \frac{\partial}{\partial z} f(x, y, z).$$

Here the symbols $\partial/\partial x, \partial/\partial y, \partial/\partial z$ are partial derivative operators.

15.3   Calculation of Partial Derivatives.   The formulas for ordinary derivative of functions $f(x)$ on $\mathscr{R}$ apply (Table I, Appendix B).

Example A.   For $f(x, y) = 5x^2y^3$, calculate $f_1(x, y)$ and $f_2(x, y)$.

Solution.   To calculate f_1 we treat y as a constant. Then

$$f_1(x, y) = \frac{\partial}{\partial x} (5y^3)x^2 = 5y^3 \frac{\partial}{\partial x} x^2 = 10xy^3.$$

In the calculation of f_2, x is a constant. Then

$$f_2(x, y) = \frac{\partial}{\partial y} 5x^2y^3 = 5x^2 \frac{\partial}{\partial y} y^3 = 15x^2y^2.$$

Example B. Calculate the partial derivatives of

$$f(x, y) = \frac{x^2 - xy - y^2}{2y^3}.$$

Solution. In calculating f_1, we hold y constant. Then $1/2y^3$ is a constant factor and

$$f_1(x, y) = \frac{1}{2y^3} \frac{\partial}{\partial x} (x^2 - xy - y^2) = \frac{1}{2y^3} (2x - y - 0) = \frac{2x - y}{2y^3},$$

valid wherever $y \neq 0$. In calculating f_2, we hold x constant but the differentiation with respect to y must follow the quotient rule Va,

$$D \frac{u}{v} = \frac{v\, Du - u\, Dv}{v^2}.$$

In this way we find

$$f_2(x, y) = \frac{2y^3 \frac{\partial}{\partial y} (x^2 - xy - y^2) - (x^2 - xy - y^2) \frac{\partial}{\partial y} 2y^3}{(2y^3)^2}$$

$$= \frac{2y^3(0 - x - 2y) - (x^2 - xy - y^2)(6y^2)}{4y^6}.$$

We may simplify this algebraically.

Example C. Calculate the three partial derivatives of

$$f(r, \theta, z) = \frac{r(2 - \cos 2\theta)}{r^2 + z^2}.$$

Solution. By the quotient formula for ordinary differentiation, when we keep θ and z constant,

$$f_1(r, \theta, z) = (2 - \cos 2\theta) \frac{(r^2 + z^2)(1) - r(2r + 0)}{(r^2 + z^2)^2}$$

$$= (2 - \cos 2\theta) \frac{z^2 - r^2}{(r^2 + z^2)^2}.$$

Then, keeping r and z constant so that $r/(r^2 + z^2)$ is a constant factor, we obtain

$$f_2(r, \theta, z) = \left(\frac{r}{r^2 + z^2} \right) (0 + 2 \sin 2\theta) = \frac{2r \sin 2\theta}{r^2 + z^2}.$$

Finally, with r and θ kept constant, which gives us a constant factor $r(2 - \cos 2\theta)$,

$$f_3(r, \theta, z) = r(2 - \cos 2\theta)(-1)(r^2 + z^2)^{-2}(2z) = \frac{-2rz(2 - \cos 2\theta)}{r^2 + z^2}.$$

This completes the problem.

15.4   Slope Interpretation of Partial Derivative. For functions of two variables we can use perspective drawings of graphs in $\mathscr{R}^3$ to interpret partial derivatives in the usual way as slopes of tangent lines to curves in the vertical sections of the graph surface (§13). We examine a particular example.

Example.   Find the slopes of tangent lines to the vertical sections of

$$f(x, y) = 12 - \frac{x^2}{4} - \frac{y^2}{9}$$

at the point $K:(4, 3, 5)$, and plot these tangent lines.

Solution.   We plot this graph and its vertical sections (Figure 9.1) and compute the partial derivative functions

$$f_1 = -\frac{x}{2} \quad \text{and} \quad f_2 = -\frac{2y}{3}.$$

Evaluating these at K, we find $f_1(4, 3) = -2$ and $f_2(4, 3) = -2$. We sketch the graph in $\mathscr{R}^3$ (Figure 9.1) and plot in the plane $x = 4$ the line KT through $(4, 3, 5)$ with slope $f_2(4, 3) = 4/-2$. This turns out to be tangent to the vertical section HJK at K. Similarly, in the plane $y = 3$ we plot the line KS through K with slope $f_2(4, 3) = 8/-4$. This is tangent to the vertical section curve KLM at K. The problem is complete.

15.5   The Rate-of-Change Interpretation of the Partial Derivative.

We recall that the ordinary derivative $f'(x)$ at x of a function $f(x)$ has not only an interpretation as the slope of the tangent line to the graph at $(x, f(x))$ but also an interpretation as the local rate of change of $f(x)$ with respect to x. In particular, when the independent variable is time, and $y = f(t)$, the derivative $f'(t)$ represents the instantaneous velocity of a point moving in the y-axis at time t. These interpretations carry over to partial derivatives of functions $f(x_1, \ldots, x_n)$ of any number of variables.

Example.   The function

$$u(x, y, t) = A \sin [k(ax + by - vt)],$$

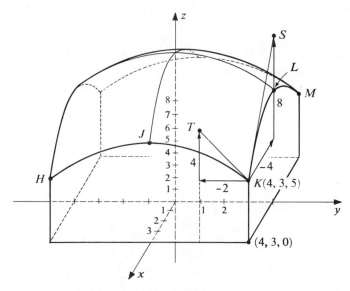

FIGURE 9.1   Partial derivatives as slopes of tangent lines, function $z = 12 - \frac{x^2}{4} - \frac{y^2}{3}$ at K: $(4, 3, 5)$.

where A, k, a, b, v are fixed numbers with $a^2 + b^2 = 1$, represents the displacement of a water surface at point $P:(x, y)$ at time t. The water surface covers the xy-plane and $u(x, y, t)$ gives the displacement of the surface above or below the xy-plane at time t. Describe the function as a moving train of water waves and interpret the partial derivatives u_1, u_2, u_3 (Figure 9.2).

Solution. We observe that with (x, y) fixed, the motion is sinusoidal as a function of time. That is, it is of the form $u = A \sin(vt + \phi)$, where $v = -kv$ and $\phi = k(ax + by)$. The height u oscillates periodically between the extremes $+|A|$ and $-|A|$ with period $T = 2\pi/kv$ seconds (I, §32.4). We may imagine a post standing in the water at the point $P:(x, y)$. Then u represents the height of the water surface on the side of this post, moving in a vertical direction up and down periodically. The partial derivative

$$u_3 = -Akv \cos[ax + by - vt]$$

represents the vertical velocity of the motion up and down of the surface. It is also sinusoidal with the same period as u. But the velocity $u_3 = 0$ when $u = \pm A$ and $|u_3|$ is a maximum when $u = 0$.

Now we hold t fixed and consider the surface as a still snapshot at the instant t. Its height is a function of x and y. We may regard the expression $\theta = k(ax + by - vt)$ as an angle. Then $u = A \sin \theta$, where θ is called the phase of the wave. We consider for the fixed t all points (x, y) that have the same phase θ. They satisfy a linear equation

$$ax + by = vt + \frac{\theta}{k}.$$

For every θ this is a straight line with unit normal $\vec{n}:(a, b)$. For example, when $\theta = \pi/2$ so that $\sin \theta = 1$ the points (x, y) lie along a crest of the wave. In general, at a fixed time the points in the wave surface that have the same phase θ all lie in a straight line and all of these lines are parallel since they have the common normal $\vec{n}:(a, b)$ (Figure 9.2). In this instantaneous static picture of the surface, we may fix y. Then the partial derivative $u_1(x, y, t)$ represents the slope of a vertical section in the plane $y = $ const. Similarly, if we fix t and x, the partial derivative $u_2(x, y, t)$ represents the slope of a vertical section of the surface in the plane $x = $ const.

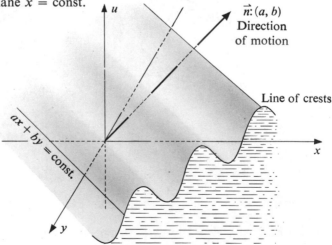

FIGURE 9.2   Instantaneous view of water wave.

Now we return to letting t vary so that we may consider the motion of a line of crests $(\theta = \pi/2)$,

$$ax + by = vt + \frac{\pi}{2k}.$$

This is the equation of a line in normal form since $|\vec{n}| = \sqrt{a^2 + b^2} = 1$ (§12.4) and the normal distance from the origin to the line is given by

$$p(t) = vt + \frac{\pi}{2k}.$$

This distance increases at the rate given by the ordinary derivative $p'(t) = v$. Hence the crests are moving in the normal direction $\vec{n}$ with scalar speed v.

In summary, the function

$$u = A \sin [ax + by - vt], \qquad \text{where} \quad a^2 + b^2 = 1$$

represents a train of parallel plane waves whose crests move in the direction $\vec{n}:(a, b)$ with speed v. At any fixed point $P:(x, y)$ the motion is periodic, up and down with velocity u_3. The period of this up-and-down motion is $2\pi/kv$. The slopes u_1 and u_2 give the steepness of the surface in the x-direction and y-direction respectively, for a fixed time and place.

15.6 Partial Derivatives of Higher Order.

Successive partial differentiations yield partial derivatives of order higher than one. We may differentiate partially twice with respect to the same variable or with respect to two different variables. For example, if we have $f(x, y) = 7x^5y^3$, the symbol f_{11} means the second-order partial derivative taken twice with respect to the first variable x. Thus in this case we have $f_1 = (7)5x^4y^3$ and

$$f_{11} = f_{xx} = \frac{\partial^2}{\partial x^2}f = (7)(5)4x^3y^3 = 140x^3y^3.$$

The symbol f_{12} indicates the partial derivative found by differentiating f first with respect to the first variable and then differentiating f_1 with respect to the second variable. In our example where $f_1 = 35x^4y^3$, we have

$$f_{12} = f_{xy} = \frac{\partial^2}{\partial x \, \partial y}f = 35x^4(3y^2) = 105x^4y^2.$$

This immediately raises the question of whether $f_{12} = f_{21}$. They do not mean the same thing but when we calculate f_{21}, we find first $f_2 = 21x^5y^2$, and

$$f_{21} = (21)(5x^4)y^2 = 105x^4y^2 = f_{12}.$$

We postpone the proof that when f_{12} and f_{21} are continuous in the domain of f, then

$$f_{12} = f_{21}, \qquad \text{or more generally} \qquad f_{ij} = f_{ji}.$$

Here f_{ij} is the second-order partial derivative of $f(x_1, x_2, \ldots, x_n)$, which results when we first differentiate f with respect to the ith variable to obtain f_i and then differentiate this partially with respect to the jth variable to obtain f_{ij}.

15.7 Exercises

1. For $f(x, y) = 3x^2 - 4xy + y^2$, calculate f_1, f_2.

2. For $f(x, y, z) = 1 - \dfrac{2xy}{z}$, calculate f_1, f_2, f_3.

3. For $f(x, y) = \sqrt{x^2 + y^2}$ calculate f_1, f_2, where $(x, y) \neq (0, 0)$.

4. For $f(r, \theta) = r(1 - \cos 2\theta)$, calculate f_1, f_2.

5. For $f(x, t) = e^{-t} \cos x$, calculate f_1, f_2.

6. For $f(\theta, \phi) = \sin 2\theta \cos 3\phi$, calculate f_1, f_2.

7. For $f(x, y) = \dfrac{x + 2y}{y - 3x}$, calculate f_1, f_2.

8. For $f(x, y, z) = \dfrac{1}{\sqrt{x^2 + y^2 + z^2}}$, calculate f_1, f_2, f_3. Exceptional points?

9. For $u(x, y) = 3x^4 - 4x^3y + 6x^2y^2$, calculate u_1, u_2.

10. For $v(x, y) = (x + y) \cos (x - y)$, calculate v_1, v_2.

11. For $u = xy + yz + zx$, show that $u_1 + u_2 + u_3 - 2(x + y + z)$.

12. For $f(x, y) = \dfrac{2y}{y - x}$, show that $f_1(1, 3) = \frac{3}{2}$ and $f_2(1, 3) = -\frac{1}{2}$.

13. For $u(\theta, \phi) = e^{-\theta} \sin (\theta + 2\phi)$, show that $f_1(0, \pi/4) = -1$, $f_2(0, \pi/4) = 0$.

14. For $f(x, y) = \sqrt{x^2 + y^2}$, calculate $f_{11}, f_{12}, f_{21}, f_{22}$.

15. For $u(x, y, z) = \dfrac{1}{\sqrt{x^2 + y^2 + z^2}}$, show that $u_{11} + u_{22} + u_{33} = 0$ if $(x, y, z) \neq (0, 0, 0)$.

16. For $u(x, y) = \ln \dfrac{1}{\sqrt{x^2 + y^2}}$, show that $\dfrac{\partial^2 u}{\partial x^2} + \dfrac{\partial^2 u}{\partial y^2} = 0$.

17. For $f(x, y) = \arctan \dfrac{y}{x}$, show that $f_1 + f_2 = -\dfrac{x + y}{x^2 + y^2}$.

18. For $f(x, y) = \arctan (y/x)$, show that $f_{12} = f_{21}$, where $x \neq 0$.

In Exercises 19–22, evaluate the two partial derivatives at the point indicated, sketch the surface, and plot the tangent lines to vertical sections through the point.

19. $z = 2x - 3y$ at $(-2, 1)$.   20. $z = x^2 + y^2$ at $(0, -1)$.

21. $z = -xy$ at $(3, 2)$.   22. $z = x^2 - y^2$ at $(-1, 2)$.

In Exercises 23–29, give physical or geometric interpretations to the indicated partial derivatives.

23. $u = 2x + 3y + 7z$ gives the total proceeds due to sale of x units of product A, y units of product B, and z units of product C. What is the meaning of u_1, u_2, u_3?

24. In Boyle's law, the volume v of a gas is proportional to the absolute temperature T and inversely proportional to the pressure p so that $v(T, p) = RT/p$, where R is a constant for the gas. What is the physical meaning of v_1 and v_2?

25. Coulomb's law says that the magnitude of force exerted on charge q_1 by a charge q_2 at a distance of r cm is given by $F(q_1, q_2, r) = kq_1 q_2/r^2$, where k is a constant. What is the meaning of F_1, F_2, F_3?

26. The compound amount A of \$$P$ invested at a nominal compound interest rate of r per year compounded k times per year for t years is a function $A(P, r, k, t)$ given by

$$A(P, r, k, t) = P\left(1 + \frac{r}{k}\right)^{kt},$$

Calculate A_1, A_2, A_3, A_4 and give a physical meaning to each partial derivative.

27. The velocity of a rocket at time t in space with no external forces acting on it is a function $v(c, M, k, t)$, where c is the speed of the exhaust gases, M is the initial mass at time $t = 0$ when $v = 0$, and k is the rate of burning fuel. The velocity function is given by

$$v(c, M, k, t) = c \ln \frac{M}{M - kt}.$$

Calculate v_1, v_2, v_3, v_4 and give a meaning to each one.

28.   In the motion of a wave (§15.5), calculate u_{33}, or u_{tt}, and give a physical meaning to it.
29.   Show that $T = Ce^{-m^2t} \sin(mx/a)$ is a solution of the partial differential equation

$$\frac{\partial T}{\partial t} = a^2 \frac{\partial^2 T}{\partial x^2},$$

where C, m, and a are constants.

16   Differentials

16.1   Differentiation of Vector Functions.

We proceed at once to the most general case, a function f that maps vectors $\vec{x}$ in some domain $\mathcal{D}$ into vectors $\vec{w}$ in some codomain. We will call it a *vector function*. The cases where either the variable $\vec{x}$ or the values $\vec{w}$, or both, are numerical variables are included. For any vector function we may consider a point $\vec{x}$ and a neighboring point $\vec{x} + \overrightarrow{\Delta x}$, both in the domain of f. We define $\overrightarrow{\Delta f}$ by $\overrightarrow{\Delta f} = f(\vec{x} + \overrightarrow{\Delta x}) - f(\vec{x})$. It is a vector in the $\vec{w}$-space. We say that f is continuous at $\vec{x}$ if for every ϵ there is a δ such that when $\|\overrightarrow{\Delta x}\| < \delta$, then $\|\overrightarrow{\Delta f}\| < \epsilon$.

We cannot write down a simple formula for all linear functions on the vector domain $\mathcal{D}$, so we define them abstractly.

DEFINITION.   A function ϕ that maps vector $\vec{u}$ in $\mathcal{D}$ into vectors $\vec{w}$ is *linear* if and only if ϕ is continuous and

$$\phi(a\vec{u} + b\vec{v}) = a\phi(\vec{u}) + b\phi(\vec{v}).$$

This is all of the notation that we need to define the differential of f at $\vec{x}$.

DEFINITION.   The linear function $\phi(\overrightarrow{dx})$ is a *differential* of f at $\vec{x}$ if ϕ is a best linear approximation of f at $\vec{x}$ in the sense that

$$\lim_{\|\overrightarrow{\Delta x}\| \to 0} \frac{\|\overrightarrow{\Delta f} - \phi(\overrightarrow{\Delta x})\|}{\|\overrightarrow{\Delta x}\|} = 0.$$

If f has a differential $\phi(\overrightarrow{dx})$ at $\vec{x}$, we say that f is differentiable at $\vec{x}$ and its differential $Df = \phi(\overrightarrow{dx})$.

This definition is phrased in the same way as the definition of the differential of a function of one real variable (§1.3). As in the simple case, for this to be a proper definition, we must show that only one, best linear approximation can exist, formally stated as follows (§16.7, Proof).

THEOREM (Uniqueness of the differential).   If f has differentials $\phi(\overrightarrow{dx})$ and $\psi(\overrightarrow{dx})$ at $\vec{x}$, then $\phi = \psi$.

We observe that the symbols $\overrightarrow{dx}, \overrightarrow{dy}, \overrightarrow{dw}, \overrightarrow{\Delta x}, \overrightarrow{\Delta y}, \overrightarrow{df}, \overrightarrow{\Delta f}$ do not denote differentials. They are all merely differences measured locally from the point where f is differentiated and they are not necessarily small. The local coordinates (dx, dy), $(\Delta x, \Delta y)$ in the xy-plane are included (§1.2).

16.2   Calculation of Differentials. We can give explicit formulas for calculating differentials in the case of number-valued functions.

THEOREM (Calculation of differentials).   If f is a number-valued function of rows $\vec{x}:(x_1, x_2, \ldots, x_n)$ that is differentiable at $\vec{x}$, then f has partial derivatives $f_1, f_2, \ldots, f_n$ at $\vec{x}$ and

$$Df = f_1 \, dx_1 + f_2 \, dx_2 + \cdots + f_n \, dx_n,$$

where all of the partial derivatives are evaluated at $\vec{x}$.

Before we prove the theorem, we consider some examples.

Example A.   Differentiate the function f given by $f(x, y, z) = xy \ln |z|$.

Solution.   Assuming for the present that f is differentiable, we calculate the partial derivatives

$$f_1 = y \ln |z|, \qquad f_2 = x \ln |z|, \qquad f_3 = \frac{xy}{z}.$$

Then

$$Df = f_1 \, dx + f_2 \, dy + f_3 \, dz$$
$$= y \ln |z| \, dx + x \ln |z| \, dy + \frac{xy}{z} \, dz.$$

Example B.   Find in local coordinates the equation of the tangent plane to $w = x^2 - xy + 3y^2$ at $P:(3, -4, 69)$.

Solution.   Assuming for the present that $f(x, y) = x^2 - xy + 3y^2$ is differentiable, we have $f_1 = 2x - y$ and $f_2 = -x + 6y$. Evaluating these partial derivatives at P, we find $f_1(3, -4) = 10$ and $f_2(3, -4) = -27$. Hence, the equation of the tangent plane is given by the differential

$$dw = 10dx - 27dy.$$

This completes the problem. If we want the equation of the tangent plane in xyw-coordinates, it is

$$w - 69 = 10(x - 3) - 27(y + 4).$$

Proof of the theorem.   We will prove the theorem for functions $f(x_1, x_2)$ of two variables. The extension to any number of variables introduces no new principles.

LEMMA.   Every number-valued linear function $\boldsymbol{\phi}$ of rows (dx_1, dx_2) has the form $\boldsymbol{\phi}(\vec{dx}) = m_1 \, dx_1 + m_2 \, dx_2$ for some numbers m_1 and m_2.

Proof.   We may express the rows $\vec{dx}$ in terms of the standard basis $\{(1, 0), (0, 1)\}$ by $\vec{dx}:(dx_1, dx_2) = dx_1 (1, 0) + dx_2 (0, 1)$. Then, by the linearity property of $\boldsymbol{\phi}$,

$$\phi(dx_1, dx_2) = dx_1\phi(1, 0) + dx_2\phi(0, 1).$$

If we let $m_1 = \phi(1, 0)$ and $m_2 = \phi(0, 1)$, the lemma is proved.

Using this lemma with the theorem on the uniqueness of the differential, we know that f has a uniquely determined differential $Df = m_1\, dx_1 + m_2\, dx_2$. By the definition of the differential, this means that

$$\lim_{\|\Delta x\| \to 0} \frac{|\Delta f - (m_1\,\Delta x_1 + m_2\,\Delta x_2)|}{\|\Delta x\|} = 0.$$

Now holding x_2 fixed so that

$$\Delta x_2 = 0, \qquad \|\overrightarrow{\Delta x}\| = \|(dx, 0)\| = |\Delta x_1|, \qquad \text{and} \qquad \Delta f = f(x_1 + \Delta x_1, x_2) - f(x_1, x_2),$$

this reduces to

$$\lim_{\Delta x_1 \to 0} \left| \frac{f(x_1 + \Delta x_1, x_2) - f(x_1, x_2)}{\Delta x_1} - m_1 \right| = 0.$$

But this says that $f_1 = m_1$. Similarly we can prove, by holding x_1 fixed, that $f_2 = m_2$. Hence we obtain the differential $Df = f_1\, dx_1 + f_2\, dx_2$, as stated in the theorem.

16.3 Condition for Differentiability.

We state, omitting the proof, the following theorem. It gives conditions under which the differentiability of f follows from the existence of the partial derivatives of f, which we can compute by ordinary numerical differentiation.

THEOREM (Differentiability). If a number-valued function $f(x_1, x_2 \ldots, x_n)$ is defined inside a disk $\mathscr{D}$ in $\mathscr{R}^n$ and has *continuous* partial derivatives $f_1, f_2, \ldots, f_n$ in $\mathscr{D}$, then f is differentiable in $\mathscr{D}$.

16.4 Use of Differentials in Approximate Calculations.

Suppose that we know the value $w = f(\vec{x})$ of the function f at $\vec{x}$ and wish to calculate its value $w + \Delta w = f(\vec{x} + \overrightarrow{\Delta x})$ at the nearby point $\vec{x} + \overrightarrow{\Delta x}$. If we know Δw, we just add Δw to w to evaluate f at $\vec{x} + \overrightarrow{\Delta x}$. Instead of calculating Δw, we can evaluate the best linear approximation $dw = Df$ at $\vec{x} + \overrightarrow{\Delta x}$ and add dw instead of Δw to get an approximation to $f(\vec{x} + \overrightarrow{\Delta x})$.

Example. Starting with the point $\vec{x}:(-2, 3, 1)$, calculate the function

$$w = f(x, y, z) = \frac{x^2}{4} - \frac{y^2}{9} + \frac{(z-1)^2}{5}$$

at the nearby point $(-1.8, 3.1, 0.9)$.

Solution. We first calculate $f(-2, 3, 1) = 0$. Next, the partial derivatives,

$$f_1 = \frac{x}{2}, \qquad f_2 = -\frac{2y}{9}, \qquad f_3 = \frac{2}{5}(z - 1),$$

are everywhere continuous. Hence f is differentiable. At $\vec{x}$ we have $f_1 = -1, f_2 = -\frac{2}{3}$, $f_3 = 0$. This gives the differential at $(-2, 3, 1)$ as

$$dw = -dx - \tfrac{2}{3}dy + 0dz.$$

The row is evaluated—

$$(\Delta x, \Delta y, \Delta z) = (-1.8, 3.1, 0.9) - (-2, 3, 1) = (0.2, 0.1, -0.1).$$

Evaluating dw with these local coordinates, we find

$$dw = -(0.2) - \tfrac{2}{3}(0.1) + 0(-0.1) = -0.267.$$

Then, to best linear approximation, our answer is

$$f(-1.8, 3.1, 0.9) = w + \Delta w = 0 - 0.267 = -0.267.$$

Actually this function is simple enough to evaluate exactly. We find by direct substitution that $f(-1.8, 3.1, 0.9) = -0.256$. The difference $|-0.267 - (-0.256)| = 0.011$ is the error of the approximation.

16.5 Operations with Differentials.

With the definition of the differential of a function of several variables closely paraphrasing the definition of the differential of a function of one variable, which is equivalent to ordinary differentiation, we expect and indeed we can prove (Exercises) that the rules for operating with differentials of functions of several variables are formally the same as those for operating with derivatives for functions of one variable. Let us write down the basic ones. If u and v are differentiable real-valued functions of several variables $(x_1, x_2, \ldots, x_n)$, then the following rules of differentiation are valid for number-valued functions.

I.   **$D(u + v) = Du + Dv$.**
II.   **$D(cu) = c\,Du$**, where c is a constant.
III.   **$D(uv) = u\,Dv + v\,Du$.**
IV.   **$Dc = 0$**, where c is a constant function.
V.   **$Du^n = nu^{n-1}\,Du$**, where n is any real number.
Va.   **$D\dfrac{u}{v} = \dfrac{u\,Dv - v\,Du}{v^2}$**, at any point where $v \neq 0$.
VI.   *(Chain rule) If f is a differentiable function of $\vec{x}$ defined on a domain that includes the values of the differentiable function u, then **$D(f \circ u) = Df \circ Du$**, where $f \circ u$ means the composite $f[u(\vec{x})]$ and **$Df \circ Du$** means the composite **$Df[Du]$** resulting from substituting Du into Df.

16.6 Exercises

In Exercises 1–8, calculate the differentials of the functions at the points indicated.

1. $y = 1 - x^2$ at the point where $x = 1$.
2. $y = 1/x$ at the point where $x = 1$.
3. $y = e^{2x}$ at the point where $x = 0$.
4. $y = x/(1 + x^2)$ at the origin.
5. $f(x_1, x_2) = x_1^2 - 3x_1x_2 + x_2^2$ at $(x_1, x_2) = (-1, 2)$.
6. $f(x_1, x_2, x_3) = \sqrt{x_1^2 + x_2^2 + x_3^2}$ at point $(x_1, x_2, x_3) = (1, 1, 1)$.
7. $f(x, y, z) = 1/\sqrt{x^2 + y^2 + z^2}$ at $(1, 1, 1)$.
8. $f(x, y, \theta) = 5\,x/y \sin \theta$ at the point $(x, y, \theta) = (3, -4, \pi/3)$.

In Exercises 9–12, using the best linear approximation, estimate the approximate change in the value of the functions in going from the first point given to the second.

9. $f(x, y) = \ln(x/y)$ from $(2, 2)$ to $(1.8, 2.3)$.
10. $\sqrt{x^2 + y^2}$ from $(3, -4)$ to $(3 + \Delta x, -4 + \Delta y)$.

---

* The chain rule is included here to make the list of differentiation formulas complete. It is taken up in Section 18.

11. $f(r, \theta, \phi) = r \sin \theta \cos \phi$ from $(4, 0, 0)$ to $(4.2, 0.1, 0.2)$.

12. $\dfrac{e(x_1^2 + x_2^2)}{x_3 x_4}$ from $(0, 0, -3, 2)$ to $(-0.1, 0.2, -2.7, 2.1)$.

13. Calculate $D(uv)$ and $D(u/v)$ if $u = 3xy$ and $v = \sqrt{x^2 + y^2}$.

14. Calculate $D(x_1^2 + x_2^2 - x_3^2)^5$.

15. At what point on the graph of $f(x, y) = x^2 - xy + 2y^2$ is $Df = 0$?

16. Find in local coordinates (dx, dy, dw) the equation of the tangent plane to $w = 2xy$ at the point where $(x, y) = (-3, 1)$.

17. Show that $f = \sqrt{xy}$, $x \geq 0$, $y \geq 0$, is not differentiable at $(0, 0)$.

18. Find the equation of the tangent plane to the sphere $x^2 + y^2 + z^2 = 9$ at the point $P:(2, -2, 1)$.

19. Find the equation of the tangent plane to the hyperbolic paraboloid $z = xy + 9$ at the point $P:(0, 0, 9)$.

20. Try to find the equation of the tangent plane to $z = \sqrt{xy}$ at the origin. State the result.

16.7 (Theoretical Section) Uniqueness of Differentiation.

We first prove a lemma on the best linear approximation to a constant function. If $f(\vec{x})$ is constant on its domain, then $\Delta f = 0$ for every $\overrightarrow{\Delta x}$ at $\vec{x}$.

LEMMA. If $\vec{\phi}(h)$ is a linear function such that

$$\lim_{\|\vec{h}\| \to 0} \frac{\|\vec{\phi}(h)\|}{\|\vec{h}\|} = 0,$$

then $\vec{\phi}(h) = \vec{0}$ for every $\vec{h}$.

Proof. Suppose that for some $\vec{h}$, $\vec{\phi}(h) \neq \vec{0}$. Then for this $\vec{h}$,

$$\frac{\|\vec{\phi}(h)\|}{\|\vec{h}\|} = m \neq 0.$$

Evaluating this ratio at any scalar multiple $c\vec{h}$, $c \neq 0$, we find that on account of the linearity of $\vec{\phi}$, the factor $|c|$ cancels out and we have

$$\frac{\|\vec{\phi}(c\vec{h})\|}{\|c\vec{h}\|} = \frac{|c|\,\|\vec{\phi}(h)\|}{|c|\,\|\vec{h}\|} = m \neq 0.$$

Passing to the limit as c approaches zero, this implies that

$$\lim_{\|c\vec{h}\| \to 0} \frac{\|\vec{\phi}(c\vec{h})\|}{\|c\vec{h}\|} = m \neq 0,$$

contrary to the hypothesis of the lemma. Hence there can be no $\vec{h}$ where $\vec{\phi}(h) \neq \vec{0}$, which is the conclusion of the lemma.

Now for any $f(\vec{x})$, suppose that there are two different linear functions $\phi(\overrightarrow{dx})$ and $\psi(\overrightarrow{dx})$ such that both

$$\lim_{\|\overrightarrow{\Delta x}\| \to 0} \frac{\|\Delta f - \phi(\overrightarrow{\Delta x})\|}{\|\overrightarrow{\Delta x}\|} = 0 \quad \text{and} \quad \lim_{\|\overrightarrow{\Delta x}\| \to 0} \frac{\|\overrightarrow{\Delta f} - \psi(\overrightarrow{\Delta x})\|}{\|\overrightarrow{\Delta x}\|} = 0.$$

The difference $[\overrightarrow{\Delta f} - \phi(\overrightarrow{\Delta x})] - [\overrightarrow{\Delta f} - \psi(\overrightarrow{\Delta x})] = \psi - \phi$ is a linear function with the property that

$$\lim_{\|\overrightarrow{\Delta x}\| \to 0} \frac{\|\psi(\overrightarrow{\Delta x}) - \phi(\overrightarrow{\Delta x})\|}{\|\overrightarrow{\Delta x}\|} = 0.$$

From the lemma, we can infer that $\psi(\overrightarrow{dx}) - \phi(\overrightarrow{dx}) = 0$. That is, $\psi = \phi$. This completes the proof of the theorem on the uniqueness of the differential (§16.1).

17 Tangents to Graphs of Relations

17.1 Tangents to Plane Curves Given by $F(x, y) = C$

Example. Find in local coordinates the equation of the tangent line to the curve $x^2 - y^2 = 9$ at the point $P:(x, y)$ and at the point where $(x, y) = (5, 4)$.

Solution. The graph of the relation $x^2 - y^2 - 9$ is the level contour for the function $F(x, y) = x^2 - y^2$ at the level $F(x, y) = 9$ (Figure 9.3). If (x, y) and $(x + \Delta x, y + \Delta y)$ are in this contour then

$$\Delta F = F(x + \Delta x, y + \Delta y) - F(x, y) = 9 - 9 = 0.$$

Hence by the lemma on best linear approximation to a constant function (§16.7), it follows that the differential for the function $F(x, y)$ restricted to the level $F(x, y) = 9$ must vanish,

$$F_1(x, y)\, dx + F_2(x, y)\, dy = 0.$$

This relation in local (dx, dy) coordinates gives the required tangent line. We calculate $F_1 = 2x$ and $F_2 = -2y$, so the required tangent line has the equation $2x\, dx - 2y\, dy = 0$. Or at the point $(5, 4)$, we find $F_1(5, 4) = 10$, $F_2(5, 4) = -8$, so the equation of the tangent relation is $10dx - 8dy = 0$. It has the slope $dy/dx = \frac{5}{4}$. We plot this line (§17.1). This completes the example.

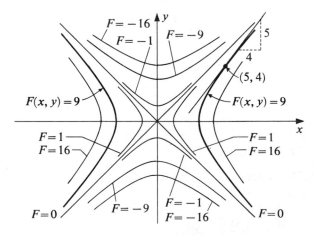

FIGURE 9.3 The relation $x^2 - y^2 = 9$ and function $F(x, y) = x^2 - y^2$.

We look again at the general point $P:(x, y)$, where the tangent relation is $2x\,dx - 2y\,dy = 0$. At any point where $F_2(x, y) \neq 0$, that is, where $y \neq 0$, we can find the slope

$$\frac{dy}{dx} = -\frac{F_1(x, y)}{F_2(x, y)} = \frac{x}{y}.$$

Even if $F_2(x, y) = 0$, however, the tangent relation

$$F_1(x, y)\,dx + F_2(x, y)\,dy = 0$$

gives the tangent line, provided that F_1 and F_2 are not both zero. In the example, at the point $(3, 0)$, $F_1(3, 0) = 6$, $F_2(3, 0) = 0$, so the tangent relation is $6\,dx + 0\,dy = 0$. This reduces to $dx = 0$, which is the local equation of the vertical line through the point. But the slope $dy/dx = \frac{6}{0}$ does not exist. Looking again at the figure, we see geometrically that this vertical line is actually the tangent line.

We summarize these results on the tangent lines to the graph of a relation $F(x, y) = C$. If $F(x, y)$ is differentiable, then at any point (x_0, y_0) in the curve the tangent relation is

$$F_1(x_0, y_0)\,dx + F_2(x_0, y_0)\,dy = 0.$$

If $F_1(x_0, y_0) = F_2(x_0, y_0) = 0$, the tangent relation describes the entire plane, but if either $F_1 \neq 0$ or $F_2 \neq 0$, the tangent relation describes the tangent line to the curve $F(x, y) = C$ at the point (x_0, y_0). If $F_2(x_0, y_0) \neq 0$, then the slope of the tangent line at the point is

$$\frac{dy}{dx} = -\frac{F_1(x_0, y_0)}{F_2(x_0, y_0)}.$$

All this does not assume an "implicit function," that is, a solution $y = f(x)$ of the relation $F(x, y) = C$.

17.2   Equation of Tangent Line in *xy*-coordinates.   If we have point $P:(x_0, y_0)$ in the relation $F(x, y) = C$, where F is differentiable at P, then the tangent relation is

$$F_1(x_0, y_0)\,dx + F_2(x_0, y_0)\,dy = 0.$$

Now let (x, y) be a variable point in the tangent line given in the original (x, y) coordinates. We recall the definition of local coordinates, which says that

$$dx = x - x_0 \qquad \text{and} \qquad dy = y - y_0.$$

Hence in *xy*-coordinates, the equation of the tangent line at $P:(x_0, y_0)$ is

$$F_1(x_0, y_0)(x - x_0) + F_2(x_0, y_0)(y - y_0) = 0.$$

Example.   Find the equation of the tangent line to the ellipse

$$\frac{x^2}{a^2} + \frac{y^2}{b^2} = 1$$

at the point (x_0, y_0) in its graph.

Solution.   With $F(x, y) = (x^2/a^2) + (y^2/b^2)$, we compute

$$F_1(x_0, y_0) = \frac{2x_0}{a^2}, \qquad F_2(x_0, y_0) = \frac{2y_0}{b^2}.$$

Hence, in local coordinates the equation of the tangent relation is

$$\frac{x_0}{a^2}\,dx + \frac{y_0}{b^2}\,dy = 0.$$

In xy-coordinates $dx = x - x_0$, $dy = y - y_0$, so this relation becomes

$$\frac{x_0}{a^2}\,(x - x_0) + \frac{y_0}{b^2}\,(y - y_0) = 0.$$

Simplifying and using the fact that (x_0, y_0) is in the curve, we find that the equation of the tangent line at P is

$$\frac{xx_0}{a^2} + \frac{yy_0}{b^2} = 1.$$

Similar calculations give equations of tangent lines to the quadratic relations in the accompanying table at the point $P : (x_0, y_0)$ on the curve.

Hyperbola: $\dfrac{x^2}{a^2} - \dfrac{y^2}{b^2} = 1.$	Tangent:	$\dfrac{xx_0}{a^2} - \dfrac{yy_0}{b^2} = 1.$
Parabola: $t^2 = 4px.$	Tangent:	$yy_0 = 2p(x + x_0).$
Quadric: $Ax^2 + Bxy + Cy^2 = k.$	Tangent:	$Axx_0 + B\left(\dfrac{xy_0 + x_0y}{2}\right) + Cyy_0 = k.$

17.3 Cusps and Double Points.

The differential style of calculus gives tangent lines where the function style may fail.

Example A. The "semicubical parabola" has the equation $y^2 = ax^3$, where $x \geqq 0$. Find the equation of the tangent at the origin $(0, 0)$.

Solution. We plot the graph (Figure 9.4). The origin clearly has unusual properties as a point of tangency. With $F(x, y) = ax^3 - y^2$, we compute $F_1 = 3ax^2$, $F_2 = -2y$. At the origin, $F_1 = 0$ and $F_2 = 0$. This gives us the tangent relation

$$0dx + 0dy = 0,$$

which is the entire plane, not a line in the plane. We can get more information if we factor $F(x, y)$ into two relations,

$$G(x, y) = (ax^3)^{1/2} - y = 0 \quad \text{and} \quad H(x, y) = (ax^3)^{1/2} + y = 0,$$

where we take the positive square root of ax^3 in each case. Then we find $G_1(x, y) = (ax)^{1/2}$ and $G_2(x, y) = -1$. At the origin, $G_1(0, 0) = 0$ and $G_2(0, 0) = -1$, so that the tangent line to the upper branch becomes

$$0dx - 1dy = 0, \quad \text{or just} \quad dy = 0.$$

Similarly for the lower branch, $H_1(0, 0) = 0$ and $H_2(0, 0) = 1$, giving us as the tangent line to the lower branch $0dx + 1dy = 0$, or just $dy = 0$. The two branches have a common tangent, the x-axis. This is a case where the tangent line intersects the curve at the point of tangency (Figure 9.4).

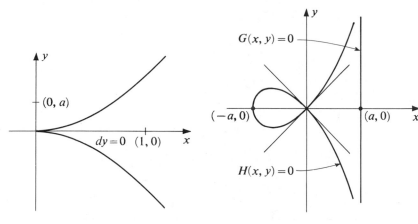

FIGURE 9.4   Semicubical parabola,
$y^2 = ax^3$, $a > 0$.

FIGURE 9.5   Strophoid,
$(a - x)y^2 = (a + x)x^2$.

A point $P:(x_0, y_0)$ on a curve $F(x, y) = C$ where $F_1(x_0, y_0) = F_2(x_0, y_0) = 0$ is called a *singular point*. A singular point P is called a *cusp* if there are two branches of the curve which have a common tangent line at P.

Example B.   Find the tangent lines to the curve $(a - x)y^2 = (a + x)x^2$ at $P:(0, 0)$.

Solution.   We plot the graph of the curve (Figure 9.5). For the function

$$F(x, y) = (a - x)y^2 - (a + x)x^2,$$

we compute

$$F_1(x, y) = -ay^2 - 2ax + 3x^2, \qquad F_2 = 2(a - x)y$$

and

$$F_1(0, 0) = 0, \; F_2(0, 0) = 0.$$

Hence the equation of the tangent relation of $F(x, y) = 0$ at $P:(0, 0)$ is $0dx + 0dy = 0$, whose graph is the entire plane and not a single line. Guided by the graph, we factor

$$F(x, y) = G(x, y)H(x, y) = (y\sqrt{a - x} - x\sqrt{a + x})(y\sqrt{a - x} + x\sqrt{a + x}).$$

Then for the factor $G(x, y)$, we find that $G_1(0, 0) = -\sqrt{a}$ and $G_2(0, 0) = +\sqrt{a}$, so that the equation of the tangent line to $G(x, y) = 0$ becomes $-\sqrt{a}\, dx + \sqrt{a}\, dy = 0$, or $dx - dy = 0$, which has the slope $dy/dx = 1$. Similarly, $H_1(0, 0) = \sqrt{a}$ and $H_2(0, 0) = \sqrt{a}$, so that the equation of the tangent relation to the other branch, $H(x, y) = 0$ at $P:(0, 0)$ is $\sqrt{a}\, dx + \sqrt{a}\, dy = 0$, or $dx + dy = 0$, which has the slope -1.

A singular point P on the graph of $F(x, y) = C$ is called a *double point* if there are two branches of the curve at P that have distinct tangent lines.

In Example B, the point $P:(0, 0)$ is a double point of the strophoid curve. We observe that cusps and double points may not exhaust all of the possible types of singular points of $F(x, y) = C$.

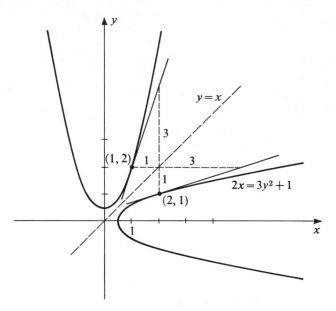

FIGURE 9.6   The relation $2y = 3x^2 + 1$ and its inverse.

17.4   Inverse Relations and their Tangents.

DEFINITION.   The relation $F^*(x, y) = C$ is the *inverse* of $F(x, y) = C$ if and only if for every (y, x) such that $F(y, x) = C$, $F^*(x, y) = F(y, x)$.

In other words, we form the inverse relation of $F(x, y)$ by interchanging x and y. Clearly, under the inversion operation, $(x, y) \to (y, x)$ and $(y, x) \to (x, y)$ so that $(F^*)^* = F$. Moreover, $(dx, dy) \to (dy, dx)$. Hence, if $A\, dx + B\, dy = 0$ is the tangent relation of $F(x, y) = C$ at (x, y), then $A\, dy + B\, dx = 0$ is the tangent relation of $F^*(x, y) = C$ at (y, x). If $A \neq 0$ and $B \neq 0$ at (x, y), then the slope of the tangent to $F(x, y) = C$ at (x, y) is $dy/dx = -A/B$. So the slope of the tangent to $F^*(x, y) = C$ at (y, x) is given by $(dy/dx)^* = -B/A$, which is the reciprocal of dy/dx. We could say, more briefly, if dy/dx is the slope of the tangent to $F(x, y) = C$ at (x, y), then dx/dy is the slope of the tangent to the inverse relation at (y, x).

Example.   Graph the relation $2y = 3x^2 + 1$ and its inverse.   Find the slope of the tangent to the relation at $(1, 2)$ and that of its inverse at $(2, 1)$.

Solution.   We plot the graphs, observing that the inversion operation $(x, y) \leftrightarrow (y, x)$ is graphically a reflection about the line $y = x$ (Figure 9.6).   With $F(x, y) = 3x^2 - 2y + 1$, we have $F_1(x, y) = 6x$, $F_2(x, y) = -2$ so that $F_1(1, 2) = 6$ and $F_2(1, 2) = -2$.   The equation of the tangent line to $F(x, y) = 0$ at $(1, 2)$ is $6dx - 2dy = 0$, whose slope is $\frac{1}{3}$. The inverse relation $F^*(x, y) = 0$ is $3y^2 - 2x + 1 = 0$, whose tangent line at $(2, 1)$ is $6dy - 2dx = 0$ and has slope $\frac{3}{1}$ at $(2, 1)$.   This completes the problem.

We observe that the inverse relation may be solved for y, and although it is not a function, it is the union of two functions $y = \frac{1}{3}\sqrt{2x - 1}$ and $y = -\frac{1}{3}\sqrt{2x - 1}$, both defined for $x \leq \frac{1}{2}$.   Only the positive function includes the point $(2, 1)$.

17.5   Tangents to Relations in 3-Space

Example.   Find in local and also *xyz*-coordinates the equation of the tangent line to $F(x, y, z) = 17$, where

$$F(x, y, z) = x^2 - 3xy + 6yz + 5xz + y^2 - z^2,$$

at the point $P:(1, -2, -1)$.

Solution.   We do not attempt to construct the graph. We compute

$$F_1 = 2x - 3y + 5z, \qquad F_2 = -3x + 2y + 6z, \qquad F_3 = 5x + 6y - 2z.$$

These are continuous everywhere, so F is differentiable, and for $F(x, y, z) = 17$, the equation of the tangent plane is $DF = 0$. Evaluating F_1, F_2, F_3 at $P:(1, -2, -1)$, we find that this becomes

$$2dx - 13dy - 5dz = 0,$$

which is an equation of the tangent plane in local coordinates. In *xyz*-coordinates, $dx = x - 1, dy = y + 2, dz = z + 1$, so the equation of the tangent plane at P is

$$3(x - 1) - 13(y + 2) - 5(z + 1) = 0.$$

This completes the example.

In general, the tangent plane to the graph of the relation $F(x, y, z) = C$ at $P:(x_0, y_0, z_0)$ is

$$F_1(x_0, y_0, z_0) \, dx + F_2(x_0, y_0, z_0) \, dy + F_3(x_0, y_0, z_0) \, dz = 0.$$

This plane has the normal vector $\bar{n}:(F_1, F_2, F_3)$ evaluated at P.

17.6   *Exercises*

In Exercises 1–5, find in local coordinates the equation of the tangent line at the point indicated. Plot the tangent line.

1. $xy = 8$ at $P:(-4, -2)$.
2. $x^2y = 4(2 - y)$ at $P:(0, 2)$.
3. $x^2 - 3xy + y^2 + 5 = 0$ at $P:(2, 3)$.
4. $x^3 + y^3 - 3axy = 0$ at $P:(0, 0)$.
5. $x^{2/3} + y^{2/3} = a^{2/3}$ at $P:(0, a)$.

In Exercises 6–11, find the equation of the tangent line in *xy*-coordinates at the point indicated.

6. $y^2 = ax^3$ at $P:(x_0, y_0)$.
7. $\sqrt{x} + \sqrt{y} = \sqrt{a}$ at point P, where $y = x$.
8.   The hyperbola (§13.1), at $P:(x_0, y_0)$.
9.   The parabola (§13.1), at $P:(x_0, y_0)$.
10.   The curve $Ax^2 + Bxy + Cy^2 = k$ at $P:(x_0, y_0)$.
11.   The line $ax + by + c = 0$ at $P:(x_0, y_0)$.

In Exercises 12–16, find the tangent plane and normal vector to the surface at the point indicated.

12.   The sphere, $x^2 + y^2 + z^2 = a^2$, at $P:(x_0, y_0, z_0)$.

13.   The ellipsoid, $\dfrac{x^2}{a^2} + \dfrac{y^2}{b^2} + \dfrac{z^2}{c^2} = 1$, at $P:(x_0, y_0, z_0)$.

14. The hyperboloid, $\dfrac{x^2}{a^2} + \dfrac{y^2}{b^2} - z^2 = 1$, at $P:(x_0, y_0, z_0)$.
15. The hyperbolic paraboloid, $z - xy = 0$, at $P:(0, 0, 0)$.
16. The cylinder, $x^2 + y^2 = 25$, at $P:(-3, 4, 6)$.
17. Find the slope of the tangent to the inverse relation of $x^2 - 4x + 2y^2 = 29$ at the point $P:(-4, 3)$ in the inverse.
18. Show that at the point $P:(x, y)$ in the inverse of $y = \ln x$, $dy/dx = y$.
19. Show that at the point $P:(x, y)$ in the inverse of $y = e^x$, $dx - x\,dy = 0$.
20. Show that at the point $P:(x, y)$ in the inverse of $\int_a^x f(u)\,du - y = 0$, the tangent relation is $dx - f(y)\,dy = 0$.
21. Calculate $(d/dx)\sin^{-1} x$ by treating $\sin^{-1} x - y = 0$ as the inverse of $\sin x - y = 0$.
22. For the function $y = x^p$, in the case that p is a fraction, $p = m/n$, calculate dy/dx from the relation $x^m - y^n = 0$. Show that the result agrees with the power formula $dy/dx = px^{p-1}$.
23. Calculate $(d/dx)\sqrt{a^2 - x^2}$ from the relation $x^2 + y^2 = a^2$.
24. Calculate $(d/dx)\sqrt{u}$ from the relation $u - y^2 = 0$, where u is a function of x.

18 Chain Rule and Directional Derivative

18.1 Chain Rule for Functions of One Variable. We return to the chain rule for differentiating composite functions $f(u)$, where f and u are differentiable functions of one variable (I, §20). If $w = f(y)$ is differentiable, its differential is $dw = f'(y)\,dy$. If $y = u(x)$ is differentiable, then its differential is $dy = u'(x)\,dx$. We form the composite function of x, $w = f[u(x)]$, by substituting $y = u(x)$ into $f(y)$. Let us perform the same substitution operation on the differentials. We form the composite differential by substituting $dy = u'(x)\,dx$ in the expression $dw = f'(y)\,dy$ and get $dw = f'[u(x)]u'(x)\,dx$. This composition actually gives the right derivative of the composite $f(u)$; for, in terms of the differentials, Df and $Du = u'(x)\,dx$, it says that $Df(u) = f'(u)\,Du$. This agrees with the chain rule previously formulated, where we formed the chain $f'(u)\,Du$ by multiplying $f'(u)$ by the chain rule factor Du.

Example. Differentiate $\ln (x^2 + 1)$.

  Solution. This is the composite formed by substituting $u = x^2 + 1$ into $w = \ln u$. Differentiating, we find $du = 2x\,dx$ and $dw = (1/u)\,du$. Substituting $u = x^2 + 1$ and $du = 2x\,dx$ into this expression for dw, we find $dw = 2x\,dx/(x^2 + 1)$. This calculates the required differential by substitution instead of by multiplication by the chain rule factor $2x$.

  It turns out for multivariable calculus that the substitution procedure for forming the differential of a composite function is much simpler than chain rule formulas, which extend the idea of multiplication by a chain rule factor.

18.2 Notation for Composite Functions. Let $\vec{w} = f(\vec{y})$ be a function that maps vectors $\vec{y}$ into vectors $\vec{w}$. Then let $\vec{y} = u(\vec{x})$ be a function that maps vectors $\vec{x}$ into vectors $\vec{y}$. We require that the images $\vec{y}$ of vectors $\vec{x}$ under u be in the domain of f. In this case we say that u is composable into f, and the composite function $f \circ u$ is *composed* of f and u if and only if

$$(f \circ u)(\vec{x}) = f[u(\vec{x})].$$

18.3   General Chain Rule

THEOREM.   Let $\vec{y} = u(\vec{x})$ be a differentiable function that is composable into the differentiable function $\vec{w} = f(\vec{y})$. Then

$$Df \circ u = Df \circ Du.$$

The rule says in words, To form the differential of the composite function obtained by substituting the function u into f, substitute the differential of u into the differential of f.

We postpone the proof (§18.6) to work some examples. We have already considered an example for functions of one variable (§18.1).

Example A.   Differentiate the composite $w = x^2 - 2y^2 + 3z^2$, where $x = 2t - 4$, $y = 5t + 7$, $z = -t + 4$. Here u gives the row (x, y, z) as a function of the real variable t. The composite function is given by

$$w = f[u(t)] = (2t - 4)^2 - 2(5t + 7)^2 + 3(-t + 4)^2.$$

Solution.   We are to calculate Dw for this composite. According to the rule, we calculate

$$Du:(Dx, Dy, Dz) = (2dt, 5dt, -1dt)$$

and

$$Df = 2x\,dx - 4y\,dy + 6z\,dz.$$

Then we substitute the expression for x, y, z as functions of t and substitute for Dx, Dy, Dz into Df. This gives us the differential of the composite

$$Dw = 2(2t - 4)\,2dt - 4(5t + 7)\,5dt + 6(-t + 4)(-1)dt = (-86t - 160)dt.$$

Since in this case the composite is an ordinary number-valued function of the number variable t we can verify the result by differentiating $f[u(t)]$ directly. This gives us

$$Dw = [2(2t - 4)2 - 4(5t + 7)5 + 6(-t + 4)(-1)]dt = (-86t - 160)dt,$$

which checks.

Example B.   We consider a case in which the differentiable function $\vec{u} = g(\vec{x})$ maps plane vectors $\vec{x}:(x, y)$ into plane vectors $\vec{u}:(u, v)$ and the differentiable function $\vec{w} = f(\vec{u})$ maps vectors $\vec{u}$ into plane vectors $\vec{w}:(p, q)$. Explicitly, u and f are given by

$$u:\quad u = x - y, \qquad v = 3xy,$$

$$f:\quad p = \frac{u}{v}, \qquad q = \frac{v}{u}.$$

Calculate the differential of the composite $f \circ u$ at the point P, where $(x, y) = (2, -1)$.

Solution.   We compute the partial derivatives

$$u_1 = 2x, \qquad u_2 = -2y,$$
$$v_1 = 3y, \qquad v_2 = 3x.$$

Evaluating them at $(2, -1)$, we find

$$u_1(2, -1) = 4, \qquad u_2(2, -1) = 2,$$
$$v_1(2, -1) = -3, \qquad v_2(2, -1) = 6.$$

Hence at P the differential Du is given by

$$Du: \quad \begin{aligned} du &= 4dx + 2dy, \\ dv &= -3dx + 6dy. \end{aligned}$$

Similarly, we calculate the partial derivatives of p and q at the point Q, where $(u, v) = (3, -6)$, which is the image of $(2, -1)$. At (u, v) they are

$$p_1 = \frac{1}{v}, \qquad p_2 = -\frac{u}{v^2}, \qquad q_1 = -\frac{v}{u^2}, \qquad q_2 = \frac{1}{u}.$$

Then, evaluating these partial derivatives at $(u, v) = (3, -6)$, we have

$$p_1(3, -6) = -\tfrac{1}{6}, \qquad p_2(3, -6) = -\tfrac{1}{12}, \qquad q_1(3, -6) = \tfrac{2}{3}, \qquad q_2(3, -6) = \tfrac{1}{3}.$$

Hence at Q the differential Df is given by

$$Df: \quad \begin{aligned} dp &= -\tfrac{1}{6} \, du - \tfrac{1}{12} \, dv, \\ dq &= \tfrac{2}{3} \, du + \tfrac{1}{3} \, dv. \end{aligned}$$

The chain rule says that to form the differential $D(f \circ u)$ we substitute Du into Df, obtaining $Df \circ Du$. If we carry out this substitution, we get

$$D(f \circ u): \quad \begin{aligned} dp &= -\tfrac{5}{12} \, dx - \tfrac{10}{12} \, dy, \\ dq &= \tfrac{5}{3} \, dx + \tfrac{10}{3} \, dy. \end{aligned}$$

This completes the problem.

18.4 Directional Derivative.

We consider a scalar-valued function $f(x, y)$ at a point $P:(x_0, y_0)$ interior to its domain. For each unit direction $\hat{h}:(h, k)$ out of P there is a line

$$x = x_0 + sh, \qquad y = y_0 + sk,$$

where s is a real variable. We observe that when $s = 0$, then $x = x_0$, $y = y_0$. If we restrict the points (x, y) to lying on this line, the restricted function is represented by the composite $f(x_0 + sh, y_0 + sk)$. This is an ordinary number-valued function of the real variable s. Hence when we differentiate it with respect to s and then set $s = 0$, we get the numerical rate of change of f at P along the line with direction $\hat{h}$. We denote this *directional derivative* of f in the direction $\hat{h}$ by the symbol $\vec{\nabla} f \cdot \hat{h}$. The symbol $\vec{\nabla}$ is read "del." The directional derivative is given by

$$\vec{\nabla} f \cdot \hat{h} = \frac{d}{ds} f(x_0 + sh, y_0 + sk) \Big|^{s=0}.$$

We can calculate it by the chain rule to be

$$\vec{\nabla} f \cdot \hat{h} = f_1 h + f_2 k,$$

where the partial derivatives f_1 and f_2 are evaluated at P.

Example. Calculate the directional derivative of the function f given by $f(x, y) = x^2 - 3xy + y^2$ at the point $P:(-1, 3)$ in the direction $\hat{h}:(\tfrac{3}{5}, -\tfrac{4}{5})$.

Solution. The line through P in the direction $\hat{h}$ is given by the parametric equations (§11.6)

$$x = \frac{3}{5} s - 1, \qquad y = -\frac{4}{5} s + 3.$$

The differential of this row function is

$$dx = \frac{3}{5} \, ds, \qquad dy = -\frac{4}{5} \, ds.$$

The differential of f at $P:(-1, 3)$ is

$$Df = f_1 \, dx + f_2 \, dy = -11 \, dx + 9 \, dy.$$

Hence by the chain rule, when we differentiate and set $s = 0$,

$$Df(\tfrac{3}{5}s - 1, -\tfrac{4}{5}s + 3) = [-11(\tfrac{3}{5}) + 9(-\tfrac{4}{5})]ds = -\tfrac{69}{5} ds.$$

That is, the required directional derivative is given by

$$\vec{\nabla} f \cdot \vec{h} = \frac{d}{ds} f \left(\frac{3}{5} s - 1, -\frac{4}{5} s + 3 \right) \Big|^{s=0} = -\frac{69}{5}.$$

This means that in the unit direction $\vec{h}:(\tfrac{3}{5}, -\tfrac{4}{5})$ the function f is decreasing at the rate $-\tfrac{69}{5}$ per unit s.

Clearly, the directional derivative can be calculated for a function of any number of variables. For example, the directional derivative of $f(x, y, z)$ in the unit direction $\vec{h}:(h_1, h_2, h_3)$ at a point P is given by

$$\vec{\nabla} f \cdot \vec{h} = f_1 h_1 + f_2 h_2 + f_3 h_3,$$

where the partial derivatives are evaluated at P. More generally the directional derivative of $f(\vec{x})$, where $\vec{x}:(x_1, x_2 \ldots, x_n)$ is a vector in n-space, taken in the unit direction $\vec{h}:(h_1, h_2, \ldots, h_n)$, is

$$\vec{\nabla} f \cdot \vec{h} = f_1 h_1 + f_2 h_2 + \cdots + f_n h_n.$$

These formulas, which are determined by the chain rule, justify the use of the dot-product notation for the directional derivative. For in every case the directional derivative of f in the direction $\vec{h}$ is the dot product of the vector $\vec{\nabla} f:(f_1, f_2, \ldots, f_n)$ with the unit vector $\vec{h}:(h_1, h_2, \ldots, h_n)$.

18.5   Exercises

In Exercises 1–8, using the chain rule calculate the differentials of composite functions.

1. $D(x^2 + 4y^2)$, where $x = 2t - 1$, $y = -3t + 5$.
2. $D(xy - 3)$, where $x = 6s$, $y = 1/s$.
3. $D\sqrt{u^2 + v^2}$, where $u = 3s - 1$, $v = 2s^2 + 1$.
4. $D \dfrac{1}{\sqrt{x^2 + y^2 + z^2}}$, where $x = 2t$, $y = 1$, $z = 1 - t$.
5. De^x, where $x = u^2 - v^2$.
6. $D \ln y$, where $y = u/v$.
7. $F(x_1, x_2)$ is the row $(1 - x_1 x_2, x_1 + x_2)$ and $x_1 = t^2 - 1$, $x_2 = 1 - 3t$.
8. $F(x_1, x_2)$ is the row $(1 - x_1 x_2, x_1 + x_2)$ and for each (y_1, y_2), $x_1 = 2y_1 - y_2$, $x_2 = y_1 + 1$.

In Exercises 9–16, calculate the directional derivative $\vec{\nabla} F \cdot \vec{h}$ in the given direction.

9. $F(x, y) = x^2 + xy + y^2$, $P:(2, -1)$, $\vec{h}:(\sqrt{3}/2, -1/2)$.
10. $F(x, y) = x^2 - 3y^2$, $P:(1, 1)$, $\vec{h}:(1/\sqrt{5}, -2/\sqrt{5})$.
11. $F(x, y) = \sin x \cos y$, $P:(\pi/3, \pi/2)$, in the direction of $\vec{v}:(1, \sqrt{3})$. [Note that $\vec{v}$ is not a unit vector.]

12. $F(x, y)$ at $P:(x_0, y_0)$ in the direction $\vec{h}:(1, 0)$.
13. $F(x, y)$ at $P:(x_0, y_0)$ in the direction $\vec{h}:(0, 1)$.
14. $F(x, y)$ at $P:(x_0, y_0)$ in the direction $\vec{h}:(\cos \alpha, \sin \alpha)$.
15. $F(x, y) = x \cos \alpha + y \sin \alpha - p$ at $P:(x_0, y_0)$ in the direction $\vec{h}:(\cos \alpha, \sin \alpha)$.
16. $F(x, y, z) = 1/\sqrt{x^2 + y^2 + z^2}$ at $P:(-3, 0, 4)$ in the direction $\vec{h}:(h_1, h_2, h_3)$, where $\|\vec{h}\| = 1$.

17.   In Exercise 16, choose $\vec{h}$ so that $\vec{\nabla}F \cdot \vec{h}$ attains its maximum value.
18.   The surface of a hill is in the shape of the graph of $w = 225 - 9x^2 - 25y^2$. At the point $P:(4, \frac{9}{5}, 0)$ on the hill, what is the slope of the hill in the direction $\vec{h}:(-1/\sqrt{2}, -1/\sqrt{2})$? Uphill or downhill?
19.   In Exercise 18, plot contours, $w = $ const, to represent the function. Calculate $\vec{\nabla}f \cdot \vec{h}$ in a direction $\vec{h}$ tangent to one of the contours. Calculate $\vec{\nabla}f \cdot \vec{h}$ in a direction $\vec{h}$ orthogonal to a level contour line.
20.   Using projection of vectors, show for $F(x, y)$ that

$$\vec{\nabla}F \cdot \vec{h} = \sqrt{F_1^2 + F_2^2} \cos \theta,$$

where θ is an angle between the vector $\vec{\nabla}F:(F_1, F_2)$ and the unit vector $\vec{h}$.

18.6 (Theoretical) Proof of the Chain Rule.   We will first prove a lemma about linear functions.

LEMMA.   If $\boldsymbol{\phi}(\vec{h})$ is a continuous linear function, then for some positive number b, and all $\vec{h}$, $\|\vec{\phi}(\vec{h})\| \leq b\|\vec{h}\|$.

   Proof.   Since $\vec{\boldsymbol{\phi}}$ is continuous, and $\vec{\phi}(\vec{0}) = \vec{0}$, there is a ball $\{\|\vec{h}\| \leq r\}$ in which $\|\vec{\phi}(\vec{h})\| < 1$. Now let $\vec{h}$ be any nonzero vector. Since $\vec{\boldsymbol{\phi}}(\vec{h})$ is linear, we can multiply and divide by $\|\vec{h}\|/r$ and find that

$$\vec{\phi}(\vec{h}) = \frac{\|\vec{h}\|}{r} \vec{\phi}\left(\frac{r}{\|\vec{h}\|} \vec{h}\right).$$

Since $\|(r/\|\vec{h}\|)\vec{h}\| = r$, this evaluates $\vec{\phi}$ at a point where $\|\vec{\phi}(\vec{h})\| < 1$. Hence for all $\vec{h}$ we have

$$\|\vec{\phi}(\vec{h})\| \leq \frac{\|\vec{h}\|}{r}.$$

This completes the proof, with $b = 1/r$.

   We return to the proof of the chain rule, which says that $\boldsymbol{D\vec{f}} \circ \boldsymbol{D\vec{u}}$ is the differential of $\vec{f} \circ \vec{u}$. We denote by $\Delta\vec{f}$ the difference $\Delta\vec{f} = \vec{f}(\vec{y} + \Delta\vec{y}) - \vec{f}(\vec{y})$. With $\vec{y}$ fixed, $\Delta\vec{f}$ is a function of $\Delta\vec{y}$. If we substitute $\vec{y} = \vec{u}$ and $\Delta\vec{y} = \vec{u}$ into $\Delta\vec{f}$, we get, for the composite $\vec{f} \circ \vec{u}$,

$$\Delta(\vec{f} \circ \vec{u}) = f(\vec{u} + \Delta\vec{u}) - f(\vec{u}).$$

   We can define a function $\vec{p}(\nabla\vec{y})$, which gives the "relative" deviation of the graph of $\vec{f}$ from the tangent (Figure 9.7) by

$$\frac{\Delta\vec{f} - D\vec{f}(\Delta\vec{y})}{\|\Delta\vec{y}\|} = \vec{p}(\Delta\vec{y}).$$

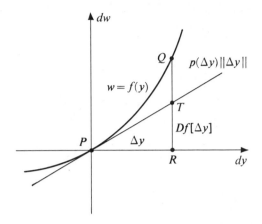

FIGURE 9.7   Deviation of graph from tangent,
$$RQ = \Delta f, \qquad RT = Df,$$
$$\text{deviation} = RQ - RT = TQ = p(\Delta y)\,\|\Delta y\|.$$
$$\text{Then } \Delta f = Df + p(\Delta y)\,\|\Delta y\|.$$

Here $D\vec{f}(\Delta \vec{y})$ means the value of $D\vec{f}$ at $\Delta \vec{y}$. We solve for $\Delta \vec{f}$ in this expression, and find that even if $\Delta \vec{y} = 0$,

$$\Delta \vec{f} = D\vec{f}(\Delta \vec{y}) + \vec{p}(\Delta \vec{y})\|\Delta \vec{y}\|.$$

Then $\vec{f}$ has $D\vec{f}$ for its differential if and only if

$$\lim_{\|\Delta \vec{y}\| \to 0} \|\vec{p}(\Delta \vec{y})\| = 0.$$

This is just a restatement of the definition of $D\vec{f}$. Similarly, $\vec{u}(\vec{x})$ has the differential $D\vec{u}$ at $\vec{x}$ if and only if

$$\Delta \vec{u} = D\vec{u} + q(\Delta \vec{q})\|\Delta \vec{x}\|,$$

and

$$\lim_{\|\Delta \vec{x}\| \to 0} \|\vec{q}(\Delta \vec{x})\| = 0.$$

Now we substitute this expression, giving $\Delta \vec{u}$, for $\Delta \vec{y}$ in that for $\Delta \vec{f}$ above and find that

$$\Delta(\vec{f} \circ \vec{u}) = D\vec{f}[D\vec{u} + \vec{q}(\Delta \vec{x})\|\Delta \vec{x}\|] + \vec{p}(\Delta \vec{u})\|D\vec{u} + \vec{q}(\Delta \vec{x})\|\Delta \vec{x}\|\|.$$

Using the linearity of Df, we see that

$$D\vec{f}[D\vec{u} + \vec{q}(\Delta \vec{x})\|\Delta \vec{x}\|] = D\vec{f}[D\vec{u}] + \|\Delta \vec{x}\|\, D\vec{f}[\vec{q}(\Delta \vec{x})].$$

Therefore, transposing the term, $D\vec{f}[D\vec{u}] = D\vec{f} \circ D\vec{u}$, and dividing both sides by $\|\Delta \vec{x}\|$, we find

$$\frac{\Delta(\vec{f} \circ \vec{u}) - D \circ \vec{f} D\vec{u}}{\|\Delta \vec{x}\|} = D\vec{f}[\vec{q}(\Delta \vec{x})] + \vec{p}(\Delta \vec{u})\left\|\frac{D\vec{u}}{\|\Delta \vec{x}\|} + \vec{q}(\Delta \vec{x})\right\|.$$

The theorem will be proved if we can show that the right-hand side approaches zero as $\|\Delta \vec{x}\| \to 0$. For in that case, $D\vec{f} \circ D\vec{u}$ is the best linear approximation to $\vec{f} \circ \vec{u}$.

Although

$$\lim_{\|\Delta \vec{x}\| \to 0} \vec{q}(\Delta \vec{x}) = 0 \qquad \text{and} \qquad \lim_{\|\Delta\|\vec{x} \to 0} \vec{p}(\Delta \vec{u}) = 0,$$

there is a difficulty. It is that the term $\|D\vec{u}\|/\|\Delta\vec{x}\|$ may not remain bounded as $\|\Delta\vec{x}\| \to 0$. But the lemma resolves this difficulty, since it says that $\|D\vec{u}\| \leq b\|\Delta\vec{x}\|$, where b is a constant. Then we see that

$$\frac{\|\Delta(\vec{f}\circ\vec{u}) - D\vec{f}\circ D\vec{u}\|}{\|\Delta\vec{x}\|} \leq \|D\vec{f}[\vec{q}(\Delta\vec{x})]\| + \|\vec{p}(\Delta\vec{u})\|[b + \|\vec{q}(\Delta\vec{x})\|].$$

Now we let $\Delta\vec{x}$ approach zero. Then since $\vec{f}$ and $\vec{u}$ are differentiable,

$$\Delta\vec{u} \to \vec{0}, \qquad \|\vec{q}(\Delta\vec{x})\| \to 0, \qquad \|\vec{p}(\Delta\vec{u})\| \to 0.$$

Moreover, the fact that $D\vec{f}$ is continuous implies that $\|D\vec{f}[\vec{q}(\Delta\vec{x})]\| \to 0$. Therefore the right member approaches zero. This completes the proof that $D\vec{f}\circ D\vec{u}$ is the differential of $\vec{f}\circ\vec{u}$.

19   Related Rates and Approximations

19.1   Related Rates

Example.   A highway and a railroad track intersect at right angles. A car $\frac{1}{2}$ mile from the intersection is moving towards it at 60 mi/hr and an engine 1 mile from the intersection is moving towards it at 80 mi/hr. At what rate is the distance between car and engine changing?

Solution.   Let x be the function of time that represents the distance of the car from the intersection and y the function (Figure 9.8) that gives the engine's distance, and let r give the distance between them. Then, at every time,

$$x^2 + y^2 = r^2.$$

Differentiating with respect to t this composite function, we have the relation between rates

$$2xx' + 2yy' = 2rr'.$$

We know that when $x = \frac{1}{2}$, $x' = -60$, then $y = 1$, $y' = -80$. It follows from the first relation that $r = \sqrt{5}/2$. Only r' is unknown. Substituting these data into the relation of rates, we find that

$$2\left(\frac{1}{2}\right)(-60) + 2(1)(-80) = 2\frac{\sqrt{5}}{2}r'$$

or $r' = -44\sqrt{5}$ mi/hr, about -98.4 mi/hr.

The pattern may be described in general. We have functions $x(t)$, $y(t)$, $z(t)$ whose values for every time t satisfy a relation

$$F(x, y, z) = C.$$

We know the values $x(t_0)$, $y(t_0)$, $z(t_0)$, and all but one of the rates $x'(t_0)$, $y'(t_0)$, $z'(t_0)$ at this time. The problem is to find the unknown rate.

To do this we consider the composite functional relation

$$F[x(t), y(t), z(t)] = C.$$

We differentiate it with respect to t, using the chain rule. This gives us the rate relation

$$F_1x' + F_2y' + F_3z' = 0.$$

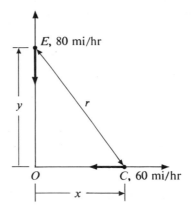

FIGURE 9.8   Car C and engine E approaching intersection O.

Only after this is done do we evaluate the functions involved at the time t_0. Using the evaluated functional relation

$$F[x(t_0), y(t_0), z(t_0)] = C$$

and the rate relation, we solve for the unknown rate.

19.2   Motion in a Circle.   In the xy-plane,

$$x = x(t), \qquad y = y(t)$$

represents a motion if $x(t)$ and $y(t)$ are functions of time t. If the moving point $P:(x(t), y(t))$ is always on the circle with center at (x_0, y_0) and radius r, then for all times t

$$[x(t) - x_0]^2 + [y(t) - y_0]^2 = r^2.$$

Differentiating with respect to t, we find a relation between the rates $x'(t)$ and $y'(t)$

$$[x(t) - x_0]x'(t) + [y(t) - y_0]y'(t) = 0.$$

This says that the radial vector given by $\vec{q}:[x(t) - x_0, y(t) - y_0]$ and the velocity given by the vector $\vec{v}:[x'(t), y'(t)]$ are always orthogonal. That is, the motion is always in the direction tangent to the circle (Figure 9.9). The linear speed is given by the norm $\|\vec{v}\| = \sqrt{x'^2 + y'^2}$. If we differentiate again, we find

$$[x(t) - x_0]x''(t) + [y(t) - y_0]y''(t) + [x'(t)]^2 + [y'(t)]^2 = 0.$$

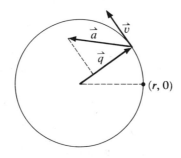

FIGURE 9.9   Motion in a circle.

This implies that the scalar projection of the acceleration $\vec{a}:(x''(t), y''(t))$ on the line of the unit radius $(1/r)\vec{q}$ is given by $-\|\vec{v}\|^2/r$. This applies whether the motion has a uniform angular rate about the center or not.

19.3   Uniform Motion in a Circle.   A motion

$$x = x(t), \qquad y = y(t)$$

in a circle $x^2 + y^2 = r^2$ is said to be *uniform* if the tangential velocity is constant in magnitude. Since the norm of the velocity vector is a constant $\|\vec{v}\|$, we have

$$x'^2 + y'^2 = \|\vec{v}\|^2,$$

and differentiation with respect to t implies that

$$x'x'' + y'y'' = 0.$$

This says that the velocity vector $\vec{v}:(x', y')$ is perpendicular to the acceleration vector with coordinates (x'', y''). Hence the projection of the acceleration vector on the tangent line is zero and we say that the tangential component of acceleration is zero.

Thus for uniform motion in a circle the velocity is all tangential with constant speed $\|\vec{v}\| = \sqrt{x'^2 + y'^2}$. The acceleration vector points inward along the radius, with norm $\|\vec{v}\|^2/r$.

19.4   Approximation of Related Quantities.   We can use the principle of best linear approximation to estimate small changes in the values of quantities that satisfy a relation of the form

$$F(x, y, z) = C.$$

The method is a variation of that of estimating the change in a function $w = f(x, y)$ due to small changes in the values of x and y (§16.3). Starting from (x_0, y_0, z_0) in the relation, the values of x and y are changed by known increments to $(x_0 + \Delta x, y_0 + \Delta y)$. The problem is to find Δz so that

$$F(x_0 + \Delta x, y_0 + \Delta y, z_0 + \Delta z) = C$$

remains true. It may be difficult to solve this equation for Δz in terms of $x_0, y_0, z_0, \Delta x, \Delta y$. We can replace F by its best linear approximation to obtain the tangent relation

$$F_1 \, dx + F_2 \, dy + F_3 \, dz = 0.$$

When $\Delta x, \Delta y, \Delta z$ are small, they satisfy this tangent relation approximately,

$$F_1 \, \Delta x + F_2 \, \Delta y + F_3 \, \Delta z = 0.$$

With evaluation of the partial derivatives F_1, F_2, F_3 at (x_0, y_0, z_0), this linear relation is easy to solve for the unknown Δz in terms of Δx and Δy.

Example.   At the point $(-1, 2)$ in the relation

$$x^2 - 3xy + 4y^2 - 6x + 2y = 33,$$

the value of y is changed to 2.1. What is the resulting change in x if the relation continues to hold?

Solution.   We replace the nonlinear relation by its best linear approximation

$$(2x - 3y - 6) \, dx + (-3x + 8y + 2) \, dy = 0.$$

If $(x, y) = (-1, 2)$ and $dy = 0.1$, the relation reduces to $-14dx + 21(0.1) = 0$, or $dx = 0.15$. Hence the approximate value of Δx is 0.15. That is, $(-0.85, 2.1)$ approximately satisfies the original relation. The method does not provide a ready estimate of the error of this approximation.

19.5   Relative Error in Relations.

The *relative change* in x from the value x to $x + \Delta x$ is $\Delta x / x$, expressed as a percentage.

Example A.   In a quantity of gas obeying the gas law

$$pv = RT,$$

where T is absolute temperature (Kelvin), p is pressure, and v is volume, the volume is increased by k percent, while T is kept constant. What is the relative change in p?

Solution.   We may use the best linear approximation given by the tangent relation

$$p \, dv + v \, dp = 0.$$

We divide by pv and find

$$\frac{dv}{v} + \frac{dp}{p} = 0.$$

Thus if $dv/v = k$ percent, then $dp/p = -k$ percent approximately. This completes the problem.

Logarithmic differentiation (I, §26.4) is useful in problems of related relative increments, since $y = \ln x$ implies that $dy = dx/x$, a relative change in x.

Example B.   If z is related to x and y by a power relation

$$z = Cx^m y^n,$$

where C is constant, what relative change is produced in z by a 2 percent increase in x and a 5 percent decrease in y?

Solution.   Using logarithm differentiation, we first take the logarithm of both members

$$\ln |z| = \ln |C| + m \ln |x| + n \ln |y|.$$

Then differentiation of this relation gives us a best linear approximation

$$\frac{dz}{z} = 0 + m \frac{dx}{x} + n \frac{dy}{y}.$$

We are given that $dx/x = 2$ percent and $dy/y = -5$ percent. Hence $dz/z = 2m - 5n$ percent.

19.6   Exercises

1. A point $P\colon(x, y)$ moves on the circle $x^2 + y^2 = 25$, not necessarily with constant velocity. At the point $(-3, 4)$ the y-coordinate is decreasing at 2 cm/sec. Find the rate of change of x.

2. The area of a circle is increasing at 4 cm²/sec. At what rate is the radius r increasing when $r = 5$?

3. A spherical raindrop of surface area S gathers moisture at a rate kS, where k is constant. Find the rate at which its radius increases.

4. A barge whose deck is 12 ft below the level of a dock is drawn to the dock by a cable attached to the deck. If the cable is pulled at the rate of 15 ft/min, how fast is the barge moving when it is 16 ft from the dock?

5. A 5 ft 9 in. man walks at 5 mi/hr away from a pinpoint light 6 ft above the ground. Show that the tip of his shadow moves at 120 mi/hr.

6. Fluid is withdrawn from a conical funnel at the rate of 1 cm³/sec. The funnel is 8 cm wide at the top and 12 cm deep. How fast is the surface falling when the fluid is 3 cm deep?

7. A spherical rubber balloon is being inflated at the rate of 6 cu in. per second (Figure 9.10).
    (a) What is the relative rate of increase of the volume when the radius is 3 in.?
    (b) What is the relative rate of increase of the volume when the radius is increasing 2 percent per second?

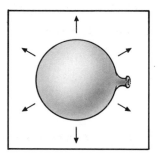

FIGURE 9.10   Exercise 7.

8. On a circle of radius 10, the radius is increased by 0.1. What is the increase in area? Find exact differential solutions.

9. A scientist measures a quantity v and finds that $v = 6.94$ with an error not exceeding 0.05. He computes A from the relation $Av^2 = 10$. Approximately what relative (percent) error is induced in A by the maximum error in v?

10. In the relation $x^2 - 3y^2 = 4$ at the point $(2\sqrt{2}, -2\sqrt{3})$, if x is increased by 2 percent what is the approximate relative change in y?

11. Calculate $\sqrt[3]{1008}$ using differential approximation.

12. Approximately what percentage change is induced in the volume of a gas under the adiabatic law $pv^\gamma = c$, when pressure is increased q percent? Here γ and c are constants.

13. In calculating the volume of the earth, if an error of $\frac{1}{2}$ mile is made in the radius of 3960 miles, approximately what error does this make in the result?

14. If a body weighs 10 pounds at the surface of the earth, find the approximate change in weight if it is carried up 10 miles.

15. In the manufacture of a steel ball bearing, the diameter can be controlled to an error of not more than 0.1 percent. What relative variation in the volume does this imply?

16. There are about 0.5×10^{24} molecules of water in a cubic inch of water. If a cubic inch of water were poured into the ocean, and the entire ocean thoroughly stirred, and then a cubic inch of water withdrawn, about how many molecules of the water previously poured in would be in it? Assume that the volume of the ocean corresponds to a sea one mile deep covering the entire earth.

17. The period T of a simple pendulum is related to its length l and acceleration g due to gravity by $T^2g = 4\pi^2l$. Write a linear relation between small changes in T, g, and l.

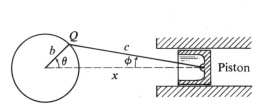

FIGURE 9.11   Exercise 18.   FIGURE 9.12   Exercise 21.

18. A piston is driven by a connecting rod QP (Figure 9.11) of length c attached to a crankshaft at Q that is rotating at the rate ω revolutions per minute. Find the linear velocity of the piston as a function of x, θ, and ω.

19. In Exercise 18, find the relation between the angular rates $d\theta/dt$ and $d\phi/dt$.

20. In Exercise 18, at what position θ is the linear velocity of the piston a maximum?

21.* The copilot in airplane A sees another airplane B at distance r flying directly towards him (Figure 9.12) at azimuth θ. If the speed of A is α mi/hr and that of B is β, what is the rate of change of θ?

22. A body at point $P:(x, y, z)$ is moving with velocity $\vec{v}:(a, b, c)$ and a body at point $P_0:(x_0, y_0, z_0)$ is moving with velocity $\vec{v}_0:(a_0, b_0, c_0)$. Find the rate of change of the distance $\|PP_0\|$.

23. For dense projectiles launched from a horizontal plane with speed v_0 and angle of elevation α, the range R is given by

$$R = \frac{v_0^2}{g} \sin 2\alpha.$$

If v_0 is increased 5 percent, what change in α will keep the range constant?

20   Gradient and Directional Derivative

20.1   Gradient.   We return to the directional derivative of the function f at the point $P:(x, y)$ in its domain. In every unit direction $\vec{h}:(h, k)$ from P the rate of change of f in the direction $\vec{h}$ is given by (§18.4)

$$(\vec{\nabla}f\cdot\vec{h}) = f_1 h + f_2 k.$$

This dot product involves a fixed vector whose components are (f_1, f_2). We give it a name.

DEFINITION.   The *gradient* of the number valued function f, at a point $P:(x, y)$ in its domain, is the vector $\vec{\nabla}f:(f_1, f_2)$, whose coordinates are the partial derivatives of f, evaluated at P.

In view of this definition, we can interpret the directional derivative as the scalar projection of the gradient $\vec{\nabla}f$ onto the line of $\vec{h}$ (Figure 9.13). We can also express this projection in terms of the cosine of an angle θ between the gradient $\vec{\nabla}f$ and $\vec{h}$ as

$$(\vec{\nabla}f\cdot\vec{h}) = \|\vec{\nabla}f\| \cos \theta.$$

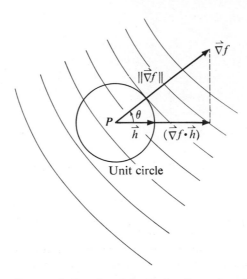

FIGURE 9.13   Gradient and directional derivative for a function represented by
level contours.

In this formula we see that the directional derivative $(\vec{\nabla} f \cdot \vec{h})$ attains its maximum value
when $\theta = 0$, and that this maximum value is $\|\vec{\nabla} f\|$. Because of its importance we state
this result as a theorem.

THEOREM.   The maximum directional derivative of f at P occurs in the direction of the
gradient $\vec{\nabla} f$, and the magnitude of this maximum directional derivative is $\|\vec{\nabla} f\|$.

   One consequence of this theorem is that the gradient is an invariant geometric object.
For the function assigns to each point P a number $f(P)$. The direction and magnitude of
the maximum rate of change of f are then determined by the distance unit in the plane
and do not depend on the particular coordinate basis in terms of which the coordinates
(x, y) are expressed. This is true even though the definition of $\vec{\nabla} f$ was originally stated in
terms of the partial derivatives (f_1, f_2) with respect to the noninvariant coordinates
(x, y).
   We easily generalize the gradient to functions $f(x_1, x_2, \ldots, x_n)$ of any number of
variables by defining the gradient as the vector

$$\vec{\nabla} f : (f_1, f_2, \ldots, f_n),$$

whose components are the partial derivatives of f at P. The theorem that the gradient
gives the maximum directional derivative of f remains true.

Example.   Find the gradient of

$$f(x, y, z) = x^2 - 4y^2 + 3z^2 - 3xy$$

at the point $\vec{a} : (1, 0, -1)$ and give it a rate-of-change interpretation.

   Solution.   We compute the partial derivatives $f_1 = 2x - 3y, f_2 = -3x - 8y, f_3 = 6z$, and evaluate them at $\vec{a}$, $f_1 = 2$, $f_2 = -3$, $f_3 = -6$. Then the gradient is the
vector $\vec{\nabla} f : (2, -3, -6)$, whose norm $\|\vec{\nabla} f\| = 7$. Hence the maximum rate of change of
f at a occurs in the unit direction $\vec{h} : (\frac{2}{7}, -\frac{3}{7}, -\frac{6}{7})$ and is 7.

20.2 Physical Interpretation of the Gradient

(1) Consider a droplet of water on a side of a hill whose height w above sea level is given at each point (x, y) by $w = f(x, y)$ (Figure 9.14). The gradient $\vec{\nabla}f$ gives the direction of steepest ascent up the hill and its negative, $-\vec{\nabla}f$, gives the direction and magnitude of steepest descent down the hill. This is the direction in which the raindrop will roll down the hill, and $\|\vec{\nabla}f\|$ determines how fast it will roll.

(2) The pressure at a point $P:(x, y, z)$ in the atmosphere is a scalar-valued function $p(x, y, z)$. The negative gradient $-\vec{\nabla}p$ at P gives the direction and magnitude of the greatest rate of decrease in pressure. This implies that the air at P is subject to a force in the direction of $-\vec{\nabla}p$ and the resulting acceleration of the air in that direction is proportional to $-\vec{\nabla}p$.

(3) Consider a particle of mass m on a line and a particle of unit mass at distance r from it (Figure 9.15). The function $U(r) = m/r$ is called the gravitational potential of the mass m. The force of attraction towards the mass m is given by the inverse-square law $-m/r^2$. We observe that

$$|\text{Force}| = -\frac{m}{r^2} = \frac{d}{dr}\,U.$$

More generally, if the mass m is located at the origin in 3-space, then the gravitational potential

$$U = \frac{m}{r} = \frac{m}{\sqrt{x^2 + y^2 + z^2}}.$$

The force attracting a unit particle at point $P:(x, y, z)$ towards the origin is the vector $\vec{\nabla}U$.

(4) Many other fields of force in 3-space can be represented by scalar-valued potential functions $U(x, y, z)$ in such a way that the vector force at the point $P:(x, y, z)$ is proportional to the gradient $\vec{\nabla}U$ of the potential function.

Hillside $w = f(x, y)$

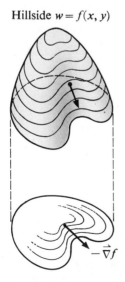

FIGURE 9.14 Contour map and gradient.   FIGURE 9.15 Attraction of a particle.

These physical examples point out the importance of the geometric invariance of the gradient, that is, its independence of the choice of coordinate basis. Droplets of water on a hillside or charged particles in a field of force respond to laws of nature that cannot know about the arbitrary choice of coordinates that we made for mathematical convenience. Our characterization of the gradient as a maximum rate of change of f has the necessary invariance to enable us to express natural *laws* in terms of it.

20.3   Gradient Normal to Level Contours.   We consider the level contours of the graph of f given by

$$f(x, y) = c.$$

The equation of the tangent line to this level curve at the point $P:(x, y)$ in local coordinates (dx, dy) at P is

$$f_1\, dx + f_2\, dy = 0.$$

The vector $d\vec{x}:(dx, dy)$ is the position vector of a point in the tangent line. So in vector notation, the equation of this tangent line can be written

$$(\vec{\nabla} f \cdot d\vec{x}) = 0.$$

This says that the gradient at P is always orthogonal to the tangent line to the level contour through P. That is, $\vec{\nabla} f$ is orthogonal to the level contour $f(x, y) = c$ itself (Figure 9.16).

Example.   For the function

$$f(x, y) = \frac{2x^2}{5} + \frac{5y^2}{8}.$$

plot the gradient at the point $\vec{a}:(3, \frac{16}{5})$ and the level contour of f through this point, as well as the tangent line.

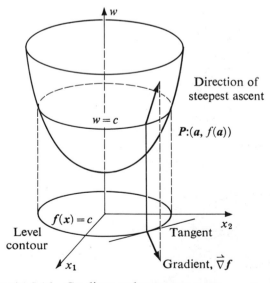

FIGURE 9.16   Gradient and steepest ascent.

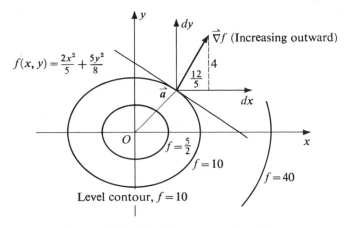

FIGURE 9.17   Level contour and gradient.

Solution.   We compute the partial derivatives $f_1 = 4x/5, f_2 = 5y/4$ and evaluate them at $\vec{a}$ and find $f_1 = 12/5, f_2 = 4$. The gradient at $\vec{a}$ is then the vector $\vec{\nabla}f:(\frac{12}{5}, 4)$. The level contour through $\vec{a}$ is the ellipse

$$\frac{x^2}{25} + \frac{y^2}{16} = 1.$$

The tangent line at $\vec{a}$ is given in local coordinates by $f_1\, dx + f_2\, dy = 0$, which in this case is $\frac{6}{25}dx + \frac{2}{5}dy = 0$, having the normal $\vec{\nabla}f:(\frac{6}{25}, \frac{2}{5})$. We plot these data (Figure 9.17).

Using vector notations $\vec{x}:(x_1, x_2, \ldots, x_n)$, we can generalize this to a function of any number of variables, with essentially the same reasoning as for functions of two variables. Since the tangent line to $f(\vec{x}) = c$ is still written $(\vec{\nabla}f \cdot d\vec{x}) = 0$, we have the following.

THEOREM.   At every point $\vec{a}$ in the domain of the differentiable function $f(\vec{x})$, the gradient $\vec{\nabla}f$ is orthogonal to the level contour $f(\vec{x}) = f(\vec{a})$, through $\vec{a}$.

20.4   Stationary Points.

As in the case of functions of one variable (I, §11.7), we can define stationary points. Consider an interior point $\vec{a}$ of the domain of $f(\vec{x})$, where the gradient $\vec{\nabla}f \neq 0$. In a neighborhood of $\vec{a}$, the gradient vector points to another point where $f(\vec{x}) > f(\vec{a})$. And the negative of the gradient $-\vec{\nabla}f$ points downhill to a nearby point where $f(\vec{x}) < f(\vec{a})$. Thus $\vec{a}$ can be neither a maximum nor a minimum point. Hence if $\vec{a}$ is an interior maximum or minimum point of f and f is differentiable there, it must be true that $\vec{\nabla}f = \vec{0}$. Points $\vec{a}$ interior to the domain of f, where $\vec{\nabla}f = 0$, are called *stationary points*. They are among the critical points for finding the maxima and minima of f.

Example.   Find the stationary points of the function f defined on the entire xy-plane by

$$f(x, y) = x^3 + 3xy^2 - 2y^3 - 12x + 5.$$

Solution.   We compute the partial derivatives

$$f_1 = 3x^2 + 3y^2 - 12 \quad\text{and}\quad f_2 = 6xy - 6y^2.$$

Since the gradient $\vec{\nabla} f:(f_1, f_2)$ is zero at stationary points, we can infer that both $f_1 = 0$ and $f_2 = 0$. Therefore we must solve the system of equations

$$3x^2 + 3y^2 - 12 = 0$$

and

$$6xy - 6y^2 = 0.$$

We find by factoring that the second equation is true if and only if $x = y$ or $y = 0$. Setting $x = y$ in the first equation, we find the solutions $(\sqrt{2}, \sqrt{2})$ and $(-\sqrt{2}, -\sqrt{2})$. Setting $y = 0$ in the first equation, we find the solutions $(2, 0)$ and $(-2, 0)$. Hence, the set of all stationary points of f is $\{(-2, 0), (-\sqrt{2}, -\sqrt{2}), (\sqrt{2}, \sqrt{2}), (2, 0)\}$. This completes the problem.

20.5 Exercises

In Exercises 1–5, calculate the indicated gradients and plot them in the domain of the functions.

1. $f(x, y) = x^2 - y^2$ at $(3, 4)$.

2. $f(x_1, x_2) = x_1^2 - 2x_1x_2 - x_2^2$ at $(-3, 5)$.

3. $f(x, y, z) = \dfrac{x^3 + y^3}{z}$ at $(1, -1, 1)$.

4. $f(x, y, z) = e^x(y + z)$ at $(0, 2, -3)$.

5. $f(x, y, z) = ax + by + cz$ at (x_0, y_0, z_0).

6. The function f defined by $f(x, y) = x^2 + y^2$ has a minimum at $(0, 0)$. Plot the level contour $f = 1$ in the xy-plane. Find gradients at intervals of $45°$ around this contour and plot them.

7. The function g defined by $g(x, y) = -x^2 - y^2$ has a maximum at $(0, 0)$. Plot the level contour $g = -1$ in the xy-plane. Find gradients at intervals of $45°$ around this contour and plot them. Compare with Exercise 6.

8. The function $h(x, y) = x^2 - y^2$ has a stationary point at $(0, 0)$. Plot level contours $h = 1$ and $h = -1$ and on these contours plot several gradients. What information does this give about whether $(0, 0)$ is a maximum or minimum point?

9. Sketch a level contour just above a minimum point of a function $f(x, y)$ and sketch in some gradient vectors at points around this contour. Do the same for a level contour just below a maximum point for $f(x, y)$.

In Exercises 10–13, find the directional derivative at the given point in the given direction by projecting the gradient onto the specified direction.

10. $f(x, y) = 2x - 3y$ at $(-3, 4)$ in the direction $(1/\sqrt{2}, -1/\sqrt{2})$.

11. $f(x, y, z) = 2x - 3y + 4z$ at $(0, 0, 0)$ in the direction of the vector $\vec{v}:(1, -2, 4)$.

12. $f(x, y, z) = y^2 - z^2$ in the direction $(1, 0, 0)$.

13. $U(x, y, z) = m/r$, where $r = \sqrt{x^2 + y^2 + z^2}$ at the point $P:(2, 1, 3)$ in the direction $\vec{h}:(1/\sqrt{3}, -1/\sqrt{3}, 1/\sqrt{3})$.

In Exercises 14–20, find the stationary points of the given functions, defined for all (x, y) except where otherwise specified.

14. $f(x, y) = 3x^2 + xy^2 - y^3$.

15. $f(x, y) = y^2 + 2x^2y - 2x^3$.

16. $f(x, y) = x^2 + 3xy - y^2 - 5x + 4y + 1$.

17. $f(x, y) = x^3 + 3xy^2 - 4y^3 - 6y^2 - 12x + 24$.

18. $f(x, y) = e^{2x}y(x^2 - y)$.

19. $f(x, y) = 2\dfrac{x^2}{y} + x + 2 \ln y$, where $y > 0$.

20. $f(x, y) = (1 - x)(1 - y)(x + y - 1)$.

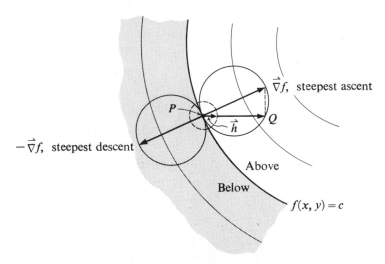

$-\vec{\nabla}f$, steepest descent

$\vec{\nabla}f$, steepest ascent

Above

Below

$f(x, y) = c$

FIGURE 9.18   Exercise 22.

21. Calculate the gradient of the potential function $U(x, y, z) = m/r$ (Section 20.4 and also Exercise 13) of a particle of mass m at the origin. Show that $\vec{\nabla}U$ is a vector that points towards the particle from any point $P:(x, y, z)$ and that $\|\vec{\nabla}U\| = m/r^2$.

22. We can represent graphically all of the directional derivatives of $f(x, y)$ at P as follows (Figure 9.18). We first construct the gradient $\vec{\nabla}f$ and its negative $-\vec{\nabla}f$. We draw circles with each of these two vectors as diameter. Any unit vector $\hat{h}$ at P we extend until it intersects the circle at Q. Show that the vector $\overrightarrow{PQ}$ represents the vector directional derivative. $\overrightarrow{PQ} = (\vec{\nabla}f \cdot \hat{h})\hat{h}$ (§12.3) of f at P. Observe how the figure represents the zero directional derivative in the direction of the level contour through P.

23. The surface of a hill is in the shape of the graph of

$$w = \frac{104}{1 + x^2 + y^2},$$

where w is the altitude in feet.

(a) Find the angle made by the steepest descent vector with the horizontal at the level $w = 52$ feet.

(b) Find the component of gravitational force that pulls a round pebble down the hill at this level.

24.* At what level is the hillside in Exercise 22 steepest?

21   Curves and Tangents

21.1   Idea of a Curve. We recall (§11.6) the use of the real-number line as a model, which we mapped faithfully into 3-space to define a line. For example, the variable point $R:(x_1, x_2, x_3)$ is in the line through the point $P:(-3, 1, -1)$, in the direction parallel to the fixed vector $\vec{a}:(2, -1, -3)$ if and only if for every member t,

$$\vec{r} = t\vec{a} + \vec{p}.$$

This vector statement says that the parametric equations of the line are

$$x_1 = 2t - 3, \qquad x_2 = -t + 1, \qquad x_3 = -3t - 1.$$

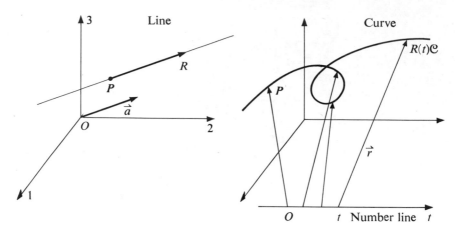

FIGURE 9.19   Line and curve.

To define a curve we again use the number line as a model (Figure 9.19) and map it continuously into 3-space, except that we do not require the mapping function to be of the first degree in t. Intuitively we may think of this mapping as taking the number line, transporting it into 3-space, deforming it so that it is no longer straight, but keeping its continuity.

DEFINITION (Curve).   Let $x_1(t)$, $x_2(t)$, $x_3(t)$ be continuous functions of the real variable t. Then the set $\mathscr{C}$ of all points $R:(x_1, x_2, x_3)$ with coordinates given by

$$x_1 = x_1(t), \qquad x_2 = x_2(t), \qquad x_3 = x_3(t)$$

is a *curve* in 3-space.

The real variable t is called *the parameter* and the set of three functions $\{x_1(t), x_2(t), x_3(t)\}$ is called a parametric representation of the curve. We will see that the same curve $\mathscr{C}$ has many different parametric representations. We observe that parametric representation is valid in 2-space or n-space.

Example.   Find a parametric representation of a curve found by winding a line around a cylinder, with progression at a constant rate along the axis of the cylinder. This is a helix.

Solution.   Let the axis of the cylinder (Figure 9.20) be the x_3-axis and let its radius be r. Then the winding may be described by the coordinates

$$x_1 = r \cos t, \qquad x_2 = r \sin t,$$

provided that it starts when $t = 0$ at a point where $x_1 = 1$ and $x_2 = 0$. The constant rate of advance along the x_3-axis is described by saying that

$$x_3 = kt,$$

where k is a constant.

Thus the helix may be described by the parametric representation

$$x_1 = r \cos t, \qquad x_2 = r \sin t, \qquad x_3 = kt.$$

This completes the problem.

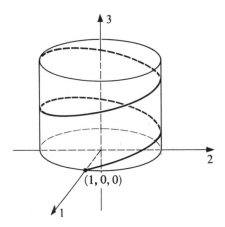

FIGURE 9.20   Helix.

21.2   Time as a Parameter for Space Curves. Frequently we have occasion to think of a curve as a path traced by a moving point P whose coordinates (x, y, z) are known at every time t. When we have such information in the form

$$x = x(t), \qquad y = y(t), \qquad z = z(t),$$

for all times t, we have not only the curve $\mathscr{C}$ traced by the moving point P but also a schedule of positions (t, x, y, z). In such a case, the parametric representation of the path $\mathscr{C}$, with time as parameter, is called a *motion*. The set of all positions (t, x, y, z) is called the *trajectory*, and the curve $\mathscr{C}$ consisting of all points $P:(x, y, z)$ occupied by the moving point is called the *orbit*.

For example, the track in space traced by a moving space capsule is its orbit, but a schedule of its positions (t, x, y, z) in that orbit is its *trajectory*. It is intuitively evident that the trajectory provides complete information about the velocity and acceleration of the space capsule, whereas the orbit does not.

A *uniform motion* is one whose orbit is a straight line and whose velocity in that line is constant.

21.3   Tangents to Space Curves. Let P be a fixed point on a curve $\mathscr{C}$. We define the tangent direction to the curve at P by the following procedure. We take a variable point Q on the curve near P and consider the unit position vector $\overrightarrow{PR}$ from P in the ray PQ. We let Q approach P along the curve and define the tangent vector $\overrightarrow{PT}$ as the limit of the unit vector $\overrightarrow{PR}$ (Figure 9.21). Guided by our study of plane curves, we may now guess a theorem that states how the derivatives of a representation determine the tangent direction.

THEOREM (Tangent to a space curve).   Let

$$\vec{x}(t):(x_1(t), \, x_2(t), \, x_3(t))$$

be a representation of a space curve $\mathscr{C}$. If these functions are differentiable at a fixed t and

$$[x_1'(t)]^2 + [x_2'(t)]^2 + [x_3'(t)]^2 \neq 0,$$

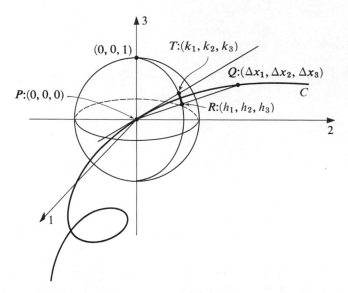

FIGURE 9.21   Tangent to a space curve, local coordinates with origin at P.

then the vector $\vec{v}:(x_1'(t),\, x_2'(t),\, x_3'(t))$ determines a direction in space that is tangent to $\mathscr{C}$ at the point $P:(x_1(t),\, x_2(t),\, x_3(t))$.

Proof.   Let the coordinates of P be given by the row $\vec{x}(t)$. A neighboring point Q on the curve having local coordinates $\Delta\vec{x}:(\Delta x_1,\, \Delta x_2,\, \Delta x_3)$ is on the curve if and only if

$$\Delta\vec{x} = \vec{x}(t + \Delta t) - \vec{x}(t).$$

Keeping both t and Δt fixed for the moment, we observe that if we multiply the row Δx by a number s we find a point on the secant ray from P to Q. One point on the ray PQ where it intersects the unit sphere about P has local coordinates $\vec{h}:(h_1, h_2, h_3)$ given by

$$\vec{h} = \frac{1}{\|\Delta\vec{x}\|}\,\Delta\vec{x},$$

or $\Delta\vec{x} = \|\Delta\vec{x}\|\vec{h}$. Putting this into the local equation of the curve above and dividing by Δt, we find that the coordinates $\vec{h}$ of the intersection R of the line and unit sphere must satisfy

$$\frac{\|\Delta\vec{x}\|}{\Delta t}\,\vec{h} = \frac{\vec{x}(t + \Delta t) - \vec{x}(t)}{\Delta t}.$$

Now we let $\Delta t \to 0$. The $\lim_{\Delta t \to 0}$ exists since the functions $\vec{x}(t)$ are differentiable. The point R then moves on the unit sphere and approaches a point $T:(k_1, k_2, k_3)$ whose coordinates $\vec{k}$ must satisfy the equation obtained from the limit as $\Delta t \to 0$; that is,

$$\sqrt{[x_1'(t)]^2 + [x_2'(t)]^2 + [x_3'(t)]^2}\,\vec{k} = \vec{x}'(t).$$

Moreover, the coordinates $\vec{k}$ of T are uniquely determined by this equation if the factor on the left is not zero. By definition, the line PT is tangent to the curve $\mathscr{C}$. This completes the proof.

COROLLARY. A representation of the tangent line to the curve $\mathscr{C}$ at the point $P:(x_1(t), x_2(t), x_3(t))$ is $d\vec{x} = \vec{x}'(t)\, dt$, where $d\vec{x}:(dx_1, dx_2, dx_3)$ is a variable point on the tangent line and dt is the parameter.

Proof. The local coordinates $d\vec{x}$ of any point on the tangent line PT may be found by multiplying the local coordinates of T by some number ds. The equation $d\vec{x} = \vec{x}'(t)\, dt$ may be written

$$dx = \frac{\vec{x}'(t)}{\|\vec{x}'(t)\|}\, \|\vec{x}'(t)\|\, dt = \vec{k}\, ds$$

if we express ds in the form $ds = \|\vec{x}'(t)\|\, dt$. This shows that the equations $d\vec{x} = \vec{x}'(t)\, dt$ are parametric equations of the tangent line, as stated in the corollary.

21.4 Velocity and Acceleration in Curvilinear Motion. The interpretation of the derivative with respect to time as velocity extends to motion in a space curve. If the moving point $R:(x, y, z)$ has coordinates given as functions of time,

$$\vec{r}(t):x = x(t),\ y = y(t),\ z = z(t),$$

then the velocity is the derivative vector $\vec{r}'(t)$,

$$\vec{v}(t):(x'(t), y'(t), x'(t)).$$

The acceleration is the second derivative vector $\vec{r}''(t)$,

$$\vec{a}(t):(x''(t), y''(t), z''(t)).$$

Example. Find the velocity and acceleration in the motion

$$x = 0, \qquad y = 2t + 1, \qquad z = -16t^2 + 80t + 5,$$

where coordinates (x, y, z) are in feet and t is in seconds. Plot the motion showing the velocity and acceleration at time when $t = 1$ sec.

Solution. We compute for any time t the velocity

$$\vec{v}(t):(0, 2, -32t + 80)$$

and the acceleration

$$\vec{a}(t):(0, 0, -32).$$

When $t = 1$, this gives $P:(0, 3, 69)$, $\vec{v}(1):(0, 2, 48)$ and $\vec{a}(1):(0, 0, -32)$. We plot the results (Figure 9.22). The orbit is in the yz-plane, since $x = 0$ for every t. We observe that the velocity when $t = 1$ sec is a vector tangent to the orbit, pointing in the direction of motion. It consists of a component of 2 ft/sec in the y-direction and 48 ft/sec upward in the z-direction. The norm of velocity, $\|\vec{v}(1)\| = \sqrt{2^2 + (48)^2} = \sqrt{2308}$, is the numerical speed in the tangential direction.

We observe that the acceleration vector $\vec{a}(1):(0, 0, -32)$ indicates in direction and magnitude the rate at which an external force is changing the velocity. In this case it is a constant vector pointing down and this determines that the motion will continue to curve downward. This example may be interpreted as the motion of a particle of mass 1 that was thrown with velocity $\vec{v}_0:(2, 80)$ at time 0 in the vertical yz-plane and thereafter moves subject to the earth's gravitational pull of -16 ft/sec².

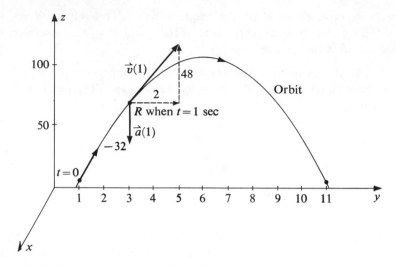

FIGURE 9.22   Velocity and acceleration in a motion, $x = 0$, $y = 2t + 1$, $z = -16t^2 + 80t + 5$.

21.5   Cycloid

Example.   A wheel of radius r rolls on a straight line. Find the parametric equations of the curve traced by a point on the rim.

Solution.   Let $P:(x, y)$ be the tracing point. We choose coordinates so that the wheel rolls on the x-axis, starting with the tracing point P at the origin (Figure 9.23). As the wheel rolls through an angle θ, measured in radians, the contact point R moves along the x-axis so that $OR = r\theta$. We look at the right triangle PQC and find that

$$x = OR - PQ = r\theta - r \sin \theta,$$
$$y = RC - QC = r - r \cos \theta.$$

Hence, the parametric equations of the cycloid curve are

$$x = r(\theta - \sin \theta), \qquad y = r(1 - \cos \theta),$$

where θ is a parameter whose range is all real numbers.. This completes the problem.

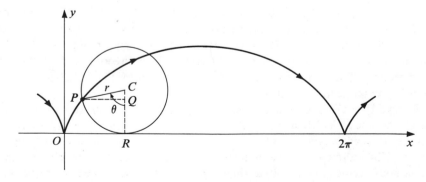

FIGURE 9.23   Cycloid.

21.6 Exercises

1. Plot the curve $x = t$, $y = -t^2 + 4t$. Find its tangent at the point where $t = 3$.
2. Plot the curve $x = r \cos \theta$, $y = r \sin \theta$. Prove that it is a circle. Find its tangent vector where $\theta = \pi/6$.
3. Plot the curve $x = 2t - 1$, $y = -6t + 3$. Show that its tangent vector $\vec{v}(t)$ is $(2, -3)$ at every point. Are all curves with constant tangent vectors straight lines?
4. Plot the orbit of the motion $x = 2 \sin^2 t$, $y = 3 \cos^2 t$, where t is time. Show that it lies on a straight line but is not a straight line.
5. Find times when the velocity is zero in Exercise 4 and points at which this occurs. Use this information to describe the motion.
6. Show that a circle of radius r with center at $(0, 0)$ can be represented by

$$x = r \cos t, \qquad y = r \sin t.$$

7. Taking t as time in Exercise 6, find the velocity and acceleration of the motion in a circle. Show that the acceleration always points to the center and that the tangential speed is a constant.
8. Find the velocity and acceleration at time $t = \pi/4$ sec for the cycloid motion

$$x = r(t - \sin t), \qquad y = r(1 - \cos t).$$

Plot these vectors.
9. Let $s(t)$ be the tangential speed in the cycloid motion of Exercise 8. Show that

$$s'(t) = 2r \left| \sin \frac{t}{2} \right|.$$

10. Plot the orbit

$$x = 3 \cos 2t, \qquad y = 5 \sin 2t.$$

Prove that it is an ellipse. Find the velocity and acceleration when $t = \pi/3$ sec. How much time is required to make one complete circuit?
11. For a motion in a helix

$$x = r \sin t, \qquad y = r \sin t, \qquad z = \frac{r}{2\pi} t,$$

find the velocity, tangential speed, and acceleration when $t = 3\pi/4$ sec.

L'Hospital's Rule.   Let

$$x = g(t), \quad y = f(t)$$

be a differential curve defined, except at $t = c$, on an open interval $c - \delta < t < c + \delta$ and having $g'(t) \neq 0$ on this deleted neighborhood of c. Suppose that

$$\lim_{t \to c} g(t) = \lim_{t \to c} f(t) = 0$$

(Figure 9.24). Then the graph is a curve on which the point (x, y) approaches $(0, 0)$ when $t \to c$. Since $g'(t) \neq 0$, this is a curve of a function, $y = F(x)$, whose tangent has slope $(d/dx) F(x) = f'(t)/g'(t)$. We observe that $\lim_{t \to c} [f(t)/g(t)] = 0/0$ is an "indeterminate form." But l'Hospital's rule says that if for some number k, or $k = +\infty$, or $k = -\infty$, it is true that

$$\lim_{t \to c} \frac{f'(t)}{g'(t)} = k, \quad \text{then} \quad \lim_{t \to c} \frac{f(t)}{g(t)} = k \quad \text{also.}$$

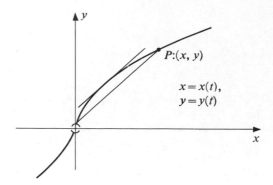

FIGURE 9.24   Exercise 12.   l'Hospital's rule.

(Proof in Exercise 18.) This evaluates the indeterminate form. If, after one application, $\lim_{t \to c} [f'(t)/g'(t)]$ is an indeterminate form, then we may apply l'Hospital's rule to it. Apply l'Hospital's rule to Exercises 12–17.

12. $\lim\limits_{t \to 0} \dfrac{\sin t}{t}.$ 13. $\lim\limits_{t \to 0} \dfrac{\sin t}{t \cos t}.$

14. $\lim\limits_{t \to \pi/2} \dfrac{\sin 2t}{\pi - 2t}.$ 15. $\lim\limits_{t \to 0} \dfrac{1 - \cos t}{2t}.$

16. $\lim\limits_{t \to 0} \dfrac{1 - \cos t}{t^3}.$ 17. $\lim\limits_{t \to 0} \dfrac{\sin t - t}{t^3}.$

18. Prove l'Hospital's rule by observing that the theorem of the mean applies to the equation of the curve in the form $y = F(x)$ when we make the curve continuous at the origin by giving it the values $g(0) = f(0) = 0$ there.

19.* An analogous rule of l'Hospital applies when $\lim_{t \to c} f(t) = \lim_{t \to c} g(t) = \infty$. Then

$$\lim_{t \to c} \frac{f(t)}{g(t)} = \frac{\infty}{\infty},$$

another indeterminate form. Prove that if $\lim_{t \to c} [f'(t)/g'(t)] = k$, then

$$\lim_{t \to c} \frac{f(t)}{g(t)} = k$$

also. This is the slope of an asymptote to the curve.

20. Show that the curve

$$x = \cos t - 3 \sin t + 2, \quad y = 2 \cos t + 2 \sin t - 4, \quad z = \cos t + 3 \sin t - 4$$

is a closed plane curve. In fact, it lies in the plane $2x - 3y + 4z = 4$ and completes a closed loop as the parameter varies from any t to $t + 2\pi$.

21. Obtain a different parametric representation of the curve in Exercise 20 by replacing t by $s + (\pi/2)$.

22. Show that

$$x = r \cos t + x_0, \qquad y = \frac{r}{\sqrt{2}} \sin t + y_0, \qquad z = \frac{r}{\sqrt{2}} \sin t + z_0$$

is a circle of radius r with center at (x_0, y_0, z_0). Find the equation of the plane of the circle.

23. Show that the curve

$$x = 3t \cos t, \qquad y = 3t \sin t, \qquad z = 3t$$

lies on the surface $x^2 + y^2 - z^2 = 0$, which is a cone. Describe the curve and sketch it.

24. Prove that the motion

$$x = r \sin t \cos 2t,$$
$$y = r \sin t \sin 2t,$$
$$z = -r \cos t,$$

lies on the sphere $x^2 + y^2 + z^2 = r^2$. Find the velocity and acceleration when $t = \pi/4$.

25. A circle of radius $r/4$ rolls inside a circle of radius r. Show that a point on the circumference of the smaller circle traces the curve

$$x = a \cos^3 \theta, \qquad y = a \sin^3 \theta.$$

Plot the curve. It is called a hypocycloid.

26. A circle of radius r rolls outside around a fixed circle of radius r. Find parametric equations of a point on the circumference of the rolling circle.

27. Solve Exercise 25 again with a rolling circle of radius $r/2$. Show that the tracing point moves back and forth on the diameter of the fixed circle.

28. Show that the motion

$$x = \sin \beta t \cos t, \qquad y = \sin \beta t \sin t, \qquad z = \cos \beta t,$$

is on a unit sphere, and find its velocity. Show that the velocity is perpendicular to the radius of the sphere to the moving point.

22   Arc Length of Curves

22.1   Arc Length. An arc of a curve (Figure 9.25) is the portion of the curve between two endpoints. It is given by a parametric representation

$$\vec{r}(t):[x_1 = r_1(t), \dots, x_n = r_n(t)], \qquad a \leq t \leq b.$$

On an arc the parameter t is restricted to an interval $[a, b]$. Thus an arc is a continuous map of a closed interval $[a, b]$ of real numbers. The length of such an arc, measured along it, is an easy notion to understand intuitively, but it is another matter to formulate a mathematical definition of arc length that is exactly computable from its parametric representation.

For simplicity, we confine our attention to smooth curves $\mathscr{C}$, that is, to curves given by a representation $\vec{r}(t)$, which has a continuous derivative $\vec{r}'(t)$. We can readily extend this to piecewise smooth curves, on which $\vec{r}'(t)$ is continuous except at a finite number of corners, where the smooth pieces are joined together.

To define the arc length of a curve $\mathscr{C}$, we approximate it by a polygon joining points on the curve. The arc length of the polygon is the sum of the lengths of its straight-line segments. These can be calculated by the distance formula (Figure 9.26). Then we proceed to the limit, by introducing more and more vertices on the curve so that the polygon approaches the curve. The arc length of the curve is obtained as the limit of the arc lengths of a sequence of approximating polygons.

In more detail, we introduce the partition points

$$a = t_0 < t_1 < t_2 < \cdots < t_n = b$$

into the interval $[a, b]$. These determine points

$$\vec{r}(a), \vec{r}(t_1), \vec{r}(t_2), \dots, \vec{r}(b)$$

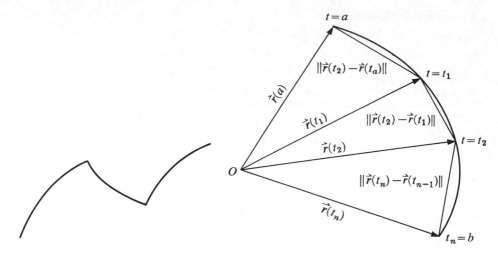

FIGURE 9.25   Arc of a curve; it is piece-wise smooth.

FIGURE 9.26   Inscribed polygon in a curve.

on the curve, which form the vertices of an approximating polygon $\mathscr{P}$. The length $L(\mathscr{P})$ of this polygon is the sum

$$L(\mathscr{P}) = \|\vec{r}(t_1) - \vec{r}(a)\| + \|\vec{r}(t_2) - \vec{r}(t_1)\| + \cdots + \|\vec{r}(b_n) - \vec{r}(t_{n-1})\|.$$

This sum can be represented as an integral. Let

$$\Delta t_i = t_i - t_{i-1} \quad \text{for} \quad i = 1, 2, 3, \ldots, n$$

and rewrite the sum $L(\mathscr{P})$ in the form

$$L(\mathscr{P}) = \frac{\|\vec{r}(t_1) - \vec{r}(a)\|}{\Delta t_1} \Delta t_1 + \frac{\|\vec{r}(t_2) - \vec{r}(t_1)\|}{\Delta t_2} \Delta t_2 + \cdots + \frac{\|\vec{r}(t_n) - \vec{r}(t_{n-1})\|}{\Delta t_n} \Delta t_n.$$

We refine the partition of the interval $[a, b]$ by introducing more and more partition points in such a way that max $\Delta t_i \to 0$. Then for the smooth curve we have

$$\lim_{\Delta t_i \to 0} \frac{\|\vec{r}(t_i) - \vec{r}(t_{i-1})\|}{\Delta t_i} = \|\vec{r}'(t_{i-1})\|,$$

and we can predict that, in the limit, the sum of the lengths of the sides of the polygon becomes the integral

$$\lim_{\max \Delta t_i \to 0} L(\mathscr{P}) = \int_a^b \|\vec{r}'(t)\| \, dt.$$

The proofs are difficult, but guided by these considerations, we bypass the theory and adopt the following definition.

DEFINITION (Arc length).   The arc length s of the smooth curve $\vec{r}(t)$, $a \leq t \leq b$, from the point $\vec{r}(a)$ to the point $\vec{r}(b)$ is given by the integral

$$s = \int_a^b \|\vec{r}'(t)\| \, dt.$$

This single formula in terms of the position vector $\vec{r}(t)$ has different forms in special cases. For a curve $\vec{r}(t):(x(t), y(t), z(t))$ in 3-space we have $\|\vec{r}'(t)\| = \sqrt{x'^2 + y'^2 + z'^2}$, so that

$$s = \int_a^b \sqrt{x'^2 + y'^2 + z'^2}\, dt.$$

For a plane curve in vector representation $\vec{r}(t):(x(t), y(t))$, it reduces to

$$s = \int_a^b \sqrt{x'^2 + y'^2}\, dt.$$

For a plane curve given by a function $y(x)$, $a \leq x \leq b$, the vector representation becomes $\vec{r}(x):(x, y(x))$, so that the arc length formula is

$$s = \int_a^b \sqrt{1 + \left(\frac{dy}{dx}\right)^2}\, dx.$$

22.2   Calculation of Arc Length

Example A.   Find the length of arc of the helix

$$\vec{r}(t):x = a \cos t, \; y = a \sin t, \; z = bt, \qquad 0 \leq t \leq 2\pi.$$

Solution.   We differentiate and find

$$\vec{r}'(t):x' = -a \sin t, \; y' = a \cos t, \; z' = b, \qquad \text{and} \qquad \|\vec{r}'(t)\| = \sqrt{a^2 + b^2}.$$

Hence,

$$s = \int_0^{2\pi} \sqrt{a^2 + b^2}\, dt = 2\pi \sqrt{a^2 + b^2}.$$

Example B.   Find the length of arc of the plane curve

$$\vec{r}(\theta):x = \cos^3 \theta, \; y = \sin^3 \theta, \qquad 0 \leq \theta \leq 2\pi.$$

Solution.   We differentiate to find

$$\vec{r}'(\theta):x' = -3a \cos^2 \theta \sin \theta, \; y' = 3a \sin^2 \theta \cos \theta,$$

then

$$\|\vec{r}'(\theta)\| = 3a \sqrt{\sin^2 \theta \cos^2 \theta (\cos^2 \theta + \sin^2 \theta)} = 3a|\sin \theta \cos \theta|.$$

Hence

$$s = \int_0^{2\pi} 3a|\sin \theta \cos \theta|\, d\theta = 4 \int_0^{\pi/2} 3a(\sin \theta \cos \theta)\, d\theta$$

$$= 12a \left. \frac{\sin^2 \theta}{2} \right|_0^{\pi/2} = 6a.$$

Here we have used the symmetry of the graph to compute the total length as 4 times the length of the arc in the first quadrant.

Example C.   Find the length of arc of the plane curve $4x^3 = 9y^2$ from $(0, 0)$ to $(4, \frac{16}{3})$.

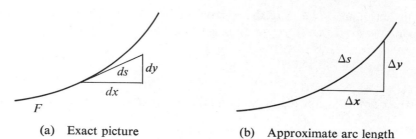

(a)   Exact picture   (b)   Approximate arc length

FIGURE 9.27   Element of arc length.

Solution.   A vector representation of the curve with parameter x is $\vec{r}(x):(x, \frac{2}{3}, x^{3/2})$. We compute $\|\vec{r}'(x)\| = \sqrt{1 + x}$. Then

$$s = \int_0^4 \sqrt{1 + x}\ dx = \frac{(1 + x)^{3/2}}{\frac{3}{2}}\bigg|_0^4 = \frac{2}{3}(5\sqrt{5} - 1) \text{ length units.}$$

22.3   Local Coordinates.   We consider plane curves $\vec{r}(t):(x(t), y(t))$. Then

$$\left(\frac{ds}{dt}\right)^2 = \left(\frac{dx}{dt}\right)^2 + \left(\frac{dy}{dt}\right)^2.$$

This can be written as a relation in local coordinates, $dx = x'\ dt$ and $dy = y'\ dt$. It says that

$$(ds)^2 = (dx)^2 + (dy)^2.$$

The exact picture of this (Figure 9.27(a)) is a right triangle with hypotenuse ds laid along the tangent to the curve and with dx and dy as sides parallel to the axes. It is convenient to think of the arc itself, for small Δs, forming the hypotenuse, with sides Δx and Δy (Figure 9.27(b)).

22.4   Arc Length as Parameter.   If we calculate the arc length of a smooth curve $\vec{r}(t)$ from a fixed point $\vec{r}(a)$ to a variable point $\vec{r}(t)$, the arc length is a function $s(t)$, whose value at t is given by

$$s(t) = \int_a^t \|\vec{r}'(u)\|\ du.$$

If we differentiate the integral with respect to the variable upper limit t, we find (I, §21.7) that

$$\frac{ds}{dt} = \|\vec{r}'(t)\|.$$

When $\|\vec{r}'(t)\| \neq 0$, this implies that $ds/dt > 0$. Thus the strictly increasing function $s = s(t)$ can be solved for t in terms of s, $t = t(s)$. If we substitute the function $t(s)$ for t in $\vec{r}(t)$, we obtain a representation of the curve $\vec{r}[t(s)]$, in which arc length is the parameter. The parameter t is arc length if and only if $\|\vec{r}'(t)\| = 1$.

Example.   Represent the curve $\vec{r}(t):x = a \cos t$, $y = a \sin t$, $a > 0$, with arc length s as parameter.

Solution. We calculate $\|\vec{r}'(t)\| = \sqrt{a^2 \sin^2 t + a^2 \cos^2 t} = a$. Hence $ds/dt = a$, and integrating gives $s = ta$, measured from the point $\vec{r}(0)$. Solving for t, we obtain $t = s/a$. Hence, a representation of the curve, with arc length s as parameter, is

$$\vec{r}\left(\frac{s}{a}\right) : x = a \cos \frac{s}{a}, \, y = a \sin \frac{s}{a}.$$

We may verify that $\|(d/ds)\vec{r}(s/a)\| = 1$.

22.5   Unit Tangent and Normal.

If the curve $\mathscr{C}$ is represented with arc length s as parameter and we differentiate the position function

$$\vec{r}(s) : x = x(s), \, y = y(s), \, z = z(s),$$

we have seen that the tangent vector $\vec{u}$,

$$\vec{u}(s) = \frac{d}{ds}\, \vec{r}(s) : [x'(s), \, y'(s), \, z'(s)],$$

has length 1. Then since $(\vec{u} \cdot \vec{u}) = x'^2 + y'^2 + z'^2 = 1$, if we differentiate $(\vec{u} \cdot \vec{u})$ with respect to s, we find that $(\vec{u} \cdot (d\vec{u}/ds)) = 0$. Hence $(d/ds)\vec{u}$ is a vector orthogonal to $\vec{u}$. We define the unit normal vector $\vec{n}$ by

$$\vec{n} = \frac{\vec{u}'(s)}{\|\vec{u}'(s)\|}.$$

Introducing the symbol $\kappa = \|\vec{u}'(s)\|$, we can write this

$$\frac{d}{ds}\, \vec{u} = \kappa \vec{n}.$$

The scalar κ is called the *curvature* of $\mathscr{C}$ at the point $\vec{r}(s)$. By its definition it measures the rate of change of the tangent vector $\vec{u}$ per unit arc length on the curve.

Example A. Calculate the unit tangent vector, normal vector, and curvature of the helix,

$$\vec{r}(t) : x = a \cos t, \, y = a \sin t, \, z = bt.$$

Solution. We have calculated $ds/dt = \|\vec{r}'(t)\| = \sqrt{a^2 + b^2}$. Hence $dt/ds = 1/\sqrt{a^2 + b^2}$. Now the unit tangent vector is

$$\vec{u}(t) : \left(-\frac{a}{\sqrt{a^2 + b^2}} \sin t, \, \frac{a}{\sqrt{a^2 + b^2}} \cos t, \, \frac{b}{\sqrt{a^2 + b^2}}\right).$$

Then

$$\frac{d\vec{u}}{ds} = \frac{d\vec{u}}{dt} \cdot \frac{dt}{ds} = \frac{d\vec{u}}{dt} \frac{1}{\sqrt{a^2 + b^2}} : \left(-\frac{a}{a^2 + b^2} \cos t, \, \frac{-a}{a^2 + b^2} \sin t, \, 0\right),$$

and

$$\kappa = \left\|\frac{d\vec{u}}{ds}\right\| = \frac{a}{a^2 + b^2}.$$

Hence the unit normal vector is $\vec{n} : (-\cos t, \, -\sin t, \, 0)$.

Example B. Calculate the curvature of the smooth plane curve $y = y(x)$.

Solution. The curve is traced by the position vector $\vec{r}(x):(x, y(x))$. Its derivative is $\vec{r}'(x):(1, dy/dx)$ and

$$\|\vec{r}'(x)\| = \sqrt{1 + \left(\frac{dy}{dx}\right)} = \frac{ds}{dx}.$$

Hence the unit tangent vector is

$$\vec{u}(x) : \left(\frac{1}{\sqrt{1 + (dy/dx)^2}}, \frac{dy/dx}{\sqrt{1 + (dy/dx)^2}}\right).$$

We calculate $d\vec{u}/ds$, the chain rule, $d\vec{u}/ds = (d\vec{u}/dx)(dx/ds)$. Doing this, we find

$$\frac{d\vec{u}}{ds} : \frac{d^2y/dx^2}{[1 + (dy/dx)^2]^2} \left(-\frac{dy}{dx}, 1\right).$$

And the curvature κ is given by

$$\kappa = \left\|\frac{d\vec{u}}{ds}\right\| = \frac{|d^2y/dx^2|}{[1 + (dy/dx)^2]^{3/2}}.$$

22.6 Exercises

In Exercises 1–6, find the length of the arcs.

1. $\vec{r}(t):x = t^2, y = 3, z = 2, 0 \leq t \leq 2.$
2. $\vec{r}(t):x = t^2, y = 3t^2, z = -t^2, -1 \leq t \leq 4.$
3. $y = x^{3/2}, 0 \leq x \leq 9.$
4. Cycloid: $x = a(t - \sin t), y = a(1 - \cos t)$, one arch.
5. $\vec{r}(t):x = \sin 2t, y = \cos 2t, z = -4t$, one time around.
6. $y = (x^3/3) + (1/4x), 1 \leq x \leq 4.$

7. Compute the curvature of the circle, $x = r \cos t, y = r \sin t$, showing that at every point $\kappa = 1/r$. This leads us to define the radius of curvature ρ for any curve to be $\rho = 1/\kappa$, so that for the circle $\rho = r$.

In Exercises 8–14, find the unit tangent vector $\vec{u}$, the unit normal $\vec{n}$ and the curvature κ at the point indicated.

8. $x = 2 \cos t, y = 2 \sin t$, at point where $t = \pi/4.$
9. The cycloid $x = a(t - \sin t), y = a(1 - \cos t)$, where $t = \pi/2.$
10. The curve $y = e^x$ at the point where $x = 0.$
11. The parabola $y^2 = 4x$ at the point $(1, 2).$
12. The catenary $y = \frac{1}{2}(e^x + e^{-x})$ at $(0, 1).$
13. The curve $x = a \cos^3 t, y = a \sin^3 t$, at a general point.
14. $\vec{r}(t):x = t \cos t, y = t \sin t, z = t$, where $t = 0.$

15. Show that the curve in Exercise 14 spirals around on the cone $z^2 = x^2 + y^2$, rising as it goes around.
16. (a) At what rate in radians per unit length of arc does the unit tangent turn in going around the circle, $x = r \cos t, y = r \sin t$?
    (b) On any curve of curvature κ?
17. Does the curvature depend upon the unit of length in which x, y, z, and s are measured?
18. Plot a circle having the same tangent and the same curvature as the curve $y = e^x$ at the point $(0, 1)$. Plot the curve also. This is the best approximating circle to the curve at the point.

19. If a wheel of radius $a = 1$ ft rolls along a level track at 1 rev/sec, in what direction is a point on the circumference moving when the point is 1 ft above the ground, going up? With what linear speed is it moving?

20.* In Exercise 19, a speck of mud flies off at the instant when $y = 1$. Where does it fall to the ground?

21.* For an electric transmission cable hanging between fixed points $(-a, h)$ and (a, h) in the form

$$y = \frac{a}{2}\left(\exp\frac{x}{a} + \exp -\frac{x}{a}\right),$$

the length increases 3 percent, due to heat expansion. Find the relative change in y at the minimum point.

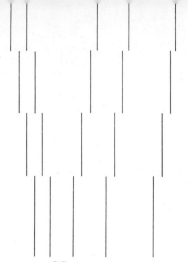

Differential Equations

CHAPTER 10

23 Inverse of Differentiation

23.1 Cross-Derivative Test. For functions of one variable, $F(x)$, we have considered the problem of both differentiation and its inverse. That is, given $F(x)$, we have calculated its differential $DF = F'(x)\, dx$, and conversely, given $f(x)\, dx$, we have developed methods of integration to produce a function $F(x)$ with differential $DF(x) = f(x)\, dx$. For example, if $f(x) = \sqrt{1 - x^2}$ on the interval $[-1, 1]$, the function

$$F(x) = \tfrac{1}{2}[x\sqrt{1 - x^2} - \arcsin x] + C$$

has $DF = f(x)\, dx$.

For a number-valued function of two variables, $F(x, y)$, differentiation produces a form in local $dx\, dy$-coordinates

$$DF(x, y) = F_1(x, y)\, dx + F_2(x, y)\, dy.$$

Conversely, the antiderivative process begins with a differential form

$$u(x, y)\, dx + v(x, y)\, dy,$$

where $u(x, y)$ and $v(x, y)$ are given functions. We seek a function $F(x, y)$ such that

$$DF(x, y) = u(x, y)\, dx + v(x, y)\, dy.$$

Example A. For the differential form $2xy\, dx + (x^2 - y^2)\, dy$, find a function $F(x, y)$ such that

$$DF(x, y) = 2xy\, dx + (x^2 - y^2)\, dy.$$

Solution. We know from the theorem on evaluation of the differential that if such a function $F(x, y)$ exists, its differential is given by

$$DF(x, y) = F_1(x, y) \, dx + F_2(x, y) \, dy.$$

Hence we set $F_1(x, y) = 2xy$ and integrate with respect to x, holding y constant. This gives $F(x, y) = x^2y + g(y)$, where $g(y)$ is some unknown function of y alone. This function $g(y)$ replaces the constant of integration, because when we differentiate it with respect to x, we get $(d/dx)g(y) = 0$. Now since $F_2(x, y) = x^2 - y^2$ must also be true, we find the partial derivative with respect to y of $F(x, y) = x^2y + g(y)$. It must satisfy the equation

$$F_2(x, y) = x^2 + \frac{d}{dy} g(y) = x^2 - y^2,$$

which simplifies to $(d/dy)g(y) = -y^2$. Integrating this as a function of y, we have

$$g(y) = -\left(\frac{y^3}{3}\right) + C.$$

Hence, we have produced the function

$$F(x, y) = x^2y - \left(\frac{y^3}{3}\right) + C.$$

We check this by differentiation and find that it does give

$$DF = 2xy \, dx + (x^2 - y^2) \, dy,$$

as required.

Example B. For the differential form $2xy \, dx - (x^2 - y^2) \, dy$, find a function $F(x, y)$ such that

$$DF(x, y) = 2xy \, dx - (x^2 - y^2) \, dy.$$

Solution. As in Example A, we first integrate the equation $F_1(x, y) = 2xy$ with respect to x alone and find that $F(x, y) = x^2y + g(y)$, where $g(y)$ is an unknown function of y only. Then we must also have by differentiation with respect to y that

$$F_2(x, y) = x^2 + \frac{d}{dy} g(y) = -x^2 + y^2.$$

This requires that $(d/dy) g(y) = -2x^2 + y^2$, which is not a function of y alone. This contradiction implies that no function $F(x, y)$ exists for which

$$DF(x, y) = 2xy \, dx - (x^2 - y^2) \, dy.$$

These examples involve only simple polynomials in x and y. For functions of one variable, every polynomial has an antiderivative that is a polynomial. These examples then show that some new principle is involved in the antiderivative process for functions of several variables that was not present for functions of one variable. The following theorem states the general facts.

THEOREM (Cross-derivative test). If the coefficient functions in the differential form $u(x, y) \, dx + v(x, y) \, dy$ are continuous and have continuous partial derivatives $u_2(x, y)$ and $v_1(x, y)$ interior to some disk in the plane, then there is a function $F(x, y)$ defined on the disk such that $DF = u(x, y)\, dx + v(x, y)\, dy$ if and only if the cross-derivative test $u_2(x, y) = v_1(x, y)$ is satisfied everywhere.

Proof.   (§23.5–§23.6).

The procedure of the examples produces an antiderivative when the cross derivative is satisfied. In Example A, the process succeeds because the cross-derivative test is satisfied and fails in Example B because the cross-derivative test is not satisfied.

23.2   Functions with zero differential

THEOREM.   If f is a differentiable function of several variables in some disk $\|\vec{x}\| < r$ and $Df = 0$ everywhere in the disk, then f is a constant.

Proof.   Consider the disk with center at $P_0:(x_0, y_0)$. Denote by $g(y)$ the values of $f(x, y)$ along the line $x = x_0$. That is, $g(y) = f(x_0, y)$ (Figure 10.1). Since $Df = 0$ implies that $f_1 = 0$ throughout the disk, the theorem on zero derivative (I, §13.4) implies that along any horizontal line with ordinate y the value of f is constant; in fact, $f(x, y) = g(y)$ along this line. Now $Df = 0$ also implies that $f_2 = 0$ and hence that $(d/dy)g(y) = 0$ along the line $x = x_0$. Hence, by the theorem on functions of one variable that have zero derivative in an interval, $g(y) = C$ (constant). Therefore for all (x, y) in the disk,

$$f(x, y) = f(x_0, y_0).$$

COROLLARY (Functions with zero partial derivative).   If $f_1(x, y) = 0$ for all (x, y) in some disk, then there is a function of y alone, $g(y)$, such that $f(x, y) = g(y)$.

Example.   Determine whether the form

$$(2xy - 3x^2)\, dx + (x^2 + y)\, dy$$

is the differential of some function $F(x, y)$, and if so, integrate it to find $F(x, y)$.

Solution.   We apply the cross-derivative test with $u = 2xy - 3x^2$ and $v = x^2 + y$. Since for all (x, y), $u_2 = 2x = v_1$, the cross-derivative test is satisfied and an antiderivative exists. Set $F_1(x, y) = 2xy - 3x^2$ and integrate with respect to x. We find $F(x, y) = x^2y - x^3 + g(y)$, where $g(y)$ is an unknown function. We calculate F_2 from this and set

$$F_2 = x^2 + \frac{dg}{dy}g(y) = x^2 + y.$$

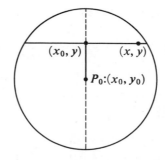

FIGURE 10.1 $Df = 0$ in a disk.

This gives us $Dg(y) = y \, dy$. Integrating this with respect to y, we find $g(y) = (y^2/2) + C$. Hence

$$F(x, y) = x^2y - x^3 + \frac{y^2}{2} + C.$$

This completes the problem.

23.3   Line Integrals.   Let $F(x, y, z)$ be a differentiable function inside some ball (or box) in 3-space. Let $\mathscr{C}$,

$$\vec{r}(t): \quad x = x_1(t), \quad y = x_2(t), \quad z = x_3(t),$$

be a smooth arc of a curve joining points

$$A:(x_1(a), x_2(a), x_3(a)) \qquad \text{and} \qquad B:(x_1(b), x_2(b), x_3(b)).$$

(See Figure 10.2). Then by the chain rule (§18.3), the composite function

$$DF \circ D\vec{r} = F_1 \, Dx_1 + F_2 \, Dx_2 + F_3 \, Dx_3 = (F_1 x_1' + F_2 x_2' + F_3 x_3') \, dt$$

is equal to $D(F \circ \vec{r})$. These are all number-valued functions of the real variable t. Hence, single-variable calculus applies and we find that the integral

$$J = \int_a^b (F_1 x_1' + F_2 x_2' + F_3 x_3') \, dt = F \circ r \Big|_a^b = F(B) - F(A),$$

where in the last terms $F(B)$ means the value of F at the point A and $F(B)$ means the value of F at B. The integral J is called the *line integral* of the gradient of F taken along the curve $\mathscr{C}$. The calculation shows that the value of the line integral J does not depend on the curve $\mathscr{C}$, which joins the endpoints A and B, but only on the values of F at A and B. We say then that the line integral J is independent of the path. We summarize these results in a theorem.

THEOREM (Line integral in a gradient field).   The line integral of the gradient $\overrightarrow{\Delta F}:(F_1, F_2, F_3)$,

$$J = \int F_1 \, Dx_1 + F_2 \, Dx_2 + F_3 \, Dx_3,$$

taken along any smooth curve $\mathscr{C}$ joining two points A and B, is independent of the path $\mathscr{C}$ and is equal to $F(B) - F(A)$.

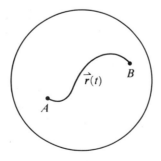

FIGURE 10.2   Path from A to B.

23.4 Exercises

In Exercises 1–4, find an unknown function $F(x, y)$ satisfying the given conditions for all (x, y) in the plane.

1. $F_1(x, y) = 0$,	$F_2(0, y) = y$,	$F(1, 1) = 3$.
2. $F_1(x, y) = 0$,	$F_2(0, y) = \cos y$,	$F(0, \pi/2) = 1$.
3. $F_2(x, y) = x^2 + 1$,	$F_2(x, 3) = 0$,	$F(-1, 3) = \frac{1}{2}$.
4. $F_1(x, y) = \sin(x/2)$,	$F_2(x, y) = 0$,	$F(\pi, -1) = a$.

In Exercises 5–12, a differential form $u(x, y)\, dx + v(x, y)\, dy$ is given. For those that satisfy the cross-derivative test, find the antiderivative $F(x, y)$ such that $DF = u\, dx + v\, dy$. In other words, integrate the forms for which integration is possible.

5. $(x - 2y)\, dx + 2(y - x)\, dy$.   6. $(x - y)\, dx + (y + x)\, dy$.

7. $(x - y)\, dx + (y - x)\, dy$.   8. $(2xy + y)\, dx + (x^2 + x)\, dy$.

9. $\dfrac{1}{y}\, dx - \dfrac{x}{y^2}\, dy$.   10. $2xy\, dx + (y^2 - x^2)\, dy$.

11. $2xy\, dx + (x^2 + y^2)\, dy$.   12. $(xy^2 + x + 2xy)\, dx + (1 + x^2 + x^2 y)\, dy$.

13. Find $F(x, y, z)$ such that $DF = 0$ and $F(1, -1, 3) = 2$.

14. Find $F(x, y, z)$ if $F_1(x, y, z) = 0$ for all (x, y, z), $F_2(x, y, z) = 2yz$, and $F_3 = y^2 + 2z$.

15. Find $f(x, y, z)$ if

$$Df = (2y + 6z)\, dx + (2x - 3z)\, dy - (3y - 6x)\, dz.$$

16. Although $y\, dx - 2x\, dy = 0$ does not satisfy the cross-derivative test, show that if we multiply the equation by $1/xy$, it does.

17. From the fact $F_{12} = F_{21}$, $F_{23} = F_{32}$, $F_{31} = F_{13}$, derive a cross-derivative test that is necessary if

$$DF = u(x, y, z)\, dx + v(x, y, z)\, dy + w(x, y, z)\, dz.$$

18. We have a function $G(x, y, z) = \sqrt{x^2 + y^2}$ and we know that for an unknown function $F(x, y, z)$,

$$D(F - G) = 0 \quad \text{and} \quad F(1, 1, 3) = \sqrt{2}.$$

Find $F(x, y, z)$.

In Exercises 19–22, calculate the line integrals of the gradient fields over the arcs indicated.

19. (Exercise 5) $\int (x - 2y)\, Dx + 2(y - x)\, Dy$ along the curve $x = \cos t$, $y = \sin t$, $0 \leq t \leq \pi$.

20. The line integral of Exercise 17 along the arc $x = \cos t$, $y = \sin t$, $0 \leq t \leq 2\pi$.

21. (Exercise 7) $\int (x - y)\, Dx + (y - x)\, Dy$ along any smooth curve $\dot{r}(t):(x(t), y(t))$ joining points $A:(-1, 3)$ and $B:(2, 0)$.

22. The line integral of Exercise 21 around any smooth closed curve in the plane.

23.5 Cross-Derivative Test

Proof of theorem. Let $O:(a, b)$ be the center of a disk in which the continuous functions $u(x, y)$ and $v(x, y)$ have continuous partial derivative $u_2(x, y) = v_1(x, y)$. We shall prove that there is a function $F(x, y)$ such that $F_1 = u$ and $F_2 = v$ so that $DF = u\, dx + v\, dy$. We define F by

$$F(x, y) = \int_a^x u(s, b)\, ds + \int_b^y v(x, t)\, dt.$$

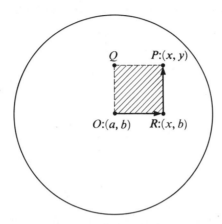

FIGURE 10.3   Function defined by line integral.

We recognize this as an integral over the line OR plus an integral over the line RP (Figure 10.3). We easily compute from the theorem

$$\frac{d}{dx} \int_a^x f(s)\,ds = f(x)$$

when f is continuous (I, §21.7), that $F_2(x, y) = 0 + v(x, y)$. To compute F_1, we must transform the expression for F. First, by subtracting and adding the same term, we find that

$$F(x, y) = \int_a^x u(s, b)\,ds + \int_b^y [v(x, t) - v(a, t)]\,dt + \int_b^y v(a, t)\,dt.$$

In the middle term of this expression we use the fundamental theorem of calculus, then the hypothesis $u_2 = v_1$, and finally the theorem on the order of repeated integrals to write

$$\int_b^y [v(x, t) - v(a, t)]\,dt = \int_a^y \left[\int_a^x v_1(s, t)\,ds \right] dt$$
$$= \int_b^y \int_a^x u_2(s, t)\,ds\,dt$$
$$= \int_a^x \int_b^y u_2(s, t)\,dt\,ds$$
$$= \int_a^x [u(s, y) - u(s, b)]\,ds.$$

Then with this replacement for the middle term, the expression for F becomes

$$F(x, y) = \int_a^x u(s, b)\,ds + \int_a^x [u(s, y) - u(s, b)]\,ds + \int_a^x v(a, t)\,dt.$$

Then we can compute the partial derivative

$$F_1(x, y) = u(x, b) + [u(x, y) - u(x, b)] + 0 = u(x, y),$$

as required. This completes the proof.

23.6   Order of Partial Differentiation

THEOREM.   If $F(x, y)$ has a continuous partial derivative $F_{12}(x, y)$ in a neighborhood of $O:(a, b)$, then it has the partial derivative $F_{21}(x, y) = F_{12}(x, y)$.

Proof. We consider the repeated integral and integrate it,

$$\int_a^x \int_b^y F_{12}(s, t) \, dt \, ds = \int_a^x [F_1(s, y) - F_1(s, b)] \, ds$$
$$= F(x, y) - F(a, y) - F(x, b) + F(a, b).$$

If we reverse the order of the repeated integral, this says that

$$F(x, y) - F(a, y) - F(x, b) + F(a, b) = \int_b^y \int_a^x F_{12}(s, t) \, ds \, dt.$$

Now computing the partial derivative $\partial^2/(\partial y \, \partial x)$ of this, we find that

$$F_{21}(x, y) = F_{12}(x, y).$$

COROLLARY. If there is a repeatedly differentiable function $F(x, y)$ such that $DF = u(x, y) \, dx + v(x, y) \, dy$, then $u_2 = v_1$.

Proof. The statement means that $F_1 = u$ and $F_2 = v$. Hence, $F_{12} = u_2$ and $F_{21} = v_1$. Since $F_{12} = F_{21}$, this implies that $u_2 = v_1$. That is, the cross-derivative test is necessary for $u(x, y) \, dx + v(x, y) \, dy$ to be the differential of some function.

23.7 Differentiation under the Integral Sign

THEOREM. If $f(x, y)$ and its partial derivative $(\partial/\partial y)f(x, y)$ are continuous in a disk with center at $O:(a, b)$ and including $P:(x, y)$, then

$$\frac{\partial}{\partial y} \int_a^x f(s, y) \, ds = \int_a^x \frac{\partial}{\partial y} f(s, y) \, ds.$$

Proof. With (x, y) fixed for the moment, let

$$F(x, y) = \int_a^x f(s, y) \, ds$$
$$= \int_a^x [f(s, y) - f(s, b)] \, ds + \int_a^x f(s, b) \, ds.$$

Applying the fundamental theorem of calculus to the bracketed term, we have

$$F(x, y) = \int_a^x \int_b^y f(s, t) \, dt \, ds + \int_a^x f(s, b) \, ds,$$

or, on changing the order of the repeated integral,

$$F(x, y) = \int_b^y \int_a^x \frac{\partial}{\partial y} f(s, t) \, ds \, dt + \int_a^x f(s, b) \, ds.$$

Now if we differentiate F partially with respect to y, we have

$$\frac{\partial F}{\partial y}(x, y) = \int_a^x \frac{\partial}{\partial y} f(s, y) \, ds + 0,$$

which is the conclusion of the theorem.

24   Integrals of Differential Equations

24.1   Tangent Relations in Local Coordinates.   We have seen that the contour lines $F(x, y) = C$ of a surface $z = F(x, y)$ have tangent lines at any point P with fixed coordinates (x, y) given in variable local coordinates (dx, dy) by the linear relation

$$F_1(x, y)\, dx + F_2(x, y)\, dy = 0 \qquad (17)$$

Now, letting (x, y) vary, we observe that this relation assigns to every point $P:(x, y)$ a line through P, where the variable coordinates on the line are (dx, dy). We may represent this situation graphically if we choose a representative distribution of points $P:(x, y)$ in the plane and through each of them plot a short portion of the line through P given by the linear relation in (dx, dy). We call such a set of lines a *direction map* in the plane (Figure 10.4).

Example.   Plot the direction map

$$xy\, dx + dy = 0$$

defined for all (x, y) in the plane.

Solution.   We need not know that there is a function $F(x, y)$ such that $F_1(x, y) = xy$ and $F_2(x, y) = 1$. We proceed directly and observe that the linear relation says that the slope of the line through the point $P:(x, y)$ is given by

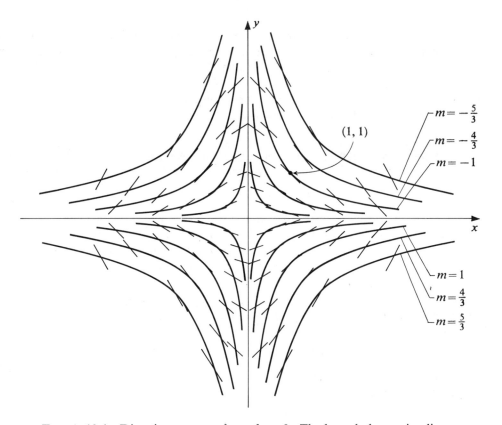

FIGURE 10.4   Direction map $xy\, dx + dy = 0$. The hyperbolas are isoclines.

$$\frac{dy}{dx} = -xy.$$

In particular, the direction may give us a line through the point $(3, \frac{1}{2})$ whose slope $dy/dx = -\frac{3}{2}$. We plot the point and a small segment of the line through it (Figure 10.5). To obtain a graphic representation of the direction map as a whole we carry out this procedure for a number of points well distributed in the plane and plot all these points and their line segments in the same coordinate plane.

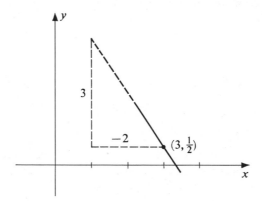

FIGURE 10.5

It is helpful to plot some of the curves in the plane along which the direction map gives the same slope m at every point. These are called *isoclines*. The isoclines for this problem are the hyperbolas $-xy = m$, where the slope m is an arbitrary constant. Then we calculate the slope at one point of a isocline and plot the directions at other points of this isocline as a series of lines parallel to the calculated one (Figure 10.4). In this problem, isoclines on which $m = 0$ are the x- and y-axes.

24.2   Tangent Relations as Differential Equations. Any relation in local coordinates (dx, dy) of the form

$$u(x, y)\, dx + v(x, y)\, dy = 0$$

determines a direction map in a region where $u(x, y)$ and $v(x, y)$ are defined, except where $u(x, y) = v(x, y) = 0$. Our problem is to find a function $F(x, y)$ such that $F_1(x, y) = u(x, y)$ and $F_2(x, y) = v(x, y)$. If this can be done, then the differential equation $u\, dx + v\, dy = 0$ is the tangent relation of the contour lines $F(x, y) = C$ of the function $z = F(x, y)$.

Before we try to solve the differential equation $u(x, y)\, dx + v(x, y)\, dy = 0$, we observe that we can form an equivalent relation by multiplying by any nonzero function $r(x, y)$. For the linear relation

$$r(x, y)u(x, y)\, dx + r(x, y)v(x, y)\, dt = 0$$

is true if and only if the relation $u\, dx + v\, dy = 0$ is true. The two relations have the same direction map. For the slope has the value,

$$\frac{dy}{dx} = -\frac{ru}{rv} = -\frac{u}{v},$$

as in the original relation. We say that the relations

$$u(x, y)\, dx + v(x, y)\, dy = 0 \qquad \text{and} \qquad ru\, dx + rv\, dy = 0$$

are equivalent if and only if their direction maps are the same.

Example.   Solve the differential equation $xy\, dx + dy = 0$.

   Solution.   This does not satisfy the cross-derivative test, $u_2 = v_1$; for $u_2 = x$ and $v_1 = 0$. Hence, there is no function $F(x, y)$ such that $F_1 = xy$ and $F_2 y = 1$. However, we divide by y and the relation becomes the equivalent one

$$x\, dx + \frac{1}{y}\, dy = 0,$$

in which the coefficient of dx is a function of y alone, and the coefficient of dy is a function of y alone. This now satisfies the cross-derivative test $u_2 = v_1 = 0$. It integrates at once to give us

$$\frac{x^2}{2} + \ln |y| = C.$$

The tangent relation of these curves is equivalent to the given relation, $xy\, dx + 1\, dy = 0$. The relation $(x^2/2) + \ln |y| = C$ is called an integral of the differential equation. If we solve it by y as a function of x, we get

$$y = \exp\left(-\frac{x^2}{2} + C\right).$$

and for every value of C this function satisfies the differential equation

$$\frac{dy}{dx} = -xy,$$

as we may verify by differentiating it and substituting into the equation. It is called a *solution* of the differential equation. We plot the graph of this solution for the case $C = 0$ (Figure 10.6). This completes the problem.

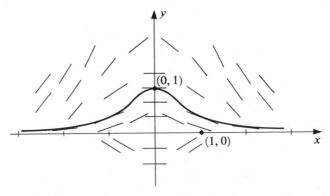

FIGURE 10.6   Solution of the differential equation $xy\, dx + dy = 0$. Solution containing (0,1) is $y = \exp(-x^2/2)$. (Compare Figure 10.5.)

In general, a relation of the form

$$u(x, y)\, dx + v(x, y)\, dy = 0$$

is a differential equation. A function $F(x, y)$ for which the level curves $F(x, y) = C$ have tangent relations equivalent to $u(x, y)\, dx + v(x, y)\, dy = 0$ is called an *integral*. Functions $y(x)$ that satisfy the differential equation in the form

$$u(x, y) + v(x, y)\frac{dy}{dx} = 0$$

are called *solutions*. Solutions may be found from integrals $F(x, y) = C$ or any equivalent relation by solving the integral for y as a function of x. Any factor $r(x, y)$ is called an *integrating factor* if it transforms the differential equation

$$u(x, y)\, dx + v(x, y)\, dy = 0$$

into an equivalent one

$$r(x, y)u(x, y)\, dx + r(x, y)v(x, y)\, dy = 0$$

for which there is a function $F(x, y)$ having $F_1(x, y) = ru$ and $F_2(x, y) = rv$.

24.3   Separation of Variables. If a differential equation can be multiplied by a suitable integrating factor to reduce it to the form

$$u(x)\, dx + v(y)\, dy = 0,$$

it is said to have *variables separated*. In practice, we integrate both sides to get

$$\int u(x)\, dx = -\int v(y)\, dy.$$

The left member is a function of x and the right member is a function of y. Hence, to be equal, the two integrals differ at most by a constant so that the relation

$$\int u(x)\, dx + \int v(y)\, dy = C$$

is an integral.

Alternatively, we can reason as we did for the problem of inverse of differentiation

$$DF = u(x, y)\, dx + v(x, y)\, dy.$$

We set $F_1(x, y) = u(x)$ and integrate with respect to x. This gives

$$F(x, y) = \int u(x)\, dx + g(y).$$

Then we differentiate partially with respect to y to find $F_2(x, y) = 0 + g'(y)$ and set

$$F_2(x, y) = g'(y) = v(y).$$

This then integrates to give $g(y) = \int v(y)\, dy + C$. Hence

$$F(x, y) = \int u(x)\, dx + \int v(y)\, dy + C.$$

Thus

$$\int u(x)\, dx + v(y) \int dy = -C$$

is an integral of the differential equation.

We consider another example.

Example. Find the solution of the differential equation

$$2x(y + 1)\, dx - y\, dy = 0$$

that satisfies the initial condition $y = -2$ when $x = 0$.

Solution. We separate variables by multiplying by $1/(y + 1)$. This gives us the equivalent relation with variables separated

$$2x\, dx - \frac{y}{y + 1}\, dy = 0.$$

To integrate the fraction in the second term, we must first carry out one division so that the degree of the numerator becomes less than the degree of the denominator. When we do this we have the equivalent relation

$$2x\, dx - \left(1 - \frac{1}{y + 1}\right) dy = 0.$$

We integrate and find that

$$x^2 - y + \ln |y + 1| = C$$

is an integral. We can determine the constant in this by use of the initial condition $(0, -2)$

$$0 + 2 + \ln |-1| = C, \quad \text{or} \quad C = 2.$$

The integral, with C determined, then becomes

$$x^2 - y + \ln |y + 1| = 2.$$

To find a solution, as defined, we must solve this for y as a function of x. This is complicated, so we find it convenient to let the integral serve instead of a solution $y = f(x)$.

24.4 Exercises

In Exercises 1–4, construct the direction maps of the given relations.

1. $y\, dx + x\, dy = 0$,   2. $y\, dx - dy = 0$.

3. $(x - 1)\, dx + (y + 1)\, dy = 0$.   4. $2y\, dx - x\, dy = 0$.

In Exercises 5–8, plot the curves $F(x, y) = C$ and find the differential equation of their tangent directions. Use the plotted curves to plot the direction maps of the tangent relations.

5. $F(x, y) = x^2 + y^2$.   6. $F(x, y) = y - x$.

7. $F(x, y) = \dfrac{x^2 + y^2}{2(x + y)}$.   8. $F(x, y) = xy$.

9. Find an integral for the differential equation of Exercise 1.
10. Find an integral for the differential equation of Exercise 2.

11. Find the solutions $y(x)$ of the differential equation in Exercise 2 using the integral in Exercise 10.

12. Solve the differential equation of Exercise 3.

13. For the direction map given by the linear relation, Exercise 3, find the locus of points (x, y) (a) where the direction is vertical; (b) where it is horizontal.

14. Find the integrals of the differential equation in Exercise 4 and plot their graphs, all in the same coordinate plane.

In Exercises 15–20, find integrals of the differential equations by separation of variables.

15. $y\,dx - 2x\,dy = 0.$ 16. $(1 - y^2)\,dx - y\,dy = 0.$

17. $\dfrac{dy}{dx} = \dfrac{3x}{1 - y^2}.$ 18. $\dfrac{dy}{dx} = \dfrac{x}{y}.$

19. $(1 + x)\dfrac{dy}{dx} = y^3.$ 20. $\dfrac{dy}{dx} = xy^2.$

21. In Exercise 16, determine C if $y = \frac{1}{2}$ when $x = 0$, and plot the graph. Show that $|y| \to 1$ when $x \to \infty$.

22. In Exercise 16, find dy/dx from the differential equation when $y = \frac{1}{3}$. Plot the tangent line to the solution curve that passes through $(1, \frac{1}{3})$.

23. Find the tangent relation of

$$\sqrt{x^2 + y^2} - \arctan\frac{y}{x} = C$$

and plot its direction map; observe that it has the characteristics of a flow into a vortex.

A function $u(x, y)$ is said to be homogeneous of degree k in x and y if $u(tx, ty) = t^k u(x, y)$. If $u(x, y)$ and $v(x, y)$ are homogeneous of the same degree, then the substitution $y = vx$, $dy = v\,dx + x\,dv$ transforms the differential equation $u(x, y)\,dx + v(x, y)\,dy = 0$ into one in x and v whose variables are separable. Use this device to solve the differential equations 24–28.

24. $xy\,dx - (x^2 + y^2)\,dy = 0.$ 25. $(x - 2y)\,dx + (2x + y)\,dy = 0.$

26. $(x^2 + y^2)\,dx + xy\,dy = 0.$ 27. $y^2\,dx + x(y - 4x)\,dy = 0.$

28. $(y - \sqrt{x^2 + y^2})\,dx - x\,dy = 0$ and $y = 1$ where $x = \sqrt{3}$.

25   Level Contours and Stream Lines

25.1   Orthogonal Nets

Example.   Find the equations of curves that intersect the family of all the curves,

$$\frac{x^2 + y^2 + 1}{x} = 2c,$$

orthogonally, that is, so that at every intersection the tangents are perpendicular.

Solution.   We plot the family of curves, finding them to be circles $(x - c)^2 + y^2 = c^2 - 1$ for all constants c such that $|c| \geq 1$. When $|c| < 1$, no points (x, y) satisfy the equation (Figure 10.7). Each value of c, $|c| \geq 1$, determines a curve of the family and

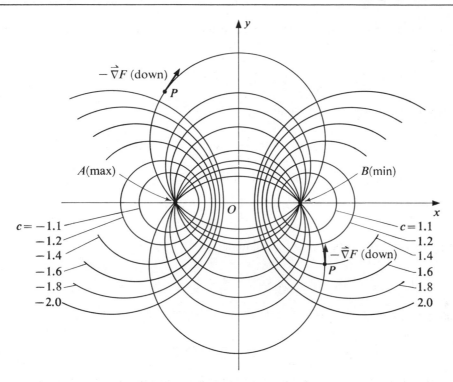

FIGURE 10.7· Orthogonal net of circles.

one and only one curve of the family passes through each point (x, y) except for points where $x = 0$. For example, the curve through $(1, 2)$ has $c = \frac{6}{4}$ and is the circle

$$\left(x - \frac{3}{2}\right)^2 + y^2 = \left(\frac{\sqrt{5}}{2}\right)^2 .$$

The tangent lines to these curves at the point $P:(x, y)$ are given in local (dx, dy) coordinates by differentiating the equation of the curve. We find them to be

$$(x^2 - y^2 - 1)\, dx + 2xy\, dy = 0$$

after removing the factor $1/x^2$. The slope of the tangent line at $P:(x, y)$ is

$$\frac{dy}{dx} = -\frac{x^2 - y^2 - 1}{2xy} .$$

and the slope of the perpendicular line through P is

$$\frac{dy}{dx} = \frac{2xy}{x^2 - y^2 - 1} .$$

Hence, the orthogonal curves must satisfy the differential equation

$$2xy\, dx - (x^2 - y^2 - 1)\, dy = 0.$$

This does not satisfy the cross-derivative test, but recalling that we have removed the factor $1/x^2$, we now multiply by $1/y^2$ and find that the equivalent direction map

$$\frac{2x}{y}\, dx - \frac{(x^2 - y^2 - 1)}{y^2}\, dy = 0$$

does satisfy the cross-derivative test. Therefore, it is the differential, $F_1\,dx + F_2\,dy$, of some function defined except where $y = 0$. We set

$$F_1 = \frac{2x}{y}$$

and integrate with respect to x, obtaining

$$F = \frac{x^2}{y} + g(y),$$

where $g(y)$ is some unknown function of y alone. Then

$$F_2 = -\frac{x^2}{y^2} + \frac{d}{dy}g(y).$$

Equating this to the coefficient of dy, we find

$$\frac{d}{dy}g(y) = 1 + \frac{1}{y^2}.$$

Hence $g(y) = y - (1/y) - 2C$ and this gives us the integral

$$\frac{x^2 + y^2 - 1}{y} = 2C$$

of the differential equation of the orthogonal curves. The equation being simplified, the curves turn out to be the circles

$$x^2 + y^2 - 2Cy = 1,$$

all of which pass through the points $(-1, 0)$ and $(1, 0)$ and have centers at $(0, C)$ on the y-axis (Figure 10.7).

25.2   Stream Lines. We now look at the preceding example in a different way. We consider the function given by

$$F(x, y) = \frac{x^2 + y^2 + 1}{x}, \qquad x \neq 0.$$

The level contour lines $z = 2c$ are the circles with centers on the x-axis (Figure 10.7). The graph is in two disconnected sheets, one approaching $+\infty$ as $x \to 0+$. There is a minimum at $B:(1, 0)$ and a maximum at $A:(-1, 0)$. At any point $P:(x, y)$, the gradient is

$$\vec{\nabla}F:(F_1, F_2) = \left(\frac{x^2 - y^2 - 1}{x^2}, \frac{2y}{x}\right).$$

This points in a direction orthogonal to the tangent $F_1\,dx + F_2\,dy = 0$ of the level contour $F(x, y) = 2c$ at P. Hence, the gradient vector is tangent to the curve of the orthogonal family at P. Moreover, the gradient gives the maximum rate of change of F at P both in magnitude and direction.

Now if we think of the surface $z = F(x, y)$ as a hillside, any droplet of water at P will move downhill in the direction $-\vec{\nabla}F$, the direction of steepest descent. On the right-hand half of the surface, where $x > 0$, this vector $-\vec{\nabla}F$ points down towards the minimum point at B. Droplets of water on the surface will all flow along the orthogonal circles

$$x^2 + y^2 - 2Cy = 1$$

towards B. On the left-hand side, where $x < 0$, the vector $-\vec{\nabla}F$ points downhill away from the maximum point A and droplets of water on this surface will flow along the circles orthogonal to the level contours.

Summarizing, for a function $z = F(x, y)$, the level contours, $F(x, y) = c$, are the integral curves of the differential equation

$$F_1 \, dx + F_2 \, dy = 0.$$

The curves of quickest descent across these level contours are the integral curves of the orthogonal family,

$$F_2 \, dx - F_1 \, dy = 0.$$

They are called the *stream lines* for the function. The stream lines are orthogonal to the level curves for every differential function $F(x, y)$ and together the level curves and stream lines always form an orthogonal net.

25.3   Integrating Factors.   In the problem of the orthogonal net of circles (I, §19.1) we encountered a differential equation

$$2xy \, dx - (x^2 - y^2 - 1) \, dy = 0$$

that did not satisfy the cross-derivative test and hence had no integral $F(x, y) = C$. But we found that by multiplying by $1/y^2$, we obtained an equivalent direction map

$$\frac{2x}{y} \, dx - \frac{(x^2 - y^2 - 1)}{y^2} \, dy = 0$$

that did satisfy the cross-derivative test. It had an integral

$$\frac{x^2 + y^2 - 1}{y} = C.$$

DEFINITION.   For a differential form

$$u(x, y) \, dx + v(x, y) \, dy,$$

a function $r(x, y)$ is called an *integrating factor* if there is a function $F(x, y)$ such that

$$DF = r(x, y)u(x, y) \, dx + r(x, y)v(x, y) \, dy.$$

The necessity of finding an appropriate integrating factor to make a differential form into a chain for integration has occurred in single-variable calculus. For example, we introduce the integrating factor 2 to complete the chain in the integral

$$\int (2x - 3)^{1/3} \, Dx = \frac{1}{2} \int (2x - 3)^{1/3} 2 \, Dx = \frac{3}{8}(2x - 3)^{4/3} + C.$$

Here we are limited to introducing a constant factor so that the compensating reciprocal factor $\frac{1}{2}$ can be applied outside the integral sign. But when we have a differential relation

$$u(x, y) \, dx + v(x, y) \, dy = 0,$$

we can introduce an integrating factor that is a function $r(x, y)$ if we can find it. We will limit ourselves to a few that can be found by inspection. We recall a few formulas for differentiation

$$\boldsymbol{D}xy = x\,dy + y\,dx, \qquad \boldsymbol{D}\left(\frac{x}{y}\right) = \frac{y\,dx - x\,dy}{y^2}.$$

$$\boldsymbol{D}\left(\frac{x}{y}\right) = \frac{x\,dy - y\,dx}{x^2}, \qquad \boldsymbol{D}\arctan\left(\frac{y}{x}\right) = \frac{x\,dy - y\,dx}{x^2 + y^2},$$

Example.   Find the orthogonal trajectories of the family of curves $y(x^2 + c) + 2 = 0$.

Solution.   We find the differential equation of the family of curves, by eliminating c from the equation of the family and the tangent relation, $(x^2 + c)\,dy + 2xy\,dx = 0$. The elimination of c gives us the differential equation

$$-\frac{dy}{y} + xy\,dx = 0.$$

The orthogonal trajectories then satisfy the differential equation

$$\frac{1}{y}\,dx + xy\,dy = 0.$$

The cross-derivative test reveals that this is not the differential of any function $F(x, y)$. However, by inspection we determine that y/x is an integrating factor. Multiplying by y/x, we find the equivalent differential equation

$$\frac{dx}{x} + y^2\,dy = 0,$$

with variables separated. Its integral is

$$\ln |x| + \frac{y^3}{3} = k, \quad x \neq 0.$$

This is the equation of the family of orthogonal trajectories.

25.4   Exercises

In Exercises 1–8, find the orthogonal trajectories of the given families of curves.

1.   The lines $x - y = c$. Plot.
2.   The circles $x^2 + y^2 = c^2$. Plot.
3.   The hyperbolas $x^2 - y^2 = c$. Plot.
4.   The hyperbolas $x^2 - y^2 = 2cx$.
5.   The ellipses $2x^2 + y^2 = b^2$.
6.   The circles $x^2 + y^2 = 2cx$. Plot.
7.   The parabolas $y^2 = 4ax$.
8.   The exponential curves $y = ce^{-x}$. Plot.
9.   For the function $F(x, y) = x^2 + y^2$, find the level contours and stream lines. Plot. (See Exercise 2.)
10.   For the potential function of a point charge at the origin

$$F(x, y) = \frac{k}{\sqrt{x^2 + y^2}},$$

find the level contours (equipotential curves) and stream lines.

11. Stream lines of a scalar field $F(x, y)$ are the solutions of the vector differential equation $d\vec{r}/dt = -\vec{\nabla}F$ (§25.3). Eliminate the parameter t and show that this equation is the same as that of the orthogonal family of the level contours $F(x, y) = C$.

12. The method of Exercise 11 has merit in finding the stream lines of a function $F(x, y, z)$. Find the stream lines of the function

$$F(x, y, z) = x^2 + y^2 + z^2.$$

13. For a hill in the shape of the surface

$$w = \frac{1}{1 + x^2 + 4y^2},$$

find the stream lines along which water flows. Plot them in the xy-plane.

In Exercises 14–18, use integrating factors to find an integral of the differential equation.

14. $y\, dx - (x + y)\, dy = 0.$ 15. $x\, dy - (y + x^2)\, dx = 0.$
16. $(x^2 + y^2)\, dx + y\, dx - x\, dy = 0.$ 17. $x\, dx + (y - \sqrt{x^2 + y^2})\, dy = 0.$
18. $y\, dx + (y^3 - x)\, dy = 0.$

19. Show that the family of parabolas $y^2 = 2cx + c^2$ is self-orthogonal.
20.* For the surface $w = \sqrt{1 - x^2} - y^2$, find the curves that intersect the level contours at a constant angle α. These are the paths on the earth traced by following a fixed direction.

26   Applications of Differential Equations in Dynamic Processes

26.1   Growth Equation. We proceed by examples, starting with the familiar differential equation of growth (I, §29).

Example. The value $\$A$ of an investment at time t years grows at a rate proportional to its value at that time. The factor of proportionality r is the compound interest rate (converted continuously). Find the amount after t years of an investment whose principal sum was $\$P$ at time $t = 0$.

Solution by Separation of Variables.   The statement of the growth law says that

$$\frac{dA}{dt} = rA.$$

We may write this relation in the form

$$\frac{dA}{A} = r\, dt,$$

where the variables are now separated. Integrating both sides by use of the theorem on functions with zero partial derivative (§23.2), we find that

$$\ln |A| = rt + C,$$

where C is an arbitrary constant. But we know that when $t = 0$, $A = P$. Hence, since both P and A are positive, $\ln P = C$. Thus $\ln A = rt + \ln P$, which gives us

$$\ln \frac{A}{P} = rt.$$

This is an integral of the differential equation. We get a solution expressing A as a function of t by solving this for A. The result is

$$A = Pe^{rt}.$$

This completes the problem.

26.2 Growth to Definitive Size. A possible mathematical model for the growth of a plant assumes that the rate of growth is proportional to the product of the present size y and the difference $h - y$ between its size now and its mature size h.

Example. Assuming this model, find the growth curve for a stalk of corn that will eventually grow to 10 ft at maturity, was 1 ft high at time 0, and was 2 ft high 8 days later.

Solution. Let p be the fraction (percentage) of mature height attained in t days. Then, according to the assumed model,

$$\frac{dp}{dt} = kp(1 - p),$$

where k is an unknown constant. By separation of variables this relation is equivalent to

$$\frac{dp}{p(1 - p)} = k \, dt.$$

We readily verify that the fraction on the left can be written as a sum of partial fractions

$$\frac{1}{p(1 - p)} = \frac{1}{p} + \frac{1}{1 - p},$$

which enables us to rewrite the differential equation as

$$\frac{dp}{p} + \frac{dp}{1 - p} = k \, dt.$$

We form the antiderivative of each side and obtain the integral

$$\ln |p| - \ln |1 - p| = kt + C,$$

or

$$\ln \frac{p}{1 - p} = kt + C.$$

It remains to determine the values of k and C from the given fact that $p = \frac{1}{10}$ when $t = 0$ and $p = \frac{2}{10}$ when $t = 8$. The first condition gives $C = \ln \frac{1}{9}$ or $C = -\ln 9$. The second says that

$$\ln \frac{\frac{2}{10}}{1 - \frac{2}{10}} = 8k - \ln 9.$$

We simplify and solve for k and find that

$$k = \tfrac{1}{8} \ln \tfrac{9}{4} = 0.102 \text{ approximately.}$$

Hence our integral of the differential equation becomes

$$\ln \frac{p}{1 - p} = \frac{t}{8} \ln \frac{9}{4} - \ln 9 \approx 0.102t - 2.197.$$

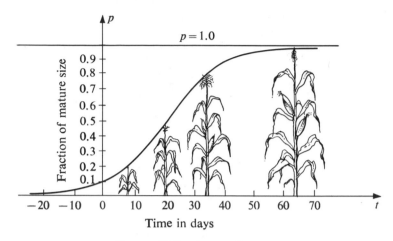

FIGURE 10.8   A growth curve.

To plot the graph of the solution we may plot the graph of this integral, which is the same. We substitute values of p and compute t as follows.

p	0.01	0.05	0.10	0.20	0.50	0.80	0.90	0.95	0.99
t	-23.5	-7.3	0.0	7.9	21.5	34.2	42.9	50.0	66.5

We plot the graph of the growth curve from this table (Figure 10.8). We observe that in this model, growth is slow at the beginning and also slow as the plant approaches mature size. It is most rapid and nearly uniform at an intermediate stage. This completes the problem.

26.3   Bimolecular Chemical Reactions.   We have seen (I, §29) that the growth or decay of a chemical substance that forms or disintegrates as a result of chance events in a single molecule is governed by the differential equation of exponential growth (§26.1). Let us now consider an example in which two different substances combine to form a third.

Example.   Two chemical substances A and B are initially present in a solution in known concentrations, a and b (Figure 10.9). The molecules of A and B combine in pairs to form a new substance AB, which is not present at the start. Assuming that the probability of forming a molecule of AB is proportional to the probability that a molecule of A will encounter a molecule of B, find a relation that represents the concentration of AB at any time t (Figure 10.9).

Solution.   It is important that the amounts be measured in mols so that the *number of molecules* of A, B, or AB present in the common unit of mass will be proportional to their concentrations. Let x be the concentration of AB at time t. Then the concentration of A is $a - x$ and that of B is $b - x$, since each molecule of AB was formed by taking away one of A and one of B. The probability that a molecule of A will encounter a

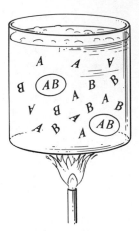

FIGURE 10.9 Bimolecular reaction process.

molecule of B in the solution is then proportional to the product of concentrations. That is, the rate dx/dt of formation of AB is given by the differential equation

$$\frac{dx}{dt} = k(a - x)(b - x),$$

where k is a constant and $x = 0$ when $t = 0$.

We separate the variables and replace this relation between the local coordinates dt and dx by the equivalent relation

$$\frac{dx}{(a - x)(b - x)} = k \, dt.$$

We find algebraically that if we assume that $a > b$ for the sake of definiteness,

$$\frac{1}{(a - x)(b - x)} = \frac{1}{a - b}\left[\frac{1}{b - x} - \frac{1}{a - x}\right].$$

Then the differential equation can be written

$$\frac{dx}{b - x} - \frac{dx}{a - x} = (a - b)k \, dt.$$

This has the integral

$$\ln \frac{b - x}{a - x} = -(a - b)kt + \ln \frac{b}{a}.$$

Simplifying, we find that this can be written

$$\frac{b - x}{a - x} = \frac{b}{a} \exp\left[-(a - b)kt\right].$$

This relation enables us to calculate x for any time t, and hence it is a solution of the problem.

26.4 Exercises

In the following exercises, set up a differential equation with initial conditions to represent the process described, and solve the equation.

1. While there is plenty of food, a culture of yeast increases at a rate proportional to the amount present, doubling in 2 hours. Express the amount of yeast as a function of time.

2. A chemical A is present in a solution initially at concentration x. It breaks down to form B at a rate proportional to the concentration of unconverted A. Half of A is converted in 20 minutes. Express the concentration y of B as a function of time. Plot it.

3. A chemical A reacts spontaneously to form B at a rate proportional to the square of the concentration x of A. Initially the concentration of A was a_0 and that of B was 0. Express the concentration y of B as a function of time. Plot this relation.

4. A hot body cools at a rate proportional to the difference between its temperature θ and that of the surrounding atmosphere, θ_0. Show that $\theta - \theta_0 = Ce^{-kt}$. What does the constant C represent? The constant k?

5. One ml of water at 90° inside a capsule is immersed in 100 ml of well-stirred water initally at 10°. Express the temperature of the bath as a function of time.

6. If corn (§26.2) reaches $\frac{1}{10}$ maturity height in 20 days and $\frac{2}{10}$ maturity height in 30 days, how many days will be required to reach $\frac{9}{10}$ maturity height?

Exercises 7–12 refer to the bimolecular reaction (§26.3).

7. For the bimolecular reaction, carry out the integration and simplify the result to obtain the final integral

$$\frac{b - x}{a - x} = \frac{b}{a} \exp\left[-(a - b)kt\right].$$

8. Prove that if $a > b$, then the reaction continues until $x = b$, that is, until all molecules of B are used up. In this respect the mathematical solution agrees with what we expect intuitively.

9. Plot the graph of the concentration of the product substance AB if $a = 2 + 10^{-3}$ and $b = 10^{-3}$.

10. At the start \$$P$ is invested to grow at the rate r according to the compound interest law, $dA/dt = kA$. In addition, new capital is invested in the fund continuously at the rate \$$R$ per year. Find the accumulated amount after t years.

11. If \$100 grows to \$200 in 16 years at compound interest, what is the growth rate?

12. A chemical substance A is changing into B at the rate rx, where x is the concentration of A. At the same time B is changing back to A at a rate sy, where y is the concentration of B. What is the concentration of B as a function of time if the initial concentrations are x_0 and y_0. Show that

$$\lim_{t \to \infty} y(t) = \frac{r}{r + s}(x_0 + y_0)$$

and that

$$x + y = \frac{r}{r + s}(x_0 + y_0).$$

13. A population $y(t)$ has an annual birth rate β and death rate δ. Find the function $y(t)$, starting with $y(0) = N$. What is the growth rate of the population?

14. In a restricted environment, the growth rate k in the growth equation $dy/dt = ky$, is not constant but decreases with increasing y according to the formula $k = \alpha - \beta y$, where α and β are constants. Solve the restricted growth equation and find the limiting population size as t increases.

15. In a chemical reaction, two molecules of substance A combine at a rate proportional to the square of the concentration of unconverted A. If the initial concentration of A is $x(0) = x_0$, find the concentration $x(t)$ of A as a function of t. Plot its graph.

16.* In the population $y(t)$ of an animal species, the fraction of females remains one half. The reproduction rate is proportional to the number of encounters between individuals of different sex, that is, to y^2. The death rate is proportional to y. These facts are expressed in the growth law $dy/dt = \beta y^2 - \delta y$, with $y(0) = N$. Solve for $y(t)$ and sketch the solution function showing the cases

$$\frac{\beta}{\delta} < \frac{1}{N}, \qquad \frac{\beta}{\delta} = \frac{1}{N}, \qquad \frac{\beta}{\delta} > \frac{1}{N}.$$

27 Differential Equations of Motion

27.1 Law of Motion of a Particle. A particle consists of a point-mass m and a vector function of time $\vec{x}(t)$, which for any time t gives the position of the particle. The particle traces a curve, as t varies, in which its velocity is the vector $\vec{v}$ and its momentum is the vector $m\vec{v}$. The particle moves in response to the forces $\vec{F}$ that act upon it. According to the law of motion, the rate of change of momentum is proportional to the force acting on the particle,

$$\frac{d}{dt}(m\vec{v}) = k\vec{F}.$$

When m is a constant, and when the units are fixed so that $k = 1$, this reduces to

$$m\frac{d^2\vec{x}}{dt^2} = \vec{F},$$

read "mass times acceleration equals force."

We recall the familiar example of the particle that falls under its weight, $-mg$, near the earth's surface,

$$m\frac{d^2z}{dt^2} = -mg.$$

Its position z as a function of t can be found by integrating twice,

$$z = -\tfrac{1}{2}gt^2 + c_1 t + c_2.$$

We can determine the constants of integration c_1 and c_2 if we know the initial position z_0 and the initial velocity v_0 at $t = 0$. With these initial conditions the motion is given by

$$z = -\tfrac{1}{2}gt^2 + v_0 t + z_0.$$

This is a motion in a vertical line, up and down.

Now we solve the same problem in 3-space, using vectors.

Example. Find the motion in 3-space of a 16-lb shot tossed by a shot-putter with the initial velocity $\vec{v}_0 : (0, 40, 40)$ from the initial position $\vec{x}_0 : (0, 0, 8)$.

Solution. The only force acting after the shot is released is $\vec{F} : (0, 0, -mg)$. The vector equation of motion

$$m\frac{d^2\vec{x}}{dt^2} = \vec{F}$$

resolves into three component equations

$$\frac{d^2x}{dt^2} = 0, \qquad \frac{d^2y}{dt^2} = 0, \qquad \frac{d^2z}{dt^2} = -g.$$

These can be integrated separately. Integrating each one twice, we find first $\vec{x}'$ and then $\vec{x}$.

$$x' = a_1, \qquad y' = b_1, \qquad z' = -gt + c_1,$$
$$x = a_1 t + a_2, \qquad y = b_1 t + b_2, \qquad z = -\tfrac{1}{2}gt^2 + c_1 t + c_2,$$

where $a_1, a_2, b_1, b_2, c_1, c_2$ are constants of integration. When $t = 0$, the velocity $\vec{x}'$ is $\vec{v}_0 : (0, 40, 40)$. Hence, $a_1 = 0, b_1 = 40, c_1 = 40$. Also, when $t = 0$, the initial position $\vec{x}$ is $\vec{x}_0 : (0, 0, 8)$. This gives us $a_2 = 0, b_2 = 0, c_2 = 8$. In conclusion, the motion of the shot is the trajectory whose parametric equations are

$$x = 0, \qquad y = 40t, \qquad z = -\tfrac{1}{2}gt^2 + 40t + 8.$$

This completes the problem.

27.2 Energy. We can integrate the equations of motion when the force $\vec{F}(\vec{x})$ is a function of the position $\vec{x}$ of the particle. We form the inner product of both members of the equation of motion

$$m\frac{d^2\vec{x}}{dt^2} = \vec{F}$$

The velocity vector $d\vec{x}/dt$ satisfies the differential equation

$$m\frac{d^2\vec{x}}{dt^2}\cdot\frac{d\vec{x}}{dt} = \vec{F}\cdot\frac{d\vec{x}}{dt}.$$

We find that the left member is the derivative

$$\frac{d}{dt}\left[\frac{1}{2}m\frac{d\vec{x}}{dt}\cdot\frac{d\vec{x}}{dt}\right] = \frac{d}{dt}\left[\frac{1}{2}m\|\vec{v}\|^2\right].$$

Hence we can integrate it with respect to t, and we find that on any solution $\vec{x}(t)$,

$$\frac{1}{2}m\|\vec{v}\|^2 - \int\left(\vec{F}\cdot\frac{d\vec{x}}{dt}\right)dt = C,$$

where the constant C has the units of energy. The first term is kinetic energy. The line integral $-\int_0^t (\vec{F}\cdot(d\vec{x}/dt))\, dt$ represents the work done by the particle against the force in moving from the position when $t = 0$ to the position at time t. The energy integral says that the sum of the potential and kinetic energies of the particle remains constant on the path, "conservation of energy."

27.3 Parachute Jumper

Example A. A parachute resists the fall of the parachutist with an upward force proportional to the velocity of fall (Figure 10.10). The parachute force drag is 150 lb when the downward velocity is 15 ft/sec. Find the velocity of a 200-lb parachutist who is falling vertically at 160 ft/sec when his chute opens.

FIGURE 10.10   Motion with resistance.

Solution.   We shall solve the problem with general parameters. Let W be the weight of the parachutist, β the coefficient of drag, and v_0 the initial velocity. Then the mass of the parachutist is W/g and the two forces acting on his fall are his weight $-W$ downward and the drag $-\beta v$ of the parachute upward. The equation of motion is then

$$\frac{W}{g}\frac{dv}{dt} = -W - \beta v.$$

Separating variables, we find that the differential equation can be written

$$\frac{dv}{W + \beta v} = -\frac{g}{W}\,dt.$$

The left member integrates in the form $\int (Du/u) = \ln |u| + C$ if we multiply by β to make the differential $D(W + \beta v)$. We find

$$\int \frac{\beta\,dv}{W + \beta v} = -\frac{\beta g}{W}\int dt, \qquad \text{or} \qquad \ln (W + \beta v) = -\frac{\beta g}{W}\,t + C.$$

Since $v = v_0$ when $t = 0$, we find that $C = \ln (W + \beta v_0)$. With this value of C, the integral of the differential equation becomes

$$\ln \frac{W + \beta v}{W + \beta v_0} = -\frac{\beta g}{W}\,t.$$

Inverting the function **ln**, we may write this relation in the form

$$\frac{W + \beta v}{W + \beta v_0} = e^{-(\beta g/W)t}.$$

Or solving for v, we get the solution that gives the velocity as a function of time:

$$v = -\frac{W}{\beta} + \left(v_0 + \frac{W}{\beta}\right)e^{-(\beta g/W)t}.$$

Since $\lim_{t \to \infty} e^{-(\beta g/W)t} = 0$, this tells us that the downward velocity of the parachutist approaches $-W/\beta$ as a limit, whatever the initial velocity.

With the numerical data of this problem, we know that $-\beta v = 150$ when $v = -15$, hence $\beta = 10$. Thus the 200-lb parachutist approaches the steady-fall velocity $-200/10 = -20$ ft/sec.

Example B.   In Example A, determine the rate of change of the total energy. Is the total energy constant (conservation of energy)?

Solution.   We calculate the energy integral of the differential equation of motion by first multiplying it by the integrating factor v.

$$\frac{W}{g} v \frac{dv}{dt} = -Wv - \beta v^2.$$

This can be written

$$\frac{d}{dt} \left[\frac{1}{2} \frac{W}{g} v^2 + W(x - x_0) \right] = -\beta v^2.$$

The expression in the brackets is the total energy, kinetic plus potential. This expression shows that the total energy is being reduced at the rate $-\beta v^2$ energy units per second by the drag of the parachute. The motion does not satisfy the conservation of energy. We may interpret this physically by observing that the parachute dissipates energy into the atmosphere at the rate $-\beta v^2$.

27.4   Rocket Flight

Example.   Find the equation of motion of a rocket-propelled mass $M(t)$ moving in a straight line under an external force F and a propulsion force P, due to the burning of rocket fuel, assuming that the burning gases are ejected backward through a nozzle with constant velocity c (Figure 10.11). Solve the differential equation under outer space conditions, where $F = 0$, with fuel burned at a constant rate, $dM/dt = -k$, and initial conditions $v = v_0$, $M = M_0$ when $t = 0$.

Solution.   We will use a plausible argument to formulate the differential equation of motion. Let $M(t)$ represent the (decreasing) mass of the rocket and fuel left unburned

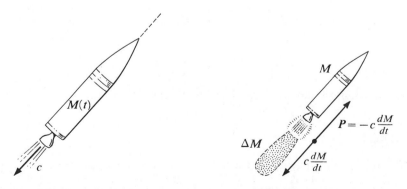

FIGURE 10.11   Motion with variable mass.   FIGURE 10.12   The rocket equation.

at time t. Let v be the velocity of the mass $M(t)$ at time t. Then the law of motion, mass × acceleration = force, says that

$$M(t)\frac{dv}{dt} = P + F.$$

The propulsion force P remains to be evaluated. We consider for the moment the body, consisting of the remaining rocket mass together with the mass ΔM of combustion gases that has been ejected in the most recent Δt seconds (Figure 10.12). The internal forces that separate these two parts of the body now consist of a pair of forces equal in magnitude but opposite in direction (Newton's third law, or in modern terminology, the conservation of linear momentum). If we calculate the rate of change of the momentum of the exhaust gases, we have the force that propels them backward at the velocity c. The ejected gas has momentum $c\,\Delta M$. The average rate of change of momentum is $c(\Delta M/\Delta t)$ and in the limit as $\Delta t \to 0$, the instantaneous rate of change of momentum is $c(dM/dt)$. This is a force opposite in sign to the force propelling the rocket mass $M(t)$, that is, $P = -c(dM/dt)$. Hence, the differential equation of rocket flight in a straight line for the remaining rocket mass $M(t)$ may be written

(1) $$M\frac{dv}{dt} + c\frac{dM}{dt} = F.$$

We now solve this differential equation under the assumption for outer space that the external force $F = 0$, and that the rate of burning fuel is constant, $dM/dt = -k$. From the constant fuel-burning rate, on integrating, we obtain

$$M(t) = M_0 - kt.$$

With $M(t)$ determined, the differential equation becomes

$$(M_0 - kt)\frac{dv}{dt} - ck = 0.$$

Considering it as a relation in the local coordinates (dt, dv), we have, by separation of variables, the equivalent relation

$$dv = \frac{ck\,dt}{M_0 - kt}.$$

On integrating, we find the integral

$$v = -c \ln (M_0 - kt) + C.$$

Since $v = v_0$ when $t = 0$, we find that $C = c \ln M_0 + v_0$. With this evaluation of C, we have

$$v = c \ln \frac{M_0}{M_0 - kt} + v_0,$$

which is both a solution and an integral of the differential equation, and completes the problem.

27.5 Exercises

1. A ball is thrown vertically upward with initial velocity 128 ft/sec. Find its velocity and height after 2, 4 and 6 sec.
2. Find the time when the ball in Exercise 1 returns to the level from which it was thrown.

3. Find the time when the 16-lb shot in the example (§27.1) strikes the ground. What is the meaning of the other time that comes out in the solution?

4. The superhuman shot-putter in the example (§27.1) easily broke the world's record. What was his distance to the point at which his shot struck the ground?

5. For a projectile released with initial velocity $\vec{v}_0$: $(v_0 \cos \alpha, 0, v_0 \sin \alpha)$ from the origin O: $(0, 0, 0)$ under the influence of a constant gravity $-mg$, find the motion. Prove that the motion is in a vertical plane and is a parabola.

6. In Exercise 5, find the range, that is, the horizontal distance from the release point to the point at which the projectile returns to the same level.

7. In Exercise 6, prove that for every v_0 and m, the maximum range is attained when $\alpha = \pi/4$.

8.* The shot-putter in Exercise 3 released his shot from a height of 8 ft, and the distance was measured on ground level. At what angle of elevation α should he release his shot to attain maximum range? Show that it is less than $\pi/4$.

9. For the parachutist of Example A (§27.3), find the velocity at which the weight of the parachutist is exactly equal to the upward pull of the parachute. Compare this velocity with the terminal velocity.

10. The parachutist of Example A (§27.3) waits to open his chute until his free-fall velocity is -160 ft/sec. Find his velocity thereafter as a function of time.

11. A mass, $m = 10$, is sliding on a horizontal plane with initial velocity 20 ft/sec. The only force affecting the motion is a force of friction $-v/10$. Find the motion. Does the mass ever actually stop? How far does it slide in the long run?

12. In space, the pull of the earth's gravity on a mass m is inversely proportional to the square of the distance r from the earth's center and it is $-mg$ when $r = R$ at the surface. Show that the equation of orbit can be written

$$v \frac{dv}{dr} = -g \frac{R^2}{r^2},$$

where $v = dr/dt$. Show that this differential equation has the integral

$$v^2 = \frac{2gR^2}{r} + C.$$

13. In Exercise 12, a particle starts straight up from the earth's surface with initial velocity v_0. Show that $C = v_0^2 - 2gR$.

14. In Exercise 12, show that if $v_0^2 - 2gR < 0$, the particle will eventually fall. Determine the minimum initial velocity that will permit the particle to escape from the earth.

15. A rocket (§27.4) reaches outer space with velocity 100 miles per minute and mass M_1. Then, burning fuel at the rate $0.1 \, M_1$ per minute, it generates a thrust of $2g \, M_1$. What is its velocity one minute later?

16. For the rocket (§27.4), set up the equation of motion for conditions near takeoff when the acceleration due to gravity, $-g$, is constant. Solve for velocity as a function of time.

17. An object weighing 1000 lb in air sinks in water, starting from rest. Two forces act on it, a buoyant force of 200 lb and a force of water resistance $80v$ lb, where v is the velocity in ft/sec. Find the distance traveled in t seconds and the limiting velocity.

18. A weight of 100 lb slides without friction on a plane inclined at angle 30 deg to the horizontal, starting with velocity 10 ft/sec up the slope. Find the motion.

19. Assuming that a free-falling parachutist encounters air resistance $-\beta v^2$ proportional to the square of the velocity, find the velocity at time t of a man who falls free after starting with zero velocity. Find the limiting velocity as $t \to \infty$. It helps to let $\beta/W = k^2$ and to use the integral

$$\int \frac{dx}{k^2 x^2 - 1} = \frac{1}{2k} \ln \left| \frac{kx - 1}{kx + 1} \right| + C.$$

20. In the theory of relativity, mass is not a constant but varies with velocity by the formula

$$m = \frac{m_0}{\sqrt{1 - (v^2/c^2)}},$$

where m_0 is the constant rest mass and c is the velocity of light. A particle starts with zero velocity and mass m_0 and moves under a constant force F. Show that its velocity approaches the velocity of light. What happens in this situation if mass is constant?

28   Models of Systems with External Input

28.1   An Integrating Factor.   The differential equation

$$\frac{dy}{dx} + p(x)y = f(x),$$

where $p(x)$ and $f(x)$ are integrable functions, does not yield to separation of variables. However, we find that if we multiply both members by the factor

$$e^{\int p(x)\, dx},$$

we can integrate it. Indeed since

$$\frac{d}{dx}\, e^{\int p(x)\, dx} = e^{\int p(x)\, dx} \frac{d}{dx} \int p(x)\, dx = p(x)e^{\int p(x)\, dx},$$

we find that

$$\frac{d}{dx}\, y e^{\int p(x)\, dx} = e^{\int p(x)\, dx} \left[\frac{dy}{dx} + p(x)y \right].$$

Hence, if we multiply both members of the equation by this integrating factor, it becomes

$$\frac{d}{dx}\left(y e^{\int p(x)\, dx} \right) = e^{\int p(x)\, dx} f(x).$$

Now integrating the equation, we find the integral

$$y e^{\int p(x)\, dx} = \int e^{\int p(x)\, dx} f(x)\, dx + C.$$

Example A.   Find the general solution of the differential equation

$$\frac{dy}{dx} + \frac{2}{x}y = x^2 + 2.$$

Solution.   This is in the form above with $p(x) = 2/x$. Hence, it has the integrating factor

$$e^{\int p(x)\, dx} = e^{\int (2/x)\, dx} = e^{2\ln x} = e^{\ln x^2} = x^2;$$

multiplying the equation by x^2, we find that

$$x^2 \frac{dy}{dx} + 2xy = x^4 + 2x^2$$

can be written

$$\frac{d}{dx} x^2 y = x^4 + 2x^2.$$

Hence, it has the integral

$$x^2 y = \frac{x^5}{5} + 2\frac{x^3}{3} + C$$

and this gives the general solution

$$y = \frac{x^3}{5} + 2\frac{x}{3} + \frac{C}{x^2}.$$

Example B.   Find the general solution of the differential equation

$$\frac{dy}{dt} + ky = A \sin \omega t,$$

where k and A are positive numbers.

Solution.   We have the integrating factor

$$e^{\int k\, dt} = e^{kt}.$$

Multiplying by this integrating factor, we have

$$\frac{d}{dt}(e^{kt}y) = Ae^{kt} \sin \omega t,$$

which can be integrated to give the integral

$$e^{kt}y = A \int e^{kt} \sin \omega t \, dt + C.$$

The right member integrates by parts to give the integral

$$e^{kt}y = A\, \frac{e^{kt}}{k^2 + \omega^2}(k \sin \omega t - \omega \cos \omega t) + C$$

and the solution

$$y = \frac{A}{k^2 + \omega^2}(k \sin \omega t - \omega \cos \omega t) + Ce^{-kt}.$$

28.2   Forced Motion

Example.   A heavy block of mass m rests on a plane where a frictional force proportional to the velocity resists the motion (Figure 10.13). An oscillating push-pull force, $f(t) = A \sin \omega t$, is applied horizontally to the block. What is the velocity at any time afterwards if $v(0) = 0$? Discuss the motion in relation to the size of the mass m, the frequency ω, and the coefficient of friction.

Solution.   The law of motion for the block says that

$$m\frac{dv}{dt} = -cv + A \sin \omega t, \quad \text{or} \quad \frac{dv}{dt} + kv = \frac{A}{m}\sin \omega t,$$

where we have put $k = c/m$ for simplicity. We multiply by the integrating factor $e^{\int k\, dt} = e^{kt}$ and find that the equation integrates (§28.1, Example B), and has the solution

$$v = \frac{1}{m}\frac{A}{(k^2 + \omega^2)}(k \sin \omega t - \omega \cos \omega t) + Ce^{-kt}.$$

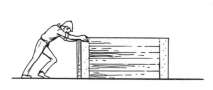

FIGURE 10.13.

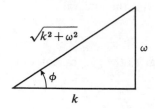

FIGURE 10.14 Phase angle.

Since $v(0) = 0$, we find that

$$C = \frac{A}{m} \frac{\omega}{k^2 + \omega^2}.$$

PHYSICAL INTERPRETATION. If we multiply and divide by $\sqrt{k^2 + \omega^2}$, we find that

$$k \sin \omega t - \omega \cos \omega t = \sqrt{k^2 + \omega^2} \left[\frac{k}{\sqrt{k^2 + \omega^2}} \sin \omega t - \frac{\omega}{\sqrt{k^2 + \omega^2}} \cos \omega t \right].$$

Then, if we define an angle ϕ so that (Figure 10.14)

$$\cos \phi = \frac{k}{\sqrt{k^2 + \omega^2}} \quad \text{and} \quad \sin \phi = \frac{\omega}{\sqrt{k^2 + \omega^2}}$$

and apply the addition theorem for the sine, we find that

$$k \sin \omega t - \omega \cos \omega t = \sqrt{k^2 + \omega^2} \sin (\omega t - \phi).$$

With this expression, the solution of the differential equation of motion becomes

$$v = \frac{1}{m} \frac{A}{\sqrt{k^2 + \omega^2}} \sin (\omega t - \phi) + Ce^{-kt}.$$

The last term is called the *transient* because it approaches zero as $t \to \infty$. The first term is called the steady state, denoted by v_∞, as in the problem of the parachute drop. We observe that the steady-state velocity is a sinusoidal oscillation with the same frequency and amplitude as the impressed force (I, §32.4)

$$\frac{1}{m} \frac{A}{\sqrt{k^2 + \omega^2}}.$$

The amplitude of the velocity response v_∞ to the impressed force decreases if m is increased; also if k is increased; also if ω is increased. Thus a heavy load or one with a high frictional drag responds with smaller velocity than a lighter or better-lubricated load does to the same impressed force. This agrees with intuition. It also agrees with intuition that a rapidly oscillating force fails to produce as big velocity oscillations as a more slowly oscillating force.

28.3 Dead Sea

Example. Rainwater flows down from the hills into a nearly flat, shallow, impervious basin with no outlet. Water is lost by evaporation. Express the process by a differential-equation model and solve it for the amount of water in the lake at time t.

Solution. Since the basin is shallow and nearly flat, the evaporation surface can be regarded as proportional to the volume of water. Hence if y represents the amount of water in the lake, the rate of evaporation is ky, for some constant k. Let $f(t)$ be the rate of rainfall expressed as an annual rate. We make no assumption about the form of this function. Then

$$\frac{dy}{dt} = -ky + f(t).$$

We shall apply the initial condition $\lim_{t \to -\infty} y = 0$, meaning physically that long ago there was no water in the lake. Multiplying by the integrating factor e^{kt} and integrating, we have

$$\int_{-\infty}^{t} \frac{d}{dt}(e^{kt}y) = \int_{-\infty}^{t} e^{kt}f(t)\, dt,$$

or

$$e^{kt}y = \int_{-\infty}^{t} e^{kt}f(t)\, dt.$$

But since

$$e^{kt} = k \int_{-\infty}^{t} e^{ks}\, ds$$

this can be written

$$y = \frac{1}{k} \frac{\int_{-\infty}^{t} e^{ks}f(s)\, ds}{\int_{-\infty}^{t} e^{ks}\, ds} = \frac{1}{k}\overline{f(t)},$$

where $\overline{f(t)}$ is the average annual rainfall computed with respect to the weight function e^{kt}. This result could almost be guessed. For example, if $k = 0.1$, that is, if the lake evaporates at the rate of $\frac{1}{10}$ of its capacity per year, then it contains at any time 10 times an average annual rainfall. We observe that the weight function e^{ks} gives maximum weight to the most recent rainfall inputs, since for the distant past, s is a large negative number and $e^{ks}f(s)$ is very small. Indeed, since the integration extends over times s before the present time t, the weight function e^{ks} has its maximum when $s = t$.

28.4 Fox and Rabbit Populations. When one species of animal lives as a predator upon another, the population of the predator at any time t and that of the prey are related in a way that cannot be described precisely by verbal analysis but, under reasonable assumptions, may be described by the solution of a differential equation. An essential part of the mathematical analysis is to set up the differential equations that serve as a model for the ecological system.

We may proceed as follows. Let x be the number of rabbits at time t, and y the number of foxes. In the presence of plenty of food and no predators, the number of rabbits increases at a rate proportional to the number of rabbits. However, with hungry foxes around, there is at the same time a loss rate equal to the number of rabbits per year taken by the foxes. It is reasonable to assume that this loss rate is proportional to xy. Hence the rate of change of the rabbit population is given by

(1)
$$\frac{dx}{dt} = ax - bxy,$$

where a and b are positive constants.

For the foxes, we may assume that it takes k rabbits per fox per year to sustain healthy life, so that y foxes will require rabbits at the rate of ky per year. We then expect that the rate of change of the fox population is proportional to the amount of excess food $bxy - ky$ available. That is,

$$\frac{dy}{dt} = m(bxy - ky).$$

Combining constants, we get the differential equation

(2)
$$\frac{dy}{dt} = cxy - py,$$

where c and p are positive constants. Our mathematical model for the variation of the two populations with time now consists of the simultaneous system of differential equations (1) and (2), together with the initial conditions $x = x_0$, $y = y_0$, when $t = 0$.

Example.   Find the relation between fox and rabbit populations, assuming this model.

Solution.   Since we are not required to find the variation of the populations with time, we may divide (2) by (1) and obtain the time-free relation

$$\frac{dy}{dx} = \frac{y(cx - p)}{x(a - by)}.$$

Separating variables, we see that this relation is equivalent to

$$\frac{a - by}{y}\, dy + \frac{p - cx}{x}\, dx = 0.$$

This has the integral (§16.4)

$$a \ln y - by + p \ln x - cx = C,$$

where

$$C = a \ln y_0 - by_0 + p \ln x_0 - cx_0.$$

Eliminating C and simplifying, we find that these statements are equivalent to

$$a \ln \frac{y}{y_0} - b(y - y_0) + p \ln \frac{x}{x_0} - c(x - x_0) = 0,$$

or

(3)
$$\ln \left(\frac{y}{y_0}\right)^a + \ln \left(\frac{x}{x_0}\right)^p = b(y - y_0) + c(x - x_0).$$

In exponential form, this says that

(3′)
$$\left(\frac{y}{y_0}\right)^a \left(\frac{x}{x_0}\right)^p = e^{b(y - y_0)} e^{c(x - x_0)}.$$

We may regard either of the relations (3) or (3′) as a mathematical solution to the problem. To check the validity of the model an ecologist would have to make enough samplings to estimate the fox and rabbit populations initially (x_0, y_0) and at four later times to determine a, b, c, p. Then he would make a number of additional determinations of (x, y) at other times to verify that they satisfied the relation (3). In the absence of any relation such as (3) to check, it would be virtually impossible to find by sampling alone the relationship between the populations.*

28.5 Exercises

In Exercises 1–7, find the general solution of the differential equations of the form $(dy/dx) + p(x)y = f(x)$ by means of an integrating factor.

1. $\dfrac{dy}{dx} - \dfrac{y}{x} = x^2.$

2. $\dfrac{dy}{dx} + \dfrac{y}{1+x} = 1.$

3. $(x^3 + 3y)\,dx - x\,dy = 0.$

4. $\dfrac{dy}{dt} + 2y = \cos t.$

5. $(y + 1)\,dx + (4x - y)\,dy = 0.$

6. $\dfrac{dv}{dr} + kr = -\dfrac{1}{r^2}.$

7. $\dfrac{dy}{dx} + 2xy = \exp(-x^2).$

8. Express the solution of the differential equation

$$\frac{dy}{dx} - xy = 1, \qquad \lim_{x \to -\infty} e^{-x^2/2}\,y(x) = 0,$$

in terms of the probability integral

$$\int_{-\infty}^{x} e^{-(x^2/2)}\,dx$$

even though this integral cannot be evaluated in terms of elementary functions.

9. A tank contains 10 gallons of brine that contains 5 lb of salt. Brine containing 2 lb of salt per gal flows into the tank at 3 gal per minute, where it is assumed to be instantly mixed. At the same time, the mixed brine flows out at the same rate. Find the amount A of salt in the tank at the end of t minutes.

10. Water containing 2 ounces of pollutant per gallon flows into a tank initially containing 10,000 gal pure water at 500 gal/min; and the thoroughly mixed water flows out at the same rate. In the tank a treatment removes 2 percent of the pollutant per minute. Find the function that gives the concentration of pollutant in the outflow.

11. We consider "continuously converted" compound interest, under which money grows at all times at an annual rate of k times the amount in the fund. If money is being deposited continuously in the fund at the rate $\$R$ per year, what is the amount s present after t years, starting from nothing? This is sometimes called a continuous annuity.

12. In Exercise 11, a continuous annuity is to be paid at the rate $\$R$ per cent for t years. If money grows continuously at the rate k per year, what is the present value of the annuity at the start?

13. A phonograph speaker is forced to move back and forth by a sinusoidal oscillating force (§28.2) representing a musical sound of frequency ω and amplitude A. The speaker has a mass m and a frictional resistance proportional to its velocity. Consider the high-fidelity question: Does the speaker respond to all frequencies with proportional amplitude of oscillation? Plot a graph of the amplitude of the response as a function of the input frequency ω, and determine which frequencies are diminished most in amplitude. Is this faithful reproduction of sound?

14. We consider the flow of blood in the aorta (Figure 10.15) under the forcing action of the heart, which pumps blood at the flow rate $A \sin \omega t$ volume units per second. At the

* This model is from J. G. Kemeny and J. L. Snell, *Mathematical Models in the Social Sciences*, (Boston: Ginn & Co., 1962).

FIGURE 10.15   Exercise 14.

same time blood flows out of the aorta at a rate $-(1/w)p(t)$, which is proportional to the pressure $p(t)$. Assuming that $p = cv$, where c is constant, show that

$$\frac{dp}{dt} + \frac{c}{w}p = cA \sin \omega t.$$

Deduce from this that the steady-state pulse in the aorta is a sinusoidal function of the same form as the heartbeat but lags behind the heartbeat. What is the effect on the volume of flow in the aorta of increasing the constant W?

15. Show from the differential equations (1) and (2) of the fox-rabbit populations that the number of foxes remains constant if and only if the number of rabbits does. Show that in this case $x = p/c$ and $y = a/b$ unless $x = y = 0$.

16. In the solution of the differential equation (§28.1, Example B)

$$\frac{dy}{dt} + ky = A \sin \omega t, \qquad y(0) = y_0,$$

show that

$$y = \int_0^t e^{-k(t-s)} A \sin \omega s \, ds + C_0 e^{-kt}.$$

17. In Exercise 16, plot the graph of the integrand as a function of s for t fixed and $0 \leq s \leq t$. Shade the regions of positive and negative area represented by the integral.

18. As in Exercise 16, show that the solution of the differential equation

$$\frac{dy}{dt} + ky = f(t), \qquad y(0) = y_0,$$

can be written

$$y = \int_0^t e^{-k(t-s)} f(s) \, ds + y_0 e^{-kt}.$$

19. In Exercise 18, observe that $t - s$ is the time interval from the time s where f is evaluated in the integral to the present time t where the solution $y(t)$ is evaluated. Describe in words and graph the process defined by the integral.

28R   Review Problems on Differential Equations

Problems 1–12 are of miscellaneous types. Integrate to find integrals, employing separation of variables or integrating factors where appropriate.

1. $dy - x^2 \, dx = 0$.

2. $x^2 \, dy - y \, dx = 0$.

3. $2x \, dx - 2y \, dy = 0$.

4. $dy + (y - e^x) \, dx = 0$.

5. $\dfrac{d^2y}{dx^2} = x^2$.

6. $\dfrac{dv}{du} = \dfrac{1 + v^2}{1 + u^2}$.

7. $x \left(\dfrac{dy}{dx}\right)^2 = 1$.

8. $\dfrac{dy}{dx} - \dfrac{2}{x}y = x^2 e^x$.

9. $(2x + y) dx + (x + 2y) dy = 0.$ 10. $x dy - y dx = 0.$

11. $x\dfrac{dy}{dx} + y = x.$ 12. $dy - f'(x) dx = 0.$

In problems 13–24, find the solution of the differential equation that satisfies the given initial condition or other requirement.

13. $y \, dy + x \, dx = 0, (x, y) = (-1, -1).$
14. $(1 - x^2) \, dy = 2dx, y = 2$ when $x = 0.$
15. $(x^2 + 4)y - 4x \, dx = 0, y(1) = 1.$

16. $\dfrac{dy}{dx} = \sin^2 ax, y(0) = 0.$

17. $e^{-x} \, dy = e^y \, dx, y(0) = \ln 2.$

18. $\dfrac{dy}{dt} + ay = E \sin \omega t, y(0) = 0.$

19. $\dfrac{dy}{dt} - y = \sin t, |y(t)|$ is bounded for all $t.$

20. $\dfrac{dy}{dt} + y = \sin t, y(t)$ is periodic.

21. $(1 - e^{-x}) \, dy = e^{-x} \, dx, \lim_{x \to \infty} y(x) = 0.$
22. $3x(xy - 2) \, dx + (x^3 + 2y) \, dy = 0.$ Find an integral for which $y = 1$ when $x = 0.$
23. $(2x - 3y) \, dx + (2y - 3x) \, dy = 0.$ Find an integral for which $x = 0$ when $y = 2.$

24. $\dfrac{dy}{dx} = \exp(2x - y), y = 0$ when $x = -1.$

25. Express the solution of the differential equation $(xy - \sin x) \, dx + x^2 \, dy = 0$ in terms of an integral of a function of $x.$
26. Find the orthogonal trajectories of the family $x \cos \alpha + y \sin \alpha = 0.$
27. Find the orthogonal trajectories of the family $xy = c.$
28. Without solving the differential equation, show that no solution of the differential equation

$$\frac{dy}{dx} = (x^2 + y^2 + 1)e^{x+y},$$

has a maximum or minimum point.
29. A 100-gal tank A initially contains 10 lb of salt, while 100-gal tank B initially contains pure water. Pure water goes into A at 2 gal/min and the overflow of A goes into B, while B overflows down the drain. Find the concentration of salt in tank B as a function of time.
30. The force of water resisting the forward motion of a boat is proportional to its velocity, and is 40 lb at 20 ft/sec. The motor can generate a thrust of 50 lb, the boat weighs 320 lb, and the passenger 160 lb. Find the maximum speed of the boat.

The pressure of the earth's atmosphere decreases to zero as we go up from sea level, but how does it decrease as a function of altitude? We may assume that the gas of the atmosphere obeys Boyle's law, $pv = c$, and the fact that the pressure at any point is equal to the weight of a column of air above a level surface of area 1. We investigate this question in Problems 31–34.

31. Let p_0 be the pressure at sea level. Show that the pressure at level z is given by

$$p(z) = p_0 - \int_0^z \sigma(y) \, dy,$$

where $\sigma(y)$ is the density at level $y.$
32. Show that the pressure is related to the density by $p = a\sigma$, where a is a constant.

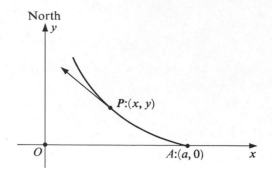

FIGURE 10.16   Pursuit curve, Exercise 35.

33. Use the result of Problem 32 in 31 to eliminate σ from the expression in 31 and show that the pressure satisfies the differential equation

$$\frac{dp}{dz} = -\frac{1}{a}p.$$

34. Solve this differential equation for $p(z)$, subject to the initial condition that $p(0) = p_0$, and find that $p = p_0 e^{-z/a}$.

35. A basset hound at $A:(a, 0)$ discovers a rabbit at O moving due north at a constant speed v (Figure 10.16). The basset gives chase, always running towards the rabbit with speed v/k, $k > 1$. Show that at any time t the arc length s of AP is given by

$$ks = y - x\frac{dy}{dx}.$$

36. In Problem 35, remembering that $ds/dx = -\sqrt{1 + (dy/dx)^2}$, since s increases as x decreases, show that the basset's curve of pursuit satisfies the differential equation

$$k\sqrt{1 + p^2} = x\frac{dp}{dx},$$

where $p = dy/dx$.

37. Integrate the differential equation in Problem 36, subject to $p = dy/dx = 0$ when $x = a$, to obtain the integral

$$\ln(p + \sqrt{1 + p^2}) = k\ln\frac{x}{a}.$$

38.* Find the equation $y = f(x)$ of the curve of pursuit.

39.* It is obvious that the basset, who with $k > 1$ is slower than the rabbit, will never catch him. Verify that the mathematical solution has this property.

40.* A greyhound with $k < 1$ starts after a rabbit under the same conditions. Find the co-ordinates of the point at which he catches the rabbit.

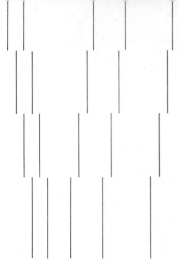

Multiple Integrals

CHAPTER 11

29 Integration over a Rectangle

29.1 Partition of a Rectangle. As a preliminary to the definition of the integral of a function, $f(x, y)$, defined on a rectangle $\mathscr{D}$ in the xy-plane, we describe a *partition p* of the rectangle into smaller rectangles, called "cells of the partition." They are no longer subintervals, as they were in the definition of the integral of $f(x)$ on $[a, b]$ (I, §19.1). We partition the intervals that form the sides of the rectangle

$$\mathscr{D} = \{a \leqq x \leqq b, c \leqq y \leqq d\}.$$

To do this, we introduce partition points $x_1, x_2, \ldots, x_m$, and $y_1, y_2, \ldots, y_n$, such that

$$a = x_1 < x_2 < \cdots < x_i < x_{i+1} < \cdots < x_m = b,$$

and

$$c = y_1 < y_2 < \cdots < y_j < y_{j+1} < \cdots < y_n = d.$$

We draw lines parallel to the axes through these partition points (Figure 11.1). This rectangular grid partitions the rectangle into smaller rectangular cells. A typical one (Figure 11.1) is $\mathscr{D}_{ij}$, where

$$\mathscr{D}_{ij} = \{x_i \leqq x < x_{i+1}, y_j \leqq y < y_{j+1}\}.$$

As a special case we include, in their adjoining cells, the points on the right and top boundaries of $\mathscr{D}$, where $x = b$ and $y = d$. Then every point (x, y) in the rectangle $\mathscr{D}$ is in one and only one cell of the partition p.

We can refine the partition p into a finer one p', each of whose cells is contained in a cell of p. To refine the partition we introduce new partitioning lines in the grid, keeping all those already used to define p.

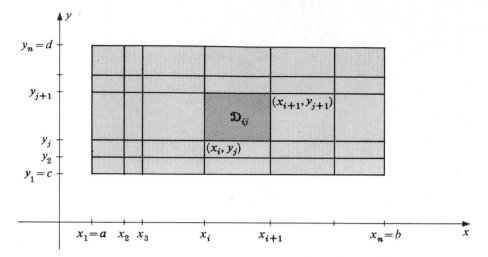

FIGURE 11.1   Partition of a rectangle $\mathscr{D}$, $\mathscr{D} = \{a \leq x \leq b, c \leq y \leq d\}$.

29.2   Definition of the Double Integral.   Let $f(x, y)$ be a bounded function on the rectangle $\mathscr{D}, \mathscr{D} = \{a \leq x \leq b, c \leq y \leq d\}$. We form a partition p of $\mathscr{D}$ into cells $\mathscr{D}_{ij}$ (Figure 11.2(a)). For every point (u, v) in the cell $\mathscr{D}_{ij}$ we define

$$s(u, v) = \inf \{f(x, y) \mid (x, y) \in \mathscr{D}_{ij}\},$$

and

$$S(u, v) = \sup \{f(x, y) \mid (x, y) \in \mathscr{D}_{ij}\}.$$

We can picture this as a constant understep and an overstep for the values of $f(x, y)$ on the cell $\mathscr{D}_{ij}$ (Figure 11.2(b)). We do this for each cell of the partition so as to define the understep function $s(u, v)$ and overstep function $S(u, v)$ throughout the rectangle $\mathscr{D}$.

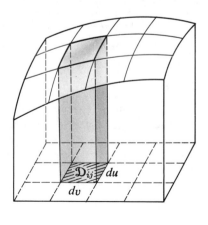

(a)   Graph and partition

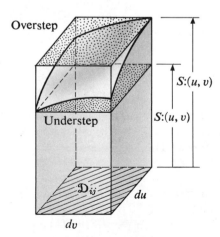

(b)   Overstep and understep

FIGURE 11.2   Integral of $f(x,y)$.

We denote by du and dv the lengths of the sides of the cell $\mathcal{D}_{ij}$, which includes the point (u, v). That is,

$$du = x_{i+1} - x_i, \qquad dv = y_{j+1} - y_j.$$

Then the area of this cell is $du\,dv$. Furthermore, the volume of the rectangular shaft below the understep in this cell is $s(u, v)\,du\,dv$ (Figure 11.2).

The sum of terms of this type for all cells of the partition is the integral of the understep function $\iint_{\mathcal{D}} s(u, v)\,du\,dv$. Similarly, the integral of the overstep function over the rectangle $\mathcal{D}$ is the sum $\iint_{\mathcal{D}} S(u, v)\,du\,dv$. For any partition ρ, there is at least one number I included between these sums, so that

$$\iint_{\mathcal{D}} s(u, v)\,du\,dv \leq I \leq \iint_{\mathcal{D}} S(u, v)\,du\,dv.$$

Now we refine the partition ρ by introducing more and more lines into the grid. If for all partitions, however fine, there is one and only one number I enclosed between the under sums and upper sums, this number I is the integral of f over the rectangle, written

$$I = \iint_{\mathcal{D}} f(x, y)\,dx\,dy.$$

29.3 Volume Interpretation of the Integral.

The definition of the double integral

$$I = \iint_{\mathcal{D}} f(x, y)\,dx\,dy$$

is constructed so that we can interpret it as the exact volume of the region below the graph of $z = f(x, y)$ for (x, y) in the rectangle (Figure 11.2(a)).

29.4 Conditions for Integrability.

For functions $f(x)$ defined on an interval $[a, b]$, we were able to show that f is integrable if it is increasing (or decreasing) in the interval $[a, b]$ (I, §9.3). The idea of an increasing function does not carry over readily to functions $f(x, y)$ of two or more variables. As a substitute for it, we can define a more general property called bounded variation.

We define the variation v_{ij} of $f(x, y)$ in the cell $\mathcal{D}_{ij}$ as the nonnegative number

$$v_{ij} = \sup f(u, v) - \inf f(u, v) = S(u, v) - s(u, v).$$

We define the variation v_p of f over the partition ρ as the sum of its variations in each cell of the partition. The function $f(x, y)$ over the rectangle $\mathcal{D}$ is *of bounded variation* if and only if there is a fixed number V such that for all partitions $v_p \leq V$.

THEOREM.   Every function of bounded variation on a rectangle is integrable.

The proof is like that for a monotone function $f(x)$ on $[a, b]$ (I, §9.3). Let A be the area of the cell of maximum area in the partition. Then the difference between the upper sum and lower sum satisfies the inequality

$$\iint_{\mathcal{D}} S(u, v)\,du\,dv - \iint_{\mathcal{D}} s(u, v)\,du\,dv \leq AV.$$

We can see this by stacking up the blocks bounded above by $S(u, v)$ and below by $s(u, v)$. We can make this arbitrarily small by refining the partition so that A approaches zero.

THEOREM.   Every continuous function $f(x, y)$ defined on a rectangle $\mathscr{D}$ is integrable.

To prove this we must first prove the lemma that for every ϵ, there is a partition p fine enough so that on each cell

$$S(u, v) - s(u, v) < \frac{\epsilon}{D},$$

where D is the area of $\mathscr{D}$, that is, $D = (b - a)(d - c)$. We omit this difficult proof, which is an extension of the same result for functions of one variable (I, §8.8). Then

$$\iint\limits_{\mathscr{D}} S(u, v) \, du \, dv - \iint\limits_{\mathscr{D}} s(u, v) \, du \, dv < \iint\limits_{\mathscr{D}} \frac{\epsilon}{D} \, du \, dv = \epsilon.$$

Hence f is integrable.

29.5   Extension to Functions of n Variables. The definition of the integral of $f(x, y)$ over a rectangle extends similarly to functions of any number of variables $w = f(x_1, x_2, \ldots, x_n)$ over a box $\mathscr{D}$ having plane boundaries, so that for constants $a_i, b_i,$

$$a_i \leqq x_i \leqq b_i, \qquad i = 1, 2, \ldots, n.$$

We represent it by the symbol

$$\iint\limits_{\mathscr{D}} \cdots \int f(x_1, x_2, \ldots, x_n) \, dx_1 \, dx_2 \cdots dx_n.$$

In particular, we have the triple integral

$$\iiint\limits_{\mathscr{D}} f(x, y, z) \, dx \, dy \, dz,$$

defined over a 3-dimensional box $\mathscr{D}$ (Figure 11.3). For integrals over a box of dimension 3, the volume interpretation (§29.3) is no longer possible, since the graph of $w = f(x, y, z)$ is in a 4-dimensional space of points (x, y, z, w). This applies whenever the dimension $n \geqq 3$. But the definition of the integral, and its mathematical properties, apply in any number of dimensions.

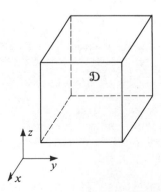

FIGURE 11.3   Domain of triple integral.

29.6   Sample Sums. For applications, or computation, we often prefer to replace the extreme upper and lower sums over the partition of $\mathscr{D}$ by sample sums obtained by evaluating f at some sample point (u_i, v_j) arbitrarily chosen in each cell $\mathscr{D}_{ij}$. Then we define a step function $h(u, v)$ whose constant values in each cell are given by the value of f at the sample point. That is,

$$h(u, v) = f(u_i, v_j)$$

for all (u, v) in $\mathscr{D}_{ij}$. Then the integral of this step function

$$\iint_{\mathscr{D}} h(u, v)\, du\, dv$$

is between the lower sum and upper sum over the partition.

Example. The function f, $f(x, y) = x + 2y - 4$, is defined over the rectangle

$$\mathscr{D}:\{0 \leq x \leq 5,\, 0 \leq y \leq 4\}.$$

We form the partition p of $\mathscr{D}$ into squares of side 1. Calculate the lower and upper sums and the sample sum obtained by evaluating f at the center of each square.

Solution. We plot $f(x, y)$ by contour lines over the rectangle (Figure 11.4). In each cell, $S(x, y) = \sup f(x, y)$ is attained at the upper right corner and $s(x, y) = \inf f(x, y)$ is attained at the lower left corner. We write these values of $s(x, y)$ and $S(x, y)$ in the lower left and upper right, respectively, of each cell. This enables us to write down the upper sum

$$
\begin{aligned}
\iint_{\mathscr{D}} S(x, y)\, dx\, dy &= 5(1) + 6(1) + 7(1) + 8(1) + 9(1) \\
&\quad + 3(1) + 4(1) + 5(1) + 6(1) + 7(1) \\
&\quad + 1(1) + 2(1) + 3(1) + 4(1) + 5(1) \\
&\quad + (-1)(1) + 0(1) + 1(1) + 2(1) + 3(1) \\
&= 80.
\end{aligned}
$$

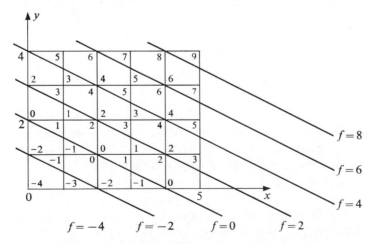

FIGURE 11.4   The domain $\mathscr{D}$ and its partition π. Each $dx = 1$, each $dy = 1$, each $dx\, dy = 1$.

The lower sum is

$$\iint_{\mathscr{D}} s(x, y)\, dx\, dy = 2(1) + 3(1) + 4(1) + 5(1) + 6(1)$$
$$+ 0(1) + 1(1) + 2(1) + 3(1) + 4(1)$$
$$+ (-2)(1) + (-1)(1) + 0(1) + 1(1) + 2(1)$$
$$+ (-4)(1) + (-3)(1) + (-2)(1) + (-1)(1) + 0(1)$$
$$= 20.$$

Hence the exact integral if it exists is somewhere between 20 and 80. The spread is large because the partition is a coarse one.

The sample sum obtained by evaluating f at the center of each cell is given by

$$\iint_{\mathscr{D}} h(x, y)\, dx\, dy = 50.$$

This is between the upper and lower sums. We should expect it to be closer to the exact integral than either of the extreme sums.

29.7 Mass Interpretation. Consider a flat rectangular plate of material whose density varies continuously from point to point, being given by a density function $\mu(x, y)$ in units of mass-per-unit-area. Then for a typical cell of a partition of $\mathscr{D}$ the $\mu(x, y)\, dx\, dy$ is an estimate of the mass of the cell determined from the density at the point (x, y) in the cell. The exact integral

$$m = \iint_{\mathscr{D}} \mu(x, y)\, dx\, dy$$

can then be interpreted as the mass of the entire plate $\mathscr{D}$.

Similarly, if $\mu(x, y, z)$ is the density of matter in a rectangular box $\mathscr{D}$, the triple integral

$$m = \iiint_{\mathscr{D}} \mu(x, y, z)\, dx\, dy\, dz$$

is the exact mass of matter in the block.

29.8 Algebraic Properties of Integrals. It is clear from the definition that multiple integrals have the same linear properties that single integrals do. That is,

$$\iint_{\mathscr{D}} (af + bg)\, dx\, dy = a \iint_{\mathscr{D}} f\, dx\, dy + b \iint_{\mathscr{D}} g\, dx\, dy,$$

where $f(x, y, z)$ and $g(x, y, z)$ are integrable functions and a and b are numbers. However, we cannot directly carry over the idea of evaluating these multiple integrals by the inverse of differentiation (fundamental theorem). For if we differentiate a function $F(x, y)$, we do not get the integrand $f(x, y)\, dx\, dy$ but find $DF = F_1\, dx + F_2\, dy$.

29.9 Exercises

Make contour maps for the following functions; find the upper sums and lower sums over the indicated partitions. Estimate the (exact) integral of the function itself.

1. $f(x, y) = x - 2y$ over the rectangle $0 \leq x \leq 3, 0 \leq y \leq 4$ with a partition into squares of side 1.

2. $f(x, y) = x^2 + y^2$ over the rectangle $0 \leq x \leq 3, 0 \leq y \leq 4$ with a partition into squares of side 1.

3. $f(x, y) = 3x + 1$ over the rectangle $0 \leq x \leq 1, 0 \leq y \leq 2$ with a partition into squares of side $\frac{1}{2}$.

4. $f(x, y, z) = 2x - 3y + 4z$ over the cube $\{|x| \leq 1, |y| \leq 1, |z| \leq 1\}$ with a partition into cubes of edge-length 1.

5. Taking the volume interpretation of the integral, use solid geometry to prove that the exact integral (§29.6, Example),

$$\iint_{\mathscr{D}} (x + 2y - 4)\, dx\, dy = 50,$$

where $\mathscr{D} = \{0 \leq x \leq 5, 0 \leq y \leq 4\}$.

6. Observe that the summation process for the integral of a step function

$$\iint_{\mathscr{D}} s(x, y)\, dx\, dy$$

can be done first by rows, keeping y constant, and then by adding the row sums (see Section 29.6, Example). Show that this idea can be expressed in symbols by

$$\iint_{\mathscr{D}} s(x, y)\, dx\, dy = \int_c^d \left[\int_a^b s(x, y)\, dx \right] dy,$$

where $\mathscr{D} = \{a \leq x \leq b, c \leq y \leq d\}$ and the inside single integral on the right is computed first, holding y constant.

7. Show that $\iint_{\mathscr{D}} xy\, dx\, dy = 0$ when computed over the square

$$\mathscr{D} = \{|x| \leq 1, |y| \leq 1\}.$$

8. Let A be the area of the rectangle

$$\mathscr{D} = \{a \leq x \leq b, c \leq y \leq d\}.$$

Let G be the maximum value of the continuous function $f(x, y)$ on $\mathscr{D}$ and let L be its minimum value. Prove that

$$LA \leq \iint_{\mathscr{D}} f(x, y)\, dx\, dy \leq GA.$$

9. Express by an integral formula the average value of the step function $s(x, y)$ on the rectangle $\mathscr{D}$ of area A.

10. If $f(x, y)$ is continuous over the rectangle $\mathscr{D}$ of area A, prove that there is some point (x_0, y_0) in $\mathscr{D}$ where

$$\iint_{\mathscr{D}} f(x, y)\, dx\, dy = f(x_0, y_0) \cdot A.$$

11. In Exercise 10, what is the average value of $f(x, y)$ on $\mathscr{D}$?

12. Compute $\iint_{\mathscr{D}} f(x, y)\, dx\, dy$ if $f(x, y) = 5$ (constant) on the rectangle

$$\mathscr{D} = \{a \leq x \leq b, c \leq y \leq d\}.$$

13. Calculate $\iint_{\mathscr{D}} (3f - 7g)\, dx\, dy$ if

$$\iint_{\mathscr{D}} f(x, y)\, dx\, dy = 5 \quad \text{and} \quad \iint_{\mathscr{D}} g(x, y)\, dx\, dy = 8.$$

14. We are to compute an approximate value of $\iint_{\mathscr{D}} (x^2 - xy + y^2)\, dx\, dy$ over the square $\mathscr{D} = \{|x| \leq 1, |y| \leq 1\}$. We partition the domain $\mathscr{D}$ into small enough squares so that in no cell does the difference (max) $-$ (min) of $x^2 - xy + y^2$ exceed 0.001. What is the largest possible error in replacing the exact integral by the integral of a step function over this partition?

15. Prove that if we partition the rectangle $\mathscr{D}$ into two rectangles $\mathscr{D}_1$ and $\mathscr{D}_2$, then

$$\iint_{\mathscr{D}} f(x, y)\, dx\, dy = \iint_{\mathscr{D}_1} f(x, y)\, dx\, dy + \iint_{\mathscr{D}_2} f(x, y)\, dx\, dy.$$

16. The density of a thin rectangular plate $\mathscr{D}$ of material is a function $\mu(x, y)$ given by $\mu(x, y) = 2xy$. Write a formula for the total mass of the plate.

17. The density of fluid in a rectangular $5 \times 5 \times 5$ tank varies with the depth z according to the function $\mu(x, y, z) = 1 - 0.2z$. Write a formula for the total mass of the fluid in the tank.

18. In the Example of §29.6 the function, $x + 2y - 4$, has the variation in the rectangle $\mathscr{D}$ as given by the difference between f at the upper right corner and lower left corner. This is $V = 13$. Find the size of square cells that will assure that any sample sum will differ from the exact integral by less than $\epsilon = 10^{-3}$.

30 Evaluation by Repeated Integral

30.1 Definition of the Repeated Integral. The idea of repeated sums is a very simple one. We can add all of the numbers in a rectangular array by first adding the

				row sums
2	3	4	1	10
				+
−1	2	3	6	10
				+
4	2	1	6	13
column sums:	5 + 7 + 8 + 13			= 33

columns, and then adding the column sums. We get the same answer by first adding the rows, and then adding the row sums. We call the method that of *repeated sums*.

The same idea applies to integrating a function $f(x, y)$ over a rectangle $\mathscr{D}:\{a \leq x \leq b, c \leq y \leq d\}$, provided that f is of bounded variation, or is continuous, so that it is integrable both over $\mathscr{D}$ and along any line in $\mathscr{D}$. We form a partition p of the rectangle $\mathscr{D}$ and consider the lower and upper sums over the rectangle.

$$\iint_{\mathscr{D}} s(x, y)\, dx\, dy \leq I \leq \iint_{\mathscr{D}} S(x, y)\, dx\, dy.$$

The terms, $s(x, y)\, dx\, dy$, of the lower sum form a rectangular array of numbers, which we can evaluate by summing with respect to x first (Figure 11.5) and then adding these x-sums.

We express this by writing

$$\iint_{\mathscr{D}} s(x, y)\, dx\, dy = \int_c^d \left(\int_a^b s(x, y)\, dx \right) dy,$$

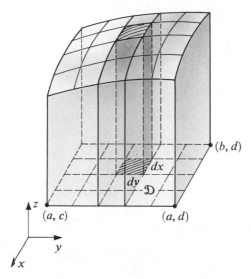

FIGURE 11.5   Repeated sum first with respect to x.

where the integral in parentheses is the sum over the column of cells between the grid lines $y = y_j$ and $y = y_{j+1}$. Similarly, for the integral of the overstep function,

$$\iint_{\mathscr{D}} S(x, y) \, du \, dv = \int_c^d \left(\int_a^d S(x, y) \, dx \right) dy.$$

The lower and upper sums in parentheses can be expected to converge, under refinement of the partition, to an exact integral

$$\int_a^b f(x, y) \, dx$$

for each fixed value of y, $c \leq y \leq d$. This integral along each line, $y = $ constant, then defines a function of y, which we can integrate between the boundaries, $y = c$ and $y = d$. The result of this repeated summation process is called the *repeated integral*

$$\int_c^d \left(\int_a^b f(x, y) \, dx \right) dy.$$

Similarly, we can first integrate along lines, $x = $ constant, and then integrate the resulting function of x. This repeated summation process gives the repeated integral

$$\int_a^b \left(\int_c^d f(x, y) \, dy \right) dx.$$

We state the result as a theorem, postponing the proof (§30.6)

THEOREM (Repeated integration).   If $f(x, y)$ is of bounded variation in the rectangle $\mathscr{D}:\{a \leqq x \leqq b, \ c \leqq y \leqq d\}$, or if f is continuous in $\mathscr{D}$, then both repeated integrals exist and are equal to the double integral of f over $\mathscr{D}$. That is,

$$\int_a^b \left(\int_c^d f(x, y) \, dy \right) dx = \iint_{\mathscr{D}} f(x, y) \, dx \, dy = \int_c^d \left(\int_a^b f(x, y) \, dx \right) dy.$$

30.2   Evaluation of the Double Integral by Means of Repeated Integrals.   The advantage of the repeated integral is that it calls for evaluating two integrals of functions of one variable and these can be found by the fundamental theorem of integral calculus (I, §19).

Example A.   Calculate the integral

$$\iint\limits_{\mathscr{D}} (x^2 - xy + y^2)\, dx\, dy$$

over the rectangle $\mathscr{D}:\{0 \leq x \leq 3,\ -1 \leq y \leq 1\}$.

Solution.   The function is continuous in the rectangle, so the integral exists. It can then be represented as a repeated integral

$$I = \int_{-1}^{1} \left(\int_{0}^{3} (x^2 - xy + y^2)\, dx \right) dy.$$

For each fixed value of y we compute the first integral (inside integral) by the fundamental theorem and then the second.

$$I = \int_{-1}^{1} \left[\frac{x^3}{3} - \frac{x^2}{2} y + xy^2 \right]_{0}^{3} dy$$

$$= \int_{-1}^{1} \left[3 - \frac{9}{2} y + 3y^2 \right] dy$$

$$= \left[3y - \frac{9}{2}\frac{y^2}{2} + 3\frac{y^3}{3} \right]_{-1}^{1}$$

$$= \left[\left(3 - \frac{9}{4} + 1 \right) - \left(-3 - \frac{9}{4} - 1 \right) \right] = 8.$$

Example B.   Calculate the integral

$$\iint\limits_{\mathscr{D}} (x + 2y - 4)\, dx\, dy$$

over the rectangle $\mathscr{D}:\{0 \leq x \leq 5,\ 0 \leq y \leq 4\}$.

Solution.   It is equal to the repeated integral

$$I = \int_{0}^{4} \int_{0}^{5} (x + 2y - 4)\, dx\, dy.$$

We omit the parentheses, understanding that the inside $\int_{0}^{5}$ and inside dx apply first. Integrating first with respect to x, then y, we get

$$I = \int_{0}^{4} \left[\frac{x^2}{2} + 2xy - 4x \right]_{0}^{5} dy$$

$$= \int_{0}^{4} \left[\frac{25}{2} + 10y - 20 \right] dy$$

$$= \left[\frac{25}{2} y\ 10\ \frac{y^2}{2} - 20y \right]_{0}^{4} = 50.$$

This is the integral (§29) we estimated from upper and lower sums. Ordinarily the average of the upper and lower sum does not give the exact integral.

Since the integral of $f(x, y)$ over $\mathcal{D}$ is equal to either of the repeated integrals, it is convenient to dispense with the distinction in notation and use the symbol

$$\int_c^d \int_a^b f(x, y)\, dx\, dy$$

for either the integral of $f(x, y)$ over $\mathcal{D}$ or the repeated integral first with respect to x, then y.

On the other hand, it is sometimes more convenient to integrate one repeated integral instead of the other.

30.3 Geometric Interpretation.

We sketch the function

$$f(x, y) = 4 - \frac{(x^2 + y^2)}{16}$$

over the rectangle $\mathcal{D} = \{0 \leq x \leq 9,\, 0 \leq y \leq 4\}$ (Figure 11.6). The multiple integral

$$V = \iint_{\mathcal{D}} f(x, y)\, dx\, dy$$

represents the volume of the region below the surface and over the rectangle $\mathcal{D}$. If we decide to compute it by repeated single integrals, first holding y fixed, we have

$$V = \int_\infty^4 \int_0^9 \left(4 - \frac{x^2 + y^2}{16}\right) dx\, dy.$$

The first integration sums the volumes of vertical shafts of base $dx\, dy$ and height $f(x, y)$ across the rectangle $\mathcal{D}$ at the fixed level y. So the result

$$S = \left[\int_0^9 \left(4 - \frac{x^2 + y^2}{16}\right) dx\right] dy$$

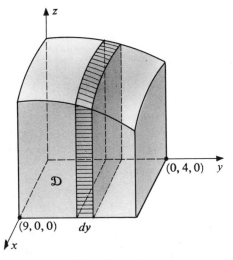

FIGURE 11.6   Slice of constant y.

represents the volume of a vertical slice of the region parallel to the x-axis and of thickness dy (Figure 11.6). We compute S.

$$S = \left[4x - \frac{(x^3/3) + y^2}{16} \right]_0^9 dy = \frac{495 + 9y^2}{16} \, dy.$$

The final integration

$$V = \int_0^4 \frac{495 + 9y^2}{16} \, dy$$

sums the volumes of these slices S for each y from 0 to 4,

$$V = \frac{1}{16} \left[495y + 9 \frac{y^3}{3} \right]_0^4 = \frac{543}{4} \text{ volume units.}$$

30.4 Average Value of a Function. The idea of the average value of a function carries over readily from functions of one variable. We define the average value of the function $f(x, y)$ over the domain $\mathscr{D}$ of area A by

$$\bar{z} = \frac{1}{A} \iint_{\mathscr{D}} f(x, y) \, dx \, dy.$$

Example. Compute the average value of $f(x, y) = x\sqrt{x^2 + y^2}$ over the rectangle $\mathscr{D}:\{0 \leq x \leq 1, -2 \leq y \leq 2\}$.

Solution. The area of the rectangle $\mathscr{D}$ is 4. The average value $\bar{z}$ of $f(x, y)$ is given by

$$\bar{z} = \frac{1}{4} \int_{-2}^2 \int_0^1 xy\sqrt{x^2 + y^2} \, dx \, dy$$

$$= \frac{1}{8} \int_{-2}^2 y \frac{[x^2 + y^2]^{3/2}}{\frac{3}{2}} \Big|_0^1 dy$$

$$= \frac{1}{12} \int_{-2}^2 [y(1 + y^2)^{3/2} - y^4] \, dy$$

$$= \frac{1}{12} \left[\frac{1}{2} \frac{(1 + y^2)^{5/2}}{\frac{5}{2}} - \frac{y^5}{5} \right]_{-2}^2 = \frac{16}{15}.$$

This gives $\bar{z}$ in the units of z, not volume units.

30.5 Exercises

In Exercises 1–8, evaluate the integrals by treating them as repeated integrals. Check by integrating in different order.

1. $\int_0^1 \int_{-1}^1 x^2 y \, dx \, dy.$

2. $\int_0^1 \int_0^1 (x^2 - y^2) \, dx \, dy.$

3. $\int_1^e \int_0^2 (x/y) \, dx \, dy.$

4. $\int_{-1}^1 \int_0^1 e^x y^2 \, dx \, dy.$

5. $\int_1^3 \int_0^2 xy\sqrt{x^2 + y^2} \, dy \, dx.$

6. $\int_0^1 \int_0^{\pi/2} y \sin xy \, dx \, dy.$

7. $\int_{-1}^1 \int_1^3 \int_0^2 (x^2 - 2xz + y^2) \, dx \, dy \, dz.$

8. $\int_{-1}^1 \int_0^1 \int_0^3 z \, e^{x+y} \, dz \, dy \, dx.$

In Exercises 9–12, the repeated integrals involve variables in the limits of integration. The order of integration must be the indicated one.

9. $\int_1^2 \int_1^x (x/y)\, dy\, dx.$ 10. $\int_0^1 \int_0^x e^x\, dy\, dx.$

11. $\int_c^d \int_a^x dy\, dx.$ 12. $\int_0^1 \int_{-a}^a ay^2\, dy\, da.$

13. Compute the volume below the surface $z = 3 - (x/3) - (y/2)$ and over the rectangle $\mathscr{D}:\{0 \le x \le \frac{9}{2}, 0 \le y \le 3\}.$

14. In Exercise 13, sketch the figure and show what the first integral represents:

$$S = \int \left[\int_0^{9/2} \left(3 - \frac{x}{3} - \frac{y}{2} \right) dx \right] dy.$$

15. In Exercise 13, if the repeated integration is performed with respect to y first, show in a figure what is represented by

$$S_1 = \int \left[\int_0^3 \left(3 - \frac{x}{3} - \frac{y}{2} \right) dy \right] dx.$$

16. Find the average value of $f(x, y) = y \sin xy$ in Exercise 6.
17. Find the average value of $f(x, y) = x$ in the rectangle

$$\mathscr{D}:\{|x| \le 3, |y| \le 5\}.$$

18. Find the average value of $f(x, y, z) = xy + y^2z^2$ over the box

$$\mathscr{D}:\{|x\{ \le 2, |y| \le 3, |z| \le 1\}.$$

19. Show that the area of the rectangle $\mathscr{D}$ is given by $A = \iint\limits_{\mathscr{D}} 1\, dx\, dy.$

20. Compute the partial derivative $F_1(x, y)$ if

$$F(x, y) = \int_a^x \int_1^y uv^2\, dv\, du.$$

21. Compute $DF(x, y, z)$ if

$$F(x, y, z) = \int_a^x \int_b^y \int_c^z uvw\, dw\, dv\, du.$$

30.6   Proof of the Theorem on Repeated Integration.

Since $f(x, y)$ is of bounded variation (or continuous) on $\mathscr{D}:\{a \le x \le b,\ c \le y \le d\}$, any function of x, $f(x, \cdot)$ obtained by fixing y is also of bounded variation (or continuous) on the interval $[a, b]$. Therefore, as a function of x, $f(x, y)$, with y held constant, is integrable over the interval $[a, b]$. The partial integral

$$\int_a^b f(x, y)\, dx$$

gives the values of a function $h(y)$ on the interval $[c, d]$. Similarly, if x is held constant, the partial integral

$$\int_c^d f(x, y)\, dy$$

exists, and gives the values of a function of x on the interval $[a, b]$.

We return to a partition $\mathscr{p}$ of the rectangle $\mathscr{D}$, and the integrals of its overstep function $S(x, y)$ and understep function $s(x, y)$. We know that since $f(x, y)$ is integrable over $\mathscr{D}$, the only number that is enclosed between the lower sum and upper sum, for every partition, is the double integral. That is, for every partition,

$$\iint\limits_{\mathscr{D}} s(x, y) \, dx \, dy \le \iint\limits_{\mathscr{D}} f(x, y) \, dx \, dy \le \iint\limits_{\mathscr{D}} S(x, y) \, dx \, dy.$$

Now suppose we fix y, so that the points (x, y) in these sums are restricted to a horizontal line that intersects a row of cells in the partition (Figure 11.7). The effect of restricting (x, y) to this line is to diminish $\sup f(x, y)$ so that in each cell $\sup f(x, y)$ along the line $y = $ const is not greater than $S(x, y)$ when (x, y) varies over the entire cell. Similarly, the $\inf f(x, y)$, with (x, y) restricted to the line $y = $ const, cannot be less than $s(x, y)$. We know that $f(x, \cdot)$, with y fixed, is integrable on $[a, b]$. It follows that under refinement of the partition of the interval $[a, b]$ by introducing more lines, $x = $ const, the exact partial integral of f satisfies the relations

$$\int_a^b s(x, y) \, dx \le \int_a^b f(x, y) \, dx \le \int_a^b S(x, y) \, dx.$$

We denote by $h(y)$ the partial integral

$$h(y) = \int_a^b f(x, y) \, dx.$$

The function $h(y)$ on $[c, d]$ is then integrable since all of its lower and upper sums lie between

$$\int_c^d \int_a^b s(x, y) \, dx \, dy \qquad \text{and} \qquad \int_c^d \int_a^b S(x, y) \, dx \, dy.$$

And we know that only one number

$$\iint\limits_{\mathscr{D}} f(x, y) \, dx \, dy$$

is enclosed between all lower and upper sums over the rectangle. Hence $h(y)$ is integrable and

$$\iint\limits_{\mathscr{D}} f(x, y) \, dx \, dy = \int_c^d h(y) \, dy = \int_c^d \left(\int_a^b f(x, y) \, dx \right) dy.$$

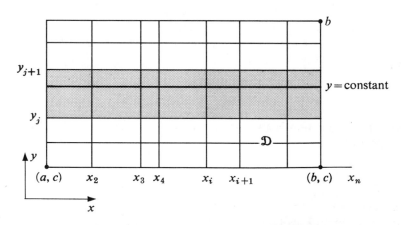

FIGURE 11.7   Line, $y = $ constant, in partition of domain $\mathscr{D}$.

By a similar proof, it can be shown that the other repeated integral exists, first with respect to y and then x, and is equal to the double integral.

31   Integration over a Region Enclosed by a Curve

31.1   Idea of Integration over a Region Bounded by a Curve.   The multiple integral

$$\iint_{\mathcal{D}} f(x, y) \, dx \, dy$$

has been defined in the case that the domain $\mathcal{D}$ is a rectangle (§29), and computed by a repeated integral (§30).   Many problems involve functions whose domain is a region bounded by a curve.

Example.   Find the volume enclosed by a sphere of radius r.

Solution: Initial Steps.   We choose a coordinate system with origin at the center of the sphere so that its equation becomes $x^2 + y^2 + z^2 = r^2$.   Then we solve for z and find two functions

$$z = +\sqrt{r^2 - (x^2 + y^2)} \qquad \text{and} \qquad z = -\sqrt{r^2 - (x^2 + y^2)},$$

both with the domain $\mathcal{D}:\{(x, y) \mid x^2 + y^2 \leqq r^2\}$.   This is a circular disk in the xy-plane. The function with the plus sign represents the hemisphere above the xy-plane (Figure 11.8).

Formally then, we expect the volume V to be given by an integral of the form

$$V = 2 \iint_{\mathcal{D}} \sqrt{r^2 - x^2 - y^2} \, dx \, dy,$$

where $\mathcal{D}$ is the circular disk in the xy-plane enclosed by the sphere.   Here the solution stops, for we have not defined multiple integrals of functions over domains bounded by a curve.

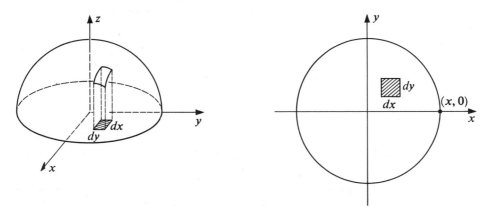

(a)   Upper   half   of   ball $x^2 + y^2 + z^2 \leqq r^2$

(b)   Domain $\mathcal{D}$ in xy-plane is the circular disk $x^2 + y^2 \leqq r^2$

FIGURE 11.8   Integration over circular disk.

31.2   Extension of the Definite Integral.   It can be proved (proof omitted) that a closed continuous curve in the plane that does not intersect itself separates the plane into two regions, one inside and one outside (Figure 11.9). Such a curve is called a simple closed curve. It is intuitively obvious, and can be proved mathematically, that a simple closed curve can be enclosed in the region bounded by a big-enough rectangle. Finally, it can be proved that a bounded function $f(x, y)$ that is defined and continuous on a rectangle $\mathscr{E}$, except at a set of points that lie on a curve of finite length, is integrable over the rectangle. The proof makes precise use of the fact that the curve of finite length occupies a region of zero area, and hence bounded discontinuities on it contribute nothing to the integral of f over the rectangle.

Now let $f(x, y)$ be a bounded function defined on the region $\mathscr{D}$ enclosed by a simple closed curve of finite length. We enclose the region $\mathscr{D}$ and its bounding curve $\mathscr{C}$ inside a rectangular $\mathscr{E}$ with sides parallel to the xy-axes (Figure 11.9). We define a function $f_E(x, y)$ over the rectangle $\mathscr{E}$ by making $f_E(x, y) = f(x, y)$ for points in $\mathscr{D}$ and $f_E(x, y) = 0$ for points in $\mathscr{E}$ outside of $\mathscr{D}$.

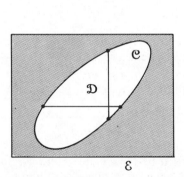
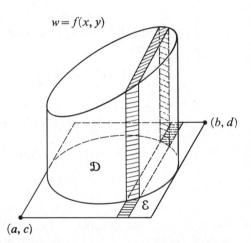

(a)   Region enclosed by a curve extended   (b)   Repeated integral over the extended
to a rectangle   domain

FIGURE 11.9   Extension of a domain to a rectangle.

DEFINITION.   The bounded function $f(x, y)$ over a region $\mathscr{D}$, enclosed by a simple closed curve of finite length, is integrable over $\mathscr{D}$ if and only if its extension $f_E(x, y)$ to a surrounding rectangle $\mathscr{E}$ is integrable, and we say that

$$\iint_{\mathscr{D}} f(x, y)\, dx\, dy = \iint_{\mathscr{E}} f_E(x, y)\, dx\, dy.$$

The basis of this definition is the fact that since $f_E(x, y) = 0$ outside of $\mathscr{D}$, the contribution of these points to the integral of $f_E(x, y)$ is also zero.

31.3   Evaluation by Repeated Integrals.   We can evaluate the double integral of $f_E(x, y)$ over the rectangle $\mathscr{E}$ by repeated integrals. Let $\mathscr{E}$ be the rectangle $\mathscr{E} = \{a \leqq x \leqq b,\ c \leqq y \leqq d\}$ (Figure 11.10). Then

$$\iint_{\mathscr{E}} f_E(x, y)\, dx\, dy = \int_a^b \left[\int_c^d f_E(x, y)\, dy \right] dx,$$

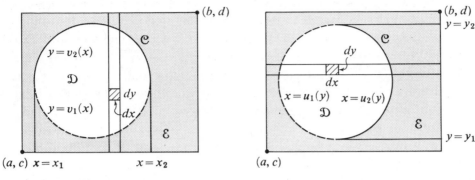

(a)   With respect to y first   (b)   With respect to x first

FIGURE 11.10   Repeated integration over extended region.

or

$$\iint_{\mathscr{E}} f_E(x, y) \, dx \, dy = \int_c^d \left[\int_a^b f_E(x, y) \, dx \right] dy.$$

We consider the first of these, integrating with respect to y first. For each x in the base of the rectangle, it carries the summation of terms of the type $f_E(x, y) \, dy \, dx$ in a vertical strip from bottom to top of the rectangle. But the portion of the strip outside of the curve contributes zero to the integral. Hence, if $y = v_2(x)$ is the function representing the upper half of the closed boundary curve $\mathscr{C}$ and if $y = v_1(x)$ represents the lower half, we see that the integration with respect to y need be carried only from $v_1(x)$ to $v_2(x)$ (Figure 11.10(a)). In symbols,

$$\left[\int_c^d f_E(x, y) \, dy \right] dx = \left[\int_{v_1(x)}^{v_2(x)} f(x, y) \, dy \right] dx.$$

Then the integration with respect to x needs to be carried only from the minimum x to the maximum x for points in $\mathscr{D}$. We denote these by the members x_1 and x_2. This gives us

$$\iint_{\mathscr{D}} f(x, y) \, dx \, dy = \int_{x_1}^{x_2} \left[\int_{v_1(x)}^{v_2(x)} f(x, y) \, dy \right] dx.$$

Similarly, integrating with respect to y first (Figure 11.10(b)), we find that

$$\iint_{\mathscr{D}} f(x, y) \, dx \, dy = \int_{y_1}^{y_2} \int_{u_1(y)}^{u_2(y)} f(x, y) \, dx \, dy,$$

where the first integration is carried in a horizontal strip from the left boundary $x = u_1(y)$ to the right boundary $x = u_2(y)$ of the domain $\mathscr{D}$. Then the integration with respect to y proceeds vertically from y_1, the minimum y for points in $\mathscr{D}$, to y_2, the maximum y for points in $\mathscr{D}$.

We observe that this technique removes the extraneous points in $\mathscr{E}$ outside of $\mathscr{D}$ that we used to put the integration on a rectangle.

31.4   Volume of the Sphere.   We return to complete the calculation of the volume V enclosed by the sphere $x^2 + y^2 + z^2 = r^2$ (§31.1), given by

$$V = 2 \iint_{\mathscr{D}} \sqrt{r^2 - x^2 - y^2} \, dx \, dy,$$

where the domain $\mathscr{D}$ was the disk $x^2 + y^2 \leqq r^2$ in the xy-plane. This integral is now defined and we can evaluate it by means of repeated single integrals. We will integrate $\sqrt{r^2 - x^2 - y^2}$ over the disk, first with respect to x holding y constant. At a fixed level y, x varies from the left boundary of the disk, where $x = -\sqrt{r^2 - y^2}$ to the right boundary, where $x = +\sqrt{r^2 - y^2}$ (Figure 11.11). We get these boundary functions by solving the equation of the circular boundary, $x^2 + y^2 = r^2$, for x in terms of y. This gives us $x = \pm\sqrt{r^2 - y^2}$. Then the integration

$$\left(\int_{-\sqrt{r^2-y^2}}^{\sqrt{r^2-y^2}} \sqrt{r^2 - x^2 - y^2}\, dx \right) dy$$

gives the volume in a slice whose base is the horizontal strip of width dy. Then when we integrate with respect to y from bottom to top, we sum these volumes of slices parallel to the x-axis to get the total volume

$$V = 2 \int_{-r}^{r} \int_{-\sqrt{a^2-y^2}}^{\sqrt{a^2-y^2}} \sqrt{z^2 - x^2 - y^2}\, dx\, dy.$$

We find in a table of integrals (Table I, Appendix B),

$$\int \sqrt{p^2 - x^2}\, dx = \frac{1}{2}\left[x\sqrt{p^2 - x^2} + p^2 \arcsin \frac{x}{p} \right]$$

Since y is fixed and $y^2 \leqq r^2$, the integrand of the volume integral is in this form with $p^2 = r^2 - y^2$. Hence

$$V = 2\left(\frac{1}{2}\right) \int_{-r}^{r} \left[x\sqrt{r^2 - y^2 - x^2} + (r^2 - y^2)\arcsin\frac{x}{\sqrt{r^2 - y^2}} \right]_{-\sqrt{r^2-y^2}}^{+\sqrt{r^2-y^2}} dy$$

$$= \int_{-r}^{r} [0 + (r^2 - y^2)(\arcsin 1 - \arcsin(-1))]\, dy$$

$$= \pi \int_{-r}^{r} (r^2 - y^2)\, dy = \pi\left[r^2 y - \frac{y^3}{3} \right]_{-r}^{r} = \frac{4}{3}\pi r^3.$$

This completes the problem.

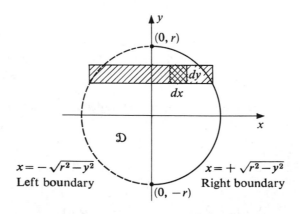

y

$(0, r)$

dy

dx

x

$\mathscr{D}$

$x = -\sqrt{r^2 - y^2}$
Left boundary

$x = +\sqrt{r^2 - y^2}$
Right boundary

$(0, -r)$

FIGURE 11.11   Repeated integration over a disk.

31.5   Area of a Plane Region.   The technique will permit us to evaluate the area of a plane region $\mathcal{D}$. The integral

$$\iint_{\mathcal{D}} 1 \, dx \, dy$$

gives the volume over $\mathcal{D}$ under a surface of constant height 1. But numerically this is $1 \times$ area of $\mathcal{D}$.

Example A.   Find by multiple integration the area of the region $\mathcal{D}$ bounded by the curves $y^2 = 6x$ and $x^2 - y^2 = 16$.

Solution.   We plot the curves and shade the region in question (Figure 11.12(a)). The required area A is given by integrating the function **1** over the region $\mathcal{D}$,

$$A = \iint_{\mathcal{D}} 1 \, dx \, dy.$$

We shall compute this by repeated integration, first in the x-direction and then in the y-direction. At every level y, a horizontal line segment stretches from the left boundary, $x = \frac{1}{6}y^2$, to the right boundary, $x = \sqrt{16 + y^2}$. We solve simultaneously for the points of intersection of the curves: $(8, 4\sqrt{3})$ and $(8, -4\sqrt{3})$. The area is then given by the repeated integral

$$A = \int_{-4\sqrt{3}}^{4\sqrt{3}} \int_{(1/6)y^2}^{\sqrt{16+y^2}} dx \, dy.$$

Since

$$\int_{(1/6)y^2}^{\sqrt{16+y^2}} dx = x \Big|_{(1/6)y^2}^{\sqrt{16+y^2}} = \sqrt{16 + y^2} - \frac{1}{6} y^2,$$

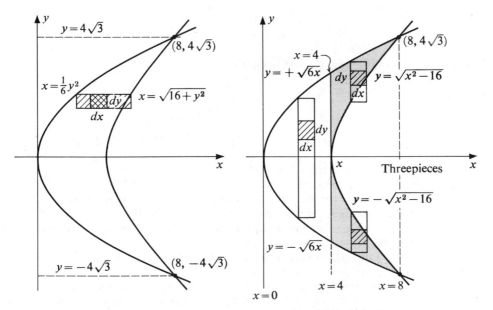

(a)   First with respect to x (b)   First with respect to y

FIGURE 11.12   Order of integration in non-convex region.

the repeated integral for A becomes

$$A = \int_{-4\sqrt{3}}^{4\sqrt{3}} \left(\sqrt{16 + y^2} - \frac{1}{6} y^2 \right) dy.$$

Using a table of integrals for $\int \sqrt{16 + y^2} \, Dy$, we find

$$A = \frac{1}{2} \left[y\sqrt{16 + y^2} + 16 \ln |y + \sqrt{16 + y^2}| - \frac{1}{9} y^3 \right]_{-4\sqrt{3}}^{4\sqrt{3}}$$

$$= \frac{224}{9} \sqrt{3} + 8 \ln (7 + 4\sqrt{3}) \text{ square units,}$$

which is the required area.

Example B. Compute the area of the region enclosed by the curves $y^2 = 6x$ and $x^2 - y^2 = 16$ by means of a repeated integral that goes first in the y-direction.

Solution. We see from Figure 11.12(b) that when $0 \leq x \leq 4$, a vertical line segment spanning the region $\mathscr{D}$ at the level x stretches from the lower branch of the parabola $y = -\sqrt{6x}$ to the upper branch $y = +\sqrt{6x}$. But when $4 \leq x \leq 8$, the vertical line segments spanning $\mathscr{D}$ break into two line segments. One stretches from $y = +\sqrt{x^2 - 16}$ to $y = \sqrt{6x}$. The other piece, below the x-axis, stretches from $y = -\sqrt{6x}$ to $y = -\sqrt{x^2 - 16}$. Hence, the area A is given by the sum of three repeated integrals

$$A = \int_0^4 \int_{-\sqrt{6x}}^{\sqrt{6x}} 1 \, dy \, dx + \int_4^8 \int_{\sqrt{x^2-16}}^{\sqrt{6x}} 1 \, dy \, dx + \int_4^8 \int_{-\sqrt{6x}}^{-\sqrt{x^2-16}} 1 \, dy \, dx.$$

This sum of repeated integrals gives the same area as that given by the repeated integral taken first in the x-direction. This calculation is not so simple as that of Example A, which is preferred.

31.6 Exercises

In Exercises 1–6, evaluate the repeated integrals.

1. $\int_0^1 \int_1^y (x + y) \, dx \, dy.$ 2. $\int_{-1}^1 \int_0^x (1 - y) \, dy \, dx.$

3. $\int_0^{2\pi} \int_0^{\cos \theta} r \, dr \, d\theta.$ 4. $\int_0^1 \int_{-\sqrt{1-v^2}}^{\sqrt{1-v^2}} v \, du \, dv.$

5. $\int_0^\infty \int_0^s e^{-(t-s)} \, dt \, ds.$ 6. $\int_0^a \int_y^{y+b} (x^2 - 2xy + y^2) \, dx \, dy.$

7. Integrate $x + y$ over the triangle bounded by the lines $y = x$, $y = 0$, $x = 1$.
8. Integrate $2xy$ over the triangle bounded by the lines $x + y = 1$, $x = 0$, $y = 0$.
9. Integrate $2xy$ over the region in the first quadrant bounded by the curve $y = 1 - x^2$. Carry out the integration with two different repeated integrals.
10. Integrate y^3 over the region bounded by the line $x = 0$ and the curve $y^2 = 1 - x$. Compute in two different ways.
11. Integrate $x^2 + 1$ over the region bounded by the curves $y = x^2 - 2$ and $y = 2 - x^2$.
12. Integrate x over the quadrant of the disk $\{x^2 + y^2 \leq 1, x \geq 0, y \geq 0\}$.
13. Integrate $x^2 - y^2$ over the region bounded by the curve $y^2 = 4x$ and the line $x = 1$.
14. Integrate $\sin^2 x$ over the region bounded by one arch of the curve $y = \cos x$ and the line $y = 0$.
15. Integrate $x^2 y^2 / \sqrt{1 - x^2}$ over the unit disk $\{x^2 + y^2 \leq 1\}$.

16. Find by a double integral the area common to the two disks

$$\{(x - 1)^2 + y^2 \leq 2\} \qquad \text{and} \qquad \{(x + 1)^2 + y^2 \leq 2\}.$$

17. Choose the more convenient repeated integral and find the area of the region bounded by $y = 4x^2$ and $y = x^2 + 3$.
18. Integrate $\sin x \cos y$ over the region bounded by $y = \arcsin x$ and the lines $x = 1$, $y = 0$.
19. Set up the repeated integral representing the volume of the sphere (§31.1), integrating first with respect to y, and compute it.
20.* Find by double integration the volume of the region bounded by the paraboloid $z = 1 - 4x^2 - y^2$ and the plane $z = 0$.

32   Generalization of the Idea of Integration

ACADEMIC NOTE.   We are here concerned more with the idea of integrating functions of several variables $x_1, x_2, \ldots, x_n$ over regions in $\mathscr{R}^n$ than with the calculation of these integrals. We want to know how to define the integrals, what they mean and what their fundamental properties are. This is necessarily somewhat abstract. For concreteness we have seen a little of the computation technique and its use of repeated single integrals for double integrals $\iint_{\mathscr{D}} f(x, y) \, dx \, dy$ over regions $\mathscr{D}$ bounded by curves in the plane (§31). When we carry this over to regions in 3-space there is essentially nothing new in the integral, but the calculation becomes technically difficult. We postpone the calculation techniques in order to proceed with the general idea first, before solving problems involving boundary surfaces.

32.1   Elements of Area and Mass in the Plane.   We recall that to define the integral

$$\iint_{\mathscr{D}} f(x, y) \, dx \, dy$$

over the rectangle domain $\mathscr{D}$ in the plane, we partitioned $\mathscr{D}$ into small rectangular cells whose sides had lengths dx and dy and hence the area $dA = dx \, dy$ (Figure 11.13). This is the formula for area in local rectangular Cartesian coordinates, dx, dy, but for other coordinates the formula for area might be different. Hence we make a first step towards generalizing the idea by replacing $dx \, dy$ by the symbol dA, which means the area of a cell of the partition however it is calculated. The symbol for the integral then becomes

$$\iint_{\mathscr{D}} f(x, y) \, dA.$$

This suggests that we generalize the summation process, defining an integral over a region by assigning to each cell some other measure of content dm, not necessarily area or volume. Then we form sample sums of terms of the type

$$f(x, y) \, dm$$

and proceed to the limit by refinement of the partition to define an exact integral

$$\iint_{\mathscr{D}} f(x, y) \, dm.$$

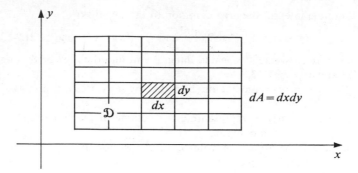

FIGURE 11.13   Element of area in xy-coordinates.

This is a conceptual integral, but to make it work we must be more specific about how the mass element dm is defined. We assume here that there is a continuous density function $\mu(x, y)$. To each cell $\mathcal{D}_{ij}$ of a partition π, this density gives the mass

$$\Delta m_{ij} = \iint_{\mathcal{D}_{ij}} \mu(x, y)\, dx\, dy = \mu(u_i, v_i)\, \Delta x_i,\, \Delta y_j,$$

where (u_i, v_j) is some point in the cell $\mathcal{D}_{ij}$ (§30.4). Then we have the sample sum

$$\sum_\pi f(x_i, y_j)\, \Delta m_{ij} = \sum_\pi f(x_i, y_j)\mu(u_i, v_j)\, \Delta x_i\, \Delta y_j.$$

The expression on the right looks like a sample sum for the function $f(x, y)\mu(x, y)$ except that the two factors are evaluated at different points in each cell. We may call upon Bliss's theorem (§32.6) to say that under refinement of the partition these sums converge to the exact integral of $f(x, y)\mu(x, y)$ so that

$$\iint_{\mathcal{D}} f(x, y)\, dm = \iint_{\mathcal{D}} f(x, y)\mu(x, y)\, dx\, dy.$$

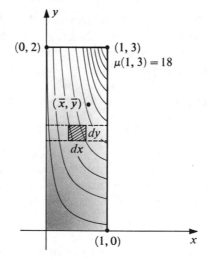

FIGURE 11.14   Mass distribution in a rectangle with density $\mu(x, y) = 0.6\, xy$. Lines of constant density are hyperbolas $0.6\, xy = $ constant. Maximum density is at the corner $(1, 3)$ and center of mass $(\bar{x}, \bar{y}) = (\tfrac{2}{3}, 2)$.

There are more general conditions under which the conceptual integral $\iint\limits_{\mathscr{D}} f(x, y)\, dm$ with respect to a measure m can be defined precisely.

Example. Find the total mass of the material in a rectangular region $\mathscr{D}:\{0 \leq x \leq 1,\ 0 \leq y \leq 3\}$ if the density of the material is proportional to the product xy and is 1.8 at $(1, 3)$ in g/cm² (Figure 11.14).

Solution. Let $\mu(x, y)$ represent the density at (x, y). Then $\mu(x, y) = kxy$. Moreover, $\mu(1, 3) = 1.8$. Hence $k = 0.6$ and $\mu(x, y) = \mathbf{0.6}xy$. Partitioning the rectangle $\mathscr{D}$ into smaller rectangular cells, we consider a typical cell of dimensions dx, dy. Its mass $dm = \mu(x, y)\, dx\, dy = 0.6xy\, dx\, dy$, where (x, y) is a suitably chosen point in the cell. Hence the total mass M is given by the exact integral

$$M = \iint\limits_{\mathscr{D}} 1 \cdot dm = \int_0^3 \int_0^1 0.6xy\, dx\, dy = 1.35\text{ g.}$$

32.2   Element of Volume and Triple Integral. We consider a function $f(x, y, z)$ defined on a rectangular box $\mathscr{D}$ in 3-space. We want to integrate $f(x, y, z)$ over the box $\mathscr{D}$. Proceeding as we did for domains in the line and in the plane, we partition the box $\mathscr{D}$ into smaller cells bounded by planes

$$x = x_i: \quad x_0 < x_1 < x_2 < \cdots < x_l,$$
$$y = y_j: \quad y_0 < y_1 < y_2 < \cdots < y_m,$$
$$z = z_k: \quad z_0 < z_1 < z_2 < \cdots < z_n,$$

(Figure 11.15). Looking at a typical cell indexed ijk, we evaluate f at some point $(\bar{x}_i, \bar{y}_j, \bar{z}_k)$ in the cell and multiply this by the volume dV of the cell. This gives us a term of the type

$$f(\bar{x}_i, \bar{y}_j, \bar{z}_k)\, dV,$$

where in Cartesian coordinates $dV = \Delta x_i\, \Delta y_j\, \Delta z_{kj}$. Then we sum all of the terms of this type over the entire box and proceed to the limit by refinement of the partition into smaller and smaller cells. The function $f(x, y, z)$ is integrable over $\mathscr{D}$ if and only if this limit procedure converges to a single number, which we denote by

$$\iiint\limits_{\mathscr{D}} f(x, y, z)\, dV,$$

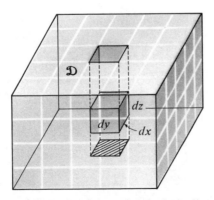

FIGURE 11.15   Partition of domain in 3-space.

where in Cartesian coordinates, $dV = dx\, dy\, dz$.

We observe that the triple integral

$$V = \iiint_{\mathscr{D}} 1\, dV$$

gives the volume V of the box itself. We also observe that for the simple box domain $\mathscr{D}$ the summation process may be carried out by holding x and y fixed and carrying a single integral from bottom to top. Then holding x constant, we may sum with respect to y. Finally we sum with respect to x.

Example. Integrate the function $f(x, y, z) = 2xy - 4yz + z^2$ over the box

$$\mathscr{D} = \{0 \leq x \leq 4,\ -2 \leq y \leq 4,\ 0 \leq z \leq 3\}.$$

Solution. The integral may be represented by a repeated triple integral

$$\iiint_{\mathscr{D}} (2xy - 4yz + z^2)\, dV = \int_0^4 \int_{-2}^4 \int_0^3 (2xy - 4yz + z^2)\, dz\, dy\, dx$$

$$= \int_0^4 \int_{-2}^4 \left[2xyz - 4y\frac{z^2}{2} + \frac{z^3}{3} \right]_0^3 dy\, dx$$

$$= \int_0^4 \int_{-2}^4 [6xy - 18y + 9]\, dy\, dx$$

$$= \int_0^4 \left[6x\frac{y^2}{2} - 18\frac{y^2}{2} + 9y \right]_{-2}^4 dx = 3 \int_0^4 (12x - 18)\, dx = 72.$$

32.3 Integration over Regions Bounded by Curved Surfaces. As for regions of the plane bounded by curves, we can integrate over domains $\mathscr{D}$ enclosed by some closed surface, like a sphere. We enclose the region $\mathscr{D}$ in a box $\mathscr{E}$ and extend the function $f(x, y, z)$ over the box by the definition $f_E(x, y, z) = f(x, y, z)$ in $\mathscr{D}$ and $f_E(x, y, z) = 0$ in $\mathscr{E}$ outside of $\mathscr{D}$ (Figure 11.16). The function is integrable over the region $\mathscr{D}$ bounded by the curved surface if its extension $f_E(x, y, z)$ is integrable over the box, and

$$\iiint_{\mathscr{D}} f(x, y, z)\, dV = \iiint_{\mathscr{E}} f_E(x, y, z)\, dV.$$

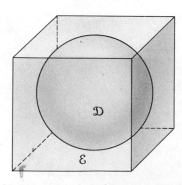

FIGURE 11.16   Curved surface enclosed in a box.

The basis for this definition is that at the points outside of $\mathscr{D}$, $f_E(x, y, z) = 0$, and hence these points make no contribution to the integral over the box. The repeated integral of f_E over the box $\mathscr{E}$ then reduces to a triple repeated integral over $\mathscr{D}$ with limits determined by the curved surface that bounds $\mathscr{D}$. We postpone this type of calculation, but let us look at one domain to see how these limits are set.

Example.   Set up a repeated triple integral equal to the integral

$$J = \iiint_{\mathscr{D}} f(x, y, z) \, dV$$

over the spherical ball $\mathscr{D} = \{x^2 + y^2 + z^2 = r^2\}$.

   Solution.   We choose to integrate in the order $dz \, dy \, dx$. If we hold x and y constant, z varies from bottom $z = -\sqrt{r^2 - x^2 - y^2}$ to top $z = +\sqrt{r^2 - x^2 - y^2}$ of the sphere. Above and below this surface, the contribution of the function f_E extended to the enclosing box is zero. Now there are only two variables (x, y) left. Only points in the disk $\{x^2 + y^2 \leq r^2\}$, $z = 0$, now contribute to the integral. Outside of this disk, $f_E = 0$. Hence y varies from the lower boundary of the circle $x^2 + y^2 = r^2$, where $y = -\sqrt{r^2 - x^2}$, to the upper boundary, where $y = +\sqrt{r^2 - x^2}$. Finally, we let x vary from its minimum to maximum value in the disk $\{x^2 + y^2 \leq r^2\}$, that is, from $x = -r$ to $x = +r$. This gives us the repeated triple integral

$$J = \int_{-r}^{r} \int_{-\sqrt{r^2 - x^2}}^{+\sqrt{r^2 - x^2}} \int_{-\sqrt{r^2 - x^2 - y^2}}^{+\sqrt{r^2 - x^2 - y^2}} f(x, y, z) \, dz \, dy \, dx.$$

Then if $f(x, y, z)$ is given by some analytic expression for which we can find an antiderivative with respect to z, and if we can also find antiderivatives with respect to y and then x of the functions that arise in the second and third stages of the calculation, we can complete the numerical calculation of the integral J.

   Alternatively, we can calculate J approximately by a numerical procedure based on the definition. Using a fine-enough partition to get a good result, we compute and add terms of the type

$$f(x_i, y_j, z_k) \, \Delta x_i \, \Delta y_j \, \Delta z_k$$

for each cell of the partition. Here it is advantageous to compute the integral over an enclosing box $\mathscr{E}$ of the function f_E extended to be zero outside of $\mathscr{D}$.

32.4   Mass Distribution.   As in the case of plane domains, we can integrate with respect to a measure of mass distributed in space. Let $\mu(x, y, z)$ be the density of a mass distribution at point $P:(x, y, z)$. Then an element of mass $dm = \mu(x, y, z) \, dV$. If $f(x, y, z)$ is defined over a box $\mathscr{D}$, then we can integrate it over the box with respect to the mass measure by forming sums of terms of the type

$$\iiint_{\mathscr{D}} f(x, y, z) \, dm = \iiint_{\mathscr{D}} f(x, y, z)[\mu(x, y, z) \, dV].$$

Example.   Find the mass of a cube of $\mathscr{D}$ edge 2, $\mathscr{D} = \{|x| \leq 1, |y| \leq 1, |z| \leq 1\}$, with center at the origin, if mass in the cube has density $\mu(x, y, z) = x^2 + y^2 + z^2$.

Solution. The mass M of the cube is given by

$$M = \iiint_{\mathcal{D}} 1 \; dm = \int_{-1}^{1} \int_{-1}^{1} \int_{-1}^{1} (x^2 + y^2 + z^2) \; dz \; dy \; dx$$

$$= \int_{-1}^{1} \int_{-1}^{1} \left[(x^2 + y^2)z + \frac{z^3}{3} \right]_{-1}^{1} dy \; dx = 2 \int_{-1}^{1} \int_{-1}^{1} \left(x^2 + y^2 + \frac{1}{3} \right) dy \; dz$$

$$= 2 \int_{-1}^{1} \left[(x^2 y + \frac{y^3}{3} + \frac{1}{3} y \right]_{-1}^{1} dx = 4 \int_{-1}^{1} \left(x^2 + \frac{2}{3} \right) dx = 8.$$

32.5 Exercises

In Exercises 1–6, integrate the repeated integrals.

1. $\int_{0}^{3} \int_{-2}^{2} \int_{0}^{1} z(x - y) \; dz \; dy \; dx$. Compute in three orders.

2. $\int_{1}^{e} \int_{0}^{3} \int_{1}^{2} \frac{xy}{z} \; dx \; dy \; dz$. Compute in three orders.

3. $\int_{0}^{5} \int_{1}^{3} \int_{0}^{1} \frac{xyz \; dy \; dx \; dz}{1 + y^2}$.

4. $\int_{0}^{a} \int_{0}^{a} \int_{-y}^{y} \sqrt{a^2 - y^2} \; dz \; dy \; dx$.

5. $\int_{-r}^{r} \int_{0}^{x} \int_{0}^{\sqrt{r^2 - x^2 - y^2}} z \; dz \; dy \; dx$.

6. $\int_{-1}^{1} \int_{0}^{y} \int_{0}^{x} ze^{-x+y} \; dz \; dx \; dy$.

In Exercises 7–12, set up triple integrals to represent the quantities described.

7. The volume of the box $\{|x| \le 3, 0 \le y \le 1, |z| \le 1\}$.
8. The volume of the region below the graph of the relation $z = 2x^2 + 5y^2$ and above the rectangle $\{0 \le x \le a, |y| \le b, z = 0\}$.
9. The mass of the material in the box $\{|x| \le a, |y| \le b, |z| \le c\}$ if the density $\mu(x, y, z) = e^z$.
10. The mass of the cube $\{|x| \le a, |y| \le a, |z| \le a\}$ in which the density $\mu(x, y, z) = 2x^2 + 3y^2 + z^2$.
11. The average value of $x^2 - y^2$ in the cube $\{|x| \le a, |y| \le a, |z| \le a\}$.
12. The average value of $z \sin x \cos y$ in the box

$$\{0 \le x \le \pi/3, 0 \le y \le \pi/2, 1 \le z \le a\}.$$

13. What volume is represented by

$$\int_{0}^{x+y+1} dz \; dy \; dx ?$$

Draw a figure to show it.

14. In Exercise 13, what volume is represented by

$$\int_{0}^{3} \int_{0}^{x+y+1} dz \; dy \; dx ?$$

Draw a figure.

15. In Exercise 13, what volume is represented by

$$V = \int_{0}^{3} \int_{0}^{3} \int_{0}^{x+y+1} dz \; dy \; dx ?$$

Integrate to find the volume.

16. What quantity is represented by

$$\frac{\iiint_{\mathcal{D}} f(x, y, z) \; dV}{\iiint_{\mathcal{D}} dV} ?$$

17. Evaluate

$$\int_{-a}^{a} \int_{0}^{1} \int_{0}^{1} z f(x, y) \, dx \, dy \, dz,$$

where $f(x, y)$ is some integrable function.

18. Interpret as a mass of a solid

$$\int_{0}^{5} \int_{1}^{3} \int_{-1}^{1} z^2 e^{-x-y} \, dx \, dy \, dz.$$

What are the boundaries of the solid? What is the density of material in the solid?

19. Explain why the following repeated integral has no meaning.

$$\int_{1}^{2} \int_{0}^{1} \int_{0}^{\sqrt{1-x^2-y^2}} dz \, dy \, dx.$$

20. What volume is represented by the integral

$$\int_{0}^{h} \int_{-r}^{r} \int_{-\sqrt{r^2-y^2}}^{\sqrt{r^2-y^2}} dx \, dy \, dz?$$

32.6   Theorem of Bliss*

THEOREM.   If functions f and g are integrable over $[a, b]$, then the sample sums over partitions π,

$$\sum_{\pi} f(x_i)g(x_i') \, \Delta x_i,$$

where f and g are evaluated at *two different* points in each cell of the partition, converge by refinement of the partition to the exact integral

$$\int_{a}^{b} f(x)g(x) \, dx.$$

Proof.   To the typical sample sum in the theorem we add and subtract $\sum_{\pi} f(x_i)g(x_i) \, \Delta x$ and find that

$$\sum_{\pi} f(x_i)g(x_i') \, \Delta x_i = \sum_{\pi} f(x_i)g(x_i) \, \Delta x_i + \left(\sum_{\pi} f(x_i)g(x_i') \, \Delta x_i - \sum_{\pi} f(x_i)g(x_i) \, \Delta x_i \right).$$

To show that the term in parentheses approaches zero, we observe that

$$\left| \sum_{\pi} f(x_i)g(x_i') \, \Delta x_i - \sum_{\pi} f(x_i)g(x_i) \, \Delta x_i \right| \leq \sum_{\pi} | f(x_i)| \, |[g(x_i) - g(x_i')] \, \Delta x_i|.$$

On the right, f is bounded, so that for some M, $|f(x)| < M$ on $[a, b]$. Then the term on the right above is less than

$$M \left| \left[\sum_{\pi} g(x_i) \, \Delta x_i \right] - \left[\sum_{\pi} g(x_i') \, \Delta x_i \right] \right|.$$

Since g is integrable, for every ϵ there is a sufficiently refined partition so that any two sample sums over it differ by less than ϵ. So the term on the right approaches zero by refinement of the partition. This implies that the sample sums with different evaluation

---

* G. A. Bliss, "A Substitute for Duhamel's Theorem," *Ann. Math.* **16** (1914–15), pp. 45–49.
   The theorem and its proof are applicable for multiple integrals with respect to any number of variables.

points for the two factors converge to the same exact integral, $\int_a^b f(x)g(x)\,dx$, as the sample sums

$$\sum_\pi f(x_i)g(x_i)\,\Delta x_i,$$

with the same evaluation point for f and g in each subinterval.

33 Average and Dispersion

33.1 Balancing Masses. We recall the principle of moments in balancing a lever (Figure 11.17(a)). A distribution of point masses $m_1, m_2, \ldots, m_n$ on a line, located at points $x_1, x_2, \ldots, x_n$, is balanced with reference to a point $\bar{x}$ if the sum of the moments is zero,

$$(x_1 - \bar{x})m_1 + (x_2 - \bar{x})m_2 + \cdots + (x_n - \bar{x})m_n = 0.$$

The products $(x_i - \bar{x})m_i$ are called the first-order moments of the masses with reference to $\bar{x}$.

We carry over this moment principle to a continuous distribution of mass in a solid body that occupies a region $\mathscr{D}$. We replace the point masses m_i by elements of mass dm (Figure 11.17(b)) and sums of moments by integrals. Specifically, the moment of the mass element dm at $P:(x, y, z)$ with reference to $\bar{x}$ is $(x - \bar{x})\,dm$. The body is balanced with respect to the plane $x = \bar{x}$ if the first-order moment of the body is zero,

$$\iiint_{\mathscr{D}} (x - \bar{x})\,dm = 0.$$

We may think of $\bar{x}$ as the x-coordinate of the center of mass. The other coordinates of the center of mass are given by the equilibrium conditions

$$\iiint_{\mathscr{D}} (y - \bar{y})\,dm = 0 \qquad \text{and} \qquad \iiint_{\mathscr{D}} (z - \bar{z})\,dm = 0.$$

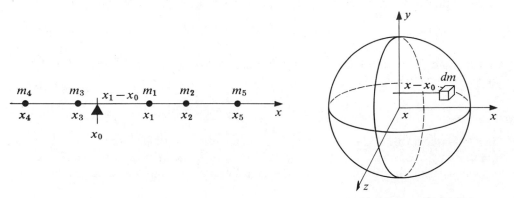

(a)   Point masses on a line   (b)   Mass elements in a solid body

FIGURE 11.17   Moments of mass.

The equilibrium conditions can be solved for $(\bar{x}, \bar{y}, \bar{z})$. For example, the equilibrium condition for $\bar{x}$ can be written

$$\iiint_{\mathscr{D}} x \, dm = \bar{x} \iiint_{\mathscr{D}} dm.$$

This gives $\bar{x}$; and $\bar{y}$ and $\bar{z}$ are found similarly. We have

$$\bar{x} = \frac{\iiint_{\mathscr{D}} x \, dm}{\iiint_{\mathscr{D}} dm}, \qquad \bar{y} = \frac{\iiint_{\mathscr{D}} y \, dm}{\iiint_{\mathscr{D}} dm}, \qquad \bar{z} = \frac{\iiint_{\mathscr{D}} z \, dm}{\iiint_{\mathscr{D}} dm}.$$

We recognize the denominators as the total mass M,

$$M = \iiint_{\mathscr{D}} dm,$$

of the body. Also we recognize these formulas as giving the weighted averages of x, y, z with respect to the mass distribution in the body.

Example.   Find the center of mass distributed in the plane domain $\mathscr{D}$ bounded by lines $x = 0$, $y = 0$, $x = 1$, $y = 3$, with density $\mu(x, y) = 0.6xy$ (§32.5).

 Solution.   The element of mass $dm = 0.6xy \, dx \, dy$. We computed the total mass M,

$$M = \iint_{\mathscr{D}} 0.6xy \, dx \, dy = 1.35.$$

The average value of x, or x-coordinate of the center of mass, is given by

$$\bar{x} = \frac{1}{M} \iint_{\mathscr{D}} x \, dm = \frac{1}{1.35} \int_0^3 \int_0^1 x(0.6xy \, dx \, dy) = \frac{2}{3}.$$

Similarly,

$$\bar{y} = \frac{1}{M} \iint_{\mathscr{D}} y \, dm = \frac{1}{1.35} \int_0^3 \int_0^1 y(0.6xy \, dx \, dy) = 2.$$

The center of mass is $(\frac{2}{3}, 2)$. We observe that it is not at the center of the rectangle $(\frac{1}{2}, \frac{3}{2})$ but located nearer the vertex $(1, 3)$, where the mass density is greatest.

33.2   Second-Order Moments.

The second-order moments of a mass distribution in the domain $\mathscr{D}$ are defined by the integrals

$$I_{x_0} = \iiint_{\mathscr{D}} (x - x_0)^2 \, dm, \qquad \text{about the plane } x = x_0,$$

$$I_{y_0} = \iiint_{\mathscr{D}} (y - y_0)^2 \, dm, \qquad \text{about the plane } y = y_0,$$

$$I_{z_0} = \iiint_{\mathscr{D}} (z - z_0)^2 \, dm, \qquad \text{about the plane } z = z_0.$$

These are called the *moments of inertia* of the mass about the planes $x = x_0$, $y = y_0$, $z = z_0$. We shall see (§33.5) that they have a physical meaning with respect to a rotating rigid mass. We consider these second-order moments now as purely mathematical objects. If we set

$$I_{x_0} = I_{y_0} = I_{z_0} = 0,$$

we do not get an equilibrium condition but have instead a requirement that all of the mass is concentrated at the point $P_0:(x_0, y_0, z_0)$. For since the integrands $(x - x_0)^2$, $(y - y_0)^2$, $(z - z_0)^2$ are all positive or zero, any positive element of mass dm for which $x - x_0 \neq 0$ will make the second-order moment I_{x_0} positive.

Now we consider second-order moments with respect to planes through the center of mass, $x = \bar{x}, y = \bar{y}, z = \bar{z}$.

$$I_{\bar{x}} = \iiint_{\mathscr{D}} (x - \bar{x})^2 \, dm.$$

We observe that a mass situated where $(x - \bar{x})^2$ is large contributes more to the second-order moment $I_{\bar{x}}$ than the same amount of mass located where $(x - \bar{x})^2$ is small. Hence a large value of $I_{\bar{x}}$ indicates that the distribution of mass is not concentrated near the equilibrium point, where $x = \bar{x}$, but is dispersed widely from the average. Conversely, if $I_{\bar{x}}, I_{\bar{y}}, I_{\bar{z}}$ are small, it is implied that the mass distribution is concentrated near the center of mass.

Example.   Find the centers of mass and moments of inertia of the mass distributions in the square $\mathscr{D} = \{|x| \leq 1, |y| \leq 1\}$ given by the density functions: (a) $\mu(x, y) = 1$; (b) $\nu(x, y) = 9x^2y^2$ (Figure 11.18). Compare them.

Solution.   We first compute the total masses M and N for the two distributions. We find $M = N = 4$. Moreover, we also find for each distribution the center of mass, $(\bar{x}, \bar{y}) = (0, 0)$. We observe that the two distributions differ in that in the first, the 4 units of mass are distributed in the square with uniform density, while in the second, the density increases from 0 at the center to 9 at the corners of the square. We compute the moments of inertia for $\mu(x, y) = 1$,

$$I_{x_0} = \int_{-1}^{1} \int_{-1}^{1} x^2(1 \, dx \, dy) = \tfrac{4}{3}. \qquad \text{Also} \qquad I_{y_0} = \tfrac{4}{3}.$$

And for $\nu(x, y) = 9x^2y^2$,

$$I_{x_0} = \int_{-1}^{1} \int_{-1}^{1} x^2(9x^2y^2 \, dx \, dy) = \tfrac{12}{5}. \qquad \text{Also} \qquad I_{y_0} = \tfrac{12}{5}.$$

The fact that $I_{x_0} = I_{y_0} = \tfrac{12}{5}$ for the second distribution is a measure of the degree to which its mass is dispersed away from the center more than for the first distribution, with the same total mass but with $I_{x_0} = I_{y_0} = \tfrac{4}{3}$.

(a)   Uniform density $\mu(x, y) = 1$ (b) $\nu(x, y) = 9x^2y^2$

FIGURE 11.18   Two distributions of mass.

33.3 Statistical Interpretation. Statistical data often take the form of a distribution of some quantity m in a line, or in a plane (or in some higher-dimensional space). For such a distribution in the plane, the average values of x and y in the distribution are given by

$$\bar{x} = \frac{\iint_{\mathscr{D}} x \, dm}{\iint_{\mathscr{D}} dm} \quad \text{and} \quad \bar{y} = \frac{\iint_{\mathscr{D}} y \, dm}{\iint_{\mathscr{D}} dm}.$$

These averages are usually called the *mean values* of x and y.

When the mean of a distribution is known, the next question is, How widely does the distributed quantity deviate from the mean values? If we look at an element dm in the distribution, its second moment

$$(x - \bar{x})^2 \, dm$$

is large if it is far from the mean point $\bar{x}$ and small if it is near the mean. It is customary to use the average value of the positive quantity $(x - \bar{x})^2$ in the distribution as a measure of its dispersion away from the mean. We have the two average values

$$\sigma_x^2 = \frac{\iint_{\mathscr{D}} (x - \bar{x})^2 \, dm}{\iint_{\mathscr{D}} dm} \quad \text{and} \quad \sigma_y^2 = \frac{\iint_{\mathscr{D}} (y - \bar{y})^2 \, dm}{\iint_{\mathscr{D}} dm}.$$

The number σ_x^2 is called the variance in x, and σ^2 is called the variance in y. σ_x itself is called the *standard deviation* in x, and σ_y the *standard deviation* in y.

Example. Find the variance and standard deviations of the rectangular distribution $\mathscr{D} = \{0 \le x \le 1, 0 \le y \le 3\}$, $\mu(x, y) = 0.6xy$ (§33.2).

Solution. We have found that $(\bar{x}, \bar{y}) = (\frac{2}{3}, 2)$ and the total mass is 1.35. Hence the variance in x is given by

$$\sigma_x^2 = \frac{1}{1.35} \int_0^3 \int_0^1 \left(x - \frac{2}{3}\right)^2 (0.6 \, dx \, dy) = \frac{4}{27}.$$

$$\sigma_y^2 = \frac{1}{1.35} \int_0^3 \int_0^1 (y - 2)^2 (0.6 \, dx \, dy) = \frac{4}{3}.$$

The standard deviations in x and y are $\sigma_x = 2/\sqrt{27}$ and $\sigma_y = 2/\sqrt{3}$. There is also defined a standard deviation of σ of the distribution

$$\sigma = \sqrt{\sigma_x^2 + \sigma_y^2} = \frac{2}{3}\sqrt{\frac{10}{3}}.$$

33.4 Exercises

1. Point masses are located at points on the x-axis indicated by their coordinates.

mass	3	4	2	1	1
x-coordinate	-2	-1	2	3	5

(a) Find $\bar{x}$, the x-coordinate of center of mass.
(b) Find the second moment I_0 with respect to the origin.
(c) Find the second moment $I_{\bar{x}}$ with respect to the mean, $\bar{x}$.

In Exercises 2–4, find the total mass of the distribution and the mean values of x and y, that is, the coordinates of the center of mass.

2. The distribution in the square $\mathscr{D} = \{|x| \leq a, |y| \leq a\}$ given by $\mu(x, y) = x^2 + y^2$.
3. The distribution in the rectangle $\mathscr{D} = \{0 \leq x \leq 1, 0 \leq y \leq 3\}$ given by $\mu(x, y) = x + y$.
4. The distribution in the triangle with vertices $(0, 0)$, $(1, 0)$, $(0, 1)$ given by $\mu(x, y) = (x + y)^2$.

5. In the distribution of Exercise 2, find the first-order moments with respect to the lines $x = 1$ and $y = 3$.

In Exercises 6–9, find the second-order moments about the lines $x = x_0$ and $y = y_0$.

6. Find I_{x_0} and I_{y_0} for the distribution in Exercise 2, with $x_0 = a$ and $y_0 = a$.
7. Find I_{x_0} and I_{y_0} for the distribution in Exercise 2, with $x_0 = 0$, $y_0 = 0$.
8. Find $I_{\bar{x}}$ and $I_{\bar{y}}$ for the distribution in Exercise 3.
9. Find $I_{\bar{x}}$ and $I_{\bar{y}}$ for the distribution in Exercise 4.

10. Find the moment of inertia about the z-axis of the homogeneous solid block, $\mathscr{D} = \{0 \leq x \leq a, 0 \leq y \leq b, 0 \leq z \leq c\}$, of density $\mu(x, y, z) = \delta$.
11. Find the moment of inertia of the plane distribution in Exercise 2 with respect to the center of mass.
12. Find the variance and standard deviation from the mean of the distribution in Exercise 2.
13. Find the variance and standard deviation from the mean of the distribution in Exercise 3.
14. Since for any distribution of mass in the region $\mathscr{D}$, $\iint_{\mathscr{D}} (x - \bar{x}) \, dm = 0$, show that for any x_0, we have the first-order moment

$$\iint_{\mathscr{D}} (x - x_0) \, dm = (\bar{x} - x_0)M,$$

where M is the total mass.

15. As in Exercise 14, show that the second-order moment has the relationship

$$I_{x_0} = \iint_{\mathscr{D}} (x - x_0)^2 \, dm = I_{\bar{x}} + (\bar{w} - x_0)^2 M.$$

16. For the distribution in the solid block $\mathscr{D} = \{|x| \leq a, |y| \leq b, |z| \leq c\}$ given by $\mu(x, y, z) = x^2$, find $M, \bar{x}, \bar{y}, \bar{z}$.
17. For the distribution in Exercise 16, find the moment of inertia with respect to the center of mass.
18. For the distribution in Exercise 16, find the variance and standard deviation from the mean.
19. For a distribution in a region $\mathscr{D}$ given by a density function $\mu(x, y, z)$, suppose that 1 is added to every coordinate. How is the mean affected? How is the standard deviation of each coordinate affected? Prove the conclusion.
20. Find the kinetic energy of rotation of the circular plate $\{x^2 + y^2 \leq a^2\}$ with constant density μ if it rotates about the z-axis at the rate of 2 revolutions per second (see §33.5).
21. Find the rotation θ as a function of time if the circular plate $\mathscr{D} = \{x^2 + y^2 \leq 4\}$ with constant density $\mu = 5$ is subjected to a constant pulling force of 10 pounds applied by a belt around its perimeter (§33.5).
22. An important property of the second-order moment is given by the answer to the following problem. For a distribution of the mass in the region $\mathscr{D}$, find the number s so that $\iint_{\mathscr{D}} (x - s)^2 \, dm$ is a minimum.

23. Show that the coordinates of the mean of a distribution and its standard deviation σ are relatively invariant, that is, that they are homogeneous physical quantities (§9).

33.5   Physical Interpretation of Moment of Inertia. We consider a single particle of mass m revolving about the z-axis in a circle of radius r (Figure 11.19). The distance s along the arc through which it has moved from the start is a function of time. Its kinetic energy is given by

$$\frac{1}{2} mv^2 = \frac{1}{2} m \left(\frac{ds}{dt}\right)^2.$$

The law of motion for it says that

$$m \frac{d^2s}{dt^2} = F,$$

where F is the force acting upon the particle in the tangential direction.

For rotation it is desirable to represent the position of the particle by the angle θ through which the vector $\overrightarrow{OP}$ has turned. Since $s = r\theta$, $ds/dt = r(d\theta/dt)$, the kinetic energy may be expressed in terms of θ by

$$\frac{1}{2} m \left(\frac{ds}{dt}\right)^2 = \frac{1}{2} mr^2 \left(\frac{d\theta}{dt}\right)^2 = \frac{1}{2} I \left(\frac{d\theta}{dt}\right)^2,$$

where $I = r^2 m$. Moreover, the law of motion becomes in angular coordinates

$$mr \frac{d^2\theta}{dt^2} = F.$$

If we multiply both sides by r, this becomes

$$I \frac{d^2\theta}{dt^2} = rF = T,$$

where T is called the *torque*, or first-order moment of force. Thus the formulas for motion in angular coordinates are in the same form as those for distance coordinates if we replace the mass m by the moment of inertia, $I = r^2 m$, and replace the force F by the torque, $T = rF$.

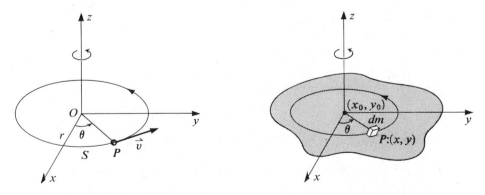

(a)   Revolving particle   (b)   Revolving rigid body

FIGURE 11.19   Moment of inertia in revolutions.

Now we consider a distribution of mass in a flat rigid sheet, revolving in a circle about the line $x = x_0$, $y = y_0$ (Figure 11.19(b)). Each mass element dm at $P:(x, y)$ revolves in a radius $\sqrt{(x - x_0)^2 + (y - y_0)^2}$ and hence has a kinetic energy

$$\frac{1}{2} dm[(x - x_0)^2 + (y - y_0)^2]\left(\frac{d\theta}{dt}\right)^2 = \frac{1}{2} I \left(\frac{d\theta}{dt}\right)^2,$$

if we define

$$I = r^2\, dm = [(x - x_0)^2 + (y - y_0)^2]\, dm.$$

The kinetic energy of a body is the sum of the kinetic energies of its particles, which, in the limit, is the integral

$$\frac{1}{2} \iint_{\mathscr{D}} [(x - x_0)^2 + (y - y_0)^2]\left(\frac{d\theta}{dt}\right)^2 dm.$$

Since $d\theta/dt$ is the same for all particles in the rigidly rotating body, it may be removed outside the integral as a factor, and we express the kinetic energy as

$$\frac{1}{2}\left(\frac{d\theta}{dt}\right)^2 \iint_{\mathscr{D}} [(x - x_0)^2 + (y - y_0)^2]\, dm = \frac{1}{2} I_0 \left(\frac{d\theta}{dt}\right)^2,$$

where

$$I = \iint_{\mathscr{D}} [(x - x_0)^2 + (y - y_0)^2]\, dm$$

is the moment of inertia about the axis $x = x_0$, $y = y_0$. For purposes of calculation this breaks into two parts,

$$I_x = \iint_{\mathscr{D}} (x - x_0)^2\, dm \quad \text{and} \quad I_y = \iint_{\mathscr{D}} (y - y_0)^2\, dm,$$

whose sum $I_x + I_y = I$.

We may proceed in a similar manner to verify that the equation of motion in an angular coordinate extends to the rotation of a solid body, with moment of inertia replacing force and with torque, the first moment of force, replacing force. We summarize by tabulating the analogous formulas.

	Kinetic energy	Equation of motion
Translational motion	$\frac{1}{2} m \left(\frac{ds}{dt}\right)^2$	$m \frac{d^2 s}{dt^2} = F$
Rotation	$\frac{1}{2} I \left(\frac{d\theta}{dt}\right)^2$	$I \frac{d^2 \theta}{dt^2} = T$

Example.   Find the kinetic energies of rotation when the two squares of Section 33.2 are rotated in the plane at one revolution per second about the center of mass.

Solution. For the square with uniformly distributed mass $\mu(x, y) = 1$, we found that the moment of inertia $I = I_{x_0} + I_{y_0} = \frac{8}{3}$. Hence its kinetic energy is given by

$$E = \frac{1}{2} I \left(\frac{d\theta}{dt}\right)^2 = \frac{1}{2} \left(\frac{8}{3}\right) (2\pi)^2 = \frac{16}{3} \pi^2 \text{ energy units.}$$

For the square with mass density $\nu(x, y) = 9x^2y^2$, we found that $I = I_{x_0} + I_{y_0} = \frac{24}{5}$. Hence, when it is rotated at 1 rps, its kinetic energy is given by

$$E = \frac{1}{2} I \left(\frac{d\theta}{dt}\right)^2 = \frac{1}{2} \left(\frac{24}{5}\right) (2\pi)^2 = \frac{48}{5} \pi^2 \text{ energy units.}$$

Hence if the two squares of the same size, same mass, and same center of gravity are rotated at the same speed, the second one would require nearly twice as much brake energy to stop it because its mass is more concentrated at a distance from the axis of rotation.

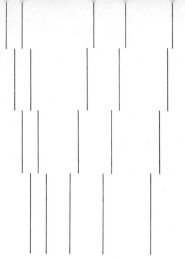

Polar Coordinates

CHAPTER 12

34 Polar Coordinates

34.1 Polar Coordinate System in the Plane. A polar coordinate system is one alternative to the use of rectangular (Cartesian) coordinates to locate points in the plane. We begin with a frame of reference that consists of a fixed point O, called the *pole*, and a fixed ray $\overrightarrow{OX}$ from O, called the *initial line* (Figure 12.1(a)). Polar coordinates (r, θ) of a point P in the plane consist of an angle θ, which the ray $\overrightarrow{OP}$ makes with the initial line, and r, the directed distance from O to P along the radius vector $\overrightarrow{OP}$. We require* that $r \geqq 0$.

Example. Plot points with polar coordinates $P:(2, \pi/3)$, $Q:(2, -5\pi/3)$.

Solution. To plot P (Figure 12.1(b)), we rotate a ray through $+\pi/3$ radians from the initial line and measure 2 units from O to locate P on this terminal ray. We observe that points P and Q are the same, though located by different pairs of polar coordinates.

The preceding example shows that although polar coordinates locate a point uniquely, a point in the plane does not determine its polar coordinates uniquely. Any point P has infinitely many pairs of polar coordinates. If (r, θ) are polar coordinates of P, then P also has the polar coordinates $(r, \theta + 2m\pi)$, where m is any integer (positive or negative). This multiplicity of coordinates for a single point is not possible in a Cartesian coordinate system and is one difficulty to be reckoned with in using polar coordinates. However, some curves have simpler equations in polar coordinates than in Cartesian. In general,

---

* Some books admit negative r coordinates. This produces different graphs for some equations in polar coordinates.

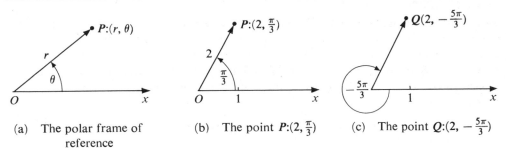

(a)   The polar frame of
reference

(b)   The point P:$(2, \frac{\pi}{3})$

(c)   The point Q:$(2, -\frac{5\pi}{3})$

FIGURE 12.1   Polar coordinates.

polar coordinates work well in configurations generated by rotation of a radius, while Cartesian coordinates work best in straight-line configurations.

34.2   Graphs in Polar Coordinates.   The coordinate lines $r =$ constant and $\theta =$ constant in polar coordinates form a grid of concentric circles and rays orthogonal to them (Figure 12.2). We may use such a polar grid to construct a polar graph of the function $r = \cos \theta$. We calculate a table of coordinate pairs (r, θ) by assigning values to θ, $-\pi/2 \leqq \theta \leqq \pi/2$, and computing r for each θ. This gives us a table of points.

θ	$\frac{2\pi}{3}$	$-\frac{\pi}{2}$	$-\frac{\pi}{3}$	$-\frac{\pi}{6}$	$\frac{0}{0}$	$\frac{\pi}{6}$	$\frac{\pi}{3}$	$\frac{\pi}{2}$	$\frac{2\pi}{3}$
r	~~$\frac{1}{2}$~~	0	$\frac{1}{2}$	$\frac{\sqrt{3}}{2}$	1	$\frac{\sqrt{3}}{2}$	$\frac{1}{2}$	0	~~$\frac{1}{2}$~~

We plot these points in a polar coordinate system (Figure 12.2(b)). The negative values of r when $\pi/2 < |\theta| < 3\pi/2$ are excluded.

We also observe that the same function $r = \cos \theta$ or $y = \cos x$ has an entirely different graph in polar coordinates from its graph in Cartesian (x, y) coordinates.

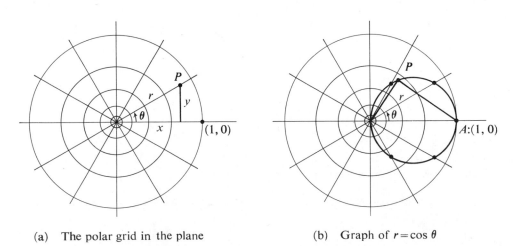

(a)   The polar grid in the plane

(b)   Graph of $r = \cos \theta$

FIGURE 12.2   Graphs in polar coordinates.

34.3   Transformations between Polar and Cartesian Coordinates.   We install both a polar and a Cartesian coordinate system in the plane in such a way that the pole coincides with the Cartesian origin, and the initial ray with the positive ray of the x-axis. We choose equal units; then simple relations may be expressed between the polar Cartesian coordinates of the same point. Indeed simple trigonometry gives us (Figure 12.2(a))

$$x = r \cos \theta \qquad \text{and} \qquad y = r \sin \theta,$$

$$r^2 = x^2 + y^2 \qquad \text{and} \qquad \theta = \arctan \frac{y}{x}, \qquad x \neq 0.$$

Example.   Express in rectangular coordinates the equation $r = \cos \theta$.

Solution.   If $r \neq 0$, the equation $r = \cos \theta$ is true if and only if $r^2 = r \cos \theta$, or $x^2 + y^2 = x$. The Cartesian equation, $x^2 + y^2 = x$, is that of a circle with center at the point where $x = \frac{1}{2}$ and $y = 0$, and radius $\frac{1}{2}$. Therefore, the polar graph of $r = \cos \theta$ (Figure 12.2(b)) is the same circle.

We could have recognized that the graph of $r = \cos \theta$ was a circle directly in terms of polar coordinates by using facts from plane geometry. For the point P traces a circle with diameter OA if and only if the angle $\angle OPA$ is a right angle. In that case the projection of the diameter OA onto the ray OP is given by $r = \text{proj}\,(OA) = 1 \cdot \cos \theta$. Hence P traces a circle with diameter $|OA| = 1$ if and only if $r = \cos \theta$.

34.4   Normal Form of Equation of a Line.   We now investigate the equation of a straight line in polar coordinates. Considering the geometry of polar coordinate systems, it is natural to determine a line L in the plane by drawing a normal $\overrightarrow{ON}$ (Figure 12.3) perpendicular to it and giving the normal intercept, $p = |ON|$, and the angle of inclination α of the normal, measured from the initial line in the same way that θ is measured. The two parameters, p and α, then determine the line L uniquely. We observe that, as defined, $p \leqq 0$.

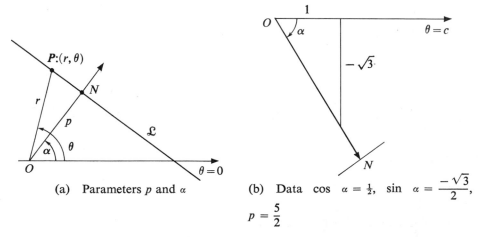

(a)   Parameters p and α

(b)   Data $\cos\ \alpha = \frac{1}{2}$, $\sin\ \alpha = \dfrac{-\sqrt{3}}{2}$,

$p = \dfrac{5}{2}$

FIGURE 12.3   Polar normal.

Now let $P:(r, \theta)$ be any point in the plane. It is clear from the right triangle ONP that the point P is in the line if and only if

$$r \cos (\theta - \alpha) = p, \qquad |\theta - \alpha| \leq \frac{\pi}{2}.$$

This is the polar-normal equation of a line.

34.5   Polar Equation of Conics.   We may define the conics (ellipse, parabola, and hyperbola) by the following variant of the classical definition.

DEFINITION.   Let O be a fixed point (focus) in the plane and $\mathscr{C}$ a circle of radius k with center at O, and L a line through O. Then the point P (Figure 12.4(a)) traces a *conic* with eccentricity e, $e \geq 0$, if and only if the radial distance QP from the circle $\mathscr{C}$ to P is related to the perpendicular distance MP from L to P by

$$QP = e(MP).$$

We now derive an equation of the general conic in polar coordinates, choosing the focus O as the pole and the initial ray perpendicular to the line L. Let the coordinates of the tracing point P be (r, θ). Then since

$$QP = r - k \qquad \text{and} \qquad MP = r \cos \theta,$$

P is on the conic if and only if

$$r - k = er \cos \theta,$$

or

$$r = \frac{k}{1 - e \cos \theta}.$$

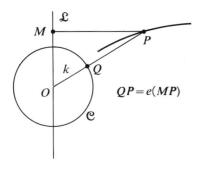

$$QP = e(MP)$$

(a)   The definition

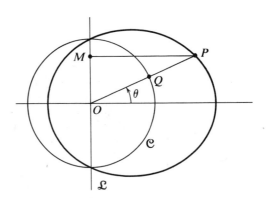

(b)   Ellipse with $k = 4$ and $e = \frac{1}{2}$

FIGURE 12.4   Conic $r = \dfrac{k}{1 - e \cos \theta}$.

This is the polar equation of a conic. We define the particular conics by conditions on e.

If $e = 0$, the conic is a *circle* (the circle $\mathscr{C}$).
If $0 < e < 1$, the conic is an *ellipse*.
If $e = 1$, the conic is a *parabola*.
If $1 < e$, the conic is a *hyperbola*.

We plot the ellipse with $k = 4$ and $e = \frac{1}{2}$ (Figure 12.4(b)).

34.6 Area in Polar Coordinates. There is of course nothing new in differentiating and integrating functions $r = f(\boldsymbol{\theta})$ when the pairs (r, θ) are interpreted as polar coordinates. However, we cannot expect that the derivative $f'(\theta)$ will represent the slope of the tangent line to the polar graph, or that $\int_a^b f(\theta)\, d\theta$ will represent an area bounded by the curve between the rays where $\theta = a$ and $\theta = b$. Let us first examine the exact area bounded by a curve in polar coordinates.

Let $\boldsymbol{f(\theta)}$ be an integrable function on the interval $[a, b]$. Then, just as in rectangular coordinates, we can partition the interval $[a, b]$ into a finite number of subintervals by partition points $a < \theta_1 < \theta_2 < \cdots < \theta_i < \cdots < \theta_n = b$. In each subinterval we can replace $f(\theta)$ by a constant value of f for some number $\bar{\theta}$ in the subinterval. The result is an approximating step function $\boldsymbol{q(\theta)}$ to $\boldsymbol{f(\theta)}$, whose graph is a series of circular arcs of radii $q(\bar{\theta})$ (Figure 12.5). A circular sector, with central angle $d\theta = \theta_i - \theta_{i-1}$ in radians and radius r, has area $\frac{1}{2}r^2\, d\theta$, or $\frac{1}{2}[q(\bar{\theta})]^2\, d\theta$. The sample sum over the partition of such areas (shaded in Figure 12.5) is given by the integral of a step function

$$\int_a^b \tfrac{1}{2}[q(\theta)]^2\, d\theta.$$

We convert this into an exact integral

$$A = \tfrac{1}{2} \int_a^b [f(\theta)]^2\, d\theta,$$

applying the definition of the integral as a limiting sum obtained by ultimate refinement of the partition (I, §19.1). And we conclude that the exact area A bounded by the curve $r = \boldsymbol{f(\theta)}$ between rays $\theta = a$ and $\theta = b$ is given by the integral $A = \int_a^b \tfrac{1}{2}r^2\, d\theta$.

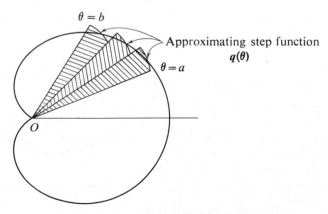

FIGURE 12.5   Area of cardioid $r = k(1 + \cos \theta)$.

Example.   (Area of the cardioid.) The curve whose polar equation is $r = k(1 + \cos \theta)$ is called a *cardioid*. It is a closed curve in which one loop is completed when θ increases from θ to 2π. Find the area enclosed by the cardioid.

Solution.   The area A of the cardioid (Figure 12.5) is given by

$$A = \frac{1}{2} \int_0^{2\pi} [k(1 + \cos \theta)]^2 \, d\theta = \frac{k^2}{2} \int_0^{2\pi} [1 + 2\cos \theta + \cos^2 \theta] \, d\theta$$

$$= \frac{k^2}{2} \int_0^{2\pi} \left[1 + 2\cos \theta \frac{1 + \cos 2\theta}{2} \right] d\theta$$

$$= \frac{k^2}{2} \left[\frac{3\theta}{2} + 2\sin \theta + \frac{1}{4} \sin 2\theta \right]_0^{2\pi} = \frac{3\pi}{2} k^2 \text{ area units.}$$

This completes the problem.

34.7 Exercises

1. The following are pairs (r, θ) of polar coordinates. Plot the points in a polar coordinate system.

(a) $\left(2, \frac{\pi}{3}\right)$;   (b) $\left(3, -\frac{\pi}{2}\right)$;   (c) $(3, 0)$;   (d) $(0, 2\pi)$;

(e) $\left(3, \frac{5\pi}{4}\right)$;   (f) $(1, 4\pi)$;   (g) $\left(2, -\frac{7\pi}{4}\right)$;   (h) $(1, -2\pi)$.

2. Find all of the pairs of polar coordinates of the point $P:\left(1, \frac{\pi}{2}\right)$.

In Exercises 3–14, plot the graphs of the relations in a polar coordinate system.

3. $r = 4\sin \theta$.	4. $r = \sin 2\theta$.
5. $r = \dfrac{8}{1 - \cos \theta}$.	6. $r = \dfrac{8}{2 - \cos \theta}$.
7. $r = \sin^2 \theta$.	8. $r = 4(1 + \sin \theta)$.
9. $r^2 = \cos 2\theta$.	10. $r \sin \theta = 5$.
11. $r \cos (\theta - (\pi/2)) = 2$.	12. $r = 0.2\theta$.
13. $r = 1/\theta$.	14. $r = 4/\theta^2$.

In Exercises 15–20, transform the equations from rectangular coordinates to equivalent relations in polar coordinates.

15. $x^2 + y^2 = 4$.	16. $x = 4$.
17. $y = 3x$.	18. $y = x^2$.
19. $y^2 = 4(x + 1)$.	20. $x^2 + y^2 - 4x = 0$.

In Exercises 21–24, transform from polar to rectangular coordinates.

21. $r = \cos \theta$.	22. $r \cos (\theta - (\pi/2)) = 1$.
23. $\theta = \pi/3$.	24. $r = \dfrac{4}{1 - 2\cos \theta}$.

In Exercises 25–29, find the areas of the regions bounded curves in a polar coordinate system between the indicated limits.

25. One loop of $r = \sin \theta$.
26. $r = 2\theta$ from $\theta = \pi/3$ to $\theta = \pi/2$.
27. One loop of $r = \cos 2\theta$.
28. $r \sin^2 \theta/2 = 1$, $\theta = \pi$, $\theta = \pi/3$.
29. One loop of $r = \sin 3\theta$.

30. The quantity

$$u = \frac{r - k}{r}$$

is called the *relative deviation* of the point $P:(r, \theta)$ from the circle $r = k$ (§34.5). Show that for every conic, $u = e \cos \theta$, and that the relative deviation satisfies the equation of simple harmonic motion $u'' + u = 0$ (§4.3).

31. Show that in rectangular coordinates the normal form of the equation of a line becomes $x \cos \alpha + y \sin \alpha = p$.

32. Using vector notation, we recognize the vector $\vec{n}:(\cos \alpha, \sin \alpha)$ as a unit normal vector to the line. For the line $4x - 3y + 7 = 0$, find the unit normal vector.

33. For the line $ax + by + c = 0$, find the unit normal vector.

34. The classical definition of the conic has a fixed line AB and a fixed point F at distance p from it (Figure 12.6). Then the conic is the locus of points P such that $FP = eMP$. Show that this gives the equation $r = ep/(1 - e \cos \theta)$. Observe that this does not include the circle.

35. A point P moves so that the line segments AP and BP to it, from two fixed points A and B, are always perpendicular. Find a polar equation of the path of P.

36. Find the polar equation of the line through the points $(2, \pi/4)$ and $(3, 2\pi/3)$.

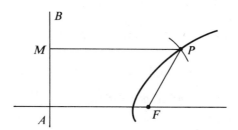

FIGURE 12.6   Directrix AB, Exercise 34.

37. Find the polar equation of the path of a point that moves so that the product of its distances from two fixed points A and B is a constant k^2. Discuss the graph in the cases: $|AB| = k$, $|AB| = 2k$, $|AB| = 4k$.

38. Plot the graph of the equation $\theta^2 - 1 = 0$ in polar coordinates.

In Exercises 39–44, find polar equations of the conics determined by the eccentricity e and radius k of the director circle $\mathscr{C}$. Identify the conic by name and sketch it. Find the areas enclosed by the circles and ellipses.

39. $e = 0, k = 4$.	40. $e = \frac{1}{2},\ k = 4$.
41. $e = 2, k = 4$.	42. $e = 0.1, k = 1$.
43. $e = 1, k = 4$.	44. $e = 3,\ k = 1$.

35   Differentiation in Polar Coordinates

35.1   Choice of Method for Differentiation in Polar Coordinates.   Many graphs are more simply described in polar coordinates than in rectangular coordinates, but the straight line is not one of them. On the other hand, the process of differentiation determines the tangent line as the best linear approximation to a curve at a point on it. Polar coordinates are not well suited to the representation of the tangent line or located

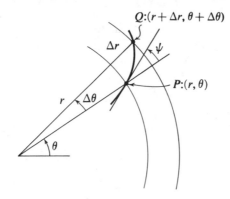

FIGURE 12.7   Differences of polar coordinates.

vectors arising from differentiation. Accordingly, we use a kind of hybrid method for differentiation of curves defined in polar coordinates. We retain the polar coordinates (r, θ) of the points P at which we differentiate but still use essentially rectangular local coordinates (dx, dy) to describe tangent lines and vectors. The numbers $(dr, d\theta)$, $(\Delta r, \Delta \theta)$ occur in differentiation with the same meaning as for Cartesian coordinates; that is, if $P:(r, \theta)$ and $Q:(r + \Delta r, \theta + \Delta \theta)$ are two points, then $(\Delta r, \Delta \theta)$ are the differences of their coordinates (Figure 12.7). In Cartesian coordinates the differences $(\Delta x, \Delta y)$ are local coordinates of Q referred to an origin at P but this is not true for the differences $(\Delta r, \Delta \theta)$ in polar coordinates. In Cartesian coordinates we reserved the symbols (dx, dy) for a point on the tangent line and we do the same for the differences $(dr, d\theta)$.

Another advantage of using Cartesian coordinates for differentiation is that we can apply all of the geometric interpretations of differentiation that we have developed.

35.2   Differential Relations. We think of a smooth curve given in parametric representation by a pair of functions $R:(r(t), \theta(t))$. These determine a rectangular coordinate representation by the relations

$$x = r \cos \theta, \qquad y = r \sin \theta.$$

Differentiating these, we find that (dx, dy) are the local coordinates of a point on the tangent line at R if and only if

$$dx = \cos \theta \, dr - r \sin \theta \, d\theta,$$

and

$$dy = \sin \theta \, dr + r \cos \theta \, d\theta.$$

Squaring and adding, we find that these imply that

$$dx^2 + dy^2 = dr^2 + r^2(d\theta)^2.$$

We know that (dx, dy) are the local coordinates of a point T on the tangent line at a distance ds from R, so that $ds^2 = dx^2 + dy^2$. But the Pythagorean relation

$$(ds)^2 = (dr)^2 + (r \, d\theta)^2$$

implies that dr and $r \, d\theta$ are also sides of a right triangle with hypotenuse ds. We can identify dr (Figure 12.8) as the difference of the r-coordinates of R and T measured along

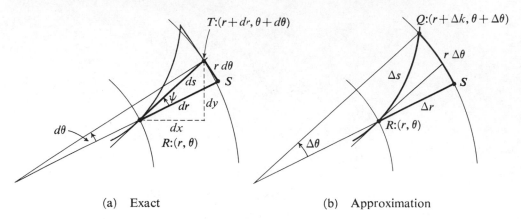

(a)   Exact   (b)   Approximation

FIGURE 12.8   Differential relations.

the ray to R. Thus in the figure, $|RS| = dr$ and $|ST| = r\, d\theta$. This is exact. No approximations are involved.

It is convenient to remember the exact differential triangle RST by the approximate "triangle" RSQ determined by a point $Q{:}(r + \Delta r,\ \theta + \Delta\theta)$ on the curve near P (Figure 12.8(b)). Regarding this figure as a right triangle, we have the approximate relation

$$(\Delta s)^2 = (\Delta r)^2 + (r\,\Delta\theta)^2.$$

Either we can use this to remember the sides of the differential triangle RST or we can use the triangle RST as an approximation to the figure RSQ.

From the differential triangle, we see that if ψ is an angle between the ray to R and the tangent line at R, then

$$\tan\psi = r\,\frac{d\theta}{dr} = \frac{r}{dr/d\theta},$$

$$\cos\psi = \frac{dr}{ds}, \quad\text{and}\quad \sin\psi = \frac{r\,d\theta}{ds}.$$

35.3   Arc Length in Polar Coordinates

Example.   Find the arc length of one loop of the closed curve $r = a(1 + \cos\theta)$.

Solution.   Since $\cos\theta$ is periodic with period 2π, the curve is a loop traced in the interval $0 \leqq \theta \leqq 2\pi$. The arc length was defined (§22.2) by the integral

$$\int_0^{2\pi} \|\vec{r}'\|\, d\theta.$$

And we have found the length of the tangent vector in polar coordinates to be given by $(ds)^2 = (dr)^2 + r^2(d\theta)^2$; that is,

$$\|\vec{r}'\| = \frac{ds}{d\theta} = \sqrt{\left(\frac{dr}{d\theta}\right)^2 + r^2}.$$

For this curve $dr/d\theta = -a\sin\theta$, and therefore

$$\frac{ds}{d\theta} = \sqrt{a^2 \sin^2\theta + a^2(1 + \cos\theta)^2} = 2a\,\sqrt{\frac{1 + \cos\theta}{2}} = 2a\left|\cos\frac{\theta}{2}\right|.$$

To avoid the complications with $|\cos (\theta/2)|$ when $\cos (\theta/2)$ becomes negative, we use symmetry to calculate half the arc in the interval $0 \leq \theta \leq \pi$.

$$\frac{s}{2} = \int_0^{\pi} 2a \cos \frac{\theta}{2} = 4a \sin \frac{\theta}{2} \Big|_0^{\pi} = 4a.$$

Hence the final result is $s = 8a$.

35.4   Examples of a Differentiation

Example A.   Find the angle between the tangent to the cardioid $r = a(1 + \cos \theta)$ and the position vector when $\theta = \pi/2$.

Solution.   From the differential triangle we have

$$\tan \psi = \frac{r}{dr/d\theta}.$$

For the cardioid this gives

$$\tan \psi = \frac{a(1 + \cos \theta)}{-a \sin \theta}$$

and when $\theta = \pi/2$,

$$\tan \psi = -1$$

whence

$$\psi = \frac{3}{4}\pi \quad \text{or} \quad -\frac{\pi}{4}.$$

Example B.   Find the slope of the cardioid $r = a(1 + \cos \theta)$ at the point where $\theta = \pi/2$.

Solution.   The slope is best described in local Cartesian coordinates by dy/dx. We have $x = r \cos \theta$, $y = r \sin \theta$, and for the cardioid these become

$$x = a(1 + \cos \theta) \cos \theta, \quad y = a(1 + \cos \theta) \sin \theta.$$

Differentiating, we find

$$dx = a(-\sin \theta + 2 \cos \theta \sin \theta) \, d\theta,$$

$$dy = a(\sin \theta + \cos^2 \theta - \sin^2 \theta) \, d\theta.$$

Evaluating these where $\theta = \pi/2$, we have $dy/dx = 1$.

Example C.   Find the curve through the point $R:(1, 0)$ that intersects all of the polar rays at the angle $\pi/6$.

Solution.   At every point on the curve, $\tan \psi = 1/\sqrt{3}$. Hence the curve satisfies the differential equation

$$r \frac{d\theta}{dr} = \frac{1}{\sqrt{3}},$$

and the initial condition $r = 1$ when $\theta = 0$. Separating variables, we have

$$\frac{dr}{r} = \sqrt{3} \, d\theta.$$

Integrating gives

$$\ln r = \sqrt{3}\,\theta + C.$$

We determine C from the initial condition and find $C = 0$. Hence the required curve is given by the relation

$$\ln r = \sqrt{3}\,\theta, \quad \text{or} \quad r = e^{\sqrt{3}\,\theta}.$$

It is called a logarithmic spiral.

35.5 Exercises

1. Find the angle between the tangent line and position vector for the curve $r = 2a\cos\theta$ at the point where $\theta = \pi/6$.
2. Find the arc length of one loop of the cardioid, $r = a(1 - \sin\theta)$.
3. Find the arc length of the curve $r = at^2$, $\theta = t$ between points where $t = 0$ and $t = \pi$.
4. Show that if β is an angle between two curves at their point of intersection, then

$$\tan\beta = \frac{\tan\psi_2 - \tan\psi_1}{1 + \tan\psi_2 \tan\psi_1}$$

where ψ_1 and ψ_2 are the angles that their tangent lines make with the position vector at point of intersection.

5. Using the result of Exercise 4, find the angle between the curves, $r = 2a\cos\theta$ and $r\cos(\theta - (\pi/4)) = \sqrt{2}\,a$, at points of intersection.
6. Using the result of Exercise 4, find the angle between the curves, $r = a(1 + \sin\theta)$ and $r = 3a\sin\theta$ at their intersections. Plot the curves.
7. Show that at every point of intersection, the parabolas

$$r = \frac{a}{1 + \cos\theta} \quad \text{and} \quad r = \frac{b}{1 - \cos\theta}$$

intersect at right angles. They form an orthogonal net.

8. Prove that at a point $R:(r, \theta)$ on the curve $r = f(\theta)$, if $r \neq 0$ and $dr/d\theta = 0$, then the tangent line is perpendicular to the polar ray to R.
9. Show that the graph of $r = a(\cos\theta + \sin\theta)$ is a circle and plot it. Find the points where the tangent lines are horizontal.
10. Find the length of the spiral $r = a\theta^2$ between points where $\theta = 0$ and $\theta = \pi$.
11. Sketch the spiral $r = ae^{\sqrt{3}\,\theta}$ (§35.4, Example C). Consider negative angles θ. How does it behave near the origin?
12. Find the length of arc of the spiral of Archimedes $r = a\theta$ for one revolution of the position vector starting at the origin.
13. Find the curve that starts from $(1, 0)$ and intersects every polar ray in an angle ψ such that $\tan\psi = r$.
14. Find the curve traced by a point $R:(r, \theta)$ that starts from $(1, \pi/2)$ and intersects every polar ray in the angle θ.
15. Show that if a family of curves satisfies the differential equation $u(r, \theta)\,dr + v(r, \theta)\,d\theta = 0$, then the orthogonal trajectories satisfy

$$v(r, \theta)\,dr - r^2 u(r, \theta)\,d\theta = 0.$$

16. Use the result of Exercise 15 to find the orthogonal trajectories of the family $r = a(1 + \cos\theta)$ for every positive value of a.
17. Find the orthogonal trajectories of $r = a\cos^2\theta$.

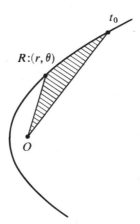

FIGURE 12.9   Exercise 18.

18. Show that on any smooth curve (Figure 12.9) $r = r(t)$, $\theta = \theta(t)$ with time t as parameter, the rate of change of the area A swept out by the position vector is

$$\frac{dA}{dt} = \frac{1}{2} r^2 \frac{d\theta}{dt},$$

even when r is not constant!

19. For the curve $r = f(\theta)$, find the slope dy/dx of the tangent line as a function of θ, as in Example B, §35.4.

36   Motion along a Curved Path

36.1   Plane Motion in Polar Coordinates.   When the position of a moving point $R:(r, \theta)$ in the plane is represented in polar coordinates, the trajectory is given by a position vector function $\vec{r}(t)$, where the parameter t is time. In such a polar representation it is appropriate to resolve the velocity and acceleration vectors into a *radial component* along the line of the position vector $\overrightarrow{OR}$ and a rotational component perpendicular to $\overrightarrow{OR}$. Let $\vec{u}_r$ be a unit vector in the direction of $\vec{r}$, and $\vec{u}_\theta$ a unit vector perpendicular to $\vec{r}$ (Figure 12.10(a)). Then, since we are treating vectors in Cartesian rather than polar representation, the velocity vector $\vec{v}$ at R is representable as the linear combination

$$\vec{v} = v_r\vec{u}_r + v_\theta\vec{u}_\theta,$$

where the scalars v_r and v_θ are called the radial and rotational components of velocity. Similarly, the acceleration vector $\vec{a}$ is representable as

$$\vec{a} = a_r\vec{u}_r + a_\theta\vec{u}_\theta.$$

To compute the coefficients in these representations, we first represent $\vec{u}_r$ and $\vec{u}_\theta$ as position vectors (Figure 12.10(b)). Since they are unit vectors, their ends lie on the unit circle. Moreover, since arc length $s = \theta$ on this circle, $d\vec{u}_r/d\theta$ is a unit tangent vector to the unit circle and is perpendicular to the position vector $\vec{u}_r$. Hence, choosing the sign to be consistent with increasing θ, we have

$$\frac{d\vec{u}_r}{d\theta} = \vec{u}_\theta.$$

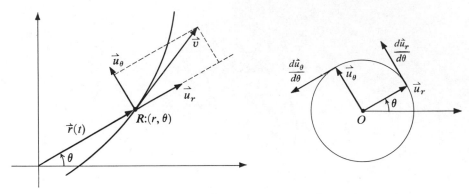

(a) Resolution of velocity (b) $\vec{u}_r$ and $\vec{u}_\theta$ as position vectors

FIGURE 12.10   Radial and rotational components.

Similarly, $d\vec{u}_\theta/d\theta$ is a unit tangent vector at the end of the position vector $\vec{u}_\theta$ and hence is parallel to $\vec{u}_r$. Now $\vec{u}_r \cdot \vec{u}_\theta = 0$ implies that

$$\vec{u}_r \cdot \frac{d\vec{u}_\theta}{d\theta} + \frac{d\vec{u}_r}{d\theta} \cdot \vec{u}_\theta = 0.$$

Substituting $d\vec{u}_r/d\theta = \vec{u}_\theta$, we find that $\vec{u}_r \cdot (d\vec{u}_\theta/d\theta) = -1$. Hence the direction of $d\vec{u}_\theta/d\theta$ is opposite that of $\vec{u}_r$, and we have

$$\frac{d\vec{u}_\theta}{d\theta} = -\vec{u}_r.$$

We now return to the position vector $\vec{r}(t)$ of the moving particle (Figure 12.10). We can express it as the scalar multiple

$$\vec{r}(t) = r\vec{u}_r,$$

where the scalar coordinate $r = |OR|$. We differentiate $\vec{r}(t)$ with respect to t and find

$$\begin{aligned}
\vec{v} = \frac{d}{dt}\,\vec{r}(t) &= \frac{d}{dt}\,r\vec{u}_r = \frac{dr}{dt}\,\vec{u}_r + r\,\frac{d\vec{u}_r}{dt} \\
&= \frac{dr}{dt}\,\vec{u}_r + r\,\frac{d\vec{u}_r}{d\theta} \cdot \frac{d\theta}{dt} \\
&= \frac{dr}{dt}\,\vec{u}_r + r\,\frac{d\theta}{dt}\,\vec{u}_\theta.
\end{aligned}$$

Differentiating again, we have the acceleration

$$\vec{a} = \frac{d^2}{dt^2}\,\vec{r}(t) = \left[\frac{d^2 r}{dt^2} - r\left(\frac{d\theta}{dt}\right)^2\right]\vec{u}_r + \left[r\,\frac{d^2\theta}{dt^2} + 2\,\frac{dr}{dt}\,\frac{d\theta}{dt}\right]\vec{u}_\theta.$$

We summarize the results in a theorem.

THEOREM (Radial and rotational components). If the position vector $\vec{r}(t)$ whose polar coordinates are (r, θ) traces a smooth curve, then the radial and rotational components of velocity v_r, v_θ and acceleration a_r, a_θ are given by

$$v_r = \frac{dr}{dt}, \qquad\qquad v_\theta = r\frac{d\theta}{dt},$$

$$a_r = \frac{d^2r}{dt^2} - r\left(\frac{d\theta}{dt}\right)^2, \qquad a_\theta = r\frac{d^2\theta}{dt^2} + 2\frac{dr}{dt}\frac{d\theta}{dt}.$$

We observe that when the position vector $\vec{r}(t)$ is described in polar coordinates, these components are not simply the first and second derivatives of r and θ.

Example. A particle moves in the curve $r = a(1 - \cos\theta)$ with angular position $\theta = 2\pi t$. Find its radial and rotational components of velocity and acceleration.

Solution. We differentiate with respect to t.

$$\frac{dr}{dt} = a\sin\theta\frac{d\theta}{dt} \qquad \text{and} \qquad \frac{d\theta}{dt} = 2\pi.$$

Hence

$$v_r = \frac{dr}{dt} = 2\pi a\sin\theta \qquad \text{and} \qquad v_\theta = r\frac{d\theta}{dt} = 2\pi a(1 - \cos\theta).$$

Then we compute $d^2\theta/dt^2 = 0$ and

$$\frac{d^2r}{dt^2} = a\cos\theta\left(\frac{d\theta}{dt}\right)^2 + a\sin\theta\frac{d^2\theta}{dt^2} = 4\pi^2a\cos\theta + 0.$$

Hence

$$a_r = \frac{d^2r}{dt^2} - r\left(\frac{d\theta}{dt}\right)^2 = 4\pi^2a\cos\theta - a(1 - \cos t)4\pi^2$$
$$= 4\pi^2a(2\cos\theta - 1),$$

and

$$a_\theta = r\frac{d^2\theta}{dt^2} + 2\frac{dr}{dt}\frac{d\theta}{dt} = 0 + 2a\sin\theta\left(\frac{d\theta}{dt}\right)^2 = 8\pi^2a\sin\theta.$$

Expressing $\vec{v}$ and $\vec{a}$ in terms of the unit radial and rotational vectors, we have (Figure 12.11)

$$\vec{v} = 2\pi a\sin\theta\vec{u}_r + 2\pi a(1 - \cos\theta)\vec{u}_\theta,$$

and

$$\vec{a} = 4\pi^2a(2\cos\theta - 1)\vec{u}_r + 8\pi^2a\sin\theta\vec{u}_\theta.$$

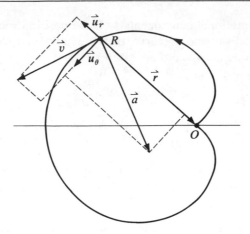

FIGURE 12.11   Motion in orbit $r = a(1 - \cos \theta)$.

36.2 Historical Commentary on Celestial Mechanics. The radial and rotational components of velocity and acceleration bring us close to the scientific objective that motivated Isaac Newton (1642–1727) to invent calculus. He wanted to use the motions of the planets as observed, organized, and recorded by Kepler (1571–1630) as evidence to test his conjectured inverse square law of gravitation. This said that the gravitational force $\vec{F}$ between two masses m_1 and m_2 acts along the line between their centers and has magnitude $\|\vec{F}\| = k(m_1 m_2/r^2)$, where k is a constant and r is the distance between their centers (Figure 12.12). Newton needed calculus for this so that he could write down and solve the equations of motion. Specifically he needed the expressions for radial and rotational components of acceleration to test the inverse square law.

Kepler, in a remarkably accurate condensation of the empirical data, had asserted that:

I.   The planets move in ellipses with sun at the focus.

II. The radius vector $\overrightarrow{OP}$ from sun to planet sweeps out equal areas in equal times.

III. The squares of the periods are proportional to the cubes of the mean diameters of the orbits.

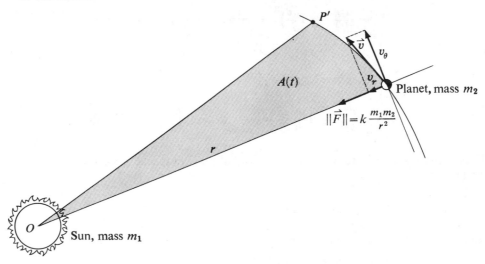

FIGURE 12.12   Newton's law of gravitation.

36.3   Motion under Inverse Square Law. The motion of a planet around the sun can be readily reduced to the following problem of one-particle motion in a plane due to a central force. A particle of mass m moves in the plane under the action of a force that is always directed towards the origin and has magnitude κ/r^2. Find the motion, starting with the position (r_0, θ_0) and velocity $\bar{v}_0$.

Solution. Since the force is directed towards the origin, the rotational component of acceleration is zero,

(1)
$$m\left(r\frac{d^2\theta}{dt^2} + 2\frac{dr}{dt}\frac{d\theta}{dt}\right) = 0.$$

The inverse square law operates on the radial component to give us the other equation of motion

(2)
$$m\left[\frac{d^2r}{dt^2} - r\left(\frac{d\theta}{dt}\right)^2\right] = -\frac{\kappa}{r^2}.$$

We observe that the rotational equation (1) can be written

$$m\frac{d}{dt}\left(r^2\frac{d\theta}{dt}\right) = 0.$$

Integrating this, we find the *area integral*

$$mr^2\frac{d\theta}{dt} = \alpha,$$

where α is the constant of integration. Here we recall that the area swept out by the position vector $\bar{r}$ is given by (§34.6)

$$A = \int_{\theta_0}^{\theta} \tfrac{1}{2}r^2\,d\theta.$$

This implies that $dA/d\theta = \tfrac{1}{2}r^2$. Hence, $dA/dt = (dA/d\theta)(d\theta/dt)$ gives us by virtue of the area integral above,

$$\frac{dA}{dt} = \frac{1}{2}r^2\frac{d\theta}{dt} = \frac{\alpha}{2m}\quad\text{(constant).}$$

This is Kepler's second law. We observe that its derivation does not depend on the inverse square law but only on the fact that the force of gravity is always directed towards the sun. So far, then, it supports only this much of Newton's conjecture.

We can make another use of the area integral to eliminate the time variable in the equation of motion for the radial component and thus obtain the orbit rather than the trajectory. The area integral implies that the derivative operators are related as follows.

$$\frac{d}{dt} = \frac{\alpha}{mr^2}\frac{d}{d\theta} \quad\text{and}\quad \frac{d^2}{dt^2} = \frac{\alpha}{mr^2}\frac{d}{dt}\left(\frac{\alpha}{mr^2}\frac{d}{d\theta}\right).$$

With this observation, we can write the equation of the radial component

$$\frac{\alpha}{r^2}\frac{d}{d\theta}\left(\frac{\alpha}{mr^2}\frac{dr}{d\theta}\right) - \frac{\alpha^2}{mr^3} = -\frac{\kappa}{r^2}.$$

This simplifies under the inverting substitution $r = 1/q$, if we observe that

$$\frac{1}{r^2}\frac{dr}{d\theta} = -\frac{d(1/r)}{d\theta} = -\frac{dq}{d\theta}.$$

The radial equation then becomes

$$-\frac{\alpha^2}{m} q^2 \left(\frac{d^2 q}{d\theta^2} + q\right) = -\kappa q^2 \qquad \text{or} \qquad \frac{d^2 q}{d\theta^2} + q = \frac{\kappa m}{\alpha^2}.$$

Now if we let $u = q - (\kappa m / \alpha^2)$, this becomes

$$\frac{d^2 u}{d\theta^2} + u = 0,$$

the equation of simple harmonic motion (I, §32.4), whose solution is

$$u = B \cos (\theta - \theta_0).$$

Here B and θ_0 are undetermined constants. We restore the variable r, recalling that $u = (1/r) - (\kappa m / \alpha^2)$. With a little simplification this gives us the solution

$$r = \frac{\alpha^2 / m\kappa}{1 + (B\alpha^2 / m\kappa) \cos (\theta - \theta_0)},$$

which we recognize as the equation of a conic (§34.5)

$$r = \frac{b}{1 + e \cos (\theta - \theta_0)},$$

with $b = \alpha^2 / m\kappa$ and $e = B\alpha^2 / m\kappa$.

This derives Kepler's first law from the inverse square law of gravitational force, at least in the case that $e < 1$, when the constants of integration are evaluated by way of the initial conditions. Actually, hyperbolic orbits, $e > 1$, are possible, though by definition planets are captive objects moving in bounded orbits, not in hyperbolic orbits. We omit the derivation of the periods of motion, which can also be deduced from the equations of motion by reinstating the time variable and computing the time for one full circuit. Kepler's third law turned out to be a rather good approximation of the exact period given by the equations of motion. Thus the conjecturing of the law of gravitation, the invention of calculus to make the necessary computations to test it, and the successful solution of the equations of motion, also due to Newton, together formed one of the great landmarks in the history of science.

36.4 Exercises

Exercises 1–11 refer to motion in a circle with center at the origin and radius ρ. The motion is called uniform if the linear speed is constant.

1. Find the radial and rotational components of velocity in uniform motion of 1 revolution per second with $\rho = 4$.
2. Find the radial and rotational components of acceleration in Exercise 1.
3. Find the radial and rotational components of acceleration in a circle of radius 4 at a time $t = 2$ under constant angular acceleration $(d^2\theta/dt^2) = 2\pi$, starting from rest at time $t = 0$.
4. A boy whirls a ball of mass 10 g at the end of a 50-cm string at 2 rps. What pull must be exerted on the string to hold the ball in the circle?
5. Show that a racing car rounding a circular track of radius ρ with linear speed $|v|$ has a radial component of acceleration $P = m|v|^2/\rho$.
6. The track exerts two forces on the car of Exercise 5. A vertical force $w = mg$ supports the weight of the car and a horizontal force P provides the radial component of acceleration. Show that if the track is banked at an angle θ to the vertical given by $\tan \theta =$

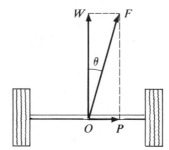

FIGURE 12.13   Exercise 6.

$|v|^2/\rho g$, the sum of the two forces will be perpendicular to the axle and hence there will be no tendency to skid sidewise (Figure 12.13). We observe that this is the correct banking angle for only one speed $|v|$.

7. The relation of the true weight of a particle of mass m at the earth's surface is expressed as $w = mg$. This is the weight a balance would register at the north or south pole, but at the equator, due to the rotation of the earth, the radial acceleration lifts up on the weight and makes the apparent weight W less than the true weight. Show that

$$W = w\left[1 - \frac{\rho(d\theta/dt)^2}{g}\right],$$

where for one revolution per day, we find $d\theta/dt = 7.3 \times 10^{-5}$ rad/sec, radius of the earth $\rho = 6.4 \times 10^6$ meters, $g = 9.80$ meters/sec^2.

8. The differential equation of motion in a plane under a central force $\|\vec{F}\| = -k/r^2$ are

$$ma_r = -\frac{k}{r^2} \quad \text{and} \quad ma_\theta = 0.$$

Show that these differential equations may be satisfied by a uniform circular motion $r = b$ provided that there is an angular velocity

$$\frac{d\theta}{dt} = \frac{k}{mb^3} = \frac{g\rho^2}{b^3}.$$

9. An astronaut is in orbit around the earth in uniform circular motion with his rocket motors turned off. He makes four revolutions per day. How high above the earth is he? Data: See Exercise 7.

10. An astronaut is attempting to maneuver his spacecraft to rendezvous with an unmanned vehicle that is moving in circular orbit with constant speed, making 4 revolutions per day. He finds himself in the same orbit with the same speed but $\frac{1}{10}$ radian behind (Figure 12.14). If he attempts to catch up by turning on his motor to increase his angular acceleration, what will happen?

11. In Exercise 10, the astronaut wishes to dock with the unmanned spacecraft at a time 24 minutes away. Write differential equations that describe a program to accomplish this action using a controlled thrust, $\vec{F}(t)$.

12. The earth orbits the sun in an ellipse with the sun at the focus, moving under the force of the sun's gravitational attraction. Find at what point in its orbit the earth's speed is maximum and at what point it is minimum.

13. In the spiral motion $r = at$, $\theta = \omega t$, find the velocity, acceleration, and speed.

In Exercises 14–17, find the radial and rotational components of velocity and acceleration in the motion.

14. $r = a \sin 2\theta$, $\theta = t^2$.

15. $r = e^{a\theta}$, $\theta = \ln t$, where $|t| > 0$.

16. $r = a(1 + \sin \theta)$, $\theta = 1 - e^{-t}$.

17. $r = e^{\alpha t} + e^{-\alpha t}$, $\theta = t$.

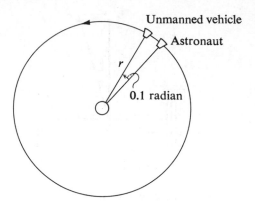

FIGURE 12.14   Exercise 10.

18. Use the results of Exercise 8 to deduce Kepler's third law for *circular* orbits. That is, find the period T and show that it satisfies the relation

$$\frac{T^2}{b^3} = 4\pi^2 \frac{m}{\kappa} \text{ (constant).}$$

19. Show that in any motion in the plane

$$\frac{1}{2} m \left[\left(\frac{dr}{dt} \right)^2 + r^2 \left(\frac{d\theta}{dt} \right)^2 \right] = \frac{1}{2} m |v|^2 \text{ (kinetic energy).}$$

20.* Prove that the equations of planetary motion in the plane (1) and (2) have the integral

$$\frac{1}{2} m \left[\left(\frac{dr}{dt} \right)^2 + r^2 \left(\frac{d\theta}{dt} \right)^2 \right] + \left(-\frac{\kappa}{r} \right) = E \text{ (constant).}$$

What is the physical meaning of this result?

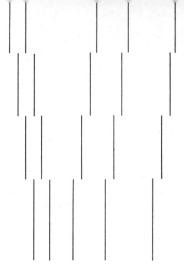

Infinite Series

CHAPTER 13

37 Infinite Series

37.1   Idea of an Infinite Series.   If we "add" the infinite sequence (I, §7) of numbers

$$1, \frac{1}{2}, \frac{1}{4}, \ldots, \frac{1}{2^n}, \ldots,$$

we obtain the formal sum

$$1 + \frac{1}{2} + \frac{1}{4} + \cdots + \frac{1}{2^n} + \cdots,$$

which has no last term. Actually, it is humanly impossible to add infinitely many terms. Yet a simple geometric argument of ancient origin shows that it has meaning to say that the sum of these terms is 2 (Figure 13.1). For we may take the interval of length 2 and bisect it, so that the left-hand half has length 1. Then we add half of the remainder to it, then half of the distance to 2 again, and so on. The sums obtained in this way form an infinite sequence that approaches 2 as a limit.

What we did here was to compute *finite* sums

$$s_1 = 1,\ s_2 = 1 + \frac{1}{2},\ s_3 = 1 + \frac{1}{2} + \frac{1}{4}, \ldots,$$

FIGURE 13.1   Geometric series.

which form an infinite sequence $\{s_n\}$. Then we said that the "sum" of infinitely many terms was given by the sequence limit, $s = \lim_{n \to \infty} s_n$.

In general, a formal sum of an infinite sequence of terms

$$a_1 + a_2 + a_3 + \cdots + a_n + \cdots$$

is called an infinite series. A shorter notation for it is $\sum_{k=1}^{\infty} a_k$, where Greek capital sigma means "sum." The notations in smaller type accompanying the letter $\sum$ (either above and below or immediately to the right) mean that the index k runs over the integers from 1 to ∞, that is, with no last element.

Since we cannot add infinitely many terms, we form the sequence of *partial sums* $\{s_n\}$,

$$s_1 = a_1,$$

$$s_2 = a_1 + a_2,$$

$$\cdot \ \cdot \ \cdot \ \cdot \ \cdot \ \cdot \ \cdot$$

$$s_n = a_1 + a_2 + \cdots + a_n = \sum_{k=1}^{n} a_k,$$

$$\cdot \ \cdot \ \cdot \ \cdot \ \cdot \ \cdot \ \cdot .$$

We define the sum of the series by means of the infinite sequence of partial sums $\{s_n\}$ as follows.

DEFINITION.   The infinite series of real numbers

$$\sum_{k=1}^{\infty} a_k = a_1 + a_2 + \cdots + a_n + \cdots$$

is *convergent* if and only if the sequence of partial sums $\{s_n\}$ is convergent (I, §71). The *sum s*,

$$s = \sum_{k=1}^{\infty} a_k = a_1 + a_2 + \cdots + a_n + \cdots,$$

of a convergent infinite series is the limit of the sequence of partial sums, $s = \lim_{n \to \infty} s_n$.

DEFINITION.   An infinite series $\sum_{k=1}^{\infty} a_k$ *diverges* if and only if its sequence $\{s_n\}$ of partial sums diverges.

Thus the theory of convergence of infinite series introduces nothing new; it is all covered by the theory of convergence of the infinite sequences of partial sums (I, §7).

37.2 A Divergent Series.   It is easy to see that the infinite series

$$1 + 1 + 1 + \cdots + 1 + \cdots$$

diverges. For its sequence of partial sums is given by $s_n = n$, and this unbounded sequence diverges. This leads to guessing the following theorem.

THEOREM.   For every convergent infinite series $\sum_{k=1}^{\infty} a_k$, it must be true that $\lim_{k \to \infty} a_k = 0$.

Proof.   We recall that an infinite sequence $\{s_n\}$ is convergent when for every ϵ there is an N such that if $m > N$ and $n > N$, then $|s_m - s_n| > \epsilon$. In particular, we may take $m = n + 1$. Then

$$|s_m - s_n| = |a_{n+1}| < \epsilon.$$

This shows that for every ϵ there exists an N such that $k > N$ implies

$$|a_k| = |a_k - 0| < \epsilon.$$

That is, $\lim_{k \to \infty} a_k = 0$, which is the conclusion of the theorem.

In this proof, the specialization $m = n + 1$ prevents us from concluding that the series is convergent if $a_k \to 0$.

Example (Harmonic series).   Investigate for convergence the harmonic series

$$\sum_{k=1}^{\infty} \frac{1}{k} = 1 + \frac{1}{2} + \frac{1}{3} + \cdots + \frac{1}{n} + \cdots.$$

Solution.   Although $a_k = 1/k$ and $a_k \to 0$, we may group the terms after the first one as follows,

$$1 + (\tfrac{1}{2}) + (\tfrac{1}{3} + \tfrac{1}{4}) + (\tfrac{1}{5} + \tfrac{1}{6} + \tfrac{1}{7} + \tfrac{1}{8}) + \cdots,$$

so that the kth parenthesis contains 2^{k-1} terms. In these parentheses we replace every term by the last term of the parenthesis, to obtain a smaller comparison series

$$1 + (\tfrac{1}{2}) + (\tfrac{1}{4} + \tfrac{1}{4}) + (\tfrac{1}{8} + \tfrac{1}{8} + \tfrac{1}{8} + \tfrac{1}{8}) + (\tfrac{1}{16} + \cdots) + \cdots.$$

Let S_n denote the partial sum of n terms of the comparison se   It was constructed so that for every n, $S_n \leqq s_n$. But the sequence $\{S_n\}$ is unbounded, s we can see by adding the terms in parentheses of the comparison series

$$1 + (\tfrac{1}{2}) + (\tfrac{1}{2}) + (\tfrac{1}{2}) + (\tfrac{1}{2}) + \cdots.$$

Since $\{S_n\}$ diverges, the larger sequence of the partial sums of the harmonic series diverges also.

Example (*p*-series). The series

$$1 + \frac{1}{2^p} + \frac{1}{3^p} + \cdots + \frac{1}{n^p} + \cdots$$

converges if $p > 1$ and diverges if $0 \leqq p \leqq 1$.

Proof.   Exercises.

37.3   Geometric Progression.   We recall the geometric progression

$$a, ar, ar^2, ar^3, \ldots, ar^{n-1}, \ldots,$$

where $a \neq 0$ and r is a constant. If we form the sum s_n of n terms, we have

$$s_n = a + ar + ar^2 + \cdots + ar^{n-1}.$$

Also,

$$rs_n = \qquad ar + ar^2 + \cdots + ar^{n-1} + ar^n.$$

Subtracting, we find that all except the first and last terms cancel to give us

$$(1 - r)s_n = a - ar^n.$$

Then if, $r \neq 1$, we have the formula for the sum of n terms of the geometric progression,

$$s_n = \frac{a - ar^n}{1 - r}.$$

37.4 Infinite Geometric Series. We seek to determine whether the infinite series

$$\sum_{k=0}^{\infty} ar^k = a + ar + ar^2 + \cdots + ar^{n-1} + \cdots$$

converges. By definition, the series converges if and only if the sequence $\{s_n\}$ of partial sums converges. Using the formula for the sum of n terms of a geometric progression (§37.3), we try to evaluate the limit

$$s = \lim_{n \to \infty} s_n = \lim_{n \to \infty} \left[\frac{a}{1-r} - \left(\frac{a}{1-r} \right) r^n \right]$$

$$= \frac{a}{1-r} - \frac{a}{1-r} \lim_{n \to \infty} r^n.$$

We know (I, §7.5) that $\lim_{n \to \infty} r^n = 0$ if $|r| < 1$. Hence, if $|r| < 1$, the geometric series converges and has the sum $s = a/(1-r)$. If $r = 1$, the sum formula is obviously invalid because it calls for division by zero. In fact, the geometric series with $r = 1$ is

$$a + a + a + a + \cdots,$$

which diverges whenever $a \neq 0$. If $r = -1$, the geometric series is

$$a - a + a - a + \cdots,$$

whose partial sums s_n oscillate between a and 0. Hence this series diverges also, provided that $a \neq 0$.

37.5 Absolute Convergence. It is often difficult to work with series

$$\sum_{k=1}^{\infty} a_k = a_1 + a_2 + a_3 + \cdots + a_n + \cdots$$

that have both positive and negative terms. To eliminate this difficulty, we may consider the series of absolute values

$$\sum_{k=1}^{\infty} |a_k| = |a_1| + |a_2| + |a_3| + \cdots + |a_n| + \cdots.$$

The sequence of partial sums $\{S_n\}$ of the absolute value series is monotonically increasing. This makes it simpler. We readily guess the following theorem.

THEOREM (Absolutely convergent series). For any series $\sum_{k=1}^{\infty} a_k$, if the series of absolute values $\sum_{k=1}^{\infty} |a_k|$ converges to a sum S, then the original series also converges to a sum $s = \sum_{k=1}^{\infty} a_k$, and $|s| \leq S$.

Proof. Since the series of absolute values converges, its sequence of partial sums $\{S_n\}$ converges. Then, by the theorem on the existence of the limit (I, §7.1), for every ϵ there is an N such that if $n > N$ and $m > N$, then $|S_m - S_n| < \epsilon$. But for the series $\sum_{k=1}^{\infty} a_k$, we know that

$$|s_m - s_n| \leq |S_n - S_n|.$$

For

$$|s_m - s_n| = |a_{n+1} + \cdots + a_m| \leq |a_{n+1}| + \cdots + |a_n| = |S_m - S_n|.$$

Hence, by the sufficient condition of the theorem on the existence of the limit, the sequence $\{s_n\}$ converges also.

In view of this result, we say that a series $\sum_{k=1}^{\infty} a_k$ is *absolutely convergent* if the series $\sum_{k=1}^{\infty} |a_k|$ converges.

37.6   Use of Geometric Series as a Comparison Series to Test Convergence

For a series $\sum_{k=1}^{\infty} a_k$, suppose we can show that the absolute values of its terms after some stage N are all less than the corresponding term of a convergent geometric series, that is, that

$$|a_k| \leqq ar^k, \qquad 0 < r < 1.$$

Then every partial sum has the relation

$$S_n = \sum_{k=1}^{n} |a_k| < \frac{a - ar^{n+1}}{1 - r} < \frac{a}{1 - r}.$$

Since the sequence S_n is monotonically increasing and bounded above by the constant $a/(1 - r)$, it converges. Hence, the original series converges.

Example. Test for convergence

$$1 + \frac{1}{2^2} + \frac{1}{3^2} + \cdots + \frac{1}{n^2} + \cdots.$$

Solution. We may group terms as follows:

$$1 + \left(\frac{1}{2^2} + \frac{1}{3^2}\right) + \left(\frac{1}{4^2} + \frac{1}{5^2} + \frac{1}{6^2} + \frac{1}{7^2}\right) + \cdots,$$

where the kth parenthesis has 2^k terms. Then if in each parenthesis we replace each term by the first term of the parenthesis, we form the larger sum

$$1 + \left(\frac{1}{2^2} + \frac{1}{2^2}\right) + \left(\frac{1}{4^2} + \frac{1}{4^2} + \frac{1}{4^2} + \frac{1}{4^2}\right) + \cdots$$

$$= 1 + \frac{2}{4} + \frac{4}{16} + \cdots = 1 + \left(\frac{1}{2}\right) + \left(\frac{1}{2}\right)^2 + \left(\frac{1}{2}\right)^3 + \cdots.$$

This series of larger positive terms is a convergent geometric series. Hence, the smaller original series converges.

The same idea can be applied by comparing the ratio $|a_{n+1}|/|a_n|$ of any term to its predecessor with the ratio r of a convergent geometric series.

THEOREM (Cauchy's ratio test).   In the series $\sum_{k=1}^{\infty} a_k$, if

$$\lim_{n \to \infty} \frac{|a_{n+1}|}{|a_n|} = \rho,$$

then the series converges if $\rho < 1$ and diverges if $\rho > 1$. The test is inconclusive if $\rho = 1$.

Proof. By the definition of

$$\lim_{n \to \infty} \frac{|a_{n+1}|}{|a_n|} = \rho,$$

it follows that if r is chosen so that $\rho < r < 1$, then there exists an N such that $n \geqq N$ implies

$$\frac{|a_{n+1}|}{|a_n|} < r, \quad \text{or} \quad |a_{n+1}| \leqq |a_n| r.$$

For we may take $\epsilon = r - \rho$. Then after some stage N the terms of $\sum_{k=1}^{\infty} |a_k|$ are less than those of the convergent geometric series

$$a_N + a_N r + a_N r^2 + \cdots.$$

This shows that the series $\sum_{k=1}^{\infty} |a_k|$ converges, and therefore the original series $\sum_{k=1}^{\infty} a_k$ does also.

Similarly, if $\rho > 1$, there is a number r such that $1 < r < \rho$. Then after some stage N, we shall have $|a_{n+1}|/|a_n| > r$. That is, the terms $|a_n|$ exceed those of the divergent geometric sequence

$$|a_N|, |a_N|r, |a_N|r^2, \ldots.$$

Hence the series $\sum_{k=1}^{\infty} a_k$ diverges, since $\lim_{k \to \infty} a_k \neq 0$.

We shall prove that the test is indeterminate when $\rho = 1$ by examples. Consider the divergent harmonic series (§37.2). If we apply the ratio test, we find

$$\lim_{n \to \infty} \frac{1/(n+1)}{(1/n)} = \lim_{n \to \infty} \frac{n}{n+1} = \lim_{n \to \infty} \frac{1}{1+(1/n)} = 1.$$

On the other hand, the series

$$1 + \frac{1}{2^2} + \frac{1}{3^2} + \frac{1}{4^2} + \cdots + \frac{1}{n^2} + \cdots$$

converges (see above). Cauchy's ratio test gives

$$\lim_{n \to \infty} \frac{|a_{n+1}|}{|a_k|} = \lim_{n \to \infty} \frac{n^2}{(n+1)^2}$$

$$= \lim_{n \to \infty} \frac{n^2}{n^2 + 2n + 1} = \lim_{n \to \infty} \frac{1}{1 + (2/n) + (1/n^2)} = 1.$$

Another version of comparison with the geometric series is given by the following theorem.

THEOREM (Cauchy's root test).   For the series $\sum_{k=1}^{\infty} a_k$, if

$$\lim_{n \to \infty} \sqrt[n]{|a_n|} = \rho,$$

then the series converges if $\rho < 1$, diverges if $\rho > 1$. The test is indeterminate if $\rho = 1$.

Proof.   (Exercises).

37.7 Exercises

In Exercises 1–4, compute the indicated partial sums.

1. $\sum_{k=0}^{5} a(\tfrac{1}{3})^k.$

2. $\sum_{k=0}^{11} (2k + 3).$

3. $\sum_{k=1}^{n} (-1)^{k-1} k.$

4. $\sum_{k=1}^{n} (2k - 1) = n^2.$

5. Show that the binomial expansion for $(x + y)^n$ can be written

$$(x + y)^n = \sum_{k=0}^{n} \binom{n}{k} x^{n-k} y^k,$$

where

$$\binom{n}{k} = \frac{n!}{k! \, (n - k)!}.$$

6. Find the exact sum of

$$3 + \frac{3}{5} + \frac{3}{25} + \cdots + \frac{3}{5^{n-1}} + \cdots.$$

In Exercises 7–16, find whether the infinite series converges, and where possible, find the exact sum. In each case write out several terms of the series.

7. $\displaystyle\sum_{k=0}^{\infty} \left(\frac{1}{3}\right)^k.$

8. $\displaystyle\sum_{k=0}^{\infty} x^k.$

9. $\displaystyle\sum_{k=1}^{\infty} (-1)^{k-1} \frac{1}{k^2}.$

10. $\displaystyle\sum_{k=1}^{\infty} e^{-k}.$

11. $\displaystyle\sum_{k=1}^{\infty} \frac{1}{10k}.$

12. $\displaystyle\sum_{k=0}^{\infty} \left(\frac{1}{1+i}\right)^k \cdot i > 0.$

13. $\displaystyle\sum_{k=1}^{\infty} \frac{k}{k+1}.$

14. $\displaystyle\sum_{k=1}^{\infty} \frac{1}{2k-1}.$

15. $\displaystyle\sum_{k=1}^{\infty} \frac{5^k}{k!}.$

16. $\displaystyle\sum_{k=1}^{\infty} \frac{k}{3^k}.$

17. A principal sum $\$P$ invested at compound interest rate r for n years accumulates to $\$A$, where $A = P(1 + r)^n$. Hence the present value P of an amount A due n years hence is given by $P = A(1 + r)^{-n}$. An infinite sequence of payments of $\$1$ is due at the end of each year forever, beginning at the end of this year. Under 4% compound interest, what is the sum of the present values of all of the future payments?

18. A nonterminating repeating decimal can be represented as the sum of an infinite series thus:

$$0.8141414 \cdots = 0.8 + 0.014 + 0.00014 + 0.0000014 + \cdots.$$

Find the sum of this series to determine the common fraction represented by this decimal. Show that every repeating decimal represents a common fraction, i.e., rational number.

19. (Paradox of Zeno, c. 500 B.C.) "If the tortoise has the start of Achilles, Achilles can never come up with the tortoise; for, while Achilles traverses the distance (S) from his starting point to the starting point of the tortoise, the tortoise advances a certain distance, and while Achilles is traversing this distance the tortoise makes a further advance, and so on ad infinitum. Consequently, Achilles may run ad infinitum without overtaking the tortoise."

Resolve this paradox in terms of the sum of a convergent infinite series. In particular, if Achilles could run 10 times as fast as the tortoise and could traverse the distance S in 8 seconds, when did he catch the tortoise?

20. A super-ball bounces back after each time it strikes the floor to $\frac{9}{10}$ of the preceding height. If it is initially dropped from a height of 16 ft, when does it stop bouncing (I, §18.4)?

21. Use the groupings

$$1 + \left(\frac{1}{2^p} + \frac{1}{3^p}\right) + \left(\frac{1}{4^p} + \frac{1}{5^p} + \frac{1}{6^p} + \frac{1}{7^p}\right) + \cdots$$

$$\leq 1 + \left(\frac{1}{2^p} + \frac{1}{2^p}\right) + \left(\frac{1}{4^p} + \frac{1}{4^p} + \frac{1}{4^p} + \frac{1}{4^p}\right) + \cdots$$

to investigate the p-series (§37.2) for convergence.

22. Prove the theorem on Cauchy's root test (§37.6).

38 Power Series

38.1 Power Series and the Interval of Convergence.

A series in the form of a sum of powers of x,

$$\sum_{k=0}^{\infty} a_k x^k = a_0 + a_1 x + a_2 x^2 + \cdots + a_n x^n + \cdots,$$

is called a *power series*. It is like an endless polynomial, written in ascending powers. To examine it for convergence, we apply Cauchy's ratio test to the series of absolute values

$$|a_0| + |a_1|\,|x| + |a_2|\,|x|^2 + \cdots + |a_n|\,|x|^n + \cdots.$$

If

$$\lim_{k \to \infty} \frac{|a_{k+1}|\,|x|^{k+1}}{|a_k|\,|x|^k} = \lim_{k \to \infty} \frac{|a_{k+1}|}{|a_k|}\,|x| = \alpha|x|,$$

then the series of absolute values converges if $\alpha|x| < 1$. It is not convergent if $\alpha|x| > 1$. The test affords no information if $|x| = 1/\alpha$. Let $R = 1/\alpha$ if $\alpha \neq 0$, and $R = +\infty$ if $\alpha = 0$. Then the original power series surely converges in the open interval, $-R < x < R$. This is called the *interval of convergence*, and R is called the *radius of convergence*. The series may converge when $x = R$ or $x = -R$ but it surely does not converge if $|x| > R$.

Example.   Find the interval of convergence of the power series

$$1 + \sum_{k=1}^{\infty} (-1)^k \frac{x^k}{k} = 1 - \frac{x}{1} + \frac{x^2}{2} - \frac{x^3}{3} + \cdots + (-1)^{n+1} \frac{x^n}{n} + \cdots.$$

Solution.   We find the limit of the ratio

$$\lim_{k \to \infty} \frac{[1/(k+1)]x^{k+1}}{(1/k)|x|^k} = \lim_{k \to \infty} \frac{k}{k+1}\,|x| = |x| \lim_{k \to \infty} \frac{1}{1 + (1/k)} = |x|.$$

Hence, by the ratio test, the series converges if $|x| < 1$ and does not converge if $|x| > 1$. When $x = 1$, the series becomes the alternating series

$$1 - \frac{1}{2} + \frac{1}{3} + \cdots + (-1)^k \frac{1}{k} + \cdots,$$

which converges (I, §7.7, Exercise 12). When $x = -1$, the series becomes the harmonic series, which does not converge. Hence the interval of convergence is $\{-1 < x \leq 1\}$. This completes the problem. It is customary to represent this graphically by plotting the interval of convergence in heavy black on the x-axis (Figure 13.2). We note the exclusion of -1 from the interval of convergence.

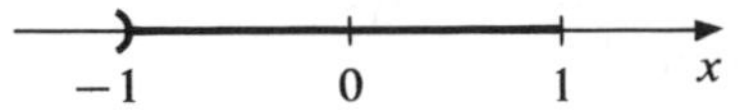

FIGURE 13.2   Interval of convergence.

We observe that the interval of convergence of a power series in powers of x always has a center where $x = 0$. A power series whose convergence interval has a different center, where $x = a$, can be written in powers of $x - a$, that is,

$$\sum_{k=0}^{\infty} a_k(x - a)^k.$$

This adaptation is easy, so we will continue to study the simpler form $\sum a_k x^k$.

38.2   Functions Represented by Power Series. If the power series $\sum_{k=0}^{\infty} a_k x^k$ converges in the closed interval $-\rho \leq x \leq \rho$, its sum is a function of x defined on the interval $[-\rho, \rho]$. We write

$$f(x) = a_0 + a_1 x + a_2 x^2 + \cdots + a_n x^n + \cdots,$$

and say that $f(x)$ is represented by the power series.

It is only a short logical step from the idea of Taylor's polynomial, with remainder (I, §14.4), for $f(x)$ to the idea of the power series that represents $f(x)$. We recall that Taylor's theorem says that a repeatedly differentiable function is given in a neighborhood of the origin by

$$f(x) = f(0) + \frac{f'(0)}{1!} x + \frac{f''(0)}{2!} x^2 + \cdots + \frac{f^{(n)}(0)}{n!} x^n + R_{n+1}(x).$$

The remainder term

$$R_{n+1}(x) = \frac{f^{(n+1)}(z)}{(n + 1)!} x^{n+1},$$

evaluated at some unspecified point z between 0 and x. If we know $f(x)$ in advance and can prove that

$$\lim_{n \to \infty} R_{n+1}(x) = 0,$$

then we say that $f(x)$ is represented by the *infinite* Taylor's series

$$f(x) = f(0) + \frac{f'(0)}{1!} x + \frac{f''(0)}{2!} x^2 + \cdots + \frac{f^{(n)}(0)}{n!} x^n + \cdots.$$

Example.   Represent **sin** x by an infinite power series and find its interval of convergence to the sum sin x.

Solution.   We calculate by Taylor's theorem (I, §34) that for an odd number n,

$$\sin x = x - \frac{x^3}{3!} + \frac{x^5}{5!} + \cdots + (-1)^{(n-1)/2} \frac{x^n}{n!} + R_{n+1}(x),$$

where

$$R_{n+1}(x) = (-1)^{(n+1)/2} \cos z \frac{x^{n+1}}{(n + 1)!}.$$

We do not know what z is, but we know that whatever it is, $|\cos z| \leqq 1$. Therefore we have the estimate of the remainder,

$$|R_{n+1}(x)| \leqq \frac{|x|^{n+1}}{(n+1)!}.$$

To see that $\lim_{n\to\infty} R_{n+1}(x) = 0$, we may fix $x = x_0$ and take M larger than $|x_0|$. Then if $n > M$ and $|x| \leqq |x_0|$,

$$|R_{n+1}(x)| \leqq \frac{|x|^{n+1}}{(n+1)!} \leqq \left|\frac{x_0}{M}\right|^{n+1}.$$

The term on the right is of the form r^{n+1}, where $r = |x_0/M| < 1$. Since we know that $\lim_{n\to\infty} r^n = 0$ when $r < 1$ (I, §7.5), it follows that

$$\lim_{n\to\infty} R_{n+1}(x) = 0.$$

That is, the Taylor's series for $\sin x$ converges to the sum $\sin x$ for every x such that $|x| \leqq |x_0|$. But since we can choose x_0 as large as we please, this implies that the power series converges to the sum, $\sin x$, for every real number x. That is

$$\sin x = x - \frac{x^3}{3!} + \frac{x^5}{5!} + \cdots + (-1)^{k+1}\frac{x^{2k-1}}{(2k-1)!} + \cdots,$$

exactly, without remainder.

While this is elegant, we observe that the finite Taylor's polynomial, taken together with the estimate of the remainder, is still what we need to compute $\sin x$ with prescribed accuracy.

38.3 Geometrically Dominated Power Series.

Power series representations are especially useful when we know $f(x)$ only by its power series representation

$$f(x) = a_0 + a_1 x + a_2 x^2 + \cdots + a_n x^n + \cdots,$$

and have no knowledge of $f(x)$ that would enable us to compute the remainder of Taylor's polynomial. Can we use such power series representations to compute $f(x)$, to add, multiply, and divide such functions? Can we differentiate them and integrate them term by term as if they were polynomials? To make this work, and obtain an estimate of the remainder after n terms, we introduce the idea of geometrically dominated power series.

DEFINITION.   A power series

$$a_0 + a_1 x + a_2 x^2 + \cdots + a_n x^n + \cdots$$

is said to be *geometrically dominated* in the closed interval $[-\rho, \rho]$ if and only if there is some geometric series

$$a + ar + ar^2 + \cdots + ar^n + \cdots$$

of constant terms, with $0 < r < 1$, such that for every x in $[-\rho, \rho]$ and every power n,

$$|a_n x^n| \leqq ar^n.$$

That is, term by term, the absolute values of the terms of the power series are less than those of a convergent geometric series. If the power series is geometrically dominated, then for the remainder, $R_{n+1}(x)$,

$$|R_{n+1}(x)| \leqq ar^{n+1} + ar^{n+2} + \cdots = \frac{ar^{n+1}}{1-r},$$

and $R_{n+1}(x) \to 0$. Similarly for the sum of n terms $s_n(x)$,

$$|s_n(x)| \leqq |a_0| + |a_1|\,|x| + \cdots + |a_n|\,|x|^n \leqq a + ar + \cdots + ar^n < \frac{a}{1-r}.$$

Thus, if we know the a and r of a dominating geometric series, we can estimate the remainder and set a definite upper bound on the partial sums.

THEOREM. If R is the radius of convergence of the power series $\sum_{k=0}^{\infty} a_k x^k$, and $0 < \rho < R$, then the power series is geometrically dominated in the interval $[-\rho, \rho]$.

Proof. We pick x_0 so that $0 \leqq \rho < x_0 < R$. Thus x_0 is in the interval of absolute convergence established by Cauchy's ratio test. Therefore there is an exact sum a of the series

$$a = |a_0| + |a_1|\,|x_0| + \cdots + |a_n|\,|x_0|^n + \cdots.$$

Since the terms are all positive, it follows that for each one separately, $|a_n|\,|x_0|^n < a$. Now with this number a and x in the closed interval $[-\rho, \rho]$, we consider

$$|a_n|\,|x|^n \leqq |a_n|\rho^n = |a_n| \left|\frac{\rho}{x_0}\right|^n |x_0|^n \leqq a \left|\frac{\rho}{x_0}\right|^n.$$

This shows that we may choose $r = |\rho/x_0|$, $0 < r < 1$, and for every n, $|a_n|\,|x^n| < ar^n$. That is, the series $\sum_{k=0}^{\infty} a_k x^k$ is geometrically dominated in the interval $[-\rho, \rho]$.

This theorem shows that not much generality is lost by restricting our power series to those that are geometrically dominated in an interval $[-\rho, \rho]$. For the interval $[-\rho, \rho]$ in which the series is geometrically dominated may be made to include any point where the series converges, except possibly $x = -R$ and $x = R$ at the ends of the interval of convergence.

We observe that the proof is valid for any number a such that for every n, $|a_n|\,|x_0|^n < a$.

38.4   Exercises

In Exercises 1–6, find the interval of convergence of the power series given and plot it.

1. $1 + 2x + 4x^2 + \cdots + 2^k x^k + \cdots$.
2. $x + x^4 + x^9 + x^{16} + \cdots + x^{k^2} + \cdots$.
3. $1 - x + x^2 - x^4 + \cdots + (-1)^k x^k + \cdots$.
4. $x + \dfrac{x^2}{\sqrt{2}} + \dfrac{x^3}{\sqrt{3}} + \cdots + \dfrac{x^k}{\sqrt{k}} + \cdots$.
5. $\displaystyle\sum_{k=1}^{\infty} \frac{(10x)^k}{k!}$.   6. $\displaystyle\sum_{k=0}^{\infty} (-1)^k \frac{x^{2k}}{(2k)!}$.

In Exercises 7–10, expand in a Taylor's series in powers of x and determine the interval of convergence.

7. e^{2x}.   8. $\cos x$.
9. $\ln(1 + x)$.   10. $1/(1 + x^2)$. (*Hint:* Divide.)

11. A power series $\sum_{k=0}^{\infty} a_k x^k$ is geometrically dominated in the interval $[-1, 1]$ by $a + ar + ar^2 + \cdots + ar^n + \cdots$, where $a = 10$, $r = \frac{1}{2}$. How many terms will assure that $s_n(x)$ gives the sum of the series with error less than $\epsilon = 0.001$?

12. Construct a geometric series that geometrically dominates the series

$$1 + \frac{x}{1} + \frac{x^2}{2} + \frac{x^3}{3} + \cdots + \frac{x^n}{n} + \cdots$$

in the interval $[-\frac{1}{2}, \frac{1}{2}]$.

13. In what intervals $[-\rho, \rho]$ is the series of Exercise 12 geometrically dominated?

14. Find the intervals in which the series of Exercise 6 is geometrically dominated.

15. The series $\sum_{k=0}^{\infty} a_k x^k$ has each coefficient determined by flipping a coin, $a_k = +1$ if the coin comes up heads, $a_k = -1$ if tails. Find the interval of convergence.

16. Find the interval of convergence of $\sum_{k=0}^{\infty} (x + 2)^k$.

17. Find the interval of convergence of

$$\sum_{k=0}^{\infty} \frac{(x - 1)^k}{2^k(k + 1)}.$$

18.* Find the interval of convergence of the binomial series

$$\sum_{k=0}^{\infty} \binom{n}{k} x^k.$$

39 Operations with Power Series

39.1 Relation to Polynomial Operations.
Intuitively, power series are just extended polynomials, written in ascending powers of x. It is plausible, then, that we can add, multiply, divide, differentiate, and integrate power series term by term, just as if they were polynomials. This intuition can be proved correct if the power series involved are all geometrically dominated (§38.3) in the same interval.

THEOREM (Operations with power series). Let $f(x)$ and $g(x)$ be two functions defined by power series

$$f(x) = a_0 + a_1 x + a_2 x^2 + \cdots + a_n x^n + \cdots,$$
$$g(x) = b_0 + b_1 x + b_2 x^2 + \cdots + b_n x^n + \cdots,$$

which are both geometrically dominated in the interval $[-\rho, \rho]$. Then the following conclusions are true.

(a) $f(x)$ and $g(x)$ are continuous on $[-\rho, \rho]$.

(b) $f(x) + g(x) = (a_0 + b_0) + (a_1 + b_1)x + \cdots + (a_n + b_n)x^n + \cdots$.

(c) The product $f(x)g(x)$ is represented by

$$f(x)g(x) = a_0 b_0 + (a_0 b_1 + a_1 b_0)x + (a_0 b_2 + a_1 b_1 + a_2 b_0)x^2 + \cdots,$$

which is formed by multiplying the series as polynomials.

(d) $f(x) = g(x)$ on $[-\rho, \rho]$ if and only if

$$a_0 = b_0, a_1 = b_1, \ldots, a_n = b_n, \ldots.$$

(e) If $g(x) \neq 0$ for every x in $[-\rho, \rho]$, then

$$\frac{f(x)}{g(x)} = \frac{a_0 + a_1 x + \cdots}{b_0 + b_1 x + \cdots} = \frac{a_0}{b_0} + \frac{a_1 b_0 - a_0 b_1}{b_0^2} x + \cdots,$$

where the quotient power series is formed by dividing the power series of f by that of g by polynomial division.

(f)   The function f is integrable on $[-\rho, \rho]$ and the integral $\int_0^x f(u)\,du$ is given by performing term-by-term integration on the power series of f,

$$\int_0^x f(u)\,du = a_0 x + a_1 \frac{x^2}{2} + a_2 \frac{x^3}{3} + \cdots + a_n \frac{x^{n+1}}{n+1} + \cdots.$$

(g)   The function f is differentiable on $[-\rho, \rho]$ and its derivative is given by term-by-term differentiation of the series for $f(x)$,

$$f'(x) = 0 + a_1 + 2a_2 x + \cdots + na_n x^{n-1} + \cdots.$$

39.2   Examples of Power Series Operations.   Before we prove the theorem, we work a few examples.

For the functions defined by power series

$$\ln(1+x) = x - \frac{x^2}{2} + \frac{x^3}{3} + \frac{x^4}{4} + \cdots + (-1)^n \frac{x^n}{n} + \cdots,$$

$$e^{-x^2} = 1 - x^2 + \frac{x^4}{2!} + \cdots + (-1)^n \frac{x^{2n}}{n!} + \cdots,$$

find power series representations for the following.

Example A. $e^{-x^2} \ln(1+x)$.

Solution.   By the ratio test the interval of convergence of the series for $\ln(1+x)$ is $-1 < x < 1$. Similarly, e^{-x^2} converges in the interval $-\infty < x < \infty$. If we take any closed interval $[-\rho, \rho]$, with $\rho < 1$, then each of the two series is geometrically dominated in this interval.

We then multiply them as polynomials.

$$
\begin{aligned}
\ln(1+x) \quad &= x - \frac{x^2}{2} + \frac{x^3}{3} \qquad\qquad\qquad - \cdots \\[2mm]
e^{-x^2} &= 1 - x^2 \qquad\quad + \frac{x^4}{2!} \qquad + \cdots \\[1mm]
\hline\\[-2mm]
&\quad\ x - \frac{x^2}{2} + \frac{x^3}{3} - \frac{x^4}{4} \\[2mm]
&\qquad\qquad\qquad\ - x^3 + \frac{x^4}{2} \qquad + \cdots \\[2mm]
&\qquad\qquad\qquad\qquad\qquad\ + \frac{x^5}{2!} + \cdots \\[1mm]
\hline\\[-2mm]
\ln(1+x)e^{-x^2} &= x - \frac{x^2}{2} - \frac{2x^3}{3} + \frac{x^4}{4} \qquad + \cdots,
\end{aligned}
$$

which converges and gives the correct product in the interval $[-\rho, \rho]$.

Example B.   Compute $\int_0^x e^{-u^2}\,du$.

Solution.

$$\int_0^x e^{-u^2} \, du = \int_0^x \left(1 - u^2 + \frac{u^4}{2!} + \cdots + (-1)^n \frac{u^{2n}}{n!} + \cdots \right) du$$

$$= x - \frac{x^3}{3} + \frac{x^5}{10} + \cdots + (-1)^n \frac{x^{2n+1}}{(2n+1)n!} + \cdots .$$

The integral power series converges and represents $\int_0^x e^{-u^2} \, du$ in any interval $[-\rho, \rho]$.

Example C.   Differentiate $\ln(1 + x)$ by means of its power series.

Solution.

$$\frac{d}{dx} \ln(1 + x) = \frac{d}{dx} \left[x - \frac{x^2}{2} + \frac{x^3}{3} - \frac{x^4}{4} + \cdots (-1)^n \frac{x^n}{n} + \cdots \right]$$

$$= 1 - x + x^2 - x^3 + \cdots (-1)^n x^n + \cdots .$$

The derived series converges in the interval $[-\rho, \rho]$ and gives the derivative of $\ln(1 + x)$, provided that $\rho < 1$.

Example D.   Find a function $f(x)$, having $f(0) = 1$, whose derivative $f'(x)$ is formed by multiplying $f(x)$ by x.

Solution.   We will seek the unknown function $f(x)$ satisfying the differential equation,

$$f'(x) = xf(x), \qquad f(0) = 1,$$

as a power series

$$f(x) = a_0 + a_1 x + a_2 x^2 + \cdots + a_n x^n + \cdots$$

that is geometrically dominated in some interval $[-\rho, \rho]$. The coefficients a_k remain to be determined. Then in the interval $[-\rho, \rho]$

$$xf(x) = a_0 x + a_1 x^2 + \cdots + a_{n+1} x^n + a_n x^{n+1} + \cdots ,$$

$$f'(x) = a_1 + 2a_2 x + 3a_3 x^2 + \cdots + (n + 1)a_{n+1} x^n + (n + 2)a_{n+2} x^{n+1} + \cdots .$$

These power series are equal if and only if the coefficients of like powers of x are equal. Since a_n is the coefficient of x^{n+1} in the series for $xf(x)$, we find that

$$(n + 2)a_{n+2} = a_n, \qquad n = 0, 1, 2, \ldots .$$

This is called a *recursive relation.* For any a_n it gives a_{n+2}.

Since we know that $a_0 = f(0) = 1$, the recursive relation gives, with $n = 0$, $2a_2 = a_0 = 1$, or $a_2 = \frac{1}{2}$. Then, with $n = 2$, it gives $4a_4 = a_2 = \frac{1}{2}$, or $a_4 = \frac{1}{8}$. With $n = 4$, it gives $a_6 = \frac{1}{48}$. Then $a_8 = \frac{1}{384}$, and so on. We can use the recursive relation to compute the odd-numbered coefficients also, if we can find a_1 to start it. By equating the constant terms in the series for $xf(x)$ and $f'(x)$ we find that $a_1 = 0$. Hence, by the recursive relation, all odd coefficients are zero, $a_3 = 0$, $a_5 = 0$, $a_7 = 0, \ldots$. Hence the solution is formally given by the power series

$$f(x) = 1 + \frac{x^2}{2} + \frac{x^4}{8} + \frac{x^6}{48} + \frac{x^8}{384} + \cdots ,$$

where $a_{n+2} = 1/(n + 2)a_n$. This recursive relation tells us that for every x,

$$\lim_{n \to \infty} \frac{|a_{n+2} x^{n+2}|}{|a_n x^n|} = \lim_{n \to \infty} \frac{1}{n + 2} |x|^2 = 0.$$

Hence, by the ratio test, the solution series converges and is geometrically dominated in every interval $[-\rho, \rho]$. This result, and the theorem on operations with power series, then justify the operations with power series that we used to compute the solution series, namely, differentiation, multiplication by x, and equating of coefficients. This completes the example.

39.3 Exercises

1. If $f(x)$ and $g(x)$ are given by power series

$$f(x) = \frac{x}{2} + \frac{x^2}{2 \cdot 3} + \frac{x^3}{3 \cdot 4} + \cdots + \frac{x^n}{n(n+1)} + \cdots,$$

and

$$g(x) = 1 - 2x + \frac{2^2 x^2}{2!} - \frac{2^3 x^3}{3!} + \cdots + (-1)^n \frac{2^n x^n}{n!} + \cdots,$$

find all intervals in which both series are geometrically dominated.

In Exercises 2–5, perform the indicated operations on the power series given in Exercise 1.

2. $\dfrac{d}{dx} f(x)$.   3. $\displaystyle\int_0^x f(u)\, du$.

4. $f(x) + g(x)$.   5. $f(x)g(x)$.

6. Solve the differential equation

$$\frac{d}{dx} f(x) = f(x), \qquad f(0) = 1,$$

for $f(x)$ as a power series. Show that the power series result is the same as $f(x) = e^x$.

7. The power series for functions $p(x)$ and $q(x)$ are both dominated by $a + ar + ar^2 + \cdots + ar^n + \cdots$, where $a = 10$ and $r = \frac{1}{2}$. How many terms of the series, with error less than 0.001, will compute (a) $p(x) + q(x)$, (b) $p(x)q(x)$?

8. For the function $p(x)$ in Exercise 7, how many terms of the integral series will compute $\int_0^x p(u)\, du$ with error less than 10^{-4}? (See the proof §39.4 (6).)

9. In Example C, §39.2, show that the derived power series,

$$D \ln (1 + x) = D \left(x - \frac{x^2}{2} + \frac{x^3}{3} + \cdots \right) = 1 - x + x^2 + \cdots + (-1)^n x^m + \cdots,$$

actually gives the derivative $D \ln (1 + x) = 1/(1 + x)$.

10. Compute a power series in powers of x for $\ln (1 + x + x^2)$. What is the interval of convergence?

11. The coefficients of the power series $a_0 + a_1 x + a_2 x^2 + \cdots$ are determined by tossing a pair of dice, so that the a_n's are randomly selected integers between 2 and 12. How many terms will compute the sum $s(\frac{1}{2})$ with error less than 10^{-3}?

39.4 (Theory)

Proof of the Theorem on Operations with Power Series. We now look in some detail at the use of dominating geometric series to validate operations with power series. If $f(x)$ is given in the interval $[-\rho, \rho]$, by a geometrically dominated power series, $\sum_{k=0}^{\infty} a_k x^k$, on $[-\rho, \rho]$, we can write $f(x) = s_n(x) + R_{n+1}(x)$, where $s_n(x)$ is the polynomial

$$s_n(x) = a_0 + a_1 x + a_2 x^2 + \cdots + a_n x^n,$$

and $R_{n+1}(x)$ is the remainder of the series,

$$R_{n+1}(x) = a_{n+1} x^{n+1} + a_{n-2} x^{n+2} + \cdots.$$

Moreover the domination by a geometric series,

$$a + ar + \cdots + ar^n + \cdots, \qquad 0 \le r < 1,$$

assures that for every x in the interval $[-\rho, \rho]$,

$$|R_{n+1}(x)| < \frac{ar^{n+1}}{1 - r}$$

(§38.3).

1. To prove that $f(x)$ is continuous at c in $[-\rho, \rho]$ we must show that for every ϵ there is a δ such that if $|x - c| < \delta$, then $|f(x) - f(c)| < \epsilon$ (§8.3). When $f(x)$ is given by the series, we can express $|f(x) - f(c)|$ as

$$
\begin{aligned}
|f(x) - f(c)| &= |s_n(x) - s_n(c) + R_{n+1}(x) - R_{n+1}(c)| \\
&\le |s_n(x) - s_n(c)| + |R_{n+1}(x)| + |R_{n+1}(c)|.
\end{aligned}
$$

Now we choose N sufficiently large so that if $n > N$,

$$|R_{n+1}(x)| < \frac{\epsilon}{4}.$$

For the same n,

$$|R_{n+1}(c)| < \frac{\epsilon}{4}.$$

Then for $n > N$ and every x in $[-\rho, \rho]$ we have

$$|f(x) - f(c)| < |s_n(x) - s_n(c)| + \frac{\epsilon}{4} + \frac{\epsilon}{4}.$$

Finally, with n fixed, $n > N$, we use the fact that the polynomial $s_n(x)$ is continuous (I, §8.3). Hence, for every positive number $\epsilon/2$, there is a δ such that when $|x - c| < \delta$, then

$$|s_n(x) - s_n(c)| < \frac{\epsilon}{2}.$$

Substituting this result into the expression for $|f(x) - f(c)|$, we find that when $|x - c| < \delta$, then

$$|f(x) - f(c)| < \frac{\epsilon}{2} + \frac{\epsilon}{4} + \frac{\epsilon}{4} = \epsilon.$$

Hence the sum function is continuous in $[-\rho, \rho]$.

2. When two functions $f(x)$ and $g(x)$ are given by geometrically dominated power series, we can find one convergent geometric series, $a + ar + ar^2 + \cdots$, that dominates both of them. Then we write

$$f(x) = (a_0 + a_1 x + \cdots + a_n x^n) + R_{n+1}(x) = s_n(x) + R_{n+1}(x),$$

and

$$g(x) = (b_0 + b_1 x + \cdots + b_n x^n) + T_{n+1}(x) = t_n(x) + T_{n+1}(x).$$

Since both series are geometrically dominated by the same geometric series, $\sum_{k=0}^{\infty} ar^k$, we have

$$|s_n(x)| \leq \frac{a}{1-r}, \qquad |R_{n+1}(x)| < \frac{ar^{n+1}}{1-r},$$

$$|t_n(x)| \leq \frac{a}{1-r}, \qquad |T_{n+1}(x)| < \frac{ar^{n+1}}{1-r}.$$

Now if we form the sum, we find that

$$f(x) + g(x) = (a_0 + b_0) + (a_1 + b_1)x + \cdots$$
$$+ (a_n + b_n)x^n + [R_{n+1}(x) + T_{n+1}(x)].$$

But for the remainder of the sum

$$|R_{n+1}(x) + T_{n+1}(x)| \leq \frac{ar^{n+1}}{1-r} + \frac{ar^{n+1}}{1-r} = \frac{2ar^{n+1}}{1-r}.$$

Hence the sum series is still geometrically dominated by a convergent geometric series, and its remainder approaches zero as $n \to \infty$. That is, we can add $f(x)$ and $g(x)$ by adding their power series term by term.

3. We consider the product of

$$f(x) = s_n(x) + R_{n+1}(x) \qquad \text{and} \qquad g(x) = t_n(x) + R_{n+1}(x).$$

Then, multiplying, we find

$$f(x)g(x) = s_n(x)t_n(x) + [R_{n+1}(x)t_n(x) + T_{n+1}(x)s_n(x) + R_{n+1}(x)T_{n+1}(x)].$$

Since both series are geometrically dominated by the same convergent geometric series $a + ar + ar^2 + \cdots$, the remainder term can be estimated in absolute value by

$$|R_{n+1}t_n + T_{n+1}s_n + R_{n+1}T_{n+1}| \leq \frac{ar^{n+1}}{1-r}\left[\frac{a}{1-r} + \frac{a}{1-r} + \frac{ar^{n+1}}{1-r}\right]$$

$$= \left[\frac{a}{1-r}\right]^2 [2 + r^{n+1}]r^{n+1}.$$

Since $r^{n+1} \to 0$ as $n \to \infty$, this remainder approaches zero. Hence the product of partial sums $\{s_n(x)t_n(x)\}$ converge to the limit $f(x)g(x)$. We omit the detail of showing that we can replace $\{s_n(x)t_n(x)\}$ by the partial sums of the product series arranged in powers of x, and find that

$$f(x)g(x) = a_0b_0 + (a_0'b_1 + a_1b_0)x + \cdots$$
$$+ (a_nb_0 + a_{n-1}b_1 + \cdots + a_0b_n)x^n + \cdots.$$

4. We shall prove that if $f(x) \equiv 0$ and it is represented by a geometrically dominated power series in the interval $[-\rho, \rho]$,

$$f(x) = a_0 + a_1x + \cdots + a_nx^n + \cdots,$$

then

$$a_0 = a_1 = \cdots = a_n = \cdots = 0.$$

For, since $f(x)$ is continuous, $\lim_{x \to 0} f(x) = f(0) = 0$. This tells us that $a_0 = 0$. Hence we have the geometrically dominated series

$$a_1x + a_2x^2 + \cdots + a_nx^n + \cdots \equiv 0$$

in $[-\rho, \rho]$. In this interval, if $x \neq 0$, this implies that

$$a_1 + a_2x + \cdots + a_nx^{n-1} + \cdots \equiv 0.$$

Again, we let $x \to 0$ for this continuous function and find that $a_1 = 0$. We repeat this argument over and over to establish the conclusion.

We use this result to "equate coefficients" in two power series

$$\sum_{k=0}^{\infty} a_k x^k = \sum_{k=0}^{\infty} b_k x^k$$

that are equal in an interval $[-\rho, \rho]$, $\rho \neq 0$. Looked at another way, it says that there can be only one series in powers x^k that represents a function $f(x)$ in an interval $[-\rho, \rho]$.

5. We omit the proof of the division theorem for power series.

6. We calculate the integral $\int_0^x f(u)\, du$ of the function

$$f(u) = a_0 + a_1u + \cdots + a_nu^n + R_{n+1}(u).$$

Since $f(u)$ is continuous in the interval $[-\rho, \rho]$, it is integrable and

$$\int_0^x f(u)\, du = a_0x + a_1\frac{x^2}{2} + \cdots + a_n\frac{x^{n+1}}{n+1} + \int_0^x R_{n+1}(u)\, du.$$

Since the series for $f(u)$ was geometrically dominated by $a + ar + \cdots + ar^n + \cdots$, it follows that (§39.3)

$$|R_{n+1}(u)| \leq \frac{ar^{n+1}}{1-r}.$$

Hence for the remainder of the integral series

$$\left| \int_0^x R_{n+1}(u)\, du \right| \leq \frac{ar^{n+1}}{1-r} \int_0^x 1\, du = \frac{ax}{1-r}(r^{n+1}).$$

Since $r^{n+1} \to 0$, it follows that the remainder

$$\left| \int_0^x R_{n+1}(u)\, du \right| \to 0.$$

This shows that term-by-term integration produces a series that converges to $\int_0^x f(u)\, du$.

7. Now we consider differentiation. If the ratio test shows that the series

$$f(x) = a_0 + a_1x + \cdots + a_nx^n + \cdots$$

is convergent in an interval $-R < x < R$, then the ratio test also shows that the derived series,

$$g(x) = a_1 + 2a_2x + \cdots + na_nx^{n-1} + \cdots,$$

has the same interval of convergence. This information tells us that the derived series has a continuous sum function $g(x)$ in any interval $[-\rho, \rho]$, where $0 < \rho < R$, but does not prove that $f'(x) = g(x)$. To prove this we apply term-by-term integration to the derived series in $[-\rho, \rho]$. This theorem tells us that since $f(0) = a_0$,

$$f(x) = \int_0^x g(x)\, dx + a_0.$$

Then, since $g(x)$ is continuous in any interval $[-\rho, \rho]$, it follows that $f'(x) = g(x)$ (I, §21.7).

Part Three

Analytic Geometry and Calculus

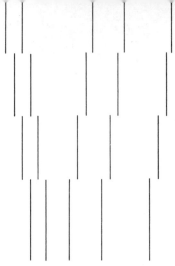

Rotation in Analytic Geometry

REVIEW AND PREVIEW.   We have seen (II, §8) that Euclidean plane geometry introduces the distance between any two points in the plane. This distance is the basis for defining perpendicular lines and proving the properties of congruent triangles. It enables us to prove the pivotal theorem of Pythagoras, which says that in any right triangle, the square of the hypotenuse is equal to the sum of the squares of the sides. When we install Cartesian coordinates in the plane, we obtain the distance between points (x_1, y_1) and (x_2, y_2) as $\sqrt{(x_2 - x_1)^2 + (y_2 - y_1)^2}$. This immediately sets up the equation of the circle (II, §8) (Figure 14.1). Also from this idea we can go to the arc length of curves (II, §22). The arc length s of the circular arc that subtends an angle θ at the center of a circle of radius r gives us a measure of the angle. It is the radian measure, $\theta = s/r$. From this we may proceed to the trigonometric functions (I, §31) defined by

$$\sin \theta = \frac{y}{r}, \qquad \cos \theta = \frac{x}{r}, \qquad \tan \theta = \frac{y}{x}.$$

We have studied the geometric and periodic properties of the trigonometric functions (I, §32). Their differentiation and integration (I; §§31, 33, 34) (II; §§3, 4, 6, 7) has enabled us to fit them into the analytic program of calculus and to compute them (I; §§34, 34R).

The Euclidean distance, generalized to spaces of higher dimension, is essential to the idea of inner product and norm for vectors, and to orthogonal projections (II; §§11, 12). Also the Euclidean distance, combined with a time coordinate, and with the inertial mass particle, is essential to the laws of motion in mechanics (II; §§21, 27, 36).

Polar coordinates in the plane (II; §§34, 35, 36) make use of the Euclidean distance r from the origin to the point R, and the angle measure θ, to locate R in the plane by its polar coordinates (r, θ). Hence differentiation and integration of functions expressed in polar coordinates can be interpreted geometrically. In particular, we defined the conics

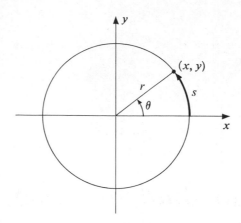

FIGURE 14.1   Rotation, angle and arc length.

in terms of polar coordinates and studied their role in the motion of planets about the sun (II; §§34.6, 36).

All of these objects of our study have been either topics of Euclidean geometry itself, or have been concerned with the interplay of calculus with Euclidean geometry. We now extend our study of Euclidean analytic geometry into some of its more technical aspects, particularly those associated with rotation about a point. This includes some further exploration of calculus in polar, cylindrical, and spherical coordinates for which sections (II; §§34, 35) are prerequisite. It includes some advanced trigonometry for which sections (I; §§31, 33, 34) and (II; §§3, 4, 6, 7) are prerequisites. This advanced trigonometry includes a new formulation of the relationship between polar coordinates, trigonometric functions, and complex numbers. It leads to a new function $e^{i\theta}$, whose values are complex numbers (III, §4).

We also study the conics (II; §34.6) in more detail (III; §§2, 3), after introducing the transformations of coordinates that effect rotations and translations of axes (III, §1). The reduction of the general second-degree equation

$$ax^2 + 2bxy + cy^2 + dx + ey + f = 0$$

to the standard forms of the conic is treated later, as a topic in linear algebra (III, §21).

1   Transformations of Euclidean Geometry

1.1   A View of Geometry.   In the study of plane geometry, without coordinates, the principle of superposition is used. For example, we say that two triangles are congruent if we can move one of the triangles in the plane until it coincides with the other.

Euclidean *analytic* geometry of the plane is the same geometry, investigated by the method of coordinates. We think of the geometric figure in the plane as fixed, and we move the coordinate system. This takes the place of the principle of superposition. The allowed transformations of coordinates are translations (keeping the axes parallel to their original directions); rotations about the fixed origin; and reorientations (which interchange x and y). These transformations of coordinates are called *Euclidean motions*. When we transform coordinates by a Euclidean motion, the coordinates of points are changed and equations of lines and other curves are changed. But distances calculated by the distance formula $\sqrt{(x_2 - x_1)^2 + (y_2 - y_1)^2}$ are not changed and angle measures are not changed. We say that they are geometric invariants.

The Euclidean motions form what is called a *group* of transformations. That is: If a Euclidean motion follows a Euclidean motion(1), the composite is a Euclidean motion; (2) the identity (stand still) motion is included; (3) each motion is invertible by another one that carries the coordinates back to their original position, that is, the two compose to form the identity.

One view of Euclidean analytic geometry is that it is the study of those properties of geometric figures that are invariant under the group of Euclidean motions. This view opens the possibility of other geometries of the plane; for we may wish to study those properties of the plane that are invariant under another group of transformations. Indeed, we have been studying another geometry of the plane. For in our vector version of analytic geometry we have allowed the replacement of a basis $\{\vec{v}_1, \vec{v}_2\}$ for position vectors of the plane by any other basis $\{\vec{w}_1, \vec{w}_2\}$, freely chosen. This establishes a group of transformations of coordinates that includes the Euclidean motions, but is larger. For in Euclidean plane geometry (vector version) we restrict bases to be a pair of perpendicular vectors $\{\vec{v}_1, \vec{v}_2\}$ of unit length, and Euclidean motions must replace one such basis by another like it. When we investigate the plane under the group of transformations effected by an unrestricted replacement of the basis, we are studying what is called *affine geometry*. The properties of straightness in lines, parallelism, intersection, tangency, and similarity are properties of affine geometry. They are also properties of Euclidean geometry.

1.2   Translations.   If the axes of coordinates (or basis vectors) are subjected to a parallel displacement that moves the origin, $x = 0, y = 0$, to a new point, $x = h, y = k$, what is the relation between the old coordinates (x, y) and coordinates (x', y') referred to the new axes? The answer is simple. We picture both sets of axes in the plane (Figure 14.2), and observe that the xy-coordinates of any point P are related to the $x'y'$-coordinates by

$$x = x' + h, \qquad y = y' + k.$$

These are called the *translation transformations*, or just translations.

Example A.   Find a translation of axes that transforms the equation of the circle $x^2 + y^2 - 4x + 6y - 1 = 0$ to the form $x'^2 + y'^2 = r^2$ for some r.

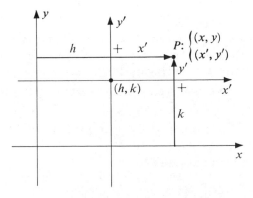

FIGURE 14.2   Translation of axes.

Solution. We have essentially solved this problem by completing the square (II, §8.3). Here we will adopt a slightly different procedure. Observing that the required form, $x'^2 + y'^2 = r^2$, has no terms of the first degree, we translate the axes and choose h and k so as to remove the linear terms in x', y'. So we substitute $x = x' + h, y = y' + k$ into the equation and find

$$x'^2 + 2hx' + h^2 + y'^2 + 2ky' + k^2 - 4x' - 4h + 6y' + 6k - 1 = 0.$$

Equating the coefficients of x' and y' (to the first power) to zero, we have

$$2h - 4 = 0 \quad \text{for } x',$$

and

$$2k + 6 = 0 \quad \text{for } y'.$$

This gives $h = 2$ and $k = -3$. With these values of h and k, the constant term,

$$h^2 + k^2 - 4h + 6k - 1 = -14.$$

Hence the equation of the circle in new coordinates is $x'^2 + y'^2 = 14$. This tells us that in the original coordinates it was a circle with center at $x = 2, y = -3$ and radius $\sqrt{14}$.

Example B. Find a translation of axes that replaces the equation of the curve

$$x^2 - 3xy + 4y^2 - 11x + 27y + 34 = 0$$

by one without first-degree terms.

Solution. We substitute $x = x' + h$ and $y = y' + k$ and equate to zero the coefficients of x' and y'. The result is the two equations

$$2h - 3k - 11 = 0,$$

and

$$-3h + 8k + 27 = 0.$$

We solve this linear system for $h = 1, k = -3$. Therefore we apply the translation $x = x' + 1, y = y' - 3$, and the equation becomes

$$x'^2 - 3x'y' + 4y'^2 = 12.$$

This completes the problem. We will return to the problem of transforming the coordinates so as to remove the term $-3x'y'$ (§21).

1.3   Rotation.

Keeping the origin fixed, we now rotate the axes through an angle θ. The first equation is, How are the coordinates (x, y) in the original system related to those (x', y') in the rotated system (Figure 14.3)? We start with x and y axes that are orthogonal and project the closed vector polygon $\overrightarrow{OQ} + \overrightarrow{QP} + \overrightarrow{PO}$ on the x-axis. The result is

$$x' \cos \theta + y' \cos \left(\frac{\pi}{2} + \theta \right) - x = 0.$$

Similarly projecting on the y-axis, we find

$$x' \cos \left(\frac{\pi}{2} - \theta \right) + y' \cos \theta - y = 0.$$

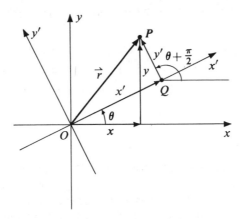

FIGURE 14.3   Rotation of axes.

Simplifying, these reduce to the equations of rotation

$$x = x' \cos \theta - y' \sin \theta,$$
$$y = x' \sin \theta + y' \cos \theta.$$

These are the substitutions that replace xy-coordinates by $x'y'$-coordinates in the rotated system. To find what the coordinates (x, y) become under the rotation, we must solve for (x', y') to obtain the mapping

$$x' = x \cos \theta + y \sin \theta,$$
$$y' = -x \sin \theta + y \cos \theta.$$

1.4   Orientation.   After the coordinate axes are located by translation and rotation, there still remains the choice of which will be the x-axis. Analogously in the vector representation of the plane, with basis $\{\vec{v}_1, \vec{v}_2\}$, we can choose to order the basis in a different way $\{\vec{v}_2, \vec{v}_1\}$. The transformation that accomplishes this is $x = y'$, $y = x'$. Including the identity $x = x'$, $y = y'$, the orientations form a group of two transformations.

1.5   Fundamental Invariants.   It is geometrically evident that translation, rotation, and orientation of axes leave distances $\sqrt{(x_2 - x_1)^2 + (y_2 - y_1)^2}$ invariant. We may verify analytically that if we substitute a translation, rotation, or orientation into this formula it becomes $\sqrt{(x_2' - x_1')^2 + (y_2' - y_1')^2}$ (Exercises). More generally, the inner product $x_1 x_2 + y_1 y_2$ is invariant under the group of Euclidean motions.

1.6   Affine Geometry.   In the plane we can find coordinates (x, y) of a position vector $\vec{r}$ with respect to any chosen basis $\{\vec{v}_1, \vec{v}_2\}$ by means of the representation $\vec{r} = x\vec{v}_1 + y\vec{v}_2$ (II, §10.4). With a different basis $\{\vec{w}_1, \vec{w}_2\}$ chosen, the same vector $\vec{r}$ has coordinates (x', y') given by $\vec{r} = x'\vec{w}_1 + y'\vec{w}_2$. We now ask, What is the relation between the coordinates (x, y) and the new coordinates (x', y')? It is easy to find, if we know the

xy-coordinates of $\vec{w}_1$ and $\vec{w}_2$ in the old basis. Let $\vec{w}_1 = a\vec{v}_1 + b\vec{v}_2$ and $\vec{w}_2 = c\vec{v}_1 + d\vec{v}_2$. Then

$$\vec{r} = x'\vec{w}_1 + y'\vec{w}_2 = x'(a\vec{v}_1 + b\vec{v}_2) + y'(c\vec{v}_1 + d\vec{v}_2)$$
$$= (ax' + cy')\vec{v}_1 + (bx_1 + dy')\vec{v}_2.$$

Equating the coefficients of $\vec{v}_1$ and $\vec{v}_2$ to those in the other representation of $\vec{r}$, $\vec{r} = x\vec{v}_1 + y\vec{v}_2$, we find

$$x = ax' + cy',$$
$$y = bx' + dy'.$$

These equations must be invertible, that is, solvable for (x', y') in terms of (x, y). The group of all such invertible transformations of the plane includes the rotations with $a = \cos\theta$, $b = \sin\theta$, $c = -\sin\theta$, $d = \cos\theta$. It also includes the changes of orientation with $a = d = 0$, $b = c = 1$. If we enlarge it to include the translations, the composite of an invertible linear transformation and a translation defines an *affine transformation*

$$x = ax' + cy' + h,$$
$$y = bx' + dy' + k.$$

Affine plane geometry is the study of properties of figures that are invariant under the group of affine transformations. It follows that all theorems of affine geometry are also theorems of Euclidean geometry; the converse is not true.

1.7 Exercises

In Exercises 1–5, perform a translation of axes so as to remove the terms of first degree, if possible.

1. $x^2 + y^2 - 4x + 8y - 6 = 0$.
2. $x^2 + xy + y^2 - 6x + 3y - 9 = 0$.
3. $y^2 = 4ax$.
4. $9x^2 + 24xy + 16y^2 + 100x - 40y + 100 = 0$.
5. $y = 3x$.

6. Verify that the slope of a pair of points (x_1, y_1) and (x_2, y_2) is invariant under translation.
7. After a translation, $x = x' + h$, $y = y' + k$, sends coordinates (x, y) to (x', y'), find the inverse translation that sends them back where they started.
8. After a rotation of axes through the angle θ sends coordinates (x, y) to (x', y'), what inverse rotation sends them back where they started?
9. In the rotation through $\theta = \pi/6$, what are the (x', y') coordinates of the point whose (x, y) coordinates are $(7, -5)$?
10. Show that the slope of two points (x_1, y_1) and (x_2, y_2) is invariant under translation but not change of orientation.
11. Show that the dot product $x_1 x_2 + y_1 y_2$ is invariant under both rotation and change of orientation; also under translation, if we treat it as $(x_1 - 0)(x_2 - 0) + (y_1 - 0)(y_2 - 0)$.
12. Rotate the axes through an angle $\theta = \pi/4$ and find what the equation $xy = 8$ becomes in $x'y'$-coordinates.
13. Rotate the axes through any angle θ and find what the equation $x^2 + y^2 = r^2$ becomes in $x'y'$-coordinates. Explain the result.
14. Verify that the distance formula $\sqrt{(x_2 - x_1)^2 + (y_2 - y_1)^2}$ is invariant under rotation of coordinates through any angle θ.
15. The line $y = (\tan\alpha)x + k$ has slope $m = \tan\alpha$, where α is the angle of inclination of the line to the x-axis. Perform a rotation transformation through angle θ. Find the slope of the line in x', y'-coordinates. Prove that it is $m' = \tan(\alpha - \theta)$.

16.   Show that the "dilation" $x = ax'$, $y = dy'$ is a special kind of affine transformation. Describe geometrically what it does to the unit square with vertices $(0, 0)$, $(1, 0)$, $(1, 1)$, $(0, 1)$ in x, y-coordinates.

17.   Apply the dilation $x = x'/a$, $y = y'/b$ to the circle $x^2 + y^2 = 1$. Is the shape of the graph unchanged?

18.   Show that an affine transformation of the real-number line is a change of scale, $x = ax' + c$. What does the requirement that the transformation be invertible imply about the coefficients a and c?

19.* A set is convex if it contains all line segments whose end-points are in the set. Prove that an affine transformation maps a convex set into a convex set.

20.   Prove that if two parallel lines are mapped by an affine transformation, they go into two straight lines that are parallel.

21.   Let the angle θ increase with time at a rate given by $\theta = 2\pi t$. Describe the motion given by the parametric representation

$$x = a \cos 2\pi t - b \sin 2\pi t,$$
$$y = a \sin 2\pi t + b \cos 2\pi t,$$

where (a, b) is a fixed point.

22.   Find the velocity of the motion in Exercise 21 and show that it also rotates. Explain in terms of the graph.

23.   Show that all theorems of affine plane geometry are also theorems of Euclidean plane geometry (§1.6).

24.   Let f, g, h be Euclidean motions of the plane. We denote by $f \circ g$ the motion that consists of g followed by f. Prove that $(f \circ g) \circ h = f \circ (g \circ h)$, or in words, that the operation $\circ$ is associative.

2   General Conic and Ellipse

2.1   History.   To the ancient Greek geometers, conic sections were the curves in which a plane intersects a cone (Figure 14.4). Euclid (born about 300 B.C.) wrote four books on conic sections, which are known only through later authors. Archimedes (c. 287–212

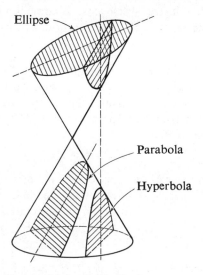

FIGURE 14.4   Conic sections.

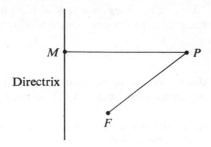

FIGURE 14.5   Focus and directrix of conic $|FP| = e|MP|$.

B.C.) found areas of ellipse and segments of the parabola by a process of integration. Apollonius of Perga (born about 261 B.C.) wrote a famous treatise on conics, which he treated as plane curves, independent of their origin as sections of a solid cone. He began with a definition in affine geometry of the plane. He gave the names *ellipse, parabola,* and *hyperbola* to the three types. Pappus of Alexandria (about A.D. 325) transmitted the work of Euclid and Apollonius and added significant theorems of his own. He defined conics in relation to a fixed point F, called the focus, and a fixed line, called the directrix. A conic is the locus of a point P (Figure 14.5) whose distance $|FP|$ from the focus is in constant ratio e to its distance from the directrix $|MP|$. That is, $|FP| = e|MP|$. The ratio e determines the shape of the conic and is called the eccentricity.

The story of the conics does not end with the ancient Greeks* but continues in modern times.

2.2   General Conic.   By a modification of the ancient definition, we may define a conic as a curve that is a perturbation from a circular path.

DEFINITION (Conic).   Through the center called the focus of a fixed basic circle, there is a fixed straight line called the transverse axis. A conic is a curve traced by a point whose radial distance from the basic circle is proportional to its perpendicular distance from the transverse axis.

The absolute value of the factor of proportionality in the definition is called the eccentricity and is denoted by e. We draw the figure with a basic circle having center at F and with the point P tracing the conic. The radial distance from the basic circle to P is $|RP|$ and the distance from the transverse axis is $|MP|$ (Figure 14.6). Then the definition of the conic can be written

$$|RP| = e|MP|.$$

Let $2k$ denote the diameter of the basic circle. (In classical terminology, $2k$ was called the "latus rectum.") To derive an equation of the conic, we choose a preliminary rectangular $x'y'$-coordinate system with origin at the focus F and x'-axis perpendicular to the transverse axis (y'-axis). Then, in terms of the coordinates of the tracing point

---

* See articles in *Encyclopedia Britannica* entitled Conic Sections, Euclid, Apollonius, Archimedes, and Pappus.

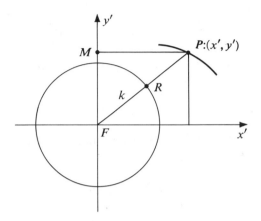

FIGURE 14.6   Conic as a locus.

$P:(x', y')$ there are two equations depending on the sign of the factor of proportionality in the definition. P is on the defined conic if and only if

$$\sqrt{x'^2 + y'^2} - k = ex' \qquad \text{(right-hand conic),}$$

or

$$\sqrt{x'^2 + y'^2} - k = -ex' \qquad \text{(left-hand conic).}$$

Squaring to eliminate the radical sign and simplifying, we find that if P is on a conic then

$$(1 - e^2)x'^2 - 2ekx' + y'^2 = k^2 \qquad \text{(right-hand conic),}$$

or

$$(1 - e^2)x'^2 + 2ekx' + y'^2 = k^2 \qquad \text{(left-hand conic).}$$

These are equations of the general conic for every value of e including the basic circle itself for $e = 0$. We observe from the symmetry, replacing x' by $-x'$, that the left-hand conic is the mirror image of the right-hand conic about the transverse axis.

If $0 \leqq e < 1$, the conic is called an ellipse. If $e = 1$, the conic is called a parabola, and if $1 < e$, the conic is called a hyperbola.

2.3   Standard Equation of the Ellipse.   For the ellipse, $e < 1$ and $1 - e^2$ is positive. Then we can rewrite the equation of the right-hand conic (§2.2) as

$$x'^2 + \frac{2ek}{1 - e^2} x' + \frac{y'^2}{1 - e^2} = \frac{k^2}{1 - e^2}.$$

Completing the square of the x'-terms, this becomes

$$\left(x' - \frac{ek}{1 - e^2}\right)^2 + \frac{y'^2}{1 - e^2} = \frac{k^2}{(1 - e^2)^2}.$$

We now choose the so-called standard coordinate system by a translation of axes

$$x = x' - \frac{ek}{1 - e^2}, \qquad y = y'.$$

We also define

$$a^2 = \frac{k^2}{(1 - e^2)^2} \quad \text{and} \quad b^2 = \frac{k^2}{1 - e^2}.$$

With these notations, the equation of the (right-hand) ellipse reduces to the standard equation

$$\frac{x^2}{a^2} + \frac{y^2}{b^2} = 1.$$

We may now compute the constants of the ellipse in terms of a and b. We observe that $(1 - e^2)a^2 = b^2$; hence $b < a$ and

$$e = \frac{\sqrt{a^2 - b^2}}{a}.$$

The focus F has x-coordinate determined by $x' = 0$,

$$x = -\frac{ek}{1 - e^2} = -ae.$$

The focus of the right-hand ellipse in standard form is $(-c, 0)$, where $c = ae = \sqrt{a^2 - b^2}$. We observe that when $x' = 0$ in the preliminary coordinates, $y' = \pm k$. Hence if we set $x = \sqrt{a^2 - b^2}$ in the standard equation, we find that $y = \pm b^2/a$. That is, $k = b^2/a$.

Turning to the left-hand ellipse, we find that it reduces to exactly the same standard equation under the translation of axes.

$$x = x' + \frac{ek}{1 - e^2}. \qquad y = y'.$$

Thus we see that the ellipse

$$\frac{x^2}{a^2} + \frac{y^2}{b^2} = 1$$

is the right-hand ellipse with respect to a focus F at $(-c, 0)$ and also the left-hand ellipse with respect to the focus F' at $(c, 0)$. This enables us to drop the distinction between left- and right-hand ellipses and speak about two foci for the same ellipse.

We summarize these results. With respect to a suitably chosen coordinate system, every ellipse has a standard equation

$$\frac{x^2}{a^2} + \frac{y^2}{b^2} = 1, \qquad \text{where} \quad 0 < b \leqq a.$$

The focal distance $c = \sqrt{a^2 - b^2}$. The ellipse has two foci with coordinates $(\pm c, 0)$. The intercepts of the ellipse on the axes are at the four points $(\pm a, 0)$, $(0, \pm b)$. The graph is symmetric to each axis and to the origin, which is the center of the ellipse. The eccentricity is $e = c/a$, and the radius of the basic circle is $k = b^2/a$. The intercepts with the two transverse axes have coordinates $(-c, \pm b^2/a)$ and $(c, \pm b^2/a)$. The axis containing the foci is called the principal axis (in this form the x-axis).

There is another standard form, found by choosing the standard coordinates with x and y interchanged,

$$\frac{y^2}{a^2} + \frac{x^2}{b^2} = 1, \qquad \text{where, as before} \quad 0 < b \leqq a.$$

The formulas for c, e, k are the same. Interchanging x- and y-coordinates in the form above, we find that the intercepts are $(0, \pm a)$, $(\pm b, 0)$ and the intercepts with the transverse axes are at $(\pm b^2/a, c)$ and $(\pm b^2/a, -c)$. The principal axis is the y-axis.

2.4   Plotting an Ellipse

Example.   For the ellipse $9x^2 + 25y^2 = 225$, reduce to standard form, find a, b, c, e, k, and plot the ellipse.

Solution.   We divide by 225 and find the standard form

$$\frac{x^2}{25} + \frac{y^2}{9} = 1, \qquad a = 5, \qquad b = 3.$$

$$c = \sqrt{a^2 - b^2} = 4. \qquad e = \frac{c}{a} = \frac{4}{5}. \qquad k = \frac{b^2}{a} = \frac{9}{5}.$$

Intercepts $(\pm 5, 0)$ and $(0, \pm 3)$. Intercepts on transverse focal axes $(4, \pm \frac{9}{5})$, $(-4, \pm \frac{9}{5})$. We plot these 8 points and the foci and sketch the ellipse (Figure 14.7).

2.5   Tangent, Normal and Area of Ellipse.

Differentiating the standard equation of the ellipse, we find that the equation of the tangent line to the ellipse at a point $P:(x, y)$ on the ellipse is given in local (dx, dy) coordinates by

$$\frac{x\,dx}{a^2} + \frac{y\,dy}{b^2} = 0.$$

The slope of the tangent line is then

$$\frac{dy}{dx} = -\frac{b^2}{a^2}\frac{x}{y}.$$

If we fix the point of tangency at $P:(x_1, y_1)$ and let (x, y) represent the variable xy-coordinates of a point T on the tangent line, then $dx = x - x_1$ and $dy = y - y_1$. The equation of the tangent line to the standard ellipse at $P:(x_1, y_1)$ becomes (Exercises)

$$\frac{xx_1}{a^2} + \frac{yy_1}{b^2} = 1.$$

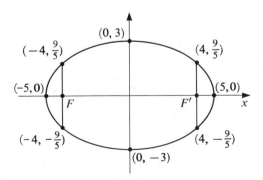

FIGURE 14.7   Ellipse.

The area of the ellipse is found by integration (Exercises) to be

$$A = \pi ab.$$

2.6 Exercises

In Exercises 1–6, reduce to standard form of the ellipse; find a, b, c, e, k, and plot the ellipse.

1. $9x^2 + 25y^2 = 900$.   2. $x^2 + 2y^2 = 2$.
3. $4x^2 + 9y^2 = 36$.   4. $4x^2 + 9y^2 = 25$.
5. $25x^2 + 9y^2 = 225$.   6. $2x^2 + y^2 = 5$.

7. Show that the equation of tangent line to the ellipse $b^2x^2 + a^2y^2 = a^2b^2$ at $P:(x_1, y_1)$ is

$$\frac{xx_1}{a^2} + \frac{yy_1}{b^2} = 1.$$

8. Set up the double integral

$$A = \iint_E dx\, dy$$

over the ellipse and compute it to show that $A = \pi ab$.

9. (a) What is the graph of $b^2x^2 + a^2y^2 = 0$?
    (b) The graph of $b^2x^2 + a^2y^2 + a^2b^2 = 0$?
10. Find the equation of the standard ellipse having $e = \frac{1}{2}$ and $k = 4$.
11. Show that for every positive value of m

$$\frac{x^2}{1 + m^2} + \frac{y^2}{m^2} = 1$$

is an ellipse with foci at $(-1, 0)$ and $(1, 0)$. Plot several ellipses in the family for different values of m.
12. Show that if P is on the ellipse with foci F_1 and F_2 then $|F_1P| + |F_2P| = 2a$.
13. Use Exercise 12 to show how to construct an ellipse using a loop of string of length $2a + 2c$ to guide the tracing point.
14. Using Exercise 13, construct an ellipse with $c = 3$ inches and $a = 4$ inches.
15. Explain how the method of Exercise 13 can be used to construct an ellipse with $e = \frac{1}{2}$ and $a = 4$.
16. In the derivation of the standard equation of the ellipse (§2.3), verify the formulas

$$e = \frac{\sqrt{a^2 - b^2}}{a}, \qquad k = \frac{b^2}{a}.$$

17. On the graph of the standard ellipse is a circle of radius a and center at y-intercept point. Show that this circle cuts the principal axes at the foci of the ellipse.
18. Find the area of the portion of the ellipse cut off by a line through a focus perpendicular to the principal axis.
19. Find the area of the portion of an ellipse cut off by a line $y = mx$ through the center.
20. Show that the normal line to the standard ellipse at the point (x_1, y_1) has the equation

$$a^2y_1x - b^2x_1y = (a^2 - b^2)x_1y_1.$$

21. Show that

$$\frac{(x - x_0)^2}{a^2} + \frac{(y - y_0)^2}{b^2} = 1$$

is an ellipse with center at (x_0, y_0).

22. Complete the square to show that $x^2 + 4y^2 + 2x - 8y + 1 = 0$ is an ellipse. Find a, b, c coordinates of foci and sketch.

23. Show that the tangent line at $P:(x_1, y_1)$ to the ellipse in Exercise 21 is given by

$$\frac{(x - x_0)(x_1 - x_0)}{a^2} + \frac{(y - y^0)(y_1 - y_0)}{b^2} = 1.$$

24. Find the area of the region enclosed by an ellipse and its basic circle.

25. Derive the general equation of the conic from the directrix definition of Pappus (§2.1) in the case that $e \neq 0$. Choose coordinates so that the focus is at the origin and the directrix is the line $x = -k/e$. Compare the result with that which we obtained (§2.2). The directrix definition does not include the circle, except as a limiting case, as $e \to 0$.

3   Hyperbola and Parabola

3.1   Standard Form of Hyperbola. We have seen (§2.2) that the $x'y'$-coordinates of a point on a right-hand conic with focus at the origin and eccentricity e satisfy an equation of the form

$$\left(x' + \frac{ek}{e^2 - 1} \right)^2 - \frac{y'^2}{e^2 - 1} = \frac{k^2}{(e^2 - 1)^2},$$

if $e \neq 1$, where the equation is rearranged to introduce $e^2 - 1$ in place of $1 - e^2$. In the hyperbola, $e > 1$, so $e^2 - 1$ is positive. As we did for the ellipse, we then transform to a new xy-coordinate system by the translation substitution

$$x = x' + \frac{ek}{e^2 - 1}, \qquad y = y'.$$

Also, since $e^2 - 1 > 0$, we may define positive numbers a and b such that

$$a^2 = \frac{k^2}{(e^2 - 1)^2}, \qquad b^2 = \frac{k^2}{e^2 - 1}.$$

With these transformations, we find that the right-hand hyperbola satisfies a standard equation of the form

$$\frac{x^2}{a^2} - \frac{y^2}{b^2} = 1.$$

Since $a^2(e^2 - 1) = b^2$, we find that $e = \sqrt{a^2 + b^2}/a$ and $k = b^2/a$. Defining c by $c = \sqrt{a^2 + b^2} = ae$, we find that the xy-coordinates of the focus of the right-hand hyperbola, corresponding to $(x', y') = (0, 0)$ become $(c, 0)$, that is $(ae, 0)$.

If we start with the left-hand conic

$$\sqrt{x'^2 + y'^2} - k = -ex',$$

assuming that $e > 1$, we find that the transformation

$$x = x' - \frac{ek}{e^2 - 1}, \qquad y = y'$$

transforms it to the same standard equation as the right-hand conic but with focus at $(-c, 0)$. However, unlike the case of the ellipse, the right- and left-hand hyperbolas do not coincide in the graph of the standard equation. If we solve the standard equation of the hyperbola for x in terms of y, we get

$$x = +\frac{a}{b}\sqrt{b^2 + y^2} \quad \text{in which} \quad x \geqq a \quad \text{for all } y,$$

or

$$x = -\frac{a}{b}\sqrt{b^2 + y^2} \quad \text{in which} \quad x \leqq -a \quad \text{for all } y.$$

The first of these separated graphs is the right-hand hyperbola and the second is the left-hand hyperbola but both satisfy $b^2x^2 - a^2y^2 = a^2b^2$, which we now say is the standard equation of the hyperbola, recognizing that it has two foci and two branches (Figure 14.8).

3.2 Intercepts and Asymptotes. We find, as for the ellipse, that the hyperbola has x-intercepts $(\pm a, 0)$ and intercepts on transverse axes through the foci $(c, \pm b^2/a)$, $(-c, \pm b^2/a)$, but when $x = 0$, $y^2 = -b^2$, which implies that there is no y-intercept of the hyperbola,

$$b^2x^2 - a^2y^2 = a^2b^2.$$

Therefore let us examine the intersections with the hyperbola of lines $y = mx$ through the origin, for varying slopes m (Figure 14.8). We substitute $y = mx$ into the equation and find that the x-coordinate of the point of intersection $P:(x_1, y_1)$ is given by

$$x_1 = \pm\frac{ab}{\sqrt{b^2 - a^2m^2}}.$$

This yields a point of intersection if and only if $b^2 - a^2m^2 > 0$, that is, if $-b/a < m < b/a$. But if $|m| > b/a$, there is no intersection point (Figure 14.8, shaded zones). There is also no x_1 when $a^2m^2 = b^2$. In this case, we observe that

$$\lim_{m \to b/a} \frac{ab}{\sqrt{b^2 - a^2m^2}} = \infty.$$

Hence, as m increases to b/a, the point of intersection P moves out along the curve without limit. We say then that the line $ax - by = 0$, and similarly the line $ax + by = 0$, are asymptotes of the graph.

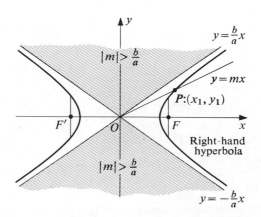

FIGURE 14.8   Asymptotes.

3.3   Alternative Standard Form.   We recognize an alternative standard form

$$\frac{y^2}{a^2} - \frac{x^2}{b^2} = 1$$

with x and y interchanged. For it, the foci are on the y-axis at $(0, \pm c)$. It has intercepts $(0, \pm a)$ and transverse intercepts at foci $(\pm b^2/a, c)$ and $(\pm b^2/a, -c)$.

In either standard form for the hyperbola, a^2 is always associated with the positive term. As opposed to the case for the ellipse, $a < b$, $a = b$, $b < a$ are all possible.

3.4   Identification and Sketch

Example.   For the hyperbola $9x^2 - 16y^2 + 144 = 0$, find a, b, e, c, k intercepts, foci, and asymptotes. Sketch the graph.

Solution.   Reducing to standard form, we find that the equation becomes

$$\frac{y^2}{9} - \frac{x^2}{16} = 1.$$

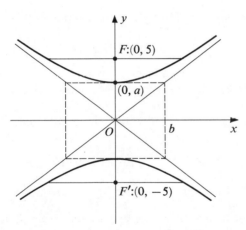

FIGURE 14.9   Hyperbola $\dfrac{y^2}{9} - \dfrac{x^2}{16} = 1$.

Hence $a = 3$ and $b = 4$. (Note that in this case $a < b$.) $e = \frac{5}{3}$, $c = 5$, $k = \frac{16}{3}$. The intercepts on the y-axis are $(0, \pm 3)$, intercepts on transverse axes through the foci $(\pm \frac{16}{3}, 5)$ and $(\pm \frac{16}{3}, -5)$. The foci are $(0, \pm 5)$. The asymptotes are $3x - 4y = 0$ and $3x + 4y = 0$. We sketch the curve by plotting the asymptotes and the six intercepts, then using the asymptotes as guides for the portions of the curve that are far from the origin (Figure 14.9).

3.5   Property of Focal Radii.   For any point P on a hyperbola, the distances $|F_1 P|$ and $|F_2 P|$ to the foci satisfy the relation (Exercises)

$$\big|\,|F_1 P| - |F_2 P|\,\big| = 2a.$$

This relation gives rise to a construction of the hyperbola (Exercises).

3.6 Parabola. The parabola is the conic with $e = 1$. In this case equation of the right-hand conic with focus at the origin (§2.2) reduces to

$$-2kx' + y'^2 = k^2.$$

We cannot complete the square to make the left member a sum or difference of two squares but we can simplify it by a substitution that eliminates the constant term. We can rewrite the equation

$$y'^2 = 2k\left(x' + \frac{k}{2}\right).$$

Hence if we translate the axes by the substitution $x = x' + (k/2)$, $y = y'$, the standard equation of the right-hand parabola becomes

$$y^2 = 2kx$$

with focus at $(k/2, 0)$. The intercepts on the transverse axis through the focus are $(k/2, \pm k)$. We may use these data to sketch the parabola. The intercept $(0, 0)$ of the parabola with its principal axis is called the vertex.

The general equation of the left-hand conic for the case that $e = 1$ reduces to

$$2kx' + y'^2 = k^2.$$

Again we translate the axes to eliminate the constant term and obtain the standard equation

$$y^2 = -2kx$$

with focus at $(-k/2, 0)$. So for the parabola, in contrast with the cases for the ellipse and the hyperbola, we have two different second-degree equations for the right- and left-hand parabolas.

Also we may interchange x and y and obtain standard equations of the parabola with focus on the y-axis,

$$x^2 = 2ky \qquad \text{with focus at} \quad (0, k/2),$$

and

$$x^2 = -2ky \qquad \text{with focus at} \quad (0, -k/2).$$

Example. Find k, coordinates of focus, and sketch the parabolas $y^2 + 4x = 0$ and $x^2 - 8y = 0$.

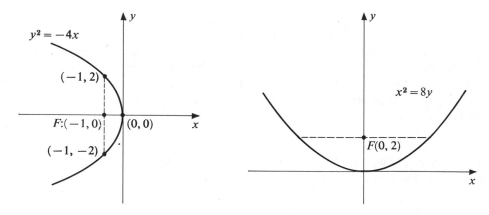

FIGURE 14.10   Parabolas.

Solution.   In standard form the first equation is $y^2 = -4x$. It is a left-hand parabola with $k = 2$ and focus at $(-1, 0)$. It has the intercepts $(0, 0)$, $(-1, 2)$ and $(-1, -2)$. We sketch the curve using these data (Figure 14.10). For the other parabola a standard equation is $x^2 = 8y$, with $k = 4$ and focus on the y-axis at $(0, 2)$. It has the intercepts $(0, 0)$, $(4, 2)$, $(-4, 2)$. These data suffice to sketch its graph (Figure 14.10).

3.7   Focusing of Parallel Rays

THEOREM.   For every point P on a parabola the focal ray FP and the ray through P parallel to the principal axis make equal angles with the tangent to the parabola at P.

Proof.   We construct the figure (Figure 14.11). We must prove that the angle α is equal to the angle β, both with vertex at $P:(x_1, y_1)$. By geometry, the angle β is equal to the angle the tangent line makes with the x-axis at T. We shall prove that the triangle TFP is isosceles. Differentiating, we find that the slope of the tangent line at P is given by

$$\frac{dy}{dx} = \frac{k}{y_1},$$

and the equation of the tangent line at P is

$$yy_1 = k(x + x_1).$$

Setting $y = 0$, we find that the x-coordinate of T is $-x_1$. Hence the line segment $|TF| = x_1 + (k/2)$. We calculate $|TP|$ by the distance formula, remembering that $y_1^2 = 2kx_1$, and find that it reduces to $|TP| = x_1 + (k/2)$ also. This proves that $\alpha = \beta$.

This theorem has applications for gathering parallel rays of light, sound, or electro-magnetic radiation, and bringing them to a focus at a point. We imagine a paraboloid surface formed by revolving the parabola about its principal axis OF. A beam MP of any radiation that strikes this surface is reflected in the direction PF through the focus. In this direction the angle of reflection is equal to the angle of incidence. In this way the reflected energy of all of the parallel beams that strike the surface are gathered at F.

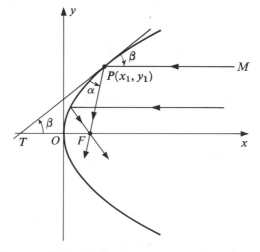

FIGURE 14.11   Reflection property of parabola.

If a receptor is placed there, it will receive energy much more intense than that of the incident radiation before being focused by the parabolic reflector.

3.8 Exercises

For the hyperbolas in Exercises 1–4, reduce to standard form, find a, b, e, c, k coordinates of foci, equations of asymptotes, intercepts on principal axis and on transerve axes through the foci, and sketch the graph.

1. $9x^2 - 16y^2 = 144$.
2. $x^2 - y^2 = 4$.
3. $9x^2 - 16y^2 + 144 = 0$.
4. $16x^2 - 9y^2 = 144$.

In Exercises 5–8, reduce to standard form, find k, coordinates of focus, intercepts on transverse focal axes, and sketch.

5. $y^2 = 4x$.
6. $y^2 = -4x$.
7. $x^2 = 9y$.
8. $x^2 + 9y = 0$.

9. Show that the tangent line to the hyperbola $b^2x^2 - a^2y^2 = a^2b^2$ at $P:(x_1, y_1)$ has the equation $b^2xx_1 - a^2yy_1 = a^2b^2$.

10. Let P be a point on a parabola and let T be the point where the tangent line at P intersects the principal axis (Figure 14.11). Show that the projection of the line segment TP on the principal axis is always bisected by the vertex.

11. Let P be a point on a parabola and let N be the point where the normal line at P intersects the principal axis. Show that the projection of the line segment PN on the principal axis has constant length, provided that P is not at the vertex.

12. Find the area of the region bounded by the parabola and a transverse axis through the focus (a result of Archimedes).

13. Find the area of the region bounded by the parabola and the basic circle.

14. Find the area of the region bounded by the circle $x^2 + y^2 = 4$ and the hyperbola $x^2 - y^2 = 1$.

15. Consider a as an arbitrary constant, and in the same coordinate plane, plot several graphs of

$$(a^2 - 9)x^2 + a^2y^2 = a^2(a^2 - 9)$$

for differing values of a including $a = 0, 1, 2, 3, 4, 5$. Show that for every value of a, except 0 and 3, the equation represents a conic with foci at $(\pm 3, 0)$. It is called a system of confocal conics.

16. In the same coordinate plane plot the graphs of

$$\frac{x^2}{a^2} - \frac{y^2}{b^2} = z$$

for several different values of z, including $z = -1, 0, 1$.

17. In Exercise 16, the hyperbolas with $z = k^2$ and $z = -k^2$ are said to be conjugate. How are they related geometrically? What are their foci? Their asymptotes? How are their eccentricities related?

18. A table lamp has a cylindrical shade with the light source inside it at the center. What is the curve marking the edge of the shadow on the wall?

19. Three LORAN stations A, B, C (Figure 14.12) receive a radio signal from a ship S in distress. By comparing the times of arrival of the signal, they find that the distances are related by $|AS| - |CS| = 800$ mi, $|AS| - |BS| = 200$ mi. Locate the ship by means of a graph.

20. In Exercise 19, set up two equations whose simultaneous solution gives the coordinates of S.

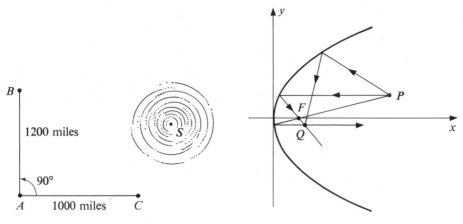

FIGURE 14.12   Exercise 13.   FIGURE 14.13   Exercise 22.

21.   What magnification, $x = cx'$, $y = cy'$, will transform the parabola $y^2 = 4ax$ into $(y')^2 = 4kx'$? Prove that thus all parabolas have the same shape, differing only in size.

22.* From a point P (Figure 14.13), not on the axis of a parabolic mirror, a light ray parallel to the axis and a light ray directly to the focus intersect at Q, after reflection. Find the coordinates of Q if those of P are (x_1, y_1).

4   Complex Numbers

4.1   Plane Vectors, Polar Coordinates, and Complex Numbers.   We recall the graphical representation of complex (imaginary) numbers such as $2 + \sqrt{-3}$. We choose an imaginary basis element $i = \sqrt{-1}$, and then every complex number z can be written in the form $x + iy$, where x and y are real numbers. x is called the *real part* and y the *imaginary part* of z. For example, the number $2 + \sqrt{-3} = 2 + i\sqrt{3}$, in which 2 is the real part and the *real* number $\sqrt{3}$ is the imaginary part. We plot pure imaginary numbers iy on the y-axis (Figure 14.14), representing $\vec{i}$ as a unit vector in the y-axis. Then every complex number $z = x + iy$ can be represented uniquely as a position vector from the origin to point $R:(x, y)$. Taking the unit vector $\vec{1}$ in the real axis and the unit vector $\vec{i}$ in the imaginary axis as the basis $\{\vec{1}, \vec{i}\}$, we can think of the complex number z as the vector $\vec{z} = x\vec{1} + y\vec{i}$. Addition of complex numbers and their multiplication by scalars

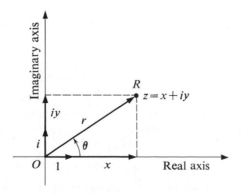

FIGURE 14.14   Complex numbers as position vectors.

(real numbers) have the same meaning as addition and multiplication of plane vectors by scalars. The absolute value of z, written $|z|$, is defined by $|z| = \sqrt{x^2 + y^2}$, which is the same as the norm of the vector $\vec{z}$, $\|\vec{z}\| = \|x\vec{1} + y\vec{i}\| = \sqrt{x^2 + y^2}$. Thus all of the algebraic structure of plane vectors is inherited by complex numbers.

Moreover, every complex number has a representation in polar coordinates. If $\vec{z} = x\vec{1} + y\vec{i}$ is the position vector $\overrightarrow{OR}$, we simply represent R in polar coordinates (r, θ). Then $|z| = r$, and

$$x = r \cos \theta, \qquad y = r \sin \theta.$$

Hence

$$\vec{z} = (r \cos \theta)\vec{1} + (r \sin \theta)\vec{i}.$$

This is usually written in complex number notation

$$z = r(\cos \theta + i \sin \theta),$$

called the *polar form*, where the angle θ is called the *argument* of z, written $\theta = \arg z$.

Thus the complex number $z = x + iy$, the vector $\vec{z} = x\vec{1} + y\vec{i}$, the Cartesian coordinates (x, y), and the polar coordinates (r, θ) each determines uniquely the position vector $\overrightarrow{OR}$. To this point these systems of representation are essentially equivalent.

4.2 Complex Multiplication and Division.

Unlike the case with plane vectors, there is a multiplication of complex numbers by complex numbers to produce a complex number product. If $z = x + iy$ and $w = u + iv$, then we multiply

$$zw = (x + iy)(u + iv) = xu + ixv + iyu + i^2 yv.$$

Here we use the fact that $i = \sqrt{-1}$ and hence $i^2 = -1$. This gives us the complex product

$$(x + iy)(u + iv) = (xu - yv) + i(xv + yu).$$

So $xu - yv$ is the real part and $xv + yu$ is the imaginary part of the complex product.

For every complex number $z = x + iy$, there is a *conjugate* complex number $\bar{z} = x - iy$. We observe that $z\bar{z} = x^2 + y^2 = |z|^2$, a nonnegative real number. This permits us to divide by a complex number z if $z \neq 0$. To divide w/z, we multiply the numerator and denominator by $\bar{z}$. This gives us

$$\frac{w}{z} = \frac{u + iv}{x + iy} = \frac{(x - iy)(u + iv)}{(x - iy)(x + iy)} = \frac{(xu + yv)}{x^2 + y^2} + \frac{(xv - yu)}{x^2 + y^2} i,$$

if $x^2 + y^2 \neq 0$, that is, if $z \neq 0$.

4.3 Exponential Property.

We consider the complex-valued function, **cis** θ = **cos** θ + **i sin** θ. For every real number θ, $|\text{cis } \theta| = 1$. Hence the position vector of cis θ traces the unit circle about the origin in the complex plane. Let us multiply

$$(\text{cis } \theta)(\text{cis } \phi) = (\cos \theta + i \sin \theta)(\cos \phi + i \sin \phi)$$
$$= (\cos \theta \cos \phi - \sin \theta \sin \phi) + i(\sin \theta \cos \phi + \cos \theta \sin \phi)$$
$$= \cos (\theta + \phi) + i \sin (\theta + \phi) = \text{cis } (\theta + \phi).$$

Thus **cis θ** has the property of an exponential function a^θ,

$$(a^\theta)(a^\phi) = a^{\theta + \phi}.$$

We obtain further evidence of the exponential character of **cis θ** if we differentiate it.

$$\frac{d}{d\theta} \text{ cis } \theta = \frac{d}{d\theta} (\cos \theta + i \sin \theta) = -\sin \theta + i \cos \theta$$

$$= i(\cos \theta + i \sin \theta) = i \text{ cis } \theta.$$

Hence **cis θ** differentiates like $(d/d\theta)e^{k\theta} = \mathbf{k}e^{k\theta}$, with $k = i$.

Observing that cis $0 = 1$, we can prove that no complex-valued function $f(\theta)$ other than **cis θ** has the property that

$$\frac{d}{d\theta} f(\theta) = if(\theta) \qquad \text{and} \qquad f(0) = 1.$$

Suppose such a function exists. Then we differentiate

$$\frac{d}{d\theta} \frac{f(\theta)}{\text{cis } \theta} = \frac{\text{cis } \theta \ (d/d\theta)f(\theta) - f(\theta)i \text{ cis } \theta}{(\text{cis } \theta)^2} = 0,$$

since $(d/d\theta)f(\theta) = if(\theta)$. Hence, $f(\theta)/\text{cis } \theta = C$, a constant, and we determine that $C = 1$ by setting $\theta = 0$. Therefore $f(\theta) = \text{cis } \theta$. This tells us that **cis θ** behaves formally like the exponential expression $e^{i\theta}$ and no other function has these properties. Hence we have a basis for a definition of $e^{i\theta}$.

DEFINITION (Complex exponential).   For every real number θ, we define $e^{i\theta}$ by

$$e^{i\theta} = \cos \theta + i \sin \theta.$$

In terms of this definition we have the following properties of the complex exponential:

$$e^{i\theta}e^{i\phi} = e^{i(\theta + \phi)},$$

$$(e^{i\theta})^k = e^{ik\theta},$$

$$e^z = e^{x + iy} = e^x e^{iy} = e^x(\cos y + i \sin y),$$

$$\frac{d}{d\theta} e^{i\theta} = ie^{i\theta},$$

$$z = re^{i\theta}, \qquad \text{where} \quad r = |z| \quad \text{and} \quad \theta = \arg z.$$

The last expression is the *polar form* of the complex number z.

4.4 Euler Relations.   We observe that since $\cos (-\theta) = \cos \theta$ and $\sin (-\theta) = -\sin \theta$,

$$e^{-i\theta} = \cos \theta - i \sin \theta = \overline{e^{i\theta}}.$$

We consider the pair of simultaneous equations in **cos θ** and **sin θ**,

$$e^{i\theta} = \cos + i \sin \theta,$$

$$e^{-i\theta} = \cos \theta - i \sin \theta.$$

We solve this system for $\cos \theta$ and $\sin \theta$ by elimination and find that

$$\cos \theta = \frac{e^{i\theta} + e^{-i\theta}}{2},$$

$$\sin \theta = \frac{e^{i\theta} - e^{-i\theta}}{2i}.$$

These two pairs of relations between $e^{i\theta}$ and the trigonometric functions are called the Euler* relations.

The properties of the complex exponential, together with the Euler relations, imply all of analytic trigonometry in a form easier to remember. We can replace **sin θ** and **cos θ** by their equivalents in complex exponentials and calculate with them.

Example A. Calculate $\int \sin^2 \theta \, d\theta$.

Solution with complex exponential.

$$\int \sin^2 \theta \, d\theta = \int \left[\frac{e^{i\theta} - e^{-i\theta}}{2i} \right]^2 d\theta = \int \frac{e^{2i\theta} - 2 + e^{-2i\theta}}{-4} \, d\theta$$

$$= -\frac{1}{4} \left[\frac{1}{2i} e^{2i\theta} - 2\theta - \frac{1}{2i} e^{-2i\theta} \right] + C$$

$$= \frac{\theta}{2} - \frac{1}{4} \frac{e^{2i\theta} - e^{-2i\theta}}{2i} + C = \frac{\theta}{2} - \frac{1}{4} \sin 2\theta + C.$$

Example B. Integrate

$$J = \int_0^{2\pi} e^{im\theta} e^{in\theta} \, d\theta,$$

where m and n are integers.

Solution. If $m \neq -n$, then

$$J = \int_0^{2\pi} e^{i(m+n)\theta} \, d\theta \, \frac{1}{i(m+n)} e^{i(m+n)\theta} \Big|_0^{2\pi} = 0,$$

since $e^{i(m+n)\theta}$ is periodic with period 2π by its definition. If $m = -n$, then

$$J = \int_0^{2\pi} 1 \, d\theta = 2\pi.$$

4.5 Extracting Roots of Complex Numbers.

To solve an equation

$$z^n = a$$

for all complex nth roots of a, we write z and a in polar form, $z = re^{i\theta}$, $a = qe^{i\phi}$. Then the equation $z^n = a$ becomes

$$r^n e^{in\theta} = qe^{i\theta}.$$

This implies that $r^n = q$ and $n\theta = \phi + 2k\pi$, where $2k\pi$ is any integral multiple of 2π. We extract the real nth root of q to find r and *divide* the angle relation by n to find all possible arguments θ.

Example. Find all cube roots of $8i$.

---

* Leonhard Euler (1707–1783) was a Swiss master of numerical analysis. Although he was not a teacher, he wrote in Latin the textbooks of algebra and calculus that became the models for modern texts in all European languages. In them, he is responsible for the standard notation $f(x)$, and also π, i and e, through his equation $e^{\pi i} + 1 = 0$.

Solution. We must solve the equation $z^3 = 8i$. We write in polar form $z = re^{i\theta}$ and $8i = 8e^{i(\pi/2)}$. Then we have

$$z^3 = r^3 e^{i(3\theta)} = 8e^{i(\pi/2)}.$$

Hence $r^3 = 8$ and $r = 2$. Also $3\theta = (\pi/2) + 2k\pi$ for any integer k. Hence $\theta = (\pi/6) + k(2\pi/3)$. Now writing $z = r(\cos\theta + i\sin\theta)$, we obtain

$$\text{for } k = 0, \qquad z_1 = 2\left(\cos\frac{\pi}{6} + i\sin\frac{\pi}{6}\right) = \sqrt{3} + i,$$

$$\text{for } k = 1, \qquad z_2 = 2\left(\cos\frac{5\pi}{6} + i\sin\frac{5\pi}{6}\right) = -\sqrt{3} + i,$$

$$\text{and for } k = 2, \qquad z_3 = 2\left(\cos\frac{9\pi}{6} + i\sin\frac{9\pi}{6}\right) = -2i.$$

For all other integers k, the solution repeats one of these. We plot the three complex cube roots of $8i$ (Figure 14.15), observing that they are distributed at the vertices of an equilateral triangle on the circle $|z| = 2$.

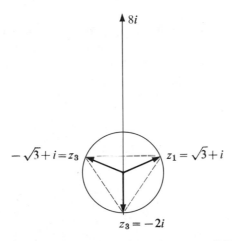

FIGURE 14.15   Complex cube roots of $8i$.

4.6 *Exercises*

1. For $z = 1 + i$, $w = 1 - 3i$, find zw, w/z, $\arg z$, $|z|$, $w\bar{w}$, $z^2 - 3z + 1$.
2. For $z = 3 - 2i$, $w = -1 + i$, find $z + w$, $z - w$, $1/z$.
3. Express the following numbers in polar form: (a) $z = 1 - i$; (b) $z = \sqrt{3} - i$; (c) $z = -2$; (d) $z = -3i$.
4. Show that if $z = x + iy$, then $z\bar{z} = x^2 + y^2$.
5. Show that $|z| = |\bar{z}|$.
6. Show that if $z = x + iy$, then $x = (z + \bar{z})/2$ and $y = (z - \bar{z})/(2i)$.
7. Prove that $\overline{zw} = \bar{z}\cdot\bar{w}$.
8. Prove that $\overline{z + w} = \bar{z} + \bar{w}$.
9. Prove that if $z = \bar{z}$, then z is a real number.
10. Express in polar form $8 - 8\sqrt{3}i$ and plot it.
11. Find all square roots of $8 - 8\sqrt{3}i$.
12. Express $\tan\theta$ in terms of $e^{i\theta}$ and $e^{-i\theta}$.
13. Prove that if $z = e^{i\theta}$, then $1/z = e^{-i\theta}$.

14. Find all cube roots of 1.
15. Find all square roots of i.
16. Integrate using complex exponentials, $\int_0^{\pi/2} \cos^2 \theta \, d\theta$.

In Exercises 17–19, use Example B, Section 4.4, to evaluate the integrals, for all integers m and n.

17. $\int_0^{2\pi} \sin mx \cos nx \, dx$.

18. $\int_0^{2\pi} \cos mx \cos nx \, dx$.

19. $\int_0^{2\pi} \cos mx \sin nx \, dx$.

20. Show that $z_1 = e^{(-2+i)t}$ and $z_2 = e^{(-2-i)t}$ are solutions of the differential equation $(d^2/dt^2)z + 4(dz/dt) + 5 = 0$.
21. In Exercise 20, show that $(z_1 + z_2)/2 = e^{-2t} \cos t$ is a solution of the differential equation. Plot the graph of this solution.
22. Show that for every complex number a the differential equation $(d/dt)y = ay$ has the solutions $y = Ce^{at}$, and only these solutions.

4.7 (Theory) Plane Numbers.

The position vectors of the plane, $\vec{u}, \vec{v}, \vec{w}, \cdots$ fill up the plane and behave like complex numbers with respect to addition, $\vec{u} + \vec{v}$, multiplication by a scalar, $c\vec{v}$, and norm, $\|\vec{u} + \vec{v}\| \leq \|\vec{u}\| + \|\vec{v}\|$. But complex numbers have a multiplication $\vec{u}\vec{v}$ to produce a complex number whereas plane vectors do not. This gives rise to the following question.

Problem.   Define a multiplication $\vec{u}\vec{v}$ in the space of plane vectors so that it contains the reals and preserves the ordinary real multiplication, and has the norm property $\|\vec{u}\vec{v}\| \leq \|\vec{u}\| \|\vec{v}\|$.

Solution.   We observe that the required norm property of multiplication is the natural analogue of the triangle law of addition. A vector space with such a multiplication is called a normed linear ring.

Since the real line is included, we take $\vec{1}$ as one basis vector and an unspecified vector $\vec{j}$, not parallel to $\vec{1}$, as the other. Then for any two vectors $\vec{u} = r\vec{1} + s\vec{j}$ and $\vec{v} = x\vec{1} + y\vec{j}$, we multiply $\vec{u}\vec{v}$ and find that the product is completely determined if we specify $\vec{j}^2$. Since it must be some vector, let $\vec{j}^2 = \alpha\vec{1} + \beta\vec{j}$, where α and β remain to be determined. Then $\vec{u}\vec{v} = (rx + sy\alpha)\vec{1} + (ry + sx + sy\beta)\vec{j}$. The norm property requires that for every r, s, x, y

$$[(rx + sy\alpha)^2 + (ry + sx + sy\beta)^2]^{1/2} \leq (r^2 + s^2)^{1/2}(x^2 + y^2)^{1/2}.$$

Squaring, we find that this is true if and only if

$$(*)s^2y^2(\alpha^2 + \beta^2 - 1) + 2rsxy(\alpha + 1) + 2sy\beta(sx + ry) \leq 0.$$

Since r, s, x, y are arbitrary, this must be true when $y = s = 1$ and $r = 0$. Then the condition (*) reduces to $\alpha^2 + \beta^2 - 1 + 2\beta x \leq 0$ for all x. If $\beta \neq 0$, this is false if $x > (1 - \alpha^2 - \beta^2)/2\beta$. Hence $\beta = 0$.

With $\beta = 0$ and $r = s = y = 1$, condition (*) now reduces to the requirement that for every x

$$(\alpha^2 - 1) + 2x(\alpha + 1) \leq 0.$$

If $a \neq -1$, this is false when $x > (1 - \alpha^2)/2(\alpha + 1)$. Hence $\alpha = -1$ is the only possible value and $\vec{j}^2 = \alpha\vec{1} + \beta\vec{j} = -\vec{1}$.

Conversely, we easily verify that with the multiplication

$$\vec{u}\vec{v} = (rx - sy)\vec{1} + (ry + sx)\vec{j},$$

determined by $\vec{j}^2 = -\vec{1}$, the plane vectors form a normed linear ring. This gives us the following theorem.

THEOREM.   The plane vectors $\vec{v} = x\vec{1} + y\vec{j}$ form a normed linear ring that includes the real number line if and only if $\vec{j}^2 = -\vec{1}$.

We observe that the multiplication of the plane vectors to give a normed ring is exactly the same as that for complex numbers. Hence the same device that produced division for complex numbers also produces division in the plane vectors. Finally (Exercises) we observe that actually $\|\vec{u}\vec{v}\| = \|\vec{u}\| \|\vec{v}\|$. Hence we have additional information about the plane numbers given by the following corollary.

COROLLARY (Plane number field).   The linear ring of plane vectors $\vec{v} = x\vec{1} + y\vec{j}$ with multiplication determined by $\vec{j}^2 = -1$ is a field of numbers admitting division $\vec{u}/\vec{v}$, if $\vec{v} \neq \vec{0}$. Moreover, $\|\vec{u}\vec{v}\| = \|\vec{u}\| \|\vec{v}\|$.

This investigation produces the only possible number system that fills the plane by making it out of the plane vectors, without introducing imaginary elements.

4.8   Problems

T1.   Show that for the normed linear ring of plane vectors $\|\vec{u}\vec{v}\| = \|\vec{u}\| \|\vec{v}\|$.

T2.   Prove that for the normed linear ring of plane vectors $\vec{u}\vec{v} = \vec{v}\vec{u}$.

T3.**(Undergraduate research question) If possible, find all multiplications that make the vectors of $\mathcal{R}^3$, $\vec{v} = x\vec{1} + y\vec{j} + z\vec{k}$ into a normed linear ring containing $\mathcal{R}^1$.

T4.**Show that there is one and only one way to define a multiplication in $\mathcal{R}^4$, $\vec{v} = t\vec{1} + x\vec{i} + y\vec{j} + z\vec{k}$ so that it becomes a normed ring. But show that $\vec{u}\vec{v} \neq \vec{v}\vec{u}$ when this is done.

5   Special Substitutions in Integration

5.1   Rationalizing Substitutions.

We consider an integrand that is a rational function of x and $(ax + b)^{1/k}$. The fractional powers of $ax + b$ may be advantageously eliminated by the substitution

$$ax + b = z^k, \qquad dx = \frac{k}{a} z^{k-1} \, dz.$$

Example.   Compute the antiderivative

$$\int \frac{x \, dx}{1 + \sqrt[3]{x + 5}}.$$

Solution.   We substitute $x + 5 = z^3$, $dx = 3z^2 \, dz$. Under this substitution,

$$\int \frac{x \, dx}{1 + \sqrt[3]{x + 5}} = \int \frac{(z^3 - 5)3z^2 \, dz}{1 + z}.$$

The substitution effects removal of the radical in the integrand and leaves us with a rational function to integrate. We combine terms in the numerator, carry out the indicated division and find that

$$\int \frac{x\,dx}{1 + \sqrt[3]{x+5}} = \int \left(3z^4 - 3z^3 + 3z^2 - 18z + 18 - \frac{18}{z+1}\right) dz$$

$$= \frac{3z^5}{5} - \frac{3z^4}{4} + \frac{3z^3}{3} - \frac{18z^2}{2} + 18z - 18 \ln|z+1| + C$$

$$= \frac{3}{5}(x+5)^{5/3} - \frac{3}{4}(x+5)^{4/3} + x + 5 - 9(x+5)^{2/3}$$

$$+ 18(x+5)^{1/3} - 18 \ln|(x+5)^{1/3} + 1| + C,$$

where in the last step we reintroduced x by the inverse substitution $z = (x+5)^{1/3}$.

5.2 Rational Functions of $\sin x$ and $\cos x$. When the integrand is a fraction whose numerator and denominator are both polynomials in $\sin x$ and $\cos x$, we can reduce it to a rational function of a variable z. From this point on we can use integration by partial fractions.

We begin with the Pythagorean relation $\sin^2 x + \cos^2 x = 1$ and observe that it is equivalent to

$$\frac{1 - \cos x}{\sin x} = \frac{\sin x}{1 + \cos x} = z, \qquad -\pi < x < \pi,$$

which we use to define the new variable z. The two forms for z imply two equations in $\sin x$ and $\cos x$,

$$z \cos x - \sin x = -z,$$

and

$$\cos x + z \sin x = 1.$$

Solving simultaneously for $\cos x$ and $\sin x$ in terms of z, we find that

$$\cos x = \frac{1 - z^2}{1 + z^2}, \qquad \sin x = \frac{2z}{1 + z^2}.$$

Differentiating either of these, we find that

$$dx = \frac{2dz}{1 + z^2}.$$

Under this substitution any integral of a rational fraction in powers of $\sin x$ and $\cos x$ becomes an integral of an ordinary algebraic rational function of z, and hence (II, §5) yields to integration by partial fractions.

Example A. Integrate

$$\int \frac{dx}{\cos x} = \int \sec x\,dx.$$

Solution.   Under the substitution,

$$\int \frac{dx}{\cos x} \rightarrow \int \frac{1+z^2}{1-z^2} \cdot \frac{2dz}{1+z^2} = \int \frac{2}{1-z^2} \, dx$$

$$= \int \left(\frac{1}{1+z} + \frac{1}{1-z} \right) dz = \ln|1+z| - \ln|1-z| + C$$

$$= \ln \left| \frac{1+z}{1-z} \right| + C.$$

Now if we return to the equations defining z, we find that we can write

$$\frac{1+z}{1-z} = \frac{1+\sin x}{\cos x} = \sec x + \tan x.$$

Hence

$$\int \frac{dx}{\cos x} = \ln \left| \frac{1+\sin x}{\cos x} \right| + C,$$

or

$$\int \sec x \, dx = \ln |\sec x + \tan x| + C.$$

Example B.   Calculate

$$J = \int \frac{\sin x}{2\cos x + \sin x \cos x} \, dx.$$

Solution.   Introducing z by the rationalizing substitution, we find that

$$J \rightarrow \int \frac{2z \, dz}{(1-z^2)(1+z^2)}.$$

The new integrand in z is a rational function of z that we can reduce to recognizable chains by expanding it in a sum of partial fractions.   Accordingly, we set

$$\frac{2z}{(1-z)(1+z)^3} = \frac{A}{1-z} + \frac{B}{1+z} + \frac{C}{(1+z)^2} + \frac{E}{(1+z)^3},$$

where we have taken account of the multiplicity of the denominator factor $(1+z)^3$. We multiply both members by $(1-z)(1+z)^3$, expand, and equate coefficients.   Upon solving the resulting system of equations for the unknowns A, B, C, E, we find $A = B = \frac{1}{4}$, $C = \frac{1}{2}$, $E = -1$.   With this expansion of the integrand into partial fractions, our integral becomes

$$\int \frac{2z \, dz}{(1-z^2)(1+z^2)} = \frac{1}{4} \int \frac{dz}{1-z} + \frac{1}{4} \int \frac{dz}{1+z} + \frac{1}{2} \int \frac{dz}{(1+z)^2} - \int \frac{dz}{(1+z)^3}$$

$$= \frac{1}{4} \ln \left| \frac{1+z}{1-z} \right| - \left(\frac{1}{2} \right) \frac{1}{1+z} + \left(\frac{1}{2} \right) \frac{1}{(1+z)^2} + C.$$

If we restore the x-variable by the substitution,

$$z = \frac{1-\cos x}{\sin x},$$

and simplify, we find that

$$\int \frac{\sin x \, dx}{2 \cos x + \sin x \cos x}$$

$$= \frac{1}{4} \ln \left| \frac{\sin x + 1 - \cos x}{\sin x - 1 + \cos x} \right| - \frac{1}{2} \frac{1 + \cos x}{1 + \sin x + \cos x} + \frac{1}{4} \frac{1 + \cos x}{1 + \sin x} + C.$$

This completes the problem.

5.3 Inverting Substitution.   If we attempt to integrate

$$\int_1^2 \frac{dx}{x\sqrt{1 - x^2}},$$

we encounter the difficulty that $\sqrt{1 - x^2}$ is not a real number for values of x in the interval $[1, 2]$. This would require that we reexamine the integral itself and we would expect to find that the correct integral is

$$J = \int_1^2 \frac{dx}{x\sqrt{x^2 - 1}},$$

but now this does not yield to the substitution $x = \sin \theta$.

We consider the inverting substitution

$$x = \frac{1}{t}, \qquad dx = -\frac{1}{t^2} \, dt.$$

When $x = 1$, then $t = 1$, and when $x = 2$, $t = \frac{1}{2}$. So the integral J becomes

$$J = \int_1^{1/2} \frac{(-1/t^2) \, dt}{(1/t)\sqrt{(1/t^2) - 1}} = \int_{1/2}^1 \frac{dt}{\sqrt{1 - t^2}} = \arcsin t \Big|_{1/2}^1 = \frac{\pi}{3}.$$

Example.   Integrate

$$J = \int \frac{Du}{\sqrt{u^2 - p^2}}.$$

Solution.   A table of integrals lists a formula for this antiderivative. To derive it directly, we recall that the substitution $u = p \cos \theta$ rationalizes the radical $\sqrt{p^2 - u^2}$. The inverting substitution above suggests that for $\sqrt{u^2 - p^2}$ we now set

$$u = \frac{p}{\cos \theta}, \qquad Du = p \frac{\sin \theta}{\cos^2 \theta} \, d\theta.$$

Then the integral J transforms into

$$\int \frac{\cos \theta}{\sin \theta} \frac{\sin \theta}{\cos^2 \theta} \, d\theta = \int \frac{d\theta}{\cos \theta} = \ln \left| \frac{1 + \sin \theta}{\cos \theta} \right| + C.$$

Using the symbolic triangle for $\cos \theta = p/u$ (Figure 14.16), we find that $\sin \theta = \sqrt{u^2 - p^2}/u$. Hence

$$J = \ln |u + \sqrt{u^2 - p^2}| + C.$$

This substitution is usually presented in the form $u = p \sec \theta$.

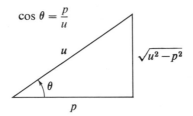

$$\cos \theta = \frac{p}{u}$$

FIGURE 14.16   Symbolic triangle.

5.4   Integrals Involving $ax^2 + bx + c$

Example.   Calculate

$$J = \int \frac{dx}{\sqrt{ax^2 + bx + c}},$$

where $a > 0$.

Solution.   We complete the square of $ax^2 + bx + c$.

$$ax^2 + bx + c = a\left(x^2 + \frac{b}{a}x + \frac{c}{a}\right) = a\left[\left(x^2 + \frac{b}{a}x + \frac{b^2}{4a^2}\right) + \left(\frac{c}{a} - \frac{b^2}{4a^2}\right)\right]$$

$$= a\left[\left(x + \frac{b}{2a}\right)^2 - \frac{b^2 - 4ac}{4a^2}\right].$$

There are three cases.

CASE I.   If $b^2 - 4ac > 0$, the integral is in the form $J = 1/\sqrt{a} \int (Du/\sqrt{u^2 - p^2})$, calculated above, with

$$u = x + \frac{b}{2a} \quad \text{and} \quad p = \frac{\sqrt{b^2 - 4ac}}{2a}, \quad \sqrt{u^2 - p^2} = \sqrt{ax^2 + bx + c}.$$

Using the result above, we have

$$J = \frac{1}{\sqrt{a}}\left(\ln \frac{|2ax + b| + \sqrt{ax^2 + bx + c}}{2a}\right) + C$$

$$= \frac{1}{\sqrt{a}} \ln \left(|2ax + b| + \sqrt{ax^2 + bx + c}\right) + C_1.$$

CASE II.   If $b^2 - 4ac < 0$, then we may set

$$u = x + \frac{b}{2a}, \quad p^2 = -\frac{b^2 - 4ac}{4a^2}, \quad \sqrt{u^2 + p^2} = \sqrt{ax^2 + bx + c},$$

and the integral J takes the form

$$J = \frac{1}{\sqrt{a}} \int \frac{Du}{\sqrt{u^2 + p^2}}.$$

In this we substitute $u = p(\sin \theta / \cos \theta) = p \tan \theta$ (II, §4.5) and find that it transforms again into

$$J \rightarrow \frac{1}{\sqrt{a}} \int \frac{d\theta}{\cos \theta} = \frac{1}{\sqrt{a}} \ln \left|\frac{1 + \sin \theta}{\cos \theta}\right|.$$

In this case the symbolic triangle is that for $\tan \theta = u/p$. It gives $\sin \theta = u/\sqrt{u^2 + p^2}$ and $\cos \theta = p/\sqrt{u^2 + p^2}$. With this we have

$$J = \frac{1}{\sqrt{a}} \ln \left(|2x + b| + \sqrt{ax^2 + bx + c} \right) + C_1$$

as in Case I.

CASE III.   If $b^2 - 4ac = 0$, then

$$\sqrt{ax^2 + bx + c} = \frac{|2ax + b|}{2\sqrt{a}}$$

and this leads directly to a logarithm form without substitution. We must observe that $|2ax + b| = 2ax + b$ when $x \geq -b/2a$, but if $x < -b/2a$, then $|2ax + b| = -(2ax + b)$. This completes the problem.

5.5 Exercises

In Exercises 1–3, verify the given integrals by means of the indicated substitutions.

1. $\int \sqrt{4x - 3} \, Dz = \frac{1}{6} \sqrt{(4x - 3)^3} + C, z^2 = 4x - 3.$

2. $\int x \sqrt{4x - 3} \, Dz = \frac{3x + 1}{20} \sqrt{(4x - 3)^3} + C, z^2 = 4x - 3.$

3. $\int \frac{x \, Dx}{\sqrt{4x - 3}} = \frac{2x + 3}{12} \sqrt{4x - 3} + C, z^2 = 4x - 3.$

4. Show that the substitution

$$\frac{1 - \cos x}{\sin x} = \frac{\sin x}{1 + \cos x} = z$$

is equivalent to $z = \tan (x/2)$. Show that there is no ambiguity of signs in this identity.

In Exercises 5–16, integrate by an appropriate substitution.

5. $\int \frac{dx}{\sin x}.$

6. $\int \frac{dx}{\cos^3 x}.$

7. $\int \frac{dx}{1 + \cos x} = \frac{1 - \cos x}{\sin x} + C.$

8. $\int \frac{\cos x \, dx}{2 - \cos x} = \frac{4}{\sqrt{3}} \arctan \left(\sqrt{3} \tan \frac{x}{2} \right) + C.$

9. $\int \frac{d\theta}{1 + \sin \theta + \cos \theta}.$

10. $\int \sqrt{\frac{1 - \cos x}{1 + \cos x}} \, dx.$

11. $\int \frac{dx}{\sqrt{1 - 4x^2}} = 1 \arcsin 2x + C, x = \frac{1}{2} \sin \theta.$

12. $\int (9 - 4x^2)^{3/2} \, dx = \frac{x}{8} (45 - 8x^2) \sqrt{9 - 4x^2} + \frac{273}{16} \arcsin \frac{2x}{3} + C.$

13. $\int \frac{\sqrt{2x^2 - 9}}{x} \, dx = \sqrt{2x^2 - 9} - 3 \arctan \frac{\sqrt{2x^2 - 9}}{3} + C.$

14. $\int \frac{dx}{\sqrt{4x^2 + 1}} = \frac{1}{2} \ln |2x + \sqrt{4x^2 + 1}| + C, 2x = \tan \theta.$

15. $\int \dfrac{\sqrt{x}\,dx}{1 + \sqrt[4]{x}} = \dfrac{4}{5}(\sqrt[4]{x})^5 - x + \dfrac{4}{3}(\sqrt[4]{x})^3 - 2\sqrt{x} + 4\sqrt[4]{x} - 4 \ln |1 + \sqrt[4]{x}| + C.$

16. $\int xe^{-\sqrt{x}}\,dx = 2e^{-\sqrt{x}}(x^{3/2} + 3x + 6x^{1/2} + 6) + C.$

In Exercises 17–20, compute the definite integrals. We recall that it is possible to change the limits of integration into appropriate ones for the new variables.

17. $\int_0^a \sqrt{a^2 - x^2}\,dx.$ 18. $\int_0^{\sqrt{3}a} \sqrt{a^2 + x^2}\,dx.$

19. $\int_0^{\pi/2} \dfrac{d\theta}{1 + \cos \theta}.$ 20. $\int_0^{5a} \sqrt{x^2 - a^2}\,dx.$

In Exercises 21–24, we will use a variant of the substitution (§5.2)

$$\frac{1 - \cos x}{\sin x} = \frac{\sin x}{1 + \cos x} = z.$$

21.   In the relation $x^2 + y^2 = 1$, defining a circle of radius 1, we observe that we may define z by

$$\frac{1 - x}{y} = \frac{y}{1 + x} = z \qquad |x| \le 1.$$

Solve for x, y, dx in terms of z to show that

$$x = \frac{1 - z^2}{1 + z^2}, \qquad y = \frac{2z}{1 + z^2}, \qquad dx = \frac{4z\,dz}{(1 + z^2)^2}.$$

22.   Use the substitution of Exercise 25 to integrate

$$\int \frac{dx}{x\sqrt{1 - x^2}}$$

without a trigonometric substitution.

23.   Use the substitution of Exercise 21 to integrate

$$\int_0^1 \sqrt{\frac{1 - x}{1 + x}}\,dx.$$

24.   Find a substitution and calculate

$$\int_1^2 \sqrt{\frac{x - 1}{x + 1}}\,dx.$$

In Exercises 25–30, integrate by completing the square and then making a substitution.

25. $\int \sqrt{x^2 + 2x}\,dx.$ 26. $\int \dfrac{dx}{\sqrt{x^2 + x + 1}}.$

27. $\int_0^1 \dfrac{dx}{x^2 + 2x + 2}.$ 28. $\int \sqrt{2x - x^2}\,dx.$

29. $\int \dfrac{dx}{x^2 - x + 2}.$ 30. $\int_0^1 \dfrac{e^x + e^{-x}}{e^{2x} + e^{-2x}}\,dx.$

31.   Find the area enclosed by the hyperbola $(x^2/9) - (y^2/16) = 1$ and the line $x = 5$.

32.   Show that the substitution for integrating rational fractions in powers of **sin** x and **cos** x (§5.2) can be expressed as $z = \tan (x/2)$. This is the customary form.

6   Advanced Trigonometric Techniques in Integration

6.1   Cotangent, Secant, and Cosecant.   It is convenient, but not necessary, to use more elaborate analytic trigonometry in calculus (§4.4). In particular, we can use the cotangent, secant, and cosecant defined by

$$\cot x = \frac{\cos x}{\sin x}, \qquad \text{where } \sin x \neq 0,$$

$$\sec x = \frac{1}{\cos x}, \qquad \text{where } \cos x \neq 0,$$

and

$$\csc x = \frac{1}{\sin x}, \qquad \text{where } \sin x \neq 0.$$

The Pythagorean identity $\sin^2 x + \cos^2 x = 1$ then implies that

$$1 + \tan^2 x = \sec^2 x \qquad \text{and} \qquad \cot^2 x + 1 = \csc^2 x.$$

We readily verify the following formulas.

XII. $D \tan u = \sec^2 u\, Du$ and $\int$ XII. $\int \sec^2 u = \tan u + C.$

XIII. $D \sec u = \sec u \tan u\, Du$ and $\int$ XIII. $\int \sec u \tan u\, Du = \sec u + C.$

$\int$ XIII$_a$. $\int \sec u\, Du = \ln |\sec u + \tan u| + C.$

We derived $\int$ XIII$_a$ in the form $\int (Du/\cos u)$ (Example A, §5.2).

6.2   Integration of Powers of Secant and Tangent.   We can use the trigonometric relation $1 + \tan^2 u = \sec^2 u$ together with $D \tan u = \sec^2 u\, Du$ and $D \sec u = \sec u \tan u\, Du$ to calculate the antiderivative $\int \tan^n u\, Du$, when n is any integer, and $\int \tan^m u \sec^n u\, Du$ or $\int \sec^n u\, Du$, when n is an even integer.

Example A.   Compute $\int \tan^4 ax\, dx.$

Solution.   We factor out one $\tan^2 ax$ factor and transform it to $\sec^2 ax$. Thus

$$\tan^4 ax = \tan^2 ax(\sec^2 ax - 1) = \tan^2 ax \sec^2 ax - \tan^2 ax.$$

Now the first term is readily made into the chain $(\tan ax)^2(\sec^2 ax\, D\, ax)$. We then replace the last term $\tan^2 ax$ by $\sec^2 ax - 1$, which also readily forms a chain. In this way we obtain

$$\int \tan^4 ax\, dx = \int \tan^2 ax(\sec^2 ax - 1)\, dx$$

$$= \int \tan^2 ax \sec^2 ax\, dx - \int \tan^2 ax\, dx$$

$$= \frac{1}{a} \int \tan^2 ax(\sec^2 ax)(a\, dx) - \frac{1}{a} \int \sec^2 ax(a\, dx) + \int dx$$

$$= \frac{1}{a} \frac{\tan^3 ax}{3} - \frac{1}{a} \tan ax + x + C.$$

This completes the problem.

Example B.   Compute the antiderivative $\int \tan^5 ax \sec^4 ax \, dx$.

Solution.

$$\int \tan^5 ax \sec^4 ax \, dx = \int \tan^6 ax \sec^2 ax \sec^2 ax \, dx$$

$$= \int \tan^5 ax(1 + \tan^2 ax) \sec^2 ax \, dx$$

$$= \int \tan^5 ax \sec^2 ax \, dx + \int \tan^7 ax \sec^2 ax \, dx.$$

We then make power-formula chains with $u = \tan ax$, $Du = \sec^2 ax(a \, dx)$ and complete the calculation as follows.

$$\int \tan^5 ax \sec^4 ax \, dx = \frac{1}{a} \int \tan^5 ax[\sec^2 ax(a \, dx)] + \frac{1}{a} \int \tan^7 ax[\sec^2 ax(a \, dx)]$$

$$= \frac{1}{a} \frac{\tan^6 ax}{6} + \frac{1}{a} \frac{\tan^8 ax}{8} + C.$$

Example C.   Find the antiderivative $\int \sec^3 x \, dx$.

Solution.   We apply integration by parts with

$$u = \sec x \qquad \text{and} \qquad Dv = \sec^2 x \, dx,$$

so that

$$Du = \sec x \tan x \, dx \qquad \text{and} \qquad v = \tan x.$$

This gives us

$$\int \sec^3 x \, dx = \sec x \tan x - \int \tan x(\sec x \tan x) \, dx$$

$$= \sec x \tan x - \int \sec x \tan^2 x \, dx$$

$$= \sec x \tan x - \int \sec x(\sec^2 x - 1) \, dx$$

$$= \sec x \tan x + \int \sec x \, Dx - \int \sec^3 x \, dx.$$

We recognize the original integral with a different coefficient. So we transpose the last term and combine it with the left member to give $2 \int \sec^3 x \, dx$. Dividing both sides by 2 and adding an arbitrary constant, we find

$$\int \sec^3 x \, dx = \tfrac{1}{2} \sec x \tan x + \tfrac{1}{2} \ln |\sec x + \tan x| + C.$$

6.3   Integrands Involving $\sqrt{a^2 - x^2}$, $\sqrt{x^2 - a^2}$, $\sqrt{x^2 + a^2}$.   We have seen (II, §4.2) that the substitution $x = a \sin \theta$, $dx = a \cos \theta \, d\theta$ into the antiderivative form $\int dx/\sqrt{a^2 - x^2}$ involving $\sqrt{a^2 - x^2}$ transforms it into a simpler one,

$$\int \frac{dx}{\sqrt{a^2 - x^2}} = \int \frac{a \cos \theta \, d\theta}{|a| \, |\cos \theta|} = \pm \int d\theta,$$

where the sign to be chosen in the last term is that of $a \cos \theta$. We integrate the last term and carry out the inverse substitution $\theta = \arcsin(x/a)$. Since $\arcsin(x/a)$ denotes the principal value, we know that $-\pi/2 \leq \arcsin(x/a) \leq \pi/2$. Hence

$$\int \frac{dx}{\sqrt{a^2 - x^2}} = \pm \int d\theta = \pm\theta + C = \arcsin\frac{x}{a} + C.$$

We consider integrands that involve $\sqrt{x^2 - a^2}$, where the substitution $x = a \sec\theta$ is effective. We have treated it before using the inverting substitution $x = 1/z$ (§5.3).

Example A. Integrate $\int \sqrt{x^2 - a^2}\, dx$.

Solution. Let $x = a \sec\theta$ so that $dx = a \sec\theta \tan\theta\, d\theta$. In the presence of this transformation we have

$$\sqrt{x^2 - a^2} = |a|\sqrt{\sec^2\theta - 1} = |a \tan\theta|.$$

Then the integral becomes

$$\int \sqrt{x^2 - a^2}\, Dx = \int |a \tan\theta|\, a \sec\theta \tan\theta\, d\theta$$

$$= \pm\, a^2 \int \sec\theta \tan^2\theta\, d\theta,$$

where the sign to be chosen is that of $\tan\theta$. We integrate this by parts, using the cyclic device (II, §7.2), and we ultimately find that

$$a^2 \int \sec^2\theta \tan^2\theta\, d\theta = \tfrac{1}{2}\, a^2[\sec\theta \tan\theta - \ln|\sec\theta + \tan\theta|] + C.$$

On restoring the x-variables by the inverse substitution $\theta = \mathbf{arcsec}\,(x/a)$, using the symbolic triangle (Figure 14.17) to find $\tan\theta = \pm\sqrt{x^2 - a^2}/a$, and taking account of the signs, we have the final result

$$\int \sqrt{x^2 - a^2}\, Dx = \tfrac{1}{2}[x\sqrt{x^2 - a^2} - a^2 \ln|x + \sqrt{x^2 - a^2}|] + C.$$

This completes the problem.

Now we consider integrands involving $\sqrt{x^2 + a^2}$. Here we may use the substitution $x = a \tan\theta$.

Example B. Calculate the antiderivative $\int \sqrt{x^2 + a^2}\, dx$ (§5.4).

Solution. We make the substitution $x = a \tan\theta$, which gives us

$$Dx = a \sec^2\theta\, D\theta \qquad \text{and} \qquad \sqrt{x^2 + a^2} = |a|\sqrt{\tan^2\theta + 1} = |a \sec\theta|.$$

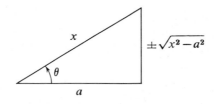

FIGURE 14.17   arcsec x/a.

Under this substitution, we have

$$\int \sqrt{x^2 + a^2}\, dx = \int |a \sec \theta|\, a \sec^2 \theta\, d\theta = \pm a^2 \int \sec^3 \theta\, d\theta,$$

where the sign is that of a. We integrate (§6.2, Example C) and find that

$$\int \sec^3 \theta\, d\theta = \tfrac{1}{2}\sec \theta \tan \theta + \tfrac{1}{2} \ln |\sec \theta + \tan \theta| + C.$$

Then, restoring this to functions of x by the inverse substitution $\theta = \arctan x/a$, taking proper account of signs, we find that

$$\int \sqrt{x^2 + a^2}\, dx = \tfrac{1}{2}[x\sqrt{x^2 + a^2} + \ln |x + \sqrt{x^2 + a^2}|] + C.$$

This completes the problem.

6.4   Use of Double-Angle Identities.   We have performed integration

$$\int \cos^m ax \sin ax\, dx$$

by reduction of power, using integration by parts (II, §4.4). An alternative procedure makes use of repeated application of the trigonometric identities

$$\cos^2 x = \tfrac{1}{2}(1 + \cos 2x) \qquad \text{and} \qquad \sin^2 x = \tfrac{1}{2}(1 - \cos 2x).$$

Example.   Compute the antiderivative

$$\int \cos^6 x \sin^2 x\, dx.$$

Solution.   We reduce the exponents by writing

$$\cos^6 x = (\cos^2 x)^3 = \left(\frac{1 + \cos 2x}{2}\right)^3$$

$$= \tfrac{1}{8}(1 + 3 \cos^2 2x + 3 \cos^2 2x + \cos^3 2x),$$

and

$$\sin^2 x = \tfrac{1}{2}(1 - \cos 2x),$$

so that

$$\cos^6 x \sin^2 x = \tfrac{1}{16}(1 + 2 \cos 2x - 2 \cos^3 2x - \cos^4 2x).$$

We continue the trigonometric reduction with appropriate transformation for the odd power, $\cos^3 2x$, and even power, $\cos^4 2x$, and obtain

$$\cos^6 x \sin^2 x = \tfrac{1}{16}[1 + 2 \cos 2x - 2(1 - \sin^2 2x) \cos 2x - \tfrac{1}{4}(1 + \cos 4x)^2]$$

$$= \tfrac{1}{16}[1 + 2 \sin^2 2x \cos 2x - \tfrac{1}{4} - \tfrac{1}{2} \cos 4x + \tfrac{1}{4} \cos^2 4x].$$

Finally, we apply one more reduction on the last term, replacing $\cos^2 4x$ by $\tfrac{1}{2}(1 + \cos 8x)$, and find

$$\cos^6 x \sin^2 x = \tfrac{1}{16}[\tfrac{3}{4} + 2 \sin^2 2x \cos 2x - \tfrac{3}{8} \cos 4x + \tfrac{1}{8} \cos 8x].$$

We can now compute the antiderivative term by term. We get

$$\int \cos^6 x \sin^2 x\, dx = \frac{1}{16}\left(\frac{3}{4} x + \frac{\sin^3 2x}{3} - \frac{3}{32} \sin 4x + \frac{1}{64} \sin 8x\right) + C.$$

This completes the problem.

6.5 Exercises

In Exercises 1–6, differentiate.

1.	$D \tan^3 2x$.	2.	$D \sec^3 u$.
3.	$D \sec 2x \tan 2x$.	4.	$D \csc x$.
5.	$D \ln |\csc x - \cot x|$.	6.	$D \operatorname{arcsec} x$.

7. Prove that $\operatorname{arcsec} x = \arccos (1/x)$, if $|x| \geq 1$.
8. Simplify $\tan (\operatorname{arcsec} x)$, where $|x| \geq 1$.
9. Simplify $\tan \frac{1}{2} (\arccos x)$, where $|x| \leq 1$.
10. Show that $D \cot x = \csc^2 x \, dx$.

In Exercises 11–16, find the indicated antiderivatives.

11. $\int \tan^3 x \, Dx$.   12. $\int \tan ax \sec^2 ax \, Dx$.

13. $\int \tan^2 x \sec x \, Dx$.   14. $\int \sec^4 ax \, Dx$.

15. $\int (\tan x / \sec x) \, Dx$.   16. $\int \csc^4 x \, Dx$.

17. Show that $\int \csc x \, Dx = \ln |\csc x - \cot x| + C$.

18. Show that $\int \csc ax \, Dx = \frac{1}{a} \ln \tan \frac{ax}{2} + C$.

19. Show that $\int \csc^3 ax \, Dx = \frac{1}{2a} \left[-\csc ax \cot ax + \ln \tan \frac{ax}{2} \right] + C$.

20. Show that $\int \tan^n ax \sec^2 ax \, Dx = \frac{1}{a(n + 1)} \tan^{n+1} ax, \ n \neq -1$.

21. Show that $\int \tan ax \sec^n ax \, Dx = \frac{1}{an} \sec^n ax, \ n \neq 0$.

In Exercises 22–25, integrate using the double-angle identities of trigonometry.

22. $\int \sin^4 x \, dx$.   23. $\int \sin^2 x \cos^2 x \, dx$.

24. $\int \cos^4 ax \, dx$.   25. $\int \sin^6 ax \, dx$.

In Exercises 26–41, integrate.

26. $\int \dfrac{dx}{\sqrt{x^2 - 4}}$.   27. $\int \dfrac{dx}{\sqrt{3 - 4x^2}}$.

28. $\int_0^a \dfrac{dx}{(x^2 + a^2)^2}$.   29. $\int \sec \theta \, (\sec \theta + \tan \theta) \, d\theta$.

30. $\int_0^a \dfrac{dx}{\sqrt{x^2 + a^2}}$.   31. $\int \operatorname{arcsec} \dfrac{1}{x} \, dx$.

32. $\int_0^\pi \sin^2 x \cos^2 x \, dx$.   33. $\int_0^1 x^2 \sqrt{1 - x^2} \, dx$.

34. $\int_0^{1/2} x^2 \sqrt{1 - 4x^2} \, dx$.   35. $\int \sin^4 3x \cos^2 3x \, dx$.

36. $\int_0^{2\pi} \sin x \cos 2x \, dx$.   37. $\int_0^1 \dfrac{\sqrt{1 - x^2}}{x^2} \, dx$ (Improper).

38. $\int_{\pi/4}^{\pi/2} \dfrac{\sec^2 x}{\tan^2 x} \, dx$ (Improper).   39. $\int t \sec t^2 \, dt$.

40. $\int \tan^3 2x \, dx$.   41. $\int \left[1 + \tan \dfrac{x}{a} \right]^2 \, dx$.

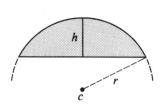

FIGURE 14.18   Exercise 50.

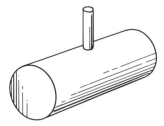

FIGURE 14.19   Exercise 52.

42.   The trigonometric substitution $u = a \sec \theta$, can be applied in some cases to integrands involving $\sqrt{ax^2 + bx + c}$. To do this, we complete the square under the radical. Observe that

$$\int \sqrt{4x^2 - 4x - 3}\ Dx = \int \sqrt{(2x - 1)^2 - 4}\ Dx,$$

and complete the integration by trigonometric substitution.

43.   Substitution of $\sec \theta$, as in Exercise 42, will be effective for $\int \sqrt{ax^2 + bx + c}\ dx$ if and only if what conditions on the signs of a and $b^2 - 4ac$ are satisfied?

44. $\int_1^2 \sqrt{2x - x^2}\ dx.$

45. $\int_0^1 \sqrt{1 - x^2} \arcsin x\ dx.$

46. $\int_0^{\sqrt{3}} \dfrac{\arctan x}{1 + x^2}\ dx.$

47. $\int \dfrac{\sqrt{x^2 - 1}}{x}\ dx.$

48. $\int_1^{-2} \dfrac{dx}{\sqrt{(1 - x)(2 + x)}}$ (Improper).

49. $\int e^{\tan x} \sec^2 x\ dx.$

50.   Find the area of a circular segment of altitude h cut from a circle of radius r by a chord (Figure 14.18).
51.   Find the coordinates of the center of area (centroid) of the segment in Exercise 50.
52.   A horizontal water tank with closed circular ends is 4 ft in diameter and is supplied by a vertical pipe in the top (Figure 14.19). If the tank is full and the water stands 8 ft above the top of the tank in the supply tank, find the total force due to fluid pressure on one end of the tank (I, §22.5).

7 Hyperbolic Functions

ACADEMIC NOTE.   We introduce the hyperbolic functions here because they are encountered frequently in the literature of applications of calculus. They are not essential to the analytic program of calculus because suitable combinations of the exponential functions e^x and e^{-x} can replace them, as we shall see. To save time, one can start with the definition at the end of §7.2.

7.1   Circular Functions.   Let us review briefly the calculus definition (I, §31.2) of the trigonometric functions **cos u** and **sin u,** and observe that, so defined, they might more appropriately be called *circular functions*, if it were not for the historical accident of their origin in the measurement of right triangles.

Starting with the unit circle $x^2 + y^2 = 1$ and a rotating vector $\overrightarrow{OP}$ from the origin to a point $P:(x, y)$ on the circle, we define the radian measure u of the rotation of $\overrightarrow{OP}$, as it rotates counterclockwise, by $u = 2A$, where A is the area of the sector MOP (Figure 14.20) swept out behind $\overrightarrow{OP}$. Then we define the circular functions **sin u, cos u, tan u**, for each u by

$$\cos u = x, \qquad \sin u = y, \qquad \tan u = \frac{y}{x}.$$

7.2   Definition of the Hyperbolic Functions.   If, instead of using the circle $x^2 + y^2 = 1$, we apply the same pattern to the hyperbola $x^2 - y^2 = 1$, we get some new-looking functions called *hyperbolic sine*, *hyperbolic cosine*, and *hyperbolic tangent*, written symbolically **sinh u, cosh u, tanh u.**

Explicitly, starting with the connected branch of the hyperbola, $x^2 - y^2 = 1$, in which $x > 0$ (Figure 14.21), and the rotating vector $\overrightarrow{OP}$ from the origin to the point $P:(x, y)$ on the hyperbola, we define the *hyperbolic radian* measure of the rotation of $\overrightarrow{OP}$ as it rotates counterclockwise. This is $u = 2A$, where A is the area of the sector MOP, swept out behind $\overrightarrow{OP}$ from its initial position $\overrightarrow{OM}$. Then we define the hyperbolic functions for each u by

$$\cosh u = x, \qquad \sinh u = y, \qquad \tanh u = \frac{y}{x}.$$

It remains to calculate the area A of the sector MOP and to use the result to express $\cosh u$, $\sinh u$, $\tanh u$ in terms of computable formulas. We solve the equation $x^2 - y^2 = 1$ for y in the case $y \geq 0$, that is, in the first quadrant, where $y = \sqrt{x^2 - 1}$, and compute the area $2A$ by integration. We find

$$u = 2A = xy - 2\int_1^x y\,dx = x\sqrt{x^2 - 1} - 2\int_1^x \sqrt{x^2 - 1}\,dx$$
$$= x\sqrt{x^2 - 1} - [x\sqrt{x^2 - 1} - \ln|x + \sqrt{x^2 - 1}|] = \ln(x + y).$$

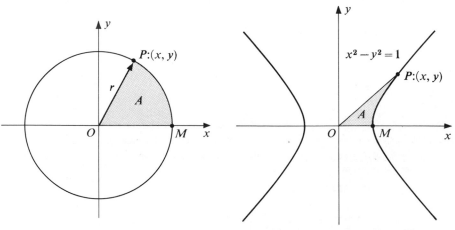

FIGURE 14.20   Circular functions.   FIGURE 14.21   Hyperbolic functions.

From the inverse of this relation we find that $x + y = e^u$.

$$e^{-u} = \frac{1}{x + y} = \frac{x - y}{x^2 - y^2} = x - y.$$

Solving these two linear equations simultaneously for x and y, recalling that $x = \cosh u$ and $y = \sinh u$, we find

$$\cosh u = \frac{e^u + e^{-u}}{2} \quad \text{and} \quad \sinh u = \frac{e^u - e^{-u}}{2}.$$

Whence is derived

$$\tanh u = \frac{e^u - e^{-u}}{e^u + e^{-u}}.$$

The domain of definition of each of these functions is the entire real line,

$$-\infty < u < +\infty$$

for as the vector $\overrightarrow{OP}$ rotates counterclockwise, $x \to +\infty$. As a function of x, u is given by

$$u = x\sqrt{x^2 - 1} - 2 \int_1^x \sqrt{x^2 - 1}\, dx.$$

The integral becomes infinite as the upper limit x becomes infinite. A similar analysis applies to the lower quadrant, where $y = -\sqrt{x^2 - 1}$, and we conclude that the domain of values of the hyperbolic radian u in this process is $-\infty < u < +\infty$. Finally, we observe that the denominator $e^u + e^{-u}$ in the formula for tanh u is nowhere zero. Hence the same domain of all real numbers applies to **tanh u** also.

Thus it turns out that the new-looking hyperbolic functions are simple combinations of the well-known functions e^u and e^{-u}. It is mathematically simpler for us to forget about the hyperbola and use these exponential formulas as definitions.

DEFINITION.   The hyperbolic functions **cosh x**, **sinh x**, and **tanh x** are defined for all real numbers x by

$$\cosh x = \frac{e^x + e^{-x}}{2}, \quad \sinh x = \frac{e^x - e^{-x}}{2}, \quad \tanh x = \frac{e^x - e^{-x}}{e^x + e^{-x}}.$$

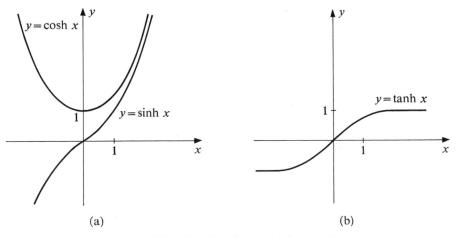

(a)   (b)

FIGURE 14.22   Graphs of hyperbolic functions.

7.3 Graphs of Hyperbolic Functions. Published tables of values of the hyperbolic functions are available. Alternatively, the functions can easily be evaluated from values of e^x and e^{-x}. These values enable us to plot the graphs of the hyperbolic functions (Figure 14.22).

7.4 Identities for Hyperbolic Functions. From their definitions in terms of exponentials we can show that for all x:

(1) $\cosh^2 x - \sinh^2 x = 1$.
(2) $\sinh(-x) = -\sinh x, \cosh(-x) = \cosh x$.
(3) $\sinh(x + y) = \sinh x \cosh y + \cosh x \sinh y$.
(4) $\cosh(x + y) = \cosh x \cosh y - \sinh x \sinh y$.
(5) $\sinh 2x = 2 \sinh x \sinh y$.
(6) $\cosh 2x = \cosh^2 x + \sinh x = 2 \cosh^2 x - 1 = 1 + 2 \sinh^2 x$.
(7) $\sinh^2 x = \dfrac{\cosh 2x - 1}{2}, \cosh^2 x = \dfrac{1 + \cosh 2x}{2}$.
(8) $e^x = \cosh x + \sinh x, e^{-x} = \cosh x - \sinh x$.

If we define $\operatorname{sech} x = \dfrac{1}{\cosh x}$, we find that

(9) $\tanh^2 x + \operatorname{sech}^2 x = 1$.

We observe the similarity of these formulas, except for certain variations of sign, with the corresponding formulas for the trigonometric functions. This is not surprising in view of the definitions from the hyperbola, $x^2 - y^2 = 1$, and the circle, $x^2 + y^2 = 1$. One obvious difference, however, is that the hyperbolic functions are not periodic.

7.5 Derivatives and Integrals of Hyperbolic Functions. By differentiating in the definitions in terms of exponentials we readily find that

XVII. $D \sinh u = \cosh u \, Du,$ $\int$ XVII. $\int \cosh u \, Du = \sinh u + C.$

XVIII. $D \cosh u = \sinh u \, Du,$ $\int$ XVIII. $\int \sinh u \, Du = \cosh u + C.$

XIX. $D \tanh u = \operatorname{sech}^2 u \, Du,$ $\int$ XIX. $\int \operatorname{sech}^2 u \, Du = \tanh u + C.$

7.6 Inverses of Hyperbolic Functions. The inverse relation of $y = \sinh x$ is $x = \sinh y$. If, in this inverse relation, each x determines one and only one y, we can solve for y in terms of x and write the solution as an inverse function $y = \sinh^{-1} x$. Let us carry this out explicitly.

We write the inverse relation

$$x = \sinh y = \frac{e^y - e^{-y}}{2} = \frac{(e^y)^2 - 1}{2e^y}$$

since for every y, $e^y \neq 0$. This is equivalent to the quadratic equation

$$(e^y)^2 + 2x(e^y) - 1 = 0$$

in the unknown e^y. Solving it, we find

$$e^y = -x + \sqrt{x^2 + 1}, \quad \text{or} \quad e^y = -x - \sqrt{x^2 + 1}.$$

Only the first gives the positive values necessary for e^y. Hence $x = \sinh y$ determines one and only one value of e^y,

$$e^y = \sqrt{x^2 + 1} - x,$$

and it is valid for all real values of x. We solve this exponential equation for y and find

$$y = \ln\left(\sqrt{x^2 + 1} - x\right).$$

Therefore the inverse of **sinh** x is a function defined on all real numbers x and it is given by

$$\mathbf{sinh}^{-1} x = \ln\left(\sqrt{x^2 + 1} - x\right).$$

It is not surprising to find the inverse function, **sinh**$^{-1}$ x, expressible in terms of the natural logarithm in view of the fact that **sinh** x is expressible in terms of the exponential.

We can differentiate this inverse function directly and find, after some simplification, that

XX. $D \, \mathbf{sinh}^{-1} u = \dfrac{1}{\sqrt{u^2 + 1}} \, Du,$ $\displaystyle\int$XX. $\displaystyle\int \dfrac{Du}{\sqrt{u^2 + 1}} = \mathbf{sinh}^{-1} u + C.$

The inverse relation of $y = \cosh x$ is $x = \cosh y$ and it is not one-to-one. As in the case of the inverse trigonometric functions, we must choose a principal value to make **cosh**$^{-1}$ x a function. We do this by choosing the solution for which $y \geq 0$ as the principal value (Exercises), which gives us the inverse function

$$\mathbf{cosh}^{-1} x = \ln\left(x + \sqrt{x^2 - 1}\right), \quad x \geq 1.$$

We find after some simplification that this yields the derivative and antiderivative

XXI. $D \, \mathbf{cosh}^{-1} u = \dfrac{1}{\sqrt{u^2 - 1}} \, Du,$ $\displaystyle\int$XXI. $\displaystyle\int \dfrac{Du}{\sqrt{u^2 - 1}} = \mathbf{cosh}^{-1} u + C,$

$$u > 1.$$

Finally, the inverse relation to $y = \tanh x$ is $x = \tanh y$, and it turns out to have a unique solution

$$\mathbf{tanh}^{-1} x = \frac{1}{2} \ln \frac{1 + x}{1 - x}, \quad |x| < 1,$$

which yields the derivative and antiderivative formulas

XXII. $D \, \mathbf{tanh}^{-1} u = \dfrac{1}{1 - u^2} \, Du,$ $\displaystyle\int$XXII. $\displaystyle\int \dfrac{Du}{1 - u^2} = \mathbf{tanh}^{-1} u + C,$

$$|u| < 1.$$

7.7   Integration by Hyperbolic Function Substitution

Example.   Calculate the integral

$$\int_a^{2a} \frac{x^2}{\sqrt{x^2 - a^2}} \, dx.$$

Solution.   The integrand becomes infinite when $x = a$. Hence this is an improper integral. We make the substitution

$$x = a \cosh z, \qquad dx = a \sinh z \, dz.$$

We also transform the limits of integration to z-limits.
   When $x = a$,

$$\cosh z = 1 \qquad \text{and} \qquad z = 0.$$

When $x = 2a$,

$$\cosh z = 2, \qquad \sinh z = \sqrt{\cosh^2 z - 1} = \sqrt{3},$$

and

$$z = \cosh^{-1} 2 = \ln (2 + \sqrt{3}).$$

Under the transformation, the antiderivative

$$\int \frac{x^2 \, dx}{\sqrt{x^2 - a^2}} = \int \frac{a^2 \cosh^2 z (a \sinh z \, dz)}{(a \sinh z)} = a^2 \int \cosh^2 z \, dz.$$

We apply integration by parts with

$$u = \cosh z, \qquad dv = \cosh z \, dz,$$
$$Du = \sinh z \, dz, \qquad v = \sinh z,$$

which gives us

$$\int \cosh^2 z \, dz = \cosh z \sinh z - \int \sinh^2 z \, dz$$
$$= \cosh z \sinh z - \int \cosh^2 z \, dz + \int 1 \, dz.$$

Transposing the term, $-\int \cosh^2 z \, dz$, we have

$$\int \cosh^2 z \, dz = \frac{1}{2} \cosh z \sinh z + \frac{z}{2} + C.$$

With this antiderivative we apply to the fundamental theorem to find that

$$\int_a^{2a} \frac{x^2 \, dx}{\sqrt{x^2 - a^2}} = a^2 \int_0^{\ln (+\sqrt{3})} \cosh^2 z \, dz = \left[\frac{1}{2} \cosh z \sinh z + z \right] \Big|_0^{\ln(2 + \sqrt{3})}$$
$$= \frac{a^2}{2} [2\sqrt{3} + \ln (2 + \sqrt{3})].$$

We observe that the transformed integral was no longer an improper one after $a \sinh z$ was cancelled from numerator and denominator. This completes the problem.

7.8 *Exercises*

1.   Using values of e^x and e^{-x}, construct the graph of $y = \cosh x$ (§7.3) and verify that $1 \leq \cosh x < +\infty$.
2.   Construct the graph of $y = \sinh x$ (§7.3).
3.   Prove from the exponential definitions the identities (1)–(7) of §7.4.
4.   Verify the formulas for $D \cosh u$, $D \sinh u$, $D \tanh u$.

In Exercises 5–16, differentiate.

5. $\sinh 3x$.
6. $a \cosh^2 (x/a)$.
7. $\tanh kx$.
8. $\operatorname{sech}^3 2x$.
9. $\ln \cosh x$.
10. $\ln \tanh x$.
11. $\cosh^2 3x - \sinh^2 3x$.
12. $\sinh^{-1} (x/4)$.
13. $\cosh (\cosh^{-1} x)$.
14. $\tanh^{-1} \cos x$.
15. $\cosh^{-1} (x/a)$.
16. $\exp (\tanh^{-1} x)$.

In Exercises 17–28, integrate in terms of hyperbolic and inverse hyperbolic functions.

17. $\displaystyle\int \cosh 2x \, dx$.
18. $\displaystyle\int \cosh 2x \sinh 2x \, dx$.

19. $\displaystyle\int \cosh^3 ax \sinh ax \, dx$.
20. $\displaystyle\int \sinh^6 kx \cosh^3 kx \, dx$.

21. $\displaystyle\int \frac{\sinh ax}{\cosh ax} \, dx$.
22. $\displaystyle\int \frac{dx}{\sqrt{4 + x^2}}$.

23. $\displaystyle\int \frac{dx}{\sqrt{x^2 - a^2}}$.
24. $\displaystyle\int \sinh^2 ax \, dx$.

25. $\displaystyle\int \tanh^2 x \, dx$.
26. $\displaystyle\int \frac{dx}{25 - x^2}$.

27. $\displaystyle\int \cosh^{-1} x \, dx$.
28. $\displaystyle\int \sqrt{x^2 - 4} \, dx$.

In Exercises 29–32, calculate the definite integrals by substitution of a hyperbolic function. Substitute $x = a \sinh z$ in $\sqrt{x^2 + a^2}$.

29. $\displaystyle\int_0^3 \frac{dx}{\sqrt{x^2 + 9}}$.
30. $\displaystyle\int_1^2 \frac{dx}{x^2 - 9}$.

31. $\displaystyle\int_4^5 \sqrt{x^2 - 16} \, dx$.
32. $\displaystyle\int_0^4 \sqrt{x^2 + 9} \, dx$.

In Exercises 33–36, evaluate the definite integrals.

33. $\displaystyle\int_y^{2a} \cosh^{-1} \frac{x}{a} \, dx$.
34. $\displaystyle\int_0^2 \frac{dx}{x^2 - 16}$.

35. $\displaystyle\int_4^5 \frac{dx}{\sqrt{x^2 - 16}}$.
36. $\displaystyle\int_0^a \frac{dx}{\sqrt{x^2 + a^2}}$.

37. Evaluate, observing that the antiderivative $\tanh^{-1} (x/4)$ does not apply:
$$\int_0^\infty \frac{dx}{x^2 - 16}.$$

38. Evaluate $\displaystyle\int_0^\infty \frac{dx}{1 + \cosh x}$.

39. Prove that for every n, $(\cosh x + \sinh x)^n = \cosh nx + \sinh nx$.

40. Show that $\cosh x - \sinh x = e^{-x}$. Show that hence the graphs of $\cosh x$ and $\sinh x$ do not intersect but if $x > 10$, $\cosh x - \sinh x < 0.00005$.

41. Prove the formulas for $D \cosh^{-1} x$, $D \sinh^{-1} x$, $D \tanh^{-1} x$.

42. A curve in the plane has the parametric equations $x = a \cosh kt$, $y = a \sinh kt$ in terms of the real parameter t. Eliminate the parameter to find the xy-equation of the curve.

43. Show that both $y = \cosh ax$ and $y = \sinh ax$ satisfy the differential equation
$$D^2 y - a^2 y = 0.$$

Prove also that
$$y = C_1 \cosh ax + C_2 \sinh ax$$

satisfies the differential equation for any pair of constants C_1 and C_2.

44. Find a solution of the differential equation in Exercise 43 for which $y = 2$ and $y' = a$ when $x = 0$.

45. As in the rationalizing substitution $z = (\sin x)/(1 + \cos x)$ (§5.2), show that if

$$z = \frac{\sinh x}{\cosh x - 1} = \frac{\cosh x + 1}{\sinh x},$$

then

$$\cosh x = \frac{1 + z^2}{1 - z^2}, \quad \sinh x = \frac{2z}{1 - z^2}, \quad \text{and} \quad Dx = \frac{2\,Dz}{1 - z^2},$$

46. Using the result of Exercise 45, show that the substitution $z = (\sinh x)/(\cosh x - 1)$ will reduce any rational function of $\cosh x$ and $\sinh x$ to integrable form. However, show that a simpler way to do this is to replace $\cosh x$ and $\sinh x$ by their exponential equivalents, and then substitute $z = e^x$.

47. Show by Taylor's theorem (I, §14.4) that

$$\cosh x = 1 + \frac{x^2}{2!} + \frac{x^4}{4!} + \cdots + \frac{x^{2k}}{(2k)!} + \frac{(\sinh z)}{(2k + 1)!}\, x^{2k+1},$$

where z is some number between 0 and x.

48. Show that the remainder term

$$R_{2k} = \frac{(\sinh z)}{(2k + 1)!}\, x^{2k+1}$$

in the Taylor's theorem for $\cosh x$ (Exercise 47) can be estimated when $x > 0$ by

$$|R_{2k}| \leqq \frac{1}{2} e^x \frac{x^{2k+1}}{(2k + 1)!},$$

which no longer contains the indefinite number z.

49. Show by Taylor's theorem that

$$\sinh x = x + \frac{x^3}{3!} + \frac{x^5}{5!} + \cdots + \frac{x^{2k+1}}{(2k + 1)!} + R_{2k+1},$$

$$R_{2k+1} = \frac{\cosh z}{(2k + 2)!}\, x^{2k+2}$$

for some number z between 0 and x.

8 Multiple Integrals in Polar Coordinates

8.1 *Problem.* Find by integration the volume of the solid figure bounded by the surface $z = f(r, \theta)$, over the domain $\mathcal{D}$ in the polar-coordinate plane, which is bounded by the rays, $\theta = \alpha$ and $\theta = \beta$, and the circles $r = a$ and $r = b$.

Solution. We sketch the figure (Figure 14.23), observing that the base is a region in the $r\theta$-plane that is like a rectangle in the xy-plane in that it is bounded by two coordinate lines (circles) on which r is constant and two coordinate lines (diverging rays) on which θ is constant. Proceeding as we did for integration over a rectangle in rectangular coordinates (II, §29), we partition the domain $\mathcal{D}$ into smaller cells. To do this, we introduce a finite number of values of r between a and b:

$$a = r_0 < r_1 < r_2 < \cdots < r_m = b;$$

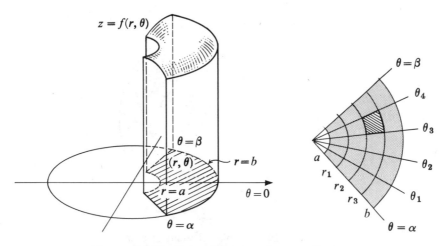

FIGURE 14.23   Integration in polar coordinates.

and a finite number of values of θ between α and β;

$$\alpha = \theta_0 < \theta_1 < \theta_2 < \cdots < \theta_n = \beta.$$

We draw the circles with radii $a, r_1, r_2, \ldots, b$ and the rays for angles $\alpha, \theta_1, \theta_2, \ldots, \beta$. These circles and rays partition the base into smaller cells D_{ij}, $i = 1, \ldots, m, j = 1, \ldots, n$, which are also rectangle-like.

In a typical cell D_{ij}, we evaluate the function at some sample point (r_i, θ_j) to get a height $f(r_i, \theta_j)$. And, representing the area of the cell D_{ij} by dA_{ij}, we find the volume of the shaft over this cell to be

$$dV_{ij} = f(r_i, \theta_j)\, dA_{ij}.$$

We then sum these small volumes over all cells of the partition of D. The resulting sum, $\sum dV_{ij}$, is an approximation to the required volume bounded by the graph of $z = f(r, \theta)$ over D.

When these sample sums converge by refinement of the partition to an exact volume, it is denoted by the double integral

$$V = \iint_{\mathscr{D}} f(r, \theta)\, dA.$$

We omit the details of using under- and overstep functions (II, §29.2) to prove that when this equation is true, the value of the integral is the same for every choice of the sample points (r_i, θ_j) in each cell of the partition.

This is only a partial solution to the stated problem. Although we have set up an integral that represents the required volume, we have not found out what the element of area dA is, and hence we are not prepared to calculate the volume V for a particular function $f(r, \theta)$ and domain D.

8.2   Element of Area in Polar Coordinates. We now seek an expression for the element of area dA in polar coordinates. We consider a typical cell of the partition bounded by the circles of radii r and $r + dr$ and rays at angles θ and $\theta + d\theta$ (Figure 14.24). The

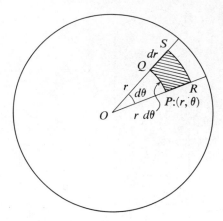

FIGURE 14.24   Polar element of area.

area dA of the cell with vertices $PQSR$ is the difference of the areas of the sectors OPQ and ORS. Since the area of a sector, having radius r and angle ϕ, is $\frac{1}{2}r^2\phi$, we find the exact area of the cell is

$$\Delta A = \frac{1}{2}(r + dr)^2 \, d\theta - \frac{1}{2}r^2 \, d\theta$$

$$= \frac{1}{2}[(r + dr) + r] \, dr \, d\theta = \left(r + \frac{dr}{2}\right) dr \, d\theta.$$

The coefficient of $dr \, d\theta$ is the mean of the radii $r + dr$ and r, so the point

$$M : (\tfrac{1}{2}[(r + dr) + r], \, \theta)$$

is in the cell. Hence, sample sums of the type

$$\sum_i \sum_j f(r_i, \, \theta_j)[r_i + \tfrac{1}{2}dr_i] \, dr_i \, d\theta_j$$

over the partitions converge to the exact integral

$$\iint\limits_{\mathscr{D}} f(r, \, \theta) \, dA.$$

However, we may regard the same sample sums as sample sums that converge under refinement to the exact integral of the function $f(r, \boldsymbol{\theta})r$,

$$\iint\limits_{\mathscr{D}} [f(r, \, \theta)r] \, dr \, d\theta.$$

In the approximating sample sums above, the factors $f(r, \boldsymbol{\theta})$ and r of the integrand are not evaluated at the same point in each cell. But the theorem of Bliss (II, §32.6) tells us that they converge to the same integral as the sample sums in which $f(r, \boldsymbol{\theta})$ and r are evaluated at the same point.

The fact that the two integrals of $f(r, \boldsymbol{\theta})$ are equal shows that we can take as the element of area in polar coordinates, for purposes of integration,

$$dA = r \, dr \, d\theta.$$

We may give a geometric interpretation to the element of area $dA = r\,dr\,d\theta$, which helps to remember it. A typical cell of a partition of $\mathcal{D}$ (Figure 14.24) has an inner arc of length $r\,d\theta$, and the side along one of the rays is dr. When dr and $d\theta$ are small, the cell is approximately a square whose area is $dA = (r\,d\theta)\,dr$. This is not exact, but we proved that integrating f with respect to this element gives the exact integral $\iint_{\mathcal{D}} f(r,\,\theta)\,dA$.

8.3   Repeated Integrals in Polar Coordinates.   Ignoring the geometric derivation, the integral

$$\iint_{\mathcal{D}} f(r,\,\theta)r\,dr\,d\theta$$

is analytically exactly the same as the integral over the rectangle $\mathcal{R}:\{a \leq x \leq b,\ \alpha \leq y \leq \beta\}$,

$$\iint_{\mathcal{R}} f(x,\,y)x\,dx\,dy.$$

This double integral permits integration by repeated single integrals (II, §30). The repeated integral can also be adapted over regions not necessarily bounded by polar coordinate lines (II, §31).

Example A.   Find the volume of the figure bounded by the graph of $z = r \sin \theta$ over the region $\mathcal{D}$ in the polar coordinate plane between the rays $\theta = \pi/6$ and $\theta = \pi/3$ inside the circle $r = a$.

Solution.   We plot the figure and the region $\mathcal{D}$ (Figure 14.25). The volume is given by the double integral

$$V = \iint_{\mathcal{D}} (r \sin \theta)r\,dr\,d\theta.$$

We shall calculate this by a repeated integral with respect to θ first,

$$V = \int_0^a \int_{\pi/6}^{\pi/3} (r \sin \theta)r\,d\theta\,dr$$

$$= \int_0^a r^2 \int_{\pi/6}^{\pi/3} \sin \theta\,d\theta\,dr = \int_0^a r^2 \Big[-\cos \theta\Big]_{\pi/6}^{\pi/3} dr$$

$$= \frac{\sqrt{3} - 1}{2} \int_0^a r^2\,dr = \frac{\sqrt{3} - 1}{6} a^3 \text{ volume units.}$$

This completes the problem.

Example B.   Use a double integral to find the area of the region between the circles $r = a \cos \theta$ and $r = 2a \cos \theta$.

Solution.   We plot the curves in polar coordinates (Figure 14.26). Each of the circles is traversed once completely as θ varies from $-\pi/2$ to $+\pi/2$. The area is given by the double integral

$$A = \iint_{\mathcal{D}} 1r\,dr\,d\theta,$$

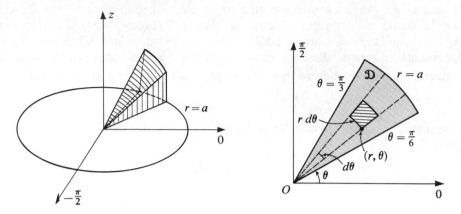

FIGURE 14.25   Repeated integrals in polar coordinates.

which we interpret as meaning the integral

$$A = \iint_{\mathscr{E}} f_E(r, \theta)r \, dr \, d\theta$$

over some domain $\mathscr{E} = \{0 \leqq r \leqq 2a, \alpha \leqq \theta \leqq \beta\}$ that contains $\mathscr{D}$. The function f_E has the value 1 in $\mathscr{D}$, and 0 outside of $\mathscr{D}$, so that only points in $\mathscr{D}$ actually contribute to the integral. For calculation we represent the area A as the repeated integral first with respect to r

$$A = \int_{-\pi/2}^{\pi/2} \int_0^{2a} f_E(r, \theta)r \, dr \, d\theta.$$

Then we cut this down to the domain $\mathscr{D}$ by observing (Figure 14.26) that only points for which $a \cos \theta \leqq r \leqq 2a \cos \theta$ contribute to the integral so that

$$A = \int_{-\pi/2}^{\pi/2} \int_{a \cos \theta}^{2a \cos \theta} r \, dr \, d\theta$$

$$= \int_{-\pi/2}^{\pi/2} \frac{r^2}{2} \Bigg|_{a \cos \theta}^{2a \cos \theta} d\theta = \frac{3a^2}{2} \int_{-\pi/2}^{\pi/2} \cos^2 \theta \, d\theta$$

$$= \frac{3a^2}{2} \left[\frac{\theta}{2} + \frac{1}{4} \sin 2\theta \right]_{-\pi/2}^{\pi/2} = \frac{3\pi a^2}{4}.$$

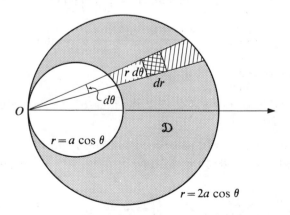

FIGURE 14.26   Example B.

We observe that the first integration sums the areas of elements $r \, dr \, d\theta$ in a wedge-strip extending from the small circle to the large one (Figure 14.26). Then the second integration with respect to θ sums the areas of all such strips from the minimum value of θ, $\theta = -\pi/2$, to the maximum $\theta = \pi/2$.

8.4 General Cases. In general, if the domain $\mathcal{D}$ is bounded by two rays $\theta = \alpha$ and $\theta = \beta$ and two curves $r = g_1(\theta)$ and $r = g_2(\theta)$ (Figure 14.27(a)), then the double integral of $f(r, \theta)$ over $\mathcal{D}$ can be represented by the repeated integral

$$\int_\alpha^\beta \int_{g_1(\theta)}^{g_2(\theta)} f(r, \theta) r \, dr \, d\theta.$$

Or, if the domain $\mathcal{D}$ is bounded by the two circles $r = a$ and $r = b$ between the curves $\theta = \phi_1(r)$ and $\theta = \phi_2(r)$ (Figure 14.27(b)), the double integral of $f(r, \theta)$ over $\mathcal{D}$ can be represented by the repeated integral

$$\int_a^b \int_{\phi_1(r)}^{\phi_2(r)} f(r, \theta) r \, d\theta \, dr.$$

When we integrate with respect to r, first the single integral

$$\int_{g_1(\theta)}^{g_2(\theta)} f(r, \theta) r \, dr \, d\theta$$

is carried over the wedge (shaded in Figure 14.27(a)) with θ and $d\theta$ fixed, going from the inner boundary $r = g_1(\theta)$ to the outer boundary $r = g_2(\theta)$. Then the second integral sums up over all such wedges the volumes resulting from the first integration.

On the other hand, when we integrate with respect to θ first, the single integral

$$\int_{\phi_1(r)}^{\phi_2(r)} f(r, \theta) r \, d\theta \, dr$$

is carried over a strip bounded by circular arcs at r and $r + dr$ (shaded in Figure 14.27(b)), going from the boundary curve $\theta = \phi_1(r)$ to the boundary curve $\theta = \phi_2(r)$. Then the second integration with respect to r sums up the volumes over these circular strips as r goes from the inner circle $r = a$ to $r = b$.

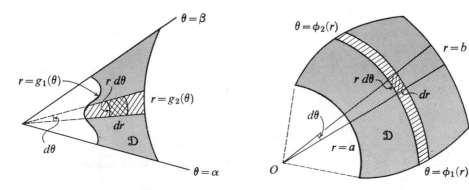

(a)   First with respect to r (b)   First with respect to θ

FIGURE 14.27   Limits of repeated integrals.

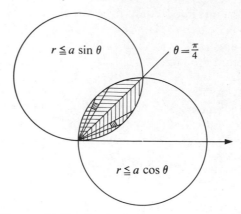

FIGURE 14.28   Region common to two disks.

8.5   Complicated Domains.   The general cases do not take care of all domains $\mathscr{D}$. The boundaries may be neither single-valued functions of r nor of θ, or at least one of the boundaries may require several formulas to represent it.   In such cases, we try to subdivide the domain $\mathscr{D}$ into a finite number of regions in which we can apply one of the general cases.

Example.   Find the area of the domain $\mathscr{D}$ common to the two disks $r \leq a \cos \theta$ and $r \leq a \sin \theta$.

Solution.   We plot the circles $r = a \cos \theta$ and $r = a \sin \theta$ and find that they intersect at the points $(0, 0)$ and $((\sqrt{2}/2)a, \pi/4)$.   For angles $0 \leq \theta \leq \pi/4$, the outer boundary with respect to r is the circle $r = a \sin \theta$, but when $\pi/4 \leq \theta < \pi/2$, it is the circle $r = a \cos \theta$. Hence we subdivide the domain $\mathscr{D}$ into two regions separated by the ray $\theta = \pi/4$ (Figure 14.28).   The area A is given by the sum of two repeated integrals taken over the two parts

$$A = \int_0^{\pi/4} \int_0^{a \sin \theta} r \, dr \, d\theta + \int_{\pi/4}^{\pi/2} \int_0^{a \cos \theta} r \, dr \, d\theta$$

$$= \frac{a^2}{2} \left(\frac{\pi}{8} - \frac{1}{4} \right) + \frac{a^2}{2} \left(\frac{\pi}{8} - \frac{1}{4} \right) = \frac{a^2}{8} (\pi - 2).$$

8.6   Exercises

In Exercises 1–6, evaluate the repeated integrals involving polar coordinates.   In each case sketch the domain over which the integral is extended.

1. $\int_0^{\pi/2} \int_0^1 \theta r \, dr \, d\theta.$

2. $\int_0^{2\pi} \int_0^a e^{r^2} r \, dr \, d\theta.$

3. $\int_0^{\pi} \int_0^{\sqrt{\sin \theta}} r \, dr \, d\theta.$

4. $\int_0^{\pi} \int_0^{a \sin \theta} \frac{r}{\sin \theta} \, dr \, d\theta.$

5. $\int_0^{\pi/2} \int_0^{\cos (\theta + \pi/4)} r \, dr \, d\theta.$

6. $\int_1^a \int_0^r r \, d\theta \, dr.$

In Exercises 7–10, find the area of the region described by a double integral with respect to polar coordinates.

7.   The area of the region inside the circle $r = a$.
8.   The area of the circle $r = a \sin \theta$.
9.   The area of the ellipse $r = a/(2 - \cos \theta)$.

10.   The area swept out by a line segment from the origin to the spiral $r = 2\theta$ in one revolution starting at $\theta = 0$.

11.   Find the volume of the region below the curve $z = r^2 + 1$ over the disk $r \leq a$.

12.   Find the volume of the region below the hemisphere $z = \sqrt{a^2 - r^2}$ and inside the circle $r = a$ in the $r\theta$-plane.

13.   Find the integral of θ^2 over the disk $r \leq a$.

14.   Find the volume below the surface $z = e^{-r^2}$ over the region inside a circle of radius a.

15.   Verify the calculation in the Example of §8.5.

16.   Find the area of the region common to the disks $r \leq 2a \sin \theta$ and $r \leq a$.

17.   Find the area of the region bounded by the spirals $r = \theta$ and $r = (\pi/2) - \theta$.

18.   Integrate $r \sin \theta$ over the region bounded by $\theta = 0, \theta = \pi/2$, and $r \cos (\theta - (\pi/4)) = \sqrt{2}$.

9   Cylindrical Coordinates and Integration over Unbounded Regions

9.1   Cylindrical Coordinates.   A coordinate system for 3-dimensional space that combines polar coordinates in the horizontal plane with one Cartesian z-coordinate is called a cylindrical coordinate system.   It locates a point P by the three coordinates (r, θ, z) (Figure 14.29).   The points where $r =$ constant form a cylinder of radius r with the z-axis for its principal axis.   The points where $\theta =$ constant are in a plane through the z-axis; and the points where $z =$ constant are in a horizontal plane.   Cylindrical coordinates simplify analysis in configurations that have a central line of symmetry.

If we choose xy-axes related in the standard way to the polar frame of reference, we see that the Cartesian coordinates (x, y, z) of a point are related to its cylindrical coordinates (r, θ, z) by

$$x = r \cos \theta, \qquad y = r \sin \theta, \qquad z = z.$$

We have also the relations $x^2 + y^2 = r^2$ and $\tan \theta = y/x$.   We are concerned with the volume element dV for purposes of integration.   The cell bounded by the horizontal planes at z and $z + dz$, the radial planes θ and $\theta + d\theta$, and the cylinders of radius r

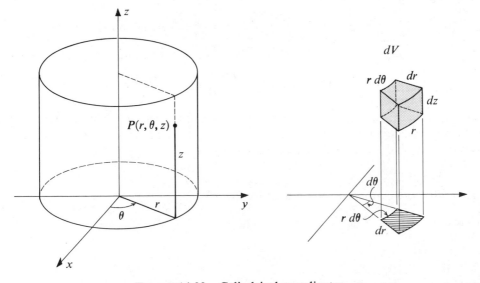

FIGURE 14.29   Cylindrical coordinates.

and $r + dr$ has a base of area $r\,dr\,d\theta$, (II, §34.6). Its altitude is dz and hence its volume is

$$dV = r\,dr\,d\theta\,dz.$$

This volume element dV is apparently not exact, but as in the case of polar coordinates (§8.2), we can show that the triple integral calculated with it gives the same result as the more complicated exact dV.

9.2   Integration in Cylindrical Coordinates

Example.   Find by triple integration the volume of the region bound by the paraboloid $z = 9 - x^2 - y^2$ and the plane $z = 0$.

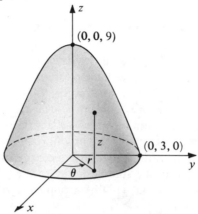

FIGURE 14.30   Volume in cylindrical coordinates.

Solution.   We install a cylindrical coordinate system. In cylindrical coordinates the surface has the equation $z = 9 - r^2$, which intersects the plane $z = 0$ in the circle $r = 3$ (Figure 14.30). Hence the volume is given by the repeated integral

$$V = \int_0^{2\pi} \int_0^3 \int_0^{9 - r^2} dz\, r\,dr\,d\theta = \tfrac{81}{2}\pi \text{ volume units.}$$

We observe that this integral could have been treated just as well as a double integral

$$V = \iint_{\mathscr{D}} (9 - r^2)\,dA = \int_0^{2\pi} \int_0^3 (9 - r^2) r\,dr\,d\theta.$$

The reduction to a double integral is the first stage of the repeated triple integral above. In the double integral the volume element $(9 - r^2)\,dA$ is the volume of a vertical shaft based at (r, θ) and having $r\,dr\,d\theta$ as the area of its base.

9.3   An Integral Extended over the Entire Plane

Example.   Define and calculate the integral

$$I = \iint_{\mathscr{R}^2} e^{-(x^2 + y^2)/2}\,dx\,dy$$

over the entire plane $\mathscr{R}^2$ (Figure 14.31(a)).

Solution.   As we defined the improper integral over an unbounded domain in the real line, we first calculate the integral over a bounded region, choosing the disk $\mathscr{C}_a = \{(x, y) \mid x^2 + y^2 \leqq a^2\}$, (Figure 14.31(b)),

$$I_a = \iint_{\mathscr{C}_a} e^{-(x^2+y^2)/2} \, dx \, dy.$$

Then the integral I over the entire plane is defined by $I = \lim_{a \to \infty} I_a$.

We find it convenient to calculate the integral I_a over the disk by use of polar coordinates (r, θ) in the plane (Figure 14.31(c)), that is, cylindrical coordinates (z, r, θ) for the volume. This requires that we replace the element of area $dx \, dy$ by $r \, dr \, d\theta$ (§8.2), and replace $e^{-(x^2+y^2)/2}$ by $e^{-r^2/2}$, since $x^2 + y^2 = r^2$. In polar coordinates, the integral I_a is represented by the repeated integral

$$I_a = \int_0^{2\pi} \int_0^a e^{-r^2/2} r \, dr \, d\theta = 2\pi(1 - e^{-(a^2/2)}).$$

Since $\lim_{a \to \infty} e^{-(a^2/2)} = 0$, this implies that

$$I = \lim_{a \to \infty} I_a = 2\pi.$$

This completes the problem. It is customary to represent the integral I over the entire xy-plane by the notation

$$I = \int_{-\infty}^{\infty} \int_{-\infty}^{\infty} e^{-(x^2+y^2)/2} \, dx \, dy = 2\pi.$$

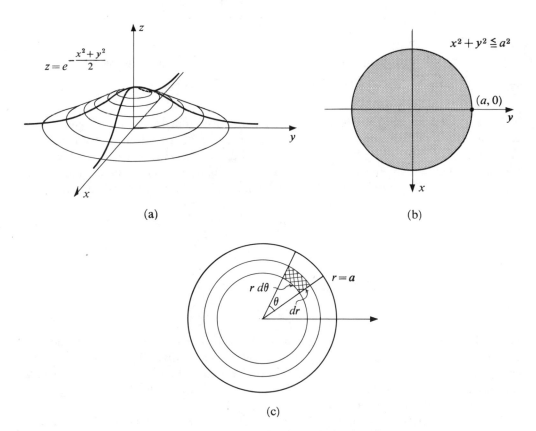

(a)

(b)

(c)

FIGURE 14.31   Integration over the entire plane.

9.4   Integration over Unbounded Domains, General Case

DEFINITION.   A system of bounded regions $\{\mathscr{C}_a\}$ in an unbounded region $\mathscr{D}$ is said to be *expanding into* $\mathscr{D}$ if (1) for every positive real number a, $\mathscr{C}_a$ is contained in $\mathscr{D}$; (2) if $a < b$, then $\mathscr{C}_a \subset \mathscr{C}_b$; (3) every point in $\mathscr{D}$ is included in some $\mathscr{C}_a$ (Figure 14.32).

DEFINITION.   If $f(x, y)$ is a nonnegative function defined on the unbounded domain $\mathscr{D}$ and $f(x, y)$ is integrable on some system of bounded regions $\mathscr{C}_a$ expanding into $\mathscr{D}$, then the integral I of $f(x, y)$ over $\mathscr{D}$ is defined by

$$I = \iint_{\mathscr{D}} f(x, y)\, dx\, dy = \lim_{a \to \infty} \iint_{\mathscr{C}_a} f(x, y)\, dx\, dy,$$

whenever this limit exists.

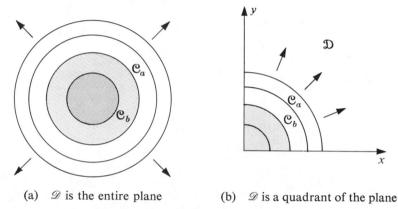

(a) $\mathscr{D}$ is the entire plane   (b) $\mathscr{D}$ is a quadrant of the plane

FIGURE 14.32   Systems of expanding domains.

For the purposes of this definition two systems of bounded domains $\mathscr{C}_a$ and $\mathscr{C}'_b$ are *equivalent* if every $\mathscr{C}_a$ is contained in some $\mathscr{C}'_b$ and every $\mathscr{C}'_b$ is contained in some $\mathscr{C}_a$. It can be proved (Theoretical Section 9.7) that the integral I of a positive function f over $\mathscr{D}$ defined by two equivalent systems of expanding domains in $\mathscr{D}$ is the same. This enables us to use circular disks or squares, whichever gives the simplest calculation of the repeated integral over $\mathscr{C}_a$, without affecting the value of the integral I over $\mathscr{D}$.

These definitions apply equally well to integration over unbounded regions in a space of n-rows $(x_1, x_2, \ldots, x_n)$.

9.5   The Probability Integral on the Real Line.   The integral

$$\int_{-\infty}^{\infty} e^{-x^2/2}\, dx$$

is called the *probability integral* because of its importance in connection with the normal distribution, $y = e^{-x^2/2}$, of the theory of probability.   The probability integral cannot be calculated directly by the fundamental theorem because there is no explicit formula that gives an antiderivative of $e^{-x^2/2}$.   We can overcome this difficulty by an ingenious method of double integration.

Returning to the integral of $f(x, y) = e^{-(x^2 + y^2)/2}$ over the xy-plane $\mathscr{R}^2$ (§9.3), we take advantage of a system of bounded squares that is equivalent to the system of circular disks expanding to $\mathcal{R}^2$. We first integrate over a square of side $2a$ with center at $(0, 0)$ (Figure 14.33) and find

$$
\begin{aligned}
J_a &= \int_{-a}^{a} \int_{-a}^{a} e^{-(x^2 + y^2)/2} \, dx \, dy \\
&= \int_{-a}^{a} e^{-y^2/2} \left(\int_{-a}^{a} e^{-x^2/2} \, dx \right) dy \\
&= \left(\int_{-a}^{a} e^{-x^2/2} \, dx \right) \left(\int_{-a}^{a} e^{-y^2/2} \, dy \right) \\
&= \left(\int_{-a}^{a} e^{-x^2/2} \, dx \right)^2 .
\end{aligned}
$$

Since the system of square domains is equivalent to the system of disks $\mathscr{C}_a$ (§9.3),

$$
\lim_{a \to \infty} J_a = \lim_{a \to \infty} I_a = 2\pi.
$$

Hence the probability integral over the real line is given by

$$
\int_{-\infty}^{\infty} e^{-x^2/2} \, dx = \sqrt{2\pi}.
$$

9.6 Exercises

In Exercises 1–6, sketch the lines, surfaces, solids given in cylindrical coordinates (r, θ, z) by the following.

1. $z^2 + r^2 = 4$.   2. $r = 4$.
3. $r = 2a \cos \theta$.   4. $0 \le z \le 2r \cos \theta, \, r \le 2a$.
5. $1 \le r \le 2, \, 0 \le z \le 3$.   6. $r \le 3, \, \pi/2 \le \theta \le \pi/2, \, -1 \le z \le 1$.

7. Using the integration in cylindrical coordinates, find the volume of the solid cone

$$
0 \le z \le 4 - 2r.
$$

8. Find by integration in cylindrical coordinates the volume of the solid $0 \le z \le 2r \cos^2 \theta$, $1 \le r \le 2$.

In Exercises 9–13, integrate the function over the indicated unbounded domain.

9. $f(x, y) = e^{-x - y}$ over the quadrant $\{x \ge 0, y \ge 0\}$.

10. $f(r, \theta) = \dfrac{1}{r^3}$ over the region outside the disk $\{r \le 1\}$.

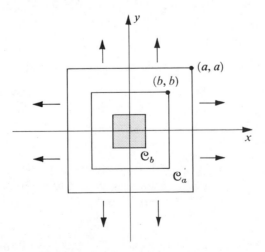

FIGURE 14.33   Square regions expanding into the entire plane.

11. $f(x, y) = \dfrac{1}{x^2 + y^2}$ outside the disk $\{x^2 + y^2 \leq 1\}$.

12. $f(x, y) = \dfrac{1}{x^2 + y^2}$ outside of the square $\{|x| \leq 1, |y| \leq 1\}$.

13. $f(x) = \dfrac{1}{1 + x^2}$ over the real line.

14. Find the volume of the region below the plane $z = 2r \cos \theta$ and above $z = r^2$.

15. Find the integral of the function $f(x, y, z) = x^2$ over the portion of the cylinder $\{x^2 + y^2 \leq a^2, 0 \leq z \leq b\}$.

16. A point R moves so that its cylindrical coordinates are given as functions of time by

$$z = 2t, \qquad r = 4, \qquad \theta = 2\pi t.$$

Find the velocity vector $\vec{v}$ and acceleration vector $\vec{a}$ at time t, and the linear speed.

17. Sketch and identify the curve in Exercise 16.

18. Show that

$$\int_0^\infty \int_0^\infty \frac{dx \, dy}{(x + y + 1)^2}$$

does not have a finite value.

19. Evaluate

$$\int_1^\infty \int_1^\infty \frac{dx \, dy}{x^2 y^2}.$$

20. Show that if a distribution of mass in the entire line has the density

$$\mu = \frac{1}{\sigma \sqrt{2\pi}} e^{-(x-a)^2/2\sigma^2},$$

then the mean x is a, and the standard deviation is σ (II, §33).

21. Show by integration that

$$\frac{1}{2\pi\sigma^2} \int_0^{2\pi} \int_0^a e^{-r^2/2\sigma^2} r \, dr \, d\theta = 1 - e^{-a^2/2\sigma^2}.$$

22. Show that

$$\int_{-a}^a e^{-x^2/2} \, dx = \sqrt{2\pi} - 2 \int_{-\infty}^a e^{-x^2/2} \, dx.$$

9.7   (Theory) Convergence of Integrals over Infinite Domains

THEOREM.   If the integral of the positive function $f(x, y)$ over the unbounded domain $\mathscr{D}$ converges with respect to some system $\{\mathscr{C}_a\}$ of bounded domains expanding to $\mathscr{D}$, then the integral converges with respect to any system $\{\mathscr{C}_b'\}$ of bounded domains expanding to $\mathscr{D}$, which is equivalent to $\{\mathscr{C}_a\}$.

   Proof.   We recall the definition of a system of bounded domains expanding to $\mathscr{D}$, and the idea of equivalence of two such systems of domains (§9.4). We consider two equivalent systems $\{\mathscr{C}_a\}$ and $\{\mathscr{C}_b'\}$ (Figure 14.34). We define the symbols I_a, I_b and I by

$$I_a = \iint_{\mathscr{C}_a} f(x, y) \, dx \, dy, \qquad I_b = \iint_{\mathscr{C}_b'} f(x, y) \, dx \, dy,$$

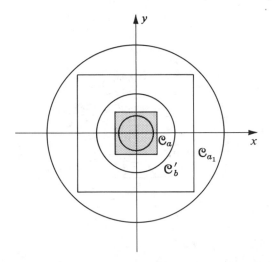

FIGURE 14.34   Equivalent systems of domains expanding into the entire plane $\mathscr{D}$. The circular disks are members of the system, $\{\mathscr{C}_a\}$. The squares are members of the system $\{\mathscr{C}_b'\}$. Every $\mathscr{C}_a$ is contained in a $\mathscr{C}_b'$ and every $\mathscr{C}_b'$ is contained in some larger $\mathscr{C}_{a_1}$.

and $I = \lim_{a \to \infty} I_a$. We have to prove that $\lim_{b \to \infty} I_b = I$ also. Since $f(x, y) \geqq 0$ and we may take $dx\, dy > 0$, the integrals I_a increase as $a \to \infty$. Since $\lim_{a \to \infty} I_a = I$, for every ϵ there is a number A such that if $a > A$, then

$$I - \epsilon < I_a \leqq I.$$

In particular, $I - \epsilon < I_A \leqq I$. But since $\{\mathscr{C}_b'\}$ is an equivalent system of domains expanding to $\mathscr{D}$, there is a domain $\mathscr{C}_B'$ that contains $\mathscr{C}_A$. Then $I - \epsilon < I_A \leqq I_B$. Moreover, every $I_b \leqq I$, since there is some $\mathscr{C}_{a_1}$ that contains $\mathscr{C}_b'$. Hence, whenever $b > B$, it is true that $I = \epsilon < I_b \leqq I$. This says that $\lim_{b \to \infty} I_b = I$. This completes the proof.

10   Geometry of the Sphere

10.1   Spherical Coordinates.   A spherical coordinate system for 3-dimensional space locates a point $R:(\rho, \theta, \phi)$ by a radial distance ρ and two angular coordinates θ and ϕ (Figure 14.35). Starting with a Cartesian xyz-coordinate system, we determine the spherical coordinates of a point R as follows. We construct a sphere with center at the origin O passing through R. The radius of this sphere is ρ. The z-axis intersects the sphere at a point Z, called the zenith, and a point N, called nadir. We then pass a plane through R containing the z-axis. This plane intersects the sphere in a half-circle ZRN, called a meridian. The angle ZOR, measured from the positive z-axis, is ϕ. The sphere intersects the xy-plane in a circle called the horizon and the meridian intersects the horizon in a point Q. We measure the angle θ as the rotation about the z-axis from the meridian of the positive x-axis to the meridian of R. Spherical coordinates simplify analysis in configurations that are symmetric with respect to a fixed center point.

If we project R perpendicularly onto the point R' in the horizontal plane, the polar coordinates of R' are (r, θ). We see that

$$r = \rho \sin \phi,$$

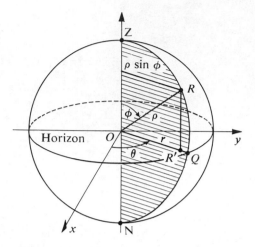

FIGURE 14.35   Spherical coordinates.

and since $x = r \cos \theta$, $y = r \sin \theta$, we find the relations between xyz-coordinates and spherical coordinates,

$$x = \rho \sin \phi \cos \theta, \qquad y = \rho \sin \phi \sin \theta, \qquad z = \rho \cos \phi.$$

The intersection of the sphere with a plane through the center is called a *great circle*.

10.2 Differentiation in Spherical Coordinates.

As in polar and cylindrical coordinates, we locate points in spherical coordinates but we use local Cartesian coordinates for vectors and tangents.

In general, if the position $\vec{r} : (\rho, \theta, \phi)$ is given at time t by functions $\rho(t)$, $\theta(t)$, $\phi(t)$, then the xyz-coordinates are functions of t determined by the relations

$$x = \rho \sin \phi \cos \theta,$$
$$y = \sin \phi \sin \theta,$$
$$z = \rho \cos \phi.$$

We consider these as composite functions with ρ, θ, ϕ expressed as functions of t by the given motion. Differentiating with respect to t, using the chain rule, we find

$$\frac{dx}{dt} = \sin \phi \cos \theta \, \frac{d\rho}{dt} - \rho \sin \phi \sin \theta \, \frac{d\theta}{dt} + \rho \cos \phi \cos \theta \, \frac{d\phi}{dt},$$
$$\frac{dy}{dt} = \sin \phi \sin \theta \, \frac{d\rho}{dt} + \rho \sin \phi \cos \theta \, \frac{d\theta}{dt} + \rho \cos \phi \sin \theta \, \frac{d\phi}{dt},$$
$$\frac{dz}{dt} = \cos \phi \, \frac{d\rho}{dt} \qquad\qquad\qquad - \rho \sin \phi \, \frac{d\phi}{dt}.$$

We can find $d\rho/dt$, $d\theta/dt$, $d\phi/dt$ from the given parametric representation of the motion and substitute them into these differential relations to find dx/dt, dy/dt, dz/dt.

Example.   Find the velocity in the object observed by a tracking station to be located at time t by

$$\rho(t) = 3\sqrt{t^2 + 1}, \qquad \tan \theta = \frac{2t - 2}{2t + 1}, \qquad \cos \phi = \frac{t + 2}{3\sqrt{t^2 + 1}}.$$

Solution.   We compute

$$\frac{d\rho}{dt} = \frac{3t}{\sqrt{t^2 + 1}}, \qquad \frac{d\theta}{dt} = \frac{6}{\sqrt{T}}, \qquad \frac{d\phi}{dt} = \frac{2t - 1}{(t^2 + 1)\sqrt{T}},$$

where $T = 8t^2 - 4t + 5$. Also

$$\cos\theta = \frac{2t + 1}{\sqrt{T}}, \qquad \sin\theta = \frac{2t - 2}{\sqrt{T}}; \qquad \cos\phi = \frac{t + 2}{3\sqrt{t^2 + 1}}, \qquad \sin\phi = \frac{\sqrt{T}}{3\sqrt{t^2 + 1}}.$$

Substituting these expressions into the formulas above, we find after some laborious algebra that

$$\frac{dx}{dt} = 2, \qquad \frac{dy}{dt} = 2, \qquad \frac{dz}{dt} = 1.$$

The object is moving in a straight line with constant velocity $\vec{v}:(2, 2, 1)$. This completes the problem.

10.3   Shortest Distances on a Sphere.   We recall (II, §22) that the derivative of arc length for a smooth curve $\vec{r}(t):[x(t), y(t), z(t)]$ is given by

$$\left(\frac{ds}{dt}\right)^2 = \left(\frac{dx}{dt}\right)^2 + \left(\frac{dy}{dt}\right)^2 + \left(\frac{dz}{dt}\right)^2.$$

We can express this in spherical coordinates from the expressions for dx/dt, dy/dt, dz/dt that we calculated above (§10.2). Squaring them, adding, and simplifying, we find that (Exercises)

$$\left(\frac{ds}{dt}\right)^2 = \left(\frac{d\rho}{dt}\right)^2 + \rho^2 \sin^2\phi \left(\frac{d\theta}{dt}\right)^2 + \rho^2 \left(\frac{d\phi}{dt}\right)^2.$$

Now we consider a curve on a sphere of radius R joining two points A and B. Then $\rho = R$ and $d\rho/dt = 0$, so that

$$s = R \int_{t_A}^{t_B} \sqrt{\sin^2\phi \left(\frac{d\theta}{dt}\right)^2 + \left(\frac{d\phi}{dt}\right)^2}\, dt.$$

We can choose the spherical coordinate system so that A and B are on the same meridian $\theta = \theta_0$ along which $d\theta/dt = 0$. Then its arc length is

$$s = R \int_{t_A}^{t_B} \left|\frac{d\phi}{dt}\right| dt = R\, |\phi_B - \phi_A|.$$

This meridian is the shortest arc joining A to B; for the contribution from the term $(d\phi/dt)^2$ is already minimal and on any other arc the term $\sin^2\phi(d\theta/dt)^2 > 0$ and hence its arc length exceeds that of the meridian. We state this result as a theorem.

THEOREM (Shortest distance on a sphere).   The shortest arc joining two points A and B on a sphere is on the great circle joining A and B.

10.4   Integration in Spherical Coordinates.   For a fixed point $R:(\rho, \theta, \phi)$, we consider the cell bounded by the spheres of radius ρ and $\rho + d\rho$, the planes at angle θ and $\theta + d\theta$, and the cones of angle ϕ and $\phi + d\phi$ (Figure 14.36). It is a boxlike cell whose edges have

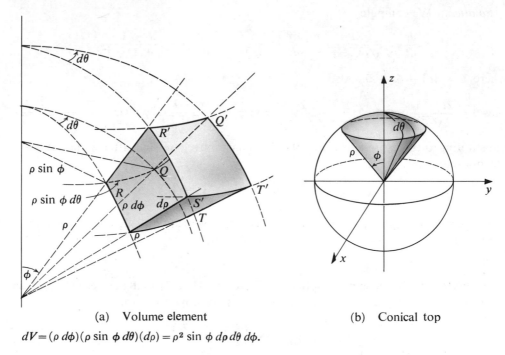

(a)   Volume element   (b)   Conical top

$dV = (\rho\, d\phi)(\rho \sin \phi\, d\theta)(d\rho) = \rho^2 \sin \phi\, d\rho\, d\theta\, d\phi.$

FIGURE 14.36   Volume element in spherical coordinates.

lengths $d\rho$, $\rho \sin \phi\, d\theta$, and $\rho d\phi$. Intuitively treating this as a rectangular box, we multiply the lengths of the edges to obtain the volume element

$$dV = \rho^2 \sin \phi\, d\rho\, d\theta\, d\phi.$$

We can prove (§10.8) that integration

$$\iiint_{\mathscr{D}} dV$$

with this volume element dV gives the exact volume of $\mathscr{D}$.

10.5   Surface Area Element.   Similarly the element of surface area of a sphere of radius R may be found by multiplying the lengths of two edges of the "square" on the sphere (Figure 14.37).   This gives us

$$dS = R^2 \sin \phi\, d\theta\, d\phi.$$

10.6   Exercises

In Exercises 1–8, describe the locus of points that satisfy the given relations in spherical coordinates (ρ, θ, ϕ).   Sketch.

1. $\rho = 4$.   2. $\theta = \pi/3$.
3. $\phi = \pi/4$.   4. $0 \leq \phi \leq \pi/4$ and $\rho \leq 1$.
5. $\rho \sin \phi = 4$.   6. $\rho = 4 \cos \phi$.
7. $1 \leq \rho \leq 2$.   8. $0 \leq \theta \leq \pi/6$ and $\rho > 1$.

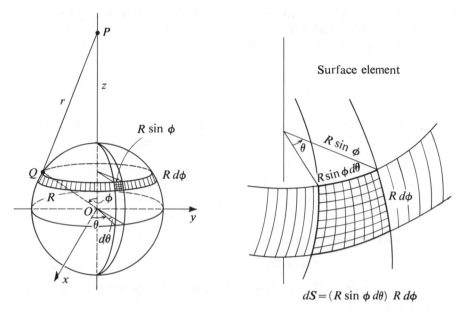

$$dS = (R \sin \phi \, d\theta) \, R \, d\phi$$

FIGURE 14.37   Spherical element of surface area.

In Exercises 9–12, transform to spherical coordinates.

9. $y = 2x$.   10. $\rho = 8 \cos \phi$.
11. $x^2 + y^2 \leqq 2x$.   12. $y^2 + z^2 = R^2$.

13. Find by integration in spherical coordinates the volume of the sphere of radius R.
14. Find by integration in spherical coordinates the volume of the top (Figure 14.36) enclosed by a sphere of radius R and cone of angle ϕ.
15. Find the volume enclosed by the surface $\rho = a(1 - \cos \phi)$.
16. Find the area on a sphere of radius R of the zone cut off by planes $z = z_0$, and $z = z_0 + h$.
17. Find the area of the zone on the cylinder $x^2 + y^2 = R^2$ cut off by the planes $z = z_0$ and $z = z_0 + h$. Draw a figure showing the regions in Exercises 16 and 17 that have equal area.
18. Find the velocity at time t of a point that moves on the sphere of radius 5 having center at O with angles $\theta = 2 \arcsin t$ and $\phi = \arccos t$.
19. Verify the calculations of dx/dt, dy/dt, dz/dt in §10.2, Example.
20. Set up the integral representing the arc length of the curve $\phi = 2\theta$ on the sphere of radius 5 from the point where $\theta = 0$ to the point where $\theta = \pi/2$. The integral does not respond to elementary integration devices.
21. In the geometry of the plane, Euclid's parallel axiom states that for any line $\mathscr{L}$ and point R, not in $\mathscr{L}$, there exists one and only one line through R parallel to $\mathscr{L}$. Show whether there is an analogous proposition that is true in the geometry of the sphere.
22.* Prove by a method like that of §10.3 that the straight line is the shortest arc connecting two points in the plane.
23. A vector $\vec{r}$ of unit length moves so that its endpoint traces a curve $\vec{r}(t)$. Prove that the velocity vector is always perpendicular to $\vec{r}$.

10.7   (Theory) Exactness of the Spherical Volume Element.   We estimated the volume element in spherical coordinates to be $dV = \rho^2 \sin \phi \, d\rho \, d\theta \, d\phi$ (§10.4). We now prove that integration with respect to this volume element gives exact volume. We begin by calculating the volume of a conical top (Figure 14.36(b)) cut from a sphere of radius ρ by a cone of angle ϕ to be $V = 2\pi(\rho^3/3)(1 - \cos \phi)$. We may do this by integration in cylindrical coordinates (§32.2, Example B). For a segment of angle $d\theta$ radians out of this top, the volume is

$$V(\rho, \phi) = d\theta \frac{\rho^3}{3}(1 - \cos \phi).$$

Using this formula for similar segments, all of angle $d\theta$, we may calculate the exact volume, ΔV, of the cell with edges dp, $pd\phi$, $p \sin \phi\, d\theta$ (Figure 14.36(b)) by

$$\begin{aligned}
\Delta V &= [V(\rho + d\rho, \phi + d\phi) - V(\rho + d\rho, \phi)] - [V(\rho, \phi + d\phi) - V(\rho, \phi)] \\
&= \tfrac{1}{3} d\theta[(\rho + d\rho)^3(1 - \cos (\phi + d\phi)) - (\rho + d\rho)^3(1 - \cos \phi)] \\
&\quad - \tfrac{1}{3} d\theta[\rho^3(1 - \cos (\phi + d\phi)) - \rho^3(1 - \cos \phi)] \\
&= \tfrac{1}{3} d\theta[(\rho + d\rho)^3 - \rho^3][\cos (\phi + d\phi) - \cos \phi].
\end{aligned}$$

We apply the mean value theorem,

$$f(x + \Delta x) - f(x) = f'(\bar{x})\, \Delta x,$$

to each of the differences in the brackets. Since $D\rho^3 = 3\rho^2\, d\rho$ and $D \cos \phi = -\sin \phi\, d\phi$, this gives us exactly

$$\Delta V = \tfrac{1}{3} d\theta[3\bar{\rho}^2\, d\rho][-\sin \bar{\phi}\, d\phi] = -\bar{\rho}^2 \sin \bar{\phi}\, d\rho\, d\theta\, d\phi,$$

where $\bar{\rho}$ is some number between ρ and $\rho + d\rho$, and $\bar{\phi}$ is a number between ϕ and $\phi + d\phi$. That is, $(\bar{\rho}, \theta, \bar{\phi})$ is some point in the cell. Now since sample sums of terms of the type $\bar{\rho}^2 \sin \bar{\phi}\, d\rho\, d\theta\, d\phi$ converge to the same exact integral as sample sums evaluated at the point (ρ, θ, ϕ) in the same cell, the sample sums with the exact volume elements converge precisely to the integral

$$\iiint_{\mathcal{D}} \rho^2 \sin \phi\, d\rho\, d\theta\, d\phi.$$

The volume element $dV = \rho^2 \sin \phi\, d\rho\, d\theta\, d\phi$ also gives the correct integral when we integrate a function $f(\rho, \theta, \phi)$ over $\mathcal{D}$. To prove this we may insert the exact volume element $\Delta V = \bar{\rho}^2 \sin \bar{\phi}\, d\rho\, d\theta\, d\phi$ into the sample sums, in which f is evaluated at a different point $f(\rho', \theta', \phi')$. We then appeal to the theorem of Bliss (II, §32.6) to find that this converges to the exact integral

$$\iiint_{\mathcal{D}} f(\rho, \theta, \phi)\rho^2 \sin \phi\, d\rho\, d\theta\, d\phi.$$

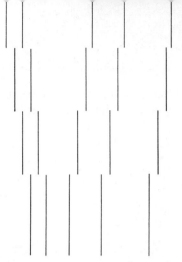

Linear Algebra

CHAPTER 15

Note: In courses that do not include linear algebra, this chapter may be omitted. One may proceed to the techniques of multiple integration, Section 30, if it is so desired.

11 Systems of Linear Equations

11.1 The Method of Elimination. We consider the problem of solving ("simultaneously") a system of linear equations.

Example A. Solve the system of linear equations

(1)
$$\begin{cases} 3x_2 - 2x_3 + 4x_4 = 1 & \text{and} \\ 2x_1 - x_2 + 3x_3 - x_4 = 3 & \text{and} \\ 6x_1 + 3x_2 + 5x_3 + 5x_4 = 5, \end{cases}$$

for *all* sets of numbers (x_1, x_2, x_3, x_4) that satisfy all of the equations.

We will use the familiar method of elimination. Let us review the operations, called *elementary row operations*, used in the method of elimination to replace the system by a simpler one that is logically equivalent to it. They are three:

A. Interchange two equations (rows) in the system;
B. Multiply an equation (row) by a nonzero number;
C. Replace any equation (row) by a new one obtained from it by adding to it a constant multiple of another equation (row) in the system.

We observe that elementary row operations are *invertible*. That is, the operation that restores the original system, after we have performed an elementary row operation on it,

583

is itself an elementary row operation. We say that two systems of linear equations are *equivalent* if and only if one can be reduced to the other by a finite sequence of elementary row operations. By examining the effect of each elementary row operation, we see that $x_1, x_2, \ldots, x_n$ is a solution of a linear system if and only if it is a solution of an equivalent system. That is, equivalent systems have exactly the same set of solutions.

Solution of Example A, using elementary row operations. We now interchange the first two equations to produce the equivalent system

(2)
$$\begin{cases} 2x_1 - x_2 + 3x_3 - x_4 = 3 & \text{and} \\ 3x_2 - 2x_3 + 4x_4 = 1 & \text{and} \\ 6x_1 + 3x_2 + 5x_3 + 5x_4 = 5. \end{cases}$$

The object of this is to bring a nonzero coefficient of x_1 into the upper left-hand corner of the array of coefficients. We now observe that if we multiply the first equation by the nonzero number $\frac{1}{2}$, we produce a new system (3) equivalent to (2),

(3)
$$\begin{cases} x_1 - \frac{1}{2}x_2 + \frac{3}{2}x_3 - \frac{1}{2}x_4 = \frac{3}{2} & \text{and} \\ 3x_2 - 2x_3 + 4x_4 = 1 & \text{and} \\ 6x_1 + 3x_2 + 5x_3 + 5x_4 = 5. \end{cases}$$

The objective of this is to bring the number 1 into the upper left-hand corner as the coefficient of x_1.

We now use the coefficient 1 in the upper left-hand corner to eliminate all x_1's except the one in the first equation. That is, we form an equivalent system in which the co-efficients of the remaining x_1's are all 0. To do this, we add -6 times the first equation to the last equation and obtain the equivalent system

(4)
$$\begin{cases} x_1 - \frac{1}{2}x_2 + \frac{3}{2}x_3 - \frac{1}{2}x_4 = \frac{3}{2} & \text{and} \\ 3x_2 - 2x_3 + 4x_4 = 1 & \text{and} \\ 6x_2 - 4x_3 + 8x_4 = -4. \end{cases}$$

We now introduce 1 as the leading coefficient in the second equation by multiplying it by $\frac{1}{3}$. This gives us the equivalent system

(5)
$$\begin{cases} x_1 - \frac{1}{2}x_2 + \frac{3}{2}x_3 - \frac{1}{2}x_4 = \frac{3}{2} & \text{and} \\ x_2 - \frac{2}{3}x_3 + \frac{4}{3}x_4 = \frac{1}{3} & \text{and} \\ 6x_2 - 4x_3 + 8x_4 = -4. \end{cases}$$

As we did for x_1, we now use the 1 that is the leading coefficient in the second equation to bring zero coefficients into the other places in the second column, the column of co-efficients of x_2. We add $\frac{1}{2}$ times the second equation to the first; and then we add -6 times the second equation to the last equation. These two operations now give us the equivalent system

(6)
$$\begin{cases} x_1 + 0 + \frac{7}{3}x_3 + \frac{1}{6}x_4 = \frac{10}{6} & \text{and} \\ 0 + x_2 - \frac{2}{3}x_3 + \frac{4}{3}x_4 = \frac{1}{3} & \text{and} \\ 0 + 0 + 0 + 0 = -6. \end{cases}$$

The system (6) is still equivalent to the system (1). Inspection of (6) reveals that no set of numbers (x_1, x_2, x_3, x_4) can satisfy it. For the last equation, $0 = -6$, is false, whatever the values of (x_1, x_2, x_3, x_4) may be. Hence the equivalent original system (1) has no solutions either. This completes the problem.

11.2 Pivoting. We observe that in the solution of a linear system, we repeatedly perform a certain set of elementary row operations that have the effect of replacing one coefficient of some x_j by 1, and then making all other coefficients of x_j equal to zero. In the rectangular array of coefficients, these are all in the jth column of the array. In the method of elimination, we apply this set of elementary row operations to the first column, then to the second column, and so on. This set of elementary row operations on the jth column can be described as follows.

(1)   Pick a nonzero coefficient a_{ij} of x_j in the ith equation.
(2)   Multiply this equation by $1/a_{ij}$ to produce $1x_j$ there.
(3)   Then add to each other equation the appropriate multiple of this new ith equation to give x_j the coefficient zero in each case.

In notation, this set of elementary row operations replaces the column of coefficients of x_j

$$\begin{bmatrix} a_{1j} \\ a_{2j} \\ \vdots \\ a_{ij} \\ \vdots \\ a_{mj} \end{bmatrix} \text{ by } \begin{bmatrix} a_{1j} - a_{1j}(1) \\ a_{2j} - a_{2j}(1) \\ \vdots \\ (1/a_{ij})a_{ij} \\ \vdots \\ a_{mj} - a_{mj}(1) \end{bmatrix} = \begin{bmatrix} 0 \\ 0 \\ \vdots \\ 1 \\ \vdots \\ 0 \end{bmatrix}$$

This set of elementary row operations, starting with a chosen nonzero coefficient a_{ij}, is called *pivoting on* a_{ij}. It is simpler to think in terms of pivoting on appropriately chosen coefficients in each column than to describe the individual elementary operations.

Example B.   Solve the following linear system, using elementary row operations in the method of elimination.

(1)
$$\begin{cases} 2x_1 - 6x_2 + 6x_3 = 5 & \text{and} \\ 3x_1 + 2x_2 - 5x_3 = 4 & \text{and} \\ 5x_1 - 3x_2 + 4x_3 = 1. \end{cases}$$

Solution.   Our tactics will be to proceed through a series of equivalent systems toward one having the leading coefficient 1 in each equation, arranged in echelon down the diagonal of the array of coefficients. Then we will reduce the coefficients off the diagonal to zeros.

We begin with the first column, that is the coefficients of x_1. We pivot on the coefficient $a_{11} = 2$ in the upper left-hand corner. This gives us the equivalent system

(2)
$$\begin{cases} x_1 - 3x_2 + 3x_3 = \frac{5}{2} & \text{and} \\ 0 + 11x_2 - 14x_3 = -\frac{7}{2} & \text{and} \\ 0 + 12x_2 - 11x_3 = -\frac{23}{2}. \end{cases}$$

Then we pivot on the coefficient $a_{22} = 11$ in the second equation. This leaves the first column unchanged and produces the equivalent system

(3)
$$\begin{cases} x_1 + 0 - \frac{9}{11}x_3 = \frac{34}{12} & \text{and} \\ 0 + x_2 - \frac{14}{11}x_3 = -\frac{7}{22} & \text{and} \\ 0 + 0 + \frac{47}{11}x_3 = -\frac{158}{22}. \end{cases}$$

Finally we pivot on the coefficient $a_{33} = \frac{47}{11}$ in the third column of system (3). This produces the system, still equivalent to (1),

(4)
$$
\begin{cases}
x_1 + 0 \ + 0 \ = \frac{16}{94} & \text{and} \\
0 \ + x_2 + 0 \ = -\frac{231}{94} & \text{and} \\
0 \ + 0 \ + x_3 = -\frac{158}{94}.
\end{cases}
$$

From (4) we can read off that the one and only one solution of the system (1) is the row of numbers $(x_1, x_2, x_3) = (\frac{16}{94}, -\frac{231}{94}, -\frac{158}{94})$. This completes the problem.

11.3   Systems with Infinitely Many Solutions

Example.   Solve the linear system

(1)
$$
\begin{cases}
\quad\ \ 3x_2 - 2x_3 + 4x_4 = 1 & \text{and} \\
2x_1 - \ x_2 + 3x_3 - \ x_4 = 3 & \text{and} \\
6x_1 + 3x_2 + 5x_3 + 5x_4 = 11.
\end{cases}
$$

Solution.   The array of coefficients of (x_1, x_2, x_3, x_4) is the same as that of a system we have already solved (§11.1). Only the last number 11 on the right is different. Applying the same sequence of elementary row operations that we did for the previous problem (§11.1), we obtain the equivalent system

(2)
$$
\begin{cases}
x_1 + 0 \ + \frac{7}{6}x_3 + \frac{1}{6}x_4 = \frac{10}{6} & \text{and} \\
0 \ + x_2 - \frac{2}{3}x_3 + \frac{4}{3}x_4 = \frac{1}{3} & \text{and} \\
0 \ + 0 \ + 0 \quad + 0 \quad = 0.
\end{cases}
$$

The last equation of the system (2) is true for every set of values of (x_1, x_2, x_3, x_4). Hence we have only to solve the system composed of the first two equations. We observe that this can be done by assigning any arbitrary values, u and v, to x_3 and x_4. We set $x_3 = u$ and $x_4 = v$. Then all solutions of the system (1) are given by the numbers

$$
\begin{aligned}
x_1 &= \tfrac{10}{6} - \tfrac{7}{6}u - \tfrac{1}{6}v, \\
x_2 &= \tfrac{1}{3} + \tfrac{2}{3}u - \tfrac{4}{3}v, \\
x_3 &= u, \\
x_4 &= v,
\end{aligned}
$$

where (u, v) ranges over all pairs of numbers. This completes the solution. For example, if $u = 0$ and $v = 1$, then one solution is $x_1 = \frac{3}{2}$, $x_2 = -1$, $x_3 = 0$, $x_4 = 1$; but a two-dimensional infinitude of other solutions can be found by assigning other values to (u, v).

These examples show that the method of reduction by elementary row operations will solve any linear system, whether it has no solutions, exactly one solution, or infinitely many solutions. Moreover, it can solve a system of linear equations whether the number of equations is the same as the number of unknowns, or not. It lends itself easily to computation and is economical in computation.

11.4   Reduced Echelon Form.   We may state a general definition of the reduced echelon from of a system of m linear equations in n unknowns.

DEFINITION.   The system,

$$a_{11}x_1 + a_{12}x_2 + \cdots + a_{1n}x_n = b_1, \quad \text{and}$$
$$a_{21}x_1 + a_{22}x_2 + \cdots + a_{2n}x_n = b_2, \quad \text{and}$$
$$\cdots \cdots \cdots \cdots \cdots \cdots$$
$$a_{m1}x_1 + a_{m2}x_2 + \cdots + a_{mn}x_n = b_m,$$

is in reduced echelon form if and only if:

(a)   The first nonzero coefficient a_{ij} in each nonzero equation (row) is 1; and
(b)   Each column that contains the leading coefficient 1 of x_j in some equation (row) has the coefficient 0 of x_j in all other equations (rows); and
(c)   Each equation (row) that has all of its coefficients $a_{ij} = 0$ is placed below every equation (row) that contains some nonzero coefficient; and
(d)   The leading coefficients 1 appear in echelon. That is, if $a_{ij} = 1$ is a leading 1 and $a_{kl} = 1$ is a leading 1, and $i < k$, then $j < l$.

In simpler words, the leading coefficients 1 are staggered downward to the right, while all other coefficients in the same column with a leading 1 are zeros. The leading coefficients usually form a diagonal of the array of coefficients, but not in every case. We omit the proof that every system is equivalent to a reduced echelon form and this is uniquely determined by the choice of coefficients for pivoting.

11.5   Exercises

In Exercises 1–8, the systems are given in reduced echelon form. For each system state precisely what the solution is.

1. $\begin{cases} x_1 + 0 = 4 \quad \text{and} \\ 0 + x_2 = 3. \end{cases}$

2. $\begin{cases} x_1 + x_2 = 1 \quad \text{and} \\ 0 + 0 = 0. \end{cases}$

3. $\begin{cases} x_1 + x_2 = 1 \quad \text{and} \\ 0 + 0 = 3. \end{cases}$

4. $\begin{cases} x_1 + x_2 = 0 \quad \text{and} \\ 0 + 0 = 0. \end{cases}$

5. $\begin{cases} x_1 + 2x_2 - x_3 = 1 \quad \text{and} \\ \quad x_2 + 3x_3 = 4. \end{cases}$

6. $\begin{cases} x_1 + 0 - x_3 = 1 \quad \text{and} \\ 0 + x_2 + 3x_3 = 4 \quad \text{and} \\ 0 + 0 + 0 = 0. \end{cases}$

7. $\begin{cases} x_1 + 0 - x_3 = 1 \quad \text{and} \\ 0 + x_2 + 3x_3 = 4 \quad \text{and} \\ 0 + 0 + 0 = 5. \end{cases}$

8. $\begin{cases} x_1 + 2x_2 + 0 = 1 \quad \text{and} \\ 0 + 0 + x_3 = 3. \end{cases}$

9.   In Exercise 8, check that the system is in reduced echelon form but does not have the leading coefficients 1 on the diagonal of the coefficient array. Show that no combinations of elementary row operations can give x_2 a leading coefficient 1.

In Exercises 10–21, find the equivalent reduced echelon form by use of elementary row operations. Use this to find *all* solutions of the system. State carefully what the set of all solutions is.

10. $\begin{cases} 2x + 3y = 5 \quad \text{and} \\ 4x + 6y = -1. \end{cases}$

11. $\begin{cases} 2x + 3y = 5 \quad \text{and} \\ 4x + 6y = 10. \end{cases}$

12. $\begin{cases} 2x + 3y - z = 1 \quad \text{and} \\ x - y + z = 4. \end{cases}$

13. $\begin{cases} 2x_1 + 3x_2 - x_3 = 0 \\ x_1 - x_2 + 4x_3 = 0. \end{cases}$

14. $\begin{cases} -x_1 + x_2 + 4x_3 = 0 \quad \text{and} \\ x_1 + 3x_2 + 8x_3 = 0 \quad \text{and} \\ \frac{1}{2}x_1 + x_2 + \frac{5}{2}x_3 = 0. \end{cases}$

15. $\begin{cases} x_1 - x_3 = 0 \quad \text{and} \\ x_2 + 3x_3 = 0. \end{cases}$

16. $\begin{cases} x_1 + 2x_2 + x_3 - x_4 = 0 \quad \text{and} \\ 3x_1 - x_3 + 5x_4 = 0. \end{cases}$

17. $\begin{cases} \frac{1}{2}x_1 + 2x_2 - 6x_3 = 0 \quad \text{and} \\ -4x_1 \qquad + 5x_3 = 0 \quad \text{and} \\ -3x_1 + 6x_2 - 13x_3 = 0 \quad \text{and} \\ -\frac{7}{3}x_1 + 2x_2 - \frac{8}{3}x_3 = 0. \end{cases}$

18. $\begin{cases} x_1 - x_2 + 2x_3 = 1 \quad \text{and} \\ 2x_1 \qquad + 2x_3 = 1 \quad \text{and} \\ x_1 - 3x_2 + 4x_3 = 2. \end{cases}$

19. $\begin{cases} -x_1 - x_2 + 3x_3 = 6 \quad \text{and} \\ 3x_1 + 2x_2 - x_3 = 4 \quad \text{and} \\ -5x_1 - 4x_2 + 7x_3 = 8. \end{cases}$

20. $\begin{cases} -x_1 - x_2 + 3x_3 = 6 \quad \text{and} \\ 3x_1 + 2x_2 - x_3 = 4 \quad \text{and} \\ 5x_1 + 4x_2 - 7x_3 = 8. \end{cases}$

21. For what values of y_1, y_2, y_3 does the following system have a solution? Write the solution explicitly in the case that it exists.

$$\begin{cases} 3x_1 - x_2 + 2x_3 = y_1 \quad \text{and} \\ x_1 + x_2 + x_3 = y_2 \quad \text{and} \\ x_1 - 3x_2 \qquad = y_3. \end{cases}$$

22.* Show that if a linear system has two different solutions, it has infinitely many.

12 Matrices

12.1 The Matrix of Coefficients in a Linear System.

We observe that when we perform elementary row operations on a system of linear equations such as

(1) $$\begin{cases} 2x_1 + 3x_2 - 6x_3 = 0 \quad \text{and} \\ 3x_1 - 5x_2 \qquad = 0, \end{cases}$$

the unknowns, x_1, x_2, x_3, are superfluous. Actually we operate only on the coefficients. So we abbreviate by writing down the *matrix* of coefficients

$$\begin{bmatrix} 2 & 3 & -6 \\ 3 & -5 & 0 \end{bmatrix}.$$

For some purposes we want the *augmented matrix*, which includes the numbers on the right-hand side of each equation. In this case the augmented matrix is

$$\begin{bmatrix} 2 & 3 & -6 & 0 \\ 3 & -5 & 0 & 0 \end{bmatrix}.$$

We recognize the horizontal *rows* of the matrix and the vertical *columns* of numbers. Each column consists of the coefficients of one unknown.

Elementary row operations are defined for matrices in the same way as for systems of linear equations (§11.2). We simply read "row" instead of "equation" in the definition. That is, elementary row operations consist of interchanging two rows, or multiplying a row by a constant c, or changing a row by adding to a constant multiple of another row. To reduce the matrix, using these operations, we pivot on some element in each column until all of the rows, or columns, are exhausted.

Let us perform elementary row operations on the augmented matrix to solve the system (1). We first pivot on the element $a_{11} = 2$ in the upper left-hand corner. This gives us the equivalent matrix

$$\begin{bmatrix} 1 & \frac{3}{2} & -3 & 0 \\ 0 & -\frac{19}{2} & 9 & 0 \end{bmatrix}.$$

Pivoting on the element $a_{22} = -\frac{19}{2}$ in this matrix we find that it is equivalent to

$$\begin{bmatrix} 1 & 0 & -\frac{30}{19} & 0 \\ 0 & 1 & -\frac{18}{19} & 0 \end{bmatrix},$$

which we say is in reduced echelon form (§11.4). If we restore the variables it says

(2)
$$\begin{cases} x_1 + 0 - \frac{30}{19}x_3 = 0 \quad \text{and} \\ 0 + x_2 - \frac{18}{19}x_3 = 0. \end{cases}$$

We see that we can assign x_3 an arbitrary value $x_3 = u$ and find x_1 and x_2 in terms of u to satisfy the system. Thus for all real numbers u the rows

$$(\tfrac{30}{19}u, \tfrac{18}{19}u, u)$$

are solutions, and these are the only solutions of (2) and therefore of (1) also.

12.2   Formal Description of an $m \times n$ Matrix.

We may display an $m \times n$ matrix A as a rectangular array of numbers, consisting of m rows and n columns.

$$A = \begin{bmatrix} a_{11} & a_{12} & \cdots & a_{1n} \\ a_{21} & a_{22} & \cdots & a_{2n} \\ \vdots & & & \vdots \\ a_{m1} & a_{m2} & \cdots & a_{mn} \end{bmatrix},$$

or more briefly,

$$A = [a_{ij}], \ i = 1, \ldots, m; j = 1, \ldots, n.$$

The number a_{ij} is called the *element* in the ith row and jth column of A. Thus a_{25} is in the second row and fifth column of a matrix A. We note that the matrix itself is denoted by capital letters, A, B, C, whereas its elements are small letters a_{ij}, b_{ij}, c_{ij}.

Two matrices A and B are *equal* if they are identical in every element. They must have the same number of rows and columns. Then $A = B$ if and only if $a_{ij} = b_{ij}, i = 1, \ldots, m; j = 1, \ldots, n$.

Among $n \times n$ square matrices, we distinguish the zero matrix O and the identity matrix I,

$$O = \begin{bmatrix} 0 & 0 & \cdots & 0 & 0 \\ 0 & 0 & \cdots & 0 & 0 \\ \vdots & & & & \vdots \\ 0 & 0 & \cdots & 0 & 0 \end{bmatrix}, \quad I = \begin{bmatrix} 1 & 0 & \cdots & 0 & 0 \\ 0 & 1 & \cdots & 0 & 0 \\ \vdots & & & & \vdots \\ 0 & 0 & \cdots & 0 & 1 \end{bmatrix}.$$

The identity matrix has 1's down the main diagonal and zeros elsewhere. That is $I = [a_{ij}]$, where $a_{ii} = 1$ and $a_{ij} = 0$ if $i \neq j$.

The zero and identity matrices are in reduced echelon form. So is the matrix

$$\begin{bmatrix} 0 & 1 & 0 & 0 & 0 & 3 \\ 0 & 0 & 1 & 2 & 0 & 5 \\ 0 & 0 & 0 & 0 & 1 & -1 \\ 0 & 0 & 0 & 0 & 0 & 0 \end{bmatrix},$$

$$\cdots\text{—diagonal}$$

which does not have its leading 1's on the main diagonal. The following matrices are not in reduced echelon form.

$$
\begin{bmatrix} 1 & 0 & 0 & 2 \\ 0 & 1 & 0 & -1 \\ 0 & 0 & 2 & 0 \\ 0 & 0 & 0 & 0 \end{bmatrix}, \quad \begin{bmatrix} 1 \\ 0 \\ 0 \\ -1 \end{bmatrix}, \quad \begin{bmatrix} 0 & 0 & 0 & 0 \\ 0 & 1 & 0 & 0 \\ 0 & 0 & 1 & 0 \\ 0 & 0 & 0 & 1 \end{bmatrix}, \quad \begin{bmatrix} 0 & 1 \\ 1 & 0 \end{bmatrix}.
$$

As for systems of linear equations, we can prove by mathematical induction (omitted) the following theorem.

THEOREM. Every $m \times n$ matrix is equivalent under elementary row operations to a matrix in reduced echelon form, uniquely determined by the pivoting elements.

12.3 **Theory of Homogeneous Linear Equations.** A *homogeneous* linear system is a linear system with the right-hand members all zero. An astonishing number of mathematical problems and proofs hinge on the solution of homogeneous linear systems. Our method of solution is by elementary row operations that lead to the reduced echelon form. This elimination method is both logically and computationally efficient. We examine it in some detail as it applies to homogeneous linear systems.

Explicitly, a homogeneous linear system is one of the form

$$
\begin{aligned}
a_{11}x_1 + a_{12}x_2 + \cdots + a_{1n}x_n &= 0, \\
a_{21}x_1 + a_{22}x_2 + \cdots + a_{2n}x_n &= 0, \\
& \cdot \cdot \cdot \cdot \\
a_{m1}x_1 + a_{m2}x_2 + \cdots + a_{mn}x_n &= 0.
\end{aligned}
$$

We may abbreviate it this way. Let A be the matrix of coefficients, $A = [a_{ij}]$, and

$$
X = \begin{bmatrix} x_1 \\ x_2 \\ \vdots \\ x_n \end{bmatrix}, \quad \text{and} \quad O = \begin{bmatrix} 0 \\ 0 \\ \vdots \\ 0 \end{bmatrix}
$$

are single-column matrices of n unknowns and n zeros. Then we write the system briefly as

$$
AX = O.
$$

We observe first that a homogeneous linear system always has a solution. For the set of values, $x_1 = 0, x_2 = 0, \ldots, x_n = 0$ always satisfies any homogeneous linear system. This zero solution is called the *trivial solution*. All other solutions are called *nontrivial*.

DEFINITION. A homogeneous linear system has *rank r* if and only if its matrix A is equivalent under elementary row operations to a reduced echelon matrix that has exactly r nonzero rows.

THEOREM ON HOMOGENEOUS LINEAR SYSTEMS. If A is an $m \times n$ matrix of rank r, then the homogeneous linear system $AX = O$ has a nontrivial solution if and only if $r < n$.

Proof.   By elementary row operations we find an equivalent homogeneous linear system $RX = O$, where R is a matrix in reduced echelon form. By hypothesis there are exactly r rows in which an unknown appears with a leading 1 as its coefficient. It follows that $n - r \geq 1$, since $r < n$. Then the remaining $n - r$ unknowns may be assigned values arbitrarily. In particular, at least one of these can be assigned a value different from zero. These will determine the values of the r unknowns that appear with leading coefficient 1 (§11.3) to yield a nontrivial solution of the system $AX = O$. This completes the proof.

COROLLARY 1.   If A is an $m \times n$ matrix and $m < n$, then the homogeneous system $AX = O$ has a nontrivial solution.

COROLLARY 2.   If A is an $n \times n$ square matrix, then the only solution of $AX = O$ is the trivial solution, if and only if A is equivalent to the identity matrix.

COROLLARY 3.   If A is an $m \times n$ matrix of rank r and $r < n$, then $n - r$ unknowns can be assigned values arbitrarily and each such assignment determines the values of the other r unknowns. This gives all solutions of homogeneous linear system $AX = O$.

12.4   Exercises

For the matrices given in Exercises 1–10, find the reduced echelon form. Use it to find the rank of the matrix and find all solutions of the homogeneous linear system $AX = O$.

1. $A = \begin{bmatrix} 2 & 3 \\ -1 & 4 \end{bmatrix}$,

2. $A = \begin{bmatrix} 2 & 3 \\ 4 & 6 \end{bmatrix}$,

3. $A = \begin{bmatrix} 2 & 3 & -1 \\ 5 & 0 & 4 \\ 3 & -3 & 5 \end{bmatrix}$,

4. $A = \begin{bmatrix} 0 & 0 & 0 & 0 & 0 \\ 4 & 6 & -1 & 2 & 3 \\ 2 & 3 & -\frac{1}{2} & 1 & \frac{3}{2} \end{bmatrix}$,

5. $A = \begin{bmatrix} 1 & 1 & 1 \\ 2 & 2 & 2 \\ 3 & 3 & 3 \\ 4 & 4 & 4 \end{bmatrix}$,

6. $A = \begin{bmatrix} 4 \\ -7 \\ 0 \\ 3 \end{bmatrix}$,

7. $A = \begin{bmatrix} 6 & 0 & 4 \\ -2 & 1 & 3 \\ 8 & 9 & 11 \end{bmatrix}$,

8. $A = \begin{bmatrix} 1 & -2 & 3 & 8 \end{bmatrix}$,

9. $A = \begin{bmatrix} \cos \theta & -\sin \theta \\ \sin \theta & \cos \theta \end{bmatrix}$,

10. $A = \begin{bmatrix} 1 & 1 & 1 & 1 \\ t_1 & t_2 & t_3 & t_4 \\ t_1^2 & t_2^2 & t_3^2 & t_4^2 \\ t_1^3 & t_2^3 & t_3^3 & t_4^3 \end{bmatrix}$,

where t_1, t_2, t_3, t_4, are four distinct numbers.

11. Find all solutions of the homogeneous systems given as $AX = O$ by the matrices $A = O$ and $A = I$.

12. Find the numbers t so that the system $AX = O$ with matrix

$$\begin{bmatrix} 1 & 2 & -3 \\ 0 & 4 & 1 \\ 2 & t & -1 \end{bmatrix}$$

shall have nontrivial solutions.

The homogeneous linear system $AX = O$, with a matrix that involves an unknown λ,

$$A = \begin{bmatrix} 2 - \lambda & 3 \\ 1 & 1 - \lambda \end{bmatrix}$$

can be solved by reducing it to one with equivalent matrix

$$R = \begin{bmatrix} 1 & 1 - \lambda \\ 0 & 3 - (2 - \lambda)(1 - \lambda) \end{bmatrix}.$$

The system $RX = O$ has nontrivial solutions only if λ satisfies the quadratic equation $3 - (2 - \lambda)(1 - \lambda) = 0$. For only in this case has R a rank $r < 2$.

13. Find the numbers λ such that the system $AX = O$ with matrix

$$\begin{bmatrix} 1 - \lambda & 2 \\ 2 & 3 - \lambda \end{bmatrix}$$

shall have nontrivial solution.

14. For what numbers λ does the system

$$\begin{cases} 2x_1 + 3x_2 = \lambda x_1 \\ 3x_1 - 4x_2 = \lambda x_2 \end{cases}$$

have nontrivial solutions (x_1, x_2)?

15. Show that the homogeneous linear system $AX = O$ with two unknowns x_1, x_2,

$$A = \begin{bmatrix} a - \lambda & 0 \\ 0 & b - \lambda \end{bmatrix}$$

has a nontrivial solution only if $\lambda = a$ or $\lambda = b$.

16. Find all of the vectors $\vec{v} = (x_1, x_2, x_3)$ that are orthogonal to both the vectors $\vec{a} = (1, -1, 2)$ and $\vec{b} = (\vec{3}, -2, 4)$.

17. Find all of the vectors $\vec{v} = (x_1, x_2, x_3)$ that are orthogonal to $\vec{a} = (1, -1, 2)$. Show that they form a plane.

18. Find all of the vectors $\vec{v} = (x_1, x_2, x_3)$ that are orthogonal to both $\vec{a} = (a_1, a_2, a_3)$ and $\vec{b} = (b_1, b_2, b_3)$, where $\vec{a}$ is not parallel to $\vec{b}$.

19. Prove that there are exactly two reduced echelon matrices

$$\begin{bmatrix} a & b \\ c & d \end{bmatrix}$$

such that $a + b + c + d = 0$. Exhibit them.

13   Dimension of a Linear Space

Intuitively, the line is one-dimensional, the plane is two-dimensional, and solid space is three-dimensional. To give these assertions a logical meaning we must define (linear) "dimension" mathematically without appeal to geometric intuition. In so doing, we find that vector spaces can have dimension 4, or 5, or n, or infinity, just as well as 1, 2, or 3. The basic tool is the solution of homogeneous linear systems.

13.1   Linear Dependence and Bases.   We recall that a finite set of vectors $\{\vec{v}_1, \vec{v}_2, \ldots, \vec{v}_n\}$ in a vector space $\mathscr{V}$ is *linearly dependent* if for some row of coefficients $(x_1, x_2, \ldots, x_n)$, not all zero, the linear combination

$$x_1\vec{v}_1 + x_2\vec{v}_2 + \cdots + x_n\vec{v}_n = \vec{0}.$$

On the other hand, the set of vectors is *linearly independent* if every linear combination of them which vanishes has $x_1 = x_2 = \cdots = x_n = 0$. The set of vectors $\mathscr{B} = \{\vec{v}_1, \vec{v}_2 \ldots, \vec{v}_n\}$ *spans* the space $\mathscr{V}$ if and only if every vector $\vec{v}$ in $\mathscr{V}$ can be expressed as a linear combination of the vectors of $\mathscr{B}$

$$\vec{v} = x_1\vec{v}_1 + x_2\vec{v}_2 + \cdots + x_n\vec{v}_n.$$

Clearly a larger set of vectors, containing the spanning set $\mathscr{B}$, also spans $\mathscr{V}$. Hence $\mathscr{V}$ has many spanning sets if it has one.

A finite set of vectors $\mathscr{B} = \{\vec{v}_1, \vec{v}_2, \ldots, \vec{v}_n\}$ is a (*finite*) *basis* for $\mathscr{V}$ if and only if $\mathscr{B}$ spans $\mathscr{V}$ and is linearly independent.

We observe that a vector space with one finite basis has many different bases. For example, if $\mathscr{B}$ is a basis, then the set $\{\vec{v}_1 - \vec{v}_2, \vec{v}_1 + \vec{v}_2, \vec{v}_3, \ldots, \vec{v}_n\}$ is a different basis. But if we can prove that every finite basis has the same number of vectors in it as any other basis, then it will be natural to say that this invariant number of vectors in a basis is the dimension of the space.

THEOREM. If a vector space $\mathscr{V}$ has the basis $\mathscr{B} = \{\vec{v}_1, \vec{v}_2, \ldots, \vec{v}_n\}$ and also the basis $\mathscr{C} = \{\vec{w}_1, \vec{w}_2, \ldots, \vec{w}_m\}$, then $m = n$.

Proof. Since $\mathscr{C}$ spans $\mathscr{V}$, each of the vectors $\vec{v}_j$ in $\mathscr{B}$ can be expressed as linear combination

$$\vec{v}_j = c_{1j}\vec{w}_1 + c_{2j}\vec{w}_2 + \cdots + c_{mj}\vec{w}_m, \qquad j = 1, 2, \ldots, n.$$

By substituting these expressions for $\vec{v}_j$ into any linear combination of the vectors in the basis $\mathscr{B}$,

$$x_1\vec{v}_1 + x_2\vec{v}_2 + \cdots + x_n\vec{v}_n,$$

and regrouping to combine the coefficients of $\vec{w}_i$, we can express it as a linear combination of the vectors $\vec{w}_i$ in the basis $\mathscr{C}$. The result is

$$\begin{aligned}
x_1\vec{v}_1 + x_2\vec{v}_2 + \cdots + x_n\vec{v}_n = {} & (c_{11}x_1 + c_{12}x_2 + \cdots + c_{1n}x_n)\vec{w}_1 \\
& + (c_{21}x_1 + c_{22}x_2 + \cdots + c_{2n}x_n)\vec{w}_2 \\
& \cdots \cdots \cdots \cdots \cdots \cdots \cdots \\
& + (c_{m1}x_1 + c_{m2}x_2 + \cdots + c_{mn}x_n)\vec{w}_m.
\end{aligned}$$

Now if $m < n$, the homogeneous linear system

$$\begin{aligned}
c_{11}x_1 + c_{12}x_2 + \cdots + c_{1n}x_n &= 0, \\
c_{21}x_1 + c_{22}x_2 + \cdots + c_{2n}x_n &= 0, \\
\cdots \cdots \cdots \cdots \cdots \cdots \cdots & \\
c_{m1}x_1 + c_{m2}x_2 + \cdots + c_{mn}x_n &= 0
\end{aligned}$$

has a nontrivial solution $(x_1, x_2, \ldots, x_n)$ (§12.3, Corollary 1). For these coefficients, not all zero

$$x_1\vec{v}_1 + x_2\vec{v}_2 + \cdots + x_n\vec{v}_n = \vec{0}.$$

Hence the vectors $\{\vec{v}_1, \vec{v}_2 \ldots, \vec{v}_n\}$ form a linearly dependent set and therefore cannot be a basis. This shows that if both $\mathscr{B}$ and $\mathscr{C}$ are bases, then it is false that $m < n$. Similarly, we can show that $n < m$ is false and hence we prove that $m = n$.

13.2   Definition of Dimension.   A vector space $\mathscr{V}$ has *dimension n* if and only if it has a basis $\{\vec{v}_1, \vec{v}_2, \ldots, \vec{v}_n\}$ with n vectors in it.

Example.   Find the dimension of the space $\mathscr{V}$ of rows (x_1, x_2, x_3) that satisfy the equations

$$2x_1 - 3x_2 + x_3 = 0 \quad \text{and}$$
$$x_1 + x_2 - x_3 = 0 \quad \text{and}$$
$$5x_1 - 5x_2 + x_3 = 0.$$

Solution.   Using elementary row operations, we find the equivalent reduced echelon form

$$x_1 + 0 - \tfrac{2}{5}x_3 = 0 \quad \text{and}$$
$$0 + x_2 - \tfrac{3}{5}x_3 = 0 \quad \text{and}$$
$$0 + 0 + 0 = 0.$$

Hence we may assign arbitrarily $x_3 = c$ and find that all solutions (x_1, x_2, x_3) are of the form

$$(x_1, x_2, x_3) = c(\tfrac{2}{5}, \tfrac{3}{5}, 1).$$

Since the singleton set $(\tfrac{2}{5}, \tfrac{3}{5}, 1)$ is linearly independent and spans the space of solutions, it is a basis. Hence the dimension of the linear space of solutions is 1.

We note that the dimension cannot be determined just by counting the coordinates (x_1, x_2, x_3) in the rows that represent points in the space. For in this case the dimension was 1, not 3. The solution space is a line imbedded in the space of all points (x_1, x_2, x_3), which is 3-dimensional because it has the basis $\{(1, 0, 0), (0, 1, 0), (0, 0, 1)\}$.

13.3   Isomorphic Representation by Rows.   We consider a vector space of dimension 2, having the real numbers as scalar multipliers. There is essentially only one such space and it may be represented by the space of rows (x_1, x_2). For let $\mathscr{V}$ be any vector space of dimension 2, having the real numbers as scalars. We choose one of its bases, $\mathscr{B}:(\vec{v}_1, \vec{v}_2)$. Then, in terms of this basis, every vector $\vec{v}$ in $\mathscr{V}$ can be represented uniquely as a linear combination

$$\vec{v} = x_1\vec{v}_1 + x_2\vec{v}_2.$$

Thus every vector $\vec{v}$ in $\mathscr{V}$ determines a row of coordinates (x_1, x_2). On the other hand, every row of coordinates, with the chosen basis, determines a unique vector $\vec{v}$. Hence there is a one-to-one pairing of all vectors $\vec{u}:(y_1, y_2)$ and $\vec{v}:(x_1, x_2)$, with their rows of coordinates. Moreover, the sum vector $\vec{u} + \vec{v}$ is paired with the sum of the row vectors $(y_1 + x_1, y_2 + x_2)$. And the scalar multiple $c\vec{v}$ is paired with the scalar multiple of the row $c(x_1, x_2) = (cx_1, cx_2)$. Thus the structure vector space cannot distinguish between the abstract space $\mathscr{V}$ of dimension 2, with real scalars, from the particular vector space $\mathscr{R}^2$ of all rows (x_1, x_2) of real numbers.

In this sense, we say that $\mathscr{R}^2$ is an *isomorphic representation* of $\mathscr{V}$, where the word *isomorphic* suggests that the two spaces have the same form. These ideas generalize at once to spaces $\mathscr{V}$ with any finite basis $\mathscr{B}:(\vec{v}_1, \vec{v}_2, \ldots, \vec{v}_n)$.

DEFINITION.   A mapping T of vector space $\mathscr{V}$ onto vector space $\mathscr{W}$ is an *isomorphism* if every vector $\vec{w}$ in $\mathscr{W}$ is the image of exactly one vector $\vec{v}$ in $\mathscr{V}$, $\vec{w} = T(\vec{v})$, $\vec{w}' = T(\vec{v}')$, and if $\vec{w} = \vec{w}' = T(\vec{v} + \vec{v}')$ and $c\vec{w} = T(c\vec{v})$.

In words, an isomorphism is a one-to-one pairing of the vectors of the spaces $\mathscr{V}$ and $\mathscr{W}$ that preserves the vector operations. If two spaces have an isomorphism, they are said to be *isomorphic*.

THEOREM.   Every vector space $\mathscr{V}$ of dimension n, with all real numbers as scalar multipliers, is isomorphic to the row space $\mathscr{R}^n$ of rows $(x_1, x_2, \ldots, x_n)$ of real numbers.

Thus the structure of vector addition and scalar multiplication can distinguish only one real vector space of dimension n. We observe, however, that the particular pairing of vectors $\vec{v}$ and rows $(x_1, \ldots, x_n)$ depends on the choice of basis. So, with different bases, the same row may be paired with many different vectors $\vec{v}$. Therefore the use of the representation by coordinates has the difficulty that it does not necessarily produce geometrically invariant statements, unless the basis itself is selected by some geometric criterion.

COROLLARY.   Every real vector $\mathscr{V}$ of dimension n is isomorphic to every other real vector space of dimension n.

13.4 Exercises

1.   Determine which of the following sets of row vectors are linearly dependent.
(a) $\{(2, 0, -1), (1, -2, 5), (5, -2, 3)\}$.
(b) $\{(2, 0, -1), (2, 0, -1), (x_1, x_2, x_3)\}$.
(c) $\{(x_1, x_2, x_3), (y_1, y_2, y_3), (0, 0, 0)\}$.

2.   Find the dimension of the space spanned by the vectors $\{(3, -1, 4, 0), (1, 4, -2, 1), (4, 3, 2, 1), (7, 2, 6, 1)\}$. Find a basis for it.

3.   Show that the vectors $\{(1, 0, 0), (0, 1, 0), (0, 0, 1)\}$ form a basis for the set of all vectors $\vec{v}:(x_1, x_2, x_3)$. This basis is called the standard basis.

4.   Show that the vectors $\{(1, 0, -1), (1, 2, 1), (0, -3, 2)\}$ form a basis for the space of all vectors $\vec{v}:(x_1, x_2, x_3)$.

5.   Express the vectors of the basis of Exercise 4 in terms of the standard basis (Exercise 3).

6.   The quadratic polynomials $ax^2 + bx + c$ form a vector space. What is its dimension?

7.   The set of all linear combinations of the functions $\sin \theta$ and $\cos \theta$ is a vector space. Find its dimension. Show that the function $\sin (\theta - \pi/3)$ is in the space by expressing it as a linear combination of the basis vectors.

8.   Show that $(3, -5, 6)$ is in the space of all vectors $\vec{v}:(x_1, x_2, x_3)$ spanned by the vectors $\{(4, -3, 1), (1, 2, -5), (7, -8, 7)\}$. Is $(4, -8, 2)$ in the space?

9.   Find the dimension of the space spanned by the rows of the matrix

$$A = \begin{bmatrix} 4 & 2 & -1 & 6 \\ -5 & 1 & 6 & 3 \\ -1 & 3 & 5 & 9 \end{bmatrix}.$$

Find the dimension of the space of solutions of the homogeneous linear system $AX = O$.

10. Let $A = [a_{ij}](i = 1, \ldots, m; j = 1, \ldots, n)$ be a matrix with m rows and n columns that has rank r. Show that the dimension of the space spanned by the rows of A is r.

11. In Exercise 10, show that the space of all solutions of the homogeneous system $AX = O$ is $n - r$.

12. Recall that two vectors are orthogonal if their inner product $(\vec{v} \cdot \vec{w}) = 0$. Prove that if neither $\vec{v}$ nor $\vec{w}$ is zero and they are orthogonal, then $\{\vec{v}, \vec{w}\}$ is a linearly independent set.

13. Show that if $\vec{v}$ and $\vec{w}$ are linearly dependent, then one is a constant multiple of the other.

14. Find the dimension and a basis for the set of all vectors $v:(x_1, x_2, x_3)$ that are orthogonal to both $(1, -3, 2)$ and $(4, -1, 5)$.

15. As in Exercise 13, show that if $\{\vec{v}_1, \vec{v}_2, \ldots, \vec{v}_n\}$ is a set of nonzero vectors that are mutually orthogonal in pairs it is linearly independent.

16. As in Exercise 6, show that the powers $\{1, x, x^2, \ldots, x^n\}$ form a basis for a vector space. What is its dimension?

17. Show that the only vector space of dimension 0 consists of the zero vector $\vec{0}$ alone.

18. Show that the set of all infinite sequences $\{a_1, a_2, a_3, \ldots, a_n, \ldots\}$ is not a finite dimensional space.

19. What is the dimension of the space of vectors in the complex plane having the basis $\{\vec{1} + 2\vec{i}, -2\vec{1} - 3\vec{i}\}$?

20. Show that if a vector $\vec{u}$ is orthogonal to both $\vec{v}$ and $\vec{w}$, it is orthogonal to every vector in the space spanned by $\{\vec{v}, \vec{w}\}$.

21. Show that if the vectors $\{\vec{v}, \vec{v}_2, \ldots, \vec{v}_m\}$ in n-dimensional space span a space of dimension r, then the set of all vectors orthogonal to all of them has dimension $n - r$.

22. Show that the set of all time-space positions $(-ct, x, y, z)$ is 4-dimensional, where c is the velocity of light and (x, y, z) are coordinates of points in ordinary 3-space. Also show that the set of positions satisfying the additional equation $x + y + z - kt = 0$ is 3-dimensional.

23. Show that the mapping

$$T:\begin{cases} x_1 - 2x_2 = y_1 \\ 2x_1 - 4x_2 = y_2 \end{cases}$$

is not an isomorphism of the space of rows (x_1, x_2) onto the space of columns $\begin{pmatrix} y_1 \\ y_2 \end{pmatrix}$ even though it preserves the vector space operations.

14 Matrix Multiplication

14.1 Definition of Matrix Multiplication. If the number of columns of the $p \times m$ matrix B,

$$B = \begin{bmatrix} b_{11} & b_{12} & \cdots & b_{1m} \\ b_{21} & b_{22} & \cdots & b_{2m} \\ \vdots & & & \vdots \\ b_{p1} & b_{p2} & \cdots & b_{pm} \end{bmatrix},$$

is the same as the number of rows of the $m \times n$ matrix A,

$$A = \begin{bmatrix} a_{11} & a_{12} & \cdots & a_{1n} \\ a_{21} & a_{22} & \cdots & a_{2n} \\ \vdots & & & \vdots \\ a_{m1} & a_{m2} & \cdots & a_{mn} \end{bmatrix},$$

then we can form the $p \times n$ matrix product $C = BA$. The element in the kth row and jth column of C is the inner product of the kth row of B with the jth column of A. That is,

$$c_{kj} = [b_{k1} b_{k2} \cdots b_{km}] \begin{bmatrix} a_{1j} \\ a_{2j} \\ \vdots \\ a_{mj} \end{bmatrix} = b_{k1} a_{1j} + b_{k2} a_{2j} + \cdots + b_{km} a_{mj};$$

$k = 1, \ldots, p, j = 1, \ldots, n.$

Example.   For the 2 × 2 matrices

$$A = \begin{bmatrix} 2 & 3 \\ -4 & 1 \end{bmatrix} \quad \text{and} \quad B = \begin{bmatrix} 5 & 2 \\ -3 & 7 \end{bmatrix},$$

compute the products AB and BA.

Solution.   To obtain the element in the first row and first column of AB we form the inner product

$$[2 \quad 3] \begin{bmatrix} 5 \\ -3 \end{bmatrix} = 2(5) + 3(-3) = 1.$$

Similarly, we compute the other elements of the product.

$$AB = \begin{bmatrix} 2 & 3 \\ -4 & 1 \end{bmatrix} \begin{bmatrix} 5 & 2 \\ -3 & 7 \end{bmatrix} = \begin{bmatrix} 2(5) + 3(-3) & 2(2) + 3(7) \\ -4(5) + 4(-3) & -3(3) + 7(1) \end{bmatrix} = \begin{bmatrix} 1 & 25 \\ -32 & -1 \end{bmatrix}.$$

$$BA = \begin{bmatrix} 5 & 2 \\ -3 & 7 \end{bmatrix} \begin{bmatrix} 2 & 3 \\ -4 & 1 \end{bmatrix} = \begin{bmatrix} 5(2) + 2(-4) & 5(3) + 2(1) \\ -3(2) + 7(-4) & -3(3) + 7(1) \end{bmatrix} = \begin{bmatrix} 2 & 17 \\ -34 & -2 \end{bmatrix}.$$

Since $AB \neq BA$, we distinguish between multiplying A by B *on the left, BA*, and multiplying A by B *on the right, AB*. This shows that, unlike the multiplication of real numbers, matrix multiplication is not commutative in every case.

14.2   Motivation of the Definition of Matrix Multiplication.

The study of matrices brings forth many different reasons for adopting the apparently arbitrary definition of matrix multiplication. Here, elementary row operations on the matrix A lead us to the definition. Since elementary row operations consist of linear combinations of the rows of a matrix, we seek a systematic procedure for representing and computing linear combinations of rows.

Example.   For the linear system $AX = Y$,

$$2x_1 + 3x_2 + 4x_3 - 5x_4 = y_1,$$
$$x_1 + 3x_2 - 7x_3 - 4x_4 = y_2,$$
$$3x_1 - 5x_2 - x_3 + x_4 = y_3,$$

form the linear combinations of rows,

$$2y_1 - y_2 + 3y_3 \quad \text{and then} \quad 5y_1 + 4y_2 - 2y_3.$$

Solution.   We express each of the rows y_1, y_2, y_3 in terms of the x's and collect like terms. This gives us

$$
\begin{aligned}
2y_1 - y_2 + 3y_3 &= [(2)(2) + (-1)(1) + (3)(3)]x_1 \\
&\quad + [2(-3) + (-1)(3) + 3(-5)]x_2 \\
&\quad + [2(4) + (-1)(-7) + 3(-1)]x_3 \\
&\quad + [2(-5) + (-1)(-4) + 3(1)]x_4 \\
&= 12x_1 - 24x_2 + 12x_3 - 3x_4.
\end{aligned}
$$

Similarly,

$$
\begin{aligned}
5y_1 + 4y_2 - 2y_3 &= [5(2) + 4(1) - 2(3)]x_1 \\
&\quad + [5(-3) + 4(3) - 2(-5)]x_2 \\
&\quad + [5(4) + 4(-7) - 2(-1)]x_3 \\
&\quad + [5(-5) + 4(-4) - 2(1)]x_4 \\
&= 8x_1 + 7x_2 - 6x_3 - 43x_4.
\end{aligned}
$$

In this we observe a pattern. The coefficient of each x is an inner product of the coefficients in the prescribed linear combination of rows with the column of coefficients of that x in the linear system. Suppressing the unknowns, we can display the matrices of coefficients to show what we have done in these two linear combinations of the rows,

$$
\begin{bmatrix} 2 & -1 & 3 \\ 5 & 4 & -2 \end{bmatrix}
\begin{bmatrix} 2 & -3 & 4 & -5 \\ 1 & 3 & -7 & -4 \\ 3 & -5 & -1 & 1 \end{bmatrix}
= \begin{bmatrix} 12 & -24 & 12 & -3 \\ 8 & 7 & -6 & -43 \end{bmatrix},
$$

or

$$
B \qquad\qquad A \qquad = \qquad C.
$$

The rows of the matrix B on the left are the coefficients in the prescribed linear combinations. The matrix A is the coefficient matrix of the original linear system. The matrix C on the right is the coefficient matrix of the new linear system that results from the prescribed linear combinations. This is an example of an operation we shall call *matrix multiplication*, denoted by

$$
BA = C.
$$

Let us observe the details of the calculation in some specific instances. To get the element -24 in the first row and second column of the product C, we take the first row of B and the second column of A and combine them as an inner product, thus,

$$
\begin{bmatrix} 2 & -1 & 3 \end{bmatrix}
\begin{bmatrix} -3 \\ 3 \\ -5 \end{bmatrix}
= (2)(-3) + (-1)(3) + (3)(-5) = -24.
$$

Similarly, to get the element -6 in the second row and third column of the product matrix, we take the inner product of the second row of B with the third column of A, thus,

$$
\begin{bmatrix} 5 & 4 & -2 \end{bmatrix}
\begin{bmatrix} 4 \\ -7 \\ -1 \end{bmatrix}
= (5)(4) + 4(-7) + (-2)(-1) = -6.
$$

14.3 Multiplying by *I* and *O*. Associativity. We have defined the $m \times m$ identity matrix I (§12.2) by

$$
I = \begin{bmatrix}
1 & 0 & 0 & \cdots & 0 \\
0 & 1 & 0 & \cdots & 0 \\
0 & 0 & 1 & \cdots & 0 \\
\vdots & \vdots & \vdots & & \vdots \\
0 & 0 & 0 & \cdots & 1
\end{bmatrix}.
$$

We observe that if A is $m \times n$, then $IA = A$. For example,

$$\begin{bmatrix} 1 & 0 \\ 0 & 1 \end{bmatrix} \begin{bmatrix} 2 & -3 & 4 \\ 6 & 0 & 2 \end{bmatrix} = \begin{bmatrix} 2 & -3 & 4 \\ 6 & 0 & 2 \end{bmatrix}.$$

To multiply the same matrix A on the right by I, it must be the $n \times n$ identity in order to match the number of columns and rows,

$$\begin{bmatrix} 2 & -3 & 4 \\ 6 & 0 & 2 \end{bmatrix} \begin{bmatrix} 1 & 0 & 0 \\ 0 & 1 & 0 \\ 0 & 0 & 1 \end{bmatrix} = \begin{bmatrix} 2 & -3 & 4 \\ 6 & 0 & 2 \end{bmatrix}.$$

that is, $AI = A$. Similarly, if we multiply on the left by the 2×2 zero matrix we have

$$\begin{bmatrix} 0 & 0 \\ 0 & 0 \end{bmatrix} \begin{bmatrix} 2 & -3 & 4 \\ 6 & 0 & 2 \end{bmatrix} = \begin{bmatrix} 0 & 0 & 0 \\ 0 & 0 & 0 \end{bmatrix}.$$

Multiplying on the right by the 3×3 zero matrix we have

$$\begin{bmatrix} 2 & -3 & 4 \\ 6 & 0 & 2 \end{bmatrix} \begin{bmatrix} 0 & 0 & 0 \\ 0 & 0 & 0 \\ 0 & 0 & 0 \end{bmatrix} = \begin{bmatrix} 0 & 0 & 0 \\ 0 & 0 & 0 \end{bmatrix}.$$

We also observe that matrices obey the associative law of multiplication

$$A(BC) = (AB)C$$

provided that the three matrices are dimensionally compatible for multiplication in this order. We postpone the proof, though it can be carried out from the definition by a rather tedious calculation.

14.4 Elementary Matrices.

We have seen that we can perform linear combinations of the rows of A by multiplying A on the left by a suitably chosen matrix B. We now set out to find the matrices E that multiply A on the left to produce specified elementary row operations. We call them *elementary matrices*.

Example A. Find the matrix E such that EA interchanges the first two rows of A.

Solution. Since $IA = A$ and $E(IA) = (EI)A = EA$, we see that E must be the identity matrix with its first two rows interchanged. For example, if A is 3×4, then we form

E from the 3×3 identity $I = \begin{bmatrix} 1 & 0 & 0 \\ 0 & 1 & 0 \\ 0 & 0 & 1 \end{bmatrix}$ by interchanging its first two rows,

$E = \begin{bmatrix} 0 & 1 & 0 \\ 1 & 0 & 0 \\ 0 & 0 & 1 \end{bmatrix}$. We readily verify that EA interchanges the first two rows of A,

$$\begin{bmatrix} 0 & 1 & 0 \\ 1 & 0 & 0 \\ 1 & 0 & 0 \end{bmatrix} \begin{bmatrix} a_{11} & a_{12} & a_{13} & a_{14} \\ a_{21} & a_{22} & a_{23} & a_{24} \\ a_{31} & a_{32} & a_{33} & a_{34} \end{bmatrix} = \begin{bmatrix} a_{21} & a_{22} & a_{23} & a_{24} \\ a_{11} & a_{12} & a_{13} & a_{14} \\ a_{31} & a_{32} & a_{33} & a_{34} \end{bmatrix}.$$

Example B. Find the elementary matrix that multiplies the second row of A by a nonzero constant c.

Solution. Again we observe that E must be formed from I by performing the required elementary operation we wish to perform on A. Thus if A is 3×4, we find that the required

$$E = \begin{bmatrix} 1 & 0 & 0 \\ 0 & c & 0 \\ 0 & 0 & 1 \end{bmatrix}.$$

We can verify that EA performs the required elementary row operation.

Example C. Find the elementary matrix E such that EA replaces the third row of A by adding to it c times the first row.

Solution. Again we have only to perform the required elementary row operation on the appropriate left identity matrix I. Thus, if A is 3×4, we add c times the first row of I to the third row and get

$$E = \begin{bmatrix} 1 & 0 & 0 \\ 0 & 1 & 0 \\ c & 0 & 1 \end{bmatrix}.$$

We readily verify that EA performs the required elementary row operation.

14.5 Exercises

In Exercises 1–12, perform the indicated matrix multiplications.

1. $\begin{bmatrix} 2 & 1 \\ 3 & -2 \end{bmatrix}\begin{bmatrix} 5 & 0 \\ 1 & 4 \end{bmatrix}.$

2. $\begin{bmatrix} 3 & 6 \\ -1 & 2 \end{bmatrix}\begin{bmatrix} 0 & 1 \\ 0 & 4 \end{bmatrix}.$

3. $\begin{bmatrix} 6 & 2 \\ -1 & 2 \end{bmatrix}^2.$

4. $\begin{bmatrix} 3 & 1 \\ 2 & -2 \end{bmatrix}\begin{bmatrix} 2 & 10 \\ 0 & 3 \end{bmatrix}.$

5. $\begin{bmatrix} 2 & 0 \\ 0 & 1 \end{bmatrix}\begin{bmatrix} -1 & 0 \\ 0 & 1 \end{bmatrix}.$

6. $\begin{bmatrix} -1 & 0 \\ 0 & 1 \end{bmatrix}\begin{bmatrix} 2 & 0 \\ 0 & 1 \end{bmatrix}.$

7. $\begin{bmatrix} 3 & 1 \\ 2 & -2 \\ 5 & 4 \end{bmatrix}\begin{bmatrix} 8 & 1 \\ 9 & 7 \end{bmatrix}.$

8. $\begin{bmatrix} 3 & -2 & 1 \\ 0 & 4 & 6 \end{bmatrix}\begin{bmatrix} 2 \\ 5 \\ -7 \end{bmatrix}.$

9. $\begin{bmatrix} 2 & 1 \\ 3 & 2 \end{bmatrix}\begin{bmatrix} 1 \\ 0 \end{bmatrix}.$

10. $\begin{bmatrix} a & b \\ c & d \end{bmatrix}\begin{bmatrix} 0 \\ 1 \end{bmatrix}.$

11. $\begin{bmatrix} 1 & 0 & 0 \\ 1 & -1 & 0 \\ 0 & 0 & 1 \end{bmatrix}\begin{bmatrix} 2 & 2 \\ 3 & 3 \\ 4 & 4 \end{bmatrix}.$

12. $\begin{bmatrix} 0 & -1 \\ 1 & 0 \end{bmatrix}^2.$

In Exercises 13–15, construct the elementary matrices E that multiply A on the left, EA, to produce the required elementary row operations. Verify the result by actual calculation, where

$$A = \begin{bmatrix} 2 & -1 & 3 & 4 \\ 6 & 0 & 1 & 2 \\ -1 & 3 & 5 & 0 \end{bmatrix}.$$

13. Interchange the first and third rows of A.
14. Multiply the third row of A by -2.
15. Replace the second row of A by adding to it -2 times the third row.
16. Which of the following indicated matrix products are possible and which are not? Explain.

$$\text{(a)} \begin{bmatrix} 2 & -1 & 3 \\ 4 & 0 & 2 \end{bmatrix} \begin{bmatrix} 1 & 0 \\ 0 & 1 \end{bmatrix}; \qquad \text{(b)} \begin{bmatrix} 1 & 0 \\ 0 & 1 \end{bmatrix} \begin{bmatrix} 2 & -1 & 3 \\ 4 & 0 & 2 \end{bmatrix}.$$

17. Form the matrix products AP_1, AP_2, AP_3, where

$$A = \begin{bmatrix} a_{11} & a_{12} & a_{13} \\ a_{21} & a_{22} & a_{23} \end{bmatrix}, \qquad P_1 = \begin{bmatrix} 1 \\ 0 \\ 0 \end{bmatrix}, \qquad P_2 = \begin{bmatrix} 0 \\ 1 \\ 0 \end{bmatrix}, \qquad P_3 = \begin{bmatrix} 0 \\ 0 \\ 1 \end{bmatrix}.$$

18. What elementary row operations on

$$A = \begin{bmatrix} a & b \\ c & d \end{bmatrix}$$

do $E_1 A$ and $E_2 A$ perform if

$$E_1 = \begin{bmatrix} 0 & 1 \\ 1 & 0 \end{bmatrix} \quad \text{and} \quad E_2 = \begin{bmatrix} 1 & 0 \\ 3 & 1 \end{bmatrix}?$$

19. In Exercise 18, what elementary row operations are performed on A by multiplying it first by E_1 on the left and then by E_2, that is $E_2 E_1 A$?
20. Find a single matrix E such that EA is the reduced echelon form of A,

$$A = \begin{bmatrix} 3 & -5 \\ 4 & 2 \end{bmatrix}.$$

21.* Let A and B be 2×2 matrices such that $BA = I$. Prove that $AB = BA$.
22. Show that, if $a_{23} \neq 0$, pivoting on a_{23} in the matrix

$$A = \begin{bmatrix} a_{11} & a_{12} & a_{13} & a_{14} \\ a_{21} & a_{22} & a_{23} & a_{24} \\ a_{31} & a_{32} & a_{33} & a_{34} \end{bmatrix}$$

can be accomplished by the single operation of multiplying A on the left by the 3×3 matrix G_{23}, where

$$G_{23} = \begin{bmatrix} 1 & -a_{13}/a_{23} & 0 \\ 0 & 1/a_{23} & 0 \\ 0 & -a_{33}/a_{23} & 1 \end{bmatrix}.$$

23. In any $m \times n$ matrix A, describe how pivoting on a_{ij} can be accomplished by multiplying on the left by a matrix G_{ij}. How is G_{ij} formed by replacing the ith column of I?
24. Show that the inverse pivoting matrix $H_{ij} = G_{ij}^{-1}$ is formed by replacing the ith column of I by the jth column of A.

15   Matrix Algebra

We next consider the algebra of $n \times n$ square matrices, with n fixed. This permits us to add $A + B$ and multiply AB or BA without restriction.

15.1   Addition of Matrices. If $A = [a_{ij}]$ and $B = [b_{ij}]$ are $n \times n$ square matrices, then the sum $A + B$ is defined by

$$A + B = [a_{ij} + b_{ij}], \qquad i, j = 1, \ldots, n.$$

In other words, we add matrices by adding corresponding elements.

We may verify that addition of $n \times n$ matrices has the following properties $\mathscr{A}_1, \ldots, \mathscr{A}_5$.

$\mathscr{A}_1$ (Closure)   For every ordered pair A, B of $n \times n$ matrices there is a unique sum $A + B$, which is an $n \times n$ matrix.

$\mathscr{A}_2$ (Associativity)   For every set of three $n \times n$ matrices,

$$A + (B + C) = (A + B) + C.$$

$\mathscr{A}_3$ (Unit for addition)   There is a matrix O such that for every A,

$$A + O = A = O + A.$$

$\mathscr{A}_4$ (Additive inverse)   For every matrix A, there is a matrix $-A$ such that

$$A + (-A) = O = (-A) + A.$$

$\mathscr{A}_5$ (Commutativity)   For every pair of $n \times n$ matrices,.

$$A + B = B + A.$$

15.2   Algebraic Properties of Multiplication. Matrix multiplication for $n \times n$ matrices has the properties $\mathscr{M}_1$ to $\mathscr{M}_3$.

$\mathscr{M}_1$ (Closure)   For every pair A, B of $n \times n$ matrices, the product AB exists as an $n \times n$ matrix.

$\mathscr{M}_2$ (Associativity)   For every three $n \times n$ matrices,

$$A(BC) = (AB)C.$$

$\mathscr{M}_3$ (Unit for multiplication)   There is an $n \times n$ matrix I such that for every A,

$$AI = A = IA.$$

We have already observed that the commutative property does not hold for matrix multiplication. We will find that even if $A \neq 0$, many square matrices A do not have an inverse, A^{-1}.

We can verify, from the definitions of addition and multiplication, that matrix multiplication is distributive to addition, and written

$$\begin{aligned}&\text{(Left)} \quad A(B + C) = AB + AC, \quad \text{and} \\ &\text{(Right)} \quad (B + C)A = BA + CA.\end{aligned}$$

15.3   Scalar Multiplication. Another type of multiplication, called *scalar multiplication*, is defined for multiplying a matrix by a number. We define

$$cA = [ca_{ij}] = Ac.$$

In words, to multiply the matrix A by the number c, we multiply every element by c. Scalar multiplication can be proved to have the properties

$$1A = A, \; 0A = O, \; bcA = b(cA), \; c(A + B) = cA + cB,$$
$$(b + c)A = bA + cA, \; c(AB) = (cA)B = A(cB).$$

15.4   Invertible Matrices

DEFINITION.   The $n \times n$ matrix A^{-1} is an *inverse* of A if and only if $A^{-1}A = I = AA^{-1}$. If $A^{-1}A = I$, then A^{-1} is a *left inverse*. Hence an inverse is both a left inverse and a right inverse.

DEFINITION.   An $n \times n$ matrix A is *invertible* if and only if it has an inverse.

First we observe that every elementary matrix E is invertible. Since multiplying A on the left by E performs an elementary row operation on A to obtain EA, we simply take E^{-1} as the elementary matrix of the row operation that reverses the first one. Then $E^1(EA) = A$, and hence $E^{-1}E = I$.

The main theoretical information about invertible matrices is included in the following theorem.

THEOREM.   If A is an $n \times n$ matrix, then the following statements are equivalent (i.e., all are true or all are false):

(a) A is invertible;
(b) A has a left inverse;
(c)   The homogeneous linear system, $AX = O$, has no nontrivial solutions;
(d) A is a product of elementary matrices.

Proof.   We will prove that $(1) \Rightarrow (2) \Rightarrow (3) \Rightarrow (4) \Rightarrow (1)$. If A is invertible, then it has a two-sided inverse A^{-1}, which is necessarily a left inverse. Hence $(1) \Rightarrow (2)$. To prove that $(2) \Rightarrow (3)$, suppose that B is a left inverse of A so that $BA = I$. Let X be any solution of $AX = O$. Then

$$X = IX = (BA)X = B(AX) = BO = O.$$

That is, $AX = O$ has only the trivial solution $X = O$. Next $(3) \Rightarrow (4)$. Since $AX = O$ has only the trivial solution, A is equivalent under elementary row operations to the identity matrix I (§12.3, Corollary 2). Hence there is a chain of elementary matrices $E_1, E_2, \ldots, E_s$ such that

$$E_s \cdots E_2 E_1 A = I.$$

Since elementary matrices are invertible, we may operate on this equation with the product $E_1^{-1}E_2^{-1} \cdots E_s^{-1}$. We find

$$(E_1^{-1}E_2^{-1} \cdots E_s^{-1})(E_s \cdots E_2 E_1)A = (E_1^{-1}E_2^{-1} \cdots E_s^{-1})I.$$

By the associative property of matrix multiplication this reduces to

$$A = E_1^{-1}E_2^{-1} \cdots E_s^{-1}.$$

Since the inverses of elementary matrices are elementary matrices, this proves (4). Finally, $(4) \Rightarrow (1)$. For if $A = E_1^{-1}E_2^{-1} \cdots E_s^{-1}$, we may take $A^{-1} = E_s \cdots E_2 E_1$, and then if we multiply A either on the left or right by this A^{-1}, we find $A^{-1}A = I = AA^{-1}$.

15.5   Computation of the Inverse.   We start with the equation

$$A = IA.$$

Then we perform a sequence of elementary row operations on the matrix A on the left so as to transform it to reduced echelon form. If this reduced echelon form has a row of zeros, then A is not invertible. If the reduced echelon form has no row of zeros, then it is the identity matrix. We perform the same elementary row operations on the first factor on the right. This then transforms into A^{-1}. To see this, we express the sequence of elementary row operations by elementary matrices. We have on the left $E_s \cdots E_2 E_1 A = I$. Applying the same chain of factors on the right, we have $I = (E_s \cdots E_2 E_1)A$. Hence

$$A^{-1} = E_s \cdots E_2 E_1.$$

Example. Calculate the inverse of

$$A = \begin{bmatrix} 3 & -2 & 6 \\ 2 & 0 & 1 \\ 4 & 6 & 5 \end{bmatrix},$$

if it exists.

Solution. We write $A = IA$ and perform the same elementary row operations on each of the pair of matrices A, I, choosing them to transform A to reduced echelon form. Starting with the pair

$$A = \begin{bmatrix} 3 & -2 & 6 \\ 2 & 0 & 1 \\ 4 & 6 & 5 \end{bmatrix}, \qquad \begin{bmatrix} 1 & 0 & 0 \\ 0 & 1 & 0 \\ 0 & 0 & 1 \end{bmatrix} = I,$$

we pivot on the element $a_{11} = 3$ of A and perform the same operations on I. This transforms the pair of matrices into

$$\begin{bmatrix} 1 & -\frac{2}{3} & 2 \\ 0 & \frac{4}{3} & -3 \\ 0 & \frac{26}{3} & 1 \end{bmatrix}, \qquad \begin{bmatrix} \frac{1}{3} & 0 & 0 \\ -\frac{2}{3} & 1 & 0 \\ -\frac{4}{3} & 0 & 1 \end{bmatrix}.$$

Next we operate on both matrices to reduce the second column of the left matrix. This gives

$$\begin{bmatrix} 1 & 0 & \frac{1}{2} \\ 0 & 1 & -\frac{9}{4} \\ 0 & 0 & \frac{41}{2} \end{bmatrix}, \qquad \begin{bmatrix} 0 & \frac{1}{2} & 0 \\ -\frac{1}{2} & \frac{3}{4} & 0 \\ \frac{9}{3} & -\frac{13}{2} & 1 \end{bmatrix}.$$

Finally, we reduce the last column of the left matrix, and perform the same operations on the right, to get

$$\begin{bmatrix} 1 & 0 & 0 \\ 0 & 1 & 0 \\ 0 & 0 & 1 \end{bmatrix}, \qquad \begin{bmatrix} -\frac{9}{6} & \frac{15}{4} & -\frac{1}{41} \\ \frac{75}{12} & -\frac{111}{8} & \frac{9}{82} \\ \frac{9}{3} & -\frac{13}{2} & \frac{2}{41} \end{bmatrix} = A^{-1}.$$

This completes the problem.

15.6 Exercises

For the matrices

$$A = \begin{bmatrix} 1 & 2 \\ -2 & 1 \end{bmatrix}, \qquad B = \begin{bmatrix} 3 & -2 \\ -6 & 4 \end{bmatrix}, \qquad I = \begin{bmatrix} 1 & 0 \\ 0 & 1 \end{bmatrix},$$

carry out the indicated algebraic operations in Exercises 1–10.

1. $A + B$.
2. $A - 2B$.
3. $2A^2 - 3A + 4I$.
4. $A - \lambda I$, where λ is an unspecified number.
5. $AB - BA$.
6. $(A - B)(A + B)$. Is it $A^2 - B^2$?
7. A^{-1}, if it exists.
8. B^{-1}, if it exists.
9. Solve for the 2×2 matrix X such that $AX = B$.
10. $A(B - 2I)$.

11. Find whether the following matrices are invertible, and if so, compute their inverses.

$$M = \begin{bmatrix} 1 & 0 & 0 \\ 2 & 1 & 0 \\ 3 & 1 & 1 \end{bmatrix}, \qquad N = \begin{bmatrix} 1 & -1 & 2 \\ 0 & 1 & 3 \\ 2 & -1 & 7 \end{bmatrix}.$$

12. Exhibit elementary matrices $E_1, E_2, \ldots, E_s$ such that $E_s \cdots E_2 E_1 = A^{-1}$, where

$$A = \begin{bmatrix} 2 & -3 \\ 4 & 1 \end{bmatrix}.$$

13. Prove that if A is invertible, then $AX = B$ if and only if $X = A^{-1}B$. Note that we may solve the equation by multiplying both sides by A^{-1}.
14. Suppose for two square matrices that $AB = O$. Does this imply that A or B is the zero matrix?
15. Prove that if both A and B are invertible, then AB is invertible and

$$(AB)^{-1} = B^{-1}A^{-1}.$$

16. Prove that if either A or B is not invertible, then AB and BA are not invertible.
17. Prove $\mathscr{A}_2$ (§15.1).
18. Prove $\mathscr{A}_3$ (§15.1).
19. Prove $\mathscr{A}_2$ (§15.2).
20. Prove the left distributive property (§15.2).
21. Find all matrices S such that for every 2×2 matrix A, $AS = SA$.
22. Show that if M and N are 2×2 diagonal matrices, then $MN = NM$. A diagonal matrix is one for which $a_{ij} = 0$ if $i \neq j$.
23. Show that all 2×2 matrices with addition and scalar multiplication form a vector space with a finite basis. What is its dimension?
24. Show that all 2×2 matrices of the form $\begin{pmatrix} x & y \\ -y & x \end{pmatrix}$ make a vector space. What is its dimension?

16   Substitution in Linear Systems

16.1   Linear Substitutions

Example. The variables (x_1, x_2) are given in terms of (y_1, y_2, y_3) by the linear system

$$\begin{aligned} x_1 &= 2y_1 - 3y_2 + 5y_3, \\ x_2 &= -y_1 - 4y_2 + y_3, \end{aligned}$$

with matrix

$$A = \begin{bmatrix} 2 & -3 & 5 \\ -1 & -4 & 1 \end{bmatrix}.$$

We substitute new variables z_1, z_2 for the y's, given by

$$y_1 = 3z_1 - z_2,$$
$$y_2 = -z_1,$$
$$y_3 = 4z_1 + 2z_2,$$

with matrix

$$B = \begin{bmatrix} 3 & -1 \\ -1 & 0 \\ 4 & 2 \end{bmatrix}.$$

Find the linear system which expresses the x's in terms of z's.

Solution. We carry out the substitution and collect coefficients of z_1 and z_2. We find that

$$x_1 = [2(3) - 3(-1) + 5(4)]z_1 + [2(-1) - 3(0) + 5(2)]z_2 = 30z_1 + 8z_2,$$
$$x_2 = [-1(3) - 4(-1) + 1(4)]z_1 + [-1(-1) - 4(0) + 1(2)]z_2 = 5z_1 + 3z_2,$$

with matrix

$$C = \begin{bmatrix} 30 & 8 \\ 5 & 3 \end{bmatrix}.$$

We observe that C could be computed by multiplying the original matrix A *on the right* by B. Thus $C = AB$. This completes the problem.

We now solve the problem in a different way, using matrix operations from the start. We observe that the original linear system with matrix A expresses a single column of x-variables as a linear combination of the column of the matrix A. The coefficients of this linear combination of columns form a single column of y-variables. In fact, if we define column vectors

$$X = \begin{bmatrix} x_1 \\ x_2 \end{bmatrix}, \qquad Y = \begin{bmatrix} y_1 \\ y_2 \\ y_3 \end{bmatrix}$$

as matrices having just one column, the linear system is given by the matrix multiplication $X = AY$.

The linear substitution that replaces the y-variables by z-variables can be expressed in the same way. If Z is a column vector defined by the matrix

$$Z = \begin{bmatrix} z_1 \\ z_2 \end{bmatrix},$$

then the substitution is simply the matrix product $Y = BZ$. Consequently, if we substitute $Y = BZ$ into the linear system $X = AY$, we get

$$X = A(BZ) = (AB)Z,$$

as we did before. The only proof necessary is to observe that matrix multiplication is associative.

Using this elegant matrix formulation of the process of linear substitutions, we can generalize it to linear systems with any finite number of variables.

THEOREM.   Let column vectors X, Y, Z be defined as the single-column matrices

$$
X = \begin{bmatrix} x_1 \\ x_2 \\ \vdots \\ x_m \end{bmatrix}, \qquad
Y = \begin{bmatrix} y_1 \\ y_2 \\ \vdots \\ y_n \end{bmatrix}, \qquad
Z = \begin{bmatrix} z_1 \\ z_2 \\ \vdots \\ z_p \end{bmatrix}.
$$

In the linear system $X = AY$ with $m \times n$ matrix A, we substitute $Y = BZ$, with $n \times p$ matrix B. The result is a new linear system $X = (AB)Z$, with $m \times p$ matrix AB.

16.2   Dual Row Operations. We recall (§14.1) that elementary row operations on a matrix A were carried out by multiplying A by a suitable matrix B *on the left* to obtain BA. This suggests a dual relationship between columns on the right and rows on the left.

Let single-row matrices X^t and Y^t be defined by

$$
X^t = [x_1, x_2], \qquad Y^t = [y_1, y_2, y_3].
$$

In general, we define the *transposed matrix* A^t from A by interchanging rows and columns.

DEFINITION.   If $A = [a_{ij}]$, then the *transposed matrix* $A^t = [a'_{ij}]$ is defined by $a'_{ij} = a_{ji}$. If A is an $m \times n$ matrix, then A^t is an $n \times m$ matrix.

From the definition, it follows that

$$
(A + B)^t = A^t + B^t, \qquad (cA)^t = cA^t, \qquad (AB)^t = B^t A^t.
$$

We note the reversal of the order of matrix multiplication in the transposed product matrix $(AB)^t$.

Now the linear system $X = AY$, written in terms of column vectors X and Y, can be written as linear combinations of row vectors X^t, Y^t by $X^t = Y^t A^t$. And the substitution $Y = BZ$ becomes $Y^t = Z^t B^t$. If we carry out this substitution, we obtain $X^t = (z^t B^t) A^t = Z^t (B^t A^t)$ as the new linear system expressed in terms of the row vectors X^t and Z^t. This establishes again that a linear combination of the rows of a matrix can be carried out by multiplying it on the left by a suitable matrix B^t.

16.3   Black Box. We have a "black box" that receives inputs $\vec{v}$ from a vector space $\mathscr{V}$ and transmits outputs $\vec{w}$ in a vector space $\mathscr{W}$. It is not known what mechanism is inside the black box (Figure 15.1). But by experimenting with measured inputs and outputs, we find that the action T of the black box on inputs is linear. That is,

$$
T(\vec{u} + \vec{v}) = T(\vec{u}), \qquad \text{and} \qquad T(c\vec{v}) = cT(\vec{v}).
$$

We determine the outputs that correspond to the basis vectors of the input space. The problem then is to construct a formula that gives the output $\vec{w}$ for every input $\vec{v}$.

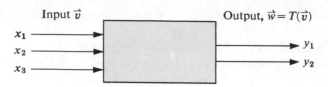

Input $\vec{v}$ Output, $\vec{w} = T(\vec{v})$

x_1 y_1

x_2 y_2

x_3

FIGURE 15.1   Black box.

Example. For a particular black box, the input space has a basis $\{\vec{v}_1, \vec{v}_2, \vec{v}_3\}$ and the output space a basis $\{\vec{w}_1, \vec{w}_2\}$. We are given that the action T of this box is linear, and on the basis inputs is given by

$$\begin{array}{ll} T(\vec{v}_1) = 2\vec{w}_1 - 3\vec{w}_2, & \\ T(\vec{v}_2) = 5\vec{w}_1 + 3\vec{w}_2, & \text{with matrix} \quad A = \begin{bmatrix} 2 & -3 \\ 5 & 3 \\ 0 & 4 \end{bmatrix}. \\ T(\vec{v}_3) = 0\vec{w}_1 + 4\vec{w}_2, & \end{array}$$

For this black box find a formula that gives, for any input $\vec{v}$, the output $\vec{w}$.

Solution.   Let

$$\vec{v} = x_1\vec{v}_1 + x_2\vec{v}_2 + x_3\vec{v}_3 \qquad \text{and} \qquad \vec{w} = y_1\vec{w}_1 + y_2\vec{w}_2.$$

The problem will be solved if we can find (y_1, y_2) when (x_1, x_2, x_3) are given.
  Since the action T is linear, we have

$$y_1\vec{w}_1 + y_2\vec{w}_2 = T(\vec{v}) = x_1 T(\vec{v}_1) + x_2 T(\vec{v}_2) + x_3 T(\vec{v}_3).$$

Substituting the given outputs $T(\vec{v}_1)$, $T(\vec{v}_2)$, $T(\vec{v}_3)$ on basis vectors, and combining terms, we find

$$y_1\vec{w}_1 + y_2\vec{w}_2 = (2x_1 + 5x_2 + 0x_3)\vec{w}_1 + (-3x_1 + 3x_2 + 4x_3)\vec{w}_2.$$

Equating coefficients of $\vec{w}_1$ and $\vec{w}_2$ (since it is a basis), we have

$$\begin{array}{l} y_1 = 2x_1 + 5x_2 + 0x_3, \\ y_2 = -3x_1 + 3x_2 + 4x_3. \end{array}$$

This solves the problem.
  It is informative to represent this solution in matrix notation, with column vectors

$$X = \begin{bmatrix} x_1 \\ x_2 \\ x_3 \end{bmatrix} \qquad \text{and} \qquad Y = \begin{bmatrix} y_1 \\ y_2 \end{bmatrix},$$

for input and output respectively. The solution in matrix notation is then $Y = A^t X$.

16.4 Exercises

In Exercises 1–3, use matrix multiplication to express $\begin{bmatrix} x_1 \\ x_2 \end{bmatrix}$ in terms of $\begin{bmatrix} z_1 \\ z_2 \end{bmatrix}$.

1. $\begin{array}{ll} x_1 = 2y_1 + 3y_2, & y_1 = -4z_1 + z_2, \\ x_2 = y_1 - y_2, & y_2 = 3z_1 - z_2. \end{array}$

2. $\begin{array}{ll} x_1 = y_1 - y_2 + y_3, & y_1 = 4z_1 - 2z_2, \\ x_2 = 2y_1 - y_3, & y_2 = z_1 + 3z_2, \\ & y_3 = 5z_1 + 2z_2, \end{array}$

3. $x_1 = 4y_1 - 3y_2 + 7y_3,$ $y_1 = 2z_1 - z_2,$
    $y_2 = z_2,$
    $y_3 = z_1.$

4. Find the transposed matrix A^t of
    $$A = \begin{bmatrix} 1 & 0 & -3 \\ 0 & 1 & 2 \end{bmatrix}.$$

5. Prove that if
    $$A = \begin{bmatrix} a_{11} & a_{12} \\ a_{21} & a_{22} \end{bmatrix} \quad \text{and} \quad B = \begin{bmatrix} b_{11} & b_{12} \\ b_{21} & b_{22} \end{bmatrix},$$
    then $(AB)^t = B^t A^t$.

In Exercises 6–12, we have black boxes whose actions are linear on inputs $(x_1, x_2, \ldots)$ to produce outputs $(y_1, y_2, \ldots)$.

6. Find a formula for the action of a black box on inputs (x_1, x_2) if $T(1, 0) = (3, -2)$ and $T(0, 1) = (1, 4)$.

7. Find a formula for the action S of a black box on inputs (x_1, x_2) if $S(1, 0) = (1, -4)$ and $S(0, 1) = (3, 5)$.

8. The output $S(x_1, x_2)$ of the black box of Exercise 7 is fed into the input of the black box of Exercise 6. Find a formula for the output of the two boxes acting in series on inputs (x_1, x_2) into the S-box.

9. In Exercise 6, what input will produce the output (a, b)?

10. Find a formula for the action of the black box on inputs (x_1, x_2, x_3) if it has the actions $T(1, 0, 0) = (3, -2, 1)$, $T(0, 1, 0) = (4, 1, -2)$, $T(0, 0, 1) = (-5, -4, 5)$. *Suggestion*: Write inputs and outputs as columns.

11. In manipulating the black box of Exercise 10, we find that many different inputs will produce the output $(0, 0, 0)$. Find all such inputs.

12. Find the formula for the action of the black box for which $T(2, -3) = (1, 4)$ and $T(4, 2) = (5, -1)$.

13. A manufacturer makes two models, Mark I and Mark II, by combining subassemblies A, B, C. Mark I requires 2 of A, 3 of B, and 5 of C. Mark II requires 6 of A, 4 of B, and 10 of C. In turn, each A uses 24 bolts, 6 plates, and 5 tubes, each B uses $(8, 3, 1)$, and each C uses $(12, 6, 4)$. Use matrix multiplication to show the numbers of bolts, plates, and tubes required for each Mark I and each Mark II.

14. In Exercise 13, how many bolts, plates, and tubes are required for an order of 100 Mark I and 120 Mark II?

15. In Exercise 13, the subassembly A costs the manufacturer \$8 to make, B costs \$3, and C costs \$5. Use matrix multiplication to express the cost of subassemblies for an order of 100 Mark I and 120 Mark II.

17   Linear Transformations

17.1   Linear Systems as Mappings.   A linear system

$$\begin{aligned} x' &= ax + by, \\ y' &= cx + dy, \end{aligned} \quad \text{with matrix} \quad [T] = \begin{bmatrix} a & b \\ c & d \end{bmatrix}$$

sends a vector $\vec{v}:(x, y)$ in the plane to a vector $T(\vec{v}):(x', y')$ in the plane. It is a kind of function T on the xy-plane to the plane. Moreover, this function has the linearity properties,

$$T(\vec{u} + \vec{v}) = T(\vec{u}) + T(\vec{v}) \quad \text{and} \quad T(c\vec{v}) = cT(\vec{v}).$$

If S is another such function having a different matrix $[S]$, we can add S and T as we add functions. That is, for each $\vec{v}$ in the plane we define $(S + T)(\vec{v}) = S(\vec{v}) + T(\vec{v})$. We regard the composition, $S \circ T$, as a multiplication, writing it simply ST. We recall that the composition, or function of a function, is defined by $(S \circ T)(\vec{v}) = T(S(\vec{v}))$. In other words, the product ST, in the order first S and then T is found by first applying S to a vector $\vec{v}$ and then applying T to $S(\vec{v})$.

We can perform these operations by matrix operations on the matrices $[S]$ and $[T]$. We find that $S + T$ is determined by the matrix sum $[S] + [T]$, that the scalar product cT corresponds to the matrix operation $c[T]$, and that the product ST corresponds to the matrix multiplication $[T][S]$, which represents composition (§16.1).

17.2   Abstract Definition of a Linear Transformation.

We observe that in this discussion the linear system and its matrix was not the essential idea. The essential idea was that of the mapping T with its linearity properties. So we define T abstractly.

DEFINITION.   A mapping T of the vector space $\mathscr{V}$ into the vector space $\mathscr{W}$ is a *linear transformation* if and only if

$$T(\vec{u} + \vec{v}) = T(\vec{u}) + T(\vec{v}) \qquad \text{and} \qquad T(c\vec{v}) = cT(\vec{v}).$$

A linear mapping T is called a *linear operator in* $\mathscr{V}$ if it maps $\mathscr{V}$ into itself.

For linear operators S, T, U, all mapping the space $\mathscr{V}$ into itself, we define algebraic operations $S + T$, cT, and ST for each vector $\vec{v}$ by

$$(S + T)(\vec{v}) = S(\vec{v}) + T(\vec{v}), \qquad (cT)(\vec{v}) = cT(\vec{v}), \qquad \text{and} \qquad ST(\vec{v}) = T(S(\vec{v})).$$

These definitions imply no knowledge of the vector space $\mathscr{V}$, no basis in it, and no matrix determining T as a linear substitution. Nevertheless, our solution of the black box problem (§16.3) leads us to expect that, at least when $\mathscr{V}$ has a finite basis, we can express T as a linear system with an $n \times n$ matrix.

THEOREM.   Let $\mathscr{T}$ be the set of all linear operators $\{S, T, U, \ldots\}$ in the vector space $\mathscr{V}$ with finite basis $\mathscr{B} = \{\vec{v}_1\,\vec{v}_2, \ldots, \vec{v}_n\}$. Then, relative to this basis, there is a one-to-one correspondence of linear operators T and $n \times n$ matrices $[T]$ such that if $S \leftrightarrow [S]$ and $T \leftrightarrow [T]$, then $S + T \leftrightarrow [S] + [T]$, $cT \leftrightarrow c[T]$, and $ST \leftrightarrow [T][S]$.

When a one-to-one correspondence between two spaces exists under which the algebraic operations of one space correspond faithfully to those in the other space, then the spaces are said to be *isomorphic*. That is, they are "of the same form." In these terms, our theorem says that matrices provide an isomorphic representation of linear operators on a finite dimensional space. The importance of the theorem lies in the fact that it enables us to use the concept of linear operators as geometric objects, without regard to any arbitrary basis. Yet, for purposes of calculation, we can go over to explicit matrix algebra. It remains only to find the matrix $[T]$ that corresponds to a linear operator T.

17.3   The Matrix of a Linear Transformation.

In solving the problem of the black box (§16.3), we derived a method for finding the explicit matrix $[T]$ of a linear transformation T that maps a vector space $\mathscr{V}$ with basis $\mathscr{B} = \{\vec{v}_1, \vec{v}_2, \ldots, \vec{v}_n\}$ into a vector space $\mathscr{W}$

with basis $\mathscr{C} = \{\vec{w}_1, \vec{w}_2, \ldots, \vec{w}_m\}$. If the action of T on the basis vectors of $\mathscr{V}$ is given in terms of the basis vectors of $\mathscr{W}$ by

$$T(\vec{v}_1) = a_{11}\vec{w}_1 + a_{12}\vec{w}_2 + \cdots + a_{1m}\vec{w}_m,$$
$$T(\vec{v}_2) = a_{21}\vec{w}_1 + a_{22}\vec{w}_2 + \cdots + a_{2m}\vec{w}_m,$$
$$\cdots\cdots\cdots\cdots\cdots\cdots\cdots\cdots\cdots\cdots$$
$$T(\vec{v}_n) = a_{n1}\vec{w}_1 + a_{n2}\vec{w}_2 + \cdots + a_{nm}\vec{w}_m,$$

with $n \times m$ matrix

$$A_T = \begin{bmatrix} a_{11} & a_{12} & \cdots & a_{1m} \\ a_{21} & a_{22} & \cdots & a_{2m} \\ \vdots & & & \vdots \\ a_{n1} & a_{n2} & \cdots & a_{nm} \end{bmatrix},$$

then the matrix $[T]$ of T is the matrix A_T transposed, $[T] = A_T^t$. Explicitly, if $\vec{v}:(x_1, x_2, \ldots, x_n)$ is any vector in $\mathscr{V}$ and $T(\vec{v}):(y_1, y_2, \ldots, y_m)$ is its image in $\mathscr{W}$ under T, then T is given by the linear system

$$y_1 = a_{11}x_1 + a_{21}x_2 + \cdots + a_{n1}x_n,$$
$$y_2 = a_{12}x_1 + a_{22}x_2 + \cdots + a_{n2}x_n,$$
$$\cdots\cdots\cdots\cdots\cdots\cdots\cdots\cdots\cdots$$
$$y_m = a_{1m}x_1 + a_{2m}x_2 + \cdots + a_{nm}x_n.$$

Example.   Find the matrix of the linear transformation T that maps the xy-plane with standard basis $\{\vec{v}_1:(1, 0), \vec{v}_2:(0, 1)\}$ into 3-space with standard basis $\{\vec{w}_1:(1, 0, 0), \vec{w}_2:(0, 1, 0), \vec{w}_3:(0, 0, 1)\}$. It is given that

$$T(\vec{v}_1) = 3\vec{w}_1 - 2\vec{w}_2 + 5\vec{w}_3,$$
$$T(\vec{v}_2) = w_1 - 8\vec{w}_3.$$

Solution.   We have the matrix

$$A_T = \begin{bmatrix} 3 & -2 & 5 \\ 1 & 0 & -8 \end{bmatrix}.$$

Its transposed matrix is the required matrix $[T]$, that is

$$[T] = \begin{bmatrix} 3 & 1 \\ -2 & 0 \\ 5 & -8 \end{bmatrix}.$$

This completes the problem. We observe that it expresses vectors $T(\vec{v}) = \vec{w}:(x', y', z')$ in terms of the components of $\vec{v}:(x, y)$ by

$$x' = 3x + y,$$
$$y' = -2x,$$
$$z' = 5x - 8y.$$

17.4   Scalar Magnification and Rotation.   A simple linear operator S in the plane, called scalar magnification, multiplies each vector $\vec{v}$ by a nonzero constant c (Figure 15.2(a)). That is, $S(\vec{v}) = c\vec{v}$. We readily find, applying S to the basis vectors $\{\vec{v}_1, \vec{v}_2\}$, that

$$S(\vec{v}_1) = c\vec{v}_1 \quad \text{and} \quad S(\vec{v}_2) = c\vec{v}_2.$$

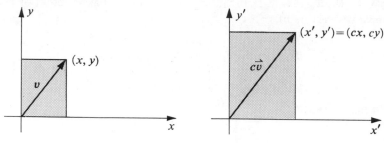

(a)   Scalar magnification by factor $c = \frac{3}{2}$

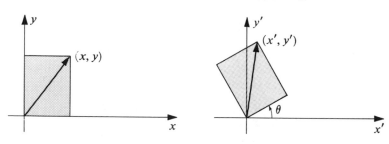

(b)   Rotation through angle θ

FIGURE 15.2   Scalar magnification and rotation.

Hence

$$A_T = \begin{bmatrix} c & 0 \\ 0 & c \end{bmatrix} \quad \text{and} \quad A_T^t = [S] = \begin{bmatrix} c & 0 \\ 0 & c \end{bmatrix}.$$

Thus the scalar magnification operator is given by

$$x' = cx, \quad y' = cy.$$

Another familiar linear operator is the rotation of the plane through an angle θ. Thus R sends the basis vector $(1, 0)$ to $(\cos \theta, \sin \theta)$. And R sends the basis vector $(0, 1)$ to $(-\sin \theta, \cos \theta)$ (Figure 15.2(b)). Hence we have

$$A_T = \begin{bmatrix} \cos \theta & \sin \theta \\ -\sin \theta & \cos \theta \end{bmatrix} \quad \text{and} \quad A_T^t = [R] = \begin{bmatrix} \cos \theta & -\sin \theta \\ \sin \theta & \cos \theta \end{bmatrix}.$$

Thus the rotation through the angle θ sends the point (x, y) to point (x', y') given by

$$x' = x \cos \theta - y \sin \theta,$$
$$y' = x \sin \theta + y \cos \theta.$$

17.5 Exercises

In Exercises 1–3, the xy-plane has the basis $\mathcal{B} = \{\vec{v}_1, \vec{v}_2\}$. The effect of a linear operator T on this basis is given for each one. Find the matrix $[T]$ of the operator and express T as a linear system of the form

$$x' = ax + by,$$
$$y' = cx + dy.$$

1. $T[\vec{v}_1] = 3\vec{v}_1 - 2\vec{v}_2, \; T(\vec{v}_2) = 5\vec{v}_1 + 7\vec{v}_2.$
2. $T(\vec{v}_1) = v_2, \; T(v_2) = -v_1.$
3. $T(\vec{v}_1): (\frac{3}{5}, -\frac{4}{5}), \; T(\vec{v}_2): (\frac{4}{5}, \frac{3}{5}).$

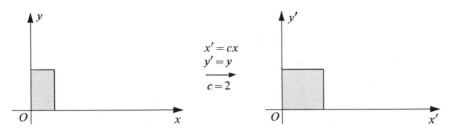

FIGURE 15.3   Dilation in x.

4. Show that the elementary row operation E_2 that multiplies the first row of a 2×2 matrix by c is given by $x' = cx, y' = y$. Plot this as a mapping of the xy-plane with $c = 2$ (Figure 15.3). It is called a *dilation* in x.

5. As in Exercise 4, the elementary row operation that interchanges the rows of a 2×2 matrix acts on column vectors $\begin{bmatrix} x \\ y \end{bmatrix}$ to produce $\begin{bmatrix} y \\ x \end{bmatrix}$. Hence it is the linear mapping $x' = y, y' = x$ with matrix $[E_1] = \begin{bmatrix} 0 & 1 \\ 1 & 0 \end{bmatrix}$. Plot it as a mapping of the xy-plane (Figure 15.4). It is called a *reflection* in the line $y = x$.

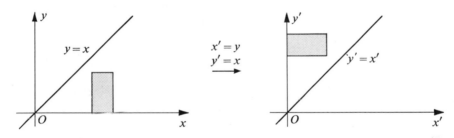

FIGURE 15.4   Reflection in line $y = x$.

6. The third type of elementary row operation E_3 replaces a row by adding to it a constant multiple of another row. Show that E_3 applied to the first row of a 2-row matrix is the linear mapping

$$\begin{aligned} x' &= x + cy, \\ y' &= y, \end{aligned} \qquad \text{with matrix} \qquad [E_3] = \begin{bmatrix} 1 & c \\ 0 & 1 \end{bmatrix}.$$

Plot it as a mapping of the xy-plane (Figure 15.5). It is called a *shear* along the x-axis.

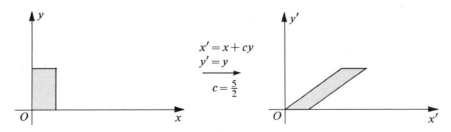

FIGURE 15.5   Shear along x-axis.

7. Show that the identity transformation I given by $I(\vec{v}) = \vec{v}$, as applied to the plane, has the matrix $[I] = \begin{bmatrix} 1 & 0 \\ 0 & 1 \end{bmatrix}$.

8. Show that the transformation of the plane that sends every vector $\vec{v}$ of the plane onto the origin $\vec{O}$, has the matrix $[O] = \begin{bmatrix} 0 & 0 \\ 0 & 0 \end{bmatrix}$.

9. Show that the projection of the plane onto the x-axis that sends every vector $\vec{v}:(x, y)$ to $P(\vec{v}) = (x, 0)$ is a linear operator that has the matrix $\begin{bmatrix} 1 & 0 \\ 0 & 0 \end{bmatrix}$. Is it invertible?

10. For the rotation transformation (§17.4), show that the rotation mapping of the xy-plane through the angle θ can be carried out by the *substitution*

$$x = x' \cos \theta + y' \sin \theta,$$
$$y = -x' \sin \theta + y' \cos \theta,$$

   whose matrix is the inverse.

11. Show graphically the effect of a scalar magnification $S(\vec{v}) = -2\vec{v}$.

12. A linear transformation D that multiplies x by constant c and y by constant d is called a *dilation*. Show that its matrix is a "diagonal" matrix $[D] = \begin{bmatrix} c & 0 \\ 0 & d \end{bmatrix}$.

13. Plot the map of the square with vertices $(0, 0)$, $(1, 0)$, $(1, 1)$, $(0, 1)$ under the dilation $x' = 2x$, $y' = \frac{1}{2}y$.

14. A linear transformation of the complex z plane is given by $w = iz$. Find the 2×2 matrix, with real entries, of this transformation with respect to the natural basis $\{\vec{1}, \vec{i}\}$.

15. The derivative operator D is a linear operator that operates on all functions of the form $a \cos x + b \sin x$ to produce another function of this form. Find the matrix of D with respect to the basis $\{\cos x, \sin x\}$.

16. The derivative operator D operates on all quadratic polynomials $ax^2 + bx + c$. Find the matrix of D with respect to the basis $\{1, x, x^2\}$. Is $[D]$ invertible?

17. The rotation matrix $R = \begin{bmatrix} \cos \theta & -\sin \theta \\ \sin \theta & \cos \theta \end{bmatrix}$ was shown in Exercise 10 to have the inverse $R^{-1} = R_t$. Show that all such 2×2 matrices are rotations or orientations (§1.4).

18. Prove that an invertible linear operator T always sends a straight line $\vec{r} = s\vec{a} + \vec{p}$, through the point $\vec{p}$ in the direction $\vec{a}$, into a straight line through point $T(\vec{p})$ in the direction $T(\vec{a})$.

19. Prove that an invertible linear operator on the plane sends intersecting lines into intersecting lines and parallel lines into parallel lines.

20. Prove that an invertible linear operator on the plane sends tangent lines into tangent lines.

21. Prove that an invertible linear operator, even though it may not preserve distances, sends midpoints of line segments into midpoints of line segments.

22.* We consider the circle $x^2 + y^2 = a^2$ with a system of parallel chords and their conjugate diameter EF, which bisects all of the chords (Figure 15.6). Parallel to these chords we have tangent lines at E and F. Map this figure into the $x'y'$-plane by the invertible linear operator $x' = x$, $y' = (b/a)x$, where $0 < b < a$. Use Exercises 19, 20, 21 to describe the image of the figure in the $x'y'$-plane. State the theorem thus proved about an ellipse, a system of parallel chords and tangent lines, and the conjugate diameter determined by them. Is the conjugate diameter perpendicular to the chords?

23. Find the matrix of the linear transformation of 3-space into the plane that sends

$$(1, 0, 0) \rightarrow (-2, 5), \qquad (0, 1, 0) \rightarrow (4, 3), \qquad (0, 0, 0) \rightarrow (6, -1).$$

24. A linear transformation T maps the real line with basis $\{\vec{1}\}$ into 3-space so that $T(1)$ is given by $(2, -3, 4)$. Find the matrix $[T]$ and express the transformation as a linear system. The image is a straight line through what point, with what direction?

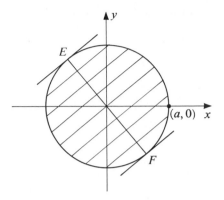

FIGURE 15.6   Exercise 22.

18   Determinants

18.1   Historical Remarks.   Determinants are functions, **det** A, that assign a number to every $n \times n$ matrix A. For example, the classical determinant of the 2×2 matrix

$$A = \begin{bmatrix} a & b \\ c & d \end{bmatrix}$$

is det $A = ad - bc$. For the 3×3 matrix

$$A = \begin{bmatrix} a_{11} & a_{12} & a_{13} \\ a_{21} & a_{22} & a_{23} \\ a_{31} & a_{32} & a_{33} \end{bmatrix},$$

$$\det A = a_{11}a_{22}a_{33} - a_{11}a_{32}a_{23} + a_{31}a_{12}a_{23} - a_{21}a_{12}a_{33} \\ + a_{21}a_{32}a_{13} - a_{31}a_{22}a_{13}.$$

The classical theory of determinants extends these combinatorial definitions to $n \times n$ matrices. When n is large, both the theory and the computation are exceedingly complicated. The recent tendency in mathematics has been to eliminate determinant techniques in favor of the simpler elementary row operations, but there remain some uses for determinants, and besides, much published mathematics requires that the reader understand determinants. So we define them, and develop a few of their general properties, using elementary row operations instead of the classical definition as a basis for the theory.

18.2   Definition.   The *determinant* is a function that assigns to every $n \times n$ matrix A a real number det A, such that

(1)   det $I = 1$, where I is the $n \times n$ identity matrix,
(2)   det A is linear function of each row,
(3)   if two rows are interchanged, the sign of det A is changed.

Property (2) requires some explanation. It says that det A is a multilinear function.* For example, if the first row of the matrix

$$A = \begin{bmatrix} a_{11} & a_{12} \\ a_{21} & a_{22} \end{bmatrix}$$

can be written as a linear combination of two pairs (u_{11}, u_{12}) and (v_{11}, v_{12}) so that

$$(a_{11}, a_{12}) = (\alpha u_{11} + \beta v_{11}, \alpha u_{12} + \beta v_{12}),$$

then

$$\det A = \alpha \det \begin{bmatrix} u_{11} & u_{12} \\ a_{21} & a_{22} \end{bmatrix} + \beta \det \begin{bmatrix} v_{11} & v_{12} \\ a_{21} & a_{22} \end{bmatrix}.$$

18.3 Effect of Elementary Row Operations. If the $n \times n$ matrix A has a row of zeros then det $A = 0$. For example, for any numbers a, b,

$$\det \begin{bmatrix} a & b \\ 0 & 0 \end{bmatrix} = 0.$$

Proof. Property (2) implies that if the row of zeros is multiplied by 2 the result is 2 det A. But since multiplying the row of zeros by 2 leaves A unchanged the result is still det A. We have, then, 2 det $A = \det A$, which implies that det $A = 0$.

If the matrix A has two identical rows, then det $A = 0$. For if we interchange the two rows, A is unchanged, but property (3) says that the sign of det A is changed. Hence we have det $A = -\det A$, which implies that det $A = 0$.

If we perform on A the elementary row operation of multiplying a row by the nonzero constant k, then the new matrix has determinant k det A. This follows immediately from property (2).

If we perform on A the elementary row operation of replacing a row by a new row formed by adding to it k times another row of A, then the determinant of the new matrix is still the same det A.

Proof. Since the ith row of the new matrix is a linear combination $1\bar{a}_i + k\bar{a}j$ of the ith row and some jth row of A, its determinant is given by

$$1 \det A + k \det A',$$

where A' is a matrix with two identical rows $\bar{a}_j$. Hence det $A' = 0$ and thus the transformed matrix has as its determinant 1 det $A + 0$.

18.4 Determinants of Elementary Matrices. We recall that elementary matrices E are formed by performing elementary row operations on the identity matrix I. Since det $I = 1$, by property (1) of the definition, and we know the effects of elementary row operations on det A, we find the following:

---

* A multilinear function is called a tensor.

(1)   If E_1 is an elementary matrix that interchanges two rows, then det $E_1 = -1$. For example,

$$\det \begin{bmatrix} 0 & 0 & 1 \\ 0 & 1 & 0 \\ 1 & 0 & 0 \end{bmatrix} = -1,$$

for it is formed from I by interchange of the first and third rows.

(2)   If E_2 is an elementary matrix formed by multiplying a row of I by k, then det $E_2 = k$.

(3)   If E_3 is an elementary matrix formed from I by adding k times another row to the ith row, then det $E_3 = 1$. For example, if we add k times the third row to the second and take the determinant, we find

$$\det \begin{bmatrix} 1 & 0 & 0 \\ 0 & 1 & k \\ 0 & 0 & 1 \end{bmatrix} = 1.$$

These arguments also show that if E is an elementary matrix, then det $E \neq 0$ and det $EA = \det E \det A$.

18.5   Calculation of Determinants.

We know (§§11.4, 12.2) that for any matrix A we can perform a finite sequence of elementary row operations to reduce it to a reduced echelon matrix R. We can represent this reduction in terms of elementary matrices $E_1, E_2, \ldots, E_k$ by writing $E_k \cdots E_2 E_1 A = R$. Hence

$$\det A = \frac{\det R}{\det E_1 \det E_2 \cdots \det E_k}.$$

The determinant, det R, can be evaluated by inspection, for R is either I, so that det $R = 1$, or $R = 1$, or R has a row of zeros, so that det $R = 0$.

This reasoning not only shows how to calculate det A but also establishes the following important theorem. (See also §15.4.)

THEOREM.   An $n \times n$ matrix A is invertible if and only if det $A \neq 0$.

In practice we perform the elementary row operations and write down the reciprocal of the factors det E_1, det $E_2, \ldots$ as we perform them. At the end we have only to multiply the product of these factors by det R.

Example.   Calculate the determinant of

$$A = \begin{bmatrix} 2 & 0 & -2 \\ 0 & 4 & 1 \\ -3 & -1 & 0 \end{bmatrix}.$$

Solution.   We multiply the first row by $\frac{1}{2}$ to place a 1 in the upper left corner. This multiplies det A by $\frac{1}{2}$, so we compensate by multiplying outside by the reciprocal of $\frac{1}{2}$, that is, by 2. Thus

$$\det A = 2 \det \begin{bmatrix} 1 & 0 & -1 \\ 0 & 4 & 1 \\ -3 & -1 & 0 \end{bmatrix}.$$

Now we add 3 times the first row to the third. This multiplies the determinant by 1. Then continuing with the reduction, we have

$$\det A = (1)(2) \det \begin{bmatrix} 1 & 0 & -1 \\ 0 & 4 & 1 \\ 0 & -1 & -3 \end{bmatrix} = (4)(1)(2) \det \begin{bmatrix} 1 & 0 & -1 \\ 0 & 1 & \frac{1}{4} \\ 0 & -1 & -3 \end{bmatrix}$$

$$= (1)(4)(1)(2) \det \begin{bmatrix} 1 & 0 & -1 \\ 0 & 1 & \frac{1}{4} \\ 0 & 0 & -\frac{11}{4} \end{bmatrix} = (-\tfrac{11}{4})(1)(4)(1)(2) \det \begin{bmatrix} 1 & 0 & -1 \\ 0 & 1 & \frac{1}{4} \\ 0 & 0 & 1 \end{bmatrix}$$

$$= (1)(1)(-\tfrac{11}{4})(1)(4)(1)(2) \det \begin{bmatrix} 1 & 0 & 0 \\ 0 & 1 & 0 \\ 0 & 0 & 1 \end{bmatrix} = -22.$$

The method is cumbersome for determinants of small matrices like this 3×3 one, but for large matrices, such as 20×20, it is much more economical than the classical definition and is easily programed for a computer.

18.6 Product Theorem. If A and B are any $n \times n$ matrices, then $\det AB = \det A \det B$.

Proof. We multiply AB on the left by a sequence $E_1, E_2, \ldots, E_k$ of elementary matrices chosen to reduce A to reduced echelon form R. Then (§18.5)

$$\det AB = \frac{\det RB}{(\det E_1) \cdots (\det E_k)}.$$

Now either $R = I$ or R has a row of zeros. If $R = I$, $\det RB = \det B$, and the product of the reciprocals of $\det E_1, \ldots, \det E_k$ is $\det A$, as in the calculations above. Hence in this case $\det AB = \det A \det B$.

If R has a row of zeros, then $\det A = 0$. Moreover, RB has a row of zeros also and hence $\det RB = 0$. Thus in the case that $\det A = 0$, it is also true that $\det AB = \det A \det B$. This completes the proof.

18.7 Row-Column Duality. We have based our theory of determinants on elementary row operations on A. These could be effected by multiplying A on the left by suitable elementary matrices E. Clearly we can make an analogous theory for elementary column operations. We may change the rows of A into the columns of A^t by transposing A. Then $A^t E^t$ performs, as elementary column operations, the same calculations that EA does for row operations. Moreover, we find that $\det E^t = \det E$. Omitting details, this leads us to calculate $\det A^t$ by elementary column operations. Thus we establish the following:

Theorem. If A is any $n \times n$ matrix, $\det A^t = \det A$.

18.8 Expansion by Minors. In the matrix $A = [a_{ij}]$, $i, j = 1, \ldots, n$, if we delete from A the row and the column occupied by a_{ij}, we obtain an $(n-1) \times (n-1)$ matrix, which we denote by M_{ij}. It is called the *minor* of a_{ij}. We calculate $\det M_{ij}$ and attach a sign to it to define the cofactor a_{ij}^* of a_{ij} thus,

$$a_{ij}^* = (-1)^{i+j} \det M_{ij}.$$

The attached signs $(-1)^{i+j}$ alternate $+1$ and -1, starting in the upper left corner with $(-1)^{1+1} = (-1)^2 = 1$. We can prove (Exercises) that we can evaluate det A in terms of the elements and cofactors of any chosen row.

THEOREM ON EXPANSION BY MINORS.   If A is an $n \times n$ matrix, then for any row number i,

$$\det A = a_{i1}a_{i1}^* + a_{i2}a_{i2}^* + \cdots + a_{in}a_{in}^*.$$

Example.   Evaluate det A if

$$A = \begin{bmatrix} 2 & 0 & -2 \\ 0 & 4 & 1 \\ -3 & -1 & 0 \end{bmatrix}$$

by minors of the second row.

Solution.   Choosing $i = 2$, we apply the theorem to say that $\det A = 0a_{21}^* + 4a_{22}^* + 1a_{23}^*$. It remains to calculate the cofactors. We find by the definition of cofactor that

$$a_{22}^* = (-1)^{2+2} \det \begin{bmatrix} 2 & -2 \\ -3 & 0 \end{bmatrix} = -6 \quad \text{and} \quad a_{23}^* = (-1)^{2+3} \det \begin{bmatrix} 2 & 0 \\ 3 & -1 \end{bmatrix} = 2.$$

Hence $\det A = 0 + 4(-6) + 1(2) = -22$ (§18.5).

18.9   Exercises

In Exercises 1–4, evaluate the determinants by reducing to reduced echelon form.

1. $\det \begin{bmatrix} 2 & 3 \\ -1 & 6 \end{bmatrix}$.

2. $\det \begin{bmatrix} 4 & -6 \\ -2 & 3 \end{bmatrix}$.

3. $\det \begin{bmatrix} 0 & 3 \\ 0 & -1 \end{bmatrix}$.

4. $\det \begin{bmatrix} 3 & 0 & -2 \\ 0 & 1 & -0 \\ 6 & 0 & -5 \end{bmatrix}$.

5.   Show by the definition (§18.2) that if $a \neq 0$,

$$\det \begin{bmatrix} a & b \\ c & d \end{bmatrix} = ab - cd.$$

6.   As in Exercise 5, show that if $a = 0$ and $c \neq 0$, then

$$\det \begin{bmatrix} a & b \\ c & d \end{bmatrix} = ab - cd.$$

7.   As in Exercise 5, show that if $a = 0$ and $c = 0$, then

$$\det \begin{bmatrix} 0 & b \\ 0 & d \end{bmatrix} = 0.$$

8.   Verify by expansion that $\det A^t = \det A$ if

$$A = \begin{bmatrix} 3 & -5 \\ 4 & 6 \end{bmatrix}.$$

9.   Prove that $\det A = 0$ if A has a column of zeros.

10. Calculate by minors of the third row and also of the second row:

$$\text{(a) } \det \begin{bmatrix} 3 & 2 & -1 \\ 4 & 0 & 5 \\ 0 & 7 & 2 \end{bmatrix}; \quad \text{(b) } \det \begin{bmatrix} 1 & 4 & -2 \\ 4 & 6 & 0 \\ -3 & -1 & 0 \end{bmatrix}.$$

11. Solve for λ the equation $\det (A - \lambda 1) = 0$, where

$$\text{(a) } A = \begin{bmatrix} 1 & 4 \\ 4 & 3 \end{bmatrix}; \quad \text{(b) } A = \begin{bmatrix} 1 & 3 & 2 \\ -1 & 4 & 6 \\ 2 & -1 & -4 \end{bmatrix}.$$

12. Show that if A is an $n \times n$ matrix, $\det (-A) = \det A$ if n is even, and $\det (-A) = -A$ if n is odd.
13. As in Exercise 12, show that if n is odd and $A^t = -A$, then $\det A = 0$.
14. Prove that if A is invertible $\det (A^{-1}) = 1/\det A$.
15. Prove that $\det (B^{-1}AB) = \det A$.
16. Prove that if $A^t A = I$, then $\det A = \pm 1$.
17. Prove without expanding that

$$\det \begin{bmatrix} x & y & 1 \\ x_1 & y_1 & 1 \\ x_2 & y_2 & 1 \end{bmatrix} = 0$$

is a first-degree equation that represents a line through points (x_1, y_1) and (x_2, y_2).

18.10 Proof of Expansion by Minors

The following problems illustrate the proof of the expansion by minors for determinants.

T1. For definiteness consider the second row of a 3×3 matrix, $A = [a_{ij}]$, $i, j = 1, 2, 3$. Since

$$(a_{21}, a_{22}, a_{23}) = a_{21}(1, 0, 0) + a_{22}(0, 1, 0) + a_{23}(0, 0, 1),$$

show that

$$\det A = a_{21} \det \begin{bmatrix} a_{11} & a_{12} & a_{13} \\ 1 & 0 & 0 \\ a_{31} & a_{32} & a_{33} \end{bmatrix} + a_{22} \det \begin{bmatrix} a_{11} & a_{12} & a_{13} \\ 0 & 1 & 0 \\ a_{31} & a_{32} & a_{33} \end{bmatrix}$$
$$+ a_{23} \det \begin{bmatrix} a_{11} & a_{12} & a_{13} \\ 0 & 0 & 1 \\ a_{31} & a_{32} & a_{33} \end{bmatrix}.$$

T2. Using T1, show by elementary row operations that zeros can be introduced into the columns containing the 1's, so that

$$\det A = a_{21} \det \begin{bmatrix} 0 & a_{12} & a_{13} \\ 1 & 0 & 0 \\ 0 & a_{32} & a_{33} \end{bmatrix} + a_{22} \det \begin{bmatrix} a_{11} & 0 & a_{13} \\ 0 & 1 & 0 \\ a_{31} & 0 & a_{33} \end{bmatrix}$$
$$+ a_{23} \det \begin{bmatrix} a_{11} & a_{12} & 0 \\ 0 & 0 & 1 \\ a_{31} & a_{32} & 0 \end{bmatrix}.$$

T3. In the first term of T_2, observe that $\begin{bmatrix} a_{12} & a_{13} \\ a_{32} & a_{33} \end{bmatrix}$ is the minor M_{23}. Perform the elementary row operations on the first and third rows that calculate det M_{23} and show that

$$\det \begin{bmatrix} 0 & a_{12} & a_{13} \\ 1 & 0 & 0 \\ 0 & a_{32} & a_{33} \end{bmatrix} = \det M_{23} \det \begin{bmatrix} 0 & 1 & 0 \\ 1 & 0 & 0 \\ 0 & 0 & 1 \end{bmatrix} = -\det M_{23}.$$

T4. As in T3, show that the det A in T2 can be represented by

$$\det A = a_{21}(-\det M_{23}) + a_{22}(+\det M_{22}) + a_{23}(-\det M_{23}).$$

This is the expansion by minors of the second row. The same procedure applies for an $n \times n$ matrix A to calculate det A by minors of any row.

19 Invariant Bases

19.1 Difficulties of Arbitrary Bases.

We have seen that in a vector space $\mathscr{V}$ we can represent a linear operator T that sends every vector $\vec{v}$, with coordinates X, to $T(\vec{v})$, with coordinates Y, by a linear system $Y = AX$ with matrix A. While this matrix representation is good for computation, it has disadvantages for physical applications. The coordinates X and Y and the elements of the matrix A all depend on the arbitrary choice of basis. Hence the representation $Y = AX$ conveys information that is meaningless, or accidental, as well as information that has physical significance. Instead of trying to sort out the invariant physical information from the static, we can use another way to attack the problem by asking whether T itself can choose its own characteristic basis. Then, if we represent T by a linear system $Y = JX$ with respect to the characteristic basis, the arbitrariness in the human choice of basis will be removed and any information derived from $Y = JX$ will have physical significance, if T does.

19.2 Change of Basis.

Let us see precisely how a change of basis affects the representation $Y = AX$ of a linear operator T. A change of basis from $\mathscr{B} = \{\vec{v}_1, \vec{v}_2, \ldots, \vec{v}_n\}$ to $\mathscr{C} = \{\vec{w}_1, \vec{w}_2, \ldots, \vec{w}_n\}$ replaces the coordinates X of a vector $\vec{v}$ by new coordinates X', and replaces the coordinates Y of its image $T(\vec{v})$ by Y'. The change of basis can be accomplished by a linear substitution, $X = BX'$, whose invertible matrix B is determined by the requirement that it map the vector of the ordered basis $\mathscr{B}$ into those of $\mathscr{C}$ (Theorem, §17.2). It is given by $X = BX'$, or $X' = B^{-1}X$. Also $Y = BY'$.

To find the matrix A' of the same mapping T in the new basis, we go from X' to Y' via $Y = AX$. Using substitution, we find that

$$Y = AX \qquad \text{becomes} \quad BY' = ABX', \qquad \text{or} \qquad Y' = (B^{-1}AB)X'.$$

So the matrix of T with respect to the new basis is $B^{-1}AB$. Two matrices A and A' are *similar* if and only if there is an invertible matrix B such that $A' = B^{-1}AB$. Similar matrices represent exactly the same transformation.

___

Example. The matrix $B = \begin{bmatrix} 2 & -1 \\ -5 & 3 \end{bmatrix}$ is invertible and has the inverse $B^{-1} = \begin{bmatrix} 3 & 1 \\ 5 & 2 \end{bmatrix}$.
The matrix $A = \begin{bmatrix} 4 & 7 \\ 0 & 6 \end{bmatrix}$ is similar to the matrix

$$A' = \begin{bmatrix} 3 & 1 \\ 5 & 2 \end{bmatrix} \begin{bmatrix} 4 & 7 \\ 0 & 6 \end{bmatrix} \begin{bmatrix} 2 & -1 \\ -5 & 3 \end{bmatrix} = \begin{bmatrix} -111 & 39 \\ -195 & 121 \end{bmatrix}.$$

This implies that A and A' produce the same mapping of the points of the plane with respect to two different bases. The coordinates (x, y) are related to the new coordinates (x', y') by the system

$$x' = 3x + y,$$
$$y' = 5x + 2y.$$

The relationship between A and A' is not apparent. If we did not know the matrix B, the fact that A and A' represent the same mapping of the planes would be obscured by the arbitrary choice of coordinates.

19.3 Invariant Lines. Suppose the linear operator T sends a line through the origin into itself, that is, the line is invariant under T. Pick any vector $\vec{v} \neq \vec{0}$ in the invariant line. Then, for some number λ, $T(\vec{v}) = \lambda \vec{v}$. Conversely, we can prove that if a number λ and a vector $\vec{v} \neq \vec{0}$ exist such that $T(\vec{v}) = \lambda \vec{v}$, then the line of $\vec{v}$ is invariant under T. For, if we pick any other vector $\vec{u} = c\vec{v}$ in the line of $\vec{v}$, we have

$$T(\vec{u}) = T(c\vec{v}) = cT(\vec{v}) = c(\lambda \vec{v}) = \lambda(c\vec{v}) = \lambda \vec{u}.$$

Thus T sends $\vec{u}$ to the vector $\lambda \vec{u}$, which is also in the line of $\vec{v}$. The same multiplier λ works for both $\vec{u}$ and $\vec{v}$, $T(\vec{v}) = \lambda \vec{v}$, and $T(\vec{u}) = \lambda \vec{u}$.

DEFINITION. For a linear operator T, the number λ is an *eigenvalue* and $\vec{v} \neq \vec{0}$ is an associated *eigenvector* if and only if $T(\vec{v}) = \lambda \vec{v}$.

We observe that the associated eigenvector of λ is not unique. For $\vec{u} = c\vec{v}$ is also an eigenvector associated with λ, if $c \neq 0$. Eigenvalues are also called *characteristic roots*, for reasons that will become apparent; and the associated *eigenvectors* are also called *characteristic vectors*. We summarize the results of this reasoning in a theorem.

THEOREM. The linear operator T has an invariant line if and only if T has an eigenvalue λ and associated eigenvector $\vec{v}$. The line of $\vec{v}$ is invariant under T.

Now in a vector space of dimension n, if it turns out that T has n linearly independent eigenvectors, these can serve as a basis, not a basis arbitrarily chosen by the mathematician, but a basis selected by T itself. The invariant lines are then coordinate axes determined by T itself.

19.4 Method of Finding Eigenvectors

Example A. Find the eigenvalues and eigenvectors of the linear operator on the plane given by

$$T: \begin{aligned} x' &= x + 2y, \\ y' &= 2x - 2y. \end{aligned}$$

Solution.   The matrix of this system is $A = \begin{bmatrix} 1 & 2 \\ 2 & -2 \end{bmatrix}$. We will represent vectors as

columns $X = \begin{bmatrix} x \\ y \end{bmatrix}$. Then λ is an eigenvalue if $AX = \lambda X$. That is,

$$x + 2y = \lambda x, \quad \text{and}$$
$$2x - 2y = \lambda y.$$

This system of linear equations simplifies to the homogeneous system

$$(1 - \lambda)x + 2y \qquad = 0,$$
$$2x + (-2 - \lambda)y = 0.$$

This system has nontrivial solutions $X \neq 0$ if and only if (Theorem, §18.5),

$$\det \begin{bmatrix} 1 - \lambda & 2 \\ 2 & -2 - \lambda \end{bmatrix} = 0.$$

Expanding the determinant, we find that eigenvalues λ must satisfy the equation

$$(1 - \lambda)(-2 - \lambda) - 4 = 0.$$

The roots of this quadratic equation are $\lambda = -3$ and $\lambda = 2$. We set $\lambda = -3$ in the linear system determining λ and find that it becomes

$$\begin{bmatrix} 1 + 3 & 2 \\ 2 & -2 + 3 \end{bmatrix} \begin{bmatrix} x \\ y \end{bmatrix} = \begin{bmatrix} 0 \\ 0 \end{bmatrix},$$

or

$$4x + 2y = 0,$$
$$2x + y = 0,$$

The reduced echelon form of this system is

$$x + \tfrac{1}{2}y = 0,$$
$$0 + 0 = 0.$$

Hence we can set $y = t$, an arbitrary number, and compute $x = -\tfrac{1}{2}t$. All solutions are then the points of the line $x = -\tfrac{1}{2}t$, $y = t$. This is the invariant line associated with the eigenvalue $\lambda = -3$. To be definite we set $t = 2$ and find that one eigenvector associated with $\lambda = -3$ is $X_1 = \begin{bmatrix} -1 \\ 2 \end{bmatrix}$.

Similarly we set $\lambda = 2$ in the equations that determined λ. They become

$$-x + 2y = 0,$$
$$2x - 4y = 0.$$

Solving this system, we find that $X_2 = \begin{bmatrix} 2 \\ 1 \end{bmatrix}$ is an eigenvector associated with $\lambda = 2$.

Example B.   In Example A find the matrix representation of T with respect to the eigenvector basis $\{X_1, X_2\}$.

Solution.   The eigenvectors form a basis because they are linearly independent and span the plane. Let X be any vector of the plane. We can express it in terms of the eigenvector basis as

$$X = xX_1 + yX_2.$$

Its image can be expressed in the same way,

$$X' = x'X_1 + y'X_2.$$

We have also

$$X' = T(X) = xT(X_1) + yT(X_2).$$

But since, by definition of eigenvector, we know that

$$T(X_1) = -3X_1 \quad \text{and} \quad T(X_2) = 2X_2,$$

the expression for $T(X)$ becomes

$$X' = T(X) = x(-3X_1) + y(2X_2).$$

Equating the two expressions for X', we have for T

$$x' = -3x + 0y,$$
$$y' = 0x + 2y.$$

With respect to the eigenvector basis, T is represented by the diagonal matrix

$$J = \begin{bmatrix} -3 & 0 \\ 0 & 2 \end{bmatrix}$$

with eigenvalues on the diagonal and zeros elsewhere.

We now consider the general case. The problem is to find the eigenvalues and eigenvectors of a linear operator T on an n-dimensional vector space. We know that with respect to some chosen basis, the operator T can be represented by a linear system $Y = AX$ with $n \times n$ matrix A. We must find λ and nonzero X such that $AX = \lambda X = \lambda IX$, or

$$(A - \lambda I)X = 0.$$

This homogeneous linear system with matrix $A - \lambda I$ has a nontrivial solution X if and only if

$$\det [A - \lambda I] = 0.$$

This is the characteristic equation whose roots are the eigenvalues λ. Since it is a polynomial equation of degree n, it has n roots, but some of them may be multiple roots. We can substitute each eigenvalue into $(A - \lambda I)X = 0$ and solve for a nonzero X. This is an eigenvector associated with λ. If there are n distinct eigenvalues $\lambda_1, \lambda_2, \ldots, \lambda_n$, they can be associated respectively with n distinct eigenvectors $X_1, X_2, \ldots, X_n$, which form a linearly independent set. Since $T(X_1) = \lambda_1 X_1, \cdots T(X_n) = \lambda_n X_n$, the matrix of the operator T is the diagonal matrix J (§17.3), where

$$J = \begin{bmatrix} \lambda_1 & 0 & 0 & \cdots & 0 \\ 0 & \lambda_2 & 0 & \cdots & 0 \\ \vdots & & & & \vdots \\ 0 & 0 & 0 & \cdots & \lambda_n \end{bmatrix}.$$

This representation of T by the canonical matrix J is both simple and independent of the choice of basis. It is not a solution of the problem in every case. The eigenvalues may be complex numbers (Exercises) or multiple roots. In the latter case, even though the eigenvalues may be real, there may not be enough eigenvectors to form a basis.*

---

* We leave consideration of the general case of multiple characteristic roots to more advanced texts.

The set of eigenvalues is called the *spectrum* of the operator. This is a name that carries over from quantum physics, where they are used extensively. The matrix J is called the *spectral representation* of T.

19.5   Exercises

Exercises 1–10 give matrices of linear operators T on the plane. For each one find the eigenvalues and a basis of eigenvectors for the plane, if it exists. Where possible, find the diagonal (spectral) representation of T with respect to the eigenvector basis.

1. $\begin{bmatrix} 1 & 1 \\ 3 & -1 \end{bmatrix}$.   2. $\begin{bmatrix} 3 & 4 \\ 4 & -3 \end{bmatrix}$.

3. $\begin{bmatrix} 5 & 2 \\ 2 & 2 \end{bmatrix}$.   4. $\begin{bmatrix} 0 & 2 \\ 2 & 3 \end{bmatrix}$.

5. $\begin{bmatrix} r & 0 \\ 0 & r \end{bmatrix}$.   6. $\begin{bmatrix} 10 & 3 \\ 3 & 2 \end{bmatrix}$.

7. $\begin{bmatrix} 2 & -3 \\ -1 & -2 \end{bmatrix}$.   8. $\begin{bmatrix} 0 & 1 \\ 0 & 0 \end{bmatrix}$.

9. $\begin{bmatrix} 0 & 1 \\ -1 & 0 \end{bmatrix}$.   10. $\begin{bmatrix} \cos\theta & -\sin\theta \\ \sin\theta & \cos\theta \end{bmatrix}$.

11. A change of basis replaces the matrix A by $A' = B^{-1}AB$ (§19.2). Show that since $B^{-1}IB = I$, $\det(A' - \lambda I) = \det(A - \lambda I)$. Hence similar matrices have the same characteristic polynomial and hence the same spectrum.

12. Using the theorem on isomorphic representation of linear operators by matrices, prove that if T has the spectral representation $J = \begin{bmatrix} \lambda_1 & 0 \\ 0 & \lambda_2 \end{bmatrix}$, then T^2 is represented by $\begin{bmatrix} \lambda_1^2 & 0 \\ 0 & \lambda_2^2 \end{bmatrix}$, and T^n by $\begin{bmatrix} \lambda_1^n & 0 \\ 0 & \lambda_2^n \end{bmatrix}$.

13. As in Exercise 12, compute the matrices of T^2, T^3, T^{-1}, $T^2 - 3T + I$, $(T - I)^{-1}$, $\sqrt{T}$.

14. Prove that every matrix of real elements

$$A = \begin{bmatrix} a & b \\ c & d \end{bmatrix}$$

that is symmetric, that is, where $A^t = A$, has real eigenvalues.

15. For the matrix A of Exercise 14, show that the characteristic equation

$$\lambda^2 - (a + d)\lambda + \det A = 0.$$

Using Exercise 11, prove that the quantities $a + d$ and $\det A$ are invariant under a change of basis. The sum of the principal diagonal elements of a matrix is called the *trace*, written tr A.

16. In Exercise 15, show that $\lambda_1 + \lambda_2 = \text{tr } A$ and $\lambda_1 \lambda_2 = \det A$.

17. A 2×2 matrix has $\det A = 4$ and tr $A = 5$. Find the eigenvalues, using Exercise 16.

18. Admitting complex numbers as scalars, find eigenvalues *and* eigenvectors for the matrix of Exercise 9,

$$A = \begin{bmatrix} 0 & 1 \\ -1 & 0 \end{bmatrix}.$$

19. For the transformation T represented by the matrix

$$A = \begin{bmatrix} 1 & 2 \\ -2 & 5 \end{bmatrix},$$

find all eigenvalues and eigenvectors. Show that no two linearly independent eigenvectors exist, even if complex scalars are admitted.

20. In Exercise 19, find a basis, including the vector $\begin{bmatrix} 1 \\ 0 \end{bmatrix}$, with the eigenvector $\begin{bmatrix} 2 \\ 2 \end{bmatrix}$, and represent the transformation by the matrix

$$J = \begin{bmatrix} \lambda_1 & 1 \\ 0 & \lambda_1 \end{bmatrix},$$

where λ_1 is the repeated eigenvalue.

21.* Show that the situation described in Exercises 19 and 20 occurs for all matrices

$$A = \begin{bmatrix} a & b \\ c & d \end{bmatrix},$$

such that $4bc = -(a - d)^2$, with the repeated eigenvalue $\lambda_1 = \frac{1}{2}(a + d)$.

20 Orthogonal Bases

20.1 Inner Products. We recall the definition of an inner product (dot product) $(\vec{u} \cdot \vec{v})$, which we defined geometrically (II, §10.6). It was related to the cosine of an angle θ between the lines of $\vec{u}$ and $\vec{v}$ by

(1) $$\cos \theta = \frac{(\vec{u} \cdot \vec{v})}{\|\vec{u}\|\, \|\vec{v}\|}.$$

In fact, norms were also defined in terms of inner products by the formula

(2) $$\|\vec{v}\| = \sqrt{(\vec{v} \cdot \vec{v})}.$$

Just as we did for vectors as elements of a linear space (II, §9), we can describe inner products by axiomatic properties, as follows.

DEFINITION. An *inner product* for a vector space with real scalars assigns to every pair of vectors $\vec{u}$, $\vec{v}$ a number $(\vec{u} \cdot \vec{v})$ such that

$\mathscr{U}_1$ $((\vec{u} + \vec{v}) \cdot \vec{w}) = (\vec{u} \cdot \vec{w}) + (\vec{v} \cdot \vec{w})$;
$\mathscr{U}_2$ $(c\vec{u} \cdot \vec{v}) = c(\vec{u} \cdot \vec{v})$;
$\mathscr{U}_3$ $(\vec{u} \cdot \vec{v}) = (\vec{v} \cdot \vec{u})$;   and
$\mathscr{U}_4$ $(\vec{v} \cdot \vec{v}) > 0$,   whenever $\vec{v} \neq \vec{0}$.

A finite-dimensional vector space with an inner product is called a *Euclidean space*. In a Euclidean space, if $\vec{v}_1$ is the position vector of P_1 and $\vec{v}_2$ is the position vector of P_2, then the distance $|P_1 P_2| = \|\vec{v}_2 - \vec{v}_1\|$. We shall see that this reduces to the familiar formula

$$|P_1 P_2| = \sqrt{(x_2 - x_1)^2 + (y_2 - y_1)^2}$$

when the coordinates (x_1, y_1), (x_2, y_2) are determined relative to a suitable basis.

DEFINITION. Two vectors, $\vec{u}$ and $\vec{v}$ are *orthogonal* if and only if $(\vec{u} \cdot \vec{v}) = 0$.

If $\|\vec{u}\| \neq 0$ and $\|\vec{v}\| \neq 0$, and $\vec{u}$ and $\vec{v}$ are orthogonal, then they are perpendicular, since this implies that $\cos \theta = 0$.

20.2   Orthonormal Bases.   A set of vectors $\{\vec{v}_1, \vec{v}_2, \ldots \vec{v}_n\}$ is said to be *orthonormal* if and only if $(\vec{v}_i \cdot \vec{v}_j) = 0$ when $i \neq j$ and for every i, $(\vec{v}_i \cdot \vec{v}_i) = 1$. In geometric language, this means that every vector in the set is perpendicular to every other one and every vector has length 1. For example, in E^3, the set $\{(1, 0, 0), (0, 1, 0), (0, 0, 1)\}$ is orthonormal (Figure 15.7).

An orthonormal set is always linearly independent. For suppose for some scalars $x_1, x_2, \ldots, x_n$ it is true that

$$x_1 \vec{v}_1 + x_2 \vec{v}_2 + \cdots + x_n \vec{v}_n = \vec{0}.$$

We form the dot product of both sides with the vector $\vec{v}_i$ out of the orthonormal set and find

$$x_1(\vec{v}_1 \cdot \vec{v}_i) + \cdots + x_n(\vec{v}_n \cdot \vec{v}_i) = 0.$$

By definition of orthonormal set, all of these dot products are zero except that $(\vec{v}_i \cdot \vec{v}_i) = 1$. Hence the equation reduces to $x_i = 0$. Since this is true for every $\vec{v}_i$, $i = 1, \ldots, n$, we see that $x_1 = x_2 = \cdots = x_n = 0$. That is, the vectors are linearly independent.

If every vector $\vec{v}$ can be represented as a linear combination of the vectors $\mathcal{B} = \{\vec{v}_1, \vec{v}_2, \ldots, \vec{v}_n\}$ of an orthonormal set, the set $\mathcal{B}$ is called an orthonormal basis for $\mathcal{V}$. The determination of coordinates is particularly simple when we have an orthonormal basis. For if

$$\vec{v} = x_1 \vec{v}_1 + x_2 \vec{v}_2 + \cdots + x_n \vec{v}_n,$$

we form the inner product of both sides with $\vec{v}_1, \vec{v}_2, \ldots, \vec{v}_n$ in succession and find that

$$x_i = (\vec{v} \cdot \vec{v}_i).$$

For the other terms are all zero, due to the orthogonality. This is called the Fourier rule for expanding $\vec{v}$ in terms of an orthonormal basis.

Example.   We consider the 3-space of rows, $\vec{v} = (x, y, z)$, $\vec{v}' = (x', y', z')$ and inner product $(\vec{v} \cdot v') = xx' + yy' + zz'$. The vectors

$$\vec{v}_1 = (\tfrac{2}{3}, -\tfrac{2}{3}, \tfrac{1}{3}), \qquad \vec{v}_2 = (\tfrac{1}{3}, \tfrac{2}{3}, \tfrac{2}{3}), \qquad \vec{v}_3 = (-\tfrac{2}{3}, -\tfrac{1}{3}, \tfrac{2}{3})$$

form an orthonormal basis. Find the coordinates of the vector $\vec{v} = (3, 12, -9)$ with respect to the orthonormal basis.

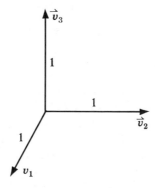

FIGURE 15.7   A set of three orthonormal vectors.

Solution. We use the Fourier rule and compute $(\vec{v} \cdot \vec{v}_1) = -9$, $(\vec{v} \cdot \vec{v}_2) = 3$, $(\vec{v} \cdot \vec{v}_3) = -12$. Hence

$$\vec{v} = -9\vec{v}_1 + 3\vec{v}_2 - 12\vec{v}_3$$

with respect to the given orthonormal basis.

20.3   Representation of Euclidean Space. We have seen (§13.3) that relative to a chosen basis $\mathscr{B} = \{\vec{v}_1, \ldots, \vec{v}_n\}$ every vector space of dimension n can be represented isomorphically by rows $(x_1, x_2, \ldots, x_n)$ of coordinates. Now we assume that there is an inner product and that the basis $\mathscr{B}$ is orthonormal. The vectors $\vec{v}$ and $\vec{w}$ have unique representations in terms of the basis

$$\vec{v} = x_1\vec{v}_1 + \cdots + x_n\vec{v}_n,$$
$$\vec{w} = y_1\vec{v}_1 + \cdots + y_n\vec{v}_n.$$

Then for the inner product

$$(\vec{v} \cdot \vec{w}) = ((x_1\vec{v}_1 + \cdots + x_n\vec{v}_n) \cdot (y_1\vec{v}_1 + \cdots + y_n\vec{v}_n))$$
$$= x_1y_1 + x_2y_2 + \cdots + x_ny_n.$$

The last expression is found by multiplying out the two vector sums, using the properties of inner products and then applying the properties $(\vec{v}_i \cdot \vec{v}_j) = 0$ if $i \neq j$ and $(\vec{v}_i \cdot \vec{v}_j) = 1$ of the orthonormal basis.

Thus if we set up the one-to-one correspondence

$$\vec{v} \leftrightarrow (x_1, x_2, \ldots, x_n)$$
$$\vec{w} \leftrightarrow (y_1, y_2, \ldots, y_n),$$

this isomorphic representation not only faithfully represents linear operations by the corresponding operations on rows but calculates inner products by the standard formula $x_1y_1 + \cdots + x_ny_n$. This procedure, however, depends upon having an orthonormal basis.

Example.   Compute $(\vec{u} \cdot \vec{v})$ if

$$\vec{u} = 3\vec{v}_1 - 2\vec{v}_2 + 5\vec{v}_3 \qquad \text{and} \qquad \vec{v} = -\vec{v}_1 + 6\vec{v}_2 - 4\vec{v}_3,$$

and $\{\vec{v}_1, \vec{v}_2, \vec{v}_3\}$ is an orthonormal basis.

Solution.   We do not need to know what the basis vectors are, provided they compose an orthonormal basis. By the isomorphic representation,

$$(\vec{u} \cdot \vec{v}) = 3(-1) - 2(6) + 5(-4) = -35.$$

20.4   Orthogonal Transformations. We know that every invertible substitution $X = BX'$ introduces a new basis for X' coordinates in place of an old basis for X coordinates. But not every such transformation of coordinates sends an orthonormal basis into a new orthonormal basis; if the inner product is invariant, however, under the transformation of coordinates, the basis will remain orthonormal. So we ask, What are the transformations $X = BX'$ that leave the inner product invariant? By this we mean that if we compute the inner product in old and new coordinates, they are equal. That is,

$$x_1y_1 + \cdots + x_ny_n = x_1'y_1' + \cdots + x_n'y_n'.$$

To answer this question we write the inner product in a new form, using matrix multiplication. Let X, Y, X', Y' be column vectors. Their transposed matrices are single-row matrices. Thus

$$X = \begin{bmatrix} x_1 \\ x_2 \\ \vdots \\ x_n \end{bmatrix} \quad \text{and} \quad X^t = [x_1 x_2 \cdots x_n].$$

Let I be the $n \times n$ identity matrix. We observe that the inner product can be written in matrix notation. For

$$X^t I Y = x_1 y_1 + \cdots + x_n y_n.$$

Now we introduce new coordinates X' and Y' by the substitution $X = BX'$ with $n \times n$ matrix B. This transforms Y too, by $Y = BY'$. Then $X^t = (X')^t B^t$ (§16.2). Substituting these into the inner product formula, we find that

$$X^t I Y = [(X')^t B^t] I [BY'] = (X')^t (B^t B) Y'.$$

The form on the right is $x'_1 y'_1 + \cdots + x'_n y'_n$ if and only if

$$B^t B = I.$$

DEFINITION. A matrix B is *orthogonal* and it defines an *orthogonal transformation* of coordinates $X = BX'$ if and only if $B^t B = I$.

Example. Find the 2×2 orthogonal matrices.

Solution. Let $B = \begin{bmatrix} a & b \\ c & d \end{bmatrix}$. Then $B^t = \begin{bmatrix} a & c \\ b & d \end{bmatrix}$ and $B^t B = I$ implies that

$$\begin{bmatrix} a^2 + c^2 & ab + cd \\ ab + cd & c^2 + d^2 \end{bmatrix} = \begin{bmatrix} 1 & 0 \\ 0 & 1 \end{bmatrix}.$$

Hence $a^2 + c^2 = 1$. By definition of $\sin \theta$ and $\cos \theta$, this implies that for some number θ, $a = \cos \theta$ and $c = \sin \theta$. Then $c^2 + d^2 = 1$ implies that $d = \pm \cos \theta$. Finally, the matrix equation implies that $ab + cd = 0$; and from this we find that if $d = \cos \theta$, then $b = -\sin \theta$. Also if $d = -\cos \theta$, then $b = \sin \theta$. Hence the 2×2 orthogonal matrices are either

$$\begin{bmatrix} \cos \theta & -\sin \theta \\ \sin \theta & \cos \theta \end{bmatrix} \quad \text{or} \quad \begin{bmatrix} \cos \theta & \sin \theta \\ \sin \theta & -\cos \theta \end{bmatrix}.$$

The first of these is the rotation transformation (§1)

$$x_1 = x'_1 \cos \theta - x'_2 \sin \theta$$
$$x_2 = x'_1 \sin \theta + x'_2 \cos \theta.$$

The second matrix gives the transformation

$$x_1 = x'_1 \cos \theta + x'_2 \sin \theta$$
$$x_2 = x'_1 \sin \theta - x'_2 \cos \theta,$$

which is not a rotation for any angle θ. But if we first reverse the positive direction on the x'_2-axis by setting

$$x'_1 = x''_1, \qquad x'_2 = -x''_2,$$

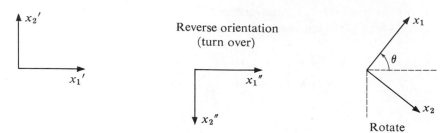

FIGURE 15.8   Reversal of orientation.

it becomes a rotation from x_1x_2-axes to $x_1''x_2''$-axes. We can picture this reversal of orientation combined with rotation if we start with x_1x_2-axes, turn the plane over, and then rotate it (Figure 15.8).

20.5   Exercises

1. Verify that the vectors
$$\vec{v}_1 = (\tfrac{3}{5}, -\tfrac{4}{5}) \quad \text{and} \quad \vec{v}_2 = (\tfrac{4}{5}, -\tfrac{3}{5})$$
form an orthonormal set.

2. Show that for every θ, the vectors
$$\vec{v}_1 = (\cos \theta, \sin \theta) \quad \text{and} \quad \vec{v}_2 = (-\sin \theta, \cos \theta)$$
form an orthonormal set.

3. Show that if the coordinates of points $P_1 : (x_1, y_1)$ and $P_2 : (x_2, y_2)$ are calculated with respect to any orthonormal basis, then we obtain the distance
$$|P_1P_2| = \sqrt{(x_2 - x_1)^2 + (y_2 - y_1)^2}.$$

4. Calculate $X^t I Y$ if
$$X = \begin{bmatrix} x_1 \\ x_2 \\ x_3 \end{bmatrix} \quad \text{and} \quad Y = \begin{bmatrix} y_1 \\ y_2 \\ y_3 \end{bmatrix}.$$

5. Find all vectors $\vec{v} = (x, y, z)$ that are orthogonal to both $\vec{v}_1 = (1, 3, -4)$ and $\vec{v}_2 = (5, 0, -2)$.

6. In Exercise 5, find all vectors of unit length that are orthogonal to both $\vec{u}$ and $\vec{v}$.

7. Express $\vec{v} = (6, -10)$ as a linear combination $x_1\vec{v}_1 + x_2\vec{v}_2$ if $\vec{v}_1 = (\tfrac{3}{5}, -\tfrac{4}{5})$, $\vec{v}_2 = (\tfrac{4}{5}, \tfrac{3}{5})$ form a basis for the plane.

8. Calculate the projection of the vector $\vec{v} = (5, -10)$ onto the line of the unit vector $\vec{u} = (\tfrac{3}{5}, -\tfrac{4}{5})$. Express the result as a vector $\vec{w}_1$ in the line of $\vec{u}$. (II, §12.3).

9. Use the result of Exercise 8 to express $\vec{v}$ as a sum, $\vec{v} = \vec{w}_1 + \vec{w}_2$, of two orthogonal vectors, one parallel to $\vec{u}$ and one perpendicular to it.

10. The vector $\vec{v}_1 = (\tfrac{2}{3}, -\tfrac{1}{3}, -\tfrac{2}{3})$ has $\|\vec{v}_1\| = 1$. Find two other vectors $\vec{v}_2$ and $\vec{v}_3$ so that $\{\vec{v}_1, \vec{v}_2, \vec{v}_3\}$ is an orthonormal set.

11. Prove that the zero vector is orthogonal to every vector $\vec{v}$.

12. Prove that $\vec{u} = \vec{0}$ if $(\vec{u} \cdot \vec{v}) = 0$ for every vector $\vec{v}$.

13. The definition of the inner point does not explicitly say that $(\vec{0} \cdot \vec{v}) = 0$. Prove it.

14. Calculate to show that $\|\vec{u} - \vec{v}\|^2 = \|\vec{u}\|^2 - 2(\vec{u} \cdot \vec{v}) + \|\vec{v}\|^2$.

15. Calculate to prove that
$$\|\vec{u} + \vec{v}\|^2 + \|\vec{u} - \vec{v}\|^2 = 2\|u\|^2 + 2\|\vec{v}\|^2.$$

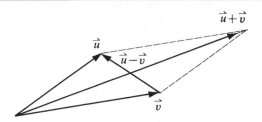

FIGURE 15.9   Exercise 15.

This is called the parallelogram theorem (Figure 15.9). Write it out in words relating the lengths of the sides to the lengths of the diagonals of any parallelogram.

16. In Exercise 15, if $\vec{u}$ is orthogonal to $\vec{v}$ then the parallelogram becomes a rectangle. Show that in this case

$$\|\vec{u} - \vec{v}\| = \|\vec{u} + \vec{v}\|.$$

17. As in Exercise 14, if $\vec{u} \neq 0$ and $\vec{v} \neq 0$, calculate

$$\left(\left(\frac{\vec{u}}{\|\vec{u}\|} - \frac{\vec{v}}{\|\vec{v}\|} \right) \cdot \left(\frac{\vec{u}}{\|\vec{u}\|} - \frac{\vec{v}}{\|\vec{v}\|} \right) \right) = 2 - 2 \frac{(\vec{u} \cdot \vec{v})}{\|\vec{u}\| \, \|\vec{v}\|}.$$

18. Use the result of Exercise 17 to prove that

$$|(\vec{u} \cdot \vec{v})| \leq \|\vec{u}\| \, \|\vec{v}\| \qquad \text{(Schwarz inequality)}.$$

19. Since

$$\|\vec{u} + \vec{v}\|^2 = ((\vec{u} + \vec{v}) \cdot (\vec{u} + \vec{v})) = \|\vec{u}\|^2 + 2(\vec{u} \cdot \vec{v}) + \|\vec{v}\|^2,$$

use the Schwarz inequality to prove that

$$\|\vec{u} + \vec{v}\| \leq \|\vec{u}\| + \|\vec{v}\| \qquad \text{(Triangle inequality)}.$$

20. Prove that the equals signs hold in the Schwarz inequality and triangle inequality if and only if $\vec{u}$ and $\vec{v}$ are parallel (and have the same direction).

21. Show that $B = \begin{bmatrix} \frac{5}{13} & \frac{12}{13} \\ -\frac{12}{13} & \frac{5}{13} \end{bmatrix}$ is an orthogonal matrix.

22. Prove that if B is an orthogonal matrix, $B^{-1} = B^t$. Use this to calculate the inverse of the matrix B in Exercise 21.

23. Show that if B is an orthogonal matrix, det $B = \pm 1$.

24. Show that if B is an orthogonal matrix, its rows (columns) form an orthonormal set.

20.6   Construction of Orthonormal Bases

In Problems T1–T7, we start with any basis $\mathscr{A} = \{\vec{u}_1, \vec{u}_2, \dots, \vec{u}_n\}$ and construct from it an orthonormal basis $\mathscr{B} = \{\vec{v}_1, \vec{v}_2, \dots, \vec{v}_n\}$. The process is called the Gram-Schmidt process.

T1. Choose $\vec{v}_1 = \vec{u}_1 / \|\vec{u}_1\|$. Show that $\{\vec{v}_1\}$ is an orthonormal set.

T2. Project $\vec{u}_2$ onto the line of $\vec{v}_1$ to obtain the vector projection $(u_2 \cdot \vec{v}_1) \vec{v}_1$. Show that $\vec{w}_2 = \vec{u}_2 - (\vec{u}_2 \cdot \vec{v}_1) \vec{v}_1$ is orthogonal to $\vec{u}_2$.

T3. Show that $\|\vec{w}_2\| \neq 0$.

T4. Choose $\vec{v}_2 = \vec{w}_2 / \|\vec{w}_2\|$. Prove that $\{\vec{v}_1, \vec{v}_2\}$ is an orthonormal set.

T5. Show that

$$\vec{w}_3 = \vec{u}_3 - (\vec{u}_3 \cdot \vec{v}_1) \vec{v}_1 - (\vec{u}_3 \cdot \vec{v}_2) \vec{v}_2$$

is orthogonal to both $\vec{v}_1$ and $\vec{v}_2$ and is not zero.

T6.   Choose $\vec{v}_3 = \vec{w}_3 / \|\vec{w}_3\|$. Then $\{\vec{v}_1, \vec{v}_2, \vec{v}_3\}$ is an orthonormal set.

T7.   Continuing this process, we obtain an orthonormal basis $\{\vec{v}_1, \ldots, \vec{v}_n\}$. Complete the necessary step of mathematical induction to show that if $\{\vec{v}_1, \ldots, \vec{v}_k\}$ is an orthonormal set and $\vec{u}_{k+1}$ is not in the space spanned by them, then

$$\vec{w}_{k+1} = \vec{u}_{k+1} - (\vec{u}_{k+1} \cdot \vec{v}_1) - \cdots - (\vec{u}_{k+1} \cdot \vec{v}_k)$$

is orthogonal to every vector $\vec{v}_1, \ldots, \vec{v}_k$ and $\vec{w}_{k+1} \neq 0$.

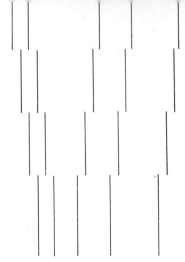

Applications of Linear Algebra

CHAPTER 16

21 Eigenvalues in Euclidean Geometry

21.1 Conics and Quadratic Forms in Two Variables. The equations of ellipse, hyperbola, and parabola (conics) in Cartesian xy-coordinates are second-degree equations. With respect to specifically chosen coordinate bases they have familiar standard equations (Figure 16.1). But with respect to a random basis, not specially chosen, the equations of conics are disguised as quadratic equations with cross-product xy-terms, as well as x^2 and y^2 terms. For example, it can be shown that $9x^2 - 6xy + 17y^2 = 288$ is an ellipse (Figure 16.2), whose equation with respect to a more geometrically chosen basis is

$$\frac{x'^2}{36} + \frac{y'^2}{16} = 1.$$

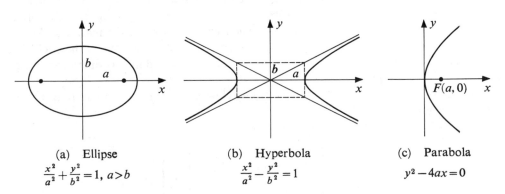

(a) Ellipse
$$\frac{x^2}{a^2} + \frac{y^2}{b^2} = 1, \ a>b$$

(b) Hyperbola
$$\frac{x^2}{a^2} - \frac{y^2}{b^2} = 1$$

(c) Parabola
$$y^2 - 4ax = 0$$

FIGURE 16.1 Standard equations of conics and their graphs.

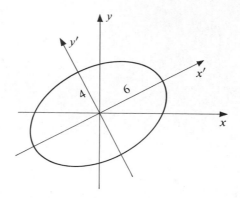

FIGURE 16.2   Ellipse $9x^2 - 6xy + 17y^2 = 288$.

In general, the second-degree terms in the equations of conics have the form

$$Q(x, y) = ax^2 + 2bxy + cy^2.$$

This is called a quadratic form in the variables. Our objective is to find a set of basis vectors with respect to which the coordinates are such that the cross-product terms drop out and the quadratic form becomes a sum of squares

$$\lambda_1 x'^2 + \lambda_2 y'^2.$$

Then we can readily identify and plot the conic.

21.2   The Matrix of a Quadratic Form.   We can introduce matrices into the representation of a quadratic form in a natural way, using matrix multiplication. We observe that

$$Q = ax^2 + 2bxy + cy^2 = [x \quad y]\begin{bmatrix} a & b \\ b & c \end{bmatrix}\begin{bmatrix} x \\ y \end{bmatrix}.$$

Or if $A = \begin{bmatrix} a & b \\ b & c \end{bmatrix}$ and $X = \begin{bmatrix} x \\ y \end{bmatrix}$, this can be written

$$Q = X^t A X.$$

We observe that the matrix A of the quadratic form is symmetric, $A^t = A$. We made it symmetric by assigning half of the coefficient $2b$ of the xy-term to the upper right corner, where it comes out as the coefficient of xy when we multiply out the matrices. We assigned the other half of $2b$ to the lower left corner, where it comes out as the coefficient of yx when we multiply the matrices. Thus from the indicated matrix operations we get $Q = (ax^2 + bxy) + (byx + cy^2)$.

The matrix formula, $Q = X^t A X$, can also be viewed as an inner product in two ways. We have seen that if

$$X = \begin{bmatrix} x_1 \\ x_2 \\ \vdots \\ x_n \end{bmatrix} \quad \text{and} \quad Y = \begin{bmatrix} y_1 \\ y_2 \\ \vdots \\ y_n \end{bmatrix}$$

are column vectors, then

$$X^t Y = [x_1 x_2 \cdots x_n] \begin{bmatrix} y_1 \\ y_2 \\ \vdots \\ y_n \end{bmatrix} = x_1 y_1 + \cdots + x_n y_n = (X \cdot Y).$$

That is, $X^t Y$ is the inner product $(X \cdot Y)$. Using this idea, we see that $Q = X^t(AX) = (X \cdot AX)$. Moreover, since $A = A^t$, we can rewrite Q as

$$Q = (X^t A^t)X = (AX)^t X = (AX \cdot X).$$

Thus, for a symmetric matrix, $(AX \cdot X) = (X \cdot AX) = Q$.

21.3   Eigenvalues and Eigenvectors.

The eigenvalues of a 2×2 symmetric matrix are always real numbers. For we readily verify that the characteristic equation of A is

$$\det [A - \lambda I] = \lambda^2 - (a + c)\lambda + (ac - b^2) = 0.$$

And it has roots given (with some simplification) by

$$\lambda = \frac{(a + c) \pm \sqrt{(a - c)^2 + 4b^2}}{2}.$$

Since the discriminant, $(a - c)^2 + 4b^2$, is a sum of squares, it is never negative. Hence the eigenvalues λ_1 and λ_2 are real numbers. Moreover $\lambda_1 \neq \lambda_2$ unless $(a - c)^2 + 4b^2 = 0$, which implies that $a = c$ and $b = 0$. Hence if $\lambda_1 = \lambda_2$, the matrix A is already in diagonal form

$$A = \begin{bmatrix} a & 0 \\ 0 & a \end{bmatrix} \quad \text{and} \quad Q = ax^2 + ay^2.$$

Hence we may consider only the case $\lambda_1 \neq \lambda_2$. We can prove a rather surprising theorem about eigenvectors associated with distinct eigenvalues of a symmetric matrix. It is just as easy to prove for $n \times n$ matrices as for 2×2.

THEOREM.   If λ_1 and λ_2 are two distinct eigenvalues of a symmetric matrix A, associated with eigenvectors X_1 and X_2, then $(X_2 \cdot X_1) = 0$.

Proof.   By definition of eigenvalue and eigenvector, $AX_1 = \lambda_1 X_1$ and $AX_2 = \lambda_2 X_2$. We form the inner product

$$(X_2 \cdot AX_1) = (X_2 \cdot \lambda_1 X_1) = \lambda_1(X_2 \cdot X_1).$$

The same inner product can be rewritten using the symmetry of A (§21.1) in the form

$$(X_2 \cdot AX_1) = (AX_2 \cdot X_1) = (\lambda_2 X_2 \cdot X_1) = \lambda_2(X_2 \cdot X_1).$$

Subtracting the second expression for $(X_2 \cdot AX_1)$ from the first, we have

$$0 = (\lambda_1 - \lambda_2)(X_2 \cdot X_1).$$

Since $\lambda_1 \neq \lambda_2$, this implies that $(X_2 \cdot X_1) = 0$. That is, the eigenvectors for two distinct eigenvalues of a symmetric matrix are always orthogonal.

21.4   Reduction of Conics to Standard Form.   We may use an eigenvector basis to reduce a quadratic equation in x and y to an equivalent one with diagonal matrix

$$J = \begin{bmatrix} \lambda_1 & 0 \\ 0 & \lambda_2 \end{bmatrix} \quad \text{so that} \quad Q = \lambda_1 x'^2 + \lambda_2 y'^2.$$

Example.   For the conic,

(1) $$9x^2 - 6xy + 17y^2 = 288,$$

introduce new coordinates (x', y') referred to an eigenvector basis. Locate the characteristic axes, identify the curve and plot.

Solution.   The quadratic form $Q = 9x^2 - 6xy + 17y^2$ has the symmetric matrix

$$A = \begin{bmatrix} 9 & -3 \\ -3 & 17 \end{bmatrix}.$$

The characteristic equation, $\det [A - \lambda I] = 0$, becomes

$$\det \begin{bmatrix} 9 - \lambda & -3 \\ -3 & 17 - \lambda \end{bmatrix} = \lambda^2 - 26\lambda + 144 = 0.$$

Factoring, we find that the eigenvalues are $\lambda_1 = 8$ and $\lambda_2 = 18$. For λ_1, the equations $[A - \lambda_1 I]X_1 = 0$; for the eigenvectors become

$$1x - 3y = 0,$$
$$-3x + 9y = 0,$$

whose reduced echelon form is

$$1x - 3y = 0,$$
$$0 + 0 = 0.$$

This tells us that all eigenvectors associated with $\lambda_1 = 8$ have the form

$$X_1 = s \begin{bmatrix} 3 \\ 1 \end{bmatrix}, \quad \text{for all real numbers } s.$$

Similarly for $\lambda_2 = 18$, the equation $[A - \lambda_2 I]X_2 = 0$ has nontrivial solutions of the form

$$X_2 = t \begin{bmatrix} -1 \\ 3 \end{bmatrix}, \quad \text{for all real numbers } t.$$

Since $\sqrt{3^2 + 1^2} = \sqrt{(-1)^2 + 3^2} = \sqrt{10}$, we may choose $s = t = 1/\sqrt{10}$ and obtain the orthonormal set of eigenvectors

(2) $$X_1 = \begin{bmatrix} \dfrac{3}{\sqrt{10}} \\ \dfrac{1}{\sqrt{10}} \end{bmatrix} \quad \text{and} \quad X_2 = \begin{bmatrix} \dfrac{-1}{\sqrt{10}} \\ \dfrac{3}{\sqrt{10}} \end{bmatrix}.$$

We plot the vectors on this orthonormal basis to locate the invariant axes (Figure 16.3(a)).

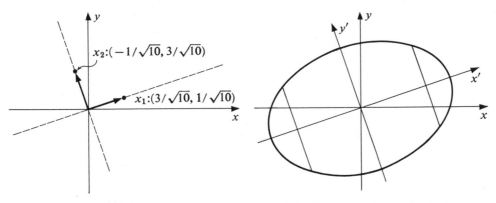

(a) Invariant axes   (b) Ellipse on eigenvector basis

FIGURE 16.3   Conic with respect to invariant axes.

Now any position vector X can be expressed in terms of the new eigenvector basis by some scalar coefficients x' and y',

$$X = x'X_1 + y'X_2.$$

Then

$$AX = x'\lambda_1 X_1 + y'\lambda_2 X_2,$$

and hence

$$Q = (X \cdot AX) = \lambda_1 x'^2 + \lambda' y'^2 = 8x'^2 + 16y'^2.$$

Thus the quadratic equation $9x^2 - 6xy + 17y^2 = 288$ becomes $8x'^2 + 18y'^2 = 288$, or

$$\frac{x'^2}{36} + \frac{y'^2}{16} = 1.$$

This is an ellipse with $a = 6$ and $b = 4$. We plot it with respect to the $x'y'$-coordinates (Figure 16.3(b)) and this gives its plot also with respect to the original xy-coordinates. This completes the problem.

21.5 The Orthogonal Transformation of Coordinates. In the preceding example, the relation between the (x, y) coordinates of a point, represented by the column matrix $X = \begin{bmatrix} x \\ y \end{bmatrix}$, were expressed in terms of the eigenvector basis $\{X_1, X_2\}$ by

$$X = x'X_1 + y'X_2.$$

We can write this out explicitly and find that

$$\begin{bmatrix} x \\ y \end{bmatrix} = x' \begin{bmatrix} \dfrac{3}{\sqrt{10}} \\ \dfrac{1}{\sqrt{10}} \end{bmatrix} + y' \begin{bmatrix} \dfrac{-1}{\sqrt{10}} \\ \dfrac{3}{\sqrt{10}} \end{bmatrix},$$

or

$$\begin{cases} x = \dfrac{3}{\sqrt{10}}\, x' - \dfrac{1}{\sqrt{10}}\, y' \\[3mm] y = \dfrac{1}{\sqrt{10}}\, x' + \dfrac{3}{\sqrt{10}}\, y' \end{cases} \qquad \text{with matrix} \qquad B = \begin{bmatrix} \dfrac{3}{\sqrt{10}} & -\dfrac{1}{\sqrt{10}} \\[3mm] \dfrac{1}{\sqrt{10}} & \dfrac{3}{\sqrt{10}} \end{bmatrix}.$$

Thus the transformation of coordinates from the original basis to the eigenvector basis is given by the substitution $X = BX'$, where B is a matrix whose columns are the eigenvectors of unit length. We readily verify that $B^t B = I$. This says that B is an orthogonal matrix (§20.4) and hence the transformation of coordinates $X = BX'$ preserves lengths. That is, the shape of the graph remains intact.

   This procedure is valid for any number of variables. For an $n \times n$ symmetric matrix A, if the distinct eigenvalues and associated unit eigenvectors are

$$\lambda_1 X_1,\ \lambda_2 X_2,\ \ldots,\ \lambda_n X_n,$$

then the position vector X can be expressed in terms of the eigenvector basis by

$$X = x_1' X_1 + x_2' X_2 + \cdots + x_n' X_n.$$

Expressing the coefficients $(x_1', x_2', \ldots, x_n')$ as a column vector X', this is a transformation of coordinates

$$X = BX'.$$

B is an $n \times n$ matrix whose columns are the eigenvectors $X_1, \ldots, X_n$. B is orthogonal, hence invertible, and it preserves distances.

21.6   The General Second-Degree Equation in x, y.

The general equation of the degree two in x, y can be written

$$ax^2 + 2bxy + cy^2 + dx + ey + f = 0.$$

Ordinarily we apply a translation

$$x = x' + h, \qquad y = y' + k$$

chosen to remove the linear terms (§1). This leaves an equation of the form

$$ax'^2 + 2bx'y' + cy'^2 = k.$$

We reduce this by going to an eigenvector basis and it becomes

$$\lambda_1 (x'')^2 + \lambda_2 (y'')^2 = k.$$

If both $\lambda_1 \neq 0$ and $\lambda_2 \neq 0$, this can be written

$$\frac{(x')^2}{(k/\lambda_1)} + \frac{(y'')^2}{(k/\lambda_2)} = 1.$$

This is the standard form of an ellipse, if λ_1 and λ_2 have the same sign, or of a hyperbola, if λ_1 and λ_2 have opposite signs. But the case where $\lambda_1 = 0$ or $\lambda_2 = 0$ is left out.

Example.   Reduce to standard form by choosing a suitable basis

$$16x^2 + 24xy + 9y^2 + 90x - 120y + 300 = 0.$$

Solution.   We try to determine h and k so that the translation $x = x' + h, y = y' + k$ will remove the linear terms. It turns out that the required linear equations have no solution (Exercises). Hence we proceed to find the eigenvector basis for the quadratic terms. The quadratic form $Q = 16x^2 + 24xy + 9y^2$ has the matrix

$$A = \begin{bmatrix} 16 & 12 \\ 12 & 9 \end{bmatrix}.$$

Its eigenvalues are $\lambda = 0$ and $\lambda = 25$. For $\lambda = 0$, we find the unit eigenvector $(\frac{3}{5}, -\frac{4}{5})$. And associated with $\lambda = 25$ we find the eigenvector $(\frac{4}{5}, \frac{3}{5})$. The substitution that transforms coordinates has a matrix having these two as columns. That is,

$$x = \tfrac{3}{5}x' + \tfrac{4}{5}y',$$
$$y = -\tfrac{4}{5}x' + \tfrac{3}{5}y'.$$

We now transform the coordinates by substituting these expressions for x and y in the original equation. The result for the quadratic terms can be written down at once, since it is $\lambda_1 x_1'^2 + \lambda_2 y'^2$. But for the linear terms we must actually carry out substitution. This gives us

$$0x'^2 + 25y'^2 + 90(\tfrac{3}{5}x' + \tfrac{4}{5}y') - 120(-\tfrac{4}{5}x' + \tfrac{3}{5}y') + 300 = 0,$$

or

$$y'^2 + 6x' + 12 = 0.$$

This is recognizable as a parabola. We can transform it to standard form $y^2 = 4ax$ by a translation

$$x' = x'' + h, \qquad y' = y'' + k.$$

We find $h = 2, k = 0$. The equation reduces to

$$y''^2 = -12x'',$$

which we plot (Figure 16.4).

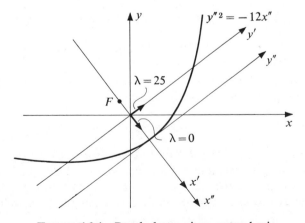

FIGURE 16.4   Parabola on eigenvector basis.

21.7 Exercises

In Exercises 1–8, use eigenvector bases to reduce the quadratic equations to standard form. Plot the invariant axes in the original coordinate system. Identify and plot the conic.

1. $2xy = 4$.
2. $9x^2 + 24xy + 16y^2 = 288$.
3. $5x^2 + 8xy + 5y^2 - 9 = 0$.
4. $3x^2 - 6xy + y^2 - 4 = 0$.
5. $x^2 + 2\sqrt{3}xy - y^2 + 1 = 0$.
6. $9x^2 + 24xy + 16y^2 + 100x - 50y + 100 = 0$.
7. $10x^2 + 6xy + 2y^2 = 44$.
8. $5x^2 + 4xy + 2y^2 - 36x - 12y - 30 = 0$.

9. In the parabola example (§21.6), carry out the translation in an attempt to remove the linear terms from the original equation. Show that it fails.
10. Show that it is not possible to remove the linear terms from

$$ax^2 + 2bxy + cy^2 + dx + cy + f = 0,$$

if $b^2 - ac = 0$.
11. In the parabola example (§21.6), find the coordinates of the focus in the original xy-coordinates.
12. In the ellipse example (§21.4), find the coordinates of the foci in xy-coordinates.
13. The quadratic equation

$$ax^2 + 2bxy + cy^2 = 4$$

has $a + c = 5$ and $b^2 - ac = -4$. Find the standard equation of the curve and sketch it.
14. The formulas $a + c$ and $b^2 - ac$ are invariant under orthogonal transformation of coordinates. Use this to prove that if the curve is an ellipse, then $b^2 - ac < 0$; if it is a parabola, then $b^2 - ac = 0$; if it is a hyperbola, then $b^2 - ac > 0$.
15. Are the converse statements to those of Exercise 14 true? Consider the quadratic

$$(2x - 3y + 4)(6x + 5y - 7) = 0.$$

Show that $b^2 - ac > 0$ but its graph is not a hyperbola. Consider also Exercise 2.

22   The Differentiation Operator

22.1   Eigenvectors of the Differentiation Operator.

The space $\mathscr{C}^\infty$ of all functions that have derivatives of all orders with respect to t on the real number line is a vector space. For if u and v are in it, $u + v$ and cu are in it. Moreover the differentiation operator d/dt is a linear operator on $\mathscr{C}^\infty$ since

$$\frac{d}{dt}(u + v) = \frac{du}{dt} + \frac{dv}{dt} \quad \text{and} \quad \frac{d}{dt}(cu) = c\frac{du}{dt}.$$

We can easily show that $\mathscr{C}^\infty$ is not a finite dimensional space, for it includes all powers $1, t, t^2, \ldots, t^n, \ldots$, and any finite set of these is linearly independent. Nevertheless we may ask if the linear operator d/dt has eigenvalues and eigenvectors, that is, if there exist numbers λ, and associated functions y, in $\mathscr{C}^\infty$ such that

$$\frac{dy}{dt} = \lambda y.$$

For any number λ we can solve this differential equation and find that $y = ce^{\lambda t}$. Hence every number λ is an eigenvalue and has associated with it the eigenvector $e^{\lambda t}$. This is true even if λ is a complex number, $\lambda = \mu + i\nu$. For we have found that the exponential is defined for complex exponents (§4.4). In fact, this was the Euler relation,

$$e^{\lambda t} = e^{(\mu + i\nu)t} = e^{\mu t}e^{i\nu t} = e^{\mu t}(\cos \nu t + i \sin \nu t).$$

We can prove that these functions $ce^{\lambda t}$ are the only eigenvectors associated with λ. For let w be another function in $\mathscr{C}^{\infty}$ for which

$$\frac{d}{dx}\,w = \lambda w.$$

Then

$$\frac{d}{dt}\,\frac{w}{e^{\lambda t}} = \frac{e^{\lambda t}(dw/dt) - w(d/dt)e^{\lambda t}}{(e^{\lambda t})^2} = \frac{e^{\lambda t}(\lambda w) - w(\lambda e^{\lambda t})}{e^{2\lambda t}} = 0.$$

Hence $w/e^{\lambda t} = c$ (constant). That is, $w = ce^{\lambda t}$, and proves the statement.

22.2   Linear Differential Equations

Example.   Solve the differential equation

$$\frac{d^2 y}{dt^2} + 2\,\frac{dy}{dt} - y = 0,$$

(a) for all solutions; (b) for the particular solution having $y(0) = 1$ and $y'(0) = 0$.

Solution.   (a) We regard this as a question that asks, What functions $y(t)$ does the linear operator

$$\frac{d^2}{dt^2} + 2\,\frac{d}{dt} - I$$

map into zero? We proceed at once to see what this linear operator does to the eigenvectors of the differentiation operator d/dt. In particular, does this operator send any eigenvector $e^{\lambda t}$ to $\mathbf{0}$? So we substitute $y = e^{\lambda t}$ where λ is any number in the spectrum of d/dt, and that is any number. We ask if we can find λ so that

$$\frac{d^2}{dt^2}\,e^{\lambda t} + 2\,\frac{d}{dt}\,e^{\lambda t} - e^{\lambda t} = 0.$$

Carrying out the differentiation, we find that this says

$$e^{\lambda t}(\lambda^2 + 2\lambda - 1) = 0.$$

Since $e^{\lambda t}$ is nowhere equal to zero, this is true if and only if

$$\lambda^2 + 2\lambda - 1 = 0.$$

We solve this quadratic equation for λ and find the roots

$$\lambda_1 = -1 + \sqrt{2} \qquad \text{and} \qquad \lambda_2 = -1 - \sqrt{2}.$$

Hence the eigenvectors

$$c_1 e^{(-1+\sqrt{2})t} \qquad \text{and} \qquad c_2 e^{(-1-\sqrt{2})t}$$

are the only ones that are solutions of the differential equation. However, since the operator $(d^2/dt^2) + 2(d/dt) - I$ is linear, any linear combination

$$y(t) = c_1 e^{(-1+\sqrt{2})t} + c_2 e^{(-1-\sqrt{2})t}$$

satisfies the differential equation.

(b) To find a particular solution for which $y(0) = 1$, $y'(0) = 0$, we differentiate the solution $y(t)$ to find

$$y'(t) = (-1 + \sqrt{2})c_1 e^{(-1+\sqrt{2})t} + (-1 - \sqrt{2})c_2 e^{(-1-\sqrt{2})t}$$

and set $t = 0$. Since $y(0) = 1$, we have

$$1 = c_1 e^0 + c_2 e^0, \qquad \text{or} \qquad c_1 + c_2 = 1.$$

Since $y'(0) = 0$, we have

$$(-1 + \sqrt{2})c_1 + (-1 - \sqrt{2})c_2 = 0.$$

We now solve simultaneously this system of two linear equations in the unknowns c_1 and c_2. The reduced echelon form is

$$c_1 + 0 = \frac{2 + \sqrt{2}}{4},$$

$$0 + c_2 = \frac{2 - \sqrt{2}}{4}.$$

Hence a particular solution of the differential equation that satisfies the given initial condition is

$$y(t) = \frac{2 + \sqrt{2}}{4} e^{(-1+\sqrt{2})t} + \frac{2 - \sqrt{2}}{4} e^{(-1-\sqrt{2})t}.$$

22.3   Imaginary Eigenvalues.   It turns out that the imaginary eigenvalues of the differentiation operator are quite important in problems of periodic oscillations. Since $e^{i\theta} = \cos\theta + i\sin\theta$, where $i = \sqrt{-1}$, we may treat the problem of simple harmonic motion by means of eigenvalues and eigenvectors.

Example A.   A mass m is driven by a spring (Figure 16.5) that exerts a force $-kx$, $k > 0$, when the mass is displaced x units from the equilibrium point. Find the motion (I, §32.3).

Solution.   The differential equation of motion is

$$m\frac{d^2x}{dt^2} = -kx.$$

We substitute $y = ce^{\lambda t}$, an eigenvector of the operator d/dt. We find that it satisfies the differential equation if and only if

$$m\lambda^2 = -k \qquad \text{or} \qquad m = \pm i\sqrt{\frac{k}{m}}.$$

If we let $\omega = \sqrt{k/m}$, this says that solutions of the differential equation are given by

$$x = c_1 e^{i\omega t} + c_2 e^{-i\omega t}.$$

Using Euler's relations (§4.4), we can write this

$$x = a\cos\omega t + b\sin\omega t.$$

It is a sinusoidal motion with period $2\pi/\omega$ seconds and amplitude $\sqrt{a^2 + b^2}$.

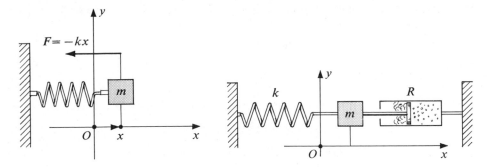

FIGURE 16.5　Undamped motion. See
Part I, Section 32.3.

FIGURE 16.6　Oscillator with resistance.

Example B.　A mass m is driven by a spring that exerts a force $-kx$ when it is displaced x units from equilibrium. There is a resistor that opposes the motion with a force $-R(dx/dt)$ proportional to the velocity, where $k > 0$, $R > 0$. Find the motion. If the mass is displaced to $x = 5$ and released with zero initial velocity, find the motion thereafter. Assume that $m = 1$, $R = 2$, $k = 5$. *Comment.* This is a spring and dashpot arrangement, like a door closer, or shock absorber (Figure 16.6).

Solution.　The forces acting on the mass are the inertial force $m(d^2 x/dt^2)$, the force of the spring $-kx$, and the force of the resistance $-R(dx/dt)$. Hence the differential equation of motion is

$$m \frac{d^2x}{dt^2} = -R \frac{dx}{dt} - kx,$$

or, with the given values of m, R, k,

$$\frac{d^2x}{dt^2} + 2 \frac{dx}{dt} + 5x = 0.$$

We find that the eigenvectors, $x = ce^{\lambda t}$, are solutions if and only if

$$\lambda^2 + 2\lambda + 5 = 0.$$

The roots of this equation are $\lambda = -1 \pm 2i$. Hence, for every value of c_1 and c_2, the linear combination

$$x = c_1 e^{(-1+2i)t} + c_2 e^{(-1-2i)t}$$

is a solution of the differential equation of motion. By use of the Euler relations, this can be written

$$x(t) = e^{-t}[(c_1 + c_2) \cos 2t + i(c_1 - c_2) \sin 2t].$$

On differentiating this, we find

$$x'(t) = e^{-t}\{[(c_1 + c_2) + 2i(c_1 - c_2)] \cos 2t + [-2(c_1 + c_2) - i(c_1 - c_2)] \sin 2t\}.$$

Applying the initial conditions $x(0) = 5$, $x'(0) = 0$, we find that $c_1 + c_2 = 5$ and $i(c_1 - c_2) = \frac{5}{2}$. Hence a motion satisfying the initial conditions is given by

$$x = e^{-t}[5 \cos 2t + \tfrac{5}{2} \sin 2t].$$

This completes the problem.

We observe that in spite of the imaginary number i that appeared, the solution is real, as physical intuition suggests it should be.

22.4   Damped Oscillation

Example.   Describe the motion given by the solution of Example B (§22.3),

$$x = e^{-t}[5 \cos 2t + \tfrac{5}{2} \sin 2t]$$

and give physical interpretations to the factors. Sketch the motion.

Solution.   The factor in brackets can be written

$$5 \cos 2t + \frac{5}{2} \sin 2t = \frac{5}{2}[2 \cos 2t + \sin 2t] = \frac{5\sqrt{5}}{2}\left[\frac{2}{\sqrt{5}} \cos 2t + \frac{1}{\sqrt{5}} \sin 2t\right]$$

$$= \frac{5\sqrt{5}}{2} \sin (2t + \phi),$$

where ϕ is a "phase" angle such that $\sin \phi = 2/\sqrt{5}$ and $\cos \phi = 1/\sqrt{5}$. Hence in the solution

$$x = \frac{5\sqrt{5}}{2} e^{-t} \sin (2t + \phi),$$

the factor $\sin (2t + \phi)$ is a periodic oscillation with period π seconds. It is multiplied by an amplitude factor $(5\sqrt{5}/2)e^{-t}$, in which $\lim_{t \to \infty} e^{-t} = 0$. Hence the amplitude of the oscillation decreases and approaches zero. We plot the graph (Figure 16.7) first by plotting $y = \sin 2t$ as a sine wave of period π and amplitude 1. Then we shift this curve to the left $-\phi$ units to graph $y = \sin (2t + \phi)$. Finally we plot the damping factor

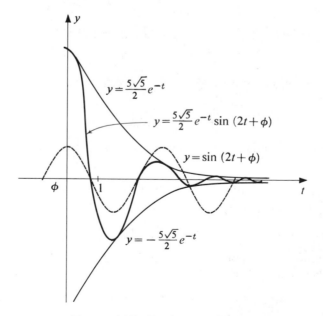

FIGURE 16.7   Damped oscillation.

$y = (5\sqrt{5}/2)e^{-t}$ and its negative mirror image. Using these ordinates as a vertical scale factor multiplying $y = \sin(2t + \phi)$, we plot the graph of $y = (5\sqrt{5}/2)e^{-t}\sin(2t + \phi)$. It oscillates, because it crosses the t axis whenever $\sin(2t + \phi)$ does. It touches the curve $y = (5\sqrt{5}/2)e^{-t}$ wherever $\sin(2t + \phi) = 1$. But, since $|\sin(2t + \phi)| \le 1$, the graph is always between the graph of the damping factor and its negative mirror image. Hence the oscillation rapidly fades out. This completes the example.

We return to the general equation of the damped oscillator, written in the form

$$mx'' + kx = -Rx'.$$

We multiply both sides by $x'\, dt$,

$$mx'x''\, dt + kxx'\, dt = -Rx'^2\, dt.$$

We may rewrite the equation in the form

$$D\left[m\frac{x'^2}{2} + k\frac{x^2}{2}\right] = -Rx'^2\, dt.$$

Hence, integration yields

$$m\frac{x'^2}{2} + k\frac{x^2}{2} = -\int_{t_0}^{t} Rx'^2\, dt + E_0.$$

This is called the energy integral. For the left-hand member is the energy of the system, that is, the sum of the kinetic energy, $\frac{1}{2}mx'^2$, and the potential energy of the stretched spring, $\frac{1}{2}kx^2$. The integral shows that the system does not obey the law of conservation of energy. When $t = t_0$, this total energy is E_0. But since $R > 0$, the integral

$$\int_{t_0}^{t} Rx'^2\, dt$$

increases with t, and it is subtracted from the total energy. This integral represents the dissipation of energy due to the resistance R (Ohm's law). As the energy loss continues, the total energy approaches zero and hence both x and x' approach zero. This is a physical view of the process that produces the exponential damping, which appears in the factor $e^{-\mu t}$, where $\mu = R/2m$, in the mathematical solution of the differential equation.

22.5   Multiple Characteristic Roots.

A special calculus formula, called the exponential shift, enables us to find a basis for the solutions of a linear differential equation, even when the characteristic roots are not distinct. We consider the differential operator $(d/dt) - \lambda I$, which operates on a function $y(t)$ to produce $(dy/dt) - \lambda y$. Repeating the application of the operator, we find

$$\left(\frac{d}{dt} - \lambda I\right)^2 y = \frac{d^2y}{dt^2} - 2\lambda\frac{dy}{dt} + \lambda^2 y.$$

Now we calculate the effect of this operator on $e^{\lambda t}z$. Since

$$\frac{d}{dt}(e^{\lambda t}z) = e^{\lambda t}\left(\frac{dz}{dt} + \lambda z\right),$$

we find that

$$\left(\frac{d}{dt} - \lambda I\right)e^{\lambda t}z = e^{\lambda t}\frac{dz}{dt}.$$

Repeating the application of the operator on $e^{\lambda t}z$ any finite number k of times, we find a formula called the *exponential shift*,

$$\left(\frac{d}{dt} - \lambda I\right)^k e^{\lambda t}z = e^{\lambda t}\frac{d^k z}{dt^k}.$$

It follows from the exponential shift, with $y = e^{\lambda t}z$, that the differential equation $(d/dt - \lambda I)^k y = 0$ is satisfied if and only if $d^k/dt^k(e^{-\lambda t}y) = 0$. By integrating this k times, we find

$$e^{-\lambda t}y = c_1 + c_2 t + c_2 t^2 + \cdots + c_k t^{k-1},$$

where $c_1, \ldots, c_k$ are arbitrary constants of integration. Hence the complete solution of the differential equation is

$$y = c_1 e^{\lambda t} + c_2 t e^{\lambda t} + \cdots + c_k t^{k-1} e^{\lambda t}.$$

Thus, although we have only one eigenvector $e^{\lambda t}$, we can construct from it k linearly independent vectors $e^{\lambda t}, te^{\lambda t}, \ldots, t^{k-1}e^{\lambda t}$, which form a basis for the linear space of all solutions.

Example. Find all solutions of the differential equation

$$4\frac{d^2 y}{dt^2} - 20\frac{dy}{dt} + 25y = 0.$$

Solution. We substitute $y = e^{\lambda t}$, the eigenvectors of d/dt and find that they satisfy the differential equation only if $4\lambda^2 - 20\lambda + 25 = 0$, or $(2\lambda - 5)^2 = 0$. This gives only one linearly independent eigenvector, $e^{(5/2)t}$, which satisfies the differential equation.

$$4\frac{d^2 y}{dt^2} - 20\frac{dy}{dt} + 25y = 4\left(\frac{d}{dt} - \frac{5}{2}I\right)^2 y = 0.$$

But the exponential shift tells us that a basis for all solutions consists of the linearly independent vectors $e^{\lambda t}$ and $te^{\lambda t}$. Hence all solutions are given by

$$y = e^{(5/2)t}(c_1 + c_2 t).$$

22.6 Exercises

In Exercises 1–6, solve the differential equations by use of eigenvalues and eigenvectors of the differentiation operator, d/dt.

1. $\dfrac{dy}{dt} = 2y.$

2. $\dfrac{d^2 y}{dt^2} + 4y = 0.$

3. $\dfrac{d^2 x}{dt^2} + \dfrac{dx}{dt} - 6x = 0.$

4. $\dfrac{d^2 x}{dt^2} - 2\dfrac{dx}{dt} = 0.$

5. Find the solution of Exercise 3, with $x(0) = 1$, $x'(0) = 0$.

6. $\dfrac{d^2 y}{dt^2} + 2\dfrac{dy}{dt} + 10y = 0$, with $y(0) = 5$ and $y'(0) = 0$.

7. Express the solution of Exercise 2 as a sum of cosine and sine terms. Determine the characteristic frequency.

8. Express the solution of Exercise 6 as a sum of cosine and sine terms multiplied by a damping factor. Find the characteristic frequency of the oscillation.

9. Show that the solution of Exercise 3 is not a sum of sine and cosine terms with exponential damping.

10. For the general equation of the damped oscillator,

$$m\frac{d^2x}{dt^2} + R\frac{dx}{dt} + kx = 0,$$

show that the solution is a damped oscillation if and only if $R^2 < 4mk$.

11. In Exercise 10, assuming that $R^2 < 4mk$, find the characteristic frequency of the oscillation. How does this frequency change as m increases? as k increases? Give a physical interpretation to these observations.

12. Plot the graph of the solution of

$$\frac{d^2x}{dt^2} + 3\frac{dx}{dt} + 2x = 0, \qquad x(0) = 2, x'(0) = 0.$$

   (a) Show that $x = 0$ at exactly one time. Find that time.
   (b) Show that $\lim_{t \to \infty} x(t) = 0$.

13. In the general equation of the damped oscillator, Exercise 10, show that $x(t)$ changes sign at most once if $R^2 \geqq 4mk$, and that in every case, $\lim_{t \to \infty} x(t) = 0$.

14. For the general damped oscillator for which $m > 0, R > 0, k > 0$, use the energy integral to prove that if $x(t_0) = x'(t_0) = 0$, then for every t, $x(t) = x'(t) = 0$.

15. From the result of Exercise 14, prove that there exists only one solution of the damped oscillator that satisfies the initial conditions $x(t_0) = x_0, x'(t_0) = v_0$.

16. Find all solutions of the differential equation

$$\frac{d^2y}{dt^2} + 4\frac{dy}{dt} + 4y = 0.$$

17. Find all solutions of the differential equation

$$\frac{d^3y}{dt^3} + 3\frac{d^2y}{dt} + 3\frac{dy}{dt} + y = 0.$$

18. Find all solutions of the damped oscillator equation

$$m\frac{d^2x}{dt^2} + R\frac{dx}{dt} + kx = 0$$

under the condition $R^2 = 4mk$. Prove that for all solutions, $\lim_{t \to \infty} x(t) = 0$ if $R > 0$.

23 Linear Differential Equations in Normal Form

23.1 Reduction to Normal Form in the Plane. A differential equation for a vector function $(x(t), y(t))$ in the plane is in normal form if the derivatives dx/dt and dy/dt are expressed as functions of the position (x, y). Thus the system of equations

$$\frac{dx}{dt} = f(x, y),$$

$$\frac{dy}{dt} = g(x, y),$$

is in normal form.

By introducing new variables, we can reduce to normal form a linear differential equation of the second order such as (§22.3)

$$\frac{d^2y}{dt^2} + 2\frac{dy}{dt} + 5y = 0.$$

We define a second function, $v(t)$, by $v = dx/dt$. Then $dv/dt = d^2y/dt^2$. The differential equation can now be rewritten as a system of two differential equations in normal form,

$$\frac{dy}{dt} = v,$$

$$\frac{dv}{dt} = -5y - 2v,$$

each of the first order. The terms on the right form a linear transformation with matrix $A = \begin{bmatrix} 0 & 1 \\ -5 & -2 \end{bmatrix}$. Let $Y = \begin{bmatrix} y \\ v \end{bmatrix}$, then the normal form of the differential equation says that

$$\frac{dY}{dt} = AY.$$

We may use this compact matrix formulation to represent in simple and compact notation a homogeneous linear system of n differential equations in n unknown functions $y_1(t), y_2(t), \ldots, y_n(t)$, whose initial values at the point where $t = 0$ are prescribed. The linear system in normal form

$$\frac{dy_1}{dt} = a_{11}y_1 + a_{12}y_2 + \cdots + a_{1n}y_n,$$

$$\frac{dy_2}{dt} = a_{21}y_1 + a_{22}y_2 + \cdots + a_{2n}y_n,$$

$$\cdots\cdots\cdots\cdots\cdots\cdots\cdots\cdots\cdots\cdots\cdots$$

$$\frac{dy_n}{dt} = a_{n1}y_1 + a_{n2}y_2 + \cdots + a_{nn}y_n,$$

has the initial values

$$y_1(0) = c_1,$$
$$y_2(0) = c_2,$$
$$\cdots\cdots\cdots\cdots$$
$$y_n(0) = c_n.$$

We let the Y and C stand for column vectors

$$Y = \begin{bmatrix} y_1 \\ y_2 \\ \vdots \\ y_n \end{bmatrix}, \qquad C = \begin{bmatrix} c_1 \\ c_2 \\ \vdots \\ c_n \end{bmatrix}$$

and let A stand for the matrix of constant coefficients

$$A = \begin{bmatrix} a_{11} & a_{12} & \cdots & a_{1n} \\ a_{21} & a_{22} & \cdots & a_{2n} \\ \vdots & & & \vdots \\ a_{n1} & a_{n2} & \cdots & a_{nn} \end{bmatrix}.$$

Then the system can be written in the brief notation,

$$\frac{dY}{dt} = AY \quad \text{with initial conditions} \quad Y(0) = C.$$

Furthermore, matrix operations permit us to perform simply the steps required to find solution functions $Y(t)$ without going back to the detailed notations.

23.2 Eigenvector Methods. To make it always possible to carry through eigenvector methods, we let scalars be the field of all complex numbers. Then eigenvalues of the differentiation operator d/dt consist of all complex numbers, real or imaginary. The associated eigenvectors are of the form $e^{\lambda t}C$, where C is a constant column vector.

The eigenvalues of the matrix A are numbers $\lambda_1, \lambda_2, \ldots, \lambda_n$ that satisfy the characteristic equation

$$\det [A - \lambda I] = 0.$$

The associated eigenvectors are nonzero column vectors Y_i, $i = 1, \ldots, n$, which satisfy the homogeneous linear equations $[A - \lambda_i I]Y_i = O$. If for the same λ and Y we can make it true that both

$$\frac{dY}{dt} = \lambda Y \quad \text{and} \quad AY = \lambda Y,$$

then we will have a solution of the differential system. So we try solutions of the form $Y = e^{\lambda t}C$ in the equation $AY = \lambda Y$.

Example.   Find solutions of the linear system (§23.1) $dY/dt = AY$, where

$$Y = \begin{bmatrix} y \\ v \end{bmatrix}, \qquad A = \begin{bmatrix} 0 & 1 \\ -5 & -2 \end{bmatrix}.$$

Solution.   The eigenvalues of A satisfy the equation

$$\det [A - I] = \det \begin{bmatrix} 0 - \lambda & 1 \\ -5 & -2 - \lambda \end{bmatrix} = 0, \quad \text{or} \quad \lambda^2 + 2\lambda + 5 = 0.$$

The eigenvalues of A are then $\lambda = -1 \pm 2i$. With these eigenvalues the eigenvectors of the differentiation operator are of the form

$$Y_1 = e^{(-1 + 2i)t}C_1, \quad \text{or} \quad Y_2 = e^{(-1 - 2i)t}C_2,$$

where C_1 and C_2 are arbitrary constant column vectors. Any linear combination $Y = aY_1 + bY_2$ is also a solution.

When $t = 0$, since $e^0 = 1$, this reduces to

$$Y_1(0) = C_1 \quad \text{and} \quad Y_2(0) = C_2.$$

The standard way to pick the initial values C_1 and C_2 so as to guarantee that the solutions will be linearly independent is to choose

$$C_1 = \begin{bmatrix} 1 \\ 0 \end{bmatrix} \quad \text{and} \quad C_2 = \begin{bmatrix} 0 \\ 1 \end{bmatrix}.$$

With this choice we will be able to show (§23.4) that *all* solutions are linear combinations $aY_1 + bY_2$.

23.3   Iteration Methods.   For large systems it is difficult to compute eigenvalues and eigenvectors. The theory is correspondingly difficult. Alternatively we can compute solutions by a direct iterative procedure, similar to Newton's method (I, §27.5).

To solve the system of linear differential equations in normal form $dY/dt = AY$, with prescribed initial conditions $Y(0) = C$, we first integrate both sides from 0 to t. We find that a continuous solution, $Y(t)$, satisfies the integral equation

$$Y(t) = C + \int_0^t A Y(s)\, ds$$

for every value of t. Here $\int_0^t A Y(s)\, ds$ means multiply the column vector $Y(s)$ on the left by the matrix A to produce a column vector $A Y(s)$, then integrate each component of $A Y(s)$ from 0 to t.

We now proceed in repeated steps to find an infinite sequence of approximate solutions $Y_0(t)$, $Y_1(t)$, ..., $Y_n(t)$, ..., which we can prove converges to an exact solution $Y(t)$.

Let $Y_0 = C$ and substitute this into the integral of the integral equation to produce $Y_1(t)$,

$$Y_1(t) = C + \int_0^t A Y_0\, ds.$$

Then substitute $Y_1(s)$ into the integral equation to produce $Y_2(t)$,

$$Y_2(t) = C + \int_0^t A Y_1(s)\, ds,$$

$$\cdots\cdots\cdots\cdots\cdots\cdots\cdots\cdots$$

$$Y_k(t) = C + \int_0^t A Y_{k-1}(s)\, ds, \qquad k = 1, 2, 3, \ldots.$$

Since $A Y_0$ is a constant vector, the first integration gives

$$Y_1(t) = C + (tA)C.$$

The integration of $A Y_1(s)$ gives, for each value of t,

$$Y_2(t) = C + (tA)C + \frac{(t^2 A^2)}{2!} C.$$

Repeating this procedure k times, we find

$$Y_k(t) = C + (tA)C + \frac{t^2 A^2}{2!} C + \cdots + \frac{t^k A^k}{k!} C$$

$$= \left[I + (tA) + \frac{(tA)^2}{2!} + \cdots + \frac{(tA)^k}{k!} \right] C.$$

Before we investigate the convergence of this sequence, let us see what it says in the special case that $n = 1$. Then the column vectors Y and C are just numbers, y and c. Also the 1×1 matrix A reduces to a single number a. Thus when $n = 1$, the linear system reduces to

$$\frac{dy}{dt} = ay, \qquad \text{with initial conditions} \quad y(0) = c.$$

We already know the exact solution of this system to be $y = (e^{ta})c$. The iteration method gives the kth approximation to the solution as

$$y_k = \left[1 + ta + \frac{(ta)^2}{2!} + \cdots + \frac{(ta)^k}{k!} \right] c.$$

We recognize the terms in brackets as the kth Taylor's polynomial in the Taylor's expansion of e^{ta} in powers of ta (I, §29.5). Hence as $k \to \infty$, the kth approximation approaches the exact solution $e^{ta}c$.

Returning to the general case, we are led to expect that in the kth approximation Y_k we have a factor that approaches an exact limit. We write it as an infinite series

$$e^{tA} = I + \frac{tA}{1!} + \frac{(tA)^r}{2!} + \cdots + \frac{(ta)^k}{k!} + \cdots.$$

We shall use the case $n = 1$ to establish the convergence of this series.

23.4   Convergence and Uniqueness of the Solution.   We define norm of a matrix $A = [a_{ij}]$ as the sum of the absolute values of all of its elements

$$\begin{aligned}
\|A\| = {} & |a_{11}| + |a_{12}| + \cdots + |a_{1n}| \\
& + |a_{21}| + |a_{22}| + \cdots + |a_{2n}| \\
& + \ \ldots\ldots\ldots\ldots\ldots\ldots\ldots \\
& + |a_{n1}| + |a_{n2}| + \cdots + |a_{nn}|.
\end{aligned}$$

This definition applies equally well to give norms of column vectors Y and C, since they are single-column matrices. We readily verify that $\|A\| \geqq 0$ and $\|A\| = 0$ if and only if $A = O$. Also

$$\|cA\| = |c|\,\|A\|, \qquad \|A + B\| \leqq \|A\| + \|B\|, \qquad \|AB\| \leqq \|A\|\,\|B\|.$$

In terms of this norm, the following theorem describes the convergence of the sequence of approximate solutions, evaluated at t,

$$Y_0(t) = C, \qquad Y_1(t) = C + \int_0^t A Y_0\, ds, \ldots, \qquad y_k(t) = C + \int_0^t A Y_{k-1}(s)\, ds, \ldots.$$

THEOREM.   There exists an exact solution $Y(t)$ of the linear system

$$Y(t) = C + \int_0^t A Y(s)\, ds.$$

The solution $Y(t)$ is the limit of the sequence of approximating solutions $\{Y_k(t)\}$ in the sense that for every ϵ and every closed interval $[0, t_1]$, there is a number N such that $k > N$ implies that $\|Y_k(t) - Y(t)\| < \epsilon$, for every t in $[0, t_1]$.

Proof Sketch.   The terms of the series

$$I + tA + \frac{(tA)^2}{2!} + \cdots + \frac{(tA)^k}{k!} + \cdots$$

are less than those of the numerical series

$$I + |t|\,\|A\| \frac{|t|^2\|A\|^2}{2!} + \cdots + \frac{|t|^k}{k!}\,\|A\|^k + \cdots,$$

which is geometrically dominated (II, §38.3) on the interval $[0, t_1]$. In fact, the series of norms converges to the function whose value at t is $e^{|t|\,\|A\|}$. Hence, by an argument similar to that for absolutely convergent series (II, §37.5), the series of matrix terms

$$I + \frac{(tA)}{1!} + \frac{(tA)^2}{2!} + \cdots$$

converges to an $n \times n$ limit matrix, which we denote by e^{tA}. Then the vector function $Y(t) = e^{tA}C$ is an exact solution of the integral equation

$$Y(t) = C + \int_0^t A Y(s) \, ds,$$

and hence of the differential system with prescribed initial values $Y(0) = C$.

THEOREM (Uniqueness of solution). The differential system $dY/dt = AY$, with initial condition $Y(0) = C$ has only one solution.

Proof. Suppose $Y(t)$ and $X(t)$ are two continuous solutions satisfying the initial conditions. Then we subtract the integral equations to find that $Y - X$ satisfies

$$Y(t) - X(t) = \int_0^t A[Y(s) - X(s)] \, ds.$$

Then

$$\| Y(t) - X(t) \| \leq \int_0^t \|A\| \, \| Y(s) - X(s) \| \, ds.$$

Since $Y(s)$ and $X(s)$ are continuous on the closed interval $[0, t_1]$, the norm $\| Y(s) - X(s) \|$ is also. Hence (I, §8.4) there is a fixed number $G = \sup_{[0, t_1]} \| Y(s) - X(s) \|$. Then

$$\| Y(s) - X(s) \| \leq \int_0^t \|A\| G \, ds = t \|A\|.$$

Repeating, with this new estimate of $\| Y(s) - X(s) \|$ in the integral, we find

$$\| Y(s) - X(s) \| \leq \int_0^t \|A\| [s \|A\|] \, ds = \frac{|t| \, \|A\|^2}{2}.$$

After k repetitions of this procedure, we find

$$\| Y(s) - X(s) \| \leq \frac{|t|^k \|A\|^k}{k!}.$$

The numbers on the right are terms of the convergent series $e^{|t| \, \|A\|}$ and hence they approach zero. This may also be verified directly from the definition of the limit. Hence $\| Y(s) - X(s) \| = 0$ and therefore $Y(s) = X(s)$ on $[0, t_1]$.

COROLLARY. The n solutions uniquely determined by initial vectors

$$E_1 = \begin{bmatrix} 1 \\ 0 \\ 0 \\ \vdots \\ 0 \end{bmatrix}, \qquad E_2 = \begin{bmatrix} 0 \\ 1 \\ 0 \\ \vdots \\ 0 \end{bmatrix}, \ldots, \qquad E_n = \begin{bmatrix} 0 \\ 0 \\ 0 \\ \vdots \\ 1 \end{bmatrix}$$

form a basis for the linear space of all continuous solutions of the system $dY/dt = AY$ on an interval $[0, t_1]$.

For the vectors $\{E_1, E_2, \ldots, E_n\}$ form a basis for the space of all initial values. Since every solution has one of these vectors for its initial value, linear combinations of these span the space of all solutions. Moreover they are linearly independent, since their initial vectors are linearly independent.

23.5

Example. Solve the system

$$\frac{dx}{dt} = 2x - 3y,$$
$$\frac{dy}{dt} = x + 2y,$$

with initial conditions $\begin{bmatrix} x \\ y \end{bmatrix} = \begin{bmatrix} 5 \\ 2 \end{bmatrix}.$

Solution.   Here

$$A = \begin{bmatrix} 2 & -3 \\ 1 & 2 \end{bmatrix}, \quad Y = \begin{bmatrix} x \\ y \end{bmatrix}, \quad C = \begin{bmatrix} 5 \\ 2 \end{bmatrix}.$$

We have

$$Y_0 = \begin{bmatrix} 5 \\ 2 \end{bmatrix}.$$

$$Y_1 = \begin{bmatrix} 5 \\ 2 \end{bmatrix} + \int_0^t \begin{bmatrix} 2 & -3 \\ 1 & 2 \end{bmatrix}\begin{bmatrix} 5 \\ 2 \end{bmatrix} ds = \begin{bmatrix} 5 \\ 2 \end{bmatrix} + \begin{bmatrix} 4t \\ 9t \end{bmatrix}.$$

$$Y_2 = \begin{bmatrix} 5 \\ 2 \end{bmatrix} + \int_0^t \begin{bmatrix} 2 & -3 \\ 1 & 2 \end{bmatrix}\left(\begin{bmatrix} 5 \\ 2 \end{bmatrix} + \begin{bmatrix} 4s \\ 9s \end{bmatrix}\right) ds$$

$$= \begin{bmatrix} 5 \\ 2 \end{bmatrix} + \begin{bmatrix} 4 \\ 9 \end{bmatrix}t + \begin{bmatrix} -19 \\ 22 \end{bmatrix}\frac{t^2}{2!}.$$

Continuing, we find that these terms converge to a unique solution,

$$Y(t) = \begin{bmatrix} 5 \\ 2 \end{bmatrix} + \begin{bmatrix} 4 \\ 9 \end{bmatrix}t + \begin{bmatrix} -19 \\ 22 \end{bmatrix}\frac{t^2}{2!} + \cdots.$$

An exact formula for the solution is

$$Y(t) = \exp t \begin{bmatrix} 2 & -3 \\ 1 & 2 \end{bmatrix}\begin{bmatrix} 5 \\ 2 \end{bmatrix}.$$

23.6   Linear Differential Equations of Higher Order. We can apply the theory of normal systems to linear differential equations of the form

(1) $$a\frac{d^2y}{dt^2} + b\frac{dy}{dt} + cy = 0, \qquad a \neq 0.$$

It is equivalent to the normal system

$$\frac{dy}{dt} = z,$$

$$\frac{dz}{dt} = \frac{c}{a}y + \frac{b}{a}z.$$

The system in normal form has a basis of two solutions determined by the initial conditions $\begin{bmatrix} 1 \\ 0 \end{bmatrix}, \begin{bmatrix} 0 \\ 1 \end{bmatrix}.$ Hence solutions of Equation (1) exist that are uniquely determined by the initial conditions $y(0) = 1, y'(0) = 0$, and $y(0) = 0, y'(0) = 1$. These two solutions form a basis for the linear space of all solutions.

Moreover, if the differential equation (1) has two eigenvector solutions with $\lambda_1 \neq \lambda_2$, $e^{\lambda_1 t}$, and $e^{\lambda_2 t}$, these form a linear basis for all solutions; they have the initial vectors

$$\begin{bmatrix} 1 \\ \lambda_1 \end{bmatrix} \quad \text{and} \quad \begin{bmatrix} 1 \\ \lambda_2 \end{bmatrix},$$

which are linearly independent. Also if $\lambda_1 = \lambda_2$, we have seen (§22.5) that $e^{\lambda_1 t}$ and $te^{\lambda_1 t}$ are solutions. These have the initial values

$$\begin{bmatrix} 1 \\ \lambda_1 \end{bmatrix} \quad \text{and} \quad \begin{bmatrix} 0 \\ 1 \end{bmatrix},$$

which are linearly independent and span the space of all initial values. Hence the solutions $e^{\lambda_1 t}$ and $te^{\lambda_1 t}$ form a basis for all solutions.

Our theory of normal systems implies that unique solutions of the nth order linear differential equation

$$a_0 \frac{d^n y}{dt^n} + a_1 \frac{d^{n-1} y}{dt^{n-1}} + \cdots + a_n y = 0, \qquad a_0 \neq 0,$$

exist, uniquely determined by initial vectors of the form $y(0), y'(0), y''(0), \ldots, y^{(n-1)}(0)$. And we can find a basis for the n-dimensional space of all solutions.

23.7 Exercises

1. Compute by iteration the solution of the differential equation $dy/dt = 2y$, with initial condition $y(0) = 5$.
2. Compute by iteration the solution of the differential equation $dy/dt = iy$, with initial condition $y(0) = 1$.
3. Compute by iteration the solution of the system

$$\frac{dy_1}{dt} = 2y_1 + y_2$$

$$\frac{dy_2}{dt} = y_1.$$

4. Find a second-order differential equation

$$a \frac{d^2 y}{dt^2} + b \frac{dy}{dt} + cy = 0$$

that is equivalent to the system in Exercise 3. Use Exercise 3 to find all solutions of Exercise 4.
5. Use the theory (§23.5) to prove that all solutions of the differential equation

$$\frac{d^2 y}{dt^2} - \frac{dy}{dt} - 6y = 0$$

are of the form $y = ae^{3t} + be^{-2t}$.
6. Find the solution of Exercise 5 that satisfies the initial conditions $y(0) = 1$, $y'(0) = 1$.
7. Treat the harmonic oscillator

$$m \frac{d^2 y}{dt^2} + ky = 0$$

by replacing the differential equation by a system of two equations in normal form. Show that if this is done by introducing the velocity $v = y'(t)$ as a new variable, then all solutions trace ellipses in the yv-plane. This is called the phase plane.

8. A matrix exponential function $Z(t) = \exp(tA)$ satisfies the differential equation

$$\frac{dZ}{dt} = AZ \qquad \text{with initial condition} \quad Z(0) = I.$$

Show that

$$Z(t) = I + tA + \frac{(tA)^2}{2!} + \cdots + \frac{(tA)^k}{k!} + \cdots.$$

In Exercises 9–11, we prove that every matrix exponential has the exponential property $Z(t + s) = Z(s)$.

9. For the matrix exponential function $Z(t)$ defined in Exercise 8, show that for every constant s,

$$\frac{d}{dt} Z(t + s) = AZ(t + s).$$

10. Show that hence for some constant matrix B,

$$Z(t + s) = Z(t)B.$$

11. Evaluate the constant matrix B by setting $t = 0$. Prove that thus

$$Z(t + s) = Z(t)Z(s), \qquad \text{where} \quad Z(t) = \exp tA.$$

24   Oscillators with External Force Applied

24.1   Nonhomogeneous Linear Differential Equations. We recall (§14.2) that we solved the homogeneous linear system $AX = 0$, where A is an $m \times n$ matrix and X is an unknown column vector, with n coordinates. Then we used this solution to solve the nonhomogeneous system $AX = Y$, where Y is a specified vector with m coordinates.

We now make the same step for linear differential equations. We proceed from the homogeneous linear differential equation

$$(1) \qquad \left(a\frac{d^2}{dt^2} + b\frac{d}{dt} + cI \right) x = 0$$

to the nonhomogeneous differential equation

$$(2) \qquad \left(a\frac{d^2}{dt^2} + b\frac{d}{dt} + cI \right) x = y,$$

where y is a specified integrable function. Just as in the case of the linear system $AX = Y$, we see that if $x_1(t)$ and $x_2(t)$ are two solutions of the nonhomogeneous equation (2), then their difference $x_2(t) - x_1(t)$ is a solution of the homogeneous system (1). Hence if we find all solutions of the homogeneous equation (1) and add to them one particular solution of the nonhomogeneous equation (2), we shall have all solutions of the nonhomogeneous equation (2).

We proceed to investigate at least one method for finding a particular solution of the nonhomogeneous equation, assuming that we have already solved the homogeneous equation (1) and have found that all of its solutions are given in the form

$$x = ue^{\lambda_1 t} + ve^{\lambda_2 t}.$$

Here λ_1 and λ_2 are distinct characteristic roots of the equation $a\lambda^2 + b\lambda + c = 0$, and u and v are constants.

24.2   Variation of Parameters. Among many methods for finding a solution of the nonhomogeneous equation, the method of variation of parameters, due to Lagrange (1736–1813), is not the easiest.* But we use it because it is quite general in its applicability and provides definite formulas for the solution.

Let the solution of the homogeneous equation be

$$x = ue^{\lambda_1 t} + ve^{\lambda_2 t}, \quad \text{where} \quad \lambda_1 \neq \lambda_2.$$

Following Lagrange, we now replace the constants (parameters) u and v by functions $u(t)$ and $v(t)$, which we will determine so that the nonhomogeneous equation

$$a\frac{d^2x}{dt^2} + b\frac{dx}{dt} + cx = y$$

is satisfied. We differentiate the solution with parameters $u(t)$ and $v(t)$ and find

$$x' = \lambda_1 ue^{\lambda_1 t} + \lambda_2 ve^{\lambda_2 t} + u'e^{\lambda_1 t} + v'e^{\lambda_2 t}.$$

We set equal to zero the terms in u' and v', that is,

$$u'e^{\lambda_1 t} + v'e^{\lambda_2 t} = 0.$$

Then we differentiate the remaining terms of x' to find

$$x'' = \lambda_1^2 ue^{\lambda_1 t} + \lambda_2^2 ve^{\lambda_2 t} + \lambda_1 u'e^{\lambda_1 t} + \lambda_2 v'e^{\lambda_2 t}.$$

We substitute x, x', x'' into the differential equation. Since the terms in which u and v were treated as constants, that is, not differentiated, satisfy the homogeneous equation, this reduces to

$$\lambda_1 u'e^{\lambda_1 t} + \lambda_2 v'e^{\lambda_2 t} = \frac{y(t)}{a} \quad \text{and} \quad u'e^{\lambda_1 t} + v'e^{\lambda_2 t} = 0,$$

as previously prescribed.

This is an algebraic linear system in the unknowns u' and v', which we can solve by the method of elimination (row reduction). We find

$$u' = \frac{1/a}{\lambda_1 - \lambda_2} e^{-\lambda_1 t}y(t) \quad \text{and} \quad v' = \frac{-1/a}{\lambda_1 - \lambda_2} e^{-\lambda_2 t}y(t).$$

Now, integrating these as derivatives of unknown functions $u(t)$ and $v(t)$, we find

$$u = \frac{1/a}{\lambda_1 - \lambda_2} \int_{t_0}^{t} e^{-\lambda_1 s}y(s)\, ds + u_0,$$

$$v = \frac{-1/a}{\lambda_1 - \lambda_2} \int_{t_0}^{t} e^{-\lambda_2 s}y(s)\, ds + v_0.$$

Here u_0 and v_0 are arbitrary constants of integration. We substitute these expressions for u and v in the expression $x = ue^{\lambda_1 t} + ve^{\lambda_2 t}$, and find the general solution

$$x = u_0 e^{\lambda_1 t} + v_0 e^{\lambda_2 t} + \frac{1/a}{\lambda_1 - \lambda_2}\left[e^{\lambda_1 t}\int_{t_0}^{t} e^{-\lambda_1 s}y(s)\, ds - e^{-\lambda_2 t}\int_{t_0}^{t} e^{-\lambda_2 s}y(s)\, ds\right],$$

$$= u_0 e^{\lambda_1 t} + v_0 e^{\lambda_2 t} + \frac{1/a}{\lambda_1 - \lambda_2}\int_{t_0}^{t} [e^{\lambda_1 (t-s)} - e^{\lambda_2 (t-s)}]y(s)\, ds.$$

---

* Textbooks on differential equations give the method of undetermined coefficients, the method of reduction of order, an infinited various series and numerical methods.

Here, since u_0 and v_0 are arbitrary constants, the first two terms constitute the general solution of the homogeneous equation, while the last term is an integral that gives one solution of the nonhomogeneous equation. This completes the solution for the case $\lambda_1 \neq \lambda_2$.

We can show (Exercises) that in case $\lambda_1 = \lambda_2 = \lambda$, the method of variation of parameters applied to the solution of the homogeneous equation

$$x = ue^{\lambda t} + vte^{\lambda t}$$

yields the solution of the nonhomogeneous equation

$$x = u_0 e^{\lambda t} + v_0 te^{\lambda t} + \frac{1}{a} \int_{t_0}^{t} e^{\lambda(t-s)} \left[1 - \frac{1}{s} \right] ds.$$

24.3   Application to Forced Oscillator. We consider an oscillator system consisting of a mass that is connected to a spring, whose motion along the x-axis is resisted by a force proportional to the velocity (Figure 16.8). An external force $y(t)$, also along the x-axis, is applied to the mass. This force is a function of time. The problem is to find how the motion $x(t)$ of the mass responds to the external force $y(t)$. We have considered problems with external input of this type for differential equations of first order (II, §28).

Example.   An oscillator has mass 1. At the point x on the x-axis, the force of the spring is $-5x$ and the motion is opposed by a force of resistance, $-2(dx/dt)$. The mass starts at time $t_0 = 0$ from rest at the point where $x = 4$ and moves under an external force $y(t)$. Find the motion.

Solution.   The differential equation of motion is

$$1 \frac{d^2x}{dt^2} + 2 \frac{dx}{dt} + 5x = y(t), \qquad \text{with } x(0) = 4 \text{ and } x'(0) = 0.$$

The associated homogeneous equation, with $y(t)$ replaced by 0, has the general solution $x = ue^{(-1+2i)t} + ve^{(-1-2i)t}$. Applying the method of variation of parameters with $y(t)$ restored, we find the solution of the original equation to be

$$x(t) = e^{-t}[ae^{2it} + be^{-2it}] + \left(\frac{1}{4} i \right) \int_0^t [e^{(-1+2i)(t-s)} - e^{(-1-2i)(t-s)}] \, y(s) ds.$$

The initial conditions $x(0) = 4$ and $x'(0) = 0$ determine that $a = 2 - i$, $b = 2 + i$. Since $e^{i\theta} = \cos\theta + i\sin\theta$, $e^{-i\theta} = \cos\theta - i\sin\theta$, this gives the required solution

$$x(t) = e^{-t}(4\cos 2t + 2\sin 2t) + \left(\frac{1}{2} \right) \int_0^t e^{-(t-s)} \sin 2(t-s) y(s) ds.$$

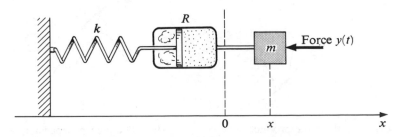

FIGURE 16.8   Forced oscillator.

The first terms represent a transient, that is, a damped oscillation independent of $y(t)$. The integral term gives the response to the external force $y(t)$. It is independent of the initial conditions and sums the product of $y(t)$ with an oscillating factor that damps out inputs of past time.

24.4 Physical Interpretation of the Solution.

We consider the general solution for the case that the characteristic equation (§24.1)

$$a\lambda^2 + b\lambda + c = 0$$

has a conjugate pair of complex roots

$$\lambda_1 = -\mu + iv, \qquad \lambda_2 = -\mu - iv,$$

with negative real part $-\mu$ and imaginary part $\pm iv$. Our example (§24.3) had

$$\lambda_1 = -\mu + iv = -1 + 2i \qquad \text{and} \qquad \lambda_2 = -\mu - iv = -1 - 2i.$$

The solution (§3) takes the form at each time t,

$$x = u_0 e^{(-\mu + iv)t} + v_0 e^{(-\mu - iv)t} + \frac{1}{2iva} \int_0^t [e^{(-\mu + iv)(t-s)} - e^{(-\mu - iv)(t-s)}]y(s)\, ds,$$

where $y(s)$ is the forcing function defined on the interval $0 \leq s \leq t$.

The first two terms can be written

$$e^{-\mu t}[u_0 e^{ivt} + v_0 e^{-ivt}].$$

This is a damped oscillation with frequency v. Its amplitude $e^{-\mu t} \to 0$ as $t \to \infty$. It is called a "transient" term in the solution.

To interpret the integral term, we observe that the domain of $y(s)$ can be extended back to $-\infty$ without changing the problem. We simply define $y(s) = 0$ when $-\infty < s < 0$. Then the integral term can be written (Exercises) as

$$\frac{1}{\mu v a} \overline{Y}, \qquad \text{where} \qquad \overline{Y} = \frac{\int_{-\infty}^t e^{\mu s} \sin v(t-s) y(s)\, ds}{\int_{-\infty}^t e^{\mu s}\, ds}.$$

The number $\overline{Y}$ is the average value of $\sin v(t-s)y(s)$ over the interval $-\infty < s \leq t$, computed with respect to the weighting factor $e^{\mu s}$. The factor $\sin v(t-s)$ changes $y(s)$ into an oscillation with frequency v whose variable amplitude at any instant is $y(s)$ (Figure 16.9). The effect of the weight factor $e^{\mu s}$ is to suppress, by exponential damping,

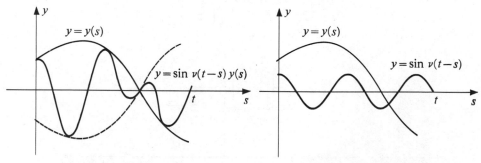

(a) Graph of the product   (b) The factors separately

FIGURE 16.9 Function with oscillating factor.

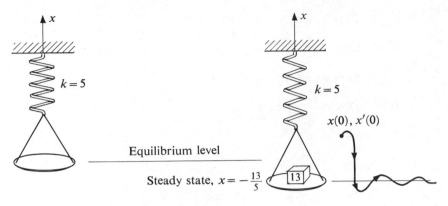

FIGURE 16.10   Weight placed on a spring scale.

the distant past values of $y(s)$, and to emphasize the most recent values of $y(s)$ in computing the average.

We compute one simple example to fine detail. Let $y(s) = 13$, a constant, for the oscillator whose equation is then (§24.3)

(1)
$$\frac{d^2x}{dt^2} + 2\frac{dx}{dt} + 5x = 13.$$

Then the general solution, satisfying the prescribed initial conditions $x(0) = 4$ and $x'(0) = 0$, is

$$x = e^{-t}(2\sin 2t + 4\cos 2t) + \tfrac{1}{2}\int_0^t e^{-(t-s)}\sin(t-s)\,13\,ds$$
$$= e^{-t}(2\sin 2t + 4\cos 2t) + \tfrac{13}{2}[-\tfrac{1}{5}e^{-(t-s)}(\sin 2(t-s) - 2\cos 2(t-s))]_0^t$$
$$= e^{-t}(2\sin 2t + 4\cos 2t) - \tfrac{13}{5} + \tfrac{13}{10}e^{-t}(\sin 2t - 2\cos 2t).$$

The first and last terms are damped oscillations with frequency 2. As t increases they rapidly approach zero. Hence $\lim_{t\to\infty} x(t) = -\tfrac{13}{5}$. This is called the "steady-state" term in the solution. This was an oscillator whose spring gave a force $F = -5x$. If a constant force $F = 13$ is applied, we find $x = -\tfrac{13}{5}$. This is the steady-state term in the solution.

We could have predicted this without any differential equations, from the behavior of a spring scale. If the spring force is $F = -5x$, then a weight of 13 units placed on it at time $t = 0$ will depress the pan to the position $x = -\tfrac{13}{5}$ (Figure 16.10). In reaching this new forced level, the pan will oscillate down and up, with a frequency that depends on the stiffness of the spring, the mass of the moving parts, and the amount of frictional resistance. But with resistance present, the oscillation will damp out and the pan will approach a new rest position at the lower level $x = -\tfrac{13}{5}$. The initial position $x(0)$, $x'(0)$ is also damped out in this motion, and has no effect on the final result or the frequency of oscillation.

24.5 Exercises

In Exercises 1–8, find the solution that satisfies the initial conditions.

1. $\dfrac{d^2x}{dt^2} + 4\dfrac{dx}{dt} + 5x = 0,\; x(0) = 7,\; x'(0) = 0.$

2. $\dfrac{d^2x}{dt^2} + 4\dfrac{dx}{dt} + 5x = y(t)$, $x(0) = 7$, $x'(0) = 0$.

3. $\dfrac{d^2x}{dt^2} + 4\dfrac{dx}{dt} + 5x = 9$, $x(0) = 7$, $x'(0) = 0$.

4. $\dfrac{d^2x}{dt^2} + 4\dfrac{dx}{dt} + 5x = \sin \omega t$, $x(0) = 7$, $x'(0) = 0$.

5. $\dfrac{d^2x}{dt^2} + 4x = y(t)$, $x(0) = x_0$, $x'(0) = v_0$.

6. $\dfrac{d^2x}{dt^2} + 4x = \sin \omega t$, $\omega \neq 2$.

7. $\dfrac{d^2x}{dt^2} + 4x = \sin 2t$.

8. $\dfrac{d^2x}{dt^2} + 3\dfrac{dx}{dt} + 2x = 13$, $x(0) = 4$, $x'(0) = 0$.

9. In Exercise 8, does the solution represent an oscillation? Describe the motion $x(t)$.

10. An oscillator with mass m, resistance R, and spring constant k has $R^2 - 4mk < 0$. In the solution of the equation

$$m\frac{d^2x}{dt^2} + R\frac{dx}{dt} + kx = F \text{ (constant):}$$

   (a) What is the frequency of the oscillation?
   (b) What is the damping coefficient μ?
   (c) What is the steady-state solution, $\lim_{t \to \infty} x(t)$?

11. In Exercise 10, discuss how increasing the value of m affects the frequency of oscillation. Repeat for the resistance R and the spring constant k.

12. Show that one solution of Exercise 6 is $x = -\sin \omega t/(\omega^2 - 4)$. Find the frequency and amplitude of the solution.

13. In Exercise 12, let ω vary, approaching 2. How does the amplitude of the oscillation $x(t)$ behave as $\omega \to 2$?

14. Show that the solution of Exercise 7 is not sinusoidal but is an oscillation. What is the frequency? What can be said about the amplitude of the oscillation?

15. Describe a spring-mass oscillator that realizes the differential equation of Exercises 7 and 14. This situation where the driving force is sinusoidal with the same frequency as the natural frequency of the oscillator, is called resonance.

16.* The oscillator

$$\frac{d^2x}{dt^2} + 2\frac{dx}{dt} + 5x = \sin \omega t$$

is driven by a sinusoidal force with frequency ω. Find a sinusoidal function that represents the steady-state term in the solution.

17.* In Exercise 16, find the amplitude of the steady-state term in the solution as a function of ω. The amplitude of the response is not proportional to that of the driving function for all frequencies ω. What frequencies ω are diminished most by the oscillator?

18. Verify the solution in §24.2.

19. Verify the formula for the weighted average $\bar{Y}$ in §24.4.

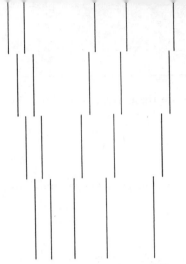

Vector Differential Calculus

CHAPTER 17

25 Taylor's Theorem

25.1 Integral Form of Mean Value Theorem. We can adapt the integral form of the mean value theorem for functions of one real variable (I, §23.2) to produce a mean value theorem for a real-valued function $f(\vec{x})$ of the vector variable $\vec{x}:(x_1, x_2, \ldots, x_n)$ defined on some disk with point $\vec{a}$ as center. Let $\vec{a} + \vec{h}$ be any other point in the domain. We connect the points $\vec{a}$ and $\vec{a} + \vec{h}$ by the straight-line segment

$$\vec{x} = \vec{a} + t\vec{h}, \qquad 0 \leq t \leq 1,$$

which includes $\vec{a}$ when $t = 0$, and $\vec{a} + \vec{h}$ when $t = 1$. We picture this for a function of two variables (Figure 17.1), representing the function $f(\vec{x})$ by contour lines. The composite function $f[\vec{a} + t\vec{h}]$ is f restricted to this line segment. It is a real-valued function of the real variable t, and hence when we differentiate it we get a numerical derived func-

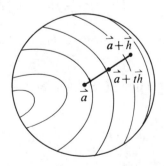

FIGURE 17.1   Function restricted to an interval.

tion $(d/dt)f[\vec{a} + t\vec{h}]$. Going the other way, we can integrate the derivative and find by the fundamental theorem that

$$f(\vec{a} + \vec{h}) - f(\vec{a}) = \int_0^1 \frac{d}{dt} f[\vec{a} + t\vec{h}] \, dt.$$

This is the integral form of the mean value theorem (I, §23.2) generalized to functions of a vector variable.

25.2   Taylor's Theorem.   We use integration by parts, $\int u \, Dv = uv - \int v \, Du$. In the integral form of the mean value theorem, we set

$$u = \frac{d}{dt} f[\vec{a} + t\vec{h}], \qquad Dv = dt,$$

$$Du = \frac{d^2}{dt^2} f[\vec{a} + t\vec{h}] \, dt, \qquad v = t + C.$$

We choose the constant C so that $v = 0$ when $t = 1$. Thus $C = -1$ and $v = -(1 - t)$. Hence, after one integration by parts, we have

$$f(\vec{a} + \vec{h}) - f(\vec{a}) = -(1 - t) \frac{d}{dt} f[\vec{a} + t\vec{h}] \bigg|_0^1 + \int_0^1 (1 - t) \frac{d^2}{dt^2} f[\vec{a} + t\vec{h}] \, dt$$

$$= \frac{d}{dt} f[\vec{a} + 0\vec{h}] + \int_0^1 (1 - t) \frac{d^2}{dt^2} f[\vec{a} + t\vec{h}] \, dt,$$

where the symbol $(d/dt)f[\vec{a} + 0\vec{h}]$ means the result of first differentiating, $(d/dt)f[\vec{a} + t\vec{h}]$, and then setting $t = 0$.

We integrate by parts again, taking

$$u = \frac{d^2}{dt^2} f[\vec{a} + t\vec{h}], \qquad Dv = (1 - t) \, dt,$$

$$Du = \frac{d^3}{dt^3} f[\vec{a} + t\vec{h}] \, dt, \qquad v = -\frac{(1 - t)^2}{2}.$$

Then the second application of integration by parts gives us

$$f(\vec{a} + \vec{h}) = f(\vec{a}) + \frac{d}{dt} f[\vec{a} + 0\vec{h}] + \frac{1}{2!} \frac{d^2}{dt^2} f[\vec{a} + 0\vec{h}]$$

$$+ \frac{1}{2!} \int_0^1 (1 - t)^2 \frac{d^3}{dt^3} f[\vec{a} + t\vec{h}] \, dt.$$

Continuing through n integrations by parts, we find the following.

TAYLOR'S THEOREM.   If the function $f(\vec{x})$ is repeatedly differentiable on the interval $[\vec{a} + t\vec{h}]$, $(0 \leq t \leq 1)$, then

$$f(\vec{a} + \vec{h}) = f(\vec{a}) + \frac{1}{1!} \frac{d}{dt} f[\vec{a} + 0\vec{h}] + \frac{1}{2!} \frac{d^2}{dt^2} f[\vec{a} + 0\vec{h}] + \frac{1}{3!} \frac{d^3}{dt^3} f[\vec{a} + 0\vec{h}]$$

$$+ \cdots + \frac{1}{n!} \frac{d^n}{dt^n} f[\vec{a} + 0\vec{h}] + R_{n+1},$$

where the remainder R_{n+1} is given by the integral

$$R_{n+1} = \frac{1}{n!} \int_0^1 (1 - t)^n \frac{d^{n+1}}{dt^{n+1}} f[\vec{a} + t\vec{h}] \, dt.$$

Example A.   Write out Taylor's theorem for a function $f(x)$ of one real variable x on the interval $[a, a + h]$.

Solution.   We can carry out the differentiation of $f[a + th]$ in this case, and it pulls out the chain rule factor h for each differentiation with respect to t (Exercises). Then Taylor's theorem with integral form of the remainder becomes

$$f(a + h) = f(a) + \frac{h}{1!} f'(a) + \cdots + \frac{h^n}{n!} f^{(n)}(a) + R_{n+1}$$

where

$$R_{n+1} = \frac{h^{n+1}}{n!} \int_0^1 (1 - t)^n f^{(n+1)}[a + th] \, dt.$$

Example B.   Write out Taylor's theorem explicitly for a function $f(x, y)$ of two variables on the line segment from (a, b) to $(a + h, b + k)$, carrying it out to terms of order 1 with R_2.

Solution.   In vector notation, we replace the pair (x, y) by the vector $\vec{x}:(x, y)$. The line joining $\vec{a}:(a, b)$ to $\vec{a} + \vec{h}:(a + h, b + k)$ has the vector equation $\vec{x} = \vec{a} + t\vec{h}$, or scalar representation,

$$x = a + th, \qquad y = b + tk, \qquad 0 \le t \le 1.$$

Then the differential $Df = f_1 \, dx + f_2 \, dy$ and the differential $D\vec{x}:(dx, dy) = (h \, dt, k \, dt)$. Hence when we substitute these expressions into Df according to the chain rule (I, §12.3), we can express the differentials of the composite $f[\vec{a} + t\vec{h}]$ in terms of partial derivatives $f_1, f_2, f_{11}, f_{12}, f_{22}$, all evaluated at $(a + th, b + tk)$. The result is, first,

$$Df[\vec{a} + t\vec{h}] = f_1 h \, dt + f_2 k \, dt.$$

Repeating the differentiation, we find

$$D^2 f[\vec{a} + t\vec{h}] = (f_{11} h^2 + 2f_{12} hk + f_{22} k^2) \, dt^2.$$

We set $t = 0$, dividing by dt to obtain the numerical derivative, in the first derivative term. We have

$$\frac{d}{dt} f(\vec{a} + 0\vec{h}) = f_1(a, b)h + f_2(a, b)k.$$

Then Taylor's theorem with remainder R_2 becomes

$$f(a + h, b + k) = f(a, b) + f_1(a, b)h + f_2(a, b)k + R_2,$$

and

$$R_2 = h^2 \int_0^1 (1 - t)f_{11} \, dt + 2hk \int_0^1 (1 - t)f_{12} \, dt + k^2 \int_0^1 (1 - t)f_{22} \, dt.$$

Example C.   Expand the function $\cos(2x - y)$ by Taylor's theorem, near the point $(a, b) = (0, 0)$. Carry out the expansion to a remainder of order 3.

Solution.   We have

$$f(\vec{a} + t\vec{h}) = f(a + th, b + tk) = \cos[2(0 + th) - (0 + tk)] = \cos(2h - k)t,$$

and

$$f(\vec{a} + 0\vec{h}) = \cos 0 = 1.$$

$$\frac{d}{dt} f(a + th) = -(2h - k)\sin(2h - k)t, \qquad \frac{d}{dt} f(a + 0h) = 0.$$

$$\frac{d^2}{dt^2} f(\vec{a} + t\vec{h}) = -(2h - k)^2 \cos(2h - k)t, \qquad \frac{d^2}{dt^2} f(\vec{a} + 0\vec{h}) = -(2h - k)^2.$$

$$\frac{d^3}{dt^3} f(\vec{a} + t\vec{h}) = (2k - k)^3 \sin(2h - k)t.$$

Hence Taylor's theorem gives us

$$\cos(2h - k) = 1 + 0 - \frac{(2h - k)^2}{2!} + \frac{(2h - k)^3}{2!} \int_0^1 (1 - t)^2 \sin(2h - k)t\, dt.$$

In this expression we observe that the first-order terms were all zero. The second-order terms make the quadratic form

$$-\frac{(2h - k)^2}{2} = -2h^2 + 2hk - \frac{k^2}{2}.$$

Since $|\sin(2h - k)t| \leq 1$, for all t, we can estimate the remainder R_3. Thus

$$|R_3| \leq \frac{|2h - k|^3}{2!} \int_0^1 (1 - t)^2 1\, dt = \frac{|2h - k|^3}{6}.$$

We observe that even for functions of two variables $f(x, y)$, even then only to a few terms, Taylor's theorem in explicit form becomes complicated. This emphasizes the simplicity of the general statement in vector notation, which tells how to compute the explicit formulas, however complicated they may be.

25.3   The Remainder R_2

Example.   Compute the average value of the partial derivative $f_{11}[a + th, b + tk]$ of $f(x, y)$ on the line segment from (a, b) to $(a + h, b + k)$ with respect to the weight function $1 - t$.

Solution.   The weighted average of f_{11} is the number (I, §23.3)

$$\bar{f}_{11} = \frac{\int_0^1 (1 - t)f_{11}[a + th, b + tk]\, dt}{\int_0^1 (1 - t)\, dt}$$

$$= 2 \int_0^1 (1 - t)f_{11}[a + th, b + tk]\, dt.$$

This completes the problem. Similar results apply for the average values $\bar{f}_{12}$ and $\bar{f}_{22}$. With these results the expression for R_2 can be simplified into

$$R_2 = \tfrac{1}{2}[\bar{f}_{11}h^2 + 2\bar{f}_{12}hk + \bar{f}_{22}k^2].$$

Thus we may regard the remainder R_2 as a quadratic form in the variables (h, k) with coefficients that are weighed average values of the second partial derivatives on the line segment from (a, b) to $(a + h, b + k)$.

25.4   A Simple Expansion.   Since Taylor's theorem for multidimensional cases becomes complicated in computation, yet is no different in basic method from that of functions of one real variable, we carry out one such example for a very simple function of one variable.

Example.   Expand $\exp x$ in powers of $x - 1$ by Taylor's theorem with remainder R_{n+1}.

Solution.   We replace x by $1 + h$, where h is a local coordinate. Then points on the interval $[1, 1 + h]$ are given by $1 + th$, $0 \le t \le 1$. By the fundamental theorem of integral calculus,

$$\exp(1 + h) - \exp(1) = \int_0^1 \frac{d}{dt} \exp(1 + th) \, dt$$

$$= \frac{h}{1} \int_0^1 \exp(1 + th) \, dt.$$

We integrate by parts with

$$u = \exp(1 + th), \qquad Dv = dt.$$
$$Du = h \exp(1 + th) \, dt, \qquad v = -(1 - t),$$

where in computing v, we chose the constant of integration so that $v(1) = 0$. Then

$$\exp(1 + h) - \exp 1 = \frac{h}{1} \left[-(1 - t) \exp(1 + th) \Big|_0^1 + h \int_0^1 (1 - t) \exp(1 + th) \, dt \right]$$

$$= \frac{h}{1} \exp 1 + \frac{h^2}{1} \int_0^1 (1 - t) \exp(1 + th) \, dt.$$

We integrate by parts again with

$$u = \exp(1 + th), \qquad Dv = (1 - t) \, dt,$$
$$Du = h \exp(1 + th) \, dt, \qquad v = -\frac{(1 - t)^2}{2}.$$

Then

$$\exp(1 + h) - \exp 1 = \frac{h}{1} \exp 1 - \frac{h^2}{2!} (1 - t)^2 \exp(1 + th) \Big|_0^1 + R_3$$

$$= \frac{h}{1} \exp 1 + \frac{h^2}{2!} \exp 1 + \frac{h^3}{2!} \Big|_0^1 (1 - t)^2 \exp(1 + th) \, dt.$$

Continuing in this way through n integrations by parts, we find

$$\exp(1 + h) - \exp 1 = \frac{h}{1} \exp 1 + \frac{h^2}{2!} \exp 1 + \cdots + \frac{h^n}{n!} \exp 1 + R_{n+1},$$

where

$$R_{n+1} = \frac{h^{n+1}}{n!} \int_0^1 (1 - t)^n \exp(1 + th) \, dt.$$

We can estimate the remainder, since $\exp(1 + th)$ is an increasing function and takes its maximum on $[0, 1]$ at $t = 1$. This maximum of $\exp(1 + th)$ is $\exp(1 + h)$. Then

$$|R_{n+1}| \le \frac{h^{n+1}}{n!} \int_0^1 (1 - t)^n \max \exp(1 + th) \, dt$$

$$= \frac{h^{n+1}}{n!} \exp(1 + h) \int_0^1 (1 - t)^n \, dt$$

$$= \frac{h^{n+1}}{n!} \exp(1 + h) \left(\frac{1}{n + 1} \right) = \frac{h^{n+1}}{(n + 1)!} \exp(1 + h).$$

Since $1/(n + 1)!$ rapidly becomes very small with increasing n, this estimate of the remainder will become very small when $|h| < 1$.

Also since when $t = 1$, $x = 1 + h$, $h = x - 1$, and exp $1 = e$, this can be written

$$e^x = e + \frac{e}{1!}(x - 1) + \frac{e}{2!}(x - 1)^2 + \cdots + \frac{e}{n!}(x - 1)^n + R_{n+1},$$

where

$$|R_{n+1}| < \frac{|x - 1|^{n+1}}{(n + 1)!}e^x.$$

25.5 Exercises

In Exercises 1–4, compute the Taylor's theorem expansion with remainder of the function $f(x)$ where x is a real variable with:

1. e^{-x} in a neighborhood of 1 with remainder R_3.
2. $\sin x$ in a neighborhood of 0 with remainder R_4.
3. x^p, p real, in a neighborhood of 1 with remainder R_3. This is the binomial series.
4. $\ln x$ in a neighborhood of 1 with remainder R_{n+1}.

5. Expand the function given by $f(x, y) = (1 - x)(1 - y)(x + y - 1)$ in a neighborhood of $(1, 1)$ with remainder R_2. Show that $R_2 = hk(h + k + 1)$.
6. In Exercise 5, show that $f(x, y) \geq f(1, 1)$ when the point $P:(x, y)$ is in the shaded $(+)$ areas of Figure 17.2.
7. In Exercises 5, 6, is the point $(1, 1)$ a maximum point or a minimum point for $f(x, y)$? Explain.
8. Expand $f(x, y) = \sin(x + y)$ near $(0, 0)$ by Taylor's theorem with remainder R_2.
9. Verify the computation of Taylor's theorem for a function $f(x)$ on the interval $[a, a + h]$, Example A, §25.2.
10. If $f(x, y, z) = xyz$, compute the Taylor's theorem expansion near $(3, 1, 4)$ with remainder R_2.
11. Show that the remainder

$$R_1 = \int_0^1 f'(a + th)h\, dt = h\bar{f}',$$

where $\bar{f}'$ is the average value of $f'(x)$ on the interval $[a, a + h]$.

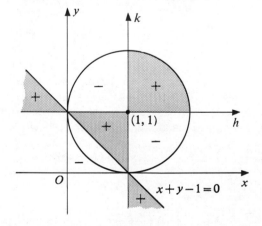

FIGURE 17.2   Exercise 6.

12. Expand $\sin x$ in powers of $x - (\pi/4)$ and estimate the remainder R_n from the fact that $|\sin x| \leq 1$.

13. From the expansion of Exercise 12, we are to compute $\sin 1$ with error less than 10^{-3}. How many terms of the expansion must be computed to assure this accuracy?

14. For the expansion of x^p in a neighborhood of 1 (Exercise 3), with $p = 0.55$, how many terms are needed to compute $x^{0.55}$ with error less than 10^{-3} when $x = 1.1$?

15. How many terms of the expansion of $\ln x$ in a neighborhood of 1 are needed to compute $\ln 2$ with error less than 10^{-3}? This is an example of a slowly converging series that we try to replace by a more efficient, rapidly converging formula.

16. Estimate the error in computing

$$\int_0^1 \exp(-x^2)\, dx,$$

if we compute it by integrating its expansion near 0 to the term in the power x^6.

17.* Show that for the function $f(x)$ on $[a, a + h]$,

$$R_3 = \frac{\overline{f^{(3)}}}{3!}\, h^3,$$

where $\overline{f^{(3)}}$ is the average value of the third derivative, $f^{(3)}(x)$, on $[a, a + h]$ with respect to the weight function $(a + h - x)^2$.

18. The derivative $(d/dt)f[\vec{a} + 0\vec{h}]$ is the directional derivative $\overrightarrow{\nabla f} \cdot \vec{h}$, if $\|\vec{h}\| = 1$ (II, §18.4). Show that whether $\|\vec{h}\| = 1$ or not,

$$\frac{d}{dt} f[\vec{a} + 0\vec{h}] = \overrightarrow{\nabla f} \cdot \vec{h},$$

where the symbol $\overrightarrow{\nabla f} \cdot \vec{h}$ is the dot product of the gradient $\overrightarrow{\nabla f}$ and the vector $\vec{h}$.

19. A remainder R_3 (§25.2, Example C), is given by

$$R_3 = \frac{(2h - k)^3}{2!} \int_0^1 (1 - t)^2 \sin(2h - k)t\, dt.$$

Show that if $|2h - k| < 1$, then $|R_3| < (2h - k)^4/24$. Recall that for small θ, $|\sin \theta| < |\theta|$ (I, §32).

26 Maximum-Minimum Problems

ACADEMIC NOTE. Modern developments such as linear programming and optimal control have revived the importance of maxima and minima that occur on the *boundary* of the domain of the function. Present-day applications also emphasize the "global" maxima over the entire domain of the function as opposed to local maxima. This presentation is devised to accommodate these current technologies without getting into them. This causes us to deviate from the concentration of the conventional calculus text on local, interior maxima and minima.

26.1 Max-Min Problems for Functions of Several Variables. We have studied a variety of methods for finding a number $x = a$ that gives a minimum (or maximum) value to a function $f(x)$ defined on some domain of real numbers (I, §§16, 25). We now consider the same problems for number-valued functions $f(\vec{x})$ of vectors $\vec{x}: (x_1, x_2, \ldots, x_n)$,

or functions of several real variables. The basic technique is to reduce the problem to a single-variable problem.

DEFINITION. A function $f(\vec{x})$ defined on the vector domain $\mathscr{D}$ has a minimum at $\vec{a}$ in $\mathscr{D}$ if and only if for every point $\vec{a} + \vec{h}$ in $\mathscr{D}$

$$f(\vec{a} + \vec{h}) - f(\vec{a}) \geqq 0.$$

Similarly, $f(\vec{x})$ has a maximum at $\vec{a}$ in $\mathscr{D}$ if for every point $\vec{a} + \vec{h}$ in $\mathscr{D}$

$$f(\vec{a} + \vec{h}) - f(\vec{a}) \leqq 0.$$

As we did for Taylor's theorem, we reduce this to a one-variable problem by fixing $\vec{a} + \vec{h}$ and restricting f to the line segment $\{\vec{x} = \vec{a} + t\vec{h}, 0 \leq t \leq 1\}$. Then the composite function $f[\vec{a} + t\vec{h}]$ is a real-valued function of the real variable t on the interval $[0, 1]$. Clearly, if $\vec{a}$ is a minimum point for $f(\vec{x})$ on its domain, then for the restricted function, $t = 0$ must give a minimum on the interval $[0, 1]$. Conversely, if $t = 0$ minimizes the restricted function $f[\vec{a} + t\vec{h}]$ for every choice of $\vec{h}$, such that $\vec{a} + \vec{h}$ is in the domain of f, then a minimizes $f(\vec{x})$ in the entire domain. We illustrate this with a graph $w = f(\vec{x})$, drawn in perspective as a disklike surface (Figure 17.3). If the point $A:(\vec{a}, f(\vec{a}))$ on this graph is to be the minimum point, then for every choice of $\vec{a} + \vec{h}$ in the domain the function $f[\vec{a} + t\vec{h}]$, $0 \leq t \leq 1$, represented by the curve AP on the surface, must have a minimum at the point $\vec{a}$ that is given by $t = 0$. We consider all such single-variable problems for every choice of $\vec{a} + \vec{h}$ in the domain of f, that is, for all vertical sections of the graph. If $\vec{a}$ gives the restricted $f[\vec{a} + t\vec{h}]$ a minimum value for every choice of $\vec{h}$, it is a minimum point of $f(\vec{x})$ on $\mathscr{D}$.

We omit separate consideration of maximum points since $\vec{a}$ is a maximum point for $f(\vec{x})$ if and only if $\vec{a}$ is a minimum point for $-f(\vec{x})$. Thus there is only an obvious change of sign to be accounted for in solving a maximum problem for $f(\vec{x})$ on $\mathscr{D}$.

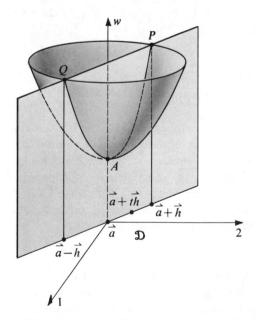

FIGURE 17.3   Minimum of restricted function $f[\vec{a} + t\vec{h}]$.

26.2   Differential Conditions for a Minimum.   We say that the unit vector $\overset{\scriptscriptstyle\vee}{h}$ at the point $\vec{a}$ in the domain $\mathscr{D}$ *leads into* the domain if some portion of the line segment $\vec{a} + t\overset{\scriptscriptstyle\vee}{h}$, $0 \leqq t < c$, is in $\mathscr{D}$. We say that $\overset{\scriptscriptstyle\vee}{h}$ is a two-way direction at point $\vec{a}$ if both $\overset{\scriptscriptstyle\vee}{h}$ and $-\overset{\scriptscriptstyle\vee}{h}$ lead into $\mathscr{D}$. We say that $\vec{a}$ is *inside* the domain $\mathscr{D}$ if every direction $\overset{\scriptscriptstyle\vee}{h}$ at $\vec{a}$ is a two-way direction (Figure 17.4).

If the function $f(\vec{x})$ on the domain $\mathscr{D}$ has a minimum at $\vec{a}$ where $f(\vec{x})$ is differentiable and the direction $\overset{\scriptscriptstyle\vee}{h}$ leads into $\mathscr{D}$ at $\vec{a}$, then the function $f[\vec{a} + t\overset{\scriptscriptstyle\vee}{h}]$ restricted to the line of $\overset{\scriptscriptstyle\vee}{h}$ has a minimum at the point where $t = 0$. Hence by single-variable arguments,

$$\frac{d}{dt}f[\vec{a} + t\overset{\scriptscriptstyle\vee}{h}]^{t=0} \geqq 0.$$

This derivative is the directional derivative $\overrightarrow{\nabla f} \cdot \overset{\scriptscriptstyle\vee}{h}$ (I, §12.4). Hence we have the following theorem.

THEOREM.   If $f(\vec{x})$ has a minimum at the point $\vec{a}$ where f is differentiable, then the directional derivative

$$\overrightarrow{\nabla f} \cdot \overset{\scriptscriptstyle\vee}{h} \geqq 0$$

for every unit direction $\overset{\scriptscriptstyle\vee}{h}$ which leads into the domain of $f(\vec{x})$.

Geometrically this theorem says that if $\vec{a}$ is a minimum point of $f(\vec{x})$ then the projection of the gradient $\overrightarrow{\nabla f}$ at $\vec{a}$ onto any direction $\overset{\scriptscriptstyle\vee}{h}$ that leads into the domain of f must be positive or zero. Intuitively, if $\vec{a}$ is a minimum, then the function cannot decrease along any ray leading into its domain from $\vec{a}$.

COROLLARY (Two-way directions at a minimum point).   If $\vec{a}$ is a minimum point of $f(\vec{x})$ and $\overset{\scriptscriptstyle\vee}{h}$ is a two-way direction at $\vec{a}$, then

$$\overrightarrow{\nabla f} \cdot \overset{\scriptscriptstyle\vee}{h} = 0.$$

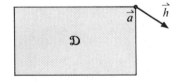

(a) $\vec{h}$ does not lead into $\mathscr{D}$.

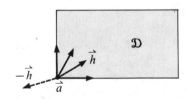

(b) $\vec{h}$ leads into $\mathscr{D}$. But $\vec{h}$ is not a 2-way direction.

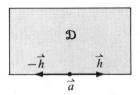

(c) $\vec{h}$ is a 2-way direction. But $\vec{a}$ is not inside $\mathscr{D}$.

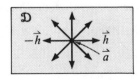

(d) $\vec{a}$ is inside $\mathscr{D}$. Every direction is a 2-way direction.

FIGURE 17.4   Point, directions and domain $\mathscr{D}$, several cases.

Proof. Since $\vec{h}$ is a two-way direction, both $\overrightarrow{\nabla f} \cdot \vec{h} \geqq 0$ and $\overrightarrow{\nabla f} \cdot (-\vec{h}) = -\overrightarrow{\nabla f} \cdot \vec{h} \geqq 0$. Hence $\overrightarrow{\nabla f} \cdot \vec{h} = 0$.

COROLLARY (Minimum at point inside the domain). If $f(\vec{x})$ has a minimum at the point $\vec{a}$ inside its domain and $f(\vec{x})$ is differentiable there, then the gradient, $\overrightarrow{\nabla f} = \vec{0}$ at $\vec{a}$.

Example. Find the minimum point of the function $f(x, y) = 2x - 3y + 4$ on the domain consisting of points (x, y) inside and on the triangle with vertices $(-1, 3)$, $(2, -5)$, $(3, 7)$.

Solution. The function is everywhere differentiable and has the constant gradient $\overrightarrow{\nabla f}: (2, -3, 4)$. Since this is nowhere zero, no minimum can occur at a point inside the triangle.

If the minimum occurs at a point on a side of the triangle between two vertices, the direction $\vec{h}$ of this side is a two-way direction and hence the constant directional derivative $\overrightarrow{\nabla f} \cdot \vec{h} = 0$ on this side. This implies that f is constant on the side. Thus if a minimum occurs at any point on the side, the same minimum obtains at the vertices that are endpoints of this side. Hence the minimum of f, if it exists, must occur at a vertex.

We evaluate f at the vertices, finding that $f(-1, 3) = -7$, $f(2, -5) = 23$, $f(3, 7) = -11$. Hence, if there is a minimum point for the function, it occurs at the point $\vec{a}: (3, 7)$ and the minimum value is -11. We may use the theorem on the values of continuous functions. Since the function $f(x, y) = 2x - 3y + 4$ is continuous on the closed triangle, it attains a minimum at some point of the domain. This completes the problem.

26.3   Use of Taylor's Theorem. We approach maximum-minimum problems with a collection of methods and principles rather than with preset formulas. One of the basic tools for investigating a point $\vec{a}$ inside the domain of f where $\overrightarrow{\nabla f} = \vec{0}$ is Taylor's theorem with remainder R_2. Since $\overrightarrow{\nabla f} \cdot \vec{h} = 0$ at $\vec{a}$, this expansion reduces to

$$f(\vec{a} + \vec{h}) - f(\vec{a}) = 0 + \int_0^1 (1 - t) \frac{d^2}{dt^2} f[\vec{a} + t\vec{h}] \, dt.$$

Then if $(d^2/dt^2)f[\vec{a} + t\vec{h}] \geqq 0$ for all vectors $\vec{h}$ such that $\vec{a} + \vec{h}$ is in the domain of f, the point $\vec{a}$ is the minimum point of f in its entire domain. We can more easily get information about the *local* minimizing property of $\vec{a}$.

THEOREM (Local minimum at a point $\vec{a}$ where the gradient is zero). If $\overrightarrow{\nabla f} = \vec{0}$ at $\vec{a}$ and if the continuous second derivative $(d^2/dt^2)f[\vec{a} + 0\vec{h}] > 0$ for all unit vectors $\vec{h}$, then for some ϵ, $\vec{a}$ minimizes f in an ϵ-neighborhood of $\vec{a}$.

Proof. We consider points $\vec{a} + t\alpha\vec{h}$ on the line segment $[\vec{a}, \vec{a} + \alpha\vec{h}]$. Then

$$f(\vec{a} + \alpha\vec{h}) - f(\vec{a}) = 0 + \int_0^1 (1 - t) \frac{d^2}{dt^2} f[a + t\alpha\vec{h}] \, dt.$$

Now if $(d^2/dt^2)f[\vec{a} + 0\vec{h}] > 0$, continuity of the second derivative for all unit vectors $\vec{h}$ implies that it remains positive on a small interval from $\vec{a}$ to $\vec{a} + \epsilon\vec{h}$. The points $\{\vec{a} + \alpha\vec{h}, 0 \leqq \alpha < \epsilon\}$ describe a disk of radius ϵ about $\vec{a}$ and in this disk $f(\vec{x}) - f(\vec{a}) \geqq 0$. That is, $\vec{a}$ minimizes f in this ϵ-neighborhood of $\vec{a}$, as stated in the theorem.

Example A. Find the minimum and maximum points of the function f given by

$$f(x, y) = \frac{8}{x} + \frac{8}{y} + xy$$

for all positive x and y.

Solution. f has the gradient

$$\overrightarrow{\nabla f} : \left(-\frac{8}{x^2} + y, \ -\frac{8}{y^2} + x \right)$$

and $\overrightarrow{\nabla f} = \vec{0}$ if and only if $x^2 y = 8$ and $xy^2 = 8$. Solving simultaneously, we find that the only point where $\overrightarrow{\nabla f} = \vec{0}$ is $\vec{a} : (2, 2)$. To expand f in a neighborhood of $\vec{a}$, we replace x by $2 + th$, y by $2 + tk$ and calculate the derivative

$$\frac{d^2}{dt^2} f[\vec{a} + t\vec{h}] = \frac{16h^2}{(2 + th)^3} + \frac{16k^2}{(2 + tk)^3} + 2hk.$$

This is easily seen to be positive if h and k have like signs but it is difficult to determine the sign of the second derivative when $2hk < 0$.

However, we observe that when $t = 0$,

$$\frac{d^2}{dt^2} f[\vec{a} + 0\vec{h}] = 2h^2 + 2hk + 2k^2 = (h + k)^2 + h^2 + k^2.$$

Therefore when (h, k) is a unit vector,

$$\frac{d^2}{dt^2} f[\vec{a} + 0\vec{h}] = (h + k)^2 + 1 \geqq 1 > 0.$$

This satisfies the conditions of the theorem and thus the point $\vec{a} : (2, 2)$ furnishes at least a local minimum for f. There are no maximum points.

Since the method is essentially the same as for functions of one variable, we apply it to a function of one variable in an easier example.

Example B. Find the maximum and minimum points of the function f given by $f(x) = 4x + (9/x)$ on the interval $\{0 < x \leqq 2\}$.

Solution. All points except $x = 2$ are inside the domain. We calculate $\overrightarrow{\nabla f} = 4 - (9/x^2)$. Then $\overrightarrow{\nabla f} = \vec{0}$ if and only if $4x^2 = 9$ or $\{x = \frac{3}{2}, x = -\frac{3}{2}\}$. Only $x = \frac{3}{2}$ is in the domain of f. The Taylor's expansion of f about this critical point is

$$f\left(\frac{3}{2} + h\right) - f\left(\frac{3}{2}\right) = 0 + \int_0^1 (1 - t) \frac{d^2}{dt^2} f\left(\frac{3}{2} + th\right) dt.$$

Calculating, we find that

$$\frac{d^2}{dt^2} f\left[\frac{3}{2} + th\right] = \frac{18}{[\frac{3}{2} + th]^3} h^2 \geqq 0$$

for all h such that $\frac{3}{2} + h$ is in the interval $0 < x \leqq 2$, since in this domain $\frac{3}{2} + th$ is positive. Hence in this case we conclude that $a = \frac{3}{2}$ minimizes f over its entire domain.

The only possible maximum point is the one where $x = 2$, but on reexamining $f(x) = 4x + (9/x)$, we see that $\lim_{x \to 0+} f(x) = \infty$. Hence there is no maximum point. This completes the problem. A graph of $y = f(x)$ illustrates this, but it is important, especially for multivariable problems, that the graph is not necessary.

26.4   Convexity.   We observe that Taylor's expansion of $w = f(\vec{x})$ at a point $\vec{a}$ where $\overrightarrow{\nabla f}$ is not necessarily zero can be written with remainder R_2,

$$f[\vec{a} + \vec{h}] - (f(\vec{a}) + \overrightarrow{\nabla f} \cdot \vec{h}) = \int_0^1 (1 - t) \frac{d^2}{dt^2} f[a + th] \, dt.$$

The equation $w = f(\vec{a}) + \overrightarrow{\nabla f} \cdot \vec{h}$ is the equation of the tangent plane in local coordinates $\vec{h}$. Hence if the second derivative $(d^2/dt^2)f[\vec{a} + 0\vec{h}] > 0$ it follows that the value $w = f(\vec{a} + \vec{h})$ of f at $\vec{h}$ is greater than the value of w in the tangent plane at $\vec{h}$. Geometrically this means that the graph of $w = f(\vec{x})$ curves upward away from its tangent plane (Figure 17.5(a)). Then if $\overrightarrow{\nabla f} = \vec{0}$, this condition implies that $\vec{a}$ minimizes $f(\vec{x})$ in a region that extends as far as $f(\vec{x})$ is above the tangent plane.

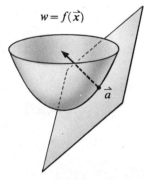

(a)   With a tangent plane   (b)   With a plane of support

FIGURE 17.5   Convexity.

This geometric condition can be generalized to include a minimum at a point where f is not differentiable, as in the positive branch of the cone $w = k\sqrt{x^2 + y^2}$ (Figure 17.5(b)) at the origin $\vec{a} = \vec{0}$.

DEFINITION.   The graph of $w = f(\vec{x})$ is *locally convex* at $\vec{a}$ if there is a plane $w = f(\vec{a}) + \vec{b} \cdot (\vec{x} - \vec{a})$ such that for all $\vec{x}$ in a neighborhood of $\vec{a}$

$$f(\vec{x}) \geq f(\vec{a}) + \vec{b}(\vec{x} - \vec{a}).$$

Such a plane is called a *plane of support* at $\vec{a}$.

We readily see that if $f(\vec{x})$ has a horizontal plane of support at $\vec{a}$ (Figure 17.5(b)), then $\vec{a}$ gives at least a local minimum to f.

26.5   Exercises

In Exercises 1–5, use the directional derivatives $(d/dt)f[a + th]$ and $(d^2/dt^2)f[a + th]$ to find the maxima and minima of the functions of one variable.

1. $f(x) = x - x^2$ on $[0, 1]$.   2. $f(x) = x^3 - 12x - 1$ on $[-5, 5]$.

3. $f(x) = x^3 - 12x - 1$ on $\mathscr{R}^1$.   4. $f(x) = \dfrac{x}{1 + x^2}$ on $\mathscr{R}^1$.

5. $f(x) = \dfrac{x^3}{x + 1}$ for all real x, $x \neq -1$.

In Exercises 6–10, find the maximum and minimum points.

6. $f(x, y) = x^2 - 2xy + 3y^2$ on $\mathscr{R}^2$.
7. $g(x, y) = x^2 + 2x - y^2 - 4y + 5$ on $\mathscr{R}^2$.
8. $p(x, y) = x^2 - xy + 2y^2 - 4x - 5y + 11$ on $\mathscr{R}^2$.
9. $f(x, y) = 3x^2 + xy^2 - y^3 + 8$ on $\mathscr{R}^2$.
10. $f(x, y) = x^3y - 3x^2 + y^2$ on the square $\{|x| \leq 1, |y| \leq 1\}$.

11. Find the maximum and minimum values of $2x + 3y + 5$ on the triangle whose vertices are $(0, 0)$, $(3, 0)$, $(0, 2)$.

12. Find the maximum and minimum values of $f(x, y) = 3\sqrt{x^2 + y^2}$ on the disk

$$\{x^2 + y^2 \leq 1\}.$$

13. Find the maximum and minimum values of $x^2 - y^2$ on the square $\{|x| \leq 1, |y| \leq 1\}$.

14. Find the maximum and minimum values of $x^2 - 4xy + 2y^2$ on the square $\{|x| + |y| \leq 1\}$.

15. Find the maximum and minimum values of $x^2 + y^2 + z^2$ in the solid pyramid whose vertices are $(1, 0, 0)$, $(0, 1, 0)$, $(0, 0, 1)$, $(2, 2, 2)$.

16. The function $f(x, y) = 2 \arctan x + \ln (1 + y^2)$ is defined on a disk $\{x^2 + y^2 \leq R^2\}$. Show that both the maximum and minimum are attained on the boundary.

17. Find the maxima and minima of $f(r, \theta) = r^2 + 2r(\sin \theta - \cos \theta)$ on $\{0 \leq \theta \leq 2\pi, 0 \leq r < \infty\}$.

18. It seems "obvious" (Figure 17.3) that, if $f(x, y)$ has a local minimum at the point $\vec{a}:(x_0, y_0)$ on every line through $\vec{a}$, then f has a minimum at $\vec{a}$ in some xy-neighborhood of $\vec{a}$. Prove that this is false. For example, show that the function given by $f(x, y) = (y - x^2)(y - x^4)$ has a local minimum at $O:(0, 0)$ on every line through O. But show that, in every neighborhood of O, there are points (x, y) where $f(x, y)$ is negative. Hence $f(0, 0) = 0$ is not a local minimum of f in xy-space.

27 Lagrange Multipliers

HISTORICAL NOTE.   J. L. Lagrange (1736–1813), Italian-French mathematician who did most of his work in Berlin, was one of the principal contributors to the development of the analytic program of calculus, following its invention by Newton (1642–1727) and Leibniz (1646–1716). His extraordinary algebraic insight enabled him to discover some of the most important methods of calculus and analytic mechanics. One of these is the method of Lagrange multipliers.

27.1   Minimum Problem with Constraints.

We consider the problem of minimizing the function $f(x, y)$ on a domain $\mathscr{D}$ in the xy-plane subject to the condition that (x, y) must satisfy some side condition $g(x, y) = c$, called a "constraint." One way to solve the problem might be to solve the equation $g(x, y) = c$ for y as a function $y(x)$ and eliminate y in the function f by substitution, so that we minimize the composite $f[x, y(x)]$. This is the naïve method (I, §25), but it is difficult in theory and practice. Following Lagrange, we may avoid the difficulty of solving the constraint equation for y. Unlike Lagrange, we shall use geometric reasoning.

We picture the domain $\mathscr{D}$ of $f(x, y)$ as a set in the xy-plane (Figure 17.6) and the constraint $g(x, y) = c$ as a curve, $\mathscr{G}$ in the domain $\mathscr{D}$. We picture a point $\vec{a}:(a, b)$ on this curve, that is a candidate for a minimum. At this point we picture the gradient $\overrightarrow{\nabla f}$

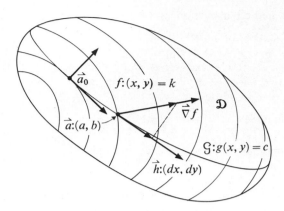

FIGURE 17.6   Constrained minimum problem.

which points in the direction of maximum increase of f, orthogonal to the contour line $f(x, y) = k$ of f through $\vec{a}$. We picture the unit tangent direction, $\vec{h}:(dx, dy)$, of $g(x, y) = c$, at $\vec{a}$. The tangent vector satisfies the equation

$$g_1 \, dx + g_2 \, dy = 0,$$

or in vector notation, $\overrightarrow{\nabla g} \cdot \vec{h} = 0$. We project the gradient of f onto the tangent line, producing a vector projection $(\overrightarrow{\nabla f} \cdot \vec{h})\vec{h}$ in the direction of the tangent line (Figure 17.6). If this vector projection of the gradient $\overrightarrow{\nabla f}$ is not zero, it gives the rate of increase of f along the tangent direction of the constraint curve $\mathcal{G}$. And if $\vec{a}$ is inside the domain $\mathcal{D}$, the oppositely directed projection, $-(\overrightarrow{\nabla f} \cdot \vec{h})\vec{h}$, points in the direction of *decreasing f* along $\mathcal{G}$. We can go along $\mathcal{G}$ in this decreasing direction to a point where the value of f is less than it is at $\vec{a}$. Hence at a minimum point, $(\overrightarrow{\nabla f} \cdot \vec{h}) = 0$, which says that the gradient of f is orthogonal to the tangent line of $\mathcal{G}$. But the gradient vector, $\overrightarrow{\nabla g}$, is also orthogonal to the tangent line of $\mathcal{G}$ at $\vec{a}$. Hence at a minimum point of the constrained problem, the gradient of f is parallel to the gradient of g. Therefore, for some nonzero number λ, if $\vec{a}$ is a minimum point, it is necessary that

$$\overrightarrow{\nabla f} - \lambda \, \overrightarrow{\nabla g} = \vec{0}.$$

The multiplier λ is called the *Lagrange multiplier.*

There is one exceptional case. That is the case when $\overrightarrow{\nabla g} = \vec{0}$. We say that the point a is *regular* if $\overrightarrow{\nabla g} \neq \vec{0}$ there. Then we may state the result of this argument as a theorem.

THEOREM (Lagrange Multiplier Rule).   If function $f(x, y)$ is differentiable in its domain $\mathcal{D}$, if the point $\vec{a}$, inside $\mathcal{D}$, is a minimum point for $f(x, y)$, subject to the constraint $g(x, y) = c$, and if $\vec{a}$ is a *regular point* for g, then for some multiplier λ,

$$\overrightarrow{\nabla f} - \lambda\overrightarrow{\nabla g} = \vec{0}$$

at the point a.

We observe that the theorem describes the minimum point as one where the tangent to the contour line of f and the tangent to $\mathcal{G}$ coincide (Point $\vec{a}_0$, Figure 17.6). Also it states a necessary condition for a minimum, which may not be sufficient to insure that $\vec{a}$ is a minimum point.

27.2   Method of Application of the Multiplier Rule.   To find a minimum point $\vec{a}$ by the multiplier rule for the minimum of $f(x, y)$ subject to the constraint $g(x, y) = c$, we need three equations. For we must determine the two coordinates (x, y) of $\vec{a}$, and the multiplier λ. The theorem gives two of these in the vector equation $\overrightarrow{\nabla f} + \lambda \overrightarrow{\nabla g} = \vec{0}$. The third equation is the constraint $g(x, y) = c$ itself.

Example.   Find the rectangle of largest area that can be inscribed in a circle of radius 4.

Solution.   We choose coordinates with origin at the center of the circle, and axes parallel to the sides of the inscribed rectangle (Figure 17.7). Let $P:(x, y)$ be a vertex of the inscribed rectangle. Then the problem is to maximize the area function

$$f(x, y) = 4xy,$$

whose domain $\mathscr{D}$ is the positive quadrant $x \geq 0$, $y \geq 0$, subject to the condition that

$$x^2 + y^2 = 16.$$

Here $g(x, y) = x^2 + y^2$. If there is a maximum point inside $\mathscr{D}$, the multiplier rule says that then for some multiplier λ, $\overrightarrow{\nabla f} - \lambda \overrightarrow{\nabla g} = \vec{0}$. Since the gradients $\overrightarrow{\nabla f}:(f_1, f_2)$, $\overrightarrow{\nabla g}:(g_1, g_2)$ each have two components, this says that

$$f_1 - \lambda g_1 = 0,$$
$$f_2 - \lambda g_2 = 0,$$

and

$$g = 16.$$

We compute $f_1 = 4y, f_2 = 4x, g_1 = 2x, g_2 = 2y$, and write the three equations explicitly,

$$4y - \lambda(2x) = 0,$$
$$4x - \lambda(2y) = 0,$$
$$x^2 + y^2 = 16,$$

and solve simultaneously for (x, y, λ). We can eliminate λ from the first two equations if we multiply the first by y, and the second by x, and subtract. The result is $y^2 - x^2 = 0$, which we solve with the last equation to give us $x = \sqrt{8}, y = \sqrt{8}$. Then we return to the first equation to find that $\lambda = 2$.

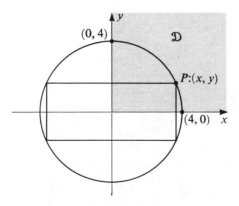

FIGURE 17.7   Rectangle inscribed in a circle.

To see that $\vec{a}:(\sqrt{8}, \sqrt{8})$ is actually the maximum point, we can reason as follows. The continuous function $f(x, y)$ on the circle arc $x^2 + y^2 = 16, x \geq 0, y \geq 0$, has a maximum and a minimum. The only possible candidate *inside* the domain is $(\sqrt{8}, \sqrt{8})$ but other critical points on the boundary of the domain are the points $(4, 0)$ and $(0, 4)$. These make the area $f(4, 0) = f(0, 4) = 0$, while $f(\sqrt{8}, \sqrt{8}) = 32$. Hence $(\sqrt{8}, \sqrt{8})$ is the maximum point and $(4, 0)$ and $(0, 4)$ are minimum points. We could also adapt the method of Taylor's theorem to prove that $\vec{a}:(\sqrt{8}, \sqrt{8})$ is the maximum point (Exercises).

27.3   Problems with More Variables and More Constraints.

We do not attempt a general theory but consider an example involving simple functions.

Example.   Find the minimum point for the function f given by $f(x, y, z) = x^2 + y^2 + z^2$ on all (x, y, z), subject to the constraints

$$x + y + z = 1, \quad \text{and}$$
$$x - y + z = 1.$$

Solution.   Denote $p(x, y, z) = x + y + z$ and $q(x, y, z) = x - y + z$. As in the simple problem, the gradient $\vec{\nabla}f$ must be orthogonal to the line of intersection of $p = 1$ and $q = 1$. But the gradients $\vec{\nabla}p$ and $\vec{\nabla}q$ are themselves orthogonal to this intersection. Moreover, any vector in the plane determined by $\vec{\nabla}p$ and $\vec{\nabla}q$ is orthogonal to the intersection of $p = 1$ and $q = 1$. All vectors in the plane of the vectors $\vec{\nabla}p$ and $\vec{\nabla}q$ are expressible in the form $\lambda \vec{\nabla}p + \mu \vec{\nabla}q$, for some multipliers λ and μ. Hence the minimum point must satisfy

$$\vec{\nabla}f - \lambda \vec{\nabla}p - \mu \vec{\nabla}q = \vec{0},$$

as well as the equation $p = 1$ and $q = 1$. This gives us five equations to determine x, y, z, λ, μ. Since $f_1 = 2x, f_2 = 2y, f_3 = 2z, p_1 = 1, p_2 = 1, p_3 = 1, q_1 = 1, q_2 = -1, q_3 = 1$, these five equations are

$$2x - \lambda - \mu = 0,$$
$$2y - \lambda + \mu = 0,$$
$$2z - \lambda - \mu = 0,$$
$$x + y + z = 1,$$
$$x - y + z = 1.$$

To solve simultaneously, we first eliminate λ from the first three equations, which results in the two equations

$$x - y - \mu = 0,$$
$$y - z + \mu = 0.$$

Then, eliminating μ, we find that

$$x - z = 0,$$

which we solve simultaneously with the last two equations of the original set. Thus we find that $x = \frac{1}{2}, y = 0, z = \frac{1}{2}$, and this enables us to determine $\lambda = \mu = \frac{1}{2}$.

To show that this is actually the minimum, we reason as follows. All points are inside the domain of f, so this is the only candidate for the minimum point. There *is* a minimum, since

$$f(x, y, z) = x^2 + y^2 + z^2 \geq 0,$$

and is continuous. At points far out along the line of intersection of $p = 1$ and $q = 1$, the function $x^2 + y^2 + z^2$ can be made as large as we please. For the same reason, there is no maximum point. Hence our candidate $\vec{a}:(\frac{1}{2}, 0, \frac{1}{2})$ is the minimum point.

27.4   Remarks on the General Theory.   To minimize a differentiable function $f(x, y, z)$ defined on a domain $\mathscr{D}$, subject to constraints $p(x, y, z) = b$ and $q(x, y, z) = c$, the multiplier rule reduces the problem of minimizing the function $F(x, y, z, \lambda, \mu)$ defined by

$$F = f - \lambda(p - b) - \mu(q - c).$$

Since $p - b = 0$ and $q - c = 0$ for any point $\vec{a}$ that satisfies the constraint equations, the gradient of F reduces to

$$\overrightarrow{\nabla F} = \overrightarrow{\nabla f} - \lambda \overrightarrow{\nabla p} - \mu \overrightarrow{\nabla q}.$$

We say that $\vec{a}$ is a regular point for the constraints if the relation

$$\lambda \overrightarrow{\nabla p} + \mu \overrightarrow{\nabla q} = \vec{0}$$

implies that $\lambda = \mu = 0$. We find that all regular points $\vec{a}$ inside the domain that minimize f, subject to the constraints, are included as the xyz-components in the set of all points (x, y, z, λ, μ) for which $\overrightarrow{\nabla F} = \vec{0}$. Thus the multiplier rule not only avoids solving the constraint equations to eliminate some of the unknowns. It actually replaces the problem of finding regular minimum points for $f(x, y, z)$, subject to the constraining equations, by another problem of minimizing the function

$$F(x, y, z, \lambda, \mu) = f - \lambda(p - b) - \mu(q - c)$$

without constraints.

To show that a point $\vec{a}$ that satisfies the constraints and for which $\overrightarrow{\nabla F} = \vec{0}$ actually gives a minimum, we may expand the function F by Taylor's theorem with remainder R_2, and examine the difference

$$F(\vec{a} + \vec{h}, \vec{\lambda} + \vec{\eta}) - F(\vec{a}, \vec{\lambda}) = f(\vec{a} + \vec{h}) - f(\vec{a})$$

to see that it is positive.

These ideas can be readily extended to minimize a function $f(x_1, x_2, \ldots, x_n)$ subject to m constraint equations, provided that $m < n$.

27.5   Exercises

In Exercises 1–13, find the maximum or minimum points, using the multiplier rule to locate all possibilities inside the domain.

1. Find the minimum distance from the origin to the line $x + 3y = 5$.
2. Find the minimum distance from the origin to the hyperbola $xy = 4$.
3. Find the area of the largest isosceles triangle that can be inscribed in a circle of radius r.
4. In Exercises 1 and 2, the positive distance is minimized if and only if the squared distance $x^2 + y^2$ is a minimum. Use this to find the minimum distance from the origin to the curve $x^2 - y^2 = 1$.
5. Find volume of the largest cylinder that can be inscribed in a sphere of radius 3.
6. Find the dimensions of the largest rectangle that can be inscribed in the graph of $x^2 + 4y^2 = 4$.

7. Find the minimizing point for $x^2 + y^2 + z^2$ subject to the constraint

$$2x + 3y - z - 6 = 0.$$

8. Find the minimizing point for $f(x, y) = x^2 + y^2$ subject to the relation $y = x^2 - 4$.
9. Find the minimum point of $f(x, y, z) = z$ on the sphere $x^2 + y^2 + (z - 1)^2 = 1$.
10. Find the maximum and minimum points for $f(x, y) = x^2 + xy + y^2$ subject to $x^2 + y^2 = 1$. Plot the graph of the constraint and plot on it the position vectors of the maximum and minimum points.
11. To find the possible maximum and minimum points of $ax^2 + 2bxy + cy^2$ subject to $x^2 + y^2 = 1$, show that the extremum points and their multiplier must satisfy the equations

$$(a - \lambda)x + by = 0,$$
$$bx + (a - \lambda)y = 0.$$

12. In Exercise 11, show that the multiplier λ must satisfy the equation

$$\det \begin{bmatrix} a - \lambda & b \\ b & a - \lambda \end{bmatrix} = 0.$$

13. Use the Taylor's theorem expansion of the function

$$F(x, y, \lambda) = 4xy - \lambda(x^2 + y^2 - 16)$$

with remainder R_2 to show that $\vec{a} : (\sqrt{8},\ \sqrt{8})$ actually gives the maximum values (Example, §27.3).

Exercises 14 and 15 present special irregularities in the multiplier method. Apply the multiplier method and find what goes wrong. Solve the problems anyway.

14. Find the point on the curve $x^3 + y^3 = 6xy$ that is at minimum distance from $P : (0, 0)$.
15. Find the point $\vec{a} : (x, y, z)$ that gives a minimum value to z, subject to $x^2 + y^2 + z^2 = 4$ and $x^2 + y^2 + (z + 1)^2 = 1$.

16. In Figure 17.8 two functions $f(x, y)$ are represented by contour lines. For each function there is a constraint $g(x, y) = c$. Use the figures to locate for each function f its minimum point in $\mathcal{D}$ subject to the condition $g(x, y) = c$.

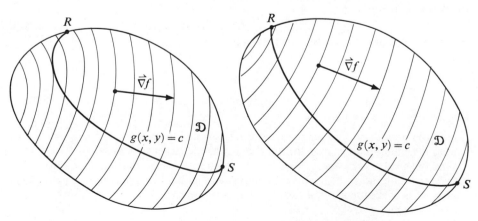

(a)   Contour lines for f and constraint   (b)   Another case

FIGURE 17.8   Exercise 16.

28 Vector Product

28.1 Definition of Vector Product. We now define, for vectors $\vec{u}$ and $\vec{v}$ in 3-space only, a vector product $\vec{u} \times \vec{v}$ that is a 3-space vector. We choose an orthonormal basis $\vec{i}:(1, 0, 0), \vec{j}:(0, 1, 0), \vec{k}:(0, 0, 1)$ and express all vectors as linear combinations of these basis vectors. Then we make the following definition.

DEFINITION.   If $\vec{u}$ and $\vec{v}$ are vectors of 3-space determined in terms of the orthonormal basis $\{\vec{i}, \vec{j}, \vec{k}\}$ by

$$\vec{u} = b_1\vec{i} + b_2\vec{j} + b_3\vec{k} \qquad \text{and} \qquad \vec{v} = c_1\vec{i} + c_2\vec{j} + c_3\vec{k},$$

then the vector ("cross") product $\vec{u} \times \vec{v}$ is defined by

$$\vec{u} \times \vec{v} = (b_2c_3 - b_3c_1)\vec{i} + (b_3c_1 - b_1c_3)\vec{j} + (b_1c_3 - b_2c_1)\vec{k}.$$

We will shortly see what this means geometrically and find a better formula for remembering it. We observe that $\vec{u} \times \vec{v} = -\vec{v} \times \vec{u}$. This shows that vector multiplication in 3-space is not commutative. This is unlike the multiplication of real numbers and unlike the multiplication of vectors of the plane (§4.7), or complex numbers. The vector cross product is not even associative. We shall see (§28.8, T8–10) that ordinarily

$$(\vec{u} \times \vec{v}) \times \vec{w} \neq \vec{u} \times (\vec{v} \times \vec{w}).$$

This is more like the operation of division in the real numbers, where ordinarily,

$$x \div y \neq y \div x, \qquad \text{and} \qquad (x \div y) \div z \neq x \div (y \div z).$$

The algebraic property that characterizes the cross product as a multiplication is the fact that it satisfies the left and right distributive laws with respect to addition.

$$\vec{u} \times (\vec{v} \times \vec{w}) = \vec{u} \times \vec{v} + \vec{u} \times \vec{w}; \qquad \text{and} \qquad (\vec{u} + \vec{v}) \times \vec{w} = \vec{u} \times \vec{w} + \vec{v} \times \vec{w}.$$

Even so, it is strictly a phenomenon of 3-space that the cross product of two vectors is another vector in the space.

By straightforward algebra we prove (Exercises) that

$$(\vec{u} + \vec{v}) \cdot \vec{u} = 0 \qquad \text{and} \qquad (\vec{u} \times \vec{v}) \cdot \vec{v} = 0.$$

Also we prove by algebraic calculation the following (Exercises).

THEOREM (Lagrange's identity).

$$\|\vec{u} \times \vec{v}\|^2 = \|\vec{u}\|^2\|\vec{v}\|^2 - (\vec{u} \cdot \vec{v})^2.$$

Then using these results we prove the following theorem.

THEOREM (Geometric interpretation of the vector product).   The vector product $\vec{u} \times \vec{v}$ of two vectors in 3-space is a vector orthogonal to both $\vec{u}$ and $\vec{v}$, having norm equal to the area of the parallelogram formed by the position vectors $\vec{u}$ and $\vec{v}$ (Figure 17.9), that is,

$$\|\vec{u} \times \vec{v}\| = \|\vec{u}\| \, \|\vec{v}\| \, |\sin \theta|.$$

Proof.   The identities $(\vec{u} \times \vec{v}) \cdot \vec{u} = 0$ and $(\vec{u} \times \vec{v}) \cdot \vec{v} = 0$ say that $\vec{u} \times \vec{v}$ is orthogonal to both $\vec{u}$ and $\vec{v}$. For the norm of $\vec{u} \times \vec{v}$ we recall (II, §12.2) that $\vec{u} \cdot \vec{v} = \|\vec{u}\| \, \|\vec{v}\| \cos \theta$. We put this result into Lagrange's identity and find that

$$\|\vec{u} \times \vec{v}\|^2 = \|\vec{u}\|^2\|\vec{v}\|^2(1 - \cos^2 \theta) = \|\vec{u}\|^2\|\vec{v}\|^2 \sin^2 \theta.$$

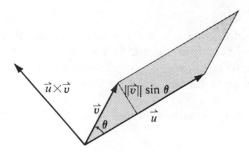

FIGURE 17.9   Vector product.

Hence $\|\vec{u} \times \vec{v}\| = \|\vec{u}\| \, \|\vec{v}\| \, |\sin\theta|$. We observe (Figure 17.9) that $\|\vec{u}\| \, |\sin\theta|$ is an altitude of the parallelogram formed by the position vectors $\vec{u}$ and $\vec{v}$, which is perpendicular to the base of length $\|\vec{v}\|$. Hence the formula gives its area. This completes the proof. Even the orientation of $\vec{u} \times \vec{v}$ is geometrically determined by the additional requirement that the sequence $\{\vec{u}, \vec{v}, \vec{u} \times \vec{v}\}$ has a right-hand orientation like any basis $\{\vec{i}, \vec{j}, \vec{k}\}$.

COROLLARY.   The vector product $\vec{u} \times \vec{v}$ is a geometric invariant. That is, it is the same vector, whatever the basis $\{\vec{i}, \vec{j}, \vec{k}\}$ chosen to calculate it. The orthogonal direction to $\vec{u}$ and $\vec{v}$, and $\sin\theta$, are both geometric invariants. Hence, so is $\vec{u} \times \vec{v}$.

28.2   Use and Interpretation of Determinant Notation. A 2×2 determinant is a number associated with the square array

$$\begin{bmatrix} a & b \\ c & d \end{bmatrix}$$

of numbers by the definition

$$\det \begin{bmatrix} a & b \\ c & d \end{bmatrix} = \begin{vmatrix} a & b \\ c & d \end{vmatrix} = ad - bc.$$

THEOREM (Area of a parallelogram).   The area A of the parallelogram formed by the position vectors $a\vec{i} + b\vec{j}$ and $c\vec{i} + d\vec{j}$ is given by

$$A = \begin{vmatrix} a & b \\ c & d \end{vmatrix}.$$

Proof.   Exercises.

Using this notation we can rewrite the product $\vec{u} \times \vec{v}$ of

$$\vec{u} = b_1\vec{i} + b_2\vec{j} + b_3\vec{k} \qquad \text{and} \qquad \vec{v} = c_1\vec{i} + c_2\vec{j} + c_3\vec{k}$$

as

$$\vec{u} \times \vec{v} = \begin{vmatrix} b_2 & b_3 \\ c_2 & c_3 \end{vmatrix} \vec{i} - \begin{vmatrix} b_1 & b_3 \\ c_1 & c_3 \end{vmatrix} \vec{j} + \begin{vmatrix} b_1 & b_2 \\ c_1 & c_2 \end{vmatrix} \vec{k}.$$

Note the alternating signs of the coefficients. We consider the rectangular array

$$\begin{matrix} + & - & + \\ \begin{bmatrix} \vec{i} & \vec{j} & \vec{k} \\ b_1 & b_2 & b_3 \\ c_1 & c_2 & c_3 \end{bmatrix} \end{matrix} .$$

The coefficient of $\vec{i}$ in $\vec{u} \times \vec{v}$ now appears as the determinant of the 2 by 2 array obtained from this by striking out the row and column in which $\vec{i}$ appears. The coefficient of $\vec{j}$ appears as *minus* the 2 by 2 determinant obtained by striking out the row and column in which $\vec{j}$ occurs. And, alternating the sign again, we find the coefficient of $\vec{k}$ to be *plus* the 2 by 2 determinant obtained by striking out the row and column of $\vec{k}$. This provides a systematic scheme for remembering the formula for $\vec{u} \times \vec{v}$.

Example.   Find the vector product

$$\vec{u} \times \vec{v} = (2\vec{i} - 3\vec{j} + \vec{k}) \times (\vec{i} - 5\vec{j} - \vec{k}).$$

Solution.   We write down the array

$$\begin{bmatrix} \vec{i} & \vec{j} & \vec{k} \\ 2 & -3 & 1 \\ 1 & -5 & -1 \end{bmatrix} .$$

By appropriately striking out rows and columns and alternating the signs of the 2 by 2 determinants, we obtain

$$\vec{u} \times \vec{v} = + \begin{vmatrix} -3 & 1 \\ -5 & -1 \end{vmatrix} \vec{i} - \begin{vmatrix} 2 & 1 \\ 1 & -1 \end{vmatrix} \vec{j} + \begin{vmatrix} 2 & -3 \\ 1 & -5 \end{vmatrix} \vec{k}$$
$$= 8\vec{i} + 3\vec{j} - 7\vec{k}.$$

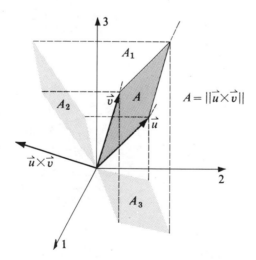

FIGURE 17.10   Components of the vector product.

We can derive more information about the coefficients in

$$\vec{u} \times \vec{v} = \begin{vmatrix} b_2 & b_3 \\ c_2 & c_3 \end{vmatrix} \vec{i} - \begin{vmatrix} b_1 & b_3 \\ c_1 & c_3 \end{vmatrix} \vec{j} + \begin{vmatrix} b_1 & b_2 \\ c_1 & c_3 \end{vmatrix} \vec{k}$$
$$= A_1 \vec{i} + A_2 \vec{j} + A_3 \vec{k}.$$

The coordinates A_1, A_2, A_3, can be interpreted as areas. Let $A = \|\vec{u} \times \vec{v}\|$ be the area of the parallelogram formed by the vectors $\vec{u}$ and $\vec{v}$ (Figure 17.10). We have at once $A = \sqrt{A_1^2 + A_2^2 + A_3^2}$. We observe that if we project the parallelogram of $\vec{u}$ and $\vec{v}$ on planes perpendicular to $\vec{i}, \vec{j}, \vec{k}$, the projected figures are parallelograms having areas A_1, A_2, A_3 respectively. (See Problem T5, §28.8).

28.3   Plane Parallel to Two Lines

Example.   Find the equation of the plane through the point $\vec{q}$ parallel to lines $\vec{r} = t\vec{a} + \vec{c}$, $\vec{r} = s\vec{b} + \vec{d}$.

Solution.   The directions of the lines are given by the vectors $\vec{a}$ and $\vec{b}$. With $\vec{a} \times \vec{b}$ as the normal vector of the required plane, the plane will be parallel to both of the lines. Let $\vec{r}:(x, y, z)$ be a variable position vector. Then $\vec{r} - \vec{q}$ is a located vector in the plane if and only if $\vec{r} - \vec{q}$ is orthogonal to $\vec{a} \times \vec{b}$. Hence the equation of the plane is

$$(\vec{a} \times \vec{b}) \cdot (\vec{r} - \vec{q}) = 0.$$

28.4   Plane through Three Points

Example.   Find the equation of the plane through points $\vec{a}:(2, 3, -2)$, $\vec{b}:(1, -1, -1)$, $\vec{c}:(1, 0, 1)$ (Figure 17.11).

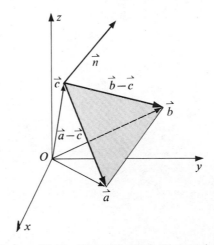

FIGURE 17.11   Plane through three points.

Solution.   The located vectors $\vec{a} - \vec{c}$:(1, 3, −3) and $\vec{b} - \vec{c}$:(0, −1, −2) are in the plane. The normal to the plane must be orthogonal to both of these. One such normal vector is $(\vec{a} - \vec{c}) \times (\vec{b} - \vec{c})$, which we find from the matrix

$$\begin{bmatrix} \vec{i} & \vec{j} & \vec{k} \\ 1 & 3 & -3 \\ 0 & -1 & -2 \end{bmatrix}$$

to be $\vec{n} = -9\vec{i} + 2\vec{j} - \vec{k}$. Let $\vec{r} = x\vec{i} + y\vec{j} + z\vec{k}$ be a variable position vector, which is in the plane if and only if $\vec{n}$ is orthogonal to the located vector $\vec{r} - \vec{c}$,

$$\vec{n} \cdot (\vec{r} - \vec{c}) = 0.$$

This is a vector equation of the plane, since $\vec{c}$ and $\vec{n}$ are known. In components it reduces to the required equation

$$-9(x - 1) + 2(y - 0) - 1(z - 1) = 0.$$

28.5   3 by 3 Determinant

DEFINITION.   We define the 3 by 3 determinant

$$\Delta = \begin{vmatrix} a_1 & a_2 & a_3 \\ b_1 & b_2 & b_3 \\ c_1 & c_2 & c_3 \end{vmatrix}$$

by $\Delta = \vec{a} \cdot (\vec{b} \times \vec{c})$ with the rows $\vec{a}$:(a_1, a_2, a_3), $\vec{b}$:(b_1, b_2, b_3), and $\vec{c}$:(c_1, c_2, c_3) of the matrix. We observe that this is formally the same process used to define the vector product $\vec{b} \times \vec{c}$ itself.

Example.   Evaluate the 3 by 3 determinant

$$\Delta = \begin{vmatrix} 2 & -1 & 3 \\ 5 & 7 & 0 \\ 0 & 4 & -4 \end{vmatrix}.$$

Solution.

$$\Delta = 2 \begin{vmatrix} 7 & 0 \\ 4 & -4 \end{vmatrix} - (-1) \begin{vmatrix} 5 & 0 \\ 0 & -4 \end{vmatrix} + 3 \begin{vmatrix} 5 & 7 \\ 0 & 4 \end{vmatrix}$$
$$= 2(-28) + (1)(-20) + 3(20) = -16.$$

28.6   Distance between Two Skew Lines

Example.   Find the minimum distance between the skew lines $\vec{p} - \vec{p}_0 = s\vec{a}$, and $\vec{q} - \vec{q}_0 = t\vec{b}$, where $\vec{p}_0, \vec{a}, \vec{q}_0, \vec{b}$ are known vectors and $\vec{p}$ and $\vec{q}$ are variable position vectors locating points in the lines.

Solution.   The minimum distance must be measured in a direction that is orthogonal to both lines, if such a direction exists. Such a direction is $\vec{a} \times \vec{b}$ and the unit direction (Figure 17.12)

$$\vec{n} = \frac{\vec{a} \times \vec{b}}{\|\vec{a} \times \vec{b}\|}.$$

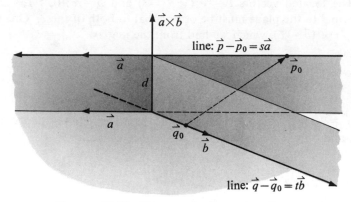

FIGURE 17.12   Distance between skew lines.

Since the lines are skew $\vec{a} \times \vec{b} \neq \vec{0}$. It is convenient to consider the line segment $\vec{p}_0 - \vec{q}_0$, which joins the point $\vec{q}_0$ in one line to $\vec{p}_0$ in the other. The scalar projection of this line segment onto $\vec{n}$ gives the required minimum distance d, which is

$$d = \left((\vec{p}_0 - \vec{q}_0) \cdot \frac{\vec{a} \times \vec{b}}{\|\vec{a} \times \vec{b}\|} \right).$$

This completes the problem. Any other pair of points $\vec{p}, \vec{q}$ on the two lines would give the same result.

28.7   Exercises

1. Compute the vector products $\vec{a} \times \vec{b}$ and $\vec{b} \times \vec{a}$ for
    (a) $\vec{a}:(3, -1, 1), \vec{b}:(1, 1, 1)$.
    (b) $\vec{a}:(2, 3, 0), \vec{b}:(0, 0, 1)$.
    (c) $\vec{a}:(2, 0, 0), \vec{b}:(3, 0, 0)$.
    (d) $\vec{a}:(x, y, z), \vec{b}:(-x, -y, -z)$.
2. Find the equation of the plane containing the position vectors $\vec{a}:(3, -1, -1)$ and $\vec{b}:(-1, 1, -3)$.
3. Find the equation of the plane containing the line $\vec{p} - \vec{p}_0 = t\vec{a}$ and parallel to the line $\vec{p} - \vec{q}_0 = s\vec{b}$.
4. Find the equation of the plane through the point $\vec{p}_0:(1, -1, -5)$, parallel to both of the vectors $\vec{b}:(3, 0, -1)$ and $\vec{c}:(1, -2, 1)$.
5. Find the equation of the line through $\vec{p}_0$ in Exercise 4 and orthogonal to both $\vec{b}$ and $\vec{c}$.
6. Find the equation of the plane containing the points $p_1:(-2, 3, -1)$, $p_2:(4, 4, -5)$, and $p_3:(1, 1, -2)$ by observing that the located vectors $\vec{p}_1 - \vec{p}_2$ and $\vec{p}_1 - \vec{p}_3$ are in the plane.
7. Find the volume of the box having edges $\vec{a}:(3, -1, -1), \vec{b}:(1, 1, -3)$, and $\vec{c}:(1, -1, -5)$.
8. Find the area of the parallelogram formed by the vectors $\vec{a}:(3, -6, 2)$ and $\vec{b}:(1, 3, -2)$.
9. For the parallelogram in Exercises 8, find the area of the projection of the parallelogram onto a plane orthogonal to $\vec{i}$.
10. Find the minimum distance between the lines

$$\vec{p} - (2\vec{i} + \vec{j} - \vec{k}) = t(\vec{i} + \vec{j} + \vec{k}) \quad \text{and} \quad \vec{p} - (4\vec{i} - 5\vec{j} + 3\vec{k}) = s(-\vec{i} + \vec{k}).$$

11. Determine whether the lines

$$x\vec{i} + y\vec{j} + z\vec{k} = t(\vec{i} + \vec{j} + \vec{k}) - (2\vec{i} + \vec{j} - \vec{k})$$
$$x\vec{i} + y\vec{j} + z\vec{k} = s(-\vec{i} + 4\vec{j} - \vec{k}) + (6\vec{j} + 3\vec{k})$$

intersect.

12. Evaluate the 3 by 3 determinants

$$\text{(a)} \begin{vmatrix} 2 & -1 & 3 \\ 0 & 4 & 2 \\ 5 & -2 & 6 \end{vmatrix}; \qquad \text{(b)} \begin{vmatrix} a_1 & a_2 & a_3 \\ b_1 & b_2 & b_3 \\ b_1 & b_2 & b_3 \end{vmatrix}.$$

13. Show that

$$\begin{vmatrix} a_1 & a_2 & a_3 \\ b_1 & b_2 & b_3 \\ c_1 & c_2 & c_3 \end{vmatrix} = a_1 b_2 c_3 + a_2 b_3 c_1 + a_3 b_1 c_2 - a_3 b_2 c_1 - a_2 b_1 c_3 - a_1 b_3 c_2.$$

14. Find the volume of the tetrahedron formed by the points $(3, 1, 0)$, $(4, 4, -1)$, $(5, 0, 3)$, $(-4, -2, 0)$.

15. Find the perpendicular distance from the plane $2x + 3y - 6z = 4$ to the point $(1, -1, 4)$.

16. Use $\vec{a} \times \vec{b}$ to find the sine of the angle θ between the vectors $\vec{a}:(1, 2, -2)$ and $\vec{b}:(-3, 4, 0)$.

17. Use vector methods to find the altitude through $P_1:(6, -1, -1)$ of the triangle formed with the other vertices $P_2:(-1, 3, -8)$ $P_3:(2, -1, 4)$.

18. Find the equation of the plane containing the line $l_1: x = 2t + 3, y = -t - 1, z = 3t + 4$ and parallel to the line $l_2: x = t, y = t + 4, z = -t + 8$.

19. For the lines in Exercise 18, find the equation of the plane containing l_2 and parallel to l_1. Use these results to find the minimum distance between the lines as the distance between the parallel planes that contain the lines.

28.8 Problems

T1. Compute a table of the nine products of $\{\vec{i}, \vec{j}, \vec{k}\}$.

T2. Use T1 to multiply $(b_1\vec{i} + b_2\vec{j} + b_3\vec{k}) \times (c_1\vec{i} + c_2\vec{j} + c_3\vec{k})$.

T3. Prove that $(\vec{u} \times \vec{v}) \cdot \vec{u} = (\vec{u} \times \vec{v}) \cdot \vec{v} = 0$.

T4. Prove Lagrange's identity (§28.1) by expanding both sides and simplifying.

T5. The area A of the parallelogram formed by the vectors $a\vec{i} + b\vec{j}$ and $c\vec{i} + d\vec{j}$ is found by subtracting the areas $2B + 2C + 2D$ from the area of the enclosing rectangle (Figure 17.13). Prove that

$$A = ac - bd.$$

T6. Prove that $\vec{a}$ and $\vec{b}$ are parallel if and only if $\vec{a} \times \vec{b} = 0$.

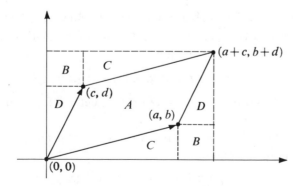

FIGURE 17.13 Problem T5.

T7.   Show that $\vec{a}$, $\vec{b}$, $\vec{c}$ are coplanar if and only if $\vec{a} \times \vec{b} = \vec{a} \times \vec{c}$.

T8.   Prove that $\vec{a} \times (\vec{b} \times \vec{c})$ is a vector in the plane of $\vec{b}$ and $\vec{c}$, hence for some numbers p and q

$$\vec{a} \times (\vec{b} \times \vec{c}) = p\vec{b} + q\vec{c}.$$

T9.   Another approach could have been used to discover the vector product of $\vec{b} = b_1\vec{i} + b_2\vec{j} + b_3\vec{k}$ and $\vec{c} = c_1\vec{i} + c_2\vec{j} + c_3\vec{k}$. The vector $\vec{p} = x\vec{i} + y\vec{j} + z\vec{k}$ is orthogonal to both $\vec{b}$ and $\vec{c}$ if and only if $\vec{b} \cdot \vec{p} = 0$ and $\vec{c} \cdot \vec{p} = 0$. Solve these two equations for (x, y, z) to obtain

$$(x, y, z) = t\left(\begin{vmatrix} b_2 & b_3 \\ c_2 & c_3 \end{vmatrix}, \ -\begin{vmatrix} b_1 & b_3 \\ c_1 & c_3 \end{vmatrix}, \ \begin{vmatrix} b_1 & b_2 \\ c_2 & c_2 \end{vmatrix} \right),$$

which leaves the scalar multiplier t to be determined.

T10.   In T9, determine that $|t| = 1$ when it is required that $\|\vec{b} \times \vec{c}\|$ is the area of the parallelogram of $\vec{b}$ and $\vec{c}$.

29   Vector Fields

NOTE: Our objective here is to establish what a vector field is, with physical interpretation, and to define and interpret the fundamental derivative operators on it. We leave the development of differentiation and integration of vector fields to more advanced texts.

29.1   **Vector Fields.**   The physical examples that gives rise to vector fields are typically velocity fields or force fields. A body of fluid, such as air or water, is moving. Fluid particles at different locations may be moving with different velocities, but at any location $\vec{r}:(x, y, z)$, there is a uniquely determined vector $\vec{v}$. Then the function $\vec{v} = \vec{F}(\vec{r})$ is a vector field. Another picture of a vector field is given by a large body, such as the earth, with its gravitational attraction. At any point r outside the earth there is for a unit mass at that point a uniquely determined force $\vec{w}$ of attraction pointed towards center of the earth. Then the function $\vec{w} = \vec{F}(\vec{r})$ is a vector (force) field. Abstractly then we define:

DEFINITION.   A vector field $\vec{w} = \vec{F}(\vec{r})$ is a function defined, and repeatedly differentiable on a domain $\mathscr{D}$ in a vector space, that assigns to each (position) vector $\vec{r}$ in $\mathscr{D}$ a vector $\vec{w}$, given by $\vec{w} = \vec{F}(\vec{r})$.

To represent a vector field $\vec{F}$ analytically we may let $\vec{r}:(x, y, z)$ be a vector in the domain $\mathscr{D}$, and choose a basis $\{\vec{i}, \vec{j}, \vec{k}\}$ in the codomain of $\vec{F}$. With respect to this basis, the vector $\vec{w} = \vec{F}(\vec{r})$ has the representation

$$\vec{w} = A(x, y, z)\vec{i} + B(x, y, z)\vec{j} + C(x, y, z)\vec{k},$$

where A, B, C are scalar-valued functions of (x, y, z) defined for points $\vec{r}:(x, y, z)$ in the domain of $\vec{F}$.

To represent a vector field $\vec{F}$ graphically, we usually plot $\vec{r}:(x, y, z)$ as a *point* in $\mathscr{D}$, and then from this point we plot the vector

$$\vec{w} = A(x, y, z)\vec{i} + B(x, y, z)\vec{j} + C(x, y, z)\vec{k},$$

as a located vector (Figure 17.14).

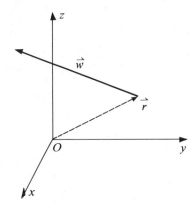

FIGURE 17.14   Representation of the vector pair $(\vec{r}, \vec{w})$.

Example A.   Graph the vector field defined in the xy-plane, $\vec{x} \neq \vec{0}$, by the function

$$\vec{w} = -\frac{x}{x^2 + y^2} \vec{i} - \frac{y}{x^2 + y^2} \vec{j} + 0\vec{k}.$$

Solution.   We observè that $\|\vec{w}\| = 1/\sqrt{x^2 + y^2}$ and that

$$\vec{w} = \frac{1}{\sqrt{x^2 + y^2}} \left(\frac{-x}{\sqrt{x^2 + y^2}} \vec{i} - \frac{y}{\sqrt{x^2 + y^2}} \vec{j} + 0\vec{k} \right).$$

The vector in the parentheses is a unit vector pointing towards the origin, and the scalar $1/\sqrt{x^2 + y^2}$ decreases as $1/r$ when we increase the distance r from the origin (Figure 17.15(a)).

Example B.   Graph the vector defined in the xy-plane for $\vec{r}: (x, y, 0)$ by

$$\vec{w} = -\omega y \vec{i} + \omega x \vec{j} + 0\vec{k}.$$

Solution.   We observe that $\|\vec{w}\| = \|\vec{r}\|$ and that $(\vec{w} \cdot \vec{r}) = 0$, so that $\vec{w}$ is orthogonal to $\vec{r}$. We plot this with $\omega = 1$ for a number of points $\vec{x}$ in the plane (Figure 17.15(b)).

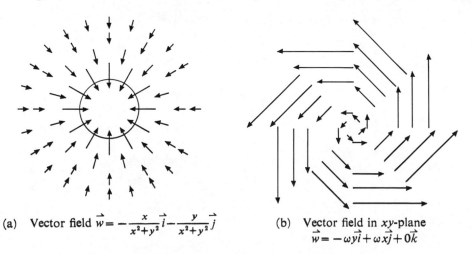

(a)   Vector field $\vec{w} = -\frac{x}{x^2+y^2} \vec{i} - \frac{y}{x^2+y^2} \vec{j}$

(b)   Vector field in xy-plane
$\vec{w} = -\omega y \vec{i} + \omega x \vec{j} + 0\vec{k}$

FIGURE 17.15   Vector fields.

29.2 Best Linear Approximation. Our definition of the differential (II, §16) applies to vector fields without change. For the vector field $\vec{w} = \vec{F}(\vec{r})$, the linear vector field $D\vec{w} = D\vec{F}$ is its best linear approximation at each point $\vec{r}$. We do not attempt to show this graphically or give it any other physical interpretation except that it is the best linear approximation to $\vec{F}$.

29.3 Divergence. We cannot interpret the operation $\vec{\nabla}\vec{F}$ for a vector field as a gradient. We define the vector derivative operator formally by

$$\vec{\nabla} = \frac{\partial}{\partial x}\vec{i} + \frac{\partial}{\partial y}\vec{j} + \frac{\partial}{\partial z}\vec{k}.$$

Then, for a vector field $\vec{F} = A\vec{i} + B\vec{j} + C\vec{k}$, the dot product $\vec{\nabla}\cdot\vec{F}$ produces the scalar,

$$\vec{\nabla}\cdot\vec{F} = \frac{\partial A}{\partial x} + \frac{\partial B}{\partial y} + \frac{\partial C}{\partial z}.$$

We call the scalar $\vec{\nabla}\cdot\vec{F}$ the *divergence* of the vector field $\vec{F}$ at $\vec{r}:(x, y, z)$, written

$$\text{div } \vec{F} = \vec{\nabla}\cdot\vec{F}.$$

To see what this means physically we first solve the following problem.

Example A. A fluid of constant density 1 is flowing through a region in xyz-space enclosed by a small cube $\mathcal{K}$ with edges of length $2dx$, $2dy$, $2dz$ parallel to the coordinate axes and center at $P_0:(x_0, y_0, z_0)$. Find the net rate of flow into and out of the cube if the velocity of flow is given by the vector field

$$\vec{v} = A(x, y, z)\vec{i} + B(x, y, z)\vec{j} + C(x, y, z)\vec{k}.$$

Solution. Consider the right face perpendicular to the y-axis at $y_0 + dy$. The flow across this face with unit normal $\vec{n}:(0, 1, 0)$ is the integral (Figure 17.16)

$$\iint (1)B(x, y_0 + dy, z)\, dx\, dz,$$

taken over the face. Similarly the flow across the opposite left face with unit normal $\vec{n}:(0, -1, 0)$ is

$$-\iint (1)B(x, y_0 - dy, z)\, dx\, dz$$

taken over the face. Thus the net flow across these two faces is the difference

$$\iint [B(x, y_0 + dy, z) - B(x, y_0 - dy, z]\, dx\, dz,$$
$$x - dx \leqq x \leqq x + dx, \qquad z - dz \leqq z \leqq z + dz.$$

We may apply the integral form of the mean value theorem (I, §23.2; III §25.1) to the difference

$$B(x, y_0 + dy, z) - B(x, y_0 - dy, z) = \int_{y_0 - dy}^{y_0 + dy} \frac{\partial B}{\partial y}\, dy.$$

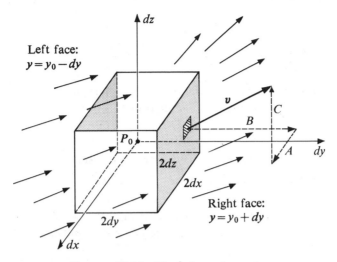

FIGURE 17.16   Flow through a cube.

Substituting this expression into the integral above, we obtain a new expression for the net flow across the two faces perpendicular to the y-axis. It is given by the repeated triple integral

$$\iiint_{\mathscr{K}} \frac{\partial B}{\partial y} \, dx \, dy \, dz$$

taken over the entire cube $\mathscr{K}$. Similar expressions give the net flow across the other two pairs of faces, and when we add them all together we get the net flow in and out of the cube as

$$\iiint_{\mathscr{K}} \left(\frac{\partial A}{\partial x} + \frac{\partial B}{\partial y} + \frac{\partial C}{\partial z} \right) dx \, dy \, dz,$$

taken over the cube. This completes the problem.

We use the result of this problem to compute the average value of the divergence over the solid cube $\mathscr{K}$. It is

$$\overline{\vec{\nabla} \cdot \vec{F}} = \frac{\iiint_{\mathscr{K}} \vec{\nabla} \cdot \vec{F} \, dx \, dy \, dz}{\iiint_{\mathscr{K}} 1 \, dx \, dy \, dz}.$$

According to the example, it is the net flow per unit volume into and out of the cube. We let the edge of the cube approach zero and this average value $\overline{\vec{\nabla} \cdot \vec{F}}$ approaches the local value $\vec{\nabla} \cdot \vec{F}$ at $\vec{r}{:}(x, y, z)$. Hence the divergence $\vec{\nabla} \cdot \vec{F}$ at $\vec{r}{:}(x, y, z)$ gives the net flow into and out of the point. It is ordinarily zero unless there is at $\vec{r}$ a "source," where mass is injected into the flow, or a "sink," where it is removed. $\vec{\nabla} \cdot \vec{F}$ is called the *flux density* at $\vec{r}$.

Example B.   Compute the divergence for the vector field (§29.1, Example A) $\vec{w} = \vec{F}(\vec{r})$ defined by

$$\vec{w} = -\frac{x}{x^2 + y^2} \, \vec{i} - \frac{y}{x^2 + y^2} \vec{j} + 0\vec{k}, \qquad \vec{x} \neq \vec{0}.$$

Solution.

$$\vec{\nabla}\cdot\vec{F} = -\frac{\partial}{\partial x}\left(\frac{x}{x^2 + y^2}\right) - \frac{\partial}{\partial y}\left(\frac{y}{x^2 + y^2}\right) + \frac{\partial}{\partial z}0 = 0, \quad \text{if} \quad \vec{r} \neq \vec{0}.$$

The divergence is undefined at the origin, where the graph indicates that there is a sink because the entire flow is directed towards the origin.

29.4   Curl.   The symbolic operator $\vec{\nabla} \times$ can be applied to a vector field $\vec{F}(\vec{r})$. It produces the vector field $\vec{\nabla} \times \vec{F}(\vec{r})$, which is called the curl $\vec{F}$.

Example.   Every point in a rigid body rotates around the z-axis at a constant angular rate ω. Find the linear velocity field of points in this body, and calculate its curl (Figure 17.17).

Solution.   The position vector at time t of a point $\vec{r}:(x, y, z)$ in this field at perpendicular distance a from the z-axis is

$$\vec{r}(t) = a \cos \omega t \vec{i} + a \sin \omega t \vec{j} + z\vec{k}.$$

Differentiating this with respect to t, we find the linear velocity field of particles located at the point $\vec{x}$

$$\begin{aligned} \vec{v} = \vec{r}'(t) &= -\omega a \sin \omega t \vec{i} + \omega a \cos \omega t \vec{j} + 0\vec{k} \\ &= -\omega y \vec{i} + \omega x \vec{j} + 0\vec{k} \end{aligned}$$

(§29.1, Example B).   Now the curl of the velocity field $\vec{r}'(t)$ is given by

$$\vec{\nabla} \times \vec{r}'(t) = \begin{vmatrix} \vec{i} & \vec{j} & \vec{k} \\ \dfrac{\partial}{\partial x} & \dfrac{\partial}{\partial y} & \dfrac{\partial}{\partial z} \\ -\omega y & +\omega x & 0 \end{vmatrix} = 2\omega\vec{k}.$$

This shows that the angular velocity ω of the rotation is given by

$$\omega = \tfrac{1}{2}\|\vec{\nabla} \times \vec{r}'(t)\|.$$

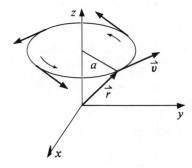

FIGURE 17.17   Rotational field.

Moreover the vector $\frac{1}{2}\vec{\nabla} \times \vec{r}'(t)$ gives not only the magnitude of the angular velocity but the axis of rotation $\vec{k}$ as well. Hence we regard the vector

$$\vec{\omega} = \tfrac{1}{2}\vec{\nabla} \times \vec{r}'(t)$$

as the *vector angular velocity* of the rigid body. This completes the problem.

Guided by this example, we generalize the meaning of $\vec{\nabla} \times \vec{F}(\vec{r})$ to any differentiable vector field. We say that curl $\vec{F}$ measures the local rotational effect of the vector velocity field $\vec{F}(\vec{r})$ at the point $\vec{r}$. A field for which

$$\text{curl } \vec{F} = \vec{0}$$

at every point is said to be *irrotational.*

29.5   Exercises

1. Plot the vector field $\vec{w} = y\vec{i} - \vec{j} + 0\vec{k}$.
2. Plot the vector field $\vec{w} = (x + y)\vec{i} + (x - y)\vec{j}$. Find its best linear approximation $D\vec{w}$ and show that it is also a vector field in local coordinates (dx, dy).
3. Calculate the divergence and curl of the vector field in Exercise 2.

In Exercises 4–11, calculate the divergence and curl. Where possible, state a physical meaning for the result obtained.

4. $\vec{F}(\vec{r}) = (x^2 - y^2)\vec{i} - 2xy\vec{j}$.   5. $\vec{F}(\vec{r}) = \dfrac{x}{x^2 + y^2}\vec{i} + \dfrac{y}{x^2 + y^2}\vec{j}$.

6. $\vec{F}(\vec{r}) = e^x \cos y\,\vec{i} - e^x \sin y\,\vec{j}$.   7. $\vec{F}(\vec{r}) = e^x \sin y\,\vec{i} - e^x \cos y\,\vec{j}$.

8. $\vec{F}(\vec{r}) = (e^y + e^{-y}) \cos x\,\vec{i} + (e^y - e^{-y}) \sin x\,\vec{j}$.
9. The vector force field in xyz-space due to a point charge at the origin, given by

$$\vec{F}(\vec{r}) = -\frac{1}{r^2}\frac{\vec{r}}{r}, \qquad \text{where} \quad r = \|\vec{r}\| = \sqrt{x^2 + y^2 + z^2}.$$

10. $\vec{F}(\vec{r}) = (y - z)\vec{i} + (z - x)\vec{j} + (x - y)\vec{k}$.
11. The gradient field $\vec{F}(\vec{r}) = \vec{\nabla}u(\vec{r})$ for the scalar-valued function $u(\vec{r})$, where
    (a) $u = x^2 - y^2$;   (b) $u = \ln(x^2 + y^2)$;
    (c) $u = \arctan \dfrac{y}{x}$;   (d) $u = e^x \cos y$.

12. Compute the curl and divergence of the gradient vector field of any repeatedly differentiable scalar field $u(\vec{r})$ in xyz-space. Show that

$$\text{div }\vec{\nabla}u = \vec{\nabla}\cdot\vec{\nabla}u = \frac{\partial^2 u}{\partial x^2} + \frac{\partial^2 u}{\partial y^2} + \frac{\partial^2 u}{\partial z^2} \quad \text{and} \quad \text{curl }\vec{\nabla}u = \vec{\nabla} \times \vec{\nabla}u = 0.$$

13. Prove that, formally, for any repeatedly differentiable scalar field $u(\vec{r})$,

$$\vec{\nabla}\cdot(\vec{\nabla}u) = (\vec{\nabla}\cdot\vec{\nabla})u = \frac{\partial^2 u}{\partial x^2} + \frac{\partial^2 u}{\partial y^2} + \frac{\partial^2 u}{\partial z^2}.$$

14. A scalar field $u(\vec{r})$ is said to satisfy Laplace's equation if $(\vec{\nabla}\cdot\vec{\nabla})u = 0$. Show that the scalar fields in Exercise 11 satisfy Laplace's equation in the plane.
15. Prove that for a vector field $\vec{F}(\vec{r})$

$$\text{div (curl }\vec{F}) = \vec{\nabla}\cdot(\vec{\nabla} \times \vec{F}) = 0.$$

16. By ordinary scalar multiplication we can multiply the vector field $\vec{F}(\vec{r})$ by the scalar-valued function $u(\vec{r})$ to produce a vector field $u\vec{F}$. Show that

$$\text{div}\,(u\vec{F}) = u\,\text{div}\,\vec{F} + \vec{\nabla}u \cdot \vec{F}.$$

17. In Exercise 16, show that

$$\text{curl}\,(u\vec{F}) = u\,\text{curl}\,\vec{F} + \vec{\nabla}u \times \vec{F}.$$

18. As in Exercises 5 and 9, prove for the vector field $r^n\vec{r}$, where n is any integer and $r = \|\vec{r}\|$, that $\text{div}\,(r^n\vec{r}) = (n + 3)r^n$.

19. For the vector field $\vec{w} = A(x, y)\vec{i} + B(x, y)\vec{j}$, show that there exists a scalar function $u(x, y)$ such that $Du = A\,dx + B\,dy$ if and only if $\text{curl}\,\vec{w} = 0$ (cross derivative test, II, §22.1).

20. Show that the vector field of Figure 17.15(b) is given by the function $\vec{F}(\vec{r}) = \omega(\vec{k} \times \vec{r})$. Replace $\vec{k}$ by any constant unit vector $\vec{a}$ in 3-space and compute $\text{curl}\,\omega(\vec{a} \times \vec{r})$. Give the result a physical interpretation.

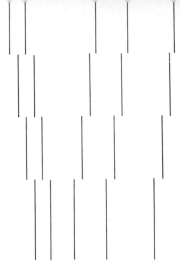

Techniques of Multiple Integration

CHAPTER 18

30 Slice Elements for Calculating Volume and Area

30.1 Parallel Slices. We consider the volume of a region in $\mathscr{R}^3$ bounded by a surface. For example, it might be the volume of water in an odd-shaped vessel up to height z (Figure 18.1). We think of this volume as being composed of thin parallel slices that we add to find the total volume. In calculation, we may find the volume $V(z)$ from the lowest level z_1 up to level z by the repeated integral

$$V(z) = \int_{z_1}^{z} \left(\iint_{\mathscr{A}(z)} dx \, dy \right) dz,$$

where the inside integral is taken over the region $\mathscr{A}(z)$ in the xy-plane, which is the projection of the horizontal section of the figure at the level z. The inside integral gives the area

$$A(z) = \iint_{\mathscr{A}(z)} dx \, dy$$

of the horizontal section at level z.

We may see this in a different way by differentiating the volume $V(z)$ with respect to z (I, §22.3). We take the limit

$$\lim_{\Delta z \to 0} \frac{V(z + \Delta z) - V(z)}{z} = \lim_{\Delta z \to 0} \frac{\Delta V}{\Delta z}.$$

ΔV is the difference in volume between levels z and $z + \Delta z$, a parallel slice of thickness Δz (Figure 18.1). Let $A_{\max}$ be the maximum area of horizontal sections in the interval $[z, z + \Delta z]$ and let $A_{\min}$ be the minimum area. Then

$$(A_{\min}) \, \Delta z \leqq \Delta V \leqq (A_{\max}) \, \Delta z.$$

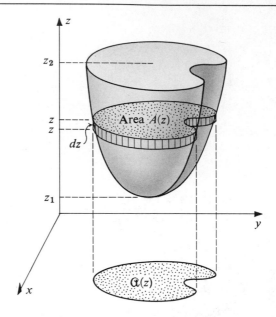

FIGURE 18.1   Volume by horizontal slices.

Dividing by Δz, which we will take to be positive, yields

$$A_{\min} \leqq \frac{\Delta V}{\Delta z} \leqq A_{\max}.$$

Now let $\Delta z \to 0$. Then, assuming that the bounding surface is smooth, we find that both $A_{\min}$ and $A_{\max}$ approach $A(z)$. Hence

$$A(z) = \frac{dV}{dz} = \frac{d}{dz} \int_{z_1}^{z} \left(\iint_{\mathscr{A}(z)} dx\, dy \right) dz = \iint_{\mathscr{A}(z)} dx\, dy.$$

So, in the last stage of the repeated integral, the calculation of the volume reduces to

$$V = \int_{z_1}^{z_2} A(z)\, dz.$$

The expression $A(z)\, dz$ is a volume element formed from a slice of area $A(z)$ and thickness dz at level z. Thus if we can use some simple geometry to calculate the areas $A(z)$, we can dispense with the first two stages of the triple repeated integral and calculate the volume V by a single integral.

Example.   Find the volume of the solid region bounded by the coordinate planes $x = 0$, $y = 0$, $z = 0$, and the surface $z^2 = 4 - x - y$.

Solution.   Planes at the fixed level z intersect the solid in right triangles whose perpendicular sides both have length $4 - z^2$ (Figure 18.2). The area $A(z) = \frac{1}{2}(4 - z^2)^2$. The volume of a slice at level z, $dV = \frac{1}{2}(4 - z^2)^2\, dz$. The solid extends from the level $z = 0$ to $z = 2$. Its volume is

$$V = \int_0^2 \tfrac{1}{2}(4 - z^2)^2\, dz = \tfrac{128}{15}.$$

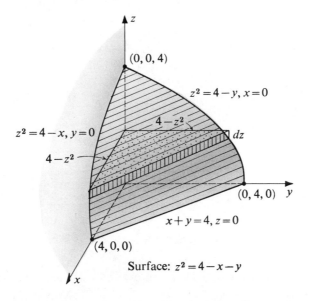

FIGURE 18.2   Solid with triangular slices.

30.2   Surfaces and Solids of Revolution.   One common situation in which we can shorten volume integration by the slice method occurs in solids that are generated by revolving a plane figure about a line in its plane.

Let $y = y(x)$, $a \leq x \leq b$ be a piecewise smooth arc in the xy-plane. We may rotate it about the x-axis to generate a surface (Figure 18.3). Each point $(x, y(x))$ on the generating arc describes a circle of radius $|y(x)|$ with center at $(x, 0, 0)$. This circle lies in the surface of revolution. With this radius, the area of the disk inside the circle is πy^2, so that the volume element $A(x)\, dx$ becomes $\pi[y(x)]^2\, dx$. Then

$$V = \int_a^b \pi y^2\, dx.$$

Example A.   Find the volume enclosed by the ellipsoid generated when the upper half of the ellipse $b^2 x^2 + a^2 y^2 = a^2 b^2$ revolves around the x-axis.

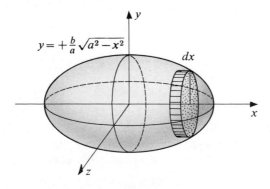

FIGURE 18.3   Solid of revolution.

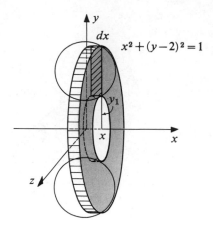

FIGURE 18.4   Solid ring with washer slice.

Solution. We solve the equation of the ellipse for y and find that the upper half of the ellipse is given by the function $y = +(b/a)\sqrt{a^2 - x^2}$. With this as a radius, we find that the volume is

$$V = \int_{-a}^{a} \pi \frac{b^2}{a^2} (a^2 - x^2)\, dx = \frac{4}{3} \pi a b^2.$$

Example B. Find the volume of the solid ring generated when the disk $x^2 + (y - 2)^2 \leqq 1$ is removed around the x-axis.

Solution. We plot the circle $x^2 + (y - 2)^2 = 1$ with center at $(0, 2)$ and radius 1 and rotate it to form the surface of the ring (Figure 18.4). Then at a constant x, we slice the ring with a plane parallel to the yz-plane. This intersects the generating circle in two points, which we find by solving for y in terms of x, $y_1 = 2 - \sqrt{1 - x^2}$, and $y_2 = 2 + \sqrt{1 - x^2}$. In the figure, we see that y_1 is the radius of the hole in a washer-shaped slice, and y_2 is the radius of the outside perimeter. The area of the washer slice,

$$A(x) = \pi y_2^2 - \pi y_1^2 = 4\pi\sqrt{1 - x^2}.$$

Hence, the volume element $dV = 4\pi\sqrt{1 - x^2}\, dx$, and

$$V = \int_{-1}^{1} 4\pi\sqrt{1 - x^2}\, dx = 4\pi(\tfrac{1}{2})[x\sqrt{1 - x^2} + \arcsin x]_{-1}^{1} = 2\pi^2.$$

30.3 Exercises

1. Find by one integration the volume of a pyramid whose base is the triangle $(1, 0)$, $(-1, 0)$, $(0, \sqrt{3})$, and whose height is 4.
2. Find by one integration the volume of wood cut from a tree of radius 2 feet by two plane cuts to the same diameter, one horizontal and one at an angle of 45 deg. Use vertical parallel of slices.

In Exercises 3–9, find by one integration, using disk element, the volume of the solid generated when the plane region is revolved around the specified axis. Draw figures for each.

3. The triangle with vertices $(0, 2)$, $(2, 0)$, $(0, 0)$ around the x-axis.
4. The region bounded by $y = x^2$, $x = 0$, $y = a$ about the y-axis.

5. The region bounded by the parabola $y = x^2$, line $y = 4$, about the y-axis.
6. The region in Exercise 4 about the x-axis.
7. The region in Exercise 4 about the line $x = -4$.
8. The disk $x^2 + y^2 \leq a^2$ about the line $x = b$, where $a < b$.
9. The intersection of the disks $x^2 + y^2 \leq 4$ and $(x - 3)^2 + y^2 \leq 4$ about the y-axis.
10. Find the volume of the solid ball of radius a, using disk element.
11. When an open reservoir of some shape holds water to depth z, the area of the surface is $A(z)$. Show that if the water is running out, the volume rate of outflow at any instant is exactly the area of the surface multiplied by the rate of fall of the surface.

31   Shells and Tubes

31.1   A Tubular Volume Element.   The volume of a solid of revolution may be computed conveniently in some cases by using a tubular volume element, or cylindrical shell. We proceed, using intuitive reasoning, and postpone the rigorous justification.

Example.   Find the volume generated by revolving about the line $x = 4$ the region above the parabola $y = x^2$ and below the line $y = 3$.

Solution.   We sketch the figure (Figure 18.5(a)) indicating the cylindrical tube swept out by a rectangle of base dx in the generating plane figure. The volume of this tube may be estimated for purposes of integration by slicing it vertically in one place and laying it out to form a rectangular slab. The dimensions of this slab give it the volume $dV = 2\pi(4 - x)(3 - y) \, dx$. Using the summation pattern to set up a formula for the exact volume, we form a limiting sum of these volume elements

$$V = \int_{-\sqrt{3}}^{\sqrt{3}} 2\pi(4 - x)(3 - y) \, dx = 2\pi \int_{-\sqrt{3}}^{\sqrt{3}} (4 - x)(3 - x^2) \, dx = 32\sqrt{3}\pi.$$

We justify this procedure by repeated integration in cylindrical coordinates. The volume of the solid $\mathscr{B}$ of revolution about the z-axis (Figure 18.5(b)) is given by the triple integral

$$V = \iiint_{\mathscr{B}} r \, dr \, d\theta \, dz.$$

From the symmetry about the z-axis, we know that the upper and lower boundary surfaces are given by functions of r alone, $z_b(r)$ and $z_a(r)$. Hence the volume can be represented by the repeated integral

$$V = \int_0^R \left(\int_0^{2\pi} \int_{z_a}^{z_b} dz \, d\theta \right) r \, dr = \int_0^R 2\pi r(z_b - z_a) \, dr = \int_0^R A(r) \, dr,$$

where $A(r) = 2\pi r(z_b - z_a)$ is the area of the cylinder of radius r and height $z_b - z_a$ in the tubular volume element $dV = 2\pi r(z_b - z_a) \, dr$, and R is the maximum radius on the generating curve.

31.2   Spherical Shell Element of Volume.   We can often shorten multiple integrations when there is symmetry with respect to a point by using the volume of a thin spherical shell as a volume element. Consider the volume of a spherical ball $\mathscr{B}$ of radius R,

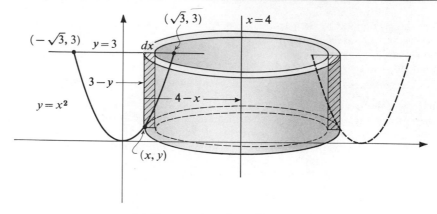

(a) For rotation about line $x = 4$

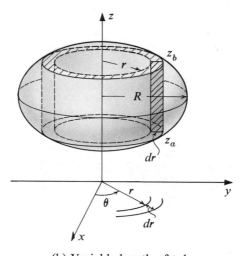

(b) Variable length of tube

FIGURE 18.5   Tubular element of volume.

described in spherical coordinates ρ, θ, ϕ (§10.1). Using the spherical volume element $dV = \rho^2 \sin \phi \, d\rho \, d\theta \, d\phi$, we find that

$$V = \iiint_{\mathscr{B}} \rho^2 \sin \phi \, d\rho \, d\theta \, d\phi = \int_0^R \rho^2 \left(\int_0^{2\pi} \int_0^{\pi} \sin \phi \, d\phi \, d\theta \right) d\rho$$

$$= \int_0^R \rho^2 (2\pi)(2) \, d\rho = \int_0^R 4\pi \rho^2 \, d\rho = \int_0^R A(\rho) \, d\rho.$$

Here $A(\rho) = 4\pi \rho^2$ is the area of the surface of a sphere of radius ρ and $A(\rho) \, d\rho$ is the approximate volume of a spherical shell of radius ρ and thickness $d\rho$. If we complete the integration we find that

$$V = \int_0^R 4\pi \rho^2 \, d\rho = 4\pi \left. \frac{\rho^3}{3} \right|_0^R = \frac{4}{3} \pi R^3,$$

which is the exact volume of the sphere.

Thus we establish the spherical shell volume element $dV = 4\pi \rho^2 \, d\rho$ (Figure 18.6).

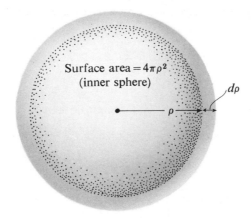

FIGURE 18.6   Spherical shell volume element.

31.3 Area of Surface of Revolution.

Let S be the area of the surface generated by revolving the piecewise smooth arc $y = y(x)$, $a \leq x \leq b$ about the x-axis. The portion of the surface between parallel planes at x and $x + \Delta x$ cuts off a portion of the arc that has a length Δs (Figure 18.7). This small arc revolves around the x-axis to form a band of area ΔS. By geometrical reasoning, if we cut the band across in the generating arc, it will open up into a strip of height Δs with base approximately $2\pi y$. Hence,

$$2\pi(y_{\min}) \, \Delta s \leqq \Delta S \leqq 2\pi(y_{\max}) \, \Delta s,$$

where $y_{\min}$ is the minimum y in the interval $[x, x + \Delta x]$ and $y_{\max}$ is the maximum. If we let $\Delta x \to 0$, so that $\Delta s \to 0$ also, we find in the limit

$$\frac{dS}{ds} = 2\pi y.$$

Therefore

$$S = \int 2\pi y \, ds + C.$$

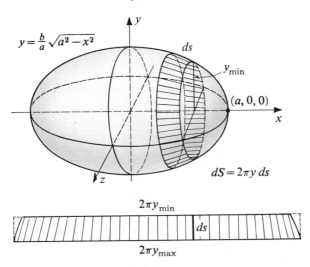

FIGURE 18.7   Surface of revolution with band element of area.

If we transform the variable in this integral to x by the substitution, $ds = \sqrt{1 + y'^2}\, dx$ (II, §22), and introduce x-limits $x = a$ and $x = b$, we find that

$$S = \int_a^b 2\pi y \sqrt{1 + y'^2}\, dx.$$

Example. Find the area of the surface of revolution generated when the upper half of the ellipse $b^2 x^2 + a^2 y^2 = a^2 b^2$ is revolved about the x-axis.

Solution. The upper half of the ellipse is given by the function

$$y = \frac{b}{a} \sqrt{a^2 - x^2}.$$

We compute ds for this and find that it is (II, §22)

$$ds = \sqrt{1 + \left(\frac{dy}{dx}\right)^2}\, dx = \frac{\sqrt{a^4 - (a^2 - b^2)x^2}}{a\sqrt{a^2 - x^2}}\, dx.$$

Then the area of the surface of revolution is

$$S = \int_{-a}^{a} 2\pi y\, ds = \int_{-a}^{a} 2\pi \frac{b}{a^2} \sqrt{a^4 - (a^2 - b^2)x^2}\, dx.$$

Here $a > b$, so the factor $a^2 - b^2 > 0$. Simplifying, we find that this reduces to

$$S = \frac{2\pi b\sqrt{a^2 - b^2}}{a^2} \int_{-a}^{a} \sqrt{p^2 - x^2}\, dx,$$

where $p^2 = a^4/(a^2 - b^2)$. This can be integrated by the substitution $x = p \sin \theta$, or by use of an extended table of integrals. The result is

$$S = \frac{2\pi b\sqrt{a^2 - b^2}}{a^2} \frac{1}{2} \left[x\sqrt{p^2 - x^2} + p^2 \arcsin \frac{x}{p} \right]_{-a}^{a}.$$

Evaluating this, we find the area to be

$$S = 2\pi \frac{b}{a} \left[ab + \frac{a^3}{\sqrt{a^2 - b^2}} \arcsin \frac{\sqrt{a^2 - b^2}}{a} \right].$$

This completes the problem.

31.4 Exercises

In Exercises 1–4, use a cylindrical tubular element to find the required volumes by a single integration.

1. The solid in Exercise 3, §30.3.
2. The solid in Exercise 4, §30.3.
3. The solid in Exercise 5, §30.3.
4. The solid in Exercise 7, §30.3.

In Exercises 5–8, find by one integration the areas of the surfaces of revolution generated when the arc revolves about the specified axis.

5. The half circle $y = \sqrt{r^2 - x^2}$, $[-r \le x \le r]$ about the x-axis.
6. The half circle in Exercise 10 about the line $y = 2r$.
7. The arc $y = x^3$, $[0 \le x \le 1]$, about the x-axis.
8. The arc of the parabola $x^2 = 4y$, $[-1 \le x \le 1]$, about the x-axis.

9. Find the volume of the solid generated when the region bounded by the curve $y^2 - x^2 = a^2$ and the lines $x = 0$, $x = a$ is revolved about the y-axis.

10. In the example concerning the area of the ellipsoid of revolution (§31.3), set $a = b$ at the place where it is correct to do so and show that the calculation gives the area of a sphere $4\pi a^2$.

11. Verify that the dimensions of the area formula for the ellipsoid of revolution (§31.3) are those of an area.

12. Find the volume and surface area of a ring generated when a circle of radius a is revolved around a line at distance b from the center, where $a < b$.

13. Find the volume of the space enclosed by revolving an equilateral triangle of side a about a line parallel to one side and at distance b from the nearest point on the triangle.

32 Moments

32.1 Integration of Functions with Respect to Disk and Shell Elements of Volume. We have seen (§§30, 31) that volumes can be calculated by single integration using disk and shell elements. We found that this occurs when we can perform by inspection the first two integrations of a repeated triple integral,

$$V = \int_a^b \left(\int\int dx\, dy \right) dz = \int_a^b A(z)\, dz,$$

by writing down the expression for the function $A(z)$ as an area. Then the volume element $dV = A(z)\, dz$ was a disk or shell element. We observe that the result of integrating with respect to x and y is to eliminate these variables from the triple integral to produce a function $A(z)$, which is a function of the one remaining variable z.

Now if we wish to use disk or shell elements to shorten the calculation of a triple integral of a function $f(x, y, z)$,

$$\iiint_{\mathscr{D}} f(x, y, z)\, dx\, dy\, dz,$$

it is *necessary that f be a function of the last integration variable only.* In this case we can integrate $f(z)$ with respect to disk or shell elements

$$\iiint_{\mathscr{D}} f(z)\, dx\, dy\, dz = \int_a^b f(z) \left(\iint_{\mathscr{A}(z)} dx\, dy \right) dz = \int_a^b f(z)\, A(z)dz = \int_{\mathscr{D}} f(z)\, dV,$$

where $dV = A(z)\, dz$ is the disk or shell element of volume. This situation occurs frequently in the calculation of moments (II, §33).

32.2 Moments. We may treat the volume dV of a cylindrical slice of a solid revolution as a volume element.

Example A. Find the center of mass of a solid hemisphere of radius a whose density at any point is proportional to the distance from the base.

Solution. We choose coordinates with the origin at the center of the sphere and xy-plane in the base (Figure 18.8(a)). The solid may be considered as a solid of revolution around the z-axis. We may take the volume element to be that of a slice of radius y and

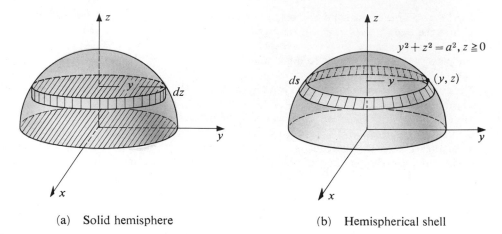

(a)   Solid hemisphere   (b)   Hemispherical shell

FIGURE 18.8   Center of mass of solid and shell.

thickness dz. Thus the volume element dV is $dV = \pi y^2\, dz$. The element of mass dm $\rho\pi y^2\, dz$, where ρ is density. The density is given as $\rho = kz$, and the generating arc is the arc of the circle $y^2 + z^2 = a^2$ in the first quadrant. The mass of the disk is then the mass element $dm = \rho\pi(a^2 - z^2)\, dz$. We can integrate $\int z\, dm$ with respect to this mass element because it has the form $dm = A(z)\, dz$. Hence (II, §33)

$$\bar{z} = \frac{\int z\, dm}{\int dm} = \frac{\int z\rho\pi y^2\, dz}{\int \rho\pi y^2\, dz}$$

$$= \frac{\int_0^a z(kz)\pi(a^2 - z^2)\, dz}{\int_0^a kz\pi(a^2 - z^2)\, dz} = \frac{8}{15}\, a.$$

The way that the property of a solid of revolution enters the calculation of $\bar{x}$ and $\bar{y}$ is that the solid is symmetric with respect to the z-axis. Hence the centroid is on the z-axis and $\bar{x} = \bar{y} = 0$.

Example B.   Find the center of mass of a homogeneous thin hemispherical shell of radius a.

Solution.   Let ρ be the density per square foot of the shell material. Then an element of mass dm can be made from the element of surface area $dS = 2\pi y\, ds$. it is $dm = \rho(2\pi y\, ds)$ (Figure 18.8(b)). Since the z-component of the center of mass is the average value of z in the distribution, we have

$$\bar{z} = \frac{\int z\, dm}{\int dm} = \frac{\int z\rho(2\pi y)\, ds}{\int \rho(2\pi y)\, ds}.$$

We compute ds from the generating circle, $y^2 + z^2 = a^2$, finding that $y = \sqrt{a^2 - z^2}$,

$$ds = \sqrt{1 + \left(\frac{dy}{dz}\right)^2}\, dz = \frac{a}{\sqrt{a^2 - z^2}}\, dz.$$

Hence, after some cancellation,

$$\bar{z} = \frac{\int_0^a z \, dz}{\int_0^a dz} = \frac{a}{2}.$$

The symmetry of the figure about the z-axis implies that $\bar{x} = \bar{y} = 0$.

32.3   Tubular Volume Element for Moment of Inertia about an Axis of Revolution.

The moment of inertia of a body $\mathscr{B}$ with respect to the z-axis is the second moment $I_z = \iiint_\mathscr{B} r^2 \, dm$. If the body is symmetric with respect to the z-axis, as a solid of revolution about the z-axis is, then the tubular volume element, $dV = 2\pi r(z_b - z_a) \, dr$, appropriately shortens the computation of the triple integral.

Example.   Find the moment of inertia of the ellipsoid generated by revolving the ellipse $b^2 y^2 + a^2 z^2 = a^2 b^2$ about its minor axis, if the material of the solid has uniform density $\delta = 1$.

We observe that the earth is such a flattened (oblate) ellipsoid with the equatorial axis as the major axis and the polar axis as the axis of revolution (Figure 18.9).

Solution.   We use cylindrical coordinates (r, θ, z). The equation of the ellipsoid is then

$$\frac{r^2}{a^2} + \frac{z^2}{b^2} = 1.$$

At a distance r from the axis of revolution, we consider a thin vertical rectangle of height $2z$ and base dr. This rectangle sweeps out a tubular volume element whose mass is

$$dm = 2\pi r(2z) \, dr = 4\pi \frac{b}{a} r \sqrt{a^2 - r^2} \, dr \, (1).$$

The moment of inertia about the z-axis is given (II, §32) by

$$I_z = \int_0^a r^2 \, dm = 4\pi \frac{b}{a} \int_0^a r^3 \sqrt{a^2 - r^2} \, dr.$$

We can integrate this by parts, $\int u \, Dv = uv - \int v \, Du$, choosing $u = r^2$ and $Dv = r\sqrt{1 - r^2} \, Dr$. The result is $I_z = \frac{3}{5} a^5$.

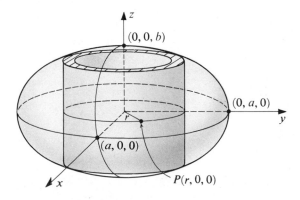

FIGURE 18.9   Oblate ellipsoid.

32.4   Potential of a Spherical Shell. We can often shorten multiple integration over a region when there is symmetry with respect to a point by using the spherical element of a surface or spherical shell.

The gravitational potential at P due to a particle of mass m at Q, where $|QP| = r$ is m/r. If the particles belong to a rigid body formed by continuously distributed mass, and dm is a mass element in the body $\mathscr{B}$, then the potential $u(x, y, z)$ at $P:(x, y, z)$ due to the entire body is found by integrating,

$$u(x, y, z) = \iiint_{\mathscr{B}} \frac{dm}{r},$$

over the region $\mathscr{B}$ occupied by the body. We can find the force of gravitational attraction of any mass at P because it is proportional to the gradient $\overrightarrow{\nabla \mu}$. From the mathematical standpoint, the problem of finding the scalar potential is just that of integrating

$$\iiint_{\mathscr{B}} \frac{dm}{r}.$$

Example. Find the gravitational potential at a point P that is at a distance z from the center of a spherical shell on which mass is distributed with uniform density σ.

Solution. We make a mass element $dm = \sigma R \sin \phi \, d\phi \, d\theta$ out of the spherical surface element (§10.4). Then we look at the mass elements $\sigma R \sin \phi \, d\phi \, d\theta$ in the shell, which are all at some fixed distance r from P (Figure 18.10). These form a ring, $\phi = \text{const}$, around the spherical shell. We can easily add the masses belonging to this ring and find the mass of the ring $dm = 2\pi \sigma R^2 \sin \phi \, d\phi$. One integration with respect to θ taken around

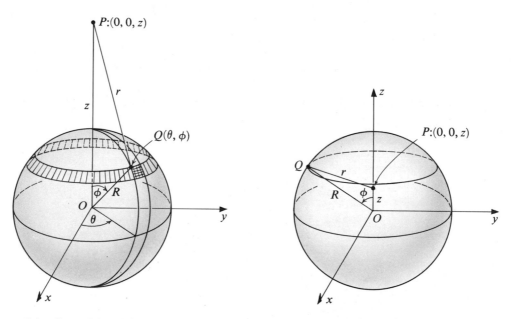

(a)   Case: *P* is outside the sphere   (b)   Case: *P* is inside the sphere

FIGURE 18.10   Potential at *P* due to a spherical shell.

the ring accomplishes this. Now with the mass of the ring as mass element, we can find the potential at P by a single integration

$$u(z) = \int_0^\pi \frac{2\pi\sigma R^2 \sin\phi \; d\phi}{r}.$$

If P is outside the sphere, as ϕ varies the distance r does also, from $z - R$ when $\phi = 0$, to $z + R$ when $\phi = \pi$. Actually, by the law of cosines, r and ϕ are related in the triangle OQP (Figure 18.10(a)) by the reasoning that

$$r^2 = z^2 + R^2 - 2zR\cos\phi.$$

If we differentiate the law of cosines statement, we find

$$r \; dr = zR \sin\phi \; d\phi.$$

We substitute this expression into the integral and find, after changing to r-limits,

$$u(z) = \int_{z-R}^{z+R} \frac{2\pi\sigma Rr \; dr}{zr}.$$

Carrying out the integration, we obtain

$$u(z) = \frac{2\pi\sigma R}{z} \int_{z-R}^{z+R} dr = \frac{4\pi\sigma R^2}{z} = \frac{M}{z},$$

where M is the total mass of the spherical shell. This completes the problem when P is outside the sphere.

If P is inside the sphere, the only difference is in the r-limits. For r varies (Figure 18.10(b)) from $R - z$ when $\phi = 0$, to $R + z$ when $\phi = \pi$. In that case the potential has the value

$$u(z) = \frac{2\pi\sigma R}{z} \int_{R-z}^{R+z} dr = 4\pi\sigma R,$$

which is a constant. This completes the problem.

Let us consider the physical meaning. When P is outside the spherical shell, we found the potential to be $u(z) = M/z$, which is the same as if the entire mass M of the shell were concentrated at the center, at distance z from P. The gradient $\overrightarrow{\nabla u} = -(M/z^2)\vec{k}$ is an inverse-square-law force directed towards the center. On the other hand, when P is inside the sphere, the potential $u(z)$ is constant and its gradient $\overrightarrow{\nabla u} = \vec{0}$. Thus a particle inside the shell behaves weightlessly. The same mathematical calculations would apply to electric potentials due to an electric charge of density σ distributed over a spherical conductor.

32.5   Potential of a Solid Ball

Example.   Find the gravitational potential $u(z)$ due to a solid ball of radius R and constant density δ at a point P outside the ball at distance z from the center.

Solution.   We may use the result of the potential due to the spherical shell, thinking of the ball as made up of layers of thin concentric shells of radius q, where q varies from 0 to R. We use a single integration with respect to q to sum all of the potentials due to the shells. We found that the potential of P due to a shell of radius q was M/z.

Here we take the mass of the shell of radius q and thickness dq to be $dm = 4\pi\delta q^2\, dq$. Therefore the potential due to the solid ball is represented by the integral

$$\phi(z) = \int_0^R \frac{4\pi\delta q^2\, dq}{z} = \frac{4\pi\delta}{z}\cdot\frac{q^3}{3}\Big|_0^R = \frac{4}{3}\,\frac{\pi R^3\delta}{z} = \frac{M}{z},$$

where M now represents the total mass of the solid ball. Once again we find that the ball produces a potential, and therefore a gravitational force, as if its entire mass M were concentrated as a particle at its center.

32.6 Exercises

In Exercises 1–4, use a disk element to complete the calculation with a single integration.

1.   Integrate z^3 over the solid cylinder $x^2 + y^2 \leq 4$, $|z| < 1$.
2.   Integrate x^2 over the ball $x^2 + y^2 + z^2 \leq 4$.
3.   Integrate $1/(1 - z)$ over the solid pyramid with vertices $(1, 0, 0)$, $(0, 1, 0)$, $(0, 0, 1)$.
4.   Integrate $z\sqrt{a^2 - z^2}$ over the solid hemisphere bounded by $x^2 + y^2 + z^2 = a^2$ and the plane $z = 0$.

In Exercises 5–8, use a cylindrical tubular volume element to complete the calculations with a single integration.

5.   Integrate $r^2 + 1$ over the solid cylinder $\{0 \leq r \leq 2, |z| \leq 1\}$.
6.   Integrate $\sqrt{x^2 + y^2}$ over the solid cylinder $\{x^2 + y^2 \leq 4, |z| \leq 1\}$.
7.   Integrate e^{-r^2} over the solid cylinder $\{0 \leq r \leq b, 0 \leq z \leq h\}$.
8.   Integrate $1/r^2$ over the hollow cylinder $\{1 \leq r \leq e, |z| \leq k\}$.
9.   Integrate $x^2 + y^2$ over the conical solid generated when the triangle $(0, 0)$, $(0, h)$, (a, h) is revolved about the y-axis.
10.   Find the average value of r in the conical solid $\{r \leq z \leq h\}$.
11.   Find the average value of z in the solid of Exercise 10.
12.   Find the center of mass of the portion of the ball $\{x^2 + y^2 + z^2 \leq 4\}$ above the plane $z = 1$.
13.   Find the center of mass of a homogeneous solid bounded by a right circular cone of height h and radius of the base R.
14.   Find the center of mass of a thin conical shell of height h and radius of the base R.
15.   Find the moment of inertia of the solid ball of radius a and mass density δ with respect to a diameter.
16.*   Find the moment of inertia of a solid ball of radius a and mass density δ when it is revolved about a line at distance q from the center.
17.   Integrate $1/r$ over the spherical ball $\{0 \leq r \leq a\}$.
18.   Find the moment of inertia with respect to the y-axis of the solid cone generated when the region inside the triangle with vertices $(0, 0)$, $(a, 0)$, $(0, b)$ is revolved about the y-axis.
19.   Find the moment of inertia with respect to its axis of symmetry of a spinning top consisting of a solid hemisphere of density δ and radius a surmounting a solid cone of altitude h (Figure 18.11).
20.   Find the moment of inertia with respect to its central axis of a hollow pipe with inside radius a, outside radius b, length l, and density δ.
21.   Find the center of mass of the top in Exercise 19.
22.   Find the potential at a point P on the z-axis of a thin hemispherical shell $x^2 + y^2 + z^2 = R^2$, $z \geq 0$, having density σ: (a) if P is on the convex side of the shell; (b) if P is on the concave side of the shell.
23.   Find the potential at a point $P:(0, 0, z)$ above a solid half ball $\{x^2 + y^2 + z^2 \leq R^2\}$, $z \geq 0$.

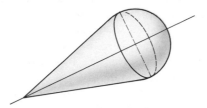

Figure 18.11   Exercise 19.

33   Techniques of Repeated Integration in $\mathscr{R}^3$

33.1   Résumé of the Method.   We return to double integration over a region enclosed by a curve (II, §31) and triple integration over regions in 3-space bounded by curved surfaces (II, §32). We used double integration over regions in which the integrand was a function $f(r, \theta)$ expressed in terms of polar coordinates (§8). And we extended these ideas to functions $f(z, r, \theta)$ defined over regions in $\mathscr{R}^3$ where points are given in cylindrical coordinates, and to functions given in spherical coordinates $f(\rho, \theta, \phi)$. The general method of applying the fundamental theorem of integral calculus was to replace the multiple integral, defined as a limiting sum,

$$\iiint_{\mathscr{D}} f(u, v, w)\, dV$$

by a sequence of integrations

$$\int_a^b \left(\int_{f(w)}^{g(w)} \left(\int_{F(v,w)}^{G(v,w)} f(u, v, w)\, du \right) dv \right) dw$$

with respect to one variable at a time. We called this a repeated integral. Each single integration of the repeated integral with respect to some variable u can be evaluated if we can find an antiderivative of the integrand viewed as a function of that variable alone.

It is easy to write down the limits of integration of the repeated integral when the domain is a rectangle or box (II, §29.32). We extended the idea to multiple integrals over domains bounded by curves and curved surfaces. To do so, we used the ideal device of enclosing the domain of integration in a box and extending the definition of the integrand function f over the enclosing box. Then we define the extended function f_E to be zero outside the domain of f. To compute a multiple integral by repeated integration with respect to one variable at a time, we must remove the extraneous extension and introduce limits of integration that are functions describing the boundary of the domain of f. In the abstract formulation above, the first integral with respect to u has limits $F(v, w)$ and $G(v, w)$, which are functions of the variables v and w yet to enter into the integration. The next integral with respect to v has limits $f(w)$ and $g(w)$, which are functions of the one remaining variable. The final integration has limits a and b, which are constants.

In practice, the problem is to choose an order of integration and then find explicit formulas for the functions that make up the limits of integration, so that it extends over the proper domain, and neither includes other points nor covers some points twice. When this is done, the problem is set up. After the required quantity is set up as a repeated

integral, the procedures are the same as for single-variable calculus. It is only more difficult to recognize the forms when several variables are present.

Finally, we recall that there are other ways to calculate a conceptual multiple integral.
Most basically, we can proceed from the definition of the integral as a limiting sum to
compute numerically a sample sum over a partition of the domain.

We can sometimes reduce the problem to a single integration by using disks or shells
as volume elements (§§30, 31, 32). Also, for some problems where the answer might
be expressed as a multiple integral, we may find it more effective to regard the unknown
object as an integral of a differential equation (II, §24).

33.2 Volume Bounded by Two Surfaces

Example A. Find the volume of the region below the graph of $z = 1 - (x^2/a^2) - (y^2/b^2)$
and above the plane $z = 0$.

Solution. The two surfaces intersect in the ellipse, $(x^2/a^2) + (y^2/b^2) = 1$, in the
xy-plane. Above this plane, where $0 < z$, we find level contours that are ellipses inside
this base up to the maximum level of z, where $z = 1$ (Figure 18.12). A suitable element
of volume is the volume of a rectangular shaft with base $dx\,dy$ and height z,

$$dV = z\,dx\,dy.$$

The total volume V is the sum of volume elements of this type taken over the domain
$\mathscr{D}:\{(x^2/a^2) + (y^2/b^2) \leq 1\}$, which is an elliptic disk. And this double integral can be
represented as a repeated integral,

$$V = \iint\limits_{\mathscr{D}} z\,dx\,dy = \int_{-a}^{a} \int_{-(b/a)\sqrt{a^2-x^2}}^{+(b/a)\sqrt{a^2-x^2}} \left(1 - \frac{x^2}{a^2} - \frac{y^2}{b^2}\right) dy\,dx.$$

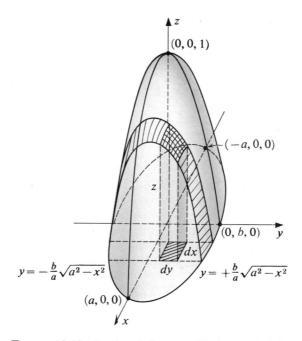

FIGURE 18.12   Region below an elliptic paraboloid.

Here we choose to hold x fixed and integrate in the y-direction from the least to the greatest y inside the disk. To find these limits of y, we solve the boundary equation $(x^2/a^2) + (y^2/b^2) = 1$ for y in terms of x, which gives us $y = -(b/a)\sqrt{a^2 - x^2}$ and $y = +(b/a)\sqrt{a^2 - x^2}$. These are the limits of the integration with respect to y. This first integration gives us the area of a typical vertical section, $x = $ constant (Figure 18.12). We multiply this area by dx to get the volume of a thin vertical slice and sum these from the minimum value, $x = -a$, to the maximum, $x - a$. This is the way we set up the integral. We could have used symmetry to calculate one-quarter of the volume and written more simply

$$V = 4 \int_0^a \int_0^{(b/a)\sqrt{a^2 - x^2}} \left(1 - \frac{x^2}{a^2} - \frac{y^2}{b^2}\right) dy\, dx.$$

Now we calculate the repeated integral. First holding x constant, we integrate

$$\int_{-(b/a)\sqrt{a^2 - x^2}}^{(b/a)\sqrt{a^2 - x^2}} \left(1 - \frac{x^2}{a^2} - \frac{y^2}{b^2}\right) dy = \left[\left(1 - \frac{x^2}{a^2}\right) y - \frac{1}{b^2}\frac{y^3}{3}\right]_{-(b/a)\sqrt{a^2 - x^2}}^{+(b/a)\sqrt{a^2 - x^2}}$$

$$= 2\left(1 - \frac{x^2}{a^2}\right)\frac{b}{a}(a^2 - x^2)^{1/2} - \frac{2}{3b^2}\frac{b^3}{a^3}(a^2 - x^2)^{3/2}$$

$$= \frac{4}{3}\frac{b}{a^3}(a^2 - x^2)^{3/2}.$$

Then the second integration with respect to x gives V,

$$V = \frac{4}{3}\frac{b}{a^3} \int_{-a}^a (a^2 - x^2)^{3/2}\, dx.$$

This integral can be calculated directly by the trigonometric substitution $x = a \sin\theta$, $dx = a \cos\theta\, d\theta$, or by appealing to a table of integrals. We find in a table of integrals

$$\int (a^2 - x^2)^{3/2}\, Dx = \frac{1}{4}\left[x(a^2 - x^2)^{3/2} + \frac{3a^2x}{2}(a^2 - x^2)^{1/2} + \frac{3a^4}{2}\arcsin\frac{x}{a}\right].$$

When we evaluate the arc sine term at the two limits we must use the principal values of arc sine on the interval $[-\pi/2, \pi/2]$. We find

$$\arcsin\frac{x}{a}\Big|_{-a}^a = \arcsin 1 - \arcsin(-1) = \frac{\pi}{2} - \left(-\frac{\pi}{2}\right) = \pi.$$

Then

$$V = \frac{4}{3}\frac{b}{a^3}\frac{1}{4}\left[0 + 0 + \frac{3a^4}{2}\pi\right] = \frac{\pi ab}{2}.$$

We now consider a more complicated example.

Example B. Find the volume cut out from a sphere $\{x^2 + y^2 + z^2 \leq a^2\}$ by a cone with vertex at the origin whose generator lines make an angle ϕ with the vertical.

Solution. We sketch the figure, shaped like a spinning top (Figure 18.13).

On account of the symmetry of the figure about the z-axis we find it advantageous to use cylindrical coordinates, that is, polar coordinates in the xy-plane. We replace $x^2 + y^2$ by r^2. Then the sphere has the equation $z = \sqrt{a^2 - r^2}$. The cone (Figure 18.13) has the equation $r/z = \tan\phi$, where ϕ is constant, or $z = r/\tan\phi$. The sphere

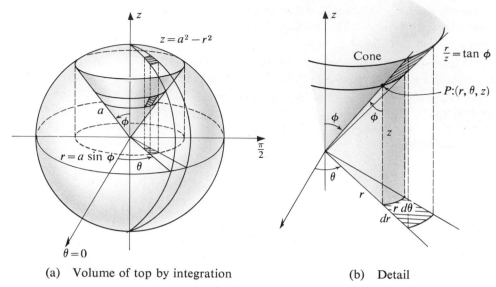

(a) Volume of top by integration   (b) Detail

FIGURE 18.13   Sphere in cylindrical coordinates.

and the cone intersect in a circle of radius $a \sin \phi$. Hence the projection of the solid onto the horizontal plane is a circular disk $r \leqq a \sin \phi$ (Figure 18.13). With this information we can write the volume integral as a repeated triple integral.

$$
\begin{aligned}
V = \iiint_{\mathscr{D}} dV &= \int_0^{2\pi} \int_0^{a \sin \phi} \int_{r/\tan \phi}^{\sqrt{a^2 - r^2}} dz \, r \, dr \, d\theta \\
&= \int_0^{2\pi} \int_0^{a \sin \phi} \left[\sqrt{a^2 - r^2} - \frac{r}{\tan \phi} \right] r \, dr \, d\theta \\
&= \int_0^{2\pi} \left[-\frac{(a^2 - r^2)^{3/2}}{3} - \frac{r^3}{3 \tan \phi} \right]_{r=0}^{r=a \sin \phi} d\theta \\
&= 2\pi \frac{a^3}{3} (1 - \cos \phi),
\end{aligned}
$$

where the last step involves some trigonometric simplification and integration with respect to θ. This is the volume of the top.

33.3 Exercises

In these exercises, a correct figure is extremely important. If possible, picture the element of integration and show the geometric effect of each successive integration as a shaft or slice or whatever it is. We start with some double integrals, for which the figures are more easily drawn (II, §31.5).

1. Find by repeated integral the area of the region enclosed by the curves $y = x^2$, $y = 2x^2 - 1$, $y = 0$, (a) integrating first in the x-direction; (b) first in the y-direction.
2. Find by repeated integral the average value of x in the region enclosed by the lines $y = 2x$, $y = 4x - 6$, $y = 0$.
3. Find by repeated integral the area enclosed between the circles $x^2 + y^2 = 4$, $x^2 + y^2 - 4y + 3 = 0$.

4. Show by a figure what geometric quantities are represented by the following two integrals:

$$\text{(a)} \quad \int_2^4 \int_2^{3x-4} dy\, dx; \quad \text{(b)} \quad \int_2^8 \int_{(y+4)/3}^4 dx\, dy.$$

5. Integrate xy over the unit disk $\{x^2 + y^2 \leq 1\}$. Can the result be calculated by integrating over the first quadrant and then multiplying this result by 4?

6. Considering the definition of the double integral and the area π of the unit disk $x^2 + y^2 \leq 1$, integrate $xy + 2$ over the unit disk by inspection.

7. Find by repeated integration the area enclosed by the curves $9y = 4x^2$, $16y = 3x^2$, $x^2 + y^2 = 25$.

8. Find by repeated integration the area of the circle $r = a \sin \theta$. Can the area be found by integrating over the first quadrant $0 \leq \theta \leq \pi/2$ and then multiplying the result by 4?

9. Find by repeated triple integration the volume of the region enclosed by the planes $x = 0$, $y = 0$, $z = 0$, $x + y + z = 2$.

10. Find the repeated triple integration in xyz-coordinates the volume of the region enclosed by the surfaces $x^2 + y^2 = 1$, $z = -1$, $z = 1$.

11. Repeat Exercise 10 using cylindrical coordinates.

12. Set up a repeated triple integral representing the volume of the region enclosed between the two spheres $x^2 + y^2 + z^2 = 1$ and $x^2 + y^2 + z^2 - 2z = 0$.

13. Choose a coordinate system in which the volume in Exercise 12 can be more easily calculated, and calculate it.

14. Find by a repeated triple integral the average values of x, y, and z in the solid half-ball bounded by the upper surface of $x^2 + y^2 + z^2 - 4x - 5 = 0$ and the plane $z = 0$.

15. Show by a figure what geometric quantities are represented by the triple repeated integrals

$$\text{(a)} \quad \int_0^6 \int_0^3 \int_0^x dy\, dx\, dz \quad \text{and} \quad \text{(b)} \quad \int_0^3 \int_y^3 \int_0^6 dz\, dx\, dy.$$

16. Invert the order of the repeated integral $\int_1^2 \int_0^{x^2-1} dy\, dx$ to express it as a repeated integral $\int \int dx\, dy$ with appropriate limits of integration.

17. A vertical column of radius 2 and height 20 contains fluid of a density that varies linearly from 1 at the top to 5 at the bottom. Find by a repeated triple integral the mass of the fluid.

18. Integrate the function $e^{-(x+y+z)}$ over the region of all points (x, y, z) in $\mathcal{R}^3$, where $x \geq 0$, $y \geq 0$, $z \geq 0$.

19. Find by a repeated triple integral the volume of the solid cylinder $x^2 + y^2 \leq a^2$ between the planes $z = x + 1$ and $z = 0$.

20. Find by a repeated triple integral the volume of the solid bounded by the surfaces $x = y^2 + 9z^2$ and $x = 18 - y^2 - 9z^2$.

21. Find the derivative

$$\frac{d}{dt} \int_a^t \int_0^x \int_0^u \sin u^2 \, dv\, du\, dx.$$

22. For the vector field given for $\vec{v} = x\vec{i} + y\vec{j} + z\vec{k}$ by $\vec{F}(\vec{v}) = \vec{v}$, integrate div $\vec{F}$ over the solid ball $\{x^2 + y^2 + z^2 \leq a^2\}$ (§29.3).

23. For the vector field given for $v = x\vec{i} + y\vec{j} + z\vec{k}$

$$\vec{F}(\vec{v}) = \frac{x}{\|\vec{v}\|}\vec{i} + \frac{y}{\|\vec{v}\|}\vec{j} + \frac{z}{\|\vec{v}\|}\vec{k}, \qquad \|\vec{v}\| \neq 0,$$

integrate div $\vec{F}$ over the ball with center at $(0, 2, 0)$ and radius 1.

34 Change of Variables in Multiple Integrals

34.1   Change of Variable in Simple Integral. We recall (II, §5.7) the procedure for changing the variable of integration in an integral of a function of one variable. If in the integral

$$I = \int_a^b f(x)\, dx$$

we substitute $x = g(u)$, which is required to be a one-to-one differentiable function, I can be calculated by integration with respect to u in

$$I = \int_\alpha^\beta f[g(u)]g'(u)\, du,$$

where $a = g(\alpha)$ and $b = g(\beta)$.

For example, if we substitute $x = a \sin u$ in the integral

$$I = \int_0^{a/\sqrt{2}} \frac{dx}{\sqrt{a^2 - x^2}},$$

we replace $dx = a \cos u\, du$ and compute new u-limits by solving $a/\sqrt{2}\, a \sin u$ for $u = \pi/4$ and $0 = a \sin u$ for $u = 0$. The integral then becomes

$$I = \int_0^{\pi/4} \frac{a \cos u\, du}{\sqrt{a^2 - a^2 \sin^2 u}} = a \int_0^{\pi/4} du = \frac{\pi}{4} a.$$

In this, observe that the *substitution* $x = g(u)$ maps the x-interval $[a, b]$ onto the u-interval $[\alpha, \beta]$ by means of $u = g^{-1}(x)$. The integration with respect to u is carried over the interval $[\alpha, \beta]$. Moreover, besides replacing $f(x)$ by $f[g(u)]$, the substitution $x = g(u)$ also replaces the element of length dx, not by du, but by $g'(u)\, du$, where $g'(u)$ was called the local scale factor.

When we apply these ideas to double integrals

$$I = \iint\limits_{\mathscr{D}} f(x, y)\, dx\, dy$$

it is easy to see that a one-to-one substitution, $x = x(u, v)$, $y = y(u, v)$, replaces $f(x, y)$ by $f[x(u, v), y(u, v)]$ and maps the domain $\mathscr{D}$ onto a domain $\mathscr{U}$ in the uv-plane, but it is not obvious what replaces the element $dx\, dy$. We may guess by analogy that $dx\, dy$ is replaced by $J(u, v)\, du\, dv$, where $J(u, v)$ is some as yet undetermined function.

34.2   Effect of Linear Substitution on the Element of Area. If we introduce new variables (u, v) by the linear substitution

$$x = au + bv,$$
$$y = cu + dv,$$

then the area of the parallelogram (§28.2) formed by the vectors $\vec{v}:(x_1, y_1)$ and $\vec{w}:(x_2, y_2)$,

$$x_1 y_2 - x_2 y_1 = (au_1 + bv_1)(cu_2 + dv_2) - (au_2 + bv_2)(cu_1 + dv_1)$$

becomes, on being simplified,

$$x_1 y_2 - x_2 y_1 = (ad - bc)(u_1 v_2 - u_2 v_1).$$

The factor $(ad - bc)$ is determinant $J = \begin{vmatrix} a & b \\ c & d \end{vmatrix}$ of the transformation. The transformation is one-to-one if and only if this determinant is not zero.

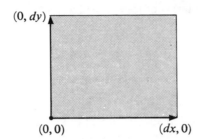

$(0, dy)$

$(0, 0)$ $(dx, 0)$

FIGURE 18.14   Element of area in local coordinates.

Going the other way, from the uv-plane to the xy-plane, we see that if the transformation is one-to-one, then every point (u, v) maps into one and only one point (x, y). This maps geometric figures in the uv-plane onto their images, which are geometric figures in the xy-plane (Figure 18.14). The linear transformation

$$x = au + bv$$
$$y = cu + dv$$

maps lines onto lines, parallel lines onto parallel lines. Moreover, every parallelogram $\mathscr{U}$ determined by the points $(0, 0)$, (u_1, v_1), (u_2, v_2), whose area is $u_1 v_2 - u_2 v_1$, maps into a parallelogram whose area $x_1 y_2 - x_2 y_1$ is J times that $\mathscr{U}$. It follows that any figure $\mathscr{U}$ in the uv-plane having an area is mapped into the xy-plane onto a figure having area J times that of $\mathscr{U}$. For example, the transformation

$$x = 2u - 3v,$$
$$y = u + 5v,$$

has determinant

$$J = \begin{vmatrix} 2 & -3 \\ 1 & 5 \end{vmatrix} = 13.$$

It maps any parallelogram $\mathscr{U}$ in the uv-plane onto an image in the xy-plane that has an area 13 times that of $\mathscr{U}$.

We observe that when the points are given in local coordinates (Figure 18.15) by $(x_1, y_1) = (dx, 0)$ and $(x_2, y_2) = (0, dy)$, this area formula reduces to $dx\, dy$.

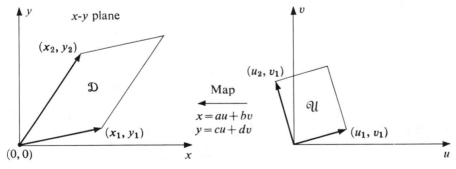

The image parallelogram $\mathscr{D}$ has area J times that of $\mathscr{U}$.

FIGURE 18.15   Linear map.

We consider linear maps in terms of local coordinates (dx, dy) and (du, dv). A one-to-one linear transformation,

$$dx = a(du) + b(dv),$$

$$dy = c(du) + d(dv),$$

makes the rectangular coordinate grid in the $dx\,dy$-plane correspond to a system of intersecting parallels in the $du\,dv$-plane. If these are taken to be a coordinate grid in the $du\,dv$-plane, then the elements of area correspond by

$$dx\,dy = J\,du\,dv,$$

where $J = \begin{vmatrix} a & b \\ c & d \end{vmatrix}.$

34.3   Transformation of Double Integrals

Problem.   Express the double integral

$$\iint_{\mathscr{D}} f(x, y)\,dx\,dy$$

as an integral in uv-coordinates that are related to the xy-coordinates by the one-to-one differentiable transformation

$$x = x(u, v), \qquad y = y(u, v).$$

   Solution.   We can map the domain of integration $\mathscr{D}$ onto a domain $\mathscr{U}$ in the uv-plane and we can transform $f(x, y)$ into the function of u and v given by $f[x(u, v), y(u, v)]$ by substitution. It remains to determine what to do with the element of area $dx\,dy$. Here we observe that the transformation relating xy-coordinates with uv-coordinates induces by differentiation a relation between the local coordinates (dx, dy) at any point (x, y) with the local coordinates (du, dv) at the corresponding point (u, v). This relation is

$$dx = \frac{\partial x}{\partial u}\,du + \frac{\partial x}{\partial v}\,dv, \qquad dy = \frac{\partial y}{\partial u}\,du + \frac{\partial y}{\partial v}\,dv.$$

The determinant of this linear transformation is

$$J = \begin{vmatrix} \dfrac{\partial x}{\partial u} & \dfrac{\partial x}{\partial v} \\ \dfrac{\partial y}{\partial u} & \dfrac{\partial y}{\partial v} \end{vmatrix}.$$

It is called the Jacobian of the original transformation $x = x(u, v)$, $y = y(u, v)$. This tells us that the element of area $dx\,dy$ at (x, y) is related to $du\,dv$ at the image point (u, v) by

$$dx\,dy = J\,du\,dv,$$

and so we use $J\,du\,dv$ as the element of area in uv-coordinates. We then guess the solution of our problem in the form of a theorem.

THEOREM. *Substitution in double integrals.* If new variables (u, v) replace (x, y) through the one-to-one differentiable transformation $x = x(u, v)$, $y = y(u, v)$, then

$$\iint_{\mathscr{D}} f(x, y)\, dx\, dy = \iint_{\mathscr{U}} f[x(u, v), y(u, v)]\,|J|\, du\, dv,$$

where $\mathscr{U}$ is the uv-image of the domain $\mathscr{D}$ and $|J|$ is the absolute value of the Jacobian

$$J = \begin{vmatrix} \dfrac{\partial x}{\partial u} & \dfrac{\partial x}{\partial v} \\[2mm] \dfrac{\partial y}{\partial u} & \dfrac{\partial y}{\partial v} \end{vmatrix}$$

of the transformation.

We omit the proof.

We use the absolute value of J so that the element of area in uv-coordinates will be positive whenever $dx\, dy$ is.

Example. Transform the integral

$$I = \iint_{\mathscr{D}} xy\, dx\, dy$$

where $\mathscr{D}$ is the circular disk bounded by $x^2 + y^2 = 4$, introducing new variables (u, v) by $x = u + 3v - 7$, $y = u - v + 5$.

Solution. The domain $\mathscr{D}$ becomes the region $\mathscr{U}$ inside the curve

$$(u + 3v - 7)^2 + (u - v + 5)^2 - 4 = 0.$$

Since

$$\frac{\partial x}{\partial u} = 1, \qquad \frac{\partial x}{\partial v} = 3, \qquad \frac{\partial y}{\partial u} = 1, \qquad \frac{\partial y}{\partial v} = -1,$$

the Jacobian J becomes

$$J = \begin{vmatrix} 1 & 3 \\ 1 & -1 \end{vmatrix} = -4, \qquad |J| = 4.$$

Hence, we replace $dx\, dy$ by $4\, du\, dv$ and the integral becomes in uv-coordinates

$$I = \iint_{\mathscr{U}} (u + 3v - 7)(u - v + 5)4\, du\, dv.$$

The original xy-coordinates are preferable for calculation. This completes the problem.

34.4 Transformation of Triple Integrals.

We state the analogous theorem for three-dimensional regions.

THEOREM. *Substitution in triple integrals.* If new variables (u, v, w) replace (x, y, z) through a one-to-one differentiable transformation

$$x = x(u, v, w), \qquad y = y(u, v, w), \qquad z = z(u, v, w),$$

then

$$\iiint\limits_{\mathscr{D}} f(x, y, z) \, dx \, dy \, dz$$

$$= \iiint\limits_{\mathscr{U}} f[x(u, v, w), y(u, v, w), z(u, v, w)]|J| \, du \, dv \, dw,$$

where $\mathscr{U}$ is the uvw-image of the domain $\mathscr{D}$ and $|J|$ is the absolute value of the Jacobian determinant

$$\begin{vmatrix} \dfrac{\partial x}{\partial u} & \dfrac{\partial x}{\partial v} & \dfrac{\partial x}{\partial w} \\[2mm] \dfrac{\partial y}{\partial u} & \dfrac{\partial y}{\partial v} & \dfrac{\partial y}{\partial w} \\[2mm] \dfrac{\partial z}{\partial u} & \dfrac{\partial z}{\partial v} & \dfrac{\partial z}{\partial w} \end{vmatrix}.$$

34.5 Exercises

1. Find the areas of the parallelograms determined by the following sets of vertices. Plot.
   (a) $(0, 0), (2, 3), (-1, 4)$.   (b) $(0, 0), (2, 4), (1, 2)$.
   (c) $(0, 0), (-1, 4), (2, 3)$.   (d) $(0, 0), (1, 0), (0, 1)$.
   (e) $(1, 4), (4, 2), (3, 4)$.   (f) $(x_1, y_1), (x_2, y_2), (x_3, y_3)$.

2. Explain the difference in sign in Exercises 1(a), 1(c).

3. When parallelograms in the uv-plane are mapped into the xy-plane by $x = 2u - 3v$, $y = 3u + 2v$, how much bigger in area are the images in the xy-plane?

4. For Exercise 3, sketch the square $\mathscr{U}$ with vertices in uv-coordinates $(0, 0), (2, 0), (2, 2)$, $(0, 2)$ and its image in the xy-plane. Verify that the area of the image is 13 times the area of $\mathscr{U}$.

5. Prove that *every* figure $\mathscr{U}$ in the uv-plane that has an area is mapped by a one-to-one transformation, $x = au + bv$, $y = cu + dv$ onto a figure in the xy-plane whose area is J times that of $\mathscr{U}$ where $J = |ad - bc|$.

In Exercises 6–9, transform the given integrals by means of the indicated transformation.

6. $\iint\limits_{\mathscr{D}} dx \, dy$, where $\mathscr{D}$ is the square $\{|x| \leq 1, |y| \leq 1\}$, under the substitution $x = u + v$, $y = u - v$.

7. $\iint\limits_{\mathscr{D}} (x - y)^2 \, dx \, dy$, where $\mathscr{D}$ is the square $\{|x| + |y| \leq 1\}$, under the substitution $x = u + v, y = u - v$.

8. $\iint\limits_{\mathscr{D}} x \, dx \, dy$, where $\mathscr{D}$ is the disk inside the circle $x^2 + y^2 = a^2$, under the transformation $x = (1/\sqrt{2})(u - v), y = (1/\sqrt{2})(u + v)$. Show that the domain $\mathscr{D}$ is unchanged.

9. $\iint\limits_{\mathscr{D}} (x + y) \, dx \, dy$, where $\mathscr{D}$ is the triangle $\{x \geq 0, y \geq 0, x + y \leq 1\}$, under the transformation $x = 2u - 3v + 1, y = -u + v - 5$.

10. Use the method of this section to show that when polar coordinates are introduced into the integral $I = \iint\limits_{\mathscr{D}} f(x, y) \, dx \, dy$, by the transformation $x = r \cos \theta, y = r \sin \theta$ the integral I becomes

$$I = \iint\limits_{\mathscr{U}} f(r \cos \theta, r \sin \theta) r \, dr \, d\theta.$$

11. Show that for every function $f(x, y)$ and every $\mathscr{D}$, the transformed integral of $I = \iint_{\mathscr{D}} f(x, y)\, dx\, dy$ into uv-coordinates by the transformation $x = 2u - 3v + 1$, $y = 4u - 6v + 7$, is zero. Explain.

In Exercises 12–13, transform the integrals into u, v, w-coordinates.

12. $\iiint_{\mathscr{D}} (xy/z)\, dx\, dy\, dz$, where $\mathscr{D}$ is the cube

$$\{|x - 2| \leq 1, |y - 2| \leq 1, |z - 2| \leq 1\},$$

under the transformation

$$x = 2u - 3v + 2w - 1, \qquad y = u + v - w + 4, \qquad z = u - 4v + w + 5.$$

13. $\iiint_{\mathscr{D}} dx\, dy\, dz$, where $\mathscr{D}$ is the spherical ball $\{x^2 + y^2 + z^2 \leq r^2\}$, under $x = u + a$, $y = v + b$, $z = w + c$. Explain the result geometrically.

14. Change the variables to cylindrical coordinates (r, θ, z) in the triple integral

$$\iiint_{\mathscr{D}} (x^2 + y^2 + z^2)\, dx\, dy\, dz,$$

where $\mathscr{D}$ is the solid cylinder $\{x^2 + y^2 \leq a^2, |z| \leq h\}$.

15. Change the variables to spherical coordinates (ρ, θ, ϕ) in the triple integral

$$\iiint \frac{1}{\sqrt{x^2 + y^2 + z^2}}\, dx\, dy\, dz,$$

where $\mathscr{D}$ is the hollow ball $\{1 \leq x^2 + y^2 + z^2 \leq 4\}$.

16. Change the variables in the double integral $I = \iint_{\mathscr{D}} xy\, dx\, dy$, where $\mathscr{D}$ is the triangle with vertices $(1, 0)$, $(0, 1)$, $(-1, 0)$, by the substitution $x = u^2 - v^2$, $y = 2uv$.

17.* Change the variables in the double integral $I = \iint_{\mathscr{D}} (x + y)\, dx\, dy$, where $\mathscr{D}$ is the disk $\{x^2 + y^2 - 2x \leq 0\}$, by the substitution

$$x = \frac{2uv^2}{u^2 + v^2}, \qquad y = \frac{2u^2v}{u^2 + v^2}.$$

35   Surfaces in Parametric Representation

ACADEMIC NOTE. We here invade the territory of advanced calculus for a brief and intuitive introduction to surfaces. Differentiation and integration in the number line $\mathscr{R}^1$, the plane $\mathscr{R}^2$, and 3-space $\mathscr{R}^3$ is the business of elementary calculus. As soon as we go to differentiation and integration of functions defined on curves, curved surfaces, or higher-dimensional manifolds, we are in the subject of advanced calculus. As a last section on elementary calculus, we look briefly at curved surfaces and integrals defined on them. We have already treated some surfaces by elementary methods but the only general representation of surfaces is the parametric one.

35.1   Idea of a Surface.
We used the real number line as a model to define a curve. A curve is a linelike set of points that is the image of the number line under a mapping of points t in the number line to points $P:(x, y, z)$ in $\mathscr{R}^3$, given by continuous functions

$$x = x(t), \qquad y = y(t), \qquad z = z(t).$$

These were called the parametric equations of the curve (II, §21).

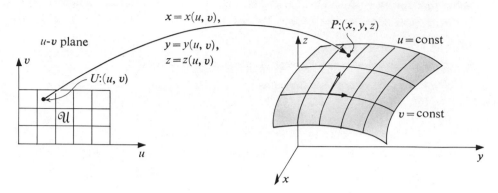

FIGURE 18.16   Surface as a mapping of a plane. For each point $U:(u, v)$ in the plane, the parametric equations determine a point $P:(x, y, z)$ in the surface.

In the same way we use the plane as a model to define a surface as a continuously deformed image of it.

DEFINITION.   A *surface* in $\mathscr{R}^3$ is the image of the plane under the continuous mapping,

$$x = x(u, v), \qquad y = y(u, v), \qquad z = z(u, v)$$

that sends points $U:(u, v)$ in the plane into points $P:(x, y, z)$ in $\mathscr{R}^3$ (Figure 18.16).

The three equations defining the mapping are called a parametric representation of the surface with two parameters (u, v).

As in the case of the curve, the continuity of the mapping sends nearby points in the plane to nearby points in the surface so that the surface is a continuous sheet of points, but a continuous mapping still allows for bending and twisting of the surface, and even bringing boundaries together to form a closed surface as the following example shows.

Example.   Describe the surface

$$x = R \sin v \cos u, \qquad y = R \sin v \sin u, \qquad z = R \cos v,$$

$\mathscr{D} = \{0 \le u \le 2\pi, 0 \le v \le \pi\}$, where R is constant.

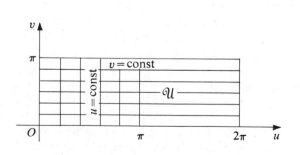
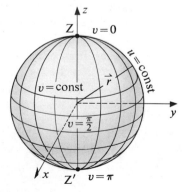

FIGURE 18.17   Sphere as a mapping. $x = R \sin v \cos u$, $y = R \sin v \sin u$, $z = R \cos v$.

Solution. We recognize this as the relation between spherical coordinates (ρ, θ, ϕ) and rectangular coordinates (x, y, z) with ρ fixed at R and (θ, ϕ) replaced by letters (u, v). Hence the surface is the sphere of radius R with center at the origin. As a mapping it maps the rectangle $\mathcal{U}$ into 3-space (Figure 18.17), and brings the edges together to close up the figure. The entire top edge of $\mathcal{U}$, $v = \pi$, maps into the south pole z', since $\sin \pi = 0$, $\cos \pi = -1$ gives $x = y = 0$, $z = -R$. Similarly, the bottom edge $v = 0$ maps into the north pole $Z:(0, 0, R)$. The left and right edges $u = 0$ and $u = 2\pi$ map into the same circle on the sphere since $\sin 0 = 2\pi$. In general the horizontal lines $v = $ constant map into parallels of latitude, and vertical lines $u = $ constant map into great circles joining north and south poles. We plot some of the orthogonal net of lines $u = c, v = k$ and their images on the sphere.

35.2   Element of Surface Area.   We now restrict our consideration of surfaces to those that have continuously differentiable representations. We shall call them *smooth* surfaces. As in the case of curves where we defined arc length for smooth curves, we can represent surface area by an integral for smooth surfaces.

We recall that for each component $x(t)$, $y(t)$, $z(t)$ of a curve, the derivative is the local scale factor that multiplies the length element dt in the line to produce the x-component of length in the curve. That is $dx = x'(t)\, dt$. Similarly, $dy = y'(t)\, dt$, $dz = z'(t)\, dt$ (I, §12.7). We have seen that, in the same way, the Jacobian is the scale factor that multiplies the element of area $du\, dv$ in the plane to produce the element of area in xy-coordinates under the mapping

$$x = x(u, v), \qquad y = y(u, v).$$

That is, the plane area element $dx\, dy$ is given by

$$dx\, dy = \begin{vmatrix} \dfrac{\partial x}{\partial u} & \dfrac{\partial x}{\partial v} \\[2mm] \dfrac{\partial y}{\partial u} & \dfrac{\partial y}{\partial v} \end{vmatrix} du\, dv.$$

We denote this Jacobian by J_{12}. Similarly, the yz-component of area and the zx-component are found by

$$dy\, dz = \begin{vmatrix} \dfrac{\partial y}{\partial u} & \dfrac{\partial y}{\partial v} \\[2mm] \dfrac{\partial z}{\partial u} & \dfrac{\partial z}{\partial v} \end{vmatrix} du\, dv \qquad \text{and} \qquad dx\, dz = \begin{vmatrix} \dfrac{\partial z}{\partial u} & \dfrac{\partial z}{\partial v} \\[2mm] \dfrac{\partial x}{\partial u} & \dfrac{\partial x}{\partial v} \end{vmatrix} du\, dv.$$

We denote these Jacobians by J_{23} and J_{31} respectively. In the sense that it is a local scale factor for areas, the Jacobian acts like the derivative for functions of one variable.

Continuing the analogy between curve and surface, we recall that when we put all three components of length together for a curve, we found that an element of length ds on the curve was found by multiplying the element of length dt on the line by $\|\vec{x}'\|$, that is, $ds = \sqrt{x'^2 + y'^2 + z'^2}\, dt$ (II, §22.3). This gave us the integral formula for arc length

$$s = \int_a^b \sqrt{x'^2 + y'^2 + z'^2}\, dt.$$

The analogous procedure for surfaces says that we find the element of area dS in the surface by multiplying the element of area $du\,dv$ in the coordinate plane by the appropriate scale factor,

$$dS = \sqrt{J_{12}^2 + J_{23}^2 + J_{31}^2}\; du\,dv.$$

This gives us an integral formula for the surface area of a portion of a surface that is the image of a region $\mathscr{U}$ in the uv-plane. It is

$$S = \iint_{\mathscr{U}} \sqrt{J_{12}^2 + J_{23}^2 + J_{31}^2}\; du\,dv.$$

For the parametric representation of the sphere of radius R

$$x = R \sin\phi \cos\theta, \qquad y = R \sin\phi \sin\theta, \qquad z = R \cos\phi,$$

we found by special spherical geometry the element of area (§10.5)

$$dS = R^2 \sin\phi\; d\phi\; d\theta.$$

If we calculate the Jacobians, treating ϕ as v and θ as u, we find (Exercises) the same result,

$$\sqrt{J_{12}^2 + J_{23}^2 + J_{31}^2} = R^2 \sin\phi.$$

The calculation by the general formula is tedious but it has the virtue of applying to any surface defined by a smooth parametric representation.

We omit the nonelementary proof of the general area formula for surfaces; indeed we have not even set down a mathematical definition of surface area. We have only constructed the surface area formula

$$S = \iint_{\mathscr{U}} \sqrt{J_{12}^2 + J_{23}^2 + J_{31}^2}\; du\,dv$$

by plausible analogy with curve theory.

Example. Find the surface area of the region $\mathscr{S}$ of the surface

$$x = r \cos\theta, \qquad y = r \sin\theta, \qquad z = 2(1 - r),$$

over the domain $\mathscr{U} = \{0 \leqq r \leqq \cos\theta, -\pi/2 \leqq \theta \leqq \pi/2\}$.

Solution. We compute the Jacobians

$$J_{12} = \begin{vmatrix} \cos\theta & -r\sin\theta \\ \sin\theta & r\cos\theta \end{vmatrix} = r, \qquad J_{23} = \begin{vmatrix} \sin\theta & r\cos\theta \\ -2 & 0 \end{vmatrix} = 2r\cos\theta,$$

$$J_{31} = \begin{vmatrix} -2 & 0 \\ \cos\theta & -r\sin\theta \end{vmatrix} = 2r\sin\theta, \qquad \text{and} \qquad \sqrt{J_{12}^2 + J_{23}^2 + J_{31}^2} = \sqrt{5}\, r.$$

Hence

$$S = \int_{-\pi/2}^{\pi/2} \int_0^{\cos\theta} \sqrt{5}\, r\, dr\, d\theta = \frac{\sqrt{5}}{2} \int_{-\pi/2}^{\pi/2} \cos^2\theta\, d\theta = \frac{\sqrt{5}\,\pi}{4}.$$

We plot the surface, which is a portion of a cone contained in the vertical cylinder $r = \cos\theta$ (Figure 18.18).

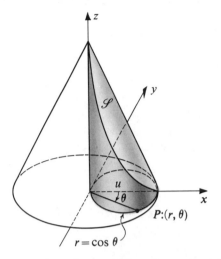

FIGURE 18.18   A portion of a conical surface cut out by a cylinder.

35.3   Use of Vector Product.   Remembering that the Jacobians J_{23}, J_{31}, J_{12}, are 2×2 determinants in the form of area components, we consider the vector $\vec{w}:(J_{23}, J_{31}, J_{12})$, whose norm $\|\vec{w}\| = \sqrt{J_{23}^2 + J_{31}^2 + J_{12}^2}$. All this so closely resembles the vector product $\vec{u} \times \vec{v}$ (§28) that we have only to identify the two vector factors.

The position vector

$$\vec{r}:[x(u, v),\ y(u, v),\ z(u\ v)]$$

traces a curve on the surface when v is held constant and only u varies. This curve $v = $ constant has a representation of the form

$$x = x(u, \cdot), \quad y = y(u, \cdot), \quad z = z(u, \cdot),$$

where we replaced the constant v by a dot. A tangent vector to the curve $v = $ constant is (Figure 18.19),

$$\frac{\partial \vec{r}}{\partial u}:\left(\frac{\partial x}{\partial u}, \frac{\partial y}{\partial u}, \frac{\partial z}{\partial u}\right).$$

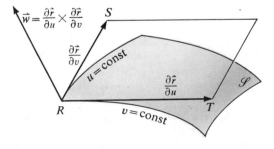

FIGURE 18.19   Surface normal and area element.

Similarly, the tangent vector to the curve $u = $ constant at the same point is

$$\frac{\partial \vec{r}}{\partial v} : \left(\frac{\partial x}{\partial v}, \frac{\partial y}{\partial v}, \frac{\partial z}{\partial v} \right).$$

Then the vector product

$$\vec{w} = \frac{\partial \vec{r}}{\partial u} \times \frac{\partial \vec{r}}{\partial v}$$

represents a vector perpendicular to the surface $\mathscr{S}$ at the point R whose position vector is $\vec{r}$. We find the components of $\vec{w}$ by the computation scheme

$$\vec{w} = \begin{vmatrix} \vec{i} & \vec{j} & \vec{k} \\ \dfrac{\partial x}{\partial u} & \dfrac{\partial y}{\partial u} & \dfrac{\partial z}{\partial u} \\ \dfrac{\partial x}{\partial v} & \dfrac{\partial y}{\partial v} & \dfrac{\partial z}{\partial v} \end{vmatrix} = J_{23}\vec{i} + J_{31}\vec{j} + J_{12}\vec{k}.$$

Hence

$$\|\vec{w}\| = \left\| \frac{\partial \vec{r}}{\partial u} \times \frac{\partial \vec{r}}{\partial v} \right\| = \sqrt{J_{23}^2 + J_{31}^2 + J_{12}^2}.$$

This is the area of the parallelogram formed by the tangent vectors $\dfrac{\partial \vec{r}}{\partial u}$ and $\dfrac{\partial \vec{r}}{\partial v}$ to the curves $v = $ constant and $u = $ constant in the surface at the point R. Hence the vector product $(\partial \vec{r}/\partial u) \times (\partial \vec{r}/\partial v)$ not only gives the area formula

$$S = \iint\limits_{\mathscr{U}} \left\| \frac{\partial \vec{r}}{\partial u} \times \frac{\partial \vec{r}}{\partial v} \right\| du\, dv$$

but gives the surface normal as well.

Example.   Find the surface normal and area element of the surface

$$\vec{r}(u, v) : (x = u \cos v, y = u \sin v, z = u^2).$$

Also find the area of the piece $\mathscr{S}$, where $\mathscr{U} = \{0 \leq u \leq 1, 0 \leq v \leq \pi/4\}$.

  Solution.   We calculate the vectors,

$$\frac{\partial \vec{r}}{\partial u} : (\cos v, \sin v, 2u) \qquad \text{and} \qquad \frac{\partial \vec{r}}{\partial v} : (-u \sin v, u \cos v, 0).$$

Then

$$\frac{\partial \vec{r}}{\partial u} \times \frac{\partial \vec{r}}{\partial v} = \begin{vmatrix} \vec{i} & \vec{j} & \vec{k} \\ \cos v & \sin v & 2u \\ -u \sin v & u \cos v & 0 \end{vmatrix} = 2u^2 \cos v\, \vec{i} + 2u^2 \sin v\, \vec{j} + u\vec{k}.$$

This is a surface normal vector at $\vec{r}(u, v)$. The area element dS is given by

$$dS = \left\| \frac{\partial \vec{r}}{\partial u} \times \frac{\partial \vec{r}}{\partial v} \right\| du\, dv = u \sqrt{4u^2 + 1}\, du\, dv.$$

We compute the area of the designated piece of surface

$$S = \int_0^{\pi/4} \int_0^1 u\sqrt{4u^2 + 1}\ du\ dv = \int_0^{\pi/4} \frac{5\sqrt{5} - 1}{12}\ dv = \frac{5\sqrt{5} - 1}{48}\ \pi.$$

35.4 Exercises

1. Find the surface normal vector $\vec{w}$ to the sphere of radius R

    $$x = R \sin \phi \cos \theta, \qquad y = R \sin \phi \sin \theta, \qquad z = R \cos \phi,$$

    at a general point determined by (u, v). By spherical geometry, this normal must be parallel to $x\vec{i} + y\vec{j} + z\vec{k}$, and the parametric representation gives the same result.
2. Derive for this sphere the element of area dS, and use this to compute the area of the sphere.
3. For the surface $x = u$, $y = v$, $z = 2u + 3v + 1$, find the surface normal at every point and the element of surface area dS. What is the surface?
4. For the surface $x = R \cos u$, $y = R \sin u$, $z = v$, find the surface normal at a general point and the element of area dS.
5. Compute the area of the portion of the surface in Exercise 4 determined by $|v| \leq h/2$ (constant). What is the surface?
6. Find the element of surface area dS for the surface $x = R \cos \theta$, $y = R \sin \theta$, $z = R \cos \phi$. Show that the surface is a zone of height $2R$ on a cylinder.
7. Find the equation of the tangent plane to the surface $x = u \cos v$, $y = u \sin v$, $z = u^2$, at the point where $(u, v) = (1, \pi/4)$ (§35.3, Example).
8. Show that the surface in Exercise 7 is generated by revolving a parabola about the z-axis and compute the area of the portion $\mathscr{U} = \{0 \leq u \leq 1,\ 0 \leq v \leq \pi/4\}$ by the method of area for surfaces of revolution (§30).
9. Compute the element of area dS for the surface $x = \sin u \cos v$, $y = \sin u \sin v$, $z = \cos 2u$. Prove that it is a surface formed by revolving a curve in the yz-plane about the z-axis. What is the generating curve?
10. In Exercise 9, compute the area of the surface that corresponds to $\mathscr{U} = \{0 \leq u \leq \pi/4,\ 0 \leq v \leq 2\pi\}$ in two ways: by integrating dS, and by the method of area of a surface of revolution (§30).
11. Generalize Exercise 9 to show that if $p(u)$ and $q(u)$ are any smooth functions $x = p(u) \cos v$, $y = p(u) \sin v$, $z = q(u)$, is a surface of revolution. Calculate dS to show that

    $$dS = |p|\sqrt{p'^2 + q'^2}\ du\ dv.$$

12. Derive from the result in Exercise 11 that a surface area element that is a function of u along $dS = 2\pi|p|\ ds$, where ds is the element of arc length on the generating curve.
13. (Archimedes' map) We consider a sphere of radius 1, $x = \sin \phi \cos \theta$, $y = \sin \phi \sin \theta$, $z = \cos \phi$, and the cylinder of radius 1 and altitude π, $x_1 = \cos \theta$, $y_1 = \sin \theta$, $z_1 = \cos \phi$. Take any point $P:(x, y, z)$ on the sphere, $|z| \neq 1$. It determines a point $U:(\theta, \phi)$ in the plane. Then U determines a point $Q:(x_1, y_1, z_1)$ on the cylinder. Prove that this mapping of the sphere on the cylinder preserves the areas of all regions.

    Figure 18.20 shows Archimedes' area-true map of the earth. It does not preserve shapes and angles, but all land regions have the correct area. For example, Alaska has the right size relative to Texas, whereas the navigator's angle-true map magnifies the area of regions near the poles out of proportion.
14. Find the area of the surface $x = u + v$, $y = u - v$ for $|u| \leq 1$, $|v| \leq 2$.
15. Find the area of the surface in Exercise 14 over the domain $\mathscr{U}$ that is the unit disk $u^2 + v^2 \leq 1$.
16. Find a parametric representation of the surface of revolution generated by revolving about the x-axis the upper half of the ellipse $b^2x^2 + a^2y^2 = a^2b^2$. Set up the integral representing its total surface area.

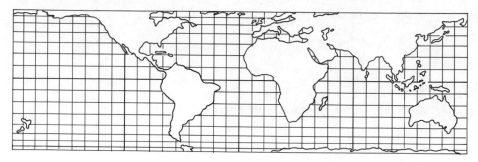

FIGURE 18.20   Exercise 13, area-true map.

17.   Find the element of area dS and total area of the cone given in parameters (r, θ) by

$$x = r \cos \theta, \qquad y = r \sin \theta, \qquad z = h\left(1 - \frac{r}{a}\right),$$

$$\{0 \leq \theta \leq 2\pi, 0 \leq r \leq a\}.$$

18.   Calculate by use of the Jacobians the element of area dS for the sphere of radius R (§35.1).

35R   Review Exercises in Multiple Integration Technique

1.   Find the volume of the spherical ball $r^2 + z^2 = R^2$, using cylindrical coordinates.
2.   Find the volume of the solid cone $0 \leq z \leq 4 - 2r$.
3.   Find the volume of the solid $\rho \leq 4$ and $\phi \leq \pi/4$, using spherical coordinates.
4.   Find the volume of the region bounded by the cylinders $r = 1$ and $r = 2$, and the planes $z = r \cos \theta$ and $z = -r \cos \theta$.
5.   Find the centroid of the solid half-cylinder $r \leq 3$, $-\pi/2 \leq \theta \leq \pi/2$, $-1 \leq z \leq 1$.
6.   Find by integration the volume of the solid spherical cap $\{kR \leq z \leq \sqrt{R^2 - r^2}\}$, where $|k| < 1$.
7.   Find the volume of the region below the plane $z = r \cos \theta$ and above the cone $z = 2r - 2$.
8.   Integrate $\iiint\limits_{\mathscr{D}} [(r^2 \cos \theta)/z]\, dV$ over the cylindrical solid $\mathscr{D} = \{1 \leq r \leq 2, 0 \leq z \leq 3\}$.
9.   Integrate $\iiint\limits_{\mathscr{S}} \rho \sin \theta \cos \phi\, dS$ over the surface of the sphere $\rho = 4$.
10.   *Theorem of Pappus* (Alexandrian Greek, c. A.D. 300). The plane region $\mathscr{A}$ of area A is revolved about the x-axis (Figure 18.21). We showed that for the volume V of the solid of revolution, $2\pi A y_{\min} \leq V \leq 2\pi A y_{\max}$. Prove Pappus' theorem, which asserts that

$$V = 2\pi A \bar{y},$$

where $\bar{y}$ is the y-coordinate of the center of area of $\mathscr{A}$.
11.   Calculate the integral

$$I = \iint\limits_{\mathscr{D}} \left(\frac{x - y}{x + y}\right)^2 dx\, dy, \quad \mathscr{D} = \{x^2 + y^2 \leq 2\}.$$

12.   Use the theorem of Pappus to calculate the volume of the solid ring generated by revolving the disk $x^2 + y^2 - 2y \leq 0$ about the x-axis.

Problems 13–21 are of mixed types, but all can be completed by a single integration with suitable volume, surface, or mass element.

13.   Find the area of the surface of revolution generated when $y = e^x$ on $[0, 1]$ is revolved about the x-axis.

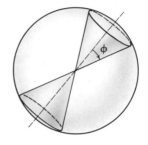

FIGURE 18.21   Problem 10.   FIGURE 18.22   Problem 17.

14. Find the area of the surface generated when one arch of the cycloid $x = a(\theta - \sin \theta)$, $y = a(1 - \cos \theta)$ is revolved about the x-axis.

15. Find the moment of inertia of a flat disk of radius a, thickness h, and density δ, when it is revolved about a line through its center perpendicular to its flat surfaces.

16. Use Problem 15 to find the moment of inertia of the solid generated when the region bounded by $y = \sin x$ on $[0, \pi]$ is revolved about the x-axis and filled with material of density δ.

17. Find the volume inside a solid ball of radius b and inside a cone with angle ϕ and vertex at the center (Figure 18.22).

18. Find the moment of inertia of a ball with respect to the center.

19. Find the center of mass of a pyramid with square base of side s and altitude h.

20. Find the work done when a spherical tank of radius 10 ft is pumped full of water weighing 62.5 lb per cu ft, from a source 50 ft below the bottom of the tank. Recall that work = force × distance.

21. Find by integration the volume generated when the region inside the square with vertices $(2, 0)$, $(3, 1)$, $(4, 0)$, $(3, -1)$ is revolved about the y-axis. Check by using the theorem of Pappus (Problem 10).

22. Find the volume common to two spheres of radii a and b that intersect in a lens-shaped solid of diameter h.

23. Find the mass of a solid ball of radius a if its density δ is a linear function of other radial distance from the center equal to 1 at the center and 2 at the surface.

24. Integrate xyz over the surface of the cube $\{|x| \leq 1, |y| \leq 1, |z| \leq 1\}$.

25. Integrate $x - y$ over the disk $\mathscr{D} = \{x^2 + y^2 - 2x \leq 0\}$.

26. Let $p(\theta, \phi)$ be the barometric pressure at each point on the surface of the earth. Let W be total weight of the earth's atmosphere. Show that

$$W = \iint_{\mathscr{S}} p(\theta, \phi)R^2 \sin \phi \, d\theta \, d\phi,$$

where R is the radius of the earth and $\mathscr{S}$ is the entire surface of the earth.

27. At a hydroelectric power plant the flow of water from the reservoir was at the rate of 100,000 cu ft/min, while at the same time the surface was falling at the rate of 0.1 ft/hr. What was the surface area of the reservoir at that time?

28. Find by double integration the area of the region bounded by the curves $y^2 = x + 4$ and $y^2 = 3x$.

29. Find the center of area of the region enclosed by the cardioid curve $r = k(1 + \cos \theta)$.

30. Integrate $\sin \theta/r$ over the region enclosed by the ellipse $r = 2/(2 - \cos \theta)$.

31. An infinitely long straight wire has a charge density of $+1$ unit charge per cm. An

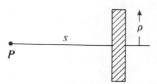

FIGURE 18.23   Problem 39.

electron with charge $-e$ is 10 cm from the wire. What is the total force of attraction acting on this electron?

32.   An infinite plane has a charge density of $+1$ unit per cm². An electron with charge $-e$ is z cm from the plane. What is the total force acting on the electron?

33.   Find the volume enclosed by the surface $x^2 + y^2 - z^2 = 1$ and the planes $z = \pm \sqrt{3}$.

34.   Find the volume of the region below the surface $z = -\frac{1}{4}xy + 9$ on $\mathscr{D}:\{|x| \leq 6, |y| \leq 6\}$ (II, Figure 8.49).

35.   A pyramid has vertices $(1, 0, 0)$, $(0, 1, 0)$, $(2, 2, 0)$, and $(3, -1, 2)$. Find by one integration $\bar{z}$, the z-coordinate of the center of volume.

36.   Find the mean and standard deviation of the distribution $\mu(x, y) + kx^2$ in the square $\mathscr{D} = \{|x| + |y| = 1\}$.

37.   Draw a figure to show the solid region whose volume

$$V = \int_0^1 \int_0^x y \, dy \, dx.$$

38.   Find the volume enclosed by the surface whose equation in spherical coordinates is $\rho = a(1 - \cos \phi)$. What does the absence of θ from the equation indicate?

39.   Show (Figure 18.23) that the potential $u(s)$ due to a thin flat disk of radius ρ and density δ at a point P that is at distance s from the center of the disk is

$$u(s) = 2\pi\delta \int_0^\rho \frac{q \, dq}{\sqrt{s^2 + q^2}} = 2\pi\delta[\sqrt{s^2 + \rho^2} - \sqrt{s^2}].$$

40.* Use the result of Problem 39 and the slice method of integration to find the potential due to a solid ball at a point outside the ball (§32.4).

41.* Use the result of Problem 39 to find the gravitational potential due to a solid ball at a point P in its interior. The result is

$$u(z) = \frac{2}{3} \pi\delta(3R^2 - z^2),$$

where M is the mass of the ball, R is its radius, and z is the coordinate of the point P on an axis with origin at the center of the ball.

42.   A hole is bored straight through the earth and a marble is dropped into it. Is it attracted to the center by an inverse-square-law attraction? Calculate the motion of the marble. (Use the result of Problem 41.)

43.* Find the volume enclosed when the line segment from $O:(0, 0, 0)$ to $P:(1, 0, 0)$ is revolved about the line $x = 2t, y = t - 1, z = -2t$.

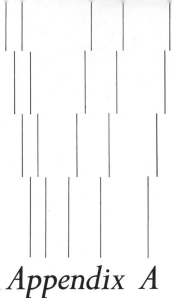

Appendix A

Review of Polynomial Algebra

ACADEMIC NOTE. Well-prepared students will find nothing new in this review section on polynomial algebra, which is not counted in the 34 sections making up the basic course in the calculus of elementary functions. For them, this section may be omitted, and used for reference. Synthetic division is useful, but not essential, in what follows.

1.1 Techniques of Polynomial Algebra. We observe that $0 + x = x$, $0x = 0$, and $1x = x$. The definition admits products of x by itself, which are called powers. Thus $(x)(x)$, written x^2, is a polynomial; so are $(x)(x)(x)$, written x^3, and all other positive integral powers $x^4, x^5, \ldots, x^n, \ldots$. By special definition we agree that $x^0 = 1$. We do *not* include in the algebra of polynomials the negative powers or fractional powers of x. Thus x^{-1}, which is $1/x$, and $x^{1/2}$, which is $\sqrt{x}$, are not polynomials.

When we multiply a power of x by a real number a, we obtain a polynomial of the form ax^n. The polynomial of the form ax^n is called a *power term*, in which a is called the *coefficient* of x^n and n is called the *exponent of the power*. We define an addition of like powers so that

$$ax^n + bx^n = (a + b)x^n$$

and a multiplication of any two power terms so that

$$(ax^m)(bx^n) = abx^{m+n}.$$

These definitions lead to the construction of the familiar polynomial forms in descending powers of x, such as $3x^4 - 12x^2 + 2x - 1$, in which there are a finite number of additions and multiplications.

Two polynomials are equal if one can be transformed into the other by applications of the rules of algebra. Every polynomial can be transformed into a unique standard form beginning with the power term that has the largest exponent with coefficient not zero and proceeding in descending powers of x. The largest exponent of any power term with nonzero coefficient in the polynomial when it is written in standard form is called the *degree* of the polynomial. Thus the following three expressions represent the same polynomial, and it is one of degree of 4:

$$2x^4 - \frac{3}{2}x^2 + 6x - 2 = 2(3x - 1) + \frac{x^2}{2}(4x^2 - 3)$$

$$= \left[\left(2x^2 - \frac{3}{2}\right)x + 6\right]x - 2.$$

The first one is in standard form.

We illustrate the three basic operations of polynomial algebra:

(1)   Polynomial addition, $(3x^2 - 6x + 1) + (-2x - 1) = 3x^2 - 8x.$
(2)   Scalar multiplication, $2(3x^2 - 6x + 1) = 6x^2 - 12x + 2.$
(3)   Polynomial multiplication, $(3x^2 - 6x + 1)(-2x - 1) = -6x^3 + 9x^2 + 4x - 1.$

These operations imply that polynomials admit a subtraction and a form of division like the division of whole numbers. We subtract by the familiar device of reducing subtraction to adding the negative of a polynomial thus:

$$(2x - 1) - (3x^2 - 6x + 1) = (2x - 1) + (-3x^2 + 6x - 1) = -3x^2 + 8x - 2.$$

Example.   Perform the division of the polynomial $f(x)$ by the polynomial $g(x)$ if $f(x) = 4x^3 - 6x + 1$ and $g(x) = -2x - 1$.

Solution.   The calculation (called the division algorithm) proceeds as follows:

$$
\begin{array}{l}
Q(x): \underline{-2x^2 + \;x \; + \frac{5}{2}} \\
\;f(x): \quad 4x^3 + 0x^2 - 6x + 1 \;\big|\; \underline{-2x - 1} : g(x) \\
\qquad\quad \underline{4x^3 + 2x^2} \\
\qquad\qquad\quad -2x^2 - 6x \\
\qquad\qquad\quad \underline{-2x^2 - \;x} \\
\qquad\qquad\qquad\qquad -5x + 1 \\
\qquad\qquad\qquad\qquad \underline{-5x - \frac{5}{2}} \\
\qquad\qquad\qquad\qquad\qquad \frac{7}{2} = R(x)
\end{array}
$$

The result is expressed by the statement

(1) $\qquad\qquad 4x^3 - 6x + 1 = (-2x - 1)(-2x^2 + x + \frac{5}{2}) + \frac{7}{2}.$

In general, the polynomial division algorithm will take a polynomial $f(x)$ and a divisor polynomial $g(x)$ that is not the zero polynomial and produce a quotient polynomial $Q(x)$ and remainder $R(x)$ such that

(2) $\qquad\qquad\qquad f(x) = g(x)Q(x) + R(x).$

The process stops and leaves $Q(x)$ and $R(x)$ uniquely determined when the degree of $R(x)$ becomes less than the degree of $g(x)$. In the example above $f(x) = 4x^3 - 6x + 1$, $g(x) = -2x - 1$, $Q(x) = -2x^2 + x + \frac{5}{2}$, and $R(x) = \frac{7}{2}$.

1.2 The Evaluation Principle. The justification of this algebra of polynomial forms lies in the fact that we can substitute any number for x in all of the polynomials involved in an operation, and the resulting numerical operations will give the same numerical result that we would find by first carrying out the polynomial operations and then substituting the number for x. The evaluation principle is often used as a check on the polynomial operations; if the polynomial operations are correct, they must give results that are correct when any number is substituted for x.

Let us evaluate $f(x)$, $g(x)$, $Q(x)$, and $R(x)$ at some number, say $x = 2$, and check the polynomial equation (2).

Polynomial	Value when $x = 2$
$f(x) = 4x^3 - 6x + 1.$	$f(2) = 4(2)^3 - 6(2) + 1 = 21.$
$g(x) = -2x - 1.$	$g(2) = -2(2) - 1 = -5.$
$Q(x) = -2x^2 + x + \frac{5}{2}.$	$Q(2) = -2(2)^2 + (2) + \frac{5}{2} = -\frac{7}{2}.$
$R(x) = \frac{7}{2}.$	$R(2) = \frac{7}{2}.$
$f(x) = g(x)Q(x) + R(x).$	$f(2) = g(2)Q(2) + R(2)$ becomes
	$21 = (-5)(-\frac{7}{2}) + \frac{7}{2} = (35 + 7)/2$
	or $21 = 21.$

Thus the result of the polynomial division $f(x) = g(x)Q(x) + R(x)$ agrees with numerical operations on the numbers obtained by evaluating the polynomials, setting $x = 2$. This is also true when we substitute any other number for x. In particular, it is true when $x = -\frac{1}{2}$, that is, when the divisor polynomial has the value 0. For the polynomial division algorithm does not require that we divide by the numerical values of $-2x - 1$.

1.3 Remainder Theorem and Factor Theorem

REMAINDER THEOREM. If the polynomial $f(x)$ is divided by $x - a$, the remainder is $f(a)$.

Proof. By the division algorithm, we can find polynomials $Q(x)$ and $R(x)$ such that $f(x) = (x - a)Q(x) + R(x)$ where $R(x)$ is of degree lower than $x - a$. This implies that $R(x)$ must be a constant R. Moreover, the evaluation principle says that when $x = a$, we have $f(a) = 0Q(a) + R$ or $f(a) = R$. This completes the proof.

DEFINITION. If $f(x)$ and $P(x)$ are polynomials then $P(x)$ is a factor of $f(x)$ if and only if there exists a polynomial $Q(x)$ such that $f(x) = P(x)Q(x)$.

COROLLARY (The Factor Theorem). If $f(x)$ is a polynomial, then $x - a$ is a factor of $f(x)$ if and only if $f(a) = 0$.

Proof. If $f(a) = 0$, then $R = 0$, and the polynomial division (2) reduces to $f(x) = (x - a)Q(x)$, which shows that $x - a$ is a factor. Conversely, if $x - a$ is a factor of $f(x)$, this means that $f(x) = (x - a)Q(x)$. Then by the evaluation principle, it must be true when $x = a$, which gives us $f(a) = (a - a)Q(a) = 0$.

We find that it is ordinarily much shorter to divide by $x - a$ to find the remainder $f(a)$ than to substitute $x = a$ directly in the polynomial to get $f(a)$.

Example. Compute $f(3)$ for the polynomial

$$f(x) = 2x^4 - 4x^3 - 3x^2 + 7x + 2$$

by use of the remainder theorem.

Solution. We carry out the division of $f(x)$ by $x - 3$.

$$
\begin{array}{l}
2x^3 + 2x^2 + 3x + 16 \\
\hline
\text{②}x^4 - 4x^3 - 3x^2 + 7x + 2 \quad\big|\,\underline{x - 3} \\
\underline{2x^4 - 6x^3} \\
\quad\; + \text{②}x^3 - 3x^2 \\
\quad\; \underline{+ 2x^3 - 6x^2} \\
\qquad\qquad + \text{③}x^2 + 7x \\
\qquad\qquad \underline{+ 3x^2 - 9x} \\
\qquad\qquad\qquad + 16x + 2 \\
\qquad\qquad\qquad \underline{+ 16x - 48} \\
\qquad\qquad\qquad\qquad + 50 \;=\; R = f(3).
\end{array}
$$

Hence the solution of the problem is $f(3) = 50$. The result can be verified by direct substitution of 3 for x.

1.4 Synthetic Division.

The substitution of a number for x by use of the division procedure can be shortened still further by synthetic division. We write down the sequence of coefficients in the polynomial $f(x)$, replacing all missing powers with coefficient 0. Then we drop out the unnecessary x's, working only with coefficients. We replace the divisor $x - a$ by the coefficient a, 3 in this case. We arrange for a blank second line of coefficients; then draw a line and provide for a third row of coefficients below it. We bring down the first coefficient, 2, into the third line. Thus we have at the start

$$
\begin{array}{l}
2 - 4 - 3 + 7 + 2 \;\big|\,3 \\
\hline
2
\end{array}
$$

Now we multiply the 2 on the third line by the divisor 3 and place the result below -4 on the second line. Then we add the numbers -4 and $+6$ in the second place and put the sum $+2$ on the third line below them. This gives us

$$
\begin{array}{l}
2 - 4 - 3 + 7 + 2 \;\big|\,3 \\
\quad\; + 6 \\
\hline
2 + 2
\end{array}
$$

We continue this process to the end, thus

$$
\begin{array}{l}
2 - 4 - 3 + 7 + 2 \;\big|\,3 \\
\quad\; + 6 + 6 + 9 + 48 \\
\hline
2 + 2 + 3 + 16 + 50
\end{array}
$$

This means that $Q(x) = 2x^3 + 2x^2 + 3x + 16$ and $R = 50$, when we restore the x's in the final line. To restore the x's in $Q(x)$ we recall that, since we are dividing a polynomial of degree 4 by a polynomial of degree 1, the quotient will be a polynomial of degree 3. Accordingly, the coefficients 2, +2, +3, +16 on the last line are the coefficients of x^3, x^2, x, 1 in descending powers.

We verify that the process computes the circled coefficients in the complete division algorithm of the example on page 522 and that these are the coefficients in the quotient. The only deviation, except for dropping out unnecessary symbols, is to add instead of subtract at each stage. This was accomplished by changing the sign of -3 in $x - 3$ and adding instead of subtracting at each stage.

Example.   Compute $Q(x)$ and R when $f(x) = x^3 - 5x$ is divided by $x + 3$. Compute $f(-3)$.

Solution.   Using synthetic division, we place -3 in the division place since $x + 3 = x - (-3)$.

$$
\begin{array}{r}
1 + 0 - 5 + 0 \underline{\,|-3} \\
- 3 + 9 - 12 \\
\hline
1 - 3 + 4 - 12.
\end{array}
$$

The quotient is $Q(x) = x^2 - 3x + 4$, $R = -12$, and $f(-3) = -12$.

1.5   The Graphs of Polynomials

Example.   Plot the graph of the function $x^3 - 5x$ (Figure A.1).

Solution.   We make a table of values $(x, f(x))$ from the equation $f(x) = x^3 - 5x$. We assign numbers to x in increasing order, say starting at -4 and increasing to $+4$. Synthetic division is helpful in the calculations. We make the table and graph (Figure A.1). To make the graph, we first plot the points representing the pairs of numbers $(x, f(x))$ in the table and then connect the points in order of increasing x by a smooth curve. That is, we connect the point where $x = -4$ to the point where $x = -3$, then the point where $x = -3$ to the point where $x = -2$, and so on. This procedure insures that our graph represents a *function*, since for every number x there then corresponds one and only one point in the graph (Figure A.2).

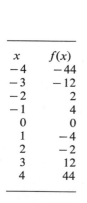

x	$f(x)$
-4	-44
-3	-12
-2	2
-1	4
0	0
1	-4
2	-2
3	12
4	44

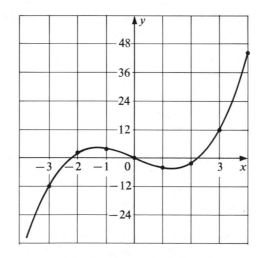

FIGURE A.1   Graph of $x^3 - 5x$.

The points

Connected incorrectly

The only possible connection for a function $f(x)$.

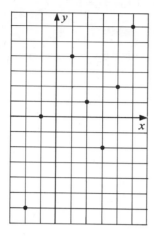

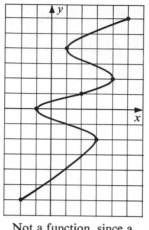

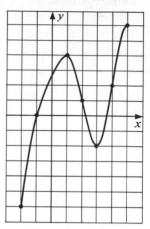

Not a function, since a value of x determines several points on the graph.

One value of $f(x)$ for each value of x.

FIGURE A.2   Connecting points on the graph of a function.

1.6 Exercises

In Exercises 1–6, carry out the indicated operations. Evaluate each polynomial at $x = 1$ and verify that the resulting numerical operations agree with the polynomial operations (evaluation principle).

1. $(2x^2 - 3x + 4) + (6x - 7)$.
2. $(x^3 - 3x + 4)/(3x^2 - 1)$.
3. $(3x^2 - 2x + 6)(x^2 - 4)$.
4. $(-x^3 + 6x - 9)/(x + 4)$.
5. $(x + 1)^2 - 6(x + 1) + 7$.
6. $(2x - 1) - (x^3 + 7x^2 - 9)$.

7. Evaluate by polynomial division: $f(-2), f(3), f(2.6)$, where $f(x) = x^3 - 5x + 7$.
8. Make a table of values of $x^3 - 3x^2 - x + 3$ and use this with the factor theorem (§1.3) to factor the polynomial into first-degree factors.

For the polynomial $f(x) = 2x^3 - 3x + 4$, carry out the divisions indicated in Exercises 9–12, using synthetic division. (Be careful to replace the missing coefficient of x^2 by 0.)

9. $\dfrac{f(x) - f(1)}{x - 1}$.

10. $\dfrac{f(x) - f(-2)}{x + 2}$.

11. $\dfrac{f(w) - f(-3)}{w + 3}$.

12. $\dfrac{f(w) - f(x)}{w - x}$.

13. Prove that if $f(x)$ is any polynomial, $f(w) - f(x)$ always divides by $w - x$ and the remainder is 0.
14. Graph the functions x^2, x^4, x^6 on the same coordinate axes. Observe that the graphs are symmetric with respect to the y-axis and that none of the functions have negative values of y.
15. Graph the functions x, x^3, x^5 on the same coordinate axes. Observe that all are increasing functions.
16. Plot the graphs of $x^2 + c$ for several values of the constant c.

In Exercises 17–24, plot the graphs of the polynomial functions. In each case use the graph to solve the equation $f(x) = 0$ by estimating the values of x where the graph crosses the x-axis. Observe that in some cases there are no roots.

17. $f(x) = 1.$ 18. $f(x) = -x^2.$
19. $f(x) = -x^3.$ 20. $f(x) = -x^2 - x - 1.$
21. $f(x) = x^3 - 7x + 2.$ 22. $f(x) = 0$ (the zero polynomial).
23. $f(x) = x^4 - 2x + 1.$ 24. $f(x) = (x - 1)^3 + 2.$

25. For the polynomial $f(x) = x^3 - 7x + 1$, find the slope of the line that joins the points on the graph where (a) $x = 1$ and $x = 3$; (b) $x = -1$ and $x = 0$; (c) $x = -2$ and $x = -1$.

26. Show that the synthetic division

$$2 - 3 + 1 - 1 + 5 \,\lfloor\underline{-2}$$
$$\underline{ - 4 + 14 - 30 + 62}$$
$$2 - 7 + 15 - 31 + 67$$

replaces the calculation

$$2(-2)^4 - 3(-2)^3 + (-2)^2 - (-2) + 5 = 67$$

by

$$\{[(2(-2) - 3)(-2) + 1](-2) - 1\}(-2) + 5 = 67.$$

27. Why is synthetic division a shorter method of computing $f(a)$ for a polynomial $f(x)$ than direct substitution? In Exercise 26, count the multiplications involved in computing $f(-2)$ by direct substitution and by synthetic division.

28. Show as in Exercise 27 that if $f(x)$ is a polynomial of degree n, then substitution ordinarily requires $(n + 1)(n + 2)/2$ multiplications to calculate $f(a)$, while synthetic division requires only n multiplications. How many multiplications are required by the two methods to compute $f(a)$ for a polynomial of degree 100?

In Exercises 29–30, divide twice by the indicated divisor to verify the given expression for the slope ratio. In each case $f(x) = x^3 - 7x + 2$.

29. $\dfrac{f(x) - f(2)}{x - 2} = 5 + (x + 4)(x - 2).$

30. $\dfrac{f(w) - f(x)}{w - x} = 3x^2 - 7 + (w + 2x)(w - x).$

31. Factor $x^n - a^n$, where a is constant.

32. Show that synthetic division replaces the general polynomial

$$a_0 x^n + a_1 x^{n+1} + a_2 x^{n-2} + a_3 x^{n-3} + \cdots + a_{n-1} x + a_n$$

by the sequence of operations

$$(\cdots((((a_0)x + a_1)x + a_2)x + a_3)x + \cdots + a_{n-1})x + a_n.$$

Describe this sequence of operations in words, beginning with: "Start with a_0 ...."

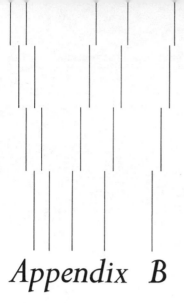

Appendix B

Table I
Derivatives and Integrals

Throughout u, v, and f are assumed to be differentiable functions. a, b, c are constants.

BASIC LIST

Derivatives	*Antiderivatives*

Linearity of derivative and antiderivative operators.

I. $D(u + v) = Du + Dv$.

$\int$ I. $\int (Du + Dv) = \int Du + \int Dv$.

II. $Dcu = c\, Du$.

$\int$ II. $\int c\, Du = c \int Du$.

Product rule.

III. $Duv = u\, Dv + v\, Du$.

Integration by parts.

$\int$ III. $\int u\, Dv = uv - \int v\, Du$.

Constant function.

IV. $Dc = 0$.

Antiderivative of zero on an interval.

$\int$ IV. $\int 0\, Du = C$ (constant), provided that the zero function is defined on a connected interval.

Power formulas.

V. $Du^n = nu^{n-1}\, Du$,

for integers n.

Power formulas.

$\int$ V. $\int u^n\, Du = \dfrac{u^{n+1}}{n+1} + C, \quad n \neq -1$,

and except where $u(x) = 0$ when $n < 0$.

Table I—*continued*

$V_p.$ $Du^p = pu^{p-1}\,Du,$ $\int V_p.$ $\int u^p\,Du = \dfrac{u^{p+1}}{p+1} + C, p \neq -1,$

valid for general real powers p, provided $u(x) > 0$ for all x.

Square root formula (Power formula with $p = \frac{1}{2}$).

$V_{1/2}.$ $D\sqrt{u} = \dfrac{Du}{2\sqrt{u}},$ provided that $\int V_{1/2}.$ $\int \sqrt{u}\,Du = \dfrac{u^{3/2}}{3/2} + C,$

$u(x) > 0.$

Quotient formula.

$V_a.$ $D\dfrac{u}{v} = \dfrac{v\,Du - u\,Dv}{v^2},$ valid where $v(x) \neq 0.$

Chain Rule.   Integration by Substitution.

VI. $Df(u) = f'(u)\,Du.$ $\int VI.$ $\int f'(u)\,Du = f(u) + C.$

Indefinite integral.   Indefinite integral.

VII. $D\int_a^u f(x)\,dx = f(u)\,Du,$ valid at $\int VII.$ An antiderivative $\int f(u)\,Du$ is given
      points of continuity of f.   by $\int_a^u f(x)\,dx$, at points where f is
                                                 continuous.

ELEMENTARY TRANSCENDENTAL FUNCTIONS

Natural logarithm.   Exceptional power formula.

VIII. $D \ln |u| = \dfrac{1}{u}\,Du,$ valid where $\int VIII.$ $\int \dfrac{1}{u}\,Du = \ln |u| + C,$ valid where

$u(x) \neq 0.$ $u(x) \neq 0.$

Other logarithms.

$VIII_a.$ $D \log_b |u| = (\log_b e)\,\dfrac{Du}{u}.$

Exponential functions.

IX. $De^u = e^u\,Du.$ $\int IX.$ $\int e^u\,Du = e^u + C.$

$IX_a.$ $Db^u = (\ln b)b^u\,Du,$ $\int IX_a.$ $\int b^u\,Du = \dfrac{b^u}{\ln b} + C,$

valid where $b > 0$ and $b \neq 1.$

Trigonometric functions.

X. $D \cos u = -\sin u\,Du.$ $\int X.$ $\int \sin u\,Du = -\cos u + C.$

XI. $D \sin u = \cos u\,Du.$ $\int XI.$ $\int \cos u\,Du = \sin u + C.$

XII. $D \tan u = \dfrac{Du}{\cos^2 u} = \sec^2 u\,Du,$ $\int XII.$ $\int \dfrac{Du}{\cos^2 u} = \tan u + C = \int \sec^2 u\,D,$

valid where $u(x) \neq \dfrac{\pi}{2}$ or $-\dfrac{\pi}{2}.$

$\int XII_a.$ $\int \dfrac{Du}{\cos u} = \ln \left| \dfrac{1 + \sin u}{\cos u} \right| + C.$

$XII_b.$ $D \cot u = \dfrac{-Du}{\sin^2 u} = -\csc^2 u\,Du,$ $\int XII_a.$ $\int \dfrac{Du}{\sin^2 u} = -\cot u + C =$

                                                  $\int \csc^2 u\,Du,$

valid where $u(x) \neq 0$ or $\pi.$

Table I—*continued*

XIII. $D \sec u = \sec u \tan u\, Du,$ $\int$ XIII. $\int \sec u \tan u\, Du = \sec u + C,$

$$\text{valid where } u(x) \neq \frac{\pi}{2} \text{ or } -\frac{\pi}{2}.$$

$\int$ XIII$_a$. $\int \sec u\, Du = \ln |\sec u + \tan u| + C.$

XIII$_b$. $D \csc u = -\csc u \cot u\, Du,$ $\int$ XIII$_b$. $\int \csc u \cot u\, Du = -\csc u + C,$

$$\text{valid where } u(x) \neq 0 \text{ or } \pi.$$

INVERSE TRIGONOMETRIC FUNCTIONS

The inverses are defined by principal values and hence are functions.

XIV. $D \arccos u = -\dfrac{Du}{\sqrt{1 - u^2}},$ $\int$ XIV. $\int \dfrac{Du}{\sqrt{1 - u^2}} = \begin{cases} -\arccos u + C, \text{ or} \\ \arcsin u + C, \end{cases}$

where $|u(x)| < 1.$ where $|u(x)| \leq 1.$

XV. $D \arcsin u = \dfrac{Du}{\sqrt{1 - u^2}},$ $\int$ XV$_a$. $\int \dfrac{Du}{\sqrt{a^2 - u^2}} = \arcsin \dfrac{u}{a} + C,$

where $|u(x)| < 1.$ where $|u| \leq |a|.$

XVI. $D \arctan u = \dfrac{Du}{1 + u^2}.$ $\int$ XVI. $\int \dfrac{Du}{1 + u^2} = \arctan u + C.$

$\int$ XVI$_a$. $\int \dfrac{Du}{a^2 + u^2} = \dfrac{1}{a} \arctan \dfrac{u}{a} + C.$

HYPERBOLIC FUNCTIONS

XVII. $D \sinh u = \cosh u\, Du.$ $\int$ XVII. $\int \cosh u\, Du = \sinh u + C.$

XVIII. $D \cosh u = \sinh u\, Du.$ $\int$ XVIII. $\int \sinh u\, Du = \cosh u + C.$

XIX. $D \tanh u = \operatorname{sech}^2 u\, Du.$ $\int$ XIX. $\int \operatorname{sech}^2 u\, Du = \tanh u + C.$

INVERSE HYPERBOLIC FUNCTIONS

XX. $D \sinh^{-1} u = \dfrac{1}{\sqrt{u^2 + 1}}\, Du.$ $\int$ XX. $\dfrac{Du}{\sqrt{u^2 + 1}} = \sinh^{-1} u + C.$

XXI. $D \cosh^{-1} u = \dfrac{1}{\sqrt{u^2 - 1}}\, Du,$ $\int$ XXI. $\int \dfrac{Du}{\sqrt{u^2 - 1}} = \cosh^{-1} u + C,$

$$|u| > 1.$$

XXII. $D \tan^{-1} u = \dfrac{1}{1 - u^2}\, Du,$ $\int$ XXII. $\int \dfrac{Du}{1 - u^2} = \tanh^{-1} u + C,$

$$|u| < 1.$$

SELECTED INTEGRATION FORMULAS THAT EXTEND THE BASIC ONES

The constant of integration is omitted. Integral powers of $(ax^2 + bx + c)^{-1}, a \neq 0.$

23a. $\displaystyle\int \dfrac{Dx}{ax^2 + bx + c} = \dfrac{1}{\sqrt{b^2 - 4ac}} \ln \dfrac{2ax + b - \sqrt{b^2 - 4ac}}{2ax + b + \sqrt{b^2 - 4ac}},$ if $b^2 - 4ac > 0.$

23b. $\displaystyle\int \dfrac{Dx}{ax^2 + bx + c} = \dfrac{2}{\sqrt{4ac - b^2}} \arctan \dfrac{2ax + b}{\sqrt{4ac - b^2}},$ if $b^2 - 4ac < 0.$

23c. $\displaystyle\int \dfrac{Dx}{ux^2 + bx + c} = -\dfrac{2}{2ax + b},$ if $b^2 - 4ac = 0.$

24. $\displaystyle\int\frac{x\,Dx}{ax^2+bx+c}=\frac{1}{2a}\ln\left|ax^2+bx+c\right|-\frac{b}{2a}\int\frac{Dx}{ax^2+bx+c}.$

25. $\displaystyle\int\frac{Dx}{(ax^2+bx+c)^{n+1}}=\frac{2ax+b}{n(4ac-b^2)(ax^2+bx+c)^n}$

$$+\frac{2(2n-1)a}{n(4ac-b^2)}\int\frac{Dx}{(ax^2+bx+c)^n},\ b^2-4ac\neq0.$$

26. $\displaystyle\int\frac{x\,Dx}{(ax^2+bx+c)^{n+1}}=\frac{-(bx+2c)}{n(4ac-b^2)(ax^2+bx+c)^n}$

$$-\frac{b(2n-1)}{n(4ac-b^2)}\int\frac{Dx}{(ax^2+bx+c)^n},\ b^2-4ac\neq0.$$

Products of integral powers of x, $\sin ax$, $\cos ax$.

27. $\displaystyle\int\sin^2 ax\,Dx=\frac{1}{2a}(ax-\sin ax\cos ax),\ a\neq0.$

28. $\displaystyle\int\frac{Dx}{\sin ax}=\int\csc ax\,Dx=\frac{1}{a}\ln\left|\tan\frac{ax}{2}\right|+\frac{1}{a}\ln\left|\frac{1-\cos x}{\sin x}\right|,\ a\neq0.$

29. $\displaystyle\int\frac{Dx}{\sin^2 ax}=-\frac{1}{a}\frac{\cos ax}{\sin ax}=\int\csc^2 ax\,Dx=-\frac{1}{a}\cot ax,\ a\neq0.$

30. $\displaystyle\int\sin^n ax\,Dx=-\frac{\sin^{n-1}ax\cos ax}{na}+\frac{n-1}{n}\int\sin^{n-2}ax\,Dx,$ if $n>0.$

$$=\frac{\sin^{n+1}ax\cos ax}{(n+1)a}+\frac{n+2}{n+1}\int\sin^{n+2}ax\,Dx,$$ if $n<0.$

31. $\displaystyle\int\cos^2 ax\,Dx=\frac{1}{2a}(ax+\sin ax\cos ax),\ a\neq0.$

32. $\displaystyle\int\frac{Dx}{\cos ax}=\int\sec ax\,Dx=\frac{1}{a}\ln\left|\tan\frac{ax}{2}+\frac{\pi}{4}\right|=\frac{1}{a}\ln\left|\frac{1+\sin x}{\cos x}\right|,\ a\neq0.$

33. $\displaystyle\int\frac{Dx}{\cos^2 ax}=\int\sec^2 ax\,Dx=\frac{1}{a}\tan ax,\ a\neq0.$

34. $\displaystyle\int\cos^n ax\,Dx=\frac{\cos^{n-1}ax\sin ax}{na}+\frac{n-1}{n}\int\cos^{n-2}ax\,Dx,$ if $n>0.$

$$=-\frac{\cos^{n+1}ax\sin ax}{(n+1)a}+\frac{n+2}{n+1}\int\cos^{n+2}ax\,Dx,$$ if $n<0.$

35. $\displaystyle\int x^n\sin ax\,Dx=-\frac{1}{a}x^n\cos ax+\frac{n}{a}\int x^{n-1}\cos ax\,Dx,$ if $n>0.$

$$=\frac{1}{n+1}x^{n+1}\sin ax-\frac{a}{n+1}\int x^{n+1}\cos as\,Dx,$$ if $n<0$ and
$$n\neq-1.$$

36. $\displaystyle\int x^n\cos ax\,Dx=\frac{1}{a}x^n\sin ax-\frac{n}{a}\int x^{n-1}\sin ax\,Dx,$ if $n>0.$

$$=\frac{1}{n+1}x^{n+1}\cos ax+\frac{a}{n+1}\int x^{n+1}\sin ax\,Dx,$$ if $n<0$ and
$$n\neq-1.$$

Other trigonometric integrals. In $\int 37-\int 53$, n is an integer.

37. $\displaystyle\int\tan^n ax\,Dx=\frac{1}{a(n-1)}\tan^{n-1}ax-\int\tan^{n-2}ax\,Dx,\ n>1.$

38. $\displaystyle\int\cot^n ax\,Dx=-\frac{1}{a(n-1)}\cot^{n-1}ax-\int\cot^{n-2}ax\,Dx,\ n>1.$

39. $\int \sec^n ax \, Dx = \dfrac{1}{a(n-1)} \dfrac{\sin ax}{\cos^{n-1} ax} + \dfrac{n-2}{n-1} \int \sec^{n-2} ax \, Dx, \, n > 1.$

40. $\int \csc^n ax \, Dx = -\dfrac{1}{a(n-1)} \dfrac{\cos ax}{\sin^{n-1} ax} + \dfrac{n-2}{n-1} \int \csc^{n-2} ax \, Dx, \, n > 1.$

41. $\int \arcsin \dfrac{x}{a} \, Dx = x \arcsin \dfrac{x}{a} + \sqrt{a^2 - x^2}, \, -\dfrac{\pi}{2} \leqq \arcsin \dfrac{x}{a} \leqq \dfrac{\pi}{2}.$

42. $\int \arccos \dfrac{x}{a} \, Dx = x \arccos \dfrac{x}{a} - \sqrt{a^2 - x^2}, \, 0 \leqq \arccos \dfrac{x}{a} \leqq \pi.$

43. $\int \arctan \dfrac{x}{a} \, Dx = x \arctan \dfrac{x}{a} - \dfrac{a}{2} \ln (a^2 + x^2), \, -\dfrac{\pi}{2} < \arctan \dfrac{x}{a} < \dfrac{\pi}{2}.$

Exponentials and logarithms.

44. $\int x^n e^{ax} \, Dx = \dfrac{1}{a} x^n e^{ax} - \dfrac{n}{a} \int x^{n-1} e^{ax} \, Dx, \, n > 0.$

45. $\int x^n \ln ax \, Dx = x^{n+1} \left[\dfrac{\ln(ax)}{n+1} - \dfrac{1}{(n+1)^2} \right], \, n \neq -1, \, ax > 0.$

46. $\int x^n (\ln ax)^m \, Dx = \dfrac{x^{n+1}}{n+1} (\ln ax)^m - \dfrac{m}{n+1} \int x^n (\ln ax)^{m-1} \, Dx, \, n \neq -1, \, ax > 0.$

Integrals involving $\sqrt{p^2 - x^2}$, $\sqrt{x^2 - p^2}$, $\sqrt{p^2 + x^2}$. For similar integrals involving $\sqrt{ax^2 + bx + c}$, complete the square to reduce it to one of the preceding.

47a. $\int \sqrt{p^2 - x^2} \, Dx = \dfrac{1}{2} \left[x\sqrt{p^2 - x^2} + p^2 \arcsin \dfrac{x}{p} \right], \, 0 \leqq |x| \leqq |p|.$

47b. $\int \sqrt{x^2 + c} \, Dx = \dfrac{2}{2} \left[x\sqrt{x^2 + c} + c \ln |x + \sqrt{x^2 + c}| \right], \, x^2 > |c| > 0.$

48. $\int \dfrac{Dx}{\sqrt{x^2 + c}} = \ln |x + \sqrt{x^2 + c}|, \, x^2 > |c|.$

49a. $\int \dfrac{Dx}{x\sqrt{p^2 + x^2}} = -\dfrac{1}{p} \ln \left| \dfrac{p + \sqrt{p^2 + x^2}}{x} \right|, \, x \neq 0.$

49b. $\int \dfrac{Dx}{x\sqrt{p^2 - x^2}} = -\dfrac{1}{p} \ln \left| \dfrac{p + \sqrt{p^2 - x^2}}{x} \right|, \, 0 < |x| < |p|.$

49c. $\int \dfrac{Dx}{x\sqrt{x^2 - p^2}} = \dfrac{1}{p} \arccos \left(\dfrac{p}{x} \right), \, \text{or} \, -\dfrac{1}{p} \arcsin \dfrac{p}{x}, \, |x| \geqq |p|.$

50. $\int \dfrac{x^n \, Dx}{\sqrt{ax^2 + c}} = \dfrac{x^{n-1}\sqrt{ax^2 + c}}{na} - \dfrac{(n-1)c}{na} \int \dfrac{x^{n-2} \, Dx}{\sqrt{ax^2 + c}}, \, ax^2 + c > 0.$

51. $\int \dfrac{Dx}{x^n \sqrt{ax^2 + c}} = -\dfrac{\sqrt{ax^2 + c}}{c(n-1)x^{n-1}} - \dfrac{(n-2)a}{(n-1)c} \int \dfrac{Dx}{x^{n-2}\sqrt{ax^2 + c}},$
$ax^2 + c > 0, \, x \neq 0, \, n > 1.$

52. $\int e^{ax} \sin bx \, Dx = \dfrac{e^{ax}}{a^2 + b^2} (a \sin bx - b \cos bx).$

53. $\int e^{ax} \sin^n bx \, Dx = \dfrac{e^{ax}(a \sin^n bx - n^b \sin^{n-1} bx \cos bx)}{a^2 + n^2 b^2}$
$\qquad\qquad + \dfrac{n(n-1)b^2}{a^2 + n^2 b^2} \int e^{ax} \sin^{n-2} bx \, Dx.$

Table II
Four-Place Common Logarithms

N L.	0	1	2	3	4	5	6	7	8	9
1.0	.0000	.0043	.0086	.0128	.0170	.0212	.0253	.0294	.0334	.0374
1.1	.0414	.0453	.0492	.0531	.0569	.0607	.0645	.0682	.0719	.0755
1.2	.0792	.0828	.0864	.0899	.0934	.0969	.1004	.1038	.1072	.1106
1.3	.1139	.1173	.1206	.1239	.1271	.1303	.1335	.1367	.1399	.1430
1.4	.1461	.1492	.1523	.1553	.1584	.1614	.1644	.1673	.1703	.1732
1.5	.1761	.1790	.1818	.1847	.1875	.1903	.1931	.1959	.1987	.2014
1.6	.2041	.2068	.2095	.2122	.2148	.2175	.2201	.2227	.2253	.2279
1.7	.2304	.2330	.2355	.2380	.2405	.2430	.2455	.2480	.2504	.2529
1.8	.2553	.2577	.2601	.2625	.2648	.2672	.2695	.2718	.2742	.2765
1.9	.2788	.2810	.2833	.2856	.2878	.2900	.2923	.2945	.2967	.2989
2.0	.3010	.3032	.3054	.3075	.3096	.3118	.3139	.3160	.3181	.3201
2.1	.3222	.3243	.3263	.3284	.3304	.3324	.3345	.3365	.3385	.3404
2.2	.3424	.3444	.3464	.3483	.3502	.3522	.3541	.3560	.3579	.3598
2.3	.3617	.3636	.3655	.3674	.3692	.3711	.3729	.3747	.3766	.3784
2.4	.3802	.3820	.3838	.3856	.3874	.3892	.3909	.3927	.3945	.3962
2.5	.3979	.3997	.4014	.4031	.4048	.4065	.4082	.4099	.4116	.4133
2.6	.4150	.4166	.4183	.4200	.4216	.4232	.4249	.4265	.4281	.4298
2.7	.4314	.4330	.4346	.4362	.4378	.4393	.4409	.4425	.4440	.4456
2.8	.4472	.4487	.4502	.4518	.4533	.4548	.4564	.4579	.4594	.4609
2.9	.4624	.4639	.4654	.4669	.4683	.4698	.4713	.4728	.4742	.4757
3.0	.4771	.4786	.4800	.4814	.4829	.4843	.4857	.4871	.4886	.4900
3.1	.4914	.4928	.4942	.4955	.4969	.4983	.4997	.5011	.5024	.5038
3.2	.5051	.5065	.5079	.5092	.5105	.5119	.5132	.5145	.5159	.5172
3.3	.5185	.5198	.5211	.5224	.5237	.5250	.5263	.5276	.5289	.5302
3.4	.5315	.5328	.5340	.5353	.5366	.5378	.5391	.5403	.5416	.5428
3.5	.5441	.5453	.5465	.5478	.5490	.5502	.5514	.5527	.5539	.5551
3.6	.5563	.5575	.5587	.5599	.5611	.5623	.5635	.5647	.5658	.5670
3.7	.5682	.5694	.5705	.5717	.5729	.5740	.5752	.5763	.5775	.5786
3.8	.5798	.5809	.5821	.5832	.5843	.5855	.5866	.5877	.5888	.5899
3.9	.5911	.5922	.5933	.5944	.5955	.5966	.5977	.5988	.5999	.6010
4.0	.6021	.6031	.6042	.6053	.6064	.6075	.6085	.6096	.6107	.6117
4.1	.6128	.6138	.6149	.6160	.6170	.6180	.6191	.6201	.6212	.6222
4.2	.6232	.6243	.6253	.6263	.6274	.6284	.6294	.6304	.6314	.6325
4.3	.6335	.6345	.6355	.6365	.6375	.6385	.6395	.6405	.6415	.6425
4.4	.6435	.6444	.6454	.6464	.6474	.6484	.6493	.6503	.6513	.6522
4.5	.6532	.6542	.6551	.6561	.6571	.6580	.6590	.6599	.6609	.6618
4.6	.6628	.6637	.6646	.6656	.6665	.6675	.6684	.6693	.6702	.6712
4.7	.6721	.6730	.6739	.6749	.6758	.6767	.6776	.6785	.6794	.6803
4.8	.6812	.6821	.6830	.6839	.6848	.6857	.6866	.6875	.6884	.6893
4.9	.6902	.6911	.6920	.6928	.6937	.6946	.6955	.6964	.6972	.6981
5.0	.6990	.6998	.7007	.7016	.7024	.7033	.7042	.7050	.7059	.7067
5.1	.7076	.7084	.7093	.7101	.7110	.7118	.7126	.7135	.7143	.7152
5.2	.7160	.7168	.7177	.7185	.7193	.7202	.7210	.7218	.7226	.7235
5.3	.7243	.7251	.7259	.7267	.7275	.7284	.7292	.7300	.7308	.7316
5.4	.7324	.7332	.7340	.7348	.7356	.7364	.7372	.7380	.7388	.7396
N L.	0	1	2	3	4	5	6	7	8	9

Table II (continued)

N L.	0	1	2	3	4	5	6	7	8	9
5.5	.7404	.7412	.7419	.7427	.7435	.7443	.7451	.7459	.7466	.7474
5.6	.7482	.7490	.7497	.7505	.7513	.7520	.7528	.7536	.7543	.7551
5.7	.7559	.7566	.7574	.7582	.7589	.7597	.7604	.7612	.7619	.7627
5.8	.7634	.7642	.7649	.7657	.7664	.7672	.7679	.7686	.7694	.7701
5.9	.7709	.7716	.7723	.7731	.7738	.7745	.7752	.7760	.7767	.7774
6.0	.7782	.7789	.7796	.7803	.7810	.7818	.7825	.7832	.7839	.7846
6.1	.7853	.7860	.7868	.7875	.7882	.7889	.7896	.7903	.7910	.7917
6.2	.7924	.7931	.7938	.7945	.7952	.7959	.7966	.7973	.7980	.7987
6.3	.7993	.8000	.8007	.8014	.8021	.8028	.8035	.8041	.8048	.8055
6.4	.8062	.8069	.8075	.8082	.8089	.8096	.8102	.8109	.8116	.8122
6.5	.8129	.8136	.8142	.8149	.8156	.8162	.8169	.8176	.8182	.8189
6.6	.8195	.8202	.8209	.8215	.8222	.8228	.8235	.8241	.8248	.8254
6.7	.8261	.8267	.8274	.8280	.8287	.8293	.8299	.8306	.8312	.8319
6.8	.8325	.8331	.8338	.8344	.8351	.8357	.8363	.8370	.8376	.8382
6.9	.8388	.8395	.8401	.8407	.8414	.8420	.8426	.8432	.8439	.8445
7.0	.8451	.8457	.8463	.8470	.8476	.8482	.8488	.8494	.8500	.8506
7.1	.8513	.8519	.8525	.8531	.8537	.8543	.8549	.8555	.8561	.8567
7.2	.8573	.8579	.8585	.8591	.8597	.8603	.8609	.8615	.8621	.8627
7.3	.8633	.8639	.8645	.8651	.8657	.8663	.8669	.8675	.8681	.8686
7.4	.8692	.8698	.8704	.8710	.8716	.8722	.8727	.8733	.8739	.8745
7.5	.8751	.8756	.8762	.8768	.8774	.8799	.8785	.8791	.8797	.8802
7.6	.8808	.8814	.8820	.8825	.8831	.8837	.8842	.8848	.8854	.8859
7.7	.8865	.8871	.8876	.8882	.8887	.8893	.8899	.8904	.8910	.8915
7.8	.8921	.8927	.8932	.8938	.8943	.8949	.8954	.8960	.8965	.8971
7.9	.8976	.8982	.8987	.8993	.8998	.9004	.9009	.9015	.9020	.9025
8.0	.9031	.9036	.9042	.9047	.9053	.9058	.9063	.9069	.9074	.9079
8.1	.9085	.9090	.9096	.9101	.9106	.9112	.9117	.9122	.9128	.9133
8.2	.9138	.9143	.9149	.9154	.9159	.9165	.9170	.9175	.9180	.9186
8.3	.9191	.9196	.9201	.9206	.9212	.9217	.9222	.9227	.9232	.9238
8.4	.9243	.9248	.9253	.9258	.9263	.9269	.9274	.9279	.9284	.9289
8.5	.9294	.9299	.9304	.9309	.9315	.9320	.9325	.9330	.9335	.9340
8.6	.9345	.9350	.9355	.9360	.9365	.9370	.9375	.9380	.9385	.9390
8.7	.9395	.9400	.9405	.9410	.9415	.9420	.9425	.9430	.9435	.9440
8.8	.9445	.9450	.9455	.9460	.9465	.9469	.9474	.9479	.9484	.9489
8.9	.9494	.9499	.9504	.9509	.9513	.9518	.9523	.9528	.9533	.9538
9.0	.9542	.9457	.9552	.9557	.9562	.9566	.9571	.9576	.9581	.9586
9.1	.9590	.9595	.9600	.9605	.9609	.9614	.9619	.9624	.9628	.9633
9.2	.9638	.9643	.9647	.9652	.9657	.9661	.9666	.9671	.9675	.9680
9.3	.9685	.9689	.9694	.9699	.9703	.9708	.9713	.9717	.9722	.9727
9.4	.9731	.9736	.9741	.9745	.9750	.9754	.9759	.9763	.9768	.9773
9.5	.9777	.9782	.9786	.9791	.9795	.9800	.9805	.9809	.9814	.9818
9.6	.9823	.9827	.9832	.9836	.9841	.9845	.9850	.9854	.9859	.9863
9.7	.9868	.9872	.9877	.9881	.9886	.9890	.9894	.9899	.9903	.9908
9.8	.9912	.9917	.9921	.9926	.9930	.9934	.9939	.9943	.9948	.9952
9.9	.9956	.9961	.9965	.9969	.9974	.9978	.9983	.9987	.9991	.9996
N L.	0	1	2	3	4	5	6	7	8	9

Table III
Values of Trigonometric Functions

Degrees	Radians	Sine	Tangent	Cotangent	Cosine		
0°	.0000	.0000	.0000		1.0000	1.5708	90°
1	.0175	.0175	.0175	57.290	.9998	1.5533	89
2	.0349	.0349	.0349	28.636	.9994	1.5359	88
3	.0524	.0523	.0524	19.081	.9986	1.5184	87
4	.0698	.0698	.0699	14.301	.9976	1.5010	86
5	.0873	.0872	.0875	11.430	.9962	1.4835	85
6	.1047	.1045	.1051	9.5144	.9945	1.4661	84
7	.1222	.1219	.1228	8.1443	.9925	1.4486	83
8	.1396	.1392	.1405	7.1154	.9903	1.4312	82
9	.1571	.1564	.1584	6.3138	.9877	1.4137	81
10	.1745	.1736	.1763	5.6713	.9848	1.3963	80
11	.1920	.1908	.1944	5.1446	.9816	1.3788	79
12	.2094	.2079	.2126	4.7046	.9781	1.3614	78
13	.2269	.2250	.2309	4.3315	.9744	1.3439	77
14	.2443	.2419	.2493	4.0108	.9703	1.3265	76
15	.2618	.2588	.2679	3.7321	.9659	1.3090	75
16	.2793	.2756	.2867	3.4874	.9613	1.2915	74
17	.2967	.2924	.3057	3.2709	.9563	1.2741	73
18	.3142	.3090	.3249	3.0777	.9511	1.2566	72
19	.3316	.3256	.3443	2.9042	.9455	1.2392	71
20	.3491	.3420	.3640	2.7475	.9397	1.2217	70
21	.3665	.3584	.3839	2.6051	.9336	1.2043	69
22	.3840	.3746	.4040	2.4751	.9272	1.1868	68
23	.4014	.3907	.4245	2.3559	.9205	1.1694	67
24	.4189	.4067	.4452	2.2460	.9135	1.1519	66
25	.4363	.4226	.4663	2.1445	.9063	1.1345	65
26	.4538	.4384	.4877	2.0503	.8988	1.1170	64
27	.4712	.4540	.5095	1.9626	.8910	1.0996	63
28	.4887	.4695	.5317	1.8807	.8829	1.0821	62
29	.5061	.4848	.5543	1.8040	.8746	1.0647	61
30	.5236	.5000	.5774	1.7321	.8660	1.0472	60
31	.5411	.5150	.6009	1.6643	.8572	1.0297	59
32	.5585	.5299	.6249	1.6003	.8480	1.0123	58
33	.5760	.5446	.6494	1.5399	.8387	.9948	57
34	.5934	.5592	.6745	1.4826	.8290	.9774	56
35	.6109	.5736	.7002	1.4281	.8192	.9599	55
36	.6283	.5878	.7265	1.3764	.8090	.9425	54
37	.6458	.6018	.7536	1.3270	.7986	.9250	53
38	.6632	.6157	.7813	1.2799	.7880	.9076	52
39	.6807	.6293	.8098	1.2349	.7771	.8901	51
40	.6981	.6428	.8391	1.1918	.7660	.8727	50
41	.7156	.6561	.8693	1.1504	.7547	.8552	49
42	.7330	.6691	.9004	1.1106	.7431	.8378	48
43	.7505	.6820	.9325	1.0724	.7314	.8203	47
44	.7679	.6947	.9657	1.0355	.7193	.8029	46
45	.7854	.7071	1.0000	1.0000	.7071	.7854	45
		Cosine	Cotangent	Tangent	Sine	Radians	Degrees

Answers to Odd-Numbered Exercises

This list *excludes* answers that are graphs or proofs, though in some of these there are brief hints. It includes answers to all of the first five exercises in each set, where they are not graphs, because in many cases these first exercises are designed to provide the student with a self-check on his reading. Also included are non-graphic, non-proof answers for all exercises in the (Review) R-sections.

Part One

I. § 1.7

7. $y = 2x + 1$.   Domain, all real numbers, x.   Range, all real numbers, y.
   $3y = 2x + 5$.   Domain, all real numbers, x.   Range, all real numbers, y.
   $y = x^2 + 1$.   Domain, all real numbers, x.   Range, all y such that $y \geq 1$.
   $y = x^3 - 5x + 1$. Domain, all real numbers, x.   Range, all real numbers, y.

9. $f(2) = 7, f(-1) = 1, f(-\frac{3}{2}) = 0, f(0) = 3, f(a) = 2a + 3$.

11. $x = 0, x = -3$.   13. $f + g = 2, fg = 1 - t^2$, both on domain of all reals, t.

21. No.   23. No.

I. § 2.5

1. $f = (1/3)(x^2 - 1)$, a polynomial on all reals x.   2. $f = x + 1$, a polynomial on all reals x.   3. $f = 1 - 1/x$, on reals, excluding $x = 0$. Not a polynomial.   4. Two solutions: $f = -1/x$, $\{x \neq 0\}$, not a polynomial, and $f = -1$ on all reals x, a polynomial.   5. No solution in the real number system.   7. $F = 2x$; 24 pounds.   9. $(1/18)(-25x^2 + 41x + 128)$ on all reals.   11. $A(x) = 2x^2 + 432/x$ on positive numbers x.

13. $T = x/20 + x, \{x \geq 0\}$.   15. $4\pi r^2 + 1200/r, \{r > 0\}$.

19. $F = -1/20r^2, \{r \neq 0\}$.   21. $A(r) = 4\pi r^2, \{r \geq 0\}$.

I. § 3.5

1. Algebraic sum of areas is -1.   2. 0 and 0.

3. (a) 60 mi, (b) 45 mi/hr.   4. $640.   5. 0.264.   11. 0.025.

21. 419.

I. § 4.7   13.   For 7: $4x - 7y + 10 = 0$.   For 8: $x + 3y = 7$.
              For 9: $5x - y = 13$.   For 10: $y = 0$.   For 11: $y = 3$.
              For 12: $x - 5y = 16$.
     15.   23/3 ft/sec.   17.   0 ft/sec.   19.   Same as the speeds over intervals.

I. § 5.8   1.   (a) True,   (b) false,   (c) false,   (d) true,   (e) true,   (f) true.
     2.   Same as the intervals: (a) $\{-1 < x < 1\}$,   (b) $\{-2 \leqq x \leqq 2\}$,
           (c) $\{-2 \leqq x < 1\}$,   (d) $\{2 < x < 3\}$.   3. $-3 < x < 5$.   4. $1 < x < 2$.
     5. $-7 < x < 3$.   7. $-0.35 < x < 0.65$.   9. $-11/6 \leqq x \leqq -9/6$.
     15.   If $a \neq 0$, $1 - 1/|a| < x < 1 + 1/|a|$. If $a = 0$, all reals x.
     17. $-8 < x < 12$.   19. x and y have the same sign.

I. § 6.5   9.   Max $y = -8$; no min exists.   11.   Min $y = 25/4$; no max exists.
     13.   Min $y = 9/2$; no max exists.   15.   Set is empty, no max or min exists.
     17.   Max $y = (4ac - b^2)/4a$; no min y exists.   19.   1/2.   25.   20¢.
     27.   H/2.

I. § 7.7   1. $1, 2, \frac{7}{3}, \frac{5}{2}, \frac{13}{5} \cdots$.   2. $s_n = 3 - 2/n$.
     3.   (a) Any N greater than $2/\epsilon$,   (b) $N = 2000$.   4. $N = 13$.
     7.   Monotone increasing and bounded above by $1 + 1$.   9. $\sigma_n \to \frac{1}{2}$.
     11.   If $|r| < 1$, $s_n \to 1/(1 - r)$.

I. § 8.6   1. $\epsilon/3$.   2.   Theory gives $\delta < 1/160$.   3.   Theory estimates
           $\delta < 10^{-5}/5$.   4.   To 11 places, since we estimate $\delta = \epsilon|a|^2/2 = 8(10^{-11})$.
     5. $\delta = \sqrt{3}\epsilon$.   7.   [1, 7].   9.   [0, 4].   11.   [1/10, 1].   13.   [63/16, 44].
     15. -2.753.   17. -0.671.   21. f is discontinuous.

I. § 9.7   1.   Understep values are: 1, 4, 9, 16; overstep values are: 4, 9, 16, 25.
           (a) 30,   (b) 54.   2.   (a) 71/2,   (b) 95/2.   3. $w = 0.01/24$.
     5. $b^3/3$ and 33.   21.   42.   25.   83/2.

I. § 10.6   1. $2x + 3$.   2. $2x - 1$.   3. $-2x$.   4.   3.   5. $3 - 2x$.
     7. $2ax + b$.   15. $-4, 0, -2$, ft/sec.   17. $7x - y + 5 = 0$,
           $8x - 4y + 5 = 0$, $x + y = 5$, $5x + y = 17$.   19.   4.   21.   5.
     23.   (a) No,   (b) No.

I. § 10.4R   1. $f(x) = 3x + 2$, uniquely determined.   2.   Step function.
     3.   (a) $3 < x < 7$,   (b) $x < -7$ or $x > -3$,
           (c) $(-1 - \sqrt{5})/2 < x < (-1 + \sqrt{5})/2$.   4. $y - 2 = (x - 3)/6$.   5.   19/3.
     6. $-5x^2/2 + 7x + c$.   7. $g = -5x^2 + 7x + 8$.   8.   (a) 50 ft/sec,
           (b) -78.8 ft/sec.
     12.   Same as line $5x - y - 2 = 0$ with point where $x = 0$ missing.
     14.   Intercept on y-axis.   16.   Down where $t < -2$ and $0 < t < 1$.
     17. $(x - 1)^3 + 9(x - 1)^2 + 12(x - 1) + 9$.   19. $x - a$ divides $f(x) - f(a)$, by
           the Factor Theorem.   20.   Max $= \frac{4}{3}$, no min.   21.   1.410.
     23. $\sqrt{2} - 1$.   24.   Trapezoidal estimate, 1.67.   25. $12x^3$.

I. § 11.8   1. $2x - 3$, Stationary points (SP): $(\frac{3}{2}, -\frac{1}{4})$.
     2. $3x^2 + 7$, SP: None.   3. $-2x + 3$, SP: $(\frac{3}{2}, -\frac{11}{12})$.
     4. $4x^3 + 2x$, SP: (0, 1).   5. $3x^2 - 7$, SP: Points where $x = \pm\sqrt{7}/3$.
     7.   (a) $2x$,   (b) $2x$,   (c) 0,   (d) Not a polynomial,   (e) Not a polynomial,
           (f) Not a polynomial.   9. $y - 1 = 6(x - 1)$.   11. $y - 1 = 4(x - 0)$.
     13. $-1 - 2x$.   15. $2z$.   17.   1.   19. $2ax + b$.
     21. $(2x - 1)^2 (x + 4) (10x + 22)$.   23. $(2x - 1)^4 (3x + 2)^6 (72x + 1)$.

I. § 12.9   1. $f(t) = 40t + c$, c = constant. $v = D_t f(t) = 40$ ft/sec.
2. $D_t g = 0$, zero acceleration.   3. $D_t y = -32t + 96$.
4.   (a) 96 ft/sec up,   (b) when $t = 3$ sec,   (c) -32 ft/sec down,
     (d) when $t = 6$ sec,   (e) -96 ft/sec down.
5.   (a) $D_t y = -gt + v_0$, velocity,   (b) $D_t v = -g$, acceleration due to gravity,
     (c) v_0,   (d) y_0.   7.   48 in.^{3}/in.   9. $D_q C' = -12 + 6q$, decreasing when
     $q > 2$, min when $q = 2$.   11. $2 + \sqrt{3}$.
13.   0.5%.   15.   (a) $\{x \mid x < -1 \text{ or } x > 3\}$,   (b) $\{x \mid -1 < x < 3\}$,
      (c) Stationary when $x = -1$ or $x = 3$.   17.   (a) -1,   (b) 2,   (c) 11.
19. $\frac{5}{9}$.   21. $1/f'(x)$.

I. § 13.5   1.   None.   2.   None.   3. $(0, -32)$.   4.   Where $u = 2$,
      $-\frac{1}{3}, \frac{16}{15}$.   5.   Where $x = -1, \frac{1}{2}, 2$.   7.   Decreasing only
      when $x < -\frac{1}{2}$.   9.   Decreasing only where $-\frac{5}{2} < t < -\frac{1}{2}$.
11.   Decreasing only where $x > -b/2a$.   13.   Decreasing only where
      $x < (5a + 3b)/8$.   23.   (b) Decrease.   25. $5x^3/3 + c$.
27. $-gt^2/2 + c$.   29. $(x^{n+1}/n + 1) + c$.   31. $n = 4$.
35.   Down between stationary points where $t = 1$ and $t = 3$.
37.   (a) $|f'(x)| > 1$,   (b) $|f'(x)| < 1$.   39. $4a$.   41.   12.

I. § 14.5   1. $f(x) = 11 + 5(x - 4) + (x - 4)^2 + 0 + \cdots$.
2. $f(t) = 6 + 32(t - 3) + 27(t - 3)^3 + (t - 3)^4 + 0 + \cdots$.
3. $f(x) = 14 + 3(t + 2) - 6(t + 2)^2 + (t + 2)^3 + 0 + \cdots$.
4. $f(w) = 3 + 30w + 120w^2 + 240w^3 + \cdots$.
5. $f(t) = t^3 + 4t^4 + 4t^5 + 0 + \cdots$.
7. $f(x) = 0 + \cdots + x^7 + 0 + \cdots$.
9. $f(x) = 1 + (x - 4) + (x - 4)^2 + (x - 4)^3$.
11. $s(t) = 2 + 2(t + 1)^2$.   13. $f(x) = x^n$.
15. $1 + 8t$ and $1 + 8t + 4t^2$.   17.   0 and 0.
19.   Coefficients of $(t - t_0)^1$ and $(t - t_0)^2$.   21. $|t - 1| < 0.526$.
25.   (a) $f'(a)$,   (b) $f^n(a)/n!$

I. § 15.5   1 to 5.   The graph is below the tangent line if $f^{(2)}(x)$ is negative; above
      if $f^{(2)}(x)$ is positive.
7.   Everywhere negative; no inflection.   9.   Negative if $x < 0$;
     inflection point $(0, 4)$.   11.   Negative when $-2 < t < 1$;
     inflection points where $t = -2$ and $t = 1$.
17. $x = x_0 + v_0(t - t_0)$.   21.   Rate of change of marginal cost.

I. § 16.4   1.   Max $(4, 7)$; min $(0, 1)$.   2.   Max $(-1, 2)$; min $(1, 0)$.
3.   Max $(-1, 1)$ and $(1, 1)$; min $(0, 0)$.   4.   Every point is both a max and a min.
5.   Max $(1, 105)$; min $(0, 8)$.   7.   Where $x = -1, \frac{3}{4}, 1$.
9.   Where $x = -1, 0, 1$.   11.   Where $x = -3, -\frac{1}{2}, 1$.
13.   Max $(-\sqrt{14}, 28\sqrt{14} + 1)$; min $(14, -28\sqrt{14} + 1)$.
15.   Max $(3, 39)$; min $(-3, -10)$.   17.   No max; min at $x = 1.345$ approx.
19.   Max $(1, 2)$ and $(-3, 2)$; min $(0, -7)$.   21.   Max $(2, 42)$; min $(1, 16)$.
23.   No max; min $(0, 4)$.   25.   No local max; local (and global) min at $\frac{5}{2}$.
27.   Local max at $\frac{1}{4}$; local min at $-\frac{1}{2}$ and 1.

I. § 17.5   1. $x^3/3 - x^2/2 + x + c$.   2. $8x + c$.
3. $x - x^2/2 + c$.   4. $x^4/4 - 7x^2/2 + 2x + c$.
5. $x^4/4 + 11x^3/3 - 7x^2/2 + 2x + c$.   7. $x^4/12 + 4x^3/3 - x^2/2 + c_1 x + c_2$.
9. $\frac{1}{2}(2x + 3)^3/8$.   11. $x^3 - x^2 + 4x$.   13. $f(x) = x^3/3 + x + c$.
15. $g(u) = u^4/4 - u^2 + 4$.   17. $f(x) = 2x^2 + 3x - 58$.

19. $x = 2/15$. 21. $625/4$. 23. 2.
25. $(b - a) + (b^2 - a^2)/2 + (b^3 - a^3)/3$. 27. $c = (3 + \sqrt{5})/2$.
29. $t = 10; 220/3$. 31. 10 ft/sec². 33. $228/3$.

I. § 18.5   1. $x = 3t^2 - 5t + 10$.   2. $x = 2t^3 - 15t^2 + 36t + 18$.
3. $x = t^3 - 6t^2 - 15t - 4$.   4. $x = 10 - 9t + 6t^2 - t^3$.
5. $x = t^3 - t^2 + t$.   7. Velocity begins to increase.
9. (a) $y = -16.1t^2 + 64.4t$,   (b) $t = 2, y = 64.4$.
11. About -191 ft/sec.   15. $\frac{1}{2}$.   17. $\frac{3}{2}$.   19. $100/g$.
21. Velocity begins to increase at $t = 4$ and after $t = 12$ remains constant $(-18/25)$.
23. $y = -(g/16)t^2 + 4t + 8$.
25. $y = (2t + 1)/4M$; inversely proportional to the total mass.

I. § 19.6   1. 8.   2. $28/3$.   3. 0.   4. $14/3$.   5. $2\sqrt{2}$.
7. $c(b^2 - a^2)/2 + d(b - a)$.   9. 6.   11. $\frac{4}{3}$.   13. $125/6$.
15. 12.   17. $\frac{8}{3}$.   19. 3.   27. $625/2$.   29. $(x^4 - 4x)/4$.
31. $x = 1 + 2\sqrt{2}$.

I. § 20.6   1. $15(3x - 4)^4$.   2. $-4(1 - 2x) - 30(1 - 2x)^2$.
3. $21(6x - 1)^6/4$.   4. 0.   5. $x^3/3 + c$.   7. $x^4/4 - 3x^2/2 + 4x + c$.
9. $(6x - 5)^{19}/114 + c$.   11. $-(7 - x)^3/3 + c$.   13. $2x + x^2/2 + c$.
15. $243/10$.   17. $1/40$.   19. $179/30$.   21. $(4/3)^7/7$.
23. $(2x - 1)^8/16 - (2x - 1)^9/288 + c$.
25. $(2x - 1)x^9/9 - x^{10}/45 + c$.
27. $(1 - x)^2 (2x - 1)^4 + (1 - x) (2x - 1)^5/5 + (2x - 1)^6/20 + c$.
31. Absolute area = 48.   33. $1031/5$.

I. § 21.9   1. $x^2/2 + x - 3/2$.   2. $x^3/3 - x - 2/3$.   3. $t^3/3 - t^2/2 + t + 11/6$.
4. $-x^2 + 3x - 2$.   5. $[(3x - 2)^6 - 2^6]/18$.   7. $(x - a)^2/2$.
9. $au^4/4 - bu^3/3 + b^4/12a^3$.   11. -4.   13. 0.
15. -104.   17. $34/4$.   19. $249/2$.
21. Net distance = $490/3$, total variation = 166.
23. $x = (1 - \sqrt{26})/2$.   25. $64/3$.   27. $7/2$.
29. $\int_a^x f = (x^2 - a^2)/3$; implies that $f(x) = 2x/3$.
31. $(-1)^n 2^{n+1} (n + 1)/(n + 2) (n + 3)$.

I. § 22.7   1. $A = \int_0^1 (x - x^2)dx$.   2. $A = \int_1^3 (-3 + 4y - y^2)dy$.
3. $A = \int_0^1 (x^2 - \sqrt{x})dx$.   4. $A = \int_1^3 (3 + 2x - x^2)dx$.
5. $A = 2\int_0^1 2(1 - x)dx$.   7. $A = \int_{-1}^1 2(1 - x^2)dx$.
9. $16/3$.   11. $343\pi/27$.   13. $7\pi/3$.
15. $\pi(2r^3/3 - hr^2 + h^3/3)$.   17. 31,250 lbs.
19. $F = -125 \int_{-10}^0 y\sqrt{y + 10}\, dy$; no antiderivative yet available.
21. $200/3$ ft-lbs.   27. $W = 62.5 \int_0^5 \pi(10y - y^2)(y + 30)dy$.
29. $W = 62.5 \int_0^6 \pi(25/4)(25y/64 + 28 + y)dy$.

I. § 23.8   1. $\frac{4}{3}$.   2. k.   3. 0.   4. (a) 0, (b) $3 - t$.
5. $\frac{2}{3}$.   7. $\bar{x} = 0, \sigma^2 = a^3/2$.   9. $(\bar{x}, \bar{y}) = (10/3, 5)$.

10. $\bar{x} = 0$. $\bar{y} = \int_0^5 y \sqrt{y}\, dy / \int_0^5 \sqrt{y}\, dy$, no antiderivative yet available.

11. $\bar{x} = 1/2$, $y = \int_0^1 y(\sqrt{y} - y)dy / \int_0^1 (\sqrt{y} - y)dy$, antiderivative not yet available.

13. $(\bar{x}, \bar{y}) = (0, 1)$.   15. $\bar{x} = 2$, $\sigma = \sqrt{37/6}$.   19. (a) $\bar{x} + x_0$, (b) $2\bar{x}$.

22. A ramp function, $(t)\, \bar{s} = \int_{t-1}^1 s(u)du$.

I. § 23.4R   1. At point $(2, 5)$ tangent line has slope 7.
  2. $y - 5 = 7(x - 2)$.   4. 0.4449 with third-order error less than 0.0154.
  5. $F = (x - 2)^4 + 4$.   6. $(2x + 3)^6 (4x + 1)/56 + c$.   8. 20/3.
  9. Step function $F = 1$ on $[-1, 1]$ and $F = -1$ on $[2, 4]$, not a constant.
  11. (a) 0, (b) 24.   12. 96, assuming $g = 32$.   13. 6, $k \approx 6h$.
  14. Increasing when $x < -3$, decreasing when $-3 < x < 3$, increasing when $3 < x$.
  15. Convex if $0 < x$; inflection at $x = 0$.   16. Assume that $C(x)$ and $S(x)$ are
  everywhere differentiable.   17. -15; decreasing.

  18. Max $(5, 152)$; min $(2, -37)$.   19. $625\int_0^5 (5 - y)(2y)dy$.

  21. None.   22. 10,198.   23. 25.   24. (a) Max $(-1, 1)$, min $(1, -3)$;
  (b) Max $(2, -2)$, min $(1, -3)$; (c) Max $(-1, 1)$, min $(1, -3)$;
  (d) No max, min $(1, -3)$; (e) No max, min $(0, -2)$.   25. $\frac{1}{2}$.
  26. $x_0 = 3$.   27. $\bar{x} = 3$, $\sigma = 2\sqrt{46/5}$.   28. $(a + b)/2$.
  29. No max; min $(2, 0)$.   30. Critical point $(0, 0)$ undecided.
  31. Local max at $-\sqrt{14}$, local min at $\sqrt{14}$.   32. Critical point at $\frac{1}{3}$, undecided.
  33. No max, no min, but local max at 1 and local min at 2.

I. § 24.5   1. $-2/x^3$, $x \neq 0$.   2. $-3/(3x + 5)^2$, $x \neq -\frac{5}{3}$.
  3. $-10/x^3 - 1/x^2 - 7 + 12x^2$, $x \neq 0$.
  4. $-2/x^3 + 6/(2x - 3)^4$, $x \neq 0$ and $x \neq \frac{3}{2}$.   5. $1/x$, $x \neq 0$.
  7. $(3x + 5 - 3x \ln 2x)/x(3x + 5)^2$, where $x > 0$.
  9. $(1 - \ln x)/x^2$, $x > 0$.   11. $-1/x(\ln x)^2$, $x > 0$.
  15. $-95/x + c$, $x \neq 0$.   17. $-cx^{-4}/(-4) + c$, $x \neq 0$.
  19. $5 \ln|x| + c$, $x \neq 0$.   21. $-\ln|1 - x| + c$, $x \neq 1$.
  23. $x^3/6 - x^2/2 + 2x - 4 \ln|2x + 4|$, $x \neq -2$.
  25. $\ln 2 + \frac{1}{2}$.   27. $-\frac{9}{2} + 14 \ln 8 + 5 \ln 5$.   29. $\ln|x + 1|$, $x \neq -1$.
  31. Line through $(1, 0)$ with slope 2.   33. No max, min where $\ln x = -1$.
  35. $(8/3) \times 10^5$ mile-pounds.   36. 7.85×10^5 mile-pounds.
  37. $(7.85 - 0.12) \times 10^5$ mile-pounds without soft landing.
  38. 3.08×10^4 to equilibrium point between earth and moon gravities.
  39. $f(t) = \ln(t/2)$.

I. § 25.3   1. $x = y = 12/2$.   2. $\frac{3}{2}$.   3. $x = y = 12$.
  4. Max where $x = (20 - 4\sqrt{7})/3$.   5. 5 and 10.   7. Square.
  9. Squares.   11. The same as the farmer.   15. $d/w = \sqrt{2}$.
  17. $h/r = 3$.   19. $v = 50$ mi/hr.   21. 8.

I. § 26.9   1. $3/(2\sqrt{3x})$.   2. Line through $(2, \sqrt{6})$ with slope $\sqrt{6}/4$.
  3. (a) $2\sqrt[3]{4x^2}/3x$, $x \neq 0$. (b) $\sqrt[3]{32/81}$.   4. $\frac{3}{2}\sqrt{3x + 5}$, $x > -\frac{5}{3}$.
  5. $-\frac{1}{2}x\sqrt{x}$, $x \neq 0$.   7. $\sqrt{2}\, x^{\sqrt{2}-1}$, $x > 0$.   9. $6(2x - 1)^{-2/5}/5$, $x > 1/2$.
  11. $10^x \ln 10$.   13. $b^{x+1} \ln b$, $b > 0$.   15. $2x$.
  17. $2x(x^2 + 1)^{-2/3}/3$.   19. $\sqrt{2x} + c$, $x > 0$.   21. 1.   23. 1.
  25. $\ln^2 x/2 + c$, $x > 0$.   27. $x(\ln^2 x - 2 \ln x + 2) + c$.
  29. $-(\ln x + 1)/x$, $x > 0$.   31. $(2 + x) \ln x - x + c$, $x > 0$.
  33. $\ln(\ln x) + c$, $x > e$.   35. $\ln^6 x/6 + c$.   37. $2/e$.   39. $(e - 1)^2/2$.
  41. $\ln 2 - \frac{3}{2}$.   43. $f(x) = \ln(x + 1) + 2 - \ln 2$, $x > -1$.

I. § 27.5   1.   1.25.   2.   0.1623.   3. -0.281.
4.   (a) 9,   (b) 26/3,   (c) 26/3.   5.   1.1924.   7.   0.002585.   9.   3.875.
11. 1.309×10^{12}.   13.   If $1 < r < 1.5$, then $x = 1.16$ with error less than 0.33.
15.   0.1119.   17. -0.784.   19.   2.94 without refinement of the partition.
21.   0.513   23. $|x^2 - \sqrt{k}| < (1/2x_1)(k - k/x_1)^2$.
25. $\ln 0.6 = -\int_{.6}^{1} dx/x$. Subintervals of length 0.1 give 5-place accuracy.

I. § 28.5   2. $-2e^{-2x}$.   3. $(-2 \ln 3)3^{-2x+1}$.   4. $(1 - x)e^{-x}$.
5. $(2x + x^2 \ln 2)2^x$.   7. $-10e^{-0.1t}$.   9. $(e^x - e^{-x})/2$.   11. $(k/2)e^{kx/2}$.
13. $\ln(e^x + 1) + c$.   15. $2^x/\ln 2 + c$.   17. $\ln(1 + e^{-x}) + c$.   19. $\frac{1}{4}$.
21.   (a) $8e^{-2t}$,   (b) velocity approaches zero,   (c) approaches a rest position x_0 but
does not reach it in finite time.   23.   (a) $v(t) = 5e^{-t/2}$, $a(t) = -\frac{5}{2}e^{-t/2}$.
(b) $s \to 10$ at $t \to \infty$, with a positive velocity which approaches zero.
25.   At $(\ln 2, \ln 2 - \frac{1}{2})$ the slope of the tangent is $\frac{3}{2}$.   27.   Max $(1, 1/e)$; no min.

I. § 29.4   1. $k = 0.0289$, $A = 21.65$.   2. $A = 25 \exp(-0.0306t)$.
3. $A \exp(-0.0248t)$.   4. $A \exp(-0.000436t)$.   5. $t = 2640$ yrs.
7. $I_0 \exp(-0.0867t)$.   9. $E = E_0 \exp(-0.231)t$.
11. $v = 100 - 100 \exp(-0.322t)$.   13. $t = 88.5$ min.
19.   About 33,000 years.   21.   4.083%.   23. $n = 12$.

I. § 30.5   1.   Convergent to $\frac{1}{2}$.   2.   Convergent to $\frac{1}{3}$.   3.   Convergent to $\frac{1}{8}$.
4.   Convergent to $\frac{1}{2}$.   5.   Convergent to 1.   7.   Convergent to $2\sqrt{2}$.
9.   Does not converge.   11.   Convergent to $1/(1 - p)$.   13.   Does not exist.
The integral is not defined.   15.   Does not converge.
17.   Does not converge.   19. $3.2 \times 10^9/R$.   21.   Escape velocity $= \sqrt{2gR} =$
39,500 ft/sec.   23. $10V/\ln 2$.

I. § 31.8   1. $6 \cos 2x$.   2. $Du/\cos^2 u$.   3. $2 \sin 2x + 2 \cos 2x$.
4. $-2 \sin x \cos x$.   5. $-2/\cos^2 2x$.   7. $\sin(x/2) \cos(x/2)$.
9. $-(2 + \cos 2x)/\sin^2 2x$.   11. $-(1/2)\cos 2x + c$.
13. $x + \cos x - \frac{3}{2} \cos 2x + c$.   15. $\ln|1 + \sin x| + c$.   19. π.
23.   If $n = 2k - 1$, $D^n \sin x = -(-1)^k \cos x$. If $n = 2k$, $D^n \sin x = (-1)^k \sin x$.
27. $3a \sec^3 ax \tan ax$.   29. $(\tan ax - ax)/a + c$.
33.   At $x = n\pi$ and $x = \pi/2 + n\pi$ for all integers n.   35. $\pm\sqrt{8}/3$.

I. § 32.5   1. $T = 2\pi/3$, $|b| = 1$.   2. $T = 4\pi$, $|b| = 3$.   3. $T = 2\pi$, $|b| = 1$.
4. $T = 4\pi$, $|b| = 3$.   5. $T = \frac{1}{6}$, $|b| = 5$.   11.   Not.
13.   Every real number T is a period.   15. $T = 2\pi$.   17.   Not.
19. $\sqrt{2} \sin(t + \pi/4)$, $T = 2\pi$, amp $= \sqrt{2}$.   21. $\sqrt{a^2 + b^2} \sin(pt + \phi)$, where
$\sin \phi = b/\sqrt{a^2 + b^2}$ and $\cos \phi = a/\sqrt{a^2 + b^2}$, $T = 2\pi/p$, amp $= \sqrt{a^2 + b^2}$.
23. $\cos(t - \pi/2)$.   25. $x = 5 \cos 2t$, $T = \pi$, amp $= 5$.   27.   None.

I. § 33.6   9. $\sqrt{3}/2$.   11.   0.   13. $a/\sqrt{a^2 + x^2}$.   15. $-\pi/4$.
19. $-2/\sqrt{1 - 4x^2}$.   21. $x^2/\sqrt{1 - x^2} + 2x \arcsin x$.   23. $4/[16 + (x + 1)^2]$.
25. $\frac{1}{2} \arcsin 2x + c$.   27. $\frac{1}{3} \arctan 3x + c$.   29. $\pi/6$.   31. $\pi/4$.
33. $2a \arccos h/a$.   35. $2\pi - 4/3$.

I. § 34.7   1.   At $(\pi, 1)$, slope $= -\sqrt{3}$.   2.   2.   5.   One term will do it.
9.   Two.   11.   Compute $\sin 0.5$ and $\cos 0.5$ to 4 places, then divide. See
Exercise 4.   13.   (a)   400 cm.   (b) $T = c\sqrt{\ell}$, independent of m and β.
15.   Max: $(\pi/2, 2)$, min: $(-\pi/2, -2)$.   17.   8 mi/sec, west.
19.   (a) $x' = -2 \sin 2t$, $y' = 2 \cos 2t$,   (b) -1 and $-\sqrt{3}$.   21. $\pi/3$.

I. § 34.4R   1. $6/(2x - 1)$, $x > \frac{1}{2}$.   2. $2(1 - \cos 2x)/\sin^2 2x$, $x \neq n\pi/2$.
3. $2/(1 - x^2)$, $|x| \neq 1$.   4. $-2x^3/\sqrt{1 - x^4}$.   5. $\sqrt{3}\, x^{\sqrt{3}-1}$.
6. $a^2/(a^2 + x^2)$.   7. $1/\sqrt{x^2 - 1}$.   8. $1/(x - 2)$.   9. The step function,
$(x + 1)/|x + 1|$.   10. $-x(x^2 + a^2)^{-3/2}$.   11. $\sqrt{\dfrac{x + 1}{x - 1}} \left(\dfrac{1}{x + 1}\right)^2$.
12. $-x \exp(-x^2/2)$.   13. $\pi/6$.   14. $\pi/2a$.   15. $\pi/2$.   16. Converges, 2.
17. Converges, 1.   18. Does not converge.   19. Slope $\frac{3}{2}$ at point $(1, 1)$.
20. Slope $-2/\sqrt{3}$ at point $(\frac{1}{2}, \pi/3)$.   21. $ey = x$, at the origin.
30. For $x = \frac{1}{2}$, $N = 5$; For $x = \frac{1}{3}$, $N = 2$.   31. For arctan $\frac{1}{2}$, use 6 terms;
for arctan $\frac{1}{3}$, use 3 terms.   32. 3.14152.   34. (b) Does not converge.
35. $\pi/2$.   38. $x = y = \sqrt{16 + (S/2)} - 4$.

I. § 35.5   1. $100 - 0.04\, y$.   2. Only half of it.   5. 800.   6. 5000.
7. 20¢.   8. 27½¢.   9. (a) Zero; (b) No; (c) Only one positive y;
(d) About 685.   10. 290,000.   11. $48.75.   12. $h = 3$.
13. Max at C; min at A.

I. § 36.3   1. Additive, $y = 2x - 4.1$.   2. Not additive.
3. Additive, $t = 2.7x - 7.77$.   4. Not additive.   5. $y = (4x + 1)/3$.
12. $w = 5x + 9y$.   13. $w = (42x + 25y)/19$.
15. (a) $u = x - 2y + 5z$, $v = -3x + 4y + 6z$; (b) If $(1, 0, 0)$ produces (a_1, b_1),
$(0, 1, 0)$ produces (a_2, b_2), and $(0, 0, 1)$ produces (a_3, b_3), then $u = a_1 x + a_2 y + a_3 z$ and $v = b_1 x + b_2 y + b_3 z$.

I. § 37.5   1. Power; $s = t^2$.   2. $v = 2.7u$.   3. $y = 0.43 \ln x$.
4. None of these.   5. $s = \exp(-r)$.   6. Approximately linear; $A = 6t + 95$.
7. $y = 0.43 \ln x$.   8. None of the laws.   9. $10^3\, y = 6.2\, x^{1.7}$.
10. Approximately $\psi = s^{0.3}$.   11. Approximately $\psi = 1.7\, s^{0.77}$.
12. Approximately $A = 5.11 \exp(-0.00154t)$.

I. § 38.5   1. When $t = 0$.   2. $\dfrac{-V}{At_1} \ln \dfrac{c - y_1}{c - y_0}$.   3. $t = 10 \ln 2$.
4. $\frac{1}{2}$.   5. c.   6. $\pi r^2 v[y(0) - y(L)]$ mass units/min.
7. Max rate: $k/4$ when $p = \frac{1}{2}$.   8. $p = 1/[1 + 10 \exp(-5t)]$.
9. $t = 4.6$ days.   10. $p(t) = 1 - \exp(-kt)$.   11. $D_t y = (A/V) - ky$.
12. A/kV.   14. $y > 160,000$.   15. $x = 10^6 \exp(-0.23t)$.
17. $y = 1 - \exp(-kt)$.   18. (a) $k = 0.232$; (b) Yes.

Part Two

II. § 1.5   1. $dy = 2x\, dx$.   2. $dy = 3x^2\, dx$.   3. $dy = m\, dx$.
4. $dy = (6x - 7)dx$.   5. $dy = 0$.   7. $dy = 15(3x - 1)^4\, dx$.
9. $dy = (12t - 1)dt$.   11. $dx = 24[1 - (2t - 1)^4]^2 (2t - 1)^3\, dt$.
13. $dr = \frac{1}{2}(1 - 2\theta)d\theta$.   15. $dx = -2dt$, $\dfrac{d^2x}{dt^2} = 0$.   17. $z\, dz = (u - 1)\, du$.
19. At $x = -1$, max $y = 7$; at $x = \sqrt{7/3}$, min $y \approx -13.95$.

II. § 2.6   1. $dx/(\sqrt{x} + x)$.   2. $-x\, dx/\sqrt{1 - x^2}$.   3. $-3x\, dx/(x^2 + 1)^{5/2}$.
4. $dx/\sqrt{1 + x^2}$.   5. $(t^2 - 1)dt/(t^2 + 1)^{5/2}$.   7. dx.   9. $x(a^2 - x^2)^{-3/2}\, dx$.
11. $15 \sin^2 5x \cos 5x\, dx$.   13. $(2 \cos 2x + 1)(\exp \sin 2x)\, dx$.
15. $dx/\sqrt{1 - x^2}$.   17. $2x f(x^2)\, dx$.   21. $(x + yy')\, dx/\sqrt{x^2 + y^2}$.
23. $(y - xy')\, dx/y^2$.   25. $a\, e^{axy}(xy' + y)dx$.   27. $(xy' - y)dx/(x^2 + y^2)$.
29. $dy = \dfrac{y}{3}\left(\dfrac{2x}{x^2 + 4} - \dfrac{10}{2x - 1}\right) dx$, $x \neq \frac{1}{2}$.   31. $\theta = 2$ rad and $r = P/4$.
33. $\pi/6$.

II. § 3.5   1. $(x^3 + 4)/6 + c$.   2. $(1 + x^2)^{3/2}/3 + c$.   3. $-\sqrt{1 - x^2} + c$.
4. $\ln\sqrt{1 + x^2} + c$.   5. $-\frac{1}{2}(1 + x^2) + c$.   7. $\frac{3}{2}(2 + 4x + 4x^2)^{3/2} + c$.
9. $\ln\sqrt{1 + y^2} + c$.   11. $x^5/5 - 2x^3 + 9x + c$.   13. 2.
15. $-\sqrt{1 - \sin 2x} + c$.   17. $(2/\sqrt{3}) \arctan(x - \frac{1}{2}) + c$.   19. π.
21. $-\sqrt{2 + x - x^2} + \frac{1}{2}\arcsin(2x - 1)/3 + c$.   23. $\frac{1}{3}\sin^3 x - \frac{1}{5}\sin^5 x + c$.
25. $|26 - 21(\pi/2)^5 - 5(\pi/2)^7|/35a$.   27. $\frac{1}{2}\cos 2x + c$.   29. $\ln \tan \theta$.
31. $(1 + \sin^2 \theta)^{3/2}/3 + c$.   33. $(1 + 2x \cos x)^4/8 + c$.   35. $\ln|\arcsin x| + c$.
37. $(1 + \exp 2x)^{3/2}/3 + c$.   39. $(2/\sqrt{3}) \arctan(2e^x + 1)/\sqrt{3} + c$.
41. $(e^2 - 1)/2e^2$.   43. $u = \ln^2 x + c$.   45. $f(x) = \pm \sqrt{2(x^3 - 1)/3}$.
47. $\frac{2}{3}(1 + \exp x)^{3/2} + c$.   49. $\arctan(\exp x) + c$.   51. $\frac{2}{3}\sin^3 x + c$.
53. $-\frac{4}{3}\cos^3 (\theta/2) + c$.   55. $\frac{1}{2}\ln(x^2 + x + 1) - (1/\sqrt{3}) \arctan(2x + 1)/\sqrt{3} + c$.
57. $-2 \cos \sqrt{x} + c$.   59. $\frac{1}{2}$.   61. $a^2/6$.   63. $\ln|x^2 + 2x + 5| +$
$\frac{1}{2}\arctan(x + 1)/2 + c$.   65. 0.   67. Convergent, $\pi/2$.   69. Does not
converge.   71. Does not converge.   73. 2π.

II. § 4.6   1. $\pi/4$.   2. $x/2 + (1/12) \sin 6x + c$.   3. $(a^2/4) (2\theta - \sin 2\theta) + c$.
5. $(-1/3) (a^2 - x^2)^{3/2} + c$.   7. $2a^5/15$.   9. $\pi/2$.   11. $\pi/2$.
13. $\frac{1}{8}a^2 [2 \sin^3 ax \cos ax + 3a(ax - \sin ax \cos ax] + c$.   15. $\pi/8$.
17. $\frac{1}{2}a^3 [ax/(a^2 + x^2) + \arctan x/a] + c$.   19. $2\sqrt{3}/27(\pi - \sqrt{3}/2)$.
21. $\frac{1}{2}\ln|x^2 + x + 1| - (1/\sqrt{3}) \arctan (2x + 1)/\sqrt{3} + c$.   27. $\ln(\sqrt{2} + 1)$.
29. $\frac{1}{8}(3\theta - 3 \sin \theta \cos \theta - 2 \sin^3 \theta \cos \theta) + c$.   31. $\sqrt{\sin x} (\frac{2}{3} \sin x -$
$\frac{2}{7}\sin^3 x) + c$.   33. Valid, but does not succeed.
37. $ab[h\sqrt{a^2 + h^2}/a^2 + \ln(h + \sqrt{a^2 + h^2}) - (\sqrt{2} + 1) - \ln(\sqrt{2} + 1)]$.

II. § 5.6   1. $\dfrac{1}{5} \ln \dfrac{x - 3}{x + 2} + c$.   2. $-1/(x - 1) + c$.   3. $\ln|x - 2| -$
$2/(x - 2) + c$.   4. $\ln\sqrt{x^2 + 1} - \arctan x + c$.
5. $\frac{3}{2}\ln(x^2 + x + 1) + (2/\sqrt{3}) \arctan (2x + 1)/\sqrt{3} + c$.
7. $x^2/2 + x - (2/\sqrt{3}) \arctan (2x - 1)/\sqrt{3} + c$.   9. $\ln\sqrt{x^2 + 1} + 1/(x^2 + 1) + c$.
11. $\ln\frac{8}{9} + \pi/4$.   15. $\pi/2 - 1$.

II. § 6.5   1. $\frac{1}{5}\ln(x - 3)/(x + 2) + c$.   3. $\frac{3}{2}\ln(x^2 + x + 1) +$
$\sqrt{3} \arctan (2x + 1)/\sqrt{3} + c$.   5. $(2x - 1)/6(x^2 - x + 1)^2 +$
$(2\sqrt{3}/9) \arctan (2x - 1)/\sqrt{3} + c$.   7. $-\dfrac{x + 12}{50(x^2 - x - 6)^2} + \dfrac{3}{(50)(25)} \dfrac{2x - 1}{x^2 - x + 1} +$
$\dfrac{6}{(50)(25)(5)} \ln \dfrac{x - 3}{x + 2} + c$.   9. $\frac{3}{2}\ln(x^2 + 1) + \frac{9}{2}\arctan x + (x + 2)/2(x^2 + 1) + c$.
11. $(13 + 2\pi)/16$.   13. $3\pi/8$.   15. $[\sqrt{2} - \ln \tan (\pi/8)]/2\pi$.
17. Does not converge.   19. $3\pi^2/4$.   21. $(2/\pi) (1 - \ln 2)$.
23. $\frac{1}{6}\tan 2x(\tan^2 2x + 1) + c$.   25. $\pi a/4 - (a/2) \ln 2$.
27. $(a^2/2)[\sqrt{2} + \ln(1 + \sqrt{2})]$.   29. $(e^{-4} - 1)/1 + \pi^2$.
31. $\dfrac{6\pi^3}{\mu^2 + 9\pi^2} \left(\dfrac{1}{\mu^2 + \pi^2}\right) (1 - e^{-2\mu})$.   33. Meaningless. The integral is undefined.

II. § 7.7   1. $\arccos x - \sqrt{1 - x^2} + c$.   3. $x \text{ arcsec } x - \ln(x + \sqrt{x^2 - 1}) + c$.
5. $[(x^2 + a^2)\arctan(x/a) - ax]/2 + c$.
7. $-\frac{1}{2}e^{-x}(\sin x + \cos x) + c$.   9. $(x - \sin x \cos x)/2 + c$.
11. $\frac{1}{2}$.   13. $(1/a^4)[(3a^2x^2 - 6)\sin ax - (a^3x^3 - 6 ax) \cos ax] + c$.
15. $(1/a^3)[a^3x^3/6 - (a^2x^2/4 - 1/8)\sin 2 ax - (ax/4)\cos 2 ax] + c$.
17. $[(x^3/5) - (6x/125)] \sin 5x - (6/725) \cos 5x + c$.   19. If $n = 2k$,
$n(2\pi)^{n-1} - n(n - 1)(n - 2)(2\pi)^{n-3} + \cdots + (-1)^k n!(2\pi)$. If $n = 2k + 1$,
there is an extra term $(-1)^{k+1}(2)$.

21. $(x - 1) - \dfrac{(x^3 - 1)}{3 \cdot 3!} + \dfrac{(x^5 - 1)}{5 \cdot 5!} + \cdots$, and $\left| R_{2k-1} \right| < \dfrac{x^k - 1}{k(2k)!}$.

23. Two methods. Integration by parts gives $\ln(x - 1) - (x/x - 1)\ln x + c$.
Or expand $\ln x$ in powers of $x - 1$ and then integrate to find
$$c + \ln(x - 1) - \frac{1}{2}\frac{(x - 1)}{1} + \frac{1}{3}\frac{(x - 1)^2}{2} - \frac{1}{4}\frac{(x - 1)^3}{3} + \cdots + \text{remainder.}$$

25. $x \ln ax - x + c$.   27. $x \ln^2 ax - 2x \ln ax + 2x + c$.

29. $\frac{1}{4}[(2x^2 - 1)\arcsin x + x\sqrt{1 - x^2}] + c$.   31. $\frac{2}{3}$.   35. $e^2 + \frac{1}{4}$.

37. $(1/8a^2)(2a^2x^2 + 2\,ax \sin 2\,ax + \cos 2\,ax) + c$.

39. Appendix B #35.   41.   Appendix B #34.

43. $\dfrac{\sin^{m+1}ax \cos^{n-1}ax}{a(m + n)} + \dfrac{n - 1}{m + n}\displaystyle\int \sin^m ax \cos^{n-2} ax \, dx$.

45. $5(\pi/2)^4 - 12(\pi/2)^2 - 6$.

II. § 7.3R   1. $\dfrac{1}{4} \ln \dfrac{x - 2}{x + 2} + c$.   2. $[2 \ln(x - 1) - 1](x - 1)^2/4 + c$.

3. $\pi a^2/2$.   4. $-\exp(-2x)(2 \sin 3x + 3 \cos 3x)/13 + c$.

5. $a^2[\sqrt{2} + \ln(1 + \sqrt{2})]$.   6. $(\pi + 2)/8$.

7. $\frac{1}{2}\ln|x^2 - x + 1| + (1/\sqrt{3})\arctan(2x - 1)/\sqrt{3} + c$.

8. $\dfrac{1}{2} \ln|x^2 - 2x - 5| + \dfrac{1}{4} \ln \left| \dfrac{x - 3}{x + 1} \right| + c$.

9. $\ln|1 - \cos x| + c$.   10. $(ax + \sin ax \cos ax)/2a$.

11. $\ln|1 + \sin x| + c$.   12. $\frac{1}{2}\ln|\sin x| + c$.   13. $[\sqrt{2} + \ln(\sqrt{2} + 1)]/2$.

14. $[(2x^2 - 1)\arcsin x + x\sqrt{1 - x^4}]/4 + c$.   15. $e^{-2x}(2x^2 + 2x + 1)/4 + c$.

16. $\ln 2$.   17. $(\ln^2 x - \frac{1}{2} \ln x + \frac{1}{5}x)x^5/4 + c$.   18. $-\cos 2\sqrt{x} + c$.

19. $\dfrac{1}{2} \ln \dfrac{e^y - 1}{e^y + 1} + c$.   20. $-(1 - x^2)^{3/2}/3 + c$.

21. $x + \frac{3}{4} \ln|x + 1| + \frac{3}{4} \ln|x^2 + 4| - \frac{1}{4} \arctan(x/2) + c$.

22. $(4 \cos^3 ax - 12 \cos ax - 3 \sin^4 ax \cos ax)/15a + c$.

23. $\cosh x + c$.   24. $-1 + 2 \ln 2$.   25. $5\pi/64$.

26. $(1/a)\ln|a - \sqrt{a^2 - x^2}|/x| + c$.   27. $(\sin^2 x)/2, \sin x > 0$.

28. $(\arcsin x + x\sqrt{1 - x^2})/2 + c$.   29. $\theta - \frac{1}{2} \cos 2\theta + c$.

30. $\arctan e^x + c$.

II. § 7.4R   1.   Max, $\frac{1}{2}$, at $x = -\pi$ and π; min, $\frac{1}{4}$, at $2\pi/3$.

2. $dy = x^x(1 + \ln x)dx, x > 0$.   3. $dy = \ln 10 \, dx/x$.

5. No, $\ln x$ is not defined at 0.   6. $\dfrac{4}{3} \dfrac{(16ab^2 + a^2)^{3/2}}{16b^2}$.

7. $ab(\sqrt{3} + \ln\sqrt{2 - \sqrt{3}})$.   8. $8[\sqrt{2} + \ln(1 + \sqrt{2})]$.

9. $\frac{2}{3}$.   10. $w \approx 138$.   12. $y = x^2$.   13. $2x$.

14. $\dfrac{1}{\sqrt{c}} \arctan \dfrac{x}{\sqrt{c}} + K$ if $c > 0$; $\dfrac{1}{2\sqrt{-c}} \ln \dfrac{x - \sqrt{-c}}{x + \sqrt{-c}}$ if $c < 0$.

15. $\pi/4 - \frac{1}{2} x \sqrt{1 - x^2} - \frac{1}{2} \arcsin x + c$.   19.   0.436.

20. Approx. 52.0.   21. $[1 - \exp(-x)]/x$.

22. $[1 - \exp(-x^2)]/x$.   23. $1/\sin 2\theta$, no maximum.

24. $-(1/x) \cos(1/x) + \sin(1/x), x \neq 0, f'(0)$ does not exist.

27. They differ by a constant, $\frac{1}{2}$.

28. If n is even, $(d^n/dx^n) \cos x = (-1)^{n/2} \cos x$;
If n is odd, $(d^n/dx^n) \cos x = (-1)^{(n+1)/2} \sin x$.

29. If n is even, $(d^n/dx^n) \sin x = (-1)^{n/2} \sin x$;
If n is odd, $(d^n/dx^n) \sin x = (-1)^{(n+1)/2} \cos x$.

30. At $3\pi/4$, max $f = \sqrt{2}$; at $-\pi/4$, min $f = -\sqrt{2}$. 31. Max, $(\pi/6, \sqrt{3}/2 + 1)$; min, $(-\pi/2, -2)$. 32. Neither exists. 33. -2 mi/sec (west).
34. $x' = 0$. 35. Maximum $A = 72/\pi$ occurs at $\theta = \pi$ where the graph has a cusp. 41. 0 if $m \ne n$, π if $m = n$.
42. $\ln x + 1/(x^2 + 1) + c$. 43. $\omega/(s^2 + \omega^2)$.
44. $\dfrac{1}{3} \ln \dfrac{e^x - 1}{e^x + 1} + c$. 45. $2(x - 2)\sqrt{x + 1}/3 + c$.
46. Does not converge. 47. $\pi/4$. 48. $x/a^2\sqrt{a^2 - x^2} + c$.
49. (a) $\dfrac{1}{\sqrt{a}} \ln|x\sqrt{a} + \sqrt{ax^2 + c}| + K$ if $a > 0$ and $x^2 > -c/a$.

   (b) $\dfrac{1}{\sqrt{-a}} \arcsin x \sqrt{\dfrac{-a}{c}}$ if $a < 0 < c$ and $x^2 < -c/a$.

   (c) Not defined if $a < 0$ and $c < 0$.
50. 3. 51. $t + \ln|1 + \sqrt{1 + \exp(-2t)}| + c$. 52. Does not converge.
53. $1/(1 - k)$ if $k < 1$; does not converge if $k \geqq 1$.

II. § 8.7 1. $\sqrt{37}$. 2. $\sqrt{370}$. 3. 5. 4. $\sqrt{(x - \sqrt{2})^2 + (y + 1)^2}$.
   5. $\sqrt{21}$. 7. $\sqrt{a^2 + b^2 + c^2}$. 11. $-1/\sqrt{37}, 6/\sqrt{37}$.
   13. $(x + 1)^2 + (y - 3)^2 = 16$. 19. $13(x + 1) + 6(y - 4) = 0$.
   21. $7(x - 2) + 5(y + 3) = 0$, $(x - 4) + 11(y + 1) = 0$, $(x + 5) + 2(y - 2) = 0$.
   23. Sphere with center at $(1, -2, -3)$ and radius $\sqrt{14}$.
   25. $(x - 2)^2 + (y - 1)^2 = 15$. 27. $(x - 3/2)^2 + (y - 1)^2 + (z - 0)^2 = 33/4$.

II. § 9.7 2. $(2, -2, 3, 10)$. 3. $(-5, 2, 0, 2)$. 4. $(-18, 9, -12, -30)$.
   5. $(-22, 6, 15, -31)$. 13. Unit free. 15. Each sample sum is a sum of unit-free ratios $\Delta x/x$. 17. Unit free. 21. L/T, L/T^2.

II. § 10.6 2. $\vec{u} + \vec{v} = 4\vec{i} - 5\vec{j}$, $\|\vec{u}\| = 5$, $\|v\| = \sqrt{2}$, $\vec{u} - \vec{v} = 2\vec{i} - 3\vec{j}$.
   $\|\vec{u} + \vec{v}\| = \sqrt{41}$, $\|\vec{u} + \vec{v}\| = \sqrt{13}$, $\vec{u} \cdot \vec{v} = 7$, $3\vec{u} = 9\vec{i} - 12\vec{j}$,
   $-4\vec{v} = -4\vec{i} + 4\vec{j}$, $\|-4\vec{v}\| = 4\sqrt{2}$, $2\vec{u} - 5\vec{v} = \vec{i} - 3\vec{j}$.
   11. (a) $\arccos(-2/\sqrt{5})$, (b) $\arccos(-3/5)$. 17. Circle: $x^2 + y^2 = 16$.
   19. Line: $ax + by = 4$.

II. § 11.8 1. $\vec{u} + \vec{v}$: $(-2, 5, 4)$, $2\vec{v}$: $(-2, 4, 0)$, $\vec{v} - \vec{w}$: $(0, -1, -4)$,
   $\|\vec{v} - \vec{w}\| = \sqrt{17}$. 2. $\vec{v} + \vec{w}$: $(4, -2, 1, 5)$, $3\vec{w}$: $(3, 0, 9, 3)$, $\|\vec{w}\| = \sqrt{11}$.
   3. $c = 1/\sqrt{11}$. 5. $\vec{v} \cdot \vec{w} = -3$, $\cos\theta = -3/2\sqrt{39}$.
   9. $x + 1 = (1)t$, $y - 3 = (-1)t$, $z - 4 = (2)t$, for all real numbers t.
   13. $x - 3 = (-2)t$, $y - 1 = (-3)t$, $z + 1 = (1)t$, for all real numbers t.
   15. $\vec{r} - \vec{p}_1 = t(\vec{p}_2 - \vec{p}_1)$. 21. (a) $\vec{n}$: (a, b). (b) Direction $\vec{a}$: $(b, -a)$, Equation, $\vec{r} - \vec{p}_0 = t\vec{a}$.

II. § 12.6 1. 5. 2. $\vec{w}_1$: $(-\frac{4}{3}, \frac{2}{3}, \frac{4}{3})$, $\vec{w}_2$: $(-\frac{2}{3}, -\frac{2}{3}, -\frac{1}{3})$.
   3. Orthogonal $\vec{w}_1$: $(-\frac{4}{3}, \frac{2}{3}, \frac{4}{3})$, parallel $\vec{w}_2$: $(-\frac{2}{3}, -\frac{2}{3}, -\frac{1}{3})$.
   4. $8/\sqrt{3}$. 5. $27/2\sqrt{13}$. 7. Normal equation: $(x/\sqrt{2}) - (y/\sqrt{2}) - (3/\sqrt{2}) = 0$, normal intercept, $p = 3/\sqrt{2}$, inclination of normal, $\alpha = -\pi/4$.
   9. One equation is $x + y + z = 1$. 11. Two parallel vectors do not determine a unique common normal. 13. 0.
   15. $\cos\theta = -52/7\sqrt{69}$. 19. $\sqrt{a_1^2 + a_2^2 + a_3^2}$. 21. Eliminate x_4.
   25. $48/\sqrt{194}$.

II. § 13.7 15. Plot the line through the point $(4, 6, -2)$ parallel to the direction $(3, 3, 3)$. 16. A cone. 17. A parabolic cylinder.
   18. A circular cylinder.

II. § 14.7   In Exercises 7–14 introduce a "slack variable" w to change the inequality involving x and y into one involving only w. Then plot contour lines, $w = $ constant, for values of w which satisfy the inequality.

7.   Lines, $x - y = w$, for $w \geqq 0$.

9.   Circles, $(x - 1)^2 + y^2 = w$, for $0 \leqq w \leqq 1$.

11.   Parabolas, $y = x^2 + c$, for $c \leqq 0$.

13.   Hyperbolas, $y^2 - x^2 = w$, for $w \geqq 0$.

II. § 15.7   1. $f_1 = 6x - 4y$, $f_2 = -4x + 2y$.

2. $f_1 = -2y/z$, $f_2 = -2x/z$, $f_3 = 2xy/z^2$, $z \neq 0$.

3. $f_1 = x/\sqrt{x^2 + y^2}$, $f_2 = y/\sqrt{x^2 + y^2}$.   4. $f_1 = 1 - \cos 2\theta$, $f_2 = 2r \sin 2\theta$.

5. $f_1 = -e^{-t} \sin x$, $f_2 = -e^{-t} \cos x$.   7. $f_1 = 7y/(y - 3x)^2$,
    $f_2 = -7x/(y - 3x)^2$.   9. $u_1 = 12x(x^2 - xy + y^2)$, $u_2 = -4x^2(x - 3y)$.

23.   Marginal revenues from products A, B, C respectively.

25.   Rates of change of force per unit change in q_1, q_2, r.

27. $v_1 = v/c$, $v_2 = -kt/M(M - kt)$, $v_3 = ct/M - kt$, $v_4 = ck/M - kt$.

II. § 16.6   1. $dy = -2\,dx$.   2. $dy = -dx$.   3. $dy = 2\,dx$.

4. $dy = dx$.   5. $Df = -8\,dx_1 + 7\,dx_2$.

7. $Df = -(1/\sqrt{27})dx_1 - (1/\sqrt{27})dx_2 - (1/\sqrt{27})dx_3$.

9. $\Delta f \approx 0.05$.   11. $\Delta f \approx 0.4$.

13.   (a) $D(uv) = \dfrac{3xy}{\sqrt{x^2 + y^2}}\left[(3x^2y + 3y^3 + x)dx + (3x^3 + 3xy^2 + y)dy\right]$.

(b) $D(u/v) = \dfrac{-3xy}{(x^2 + y^2)}\,[y\,dx + x\,dy]$.

15.   At $(0, 0)$.   19. $dz = 0$.

II. § 17.5   1. $2dx - 4dy = 0$.   2. $dy = 0$, the dx-axis.   3. $dx = 0$.

4. $0\,dx + 0\,dy = 0$, the entire plane.   5. $dy = 0$.   7. $2x + 2y = a$.

11. $ax + by + c = 0$; A line is tangent to itself.

13. $(x_0\,x/a^2) + (y_0\,y/b^2) + (z_0\,z/c^2) = 1$.   15. $dz = 0$.   17. $dy/dx = 8$, slope.

21. $dy/dx = 1/\cos y = 1/\sqrt{1 - x^2}$.

II. § 18.5   1. $-(63t + 124)dt$.   2. 0.

3. $[3(3s - 1) + (2s^2 + 1)(2s)]ds/\sqrt{(3s - 1)^2 + (2s^2 + 1)^2}$.

4. $(5t - 1)dt/\sqrt{5t^2 - 2t + 2}$.   5. $2e^{u^2 - v^2}(u\,du - u\,dv)$.

7. $((9t^2 - 2t - 3)dt, (2t + 3)dt)$ 9. $3\sqrt{3}/2$.   11. $-\frac{3}{4}$.

13. $F_2(x_0, y_0)$.   15.   1, independent of x_0, y_0 and α.

17. $(\frac{3}{5}, 0, -\frac{4}{5})$.   19. $\nabla f \cdot h = 0$ when h is the tangent direction,
    $\nabla f \cdot h = \pm 30$ when h is a normal direction.

II. § 19.6   1. $x' = -\frac{8}{3}$ cm/sec.   2. $r' = 2/5\pi$ cm/sec.   3. $r' = k$.

4. $x' = -75/8$ ft/min.   7.   (a) $\frac{1}{2}\pi\%$;   (b) 6%.   9.   1.44%.

11.   10.027, approx.   13.   0.038%, approx.   15.   0.3%.

17. $2g\,T\,dT + T^2\,dg = 4\pi^2\,d\ell$.   19. $c \cos \phi \phi' = b \cos \theta \theta'$.

21.   With both planes on fixed course, $\theta' = \alpha \sin \theta/r$.

23. $d\alpha = -0.05 \tan 2\alpha$.

II. § 20.5   1. $(6, 8)$.   2. $(-4, -16)$.   3. ∇f: $(3, 3, -2)$.

4. ∇f: $(-1, 1, 1)$.   5. (a, b, c).   9.   Near a minimum point the gradients point out; near a maximum the gradients point inward.   11. $2/\sqrt{21}$.

13. $-2/7\sqrt{42}$.   15. $(0, 0)$, $(-\frac{1}{2}, -\frac{1}{4})$.   17. $(3, 0)$, $(-\frac{6}{5}, -\frac{8}{5})$.

19. $(0, 0)$, $(-2, 8)$.

21. ∇U: $(m/r^{3/2})(-x, -y, -z)$.   23.   (a) $\pi/4$,   (b) $-1/\sqrt{2}$.   23. $w = 78$.

II. § 21.6   1.   Tangent vector a: $(1, -2)$.   2.   Tangent vector a: $(-r/2, r\sqrt{3}/2)$.
3.   Yes.   5.   Zero velocity when $t = 0$ at $(0, 3)$, when $t = \pi/2$ at $(2, 0)$,
when $t = \pi$ at $(0, 3)$, also when $t = 3\pi/2, 2\pi \cdots$. Motion is back and forth
on line segment joining $(2, 0)$ to $(0, 3)$.
7. $\vec{v}$: $(-r \sin t, r \cos t)$, $\vec{a}$: $(-r \cos t, -r \sin t) = (-x, -y)$, $\|\vec{v}\| = r$ (constant).
11. $\vec{v}$: $(-r/\sqrt{2}, -r/\sqrt{2}, r/2\pi)$; $\vec{a}$: $(r/\sqrt{2}, -r/\sqrt{2}, 0)$.   13.   1.   15.   0.
17. $-1/6$.   21.   Replaces $\cos t$ by $-\sin s$ and $\sin t$ by $\cos s$.
23.   A rising spiral on the cone.

II. § 22.6   1.   4.   2. $15\sqrt{11}$.   3. $85\sqrt{85}/27$.   4. $8a$.   5. $8\pi\sqrt{5}$.
9.   HINT: $ds/dt = 2a \sin(t/2)$, $t \neq 0, 2\pi$. $\vec{u}$: $(\sin t/2, \cos t/2)$,
$\vec{n}$: $(\cos t/2, -\sin t/2)$, $\kappa = 1/4a \sin t/2$.   11. $\vec{u}$: $(1/\sqrt{2}, 1/\sqrt{2})$;
$\vec{n}$: $(1/\sqrt{2}, -1/\sqrt{2})$, $\kappa = \sqrt{2}/8$.   13. $\vec{u}$: $(-\cos t, \sin t)$;
$\vec{n}$: $(\sin t, \cos t)$, $\kappa = (1/3a) \cos t \sin t$, where $t \neq n\pi/2$.
17.   Curvature is invariant in Euclidean geometry.   19.   Direction
u: $(1/\sqrt{2}, 1/\sqrt{2})$; speed, $2\sqrt{2}\pi$.   21.   When $b = a$ and $dS/S = 3\%$,
then $db/b = -3\%$.

II. § 23.4   1. $F = (y^2 + 5)/2$.   2. $\sin y$.
3. $F = x^2y + y - 3x^2 - \frac{5}{2}$.   4. $F = -2 \cos (x/2) + a$.
5. $F = x^2/2 - 2xy + y^2 + c$.   7. $F = (x^2 + y^2)/2 - xy + c$.
9. $F = x/y + c$.   10.   None exists.   11. $F = x^2y + y^3/3 + c$.
13. $F = 2$.   15. $F = 2xy + 6xz - 3yz + c$.   17. $u_2 = v_1, v_3 = w_2, w_1 = u_3$.
19. $\frac{1}{2}$.   21. -4.

II. § 24.4   5. $x \, dx + y \, dy = 0$.   6. $-dx + dy = 0$.   7. $y \, dx - x \, dy = 0$.
9. $xy = c$.   11. $y = C_1 e^x$ where $|C_1| = e^{-c}$.   13.   (a) $y = -1$,
(b) $x = 1$.   15. $x - \ln\sqrt{1 - y^2} = c$.   17. $y - y^3/3 = 3x^2/2 + c$.
19. $1 = cy^2 - y^2 \ln(1 + x)^2$.   21. $c = -\frac{1}{2} \ln \frac{3}{4}$; Integral can be written
$1 - y^2 = 3e^{-2x}/4$.
23.   Direction map: $(x + y/\sqrt{x^2 + y^2}) \, dx + (y - x/\sqrt{x^2 + y^2})dy = 0$.
25. $\ln\sqrt{x^2 + y^2} + 2 \arctan y/x = c$.   27. $x^2y/\sqrt{xy - 2x^2} = c$.

II. § 25.4   1. $x + y = a$.   2. $y = ax$.   3. $xy = a$.   4. $3x^2y + y^3 = c$.
5. $y^2 = ax$.   6.   Integrating factor x^2/y^2 gives $x^2 + y^2 = ay$.
7. $y^2 = ax$.   13. $y = cx^4$.   15.   Integrating factor, $1/x^2$.
17.   Integrating factor, $1/y^2$.   19.   The curves consist of two sets
$c = x + \sqrt{x^2 + y^2}$ and $c = x - \sqrt{x^2 + y^2}$, one of which is orthogonal
to the other.

II. § 26.4   1. $y = y_0 \exp 0.347t$.   2. $y = x_0(1 - \exp - 0.0347t)$.
3. $y = 1/kt$, with k undetermined.   5. $\theta = (1090/101)[1 - \exp (- kt)] +$
$10 \exp (- kt)$, where k is the rate of heat conduction.   11. $r = 0.0434$.
13. $y = N \exp (\beta - \delta)t$, and $\beta - \delta$.   15. $dx/dt = -kx^2$,
$x = x_0/(1 + kx_0 \, t^2/2)$.

II. § 27.5   1. $(v, y) = (64, 64)$; $(0, 256)$; $(-64, 64)$.   2.   8 sec.
3. $(5 + \sqrt{33})/4$.   4.   107.5.   5. $x = x_0(\cos \alpha)t, y = 0$,
$z = -gt^2/2 + v_0 (\sin \alpha)t$.   9. $v = -W/\beta$, the terminal velocity.
11.   Never stops, but total distance approaches 10 mv_0.   15.   121.
17. $y = -10t + (m/8)[1 - \exp(-80/m)t]$, Limit velocity $= -10$ ft/sec.
19.   Limit velocity $v = -1/k$.

II. § 28.5   1. $y = x^3/2 + cx$.   2. $y = (1 + x)[\ln(1 + x) + c]$.
  3. $y = x^3 (\ln x + c)$.   4. $y = \frac{2}{5} \cos t + \frac{1}{5} \sin t + c \exp(-t)$.
  5. $x = y/4 - (y + 1)20 + c(y + 1)^{-4}$.   7. $y = x + c \exp(-x^2)$.
  9. $y = 20 - 15 \exp(-\frac{3}{10})t$.   11. $S = R \exp kt - R$.
  13. The amplitude of displacement, $A/m\omega\sqrt{k^2 + \omega^2}$, decreases for higher
    frequencies, ω. Thus, the high frequency response is reduced.   19.   The
    process exponentially damps out (forgets) old input values $f(s)$ and sums the
    product $e^{-k(t-s)} f(s)$. Most recent input values have the greatest influence in
    determining $y(t)$. Thus $y(t)$ is a kind of smoothed average response to
    recent inputs.

II. § 28R   1. $y = x^3/3 + c$.   2. $\ln y + 1/x = c$.   3. $x^2 - y^2 = c$.
  4. $e^x y = e^{2x}/x + c$.   5. $y = x^4/12 + c_1 x + c_2$.
  6. $\arctan v = \arctan u + c$.   7. $(y - c)^2 = x$.
  8. $y = x^2 e^x + cx^2$.   9. $x^2 + xy + y^2 = c$.   10. $y/x = c$.
  11. $xy = x^2/2 + c$.   12. $y - f(x) = c$.   13. $xy = 1$.
  14. $y = \ln\left|\dfrac{1 + x}{1 - x}\right| + 2, x \neq 1$.   15. $y = 2 \ln(x^2 + 4) - 2 \ln 5 + 1$.
  16. $y = (2 ax - \sin 2 ax)/4a$.   17. $e^x + e^{-y} = 3/2$.
  18. $y = \dfrac{E}{a^2 + \omega^2} (a \sin \omega t - \omega \cos \omega t + c)$.   19. $y = -(\sin t + \cos t)/2$.
  20. $y = (\sin x - \cos x)/2$.   21. $y = \ln(1 - e^{-x})$.
  22. $x^3 y - 3x^2 + y^2 = 1$.   23. $x^2 - 3xy + y^2 = 4$.
  24. $e^{2x} - 2e^y = e^{-1} - 2$.   25. $y = \dfrac{1}{x} \int \dfrac{\sin x}{x} dx + \dfrac{c}{x}$.
  26. $x^2 + y^2 = r^2$.   27. $x^2 - y^2 = c$.   28. $dy/dx > 0$ everywhere.
  29. Concentration $\frac{1}{500} t \exp(-1/50)t$.   30. 25 ft/sec.
  38. $y = (a/2) [(x/a)^{1+k}/1 + k] - (a/2) [(x/a)^{1-k}/1 - k] + c, k \neq 1$.
  39. If $k > 1$, the solution function is not defined when $x = 0$.
  40. If $k < 1$, catch occurs where $x = 0$ and $y = c = ak/1 - k^2$.

II. § 29.9   1. Upper sum $= -12$, lower sum $= -48$.   2. Upper sum $= 156$,
    lower sum $= 64$.   3. Upper sum $= 26$, lower sum $= 14$.
  4. Upper sum $= 36$, lower sum $= -36$.   5. Volume $=$ area of base $\times$
    average height.   9. $\overline{f(x,y)} = \iint_\infty f(x, y)dx\, dy/A$.   11. $f = f(x_0, y_0)$.

  13. -41.   17. $M = \displaystyle\int_0^5 \int_0^5 \int_0^5 (1 - 0.2z)dx\, dy\, dz$.

II. § 30.5   1. $\frac{1}{3}$.   2. 0.   3. 2.   4. $2(e - 1)/3$.
  5. $(1/15)(13^{5/2} - 5^{5/2} + 3^5 - 1)$.   7. $136/3$.   9. $2 \ln 2 - \frac{3}{4}$.
  11. $(d + c - 2a)(d - c)/2$.   13. $-81/8$.
  15. Volume of a vertical slice parallel to yz-plane at distance x and having
    thickness dx.   17. $2/\pi(1 - 2/\pi)$.
  21. $(x/4) (z^2 - c^2)(y^2 - b^2)dx + (y/4) (z^2 - c^2)(x^2 - a^2)dy + (z/4) (y^2 - b^2)$
    $(x^2 - a^2)dz$.

II. § 31.6   1. $-\frac{1}{2}$.   2. $-\frac{1}{3}$.   3. $\pi/2$.   4. $\frac{2}{3}$.   5. ∞.
  7. $\frac{1}{2}$.   9. $\frac{1}{6}$.   11. $16\sqrt{2}/3$.   13. $-104/105$.   15. $8/75$.
  17. 4.   19. $V = 8\displaystyle\int_0^a \int_0^{\sqrt{a^2 x^2}} \sqrt{a^2 - x^2 - y^2}\, dy\, dz$.

II. § 32.5   1.   55/6.   2.   27/4.   3. $50(\sqrt{2} - 1)$.   4. $2a/3$.   5.   0.

7. $V = \int_{-3}^{3} \int_{0}^{1} \int_{-1}^{1} dz \, dy \, dx$.

9. $M = \int_{-c}^{c} \int_{-b}^{b} \int_{-a}^{a} e^z \, dx \, dy \, dz$.   11. $\vec{f} = \int_{-a}^{a} \int_{-a}^{a} \int_{-a}^{a} (x^2 - y^2) dx \, dy \, dz/8a^3$.

13.   Volume of a vertical shaft of base $dx \, dy$, located at $P: (x, y)$, and extending from the plane $z = 0$ up to the plane $z = x + y + 1$.

15.   Volume of a prism under the plane $z = x + y + 1$ over the square $\{(x, y)/0 \leq x \leq 3, 0 \leq y \leq 3\}, z = 0$.   17.   0.

19.   The integral is undefined in the domain. In this case it has imaginary values at some points.   20.   Solid cylinder whose base is the disk $\{(x, y)|x^2 + y^2 \leq r^2\}$ and whose altitude is h.

II. § 33.4   1.   (a) 2/11,   (b) 58,   (c) 647/11.   2. $M = 8a^4/3$, $\bar{x} = 0$, $\bar{y} = 0$.

3. $M = 6$, $\bar{x} = 13/24$, $\bar{y} = 45/24$.   4. $M = \frac{1}{4}$, $(\bar{x}, \bar{y}) = (\frac{2}{5}, \frac{2}{5})$.

5. $M_{x-1} = -2a^4$, $M_{y-3} = -8a^4$.   7. $56a^6/45$.   9. $I_x = I_y = 7/450$.

11. $112a^6/45$, See Ex. 7.   13. $\sigma = \sqrt{398}/24$.   17. $49M(a^2 + b^2 + c^2)/240$.

19.   New means: $(\bar{x} + 1, \bar{y} + 1, \bar{z} + 1)$, standard deviation unchanged.

21. $\theta = t^2/4\pi$, starting from rest.

II. § 34.7   2.   1, $\pi/2 + 2m\pi$, $(m = 0, \pm 1, \pm 2, \cdots)$.   15. $r = 2$.

17. $\tan \theta = 3$.   19. $r = 2/(1 - \cos \theta)$.   21. $x^2 + y^2 - x = 0$.

23. $y = \sqrt{3} \, x$.   25. $\pi/4$.   27. $\frac{1}{2}$.   29. $\pi/6$.

33. $n: (-a/\sqrt{a^2 + b^2}, -b/\sqrt{a^2 + b^2})$.   35. $r = 2a \cos \theta$, where $|AB| = 2a$, the pole is at A and B is on the initial ray.   37. $r^4 + a^4 - 2r^2a^2 \cos 2\theta = k^4$.

39.   Circle, $r = k$, $A = \pi k^2$.   41. $r = 4/(1 - 2 \cos \theta)$.   43.   Parabola, $r = 4(1 - \cos \theta)$.

II. § 35.5   1. $\psi = -\pi/3$.   2. $8a$.   3. $(a/3) [(\pi^2 + 4)^{3/2} - 8]$.

5. $\beta = -\pi/4$, where $\theta = 0$ and $\beta = \pi/4$, where $\theta = \pi/4$.

9.   (b) Where $\theta = -\pi/8, 3\pi/8$.   13.   The spiral, $r = \theta$.   17. $r^2 = b \sin \theta$.

19. $dy/dx = [f'(\theta) \sin \theta + f(\theta) \cos \theta]/ [f'(\theta) \cos \theta - f(\theta) \sin \theta]$.

II. § 36.4   1. $v_r = 0$, $v_\theta = 8\pi$.   2. $a_r = -16\pi^2$, $a_\theta = 0$.   3. $a_r = -64\pi^2$, $a_\theta = 8\pi$.   4. $F = -640\pi^2$.   9. $1.62\rho = 6450$ mi.

11. $ma_r = -k/r^2 + \vec{F} \cdot u_r$, $ma_\theta = \vec{F} \cdot u_\theta$ with boundary conditions, $(r, \theta) = (b, 0)$ when $t = 0$ and $(r, \theta) = (b, 1/10)$ when $t = 24$ min.

13. $\vec{v} = a\vec{u}_r + a\omega t \, \vec{u}_\theta$, $a = -a\omega^2 t \, \vec{a}_r + 2a\omega \, \vec{u}_\theta$.   15. $(v_r, v_\theta) = t^{a-1} (a, 1)$, $(a_r, a_\theta) = t^{a-2} (a^2 - a - 1, 2a - 1)$.   17. $(v_r, v_\theta) = (\alpha e^{\alpha t} - \alpha e^{-\alpha t}, e^{\alpha t} + e^{-\alpha t})$. $(a_r, a_\theta) = (\alpha^2 - 1) (e^{\alpha t} + e^{-\alpha t}), 2\alpha(e^{\alpha t} - e^{-\alpha t})$.

II. § 37.7   1. $121a/81$.   2.   168.   3. $-n/2$ if n is even, $(n + 1)/2$ if n is odd.

7. $\frac{3}{2}$.   9.   Converges by comparison with p-series, $p = 2$.

11.   Diverges by comparison with harmonic series.   13.   Diverges, since $\lim_{k\to\infty} k/(k + 1) \neq 0$.   15.   Converges.   17. $25.

19. $T = 8 + 8/10 + 8/10^2 + \cdots = 80/9$ sec.

II. § 38.4   1. $|x| < \frac{1}{2}$.   2. $|x| < 1$.   3. $|x| < 1$.   4. $|x| < 1$.

5.   All real numbers.   7.   All real numbers.   9. $|x| < 1$.   11. $n \geq 7$.

13.   In any closed interval, $[-\rho, \rho]$ such that $\rho < 1$.   15. $|x| < 1$.

17. $0 < x < 4$.

II. § 39.3   1.   Both geometrically dominated in $[-\rho, \rho]$ where $\rho < 1$.

2. $f'(x) = 1/2 + x/3 + x^2/4 + \cdots$ in any interval $[-\rho, \rho]$, where $\rho < 1$.

3. $\dfrac{x^2}{1 \cdot 2^2} + \dfrac{x^3}{2 \cdot 3^2} + \dfrac{x^4}{3 \cdot 4^2} + \cdots + \dfrac{x^n}{(n-1)n^2} + \cdots$, where $|x| < 1$.

4. $f(x) + g(x) = 1 - (3/2)x + (13/6)x^2 + \cdots$, where $|x| < 1$.

5. $f(x) g(x) = x/2 - (5/6)x^2 + (3/4)x^3 + \cdots$, where $|x| < 1$.

7. (a) $n > 15$, (b) $n > 19$.   11. $n > 14$.

Part Three

III. § 1.7   1. $x = x' + 2$, $y = y' - 4$.   2. $x = x' + 5$, $y = y' - 4$.

3. Not possible.   4. Not possible.   5. Not possible.

7. $x' = x - h$, $y' = y - k$.   9. $\left(\dfrac{7\sqrt{3} - 5}{2}, \dfrac{-7 - 5\sqrt{3}}{2} \right)$.

13. $x'^2 + y'^2 = r^2$.   17. Shape changed to ellipse $(x'/a)^2 + (y'/b)^2 = 1$.

21. Starting at (a, b) when $t = 0$, the point (x, y) rotates counterclockwise in the circle $x^2 + y^2 = a^2 + b^2$ at 1 revolution per second.

III. § 2.6   1. $(x^2/100) + (y^2/36) = 1$, $a = 10$, $b = 6$, $c = 8$, $e = \frac{4}{5}$, $k = \frac{32}{5}$.

2. $a = \sqrt{2}$, $b = 1$, $c = 1$, $e = 1/\sqrt{2}$, $k = 1/\sqrt{2}$.   3. $x^2/9 = y^2/4 = 1$, $a = 3$,

$b = 2$, $c = \sqrt{5}$, $e = \sqrt{5}/3$, $k = \frac{4}{3}$.   4. $\dfrac{x^2}{(25/4)} + \dfrac{y^2}{(25/9)} = 1$, $a = \frac{5}{2}$, $b = \frac{5}{3}$,

$c = 5\sqrt{5}/6$, $e = \sqrt{5}/3$, $k = \frac{10}{9}$.   5. $x^2/9 + y^2/25 = 1$, foci on y-axis,

$a = 5$, $b = 3$, $c = 4$, $e = \frac{4}{5}$, $k = \frac{9}{5}$.   9. (a) Point $(0, 0)$, (b) Empty set.

15. Compute $c = ae = 2$, then proceed as in Ex. 13.   19. Using symmetry, $A = \pi ab/2$.

III. § 3.8   1. $x^2/16 - y^2/9 = 1$, $a = 4$, $b = 3$, $c = 5$, $k = 9/4$, foci: $(\pm 5, 0)$, asymptotes: $3x \pm 4y = 0$.   2. $x^2/4 - y^2/4 = 1$, $a = b = 2$, $c = 2\sqrt{2}$, $k = 2$, foci: $(\pm 2\sqrt{2}, 0)$, asymptotes: $x \pm y = 0$, intercepts: $(\pm 2, 0)$.

3. $y^2/9 - x^2/16 = 1$, $a = 3$, $b = 4$, $c = 5$, $k = 16/3$, foci: $(0, \pm 5)$.

4. $x^2/9 - y^2/16 = 1$, $a = 3$, $b = 4$, $c = 5$, $k = 16/3$, foci: $(\pm 5, 0)$, asymptotes: $4x \pm 3y = 0$.   5. $k = 2$, focus $(1, 0)$, end-points of focal chord $(1, \pm 2)$.   7. $k = \frac{9}{2}$, focus $(0, \frac{9}{4})$, end-points of focal chord $(\pm \frac{9}{2}, \frac{9}{4})$.

11. Subnormal $|P'N| = k$.   13. $k^2(\pi/2 + 16/3)$.   17. Hyperbolas have the same asymptotes but perpendicular principal axes.   19. Intersection of two hyperbolas.   21. $c = a/k$.

III. § 4.6   1. $zw = 4 - 2i$, $w/z = -1/5 + 2i/5$, $\arg z = \pi/4$, $|z| = \sqrt{2}$, $w\overline{w} = 10$, $z^2 - 3z + 1 = -2 + 5i$.   2. $z + w = 2 - i$, $z - w = 4 - 3i$, $1/z = 3/13 + 2i/13$.   3. (a) $\sqrt{2}[\cos(-\pi/4) + i \sin(-\pi/4)]$, (b) $2[\cos(-\pi/6) + i \sin(-\pi/6)]$, (c) $2(\cos \pi + i \sin \pi)$, (d) $3[\cos(-\pi/2) + i \sin(-\pi/2)]$.   11. $4[\cos(-\pi/6) + i \sin(-\pi/6)]$ and $4[\cos(5\pi/6) + i \sin(5\pi/6)]$.   15. $\pm (1/\sqrt{2} + i/\sqrt{2})$.   17. 0.   19. 0.

III. § 5.5   5. $\ln \left| \dfrac{1 - \cos x}{\sin x} \right| + c$.   9. $\ln \left| \dfrac{1 + \sin x + \cos x}{1 + \cos x} \right| + c$.   17. $\pi a^2/4$.

19. 1.   23. $\pi/2 - 1$.   25. $(1/2)[(x + 1)\sqrt{x^2 + 2x} - \ln|(x + 1) + \sqrt{x^2 + 2x}|] + c$, where $x > 0$ or $x < -2$.   27. $\ln(2x + 1 + 2\sqrt{x^2 + x + 1}) + c$.   29. $\arctan(-\pi/4)$.   31. $\frac{4}{3}(20 - 9 \ln 3)$.

III. § 6.5   1. $6 \tan^2 2x \sec^2 2x \, dx$.   2. $3 \sec^3 u \tan u \, Du$.

3. $2(\sec^3 3x + \sec 2x \tan^2 2x)dx$.   4. $-\csc x \cot x \, dx$.   5. $\csc x \, dx$.

9. $+\sqrt{(1 - x)/(1 + x)}$.   11. $\frac{1}{2} \tan^2 x + \ln|\cos x| + c$.

13. $\frac{1}{2}(\sec x \tan x - \ln|\sec x + \tan x|) + c$.   15. $-\cos x + c$.

23. $(1/32)(4x - \sin 4x) + c$.   25. $(1/192a)(60\,ax + 48 \sin ax + 9 \sin 4\,ax - 4 \sin^3 2\,ax) + c$.   27. $\frac{1}{2} \arcsin(2x/\sqrt{3}) + c$ where $|x| \leq \sqrt{3}/2$.

29. $\tan \theta + \sec \theta + c$.   31. $\ln(1 + \sqrt{2})$.   33. $\pi/16$.
35. $(\cos 12x - 4 \cos 6x)/792 + c$.   37. Does not converge.
39. $\frac{1}{2} \ln(\sec t^2 + \tan t^2) + c$.   41. $a \tan(x/a) - a \ln|\cos (x/a)| + c$.
43. $a > 0$ and $b^2 - 4ac > 0$.   45. $(\pi^2 - 4)/16$.   47. $-\pi$.
49. $e^{\tan x} + c$.   51. $\bar{y} = 2 \sqrt{2rh - h^2}/3A$.

III. § 7.8   5. $3 \cosh 3x \, dx$.   7. $k \operatorname{sech}^2 kx \, dx$.   9. $\tanh x \, dx$.   11. 0.
13. dx.   15. $dx/\sqrt{x^2 - a^2}$.   17. $(1/2)\sinh 2x + c$.   19. $\cosh^4 ax/4a + c$.
21. $(1/a) \ln \cosh ax + c$.   23. $\cosh^{-1} (x/a) + c = \ln|x + \sqrt{x^2 - a^2}| + c_1$.
25. $x - \tanh x + c$.   27. $x \cosh^{-1} x - \sqrt{x^2 - 1} + c$.   29. $\ln(1 + \sqrt{2})$.
31. $(15 - 16 \ln 2)/2$.   33. $\ln(2 + \sqrt{3}) - \sqrt{3}$.   35. $\ln 2$.   37. $(\ln 9)/8$.

III. § 8.6   1. $\pi^2/16$.   2. $\pi(\exp a^2 - 1)$.   3. $\pi/4$.   4. a^2.   5. $(\pi - 2)/8$.
7. πa^2.   9. $2\pi \, a^2/3 \sqrt{3}$.   11. $\pi a^2(a^2 + 2)/2$.   13. $4a^2 \pi^3/3$.   17. $\pi^3/192$.

III. § 9.6   7. $16\pi/3$.   9. e^2.   11. Does not converge.   13. π.
15. $\pi a^4 b/4$.   17. Rising spiral around the cylinder, $r = 4$.   19. 1.

III. § 10.6   1. Sphere of radius 4.   2. Vertical plane through the meridian,
$\theta = \pi/3$.   3. Cone of constant latitude, $\phi = \pi/4$.   4. Conical solid inside
a sphere of radius 1.   5. Cylinder of radius 4 with axis NZ.
7. Solid spherical shell of thickness 1 and interior radius 1.   9. $\tan \theta = 2$.
11. Solid cylinder, $r \leqq 2 \cos \theta$.   13. $4\pi R^3/3$.   15. $8\pi a^3/3$.   17. $2\pi Rh$.
21. No "parallel line" (great circle) exists.

III. § 11.5   1. $(4, 3)$ is the only solution.   2. For every number u, $(1 - u, u)$
is a solution and these are the only ones.   3. No solutions exist.
4. For every number u, $(u, -u)$ is a solution and these are the only ones.
5. For every number u, $(-7 + 7u, 4 - 3u)$ is a solution and these are the only ones.
7. No solutions exist.   11. $\{(x, y) = (5/2 - 3u/2, u)$ and $u \in \Re\}$.
13. $\{(x_1, x_2, x_3)|u \in \Re$ and $(x_1, x_2, x_3) = u(11/5, -\frac{9}{5}, 1)$.
15. $\{(x_1, x_2, x_3)|u \in \Re$ and $\{x_1, x_2, x_3\} = u(\frac{5}{4}, \frac{67}{12}, 1)\}$.   17. The only solution
is $(\frac{1}{4}, \frac{5}{2}, \frac{11}{4})$.   19. $\{(x_1, x_2, x_3)|u \in \Re$ and $(x_1, x_2, x_3) = (-5u + 16,$
$8u - 22, u)\}$.   21. Solutions exist if and only if $y_3 = y_1 - 2y_2$ and in that
case the solutions are, for all u, $(x_1, x_2, x_3) = \left(\dfrac{y_1 + y_2 - 3u}{4}, \dfrac{-y_1 - 3y_2 - u}{4} \right)$.

III. § 12.4   1. The only solution is $\begin{bmatrix} 0 \\ 0 \end{bmatrix}$.   2. Rank $= 1$.   All solutions $X = \begin{bmatrix} 0 \\ u \end{bmatrix}$.

3. Rank $= 2$. Solution: For all $u \in \Re$, $X = \begin{bmatrix} -4u/5 \\ 13u/15 \\ u \end{bmatrix}$.   4. Rank $= 1$.

For all real numbers u_1, u_2, u_3, u_4, $X = \begin{bmatrix} -3u_1/2 + u_2/4 - u_3/2 - 3u_4/4 \\ u_1 \\ u_2 \\ u_3 \\ u_4 \end{bmatrix}$.

5. Rank $= 1$. For all numbers u, v, $X = \begin{bmatrix} -u - v \\ u \\ v \end{bmatrix}$.

7. Rank $= 3$. The only solution is $X = \begin{bmatrix} 0 \\ 0 \\ 0 \end{bmatrix}$.

9. Rank = 2. The only solution is $X = \begin{bmatrix} 0 \\ 0 \end{bmatrix}$.

11. For $A = O$ solutions consist of all $X = \begin{bmatrix} x \\ x_2 \\ \cdot \\ \cdot \\ \cdot \\ x_n \end{bmatrix}$. For $A = I$, only $X = \begin{bmatrix} 0 \\ 0 \\ \cdot \\ \cdot \\ \cdot \\ 0 \end{bmatrix}$.

13. $\lambda = 2 + \sqrt{5}, \lambda = 2 - \sqrt{5}$.   17. All solutions, $x_1 = u - 2v, x_2 = u, x_3 = v$, form a plane.   19. $\begin{pmatrix} 1 & -1 \\ 0 & 0 \end{pmatrix}$ or $\begin{pmatrix} 0 & 0 \\ 0 & 0 \end{pmatrix}$.

III. § 13.4   1. (b) and (c) are linearly dependent.   2. Dimension = 2.
One basis: (1, 4, −2, 0) and (0, −13, 10, −3).   4. The matrix having these vectors as rows is equivalent to the identity matrix.   5. For example, (1, 0, −1) = 1(1, 0, 0) + 0(0, 1, 0) −1(0, 0, 1).
7. 2 and $\frac{1}{2} \sin \theta + (- \sqrt{3}/2) \cos \theta$.   9. Rank = 2 implies that the row space has dimension 2. We may assign arbitrary values to x_3, x_4 so the solution space has dimension, 4 − 2 = 2.   19. 2.   23. It is not one-to-one.

III. § 14.5   1. $\begin{bmatrix} 11 & 4 \\ 13 & -8 \end{bmatrix}$.   2. $\begin{bmatrix} 0 & 27 \\ 0 & 7 \end{bmatrix}$.   3. $\begin{bmatrix} 34 & 16 \\ -8 & 2 \end{bmatrix}$.   4. $\begin{bmatrix} 6 & 33 \\ 4 & 14 \end{bmatrix}$.

5. $\begin{bmatrix} -2 & 0 \\ 0 & 1 \end{bmatrix}$.   7. $\begin{bmatrix} 33 & 10 \\ -3 & -12 \\ 76 & 33 \end{bmatrix}$.   9. $\begin{bmatrix} 2 \\ 3 \end{bmatrix}$.   11. $\begin{bmatrix} 2 & 2 \\ -1 & -1 \\ 4 & 4 \end{bmatrix}$.

13. $\begin{bmatrix} 0 & 0 & 1 \\ 0 & 1 & 0 \\ 1 & 0 & 0 \end{bmatrix}$.   15. $\begin{bmatrix} 1 & 0 & 0 \\ 0 & 1 & -2 \\ 0 & 0 & 1 \end{bmatrix}$.   17. $AP_1 = \begin{bmatrix} a_{11} \\ a_{21} \end{bmatrix}, \quad AP_2 = \begin{bmatrix} a_{12} \\ a_{22} \end{bmatrix}$,

$AP_3 = \begin{bmatrix} a_{13} \\ a_{23} \end{bmatrix}$.   19. $E_2 E_1$ first performs the operation E_1, then on the result it performs the operation E_2.

III. § 15.6   1. $\begin{bmatrix} 4 & 0 \\ -8 & 5 \end{bmatrix}$.   2. $\begin{bmatrix} -5 & 6 \\ 10 & -7 \end{bmatrix}$.   3. $\begin{bmatrix} -5 & 2 \\ -2 & -5 \end{bmatrix}$.

4. $\begin{bmatrix} 1 - \lambda & 2 \\ -2 & 1 - \lambda \end{bmatrix}$.   5. $\begin{bmatrix} -16 & 2 \\ 2 & 16 \end{bmatrix}$.   7. $\begin{bmatrix} 1/5 & -2/5 \\ 2/5 & 1/5 \end{bmatrix}$.   9. $\begin{bmatrix} 3 & -2 \\ 0 & 0 \end{bmatrix}$.

11. $M^{-1} = \begin{bmatrix} 1 & 0 & 0 \\ -2 & 1 & 0 \\ -1 & -1 & 1 \end{bmatrix}$, N^{-1} does not exist.   21. $\begin{bmatrix} c & 0 \\ 0 & c \end{bmatrix}$.
23. Dimension = 4.

III. § 16.4   1. $\begin{bmatrix} x_1 \\ x_2 \end{bmatrix} = \begin{bmatrix} 2 & 3 \\ 1 & -1 \end{bmatrix} \begin{bmatrix} -4 & 1 \\ 3 & -1 \end{bmatrix} \begin{bmatrix} z_1 \\ z_2 \end{bmatrix}$.

2. $\begin{bmatrix} x_1 \\ x_2 \end{bmatrix} = \begin{bmatrix} 1 & -1 & 1 \\ 2 & 0 & -1 \end{bmatrix} \begin{bmatrix} 4 & -2 \\ 1 & 3 \\ 5 & 2 \end{bmatrix} \begin{bmatrix} z_1 \\ z_2 \end{bmatrix}$.

3. $[x_1] = [4 \quad -3 \quad 7] \begin{bmatrix} 2 & -1 \\ 0 & 2 \\ 1 & 0 \end{bmatrix} \begin{bmatrix} z_1 \\ z_2 \end{bmatrix}$.   4. $A^t = \begin{bmatrix} 1 & 0 \\ 0 & 1 \\ -3 & 2 \end{bmatrix}$.

7. $Y = \begin{bmatrix} 1 & 3 \\ -4 & 5 \end{bmatrix} \begin{bmatrix} x_1 \\ x_2 \end{bmatrix}$.   9. $\begin{bmatrix} x_1 \\ x_2 \end{bmatrix} = \begin{bmatrix} (5a - 3b)/17 \\ (4a + b)/17 \end{bmatrix}$.

11. For all numbers u, $\begin{bmatrix} x_1 \\ x_2 \\ x_3 \end{bmatrix} = \begin{bmatrix} -17u \\ 2u \\ u \end{bmatrix}$.

13. $\begin{bmatrix} b \\ p \\ t \end{bmatrix} = \begin{bmatrix} 24 & 8 & 12 \\ 6 & 3 & 6 \\ 5 & 1 & 4 \end{bmatrix} \begin{bmatrix} 2 & 6 \\ 3 & 4 \\ 5 & 10 \end{bmatrix} \begin{bmatrix} m_1 \\ m_2 \end{bmatrix}$.   15. $[c] = \begin{bmatrix} 8 & 3 & 5 \end{bmatrix} \begin{bmatrix} 2 & 6 \\ 3 & 4 \\ 5 & 10 \end{bmatrix} \begin{bmatrix} 100 \\ 120 \end{bmatrix}$.

III. § 17.5   1. $[T] = \begin{bmatrix} 3 & 5 \\ -2 & 7 \end{bmatrix}$.   2. $[T] = \begin{bmatrix} 0 & -1 \\ 1 & 0 \end{bmatrix}$.

3. $[T] = \begin{bmatrix} \frac{3}{5} & \frac{4}{5} \\ -\frac{4}{5} & \frac{3}{5} \end{bmatrix}$.   9. This projection operator is not invertible.

15. $[D] = \begin{bmatrix} 0 & 1 \\ -1 & 0 \end{bmatrix}$.   23. $\begin{bmatrix} -2 & 4 & 6 \\ 5 & 3 & -1 \end{bmatrix}$.

III. § 18.9   1. 15.   2. 24.   3. 0.   4. -27.

11. (a) $\lambda = 2 \pm \sqrt{17}$;   (b) $\lambda = 0, \lambda = (1 \pm \sqrt{45})/2$.

III. § 19.5   In Exercises 1–9 we write the eigenvectors as columns in a matrix in the same order as the associated eigenvalues λ. Remember that any scalar multiple of an eigenvalue is an eigenvalue.

1. Eigenvalues: $\lambda = 2, -2$. Eigenvectors: $\begin{bmatrix} 1 & -1 \\ 1 & 3 \end{bmatrix}$, $J = \begin{bmatrix} 2 & 0 \\ 0 & -2 \end{bmatrix}$.

2. Eigenvalues: $\lambda = 5, -5$. Eigenvectors: $\begin{bmatrix} 2 & -1 \\ 1 & 2 \end{bmatrix}$, $J = \begin{bmatrix} 5 & 0 \\ 0 & -5 \end{bmatrix}$.

3. Eigenvalues: $\lambda = 6, 1$. Eigenvectors: $\begin{bmatrix} 2 & -1 \\ 1 & 2 \end{bmatrix}$, $J = \begin{bmatrix} 6 & 0 \\ 0 & 1 \end{bmatrix}$.

4. Eigenvalues: $\lambda = 4, -1$. Eigenvectors: $\begin{bmatrix} 1 & -2 \\ 2 & 1 \end{bmatrix}$, $J = \begin{bmatrix} 4 & 0 \\ 0 & -1 \end{bmatrix}$.

5. Eigenvalues: $\lambda = 4, 4$. Pick eigenvectors: $\begin{bmatrix} 1 & 0 \\ 0 & 1 \end{bmatrix}$, $J = \begin{bmatrix} r & 0 \\ 0 & r \end{bmatrix}$.

7. Eigenvalues: $\lambda = \sqrt{7}, -\sqrt{7}$.
   Eigenvectors: $\begin{bmatrix} 2 + \sqrt{7} & -2 + \sqrt{7} \\ 1 & 1 \end{bmatrix}$, $J = \begin{bmatrix} \sqrt{7} & 0 \\ 0 & \sqrt{7} \end{bmatrix}$.

9. Eigenvalues: $\lambda = \pm i$. No real eigenvectors. See Exercise 18.

13. $T^{-1} \leftrightarrow \begin{bmatrix} 1/\lambda_1 & 0 \\ 0 & 1/\lambda_2 \end{bmatrix}$ if $\lambda_1 \neq 0$ and $\lambda_2 \neq 0$,

$(T - I)^{-1} \leftrightarrow \begin{bmatrix} 1/(\lambda_1 - 1) & 0 \\ 0 & 1/(\lambda_2 - 1) \end{bmatrix}$, $\sqrt{T} \leftrightarrow \begin{bmatrix} \sqrt{\lambda_1} & 0 \\ 0 & \sqrt{\lambda_2} \end{bmatrix}$.

17. $\lambda = 4, 1$.   19. Repeated eigenvalues $\lambda = 3, 3$. Only eigenvector is $X = u \begin{bmatrix} 1 \\ 1 \end{bmatrix}$.   21. The only eigenvector is $t \begin{bmatrix} -2bc \\ c(a - d) \end{bmatrix}$. With an additional vector $\begin{bmatrix} 0 \\ 2c \end{bmatrix}$ to make a basis, $J = \begin{bmatrix} \frac{a + d}{2} & 1 \\ 0 & \frac{a + d}{2} \end{bmatrix}$.

III. § 20.5   4. $x_1y_1 + x_2y_2 + x_3y_3$.   5. $t(2, 18, 5)$ for all numbers t.
   7. $\vec{v} = \frac{58}{5}\vec{v}_1 - \frac{6}{5}\vec{v}_2$.   9. $\vec{w}_1 = (\frac{33}{5}, -\frac{44}{5})$, $\vec{w}_2 = (-\frac{3}{5}, -\frac{6}{5})$.
   15. The sum of the squares of two sides of a parallelogram equals the average of
        the squares of the diagonals.

III. § 21.7   1. Hyperbola, $x'^2 - y'^2 = 4$.   2. Parallel straight lines,
        $25x'^2 = 288$.   3. Ellipse, $x'^2 + 9y'^2 = 9$.
   4. Hyperbola, $(\sqrt{7} + 2)x'^2 - (\sqrt{7} - 2)y'^2 = 4$.
   5. Hyperbola, $2x'^2 - 2y'^2 = -1$.   7. Ellipse, $x'^2 + 11y'^2 = 44$.
   11. $(x, y) = (-\frac{3}{5}, \frac{4}{5})$.   13. $(x'^2/4) + (y'^2/1) = 1$.

III. § 22.6   1. $y = c\, e^{\lambda t}$.   2. $y = c_1\, e^{-2it} + c_2\, e^{2it}$.
   3. $x = c_1\, e^{-3t} + c_2\, e^{2t}$.   4. $x = c_1 + c_2\, e^{2t}$.
   5. $x = \frac{1}{5}(2e^{-3t} + 3e^{2t})$.
   7. $y = a \cos 2t + b \cos 2t$, period $= \pi$, frequency $= 1/\pi$, where
        $a = c_1 + c_2$, $b = i(c_1 - c_2)$.
   9. $x = e^{-3t}(c_1 + c_2\, e^{5t})$. When λ is real, $e^{\lambda t}$ does not convert into a
        sinusoidal function.
   11. Frequency $= \sqrt{2mk - R^2}/4\pi m$ oscillations per second. It decreases as the
        mass, m, increases, and increases as the spring stiffness, k, increases.
   13. $\lambda = -\dfrac{R}{2m} \pm \dfrac{\sqrt{R^2 - 4mk}}{2m}$, both real and negative.
   15. $y = e^{-t}(c_1 + c_2\, t + c_3\, t^2)$.

III. § 23.7   1. $y = 5[1 + 2t + (2t)^2/2! + (2t)^3/3! + \cdots]$.
   2. $y = 1 + it + (it)^2/2! + (it)^3/3! + \cdots$.
   3. $\begin{bmatrix} y_1 \\ y_2 \end{bmatrix} = \begin{bmatrix} c_1 \\ c_2 \end{bmatrix} + \dfrac{t}{1!} A \begin{bmatrix} c_1 \\ c_2 \end{bmatrix} + \dfrac{t^2}{2!} A^2 \begin{bmatrix} c_1 \\ c_2 \end{bmatrix} + \cdots$, where $A = \begin{bmatrix} 2 & 1 \\ 1 & 0 \end{bmatrix}$.
   4. (a) $d^2 y_1/dt^2 - 2\, dy_1/dt - y_1 = 0$.
        (b) $y_1 = c_1 (1 + 2t + 2t^2 + \frac{5}{3} t^3 + \cdots) + c_2 (1 + t + t^2 + \frac{5}{6} t^3 + \cdots)$.
   5. Show that this includes the unique solution with $y(0) = y_0$, $y'(0) = v_0$.
   7. The system has the integral, $v^2 + (k/m)y^2 = c_0$, whose graph is an ellipse.

III. § 24.5   1. $7e^{-2t}(\cos t + 2 \sin t)$.
   2. $7e^{-2t}(\cos t + 2 \sin t) + \displaystyle\int_0^t e^{-2(t-s)}\sin(t - s)y(s)ds$.
   3. $x = \frac{26}{5}e^{-2t}(\cos t + 2 \sin t) + \frac{9}{5}$.
   5. $x = x_0 \cos 2t + (v_0/2) \sin 2t + \frac{1}{2}\displaystyle\int_0^t \sin 2(t - s) \sin \omega s\, ds$.
   7. $x = x_0 \cos 2t + (v_0/2) \sin 2t + \frac{1}{8} \sin 2t - \frac{1}{4} t \cos 2t$.
   9. $x(t)$ does not oscillate but decreases monotonically; approaches $-13/2$.
   11. Frequency: (a) decreases as m increases (heavier mass),
        (b) decreases as the resistance R increases, increases as k
        increases (stiffer spring).
   13. Increases without limit as ω approaches resonance frequency, 2.
   15. Mass $= 1$, spring modules $= 4$, impressed force is sinusoidal with frequency
        2 per 2π seconds.
   17. Amp $= 1/\sqrt{4\omega^2 + (\omega^2 - 5)^2}$; decreases with increasing ω.

III. § 25.5
   1. $e^{-x} = e + e\,\dfrac{(x - 1)}{1!} + e\,\dfrac{(x - 1)^2}{2!} + e\,\dfrac{(x - 1)^3}{2!}\displaystyle\int_0^1 (1 - t)^2\, e^{t(x-1)}dt$.

   2. $\sin x = x - \dfrac{x^3}{3!} + \dfrac{x^4}{3!}\displaystyle\int_0^1 (1 - t)^3 \sin t\, x\, dt$.

3. $x^p = 1 + \frac{p}{1!}(x-1) + \frac{p(p-1)}{2!}(x-1)^2 + \frac{p(p-1)(p-2)}{2!}$

$\int_0^1 (1-t)^2[1+t(x-1)]^{p-3}dt.$

4. $\ln x = \frac{x-1}{1} - \frac{(x-1)^2}{2} + \cdots + (-1)^{n-1}\frac{(x-1)^n}{n} +$

$(-1)^n \frac{(x-1)^{n+1}}{n} \int_0^1 \frac{(1-t)^n}{[1+t(x-1)]^{n+1}}\,dt.$

5. $R_2 = hk(1 + h + k).$

7. No neighborhoods of $(1, 1)$ exist where $f(x, y) \geq f(1, 1)$ or where $f(x, y) \leq f(1, 1).$

13. $R_n = (1 - \pi/4)^n/\sqrt{2}\,n!.$ Since $R_4 < 0.001$, three terms suffice.

15. Estimating the integral as 1, $n = 1000$.

III. § 26.5   1.   Maximum where $x = \frac{1}{2}$ and minimum where $x = 1.$

2.   Minimum at $x = 2$ and maximum at $x = 5.$ 3.   No max, no min.

4.   Max where $x = 1$, min where $x = -1.$ 5.   No max, no min.

7.   No max, no min.   9.   No max, no min but $(27/2, -9)$ is a local minimum.   11.   Min at $(0, 0)$ and max at $(3, 0)$ and $(0, 2).$

13.   Max at $(1, 0)$ and min at $(0, 1).$ 15.   Max at $(2, 2, 2)$, min at $(\frac{1}{3}, \frac{1}{3}, \frac{1}{3}).$ 17.   No max, min at $(\sqrt{2}, -\pi/4).$

III. § 27.5   1. $\sqrt{10}/2.$ 2. $2\sqrt{2}.$ 3.   Max given by (base, alt) $= (r, 3r/2).$

4.   Min distance $= 1$, given by $(x, y) = (1, 0)$ with $\lambda = 1.$

5.   Min at $(0, \pm 3)$, max at $(\pm 3/\sqrt{2}, 3/\sqrt{2})$ with $\lambda = 6\pi/\sqrt{2}.$

7. $(\frac{6}{7}, \frac{9}{7}, -\frac{3}{7})$ with $\lambda = \frac{6}{7}.$ 9. $(0, 0, 0)$ with $\lambda = -\frac{1}{2}.$

13.   Max at $(\sqrt{8}, \sqrt{8})$ with $\lambda = 2$ since $\Delta F = -\int_0^1 (1-t)(h+k)^2\,dt < 0$ for all $(h, k).$

15.   Min at $(0, 0, -2)$ with multipliers not uniquely determined.

III. § 28.7   In every case $\vec{a} \times \vec{b} = -\vec{b} \times \vec{a}.$ 1.   (a) $\vec{a} \times \vec{b} = (-2, -2, 4),$
(b) $\vec{a} \times \vec{b} = (3, -2, 0),$ (c) $\vec{a} \times \vec{b} = (0, 0, 0),$ (d) $\vec{a} \times \vec{b} = (0, 0, 0).$

2. $4x + 10y + 2z = 0.$ 3. $(\vec{a} \times \vec{b}) \cdot (\vec{p} - \vec{p_0}) = 0.$

4. $1(x-1) + 2(y+1) + 3(z+5) = 0.$ 5. $x = t + 1, y = 2t - 1, z = 3t - 5.$

7. $18.$ 9. $6.$ 11. $d = 30/\sqrt{50} \neq 0$, hence the lines do not intersect.

15. $-25/7.$ 17. $h = 75/13 = \|(p_1 - p_2) \times (p_3 - p_2)\|/\| p_3 - p_2 \|.$

19. $d = [-2(-3+0) + 5(1+4) + 3(-4+8)]/\sqrt{38}.$

III. § 29.5   2. $Dw = (dx + dy)\,\vec{i} + (dx - dy)\,\vec{j}.$ 3. $\vec{\nabla} \cdot \vec{w} = 0, \vec{\nabla} \times \vec{w} = \vec{0}.$

4. $\vec{\nabla} \cdot \vec{F} = 0, \vec{\nabla} \times \vec{F} = \vec{0}.$ 5. $\vec{\nabla} \cdot \vec{F} = 0, \vec{\nabla} \times \vec{F} = \vec{0}.$

7. $\vec{\nabla} \cdot \vec{F} = 2e^x \sin y, \vec{\nabla} \times \vec{F} = e^x(\sin y - \cos y)\,\vec{k}.$ 9. $\vec{\nabla} \cdot \vec{F} = 0, \vec{\nabla} \times \vec{F} = \vec{0}.$

11.   (a), (b), (c), (d) $\vec{\nabla} \cdot \vec{\nabla}u = 0, \vec{\nabla} \times \vec{\nabla}u = \vec{0}.$

III. § 30.3   1. $V = \int_0^4 (1 - z/4)^2\,\sqrt{3}\,dz = 4\sqrt{3}/3.$

2. $V = \frac{1}{2}\int_{-2}^2 (4 - y^2)\,dy = 16/3.$ 3. $V = \int_0^2 \pi(2 - x)^2\,dx = 8\pi/3.$

4. $V = \int_0^a \pi x^2\,dy = \pi a^2/2.$ 5. $V = 2\int_0^4 \pi x^2\,dy = 16\pi$, counting the volume twice because it is generated one time each by two different plane regions.

7. $V = \pi \int_0^a (y + 8\sqrt{y})\,dy.$ 9. $V = 2\pi \int_0^{\sqrt{7}/2} (6\sqrt{4 - y^2} - 9)\,dy =$

$18\,\pi[\frac{4}{3}(\arcsin\sqrt{7}/4) - \sqrt{7}/4].$ 11.   Since $V = \int_0^z A(y)\,dy,$ $\frac{dV}{dt} = A(z)\,\frac{dz}{dt}.$

III. § 31.4   1. $V = \int_0^2 2\pi \, y(2 - y) \, dy = 8\pi/3$.   2. $V = \int_0^{\sqrt{a}} 2\pi \, x(a - x^2) \, dx$.

3. $V = \int_0^2 2\pi \, x(4 - x^2) \, dx$.   4. $V = \int_0^{\sqrt{a}} 2\pi(x + 4) \, (a - \sqrt{x}) \, dx$.

5. $S = \int_{-r}^r 2\pi \, y \, ds = \int_{-r}^r 2\pi \, r \, dx = 4\pi r^2$.   7. $116\pi/135$.

9. $V = \int_0^a 2\pi \, x(2y) \, dx = 4\pi \, a^3(2 \sqrt{2} - 1)/3$.

13. $\pi a^2[(9 - 4\sqrt{3}) \, a + 6(3 - \sqrt{3}) \, b]/9$.

III. § 32.6   1. $I = \int_{-1}^1 z^3(4\pi) \, dz = 0$.   2. $128\pi/15$.

3. $I = \int_0^1 \dfrac{1}{1 - z} \left(\dfrac{1 - z^2}{2} \right) dz = \dfrac{1}{4}$.

4. $I = \int_0^a z \sqrt{a^2 - z^2} \, [\pi(a^2 - z^2) \, dz] = \pi a^5/5$.   5. $I = \int_0^2 (r^2 + 1)2\pi r(2) dr = 24\pi$.

7. $I = \int_0^b e^{-r^2} 2\pi \, r \, h \, dr = \pi h(1 - e^{-b^2})$.

9. $I = \int_0^a r^2(2\pi \, r) \, (h - hr/a) \, dr = \pi a^4 \, h/10$.   11. $z = 3h/4$.

13.   On the axis of the cone at height $h/4$.

15. $I = \int r^2 \, dm = \int_0^a r^2(2\pi \, \delta \, r \sqrt{a^2 - r^2} \, dr = 14\pi \, \delta \, a^5/15$.

17. $I = \int_0^a (1/r) \, 4\pi \, r^2 \, dr = 2\pi \, a^2$.   19. $\pi a^4(28a + 3h)/30$.

21.   On the axis at distance $\bar{y} = (3a^2 - h^2)/4(a + h)$ from the center of the spherical part.   23. $u = (2\pi \, \sigma/3z) \, [(R^2 + z^2)^{3/2} + R^3 - z^3 - 3z \, R^2/2]$.

III. § 33.3   1.   (a), (b) $2(2 - \sqrt{2})/3$.   2. $\bar{x} = 2$.
3. $4 \arcsin \sqrt{15}/8 + \arcsin \sqrt{15}/4 - \sqrt{15}/2$.   4.   (a), (b). Both represent the area of the triangle $(2, 2), (4, 2), (4, 8)$.   5.   0, No.   7. $25 \arcsin (7/25)$.

9. $V = \int_{-1}^1 \int_{-\sqrt{1-x^2}}^{+\sqrt{1-x^2}} \int_{-1}^1 dz \, dy \, dx = 2\pi$.   11. $V = \int_0^{2\pi} \int_{-1}^1 \int_{-1}^1 dz \, r \, dr \, d\theta = 2\pi$.

13.   Using cylindrical coordinates, $V = 5\pi/12$.   15.   Both represent the volume of a prism of height 6 whose base is the triangle $(0, 0, 0), (3, 0, 0), (0, 0, 3)$.

17. $\delta = z/5 + 1$, $M = 240\pi$.   19. $a^2(4a/3 + \pi)$.   21. $\int_0^t \int_0^u \sin u^2 \, dv \, du$.

23.   In cylindrical coordinates with polar coordinates in the xz plane
$\int div \, \vec{F} \, dV = \int_0^{2\pi} \int_1^3 \int_0^{\sqrt{1-(\bar{y}-2)^2}} \dfrac{2}{\sqrt{r^2 + y^2}} \, r \, dr \, dy \, d\theta = 4\pi/3$.

III. § 34.5   1.   (a) 11,   (b) 0,   (c) -11,   (d) 1,   (e) 4,
(f) $(x_2 - x_1) \, (y_3 - y_1) - (x_3 - x_1) \, (y_2 - y_1)$.   2.   The sign difference is due to the order of writing down the points.   3.   13 times bigger.
5.   HINT: By definition, the area of any areable figure is the limit of a sum of rectangles.   7. $\iint_{\mathcal{D}} (x - y)^2 \, dx \, dy = \iint_{\mathcal{U}} 4v^2(2) \, du \, dv$, where $\mathcal{U}$ is the parallelogram with vertices $(u, v) = (\frac{1}{2}, \frac{1}{2}), (\frac{1}{2}, -\frac{1}{2}), (-\frac{1}{2}, -\frac{1}{2}), (-\frac{1}{2}, \frac{1}{2})$.

9. $\iint_{\mathcal{D}} (x + y) \, dx \, dy = \iint_{\mathcal{U}} (u - 2v - 4) \, (1) \, du \, dv$, where $\mathcal{U}$ is the triangle with vertices $(u, v) = (-14, -9), (-15, -10), (-17, -11)$.

11. $J = 0$ shows that the transformation maps the element of area in the uv-plane onto a figure of zero area in the xy-plane.   13. $\iiint_{\mathcal{U}} dy \, dv \, dw$ where $\mathcal{U}$ is a ball of radius r with center at $(-a, -b, -c)$. The transformation may be regarded as a translation from this center to ball centered at $(x, y, z) = (0, 0, 0)$.

15. $\iiint_{\mathcal{U}} (1/\rho)\rho^2 \sin\phi \, d\rho \, d\theta \, d\phi$, where $\mathcal{U}$ is the box,

$\{1 \leq \rho \leq 2, 0 \leq \theta \leq 2\pi, 0 \leq \phi \leq \pi\}$.   17.* $J = -4u^2v^2/(u^2 + v^2)^2$.
$J = 0$ when $u = 0$ or $v = 0$. The substitution is unsuitable for transforming the integral into uv-coordinates.

III. § 35.4   1. $\vec{w} = R \sin\phi(x\vec{i} + y\vec{j} + z\vec{k})$.   2. $S = \int_0^{2\pi}\int_0^\pi R^2 \sin\phi \, d\phi \, d\theta = 4\pi R^2$.
3. $\vec{w} = -2\vec{i} - 3\vec{j} + \vec{k}$, $dS = \sqrt{14} \, du \, dv$, a plane.
4. $\vec{w} = (R \cos u)\vec{i} + (R \sin u)\vec{j} + 0\,\vec{k}$, $dS = R \, du \, dv$.   5. $2\pi Rh$, a cylinder.
7. $-\sqrt{2}\,(x - \sqrt{2}/2) + \sqrt{2}\,(y - \sqrt{2}/2) + 1(z - 1) = 0$.
9. $dS = |\sin u \cos v|\sqrt{16 \sin^2 u - 1} \, du \, dv$. Eliminating u and v shows that the surface is formed by revolving about the z-axis the parabolic arc,
$z = 1 - 2y^2$, $-1 \leq z \leq 1$.   11. Eliminating v gives $x^2 + y^2 = p^2(u)$,
$z = q(u)$, which is the surface generated by revolving about the z-axis the curve
$x = 0$, $y = p(u)$, $z = q(u)$.   13. We compute dS for each and find that for
both $dS = |\sin\phi|d\phi \, d\theta$.   15. 2π.   17. $dS = (r/a)\sqrt{h^2 + a^2} \, dr \, d\theta$,
$S = \pi a \sqrt{h^2 + a^2}$.

III. § 35R   1. $V = \int_{-R}^R \pi r^2 \, dz = 4\pi R^3/3$.   2. $V = \int_0^2 2\pi r(4 - 2r)dr = 4\pi/3$.
3. $64\pi(2 - \sqrt{2})/3$.   4. $V = 2(28/3)$, counting both sides.
5. Cartesian coordinates are appropriate, $\bar{x} = 4/\pi$, $\bar{y} = \bar{z} = 0$.
6. Disk method is convenient, $V = (\pi R^3/3)(2 - 3k + k^3)$.   7. $4\pi/\sqrt{3}$.
8. 0.   9. 0.   11. An improper integral which does not converge. The integrand becomes infinite on the line $y = -x$.   12. $2\pi^2$.
13. $\pi[e\sqrt{e+1} + \ln(e + \sqrt{1+e^2}) - \sqrt{2} - \ln(1 + \sqrt{2})]$.   14. $128\pi a^2/3$.
15. $\pi ha^4\delta/2$.   16. $3\pi^2 \delta/16$.   17. $(4\pi a^3/3)(1 - \cos^3\phi/2)$.   18. $4\pi\delta a^5/5$.
19. On the vertical axis of symmetry at height $h/4$.   20. $62.5\pi(10^5)$ ft lbs.
21. 12π.   22. $V = \dfrac{\pi}{3}\left[2(a^3 + b^3) - \left(a^2 + \dfrac{h^2}{8}\right)\sqrt{4a^2 - h^2} - \left(b^2 + \dfrac{h^2}{8}\right)\sqrt{4b^2 - h^2}\right]$.
23. $7\pi a^3/3$.   24. 0.   25. π.   27. $A(z) = 6(10)^7$ ft^2.   28. $8\sqrt{6}/3$.
29. $3k^2\pi/2$.   30. 0.   31. $\pi e/10$.   32. The integral diverges.
33. $4\pi\sqrt{3}$.   34. $9(12)^2$.   35. $\bar{z} = 1/2$.   36. $\bar{x} = \bar{y} = 0$. $\sigma = \sqrt{105}/15$.
37. The pyramid below the plane $z = y$ whose base in the plane $z = 0$ is the triangle bounded by the lines $x = 1$, $y = 0$, $y = x$.   38. $V = 6\pi a^3/5$.
40. M/R, where M is the mass of the ball.   42. Force $= -(4\pi\delta/3)z$ and it produces simple harmonic motion with frequency $\sqrt{4\pi\delta/3m}$ in 2π sec.
43. $46\pi/243$.

Appendix A Exercises
1. $2x^2 + 3x - 3$.   2. $x^3 - 3x + 4 = \frac{1}{3}(3x^2 - 1) - (3x - 11/3)$.
3. $3x^4 - 2x^3 - 6x^2 + 8x - 24$.
4. $-x^3 + 6x - 9 = (x + 4)(-x^2 + 4x - 10) + 3$.   5. $x^2 - 4x + 8$.
7. $f(-2) = 9$, $f(3) = 19$, $f(2.6) = 11.576$.   9. $2x^2 + 2x - 1$.
11. $2w^2 - 6w + 15$.   17. No roots.   19. $x = 0$.
21. $x \approx -2.7$, $x \approx 0.4$, $x \approx 2.5$.   23. $x \approx 0.7$ and $x \approx 0.9$.
25. (a) 6, (b) -6, (c) 0.   27. 8 and 4 respectively.

Index

use of — in integration, 299
— in vector calculus, III 662
Theorem
— on differentiation of the indefinite
integral, 135
evaluation — for differential calculus, 76
fundamental — of integral calculus, 121
— on increasing functions, 85
— on interior maxima and minima, 83
— on the limit of a derivative, 271
— of the mean, 84
Rolle's —, 87
Taylor's —, 92; proof, 94
Torque, 479
Total variation, 134, 449
Trajectory, 397
Transcendental function
representation of —s by power series, 228,
509
review of —s, 226*ff*.
Transformation
group of —s, III 523
rotation —, III 524
—s of affine geometry, III 526
—s of Euclidean geometry, III 522
translation —, III 523
Translation of coordinate plane, III 523
Trapezoidal rule, 57
Triangle inequality, 328
Trigonometric functions, 200
addition theorems for —, 201
antiderivatives for —, 203
area of region below the graph —, 222
differentiation of —, 203
double angle identities of —, III 555
graphs of —, 206
inverse —, 213
tangent line to the graph of a trigonometric
function, 221
theorem on existence of —, 220
Triple integral, 469

Utility in economics, 166

Variable
dummy —, 133
Variance in statistics, 477
Variance, or σ^2, in a distribution, 150
Variation of parameters, III 656
Vector, 322*ff*.
basis for a — space, III 593
coordinates of a — with respect to
orthogonal basis, III 627
— differential calculus, III 661

geometric interpretation of — product, III
679
inner product of —s in n-space, 333
linearly independent set of —s, 325, III 593
located —, 334
norm of a —, 327
norm of a — in n-space, 333
— operations, 324
orthogonal —s, 334, III 626
parallel —s, 334
position —, 323
position — in 3-space, 332
— product, or cross product, III 679
scalar multiplication of — space, 324
—s in analytic geometry, 306
—s of higher dimensional spaces, 331
unit —, 327
Vector field, III 686
curl of a —, III 690
divergence of a —, III 688
irrotational —, III 691
Vector function, 367
Vector product in surface area, III 722
Vector space
basis for a —, III 593
finite dimensional —, III 593
isomorphic representation of a —, III 594
orthogonal basis in a —, III 626
representation of a — with inner product,
III 628
Velocity
— in curvilinear motion, 399
escape —, 199
instantaneous — in a motion, 78
Volume
— calculated by parallel slices, 141
— of a region bounded by two surfaces, III
708
— of revolution, 142
Volume element
disk —, III 695
spherical shell —, III 698
tubular —, III 697

Wave train, 364
Weight and mass, 117
Weightlessness, III 705
Work
— against gravity, 196
— done in compressing a gas, 164
— done by a force, 143

Zeno's paradox, 507